国外优秀数学著作
原 版 系 列

U0212029

多元实函数教程

Function of Several Real Variables

[美] 马丁·莫斯科维茨 (Martin Moskowitz)
[美] 福蒂奥斯·帕里奥詹尼斯 (Fotios Paliogiannis) 著

（英文）

哈尔滨工业大学出版社
HARBIN INSTITUTE OF TECHNOLOGY PRESS

黑版贸审字 08-2020-186 号

图书在版编目(CIP)数据

多元实函数教程=Function of Several Real Variables：英文/(美)马丁·莫斯科维茨 (Martin Moskowitz),(美)福蒂奥斯·帕里奥詹尼斯 (Fotios Paliogiannis)著. —哈尔滨:哈尔滨工业大学出版社,2022.9
ISBN 978-7-5767-0435-8

Ⅰ.①多… Ⅱ.①马… ②福… Ⅲ.①多元函数-英文 Ⅳ.①O174.1

中国版本图书馆 CIP 数据核字(2022)第 179382 号

DUOYUAN SHIHANSHU JIAOCHENG

W$ World Scientific

策划编辑　刘培杰　杜莹雪
责任编辑　杜莹雪　张嘉芮
封面设计　孙茵艾
出版发行　哈尔滨工业大学出版社
社　　址　哈尔滨市南岗区复华四道街 10 号　邮编 150006
传　　真　0451-86414749
网　　址　http://hitpress.hit.edu.cn
印　　刷　哈尔滨博奇印刷有限公司
开　　本　787 mm×960 mm　1/16　印张 47　字数 426 千字
版　　次　2022 年 9 月第 1 版　2022 年 9 月第 1 次印刷
书　　号　ISBN 978-7-5767-0435-8
定　　价　118.00 元

Preface and Acknowledgments

This book arose from courses in calculus in several variables which each of us has taught a number of times. Our book has the dual benefits of emphasizing both the *conceptual* and *computational* aspects of the subject, while having a modern outlook. Its primary goal is to *teach* the subject in a clear and systematic way. Although the project proved more demanding than had been anticipated, having completed the process the authors believe it to have been a worthwhile endeavor. The early chapters give a mature introduction to the classical topics, under the rubric of Calculus in Several Variables, Advanced Calculus, or Vector Analysis, topics usually covered in the junior or senior year of the undergraduate mathematics curriculum. We then turn to Ordinary Differential Equations as well as the classical Partial Differential Equations of the second order usually found in books on advanced calculus, or mathematical physics for undergraduates. Finally, there is an elementary introduction to the powerful and important subject of the Calculus of Variations.

The subject matter of this book is fundamental to numerous disciplines, and so would be of use both to students in mathematics in the early stages of graduate study as well as to advanced undergraduate mathematics, physics, chemistry, biology, engineering, or even economics majors. The book consists of eight chapters. Chapters 1 and 2 deal with the basic geometry and topology of Euclidean space. Chapter 3 deals with differential calculus and Chapters 4, 5 and 6 with integral calculus of several variables. The book has two novel features: Chapter

7 is a primer on elementary but quite important aspects of ordinary differential equations, as well as the classical partial differential equations of the second order; related to this, Chapter 8 is an in depth introduction to the Calculus of Variations which is to be viewed as analogous to the usual extremal problems in the calculus of several variables. For that reason it seems to the authors to fit with the rest of the material nicely and so is quite properly a part of it.

When needed we refer to the Appendix for backround material. These fall into four categories, countability and decimal expansions, Calculus in one variable, uniform convergence and Linear Algebra. The results on countability can be found [18], those of Calculus in one variable in [1], those concerning uniform convergence, in particular the Stone-Weierstrass theorem in [13] and [12], while those concerning Linear Algebra can be found in [14].

The instructor can select numerous options for courses on the subject. For example, Chapters 3, 4, 5 and 6 make a standard year course in calculus of several variables. Chapters 3 and 4 a first semester course while 5 and 6 a second semester course. Chapter 7 can serve as a text for a one semester course in ODE and PDE. For stronger students Chapters 7 and 8 make a semester course in ODE, PDE and Calculus of Variations. The book can also be used for selected topics in Analysis and Geometry by focusing on the stared sections.

The reader will find many examples in each section. Individual sections contain exercises directly pertinent to that section. Answers to the exercises are given at the end of each section. In addition, every chapter concludes with a set of solved exercises and proposed exercises, some with answers or hints. Throughout the book there are more challenging stared exercises to interest the more advanced reader. Stars (*) indicate sections, examples or exercises that may be omitted on a first reading. However the more ambitious reader desiring a deeper understanding of the subject will wish to master this material as well.

We wish to thank Fred Greenleaf for making available to us a set of notes in Analysis which he used at NYU. We thank Stanley Kaplan who read much of the book for content and has made many comments which have considerably improved the exposition. Of course, any errors or misstatements are our responsibility. We also wish to thank Erez Shochat for creating the figures which was very helpful as we prepared the final version of our book. Finally, we thank Anita and Kassiani for their encouragement and patience during the three years since we began this project.

New York, June 2010

Martin A. Moskowitz, Fotios C. Paliogiannis

This page intentionally left blank

Notations

We let $\mathbb{Z}$ stand for the set of integers, $\mathbb{Q}$ stand for the set of rational numbers, $\mathbb{R}$ stand for the set of real numbers and $\mathbb{C}$ stands for the set of complex numbers.

A finite sum of numbers $a_1, ..., a_m$ is written as

$$\sum_{j=1}^{m} a_j = a_1 + a_2 + \ldots + a_m.$$

A set A is often defined by some property satisfied by its elements, viz,

$$A = \{x \in X : ...\}.$$

Let A and B be sets. We write $x \in A$ to mean the element x belongs to the set A. Also, $A \subseteq B$ means that if $x \in A$, then $x \in B$. We denote their intersection by $A \cap B$, that is, the common elements of both A and B. $A \cup B$ stands for the union of A and B, that is, the elements of either. A^c denotes the complement of A in some larger universal set X, that is, the elements of X not in A, viz,

$$A^c = \{x \in X : x \notin A\}.$$

$A \setminus B$ denotes the elements in A, but not in B.

More generally, if $\{A_i : i \in I\}$ is a family of sets indexed by the set I, then $\bigcap_{i \in I} A_i$ indicates the intersection of all the A_i. Similarly, $\bigcup_{i \in I} A_i$ indicates the union of all the A_i. The Cartesian product of X and Y is the set $X \times Y = \{(x, y) : x \in X, y \in Y\}$. More generally

$$\prod_{i=1}^{n} X_i = X_1 \times \ldots \times X_n = \{(x_1, ..., x_n) : x_i \in X_i, i = 1, ..., n\}$$

If $f : X \to Y$ is a function from X to Y and $A \subseteq X$ and $B \subseteq Y$, then

$$f(A) = \{f(a) \in Y : a \in A\},$$

while

$$f^{-1}(B) = \{x \in X : f(x) \in B\}.$$

Additional notations will be introduced throughout the text.

Contents

Preface and Acknowledgments   iii

Notations   vii

1   Basic Features of Euclidean Space, $\mathbb{R}^n$ 1

   1.1   Real numbers   1

      1.1.1   Convergence of sequences of real numbers   8

   1.2 $\mathbb{R}^n$ as a vector space   15

   1.3 $\mathbb{R}^n$ as an inner product space   19

      1.3.1   The inner product and norm in $\mathbb{R}^n$ 19

      1.3.2   Orthogonality   24

      1.3.3   The cross product in $\mathbb{R}^3$ 32

   1.4 $\mathbb{R}^n$ as a metric space   38

   1.5   Convergence of sequences in $\mathbb{R}^n$ 43

   1.6   Compactness   48

   1.7   Equivalent norms (*)   52

   1.8   Solved problems for Chapter 1   54

2   Functions on Euclidean Spaces   61

   2.1   Functions from $\mathbb{R}^n$ to $\mathbb{R}^m$ 61

   2.2   Limits of functions   66

   2.3   Continuous functions   71

   2.4   Linear transformations   80

   2.5   Continuous functions on compact sets   85

   2.6   Connectedness and convexity   88

      2.6.1   Connectedness   88

      2.6.2   Path-connectedness   91

2.6.3   Convex sets   95

2.7   Solved problems for Chapter 2   97

3 Differential Calculus in Several Variables   **103**

3.1   Differentiable functions   103

3.2   Partial and directional derivatives, tangent space   109

3.3   Homogeneous functions and Euler's equation   129

3.4   The mean value theorem   131

3.5   Higher order derivatives   134

    3.5.1   The second derivative   138

3.6   Taylor's theorem   144

    3.6.1   Taylor's theorem in one variable   144

    3.6.2   Taylor's theorem in several variables   154

3.7   Maxima and minima in several variables   159

    3.7.1   Local extrema for functions in several variables .   160

    3.7.2   Degenerate critical points   172

3.8   The inverse and implicit function theorems   177

    3.8.1   The Inverse Function theorem   177

    3.8.2   The Implicit Function theorem   185

3.9   Constrained extrema, Lagrange multipliers   194

    3.9.1   Applications to economics   205

3.10 Functional dependence   211

3.11 Morse's lemma (*)   217

3.12 Solved problems for Chapter 3   221

4 Integral Calculus in Several Variables   **231**

4.1   The integral in $\mathbb{R}^n$ 231

    4.1.1   Darboux sums. Integrability condition   234

    4.1.2   The integral over a bounded set   247

4.2   Properties of multiple integrals   250

4.3   Fubini's theorem   260

    4.3.1   Center of mass, centroid, moment of inertia . . .   273

4.4   Smooth Urysohn's lemma and partition of unity (*) . .   280

4.5   Sard's theorem (*)   282

4.6   Solved problems for Chapter 4   285

**5　Change of Variables Formula, Improper
　Multiple Integrals**　　　　　　　　　　　　　　　　**293**

　5.1　Change of variables formula 293

　　　5.1.1　Change of variables; linear case 297

　　　5.1.2　Change of variables; the general case 301

　　　5.1.3　Applications, polar and spherical coordinates . . 308

　5.2　Improper multiple integrals 328

　5.3　Functions defined by integrals 346

　　　5.3.1　Functions defined by improper integrals 351

　　　5.3.2　Convolution of functions 368

　5.4　The Weierstrass approximation theorem (*) 374

　5.5　The Fourier transform (*) 376

　　　5.5.1　The Schwartz space 383

　　　5.5.2　The Fourier transform on $\mathbb{R}^n$ 389

　5.6　Solved problems for Chapter 5 393

6　Line and Surface Integrals　　　　　　　　　　　　　　**405**

　6.1　Arc-length and Line integrals 405

　　　6.1.1　Paths and curves 405

　　　6.1.2　Line integrals 409

　6.2　Conservative vector fields and Poincare's lemma 417

　6.3　Surface area and surface integrals 429

　　　6.3.1　Surface area 429

　　　6.3.2　Surface integrals 441

　6.4　Green's theorem and the divergence theorem in $\mathbb{R}^2$ . . . 450

　　　6.4.1　The divergence theorem in $\mathbb{R}^2$ 457

　6.5　The divergence and curl 462

　6.6　Stokes' theorem 466

　6.7　The divergence theorem in $\mathbb{R}^3$ 474

　6.8　Differential forms (*) 483

　6.9　Vector fields on spheres and Brouwer fixed point
　　　theorem (*) 492

　　　6.9.1　Tangential vector fields on spheres 494

　　　6.9.2　The Brouwer fixed point theorem 497

　6.10　Solved problems for Chapter 6 501

**7   Elements of Ordinary and Partial
     Differential Equations   509**
     7.1   Introduction   509
     7.2   First order differential equations   511
           7.2.1   Linear first order ODE   511
           7.2.2   Equations with variables separated   517
           7.2.3   Homogeneous equations   519
           7.2.4   Exact equations   521
     7.3   Picard's theorem (*)   525
     7.4   Second order differential equations   532
           7.4.1   Linear second order ODE with constant
                   coefficients   534
           7.4.2   Special types of second order ODE; reduction
                   of order   545
     7.5   Higher order ODE and systems of ODE   548
     7.6   Some more advanced topics in ODE (*)   564
           7.6.1   The method of Frobenius; second order
                   equations with variable coefficients   564
           7.6.2   The Hermite equation   568
     7.7   Partial differential equations   573
     7.8   Second order PDE in two variables   576
           7.8.1   Classification and general solutions   576
           7.8.2   Boundary value problems for the wave equation .   583
           7.8.3   Boundary value problems for Laplace's equation   588
           7.8.4   Boundary value problems for the heat equation .   594
           7.8.5   A note on Fourier series   598
     7.9   The Fourier transform method (*)   603
     7.10 Solved problems for Chapter 7   612

8   An Introduction to the Calculus of Variations   619
     8.1   Simple variational problems   620
           8.1.1   Some classical problems   627
           8.1.2   Sufficient conditions   636
     8.2   Generalizations   638
           8.2.1   Geodesics on a Riemannian surface   646
           8.2.2   The principle of least action   653
     8.3   Variational problems with constraints   659

8.4   Multiple integral variational problems 666

     8.4.1   Variations of double integrals 667

     8.4.2   The case of n variables 673

8.5   Solved problems for Chapter 8 682

Appendix A   Countability and Decimal Expansions   **691**

Appendix B   Calculus in One Variable   **693**

B.1   Differential calculus 693

B.2   Integral calculus 694

     B.2.1   Complex-valued functions 695

B.3   Series 695

Appendix C   Uniform Convergence   **697**

C.1   The Stone-Weierstrass theorem 698

Appendix D   Linear Algebra   **701**

Bibliography   **705**

Index   **709**

编辑手记   **717**

This page intentionally left blank

Chapter 1

Basic Features of Euclidean Space, $\mathbb{R}^n$

This chapter is foundational. The reader who is familiar with this material can go on to Chapter 2. Those who are not can either go over the chapter before going on, or can begin with the latter chapters and return to the parts of Chapter 1 as needed. Here we review basic features of Euclidean space $\mathbb{R}^n$ and prove results concerning orthogonality, convergence, completeness and compactness.

1.1 Real numbers

In this section we briefly give a review of the basic properties of real numbers. There are a number of ways to deal with the real numbers, all more or less equivalent. The way we shall do it (and perhaps the most efficient way) is by means of the *least upper bound axiom* . The *integers* are denoted by $\mathbb{Z} = \{..., -2, -1, 0, 1, 2, ...\}$ and the *positive integres* by $\mathbb{Z}_+ = \{1, 2, 3, ...\}$. The set $\mathbb{Z}_+$ is also called the set of *natural numbers* and is sometimes denoted by $\mathbb{N}$. The *principle of mathematical induction* states that: If a subset $S \subseteq \mathbb{N}$ satisfies (i) $1 \in S$ (ii) $n \in S \Rightarrow (n+1) \in S$, then $S = \mathbb{N}$. The set of *rational numbers* is denote by $\mathbb{Q} = \{\frac{m}{n} : m, n \in \mathbb{Z}, n \neq 0\}$. It was known since the times of Pythagoras that $\sqrt{2} \notin \mathbb{Q}$. Such numbers can not be written as fractions and are called *irrational numbers*. As we shall see shortly, there are infinitely many irrational numbers. The set of rationals together with set of irrationals form the set of *real numbers* $\mathbb{R}$. That is, $\mathbb{R} = \mathbb{Q} \cup \mathbb{Q}^c$. We shall assume the standard arithmetic operations and their properties with respect to the real numbers. We shall also assume the ordering of real numbers ($<, \leq, >$ and $\geq$).

Definition 1.1.1. Let S be a nonempty subset of $\mathbb{R}$.

1. A real number M is called an *upper bound* of S if $x \leq M$ for all $x \in S$. In this case S is said to be *bounded above*.

2. A real number m is called a *lower bound* of S if $m \leq x$ for all $x \in S$. In this case S is said to be *bounded below*.

3. The set S is said to be *bounded* if it is both bounded above and bounded below. Thus S is bounded if there exist real numbers m and M such that $S \subseteq [m, M]$.

4. A set which is not bounded is called *unbounded*.

5. if $M \in S$ we call M the *maximal element* of S. If $m \in S$ we call m the *minimal element* of S.

Definition 1.1.2. Let S be a nonempty bounded subset of $\mathbb{R}$.

1. If S has a least upper bound, then we call it the *supremum* of S, denoted by $\sup(S)$ (or sometimes $\mathrm{lub}(S)$).

2. If S has a greatest lower bound, then we call it the *infimum* of S, denoted by $\inf(S)$ (or sometimes $\mathrm{glb}(S)$).[1]

   The Completeness Axiom. *Every nonempty subset S of $\mathbb{R}$ which is bounded above has a least upper bound. That is, there exists a real number,* $\sup(S)$.

   If S is a set of real numbers which is bouned below, then by considering the set $-S = \{-x : x \in S\}$ we may state the completeness axiom in the alternative form as:[2] *Every nonempty set of real numbers which is bounded below has an infimum in* $\mathbb{R}$.

---

[1] lub, means *least upper bound* and glb, means *greatest lower bound*.

[2] Changing sign reverses the order in $\mathbb{R}$.

Observe that if S is bounded above, then $M = \sup(S)$ is the unique number with the following two properties:
(i) $x \leq M$ for all $x \in S$, and (ii) for every $\epsilon > 0$, there exists $x \in S$ with $x > M - \epsilon$.
In other words, (i) tells us that M is an upper bound, whereas (ii) tells us that there is no smaller upper bound. Similarly, if S is bounded bellow, then $m = \inf(S)$ is the unique number such that
(i) $m \leq x$ for all $x \in S$ and (ii) for every $\epsilon > 0$, ther exist $x \in S$ with $x < m + \epsilon$.

Note that $\sup(-S) = -\inf(S)$ and $\inf(-S) = -\sup(S)$. If S has no upper bound we shall define $\sup(S) = +\infty$, and if S has no lower bound, we shall define $\inf(S) = -\infty$.

As the following example shows the $\sup(S)$ may or may not belong to S itself. However, we shall see in Section 1.5 (Corollary 1.5.8), it always belongs to the closure of S. Similarly for $\inf(S)$.

Example 1.1.3. 1. Let $S = \{\frac{1}{n} : n = 1, 2, ...\}$. Then the $\sup(S) = 1$ and is the maximal element of S. On the other hand the $\inf(S) = 0$ and $0 \notin S$.

2. Let $a, b \in \mathbb{R}$ with $a < b$. Then

$$\sup[a, b] = \sup[a, b) = \sup(a, b] = \sup(a, b) = b$$

and

$$\inf[a, b] = \inf[a, b) = \inf(a, b] = \inf(a, b) = a.$$

Theorem 1.1.4. *(Archimedean property). Let $a, b \in \mathbb{R}$ with $a > 0$ and $b > 0$. Then there exists a positive integer n such that $na > b$.*

Proof. Suppose the statement is false. That is, there exist $a > 0$ and $b > 0$ such that $na \leq b$ for all $n \in \mathbb{Z}_+$. Then, the set $S = \{na : n \in \mathbb{Z}_+\}$ is a nonempty subset of $\mathbb{R}$ bounded above by b. By the completeness axiom S has a supremum. Let $M = \sup(S)$. Hence, $na \leq M$ for all $n \in \mathbb{Z}_+$. Therefore $(n + 1)a \leq M$, or $na \leq M - a$ for all $n \in \mathbb{Z}_+$. Thus, $M - a$ is also an upper bound of S. But $M - a < M$. This contradicts that $M = \sup(S)$. This contradiction proves the theorem. $\square$

Corollary 1.1.5. *For any positive real number b there exists $n \in \mathbb{Z}_+$ such that $n > b$. In particular, for every $\epsilon > 0$ there exists $n \in \mathbb{Z}_+$ such that $\frac{1}{n} < \epsilon$.*

Proof. Taking $a = 1$ in Theorem 1.1.4, we get $n = n1 > b$. Setting $b = \frac{1}{\epsilon}$ yields $\frac{1}{n} < \epsilon$. $\qquad\square$

Corollary 1.1.6. *For any positive real number x there exists a (unique) $m \in \mathbb{Z}_+$ such that $m - 1 \le x < m$.*

Proof. By Corollary 1.1.5, there exists a positive integer n such that $x < n$. Let m be the least such positive integer (and so unique). Then $m - 1 \le x < m$. $\qquad\square$

Theorem 1.1.7. *(Denseness of $\mathbb{Q}$). Let $a, b \in \mathbb{R}$ with $a < b$. Then there exists a rational number $r \in \mathbb{Q}$ such that $a < r < b$.*

Proof. It is enough to prove the theorem for $0 < a < b$. Since $b - a > 0$, it follows from the Archimedean property that there exist $n \in \mathbb{Z}_+$ such that $n(b - a) > 1$. That is, $nb > na + 1$. By Corollary 1.1.6, there exists $m \in \mathbb{Z}_+$ such that $m - 1 \le na < m$. Now we have $na < m \le na + 1 < nb$. Hence, $na < m < nb$. Therefore, $a < \frac{m}{n} < b$. $\qquad\square$

The following proposition tells us that there are infinitely many irrationals. In fact, there are so many that, as we prove in the next theorem, they are densely distributed on the real line.

Proposition 1.1.8. *Let $n \in \mathbb{Z}_+$ with n not a perfect square. Then $\sqrt{n} \notin \mathbb{Q}$.*

Proof. Suppose that $\sqrt{n}$ were rational. Then it can be written as $\sqrt{n} = \frac{a}{b}$ with $a, b \in \mathbb{Z}_+$ and $a + b$ be *smallest possible*. Let k be the unique positive integer such that $k < \sqrt{n} < k + 1$. Then $n = \frac{a^2}{b^2}$ implies that $a^2 - kab = nb^2 - kab$ or $\frac{a}{b} = \frac{nb - ka}{a - kb}$. Since $nb - ka < a$ and $a - kb < b$. It follows that $(nb - ka) + (a - kb) < a + b$ which contadicts our hypothesis and proves that $\sqrt{n} \notin \mathbb{Q}$. $\qquad\square$

Some other examples of irrational numbers are e, π and e^π. In Chapter 3 (Proposition 3.6.13), we shall prove the irrationality of e. In general it is not too easy to establish the irrationality of certain particular numbers, for example there is no simple proof of the irrationality of π or e^π.

Theorem 1.1.9. *(Denseness of $\mathbb{Q}^c$). Let $a, b \in \mathbb{R}$ with $a < b$. Then there exists an irrational number $\tau \in \mathbb{Q}^c$ such that $a < \tau < b$.*

Proof. Since $b - a > 0$ and $\sqrt{2} > 0$, by the Archimedean property, there exist $n \in \mathbb{Z}_+$ such that $n(b - a) > \sqrt{2}$. Therefore,

$$b > a + \frac{\sqrt{2}}{n} > a + \frac{\sqrt{2}}{2n} > a.$$

Since $\left(a + \frac{\sqrt{2}}{n}\right) - \left(a + \frac{\sqrt{2}}{2n}\right) = \frac{\sqrt{2}}{n} \notin \mathbb{Q}$, it follows that at least one of the numbers $\tau_1 = a + \frac{\sqrt{2}}{n}$ or $\tau_2 = a + \frac{\sqrt{2}}{2n}$ is irrational.   □

Definition 1.1.10. The *absolute value* of a real number x is defined by

$$|x| = \begin{cases} x & \text{when } x \geq 0, \\ -x & \text{when } x < 0. \end{cases}$$

Note that $|x|^2 = x^2$ and $|x| = \sqrt{x^2}$.

For $x, y \in \mathbb{R}$ the absolute value has the following properties.

1. $|x| \geq 0$ and $|x| = 0$ if and only if $x = 0$.

2. $|xy| = |x| \cdot |y|$

3. $|x + y| \leq |x| + |y|$ *(triangle inequality)*.

Moreover,

$$\big|\, |x| - |y| \,\big| \leq |x - y|.$$

We remark that $|x + y| = |x| + |y|$ if and only if $xy \geq 0$, and $|x + y| < |x| + |y|$ if and only if $xy < 0$. The reader is invited to verify these properties in Exercise 12.

Definition 1.1.11. Let $S \subseteq \mathbb{R}$ and $a \in \mathbb{R}$ (not necessarily in S). The point a is called an *accumulation point or limit point* of S if every open interval around a contains at least one point $x \in S$ distinct from a, that is, for every $r > 0$ we have

$$((a - r, a + r) - \{a\}) \cap S \neq \emptyset.$$

It is worthwhile remarking

1. A finite set $S = \{s_1, ..., s_m\}$ has no limit points. To see this take $r = \min\{|s_i - s_j| : i, j = 1, .., m\}$, then $((s_i - r, s_i + r) - \{s_i\}) \cap S = \emptyset$.

2. If a is a limit point of S, then every open interval $(a - r, a + r)$ around a contains infinitely many points of S; for if $(a - r, a + r)$ contained a finite number of points $s_1, ..., s_k$ of S distinct from a, then the open interval around a with radius $r_0 = \min\{|a - s_j| : j = 1, ..., k\}$ contains no point of S distinct from a, which is impossible.

The following important result characterizes the sets in $\mathbb{R}$ which have limit points. In the next section we shall extend it to $\mathbb{R}^n$.

Theorem 1.1.12. *(Bolzano-Weierstrass)*[3] *Every bounded infinite set in $\mathbb{R}$ has at least one limit point.*

Proof. Let S be a bounded infinite set in $\mathbb{R}$. Since S is bounded there is a closed interval $[a, b]$ containing S. Hence, both $m = \inf(S)$ and $M = \sup(S)$ exist. Consider the set

$$A = \{x \in \mathbb{R} : (-\infty, x) \cap S \text{ is at most finite}\}.$$

Since $m \in A$ it follows that $A \neq \emptyset$. At the same time A is bounded above by M. In fact, if $x_0 \in \mathbb{R}$ with $M < x_0$, then $x < M < x_0$ for all $x \in S$. Since S is infinite, this implies that $x_0 \notin A$. Hence, $x \leq M$ for all $x \in A$. It follows that A has a least upper bound, call it $y = \sup(A)$. We show that y is a limit point of S. Let $\epsilon > 0$ be any positive number. Then $(y + \epsilon) \notin A$ and so $(-\infty, y + \epsilon) \cap S$ is infinite. Since $y = \sup(A)$ there exists $x_1 \in A$ such that $y - \epsilon < x_1 \leq y$. Therefore $(-\infty, y - \epsilon) \cap S \subseteq (-\infty, x_1) \cap S$ and so $(-\infty, y - \epsilon) \cap S$ is at most finite. Hence, $(y - \epsilon, y + \epsilon)$ contains an infinite number of points of S. Thus y is a limit point of S. $\qquad \square$

---

[3]B. Bolzano (1781-1848). Mathematician, logician and philosopher. K. Weierstrass (1815-1897). Mathematician who made many contribution in mathematical analysis and devoted a great deal of attention to the rigorous foundation of analysis.

EXERCISES

1. Show that if r is a nonzero rational number and x is irrational, then $x \pm r$, $r - x$, rx, $\frac{r}{x}$ and $\frac{x}{r}$ are all irrational.

2. Show the least upper bound of a bounded set is unique.

3. If $\emptyset \neq A \subseteq B \subseteq \mathbb{R}$ and B is bounded, show that

$$\inf(B) \leq \inf(A) \leq \sup(A) \leq \sup(B).$$

4. Let A and B be two nonempty bounded sets of real numbers and $c \in \mathbb{R}$. If $A + B = \{a + b : a \in A, b \in B\}$ and $cA = \{ca : a \in A\}$. Show that

   (a) $\sup(A + B) = \sup(A) + \sup(B)$ and $\inf(A + B) = \inf(A) + \inf(B)$.

   (b) When $c \geq 0$, then $\sup(cA) = c\sup(A)$ and $\inf(cA) = c\inf(A)$.
   When $c < 0$, then $\sup(cA) = c\inf(A)$ and $\inf(cA) = c\sup(A)$.

5. Let $S = \left\{x + \frac{1}{x} : x > 0\right\}$. Show that $\sup(S) = \infty$ and $\inf(S) = 2$.

6. Let $S = \left\{\frac{m}{m+n} : m, n \in \mathbb{N}\right\}$. Show that $\sup(S) = 1$ and $\inf(S) = 0$.
   Hint. Fix $m = 1$ (resp. $n = 1$).

7. Let $S = \left\{x \in \mathbb{Q}_+ : x^2 < 2\right\}$. Show that $\sup(S) \notin \mathbb{Q}$. Thus, $\mathbb{Q}$ is not (order) complete.

8. Let $x \in \mathbb{R}$ and $S = \{r \in \mathbb{Q} : r < x\}$. Show that $\sup(S) = x$

9. Let $a, b \in \mathbb{R}$. Show that $|ab| \leq \frac{1}{2}(a^2 + b^2)$. Deduce that if $a \geq 0$ and $b \geq 0$, then $\sqrt{ab} \leq \frac{1}{2}(a + b)$.

10. Show that if $0 \leq a \leq b \leq c$ and $b > 0$, then $2\sqrt{\frac{a}{c}} \leq \frac{a}{b} + \frac{b}{c} \leq 1 + \frac{a}{c}$.

11. (*Bernoulli's inequality*). For any $x \in \mathbb{R}$ and $k \in \mathbb{Z}_+$, show that

$$(1 + x)^k \geq 1 + kx.$$

12. Let $x, y \in \mathbb{R}$. Show that

    (a) $|xy| = |x| \cdot |y|$.

    (b) $|x \pm y| \le |x| + |y|$.

    (c) $||x| - |y|| \le |x - y|$.

13. Let $x, y \in \mathbb{R}$. Show that

$$\max\{x, y\} = \frac{x+y+|y-x|}{2}, \quad \min\{x, y\} = \frac{x+y-|y-x|}{2}.$$

14. Let $x, y \in \mathbb{R}$ with $x \ne \pm y$. Show that $\frac{|x|}{|x+y|} + \frac{|y|}{|x-y|} \ge 1$.

15. Let $x, y \in \mathbb{R}$. Show that

$$\frac{|x+y|}{1+|x+y|} \le \frac{|x|}{1+|x|} + \frac{|y|}{1+|y|}.$$

1.1.1   Convergence of sequences of real numbers

If for every $k \in \mathbb{Z}_+$ there is an associated real number a_k, the ordered list $\{a_1, a_2, ...\}$, briefly written as $\{a_k\}_{k=1}^\infty$, is called a *sequence*. Given a real sequence $\{a_k\}_{k=1}^\infty$, the major question that one might ask about the sequence is how its terms behave for k arbitrarily large, i.e, as $k \to \infty$.

Definition 1.1.13. Let $\{a_k\}_{k=1}^\infty$ be a sequences of real numbers. The sequence is said to *converge to a* $\in \mathbb{R}$, if for every $\epsilon > 0$ there exists a positive integer N(depending on ϵ) such that $|a_k - a| < \epsilon$ whenever $k \ge N$. We write $\lim_{k\to\infty} a_k = a$ or $a_k \to a$ as $k \to \infty$.

Equivalently, $\lim_{k\to\infty} a_k = a$ if for every $\epsilon > 0$ there exists a positive integer N such that $a_k \in (a - \epsilon, a + \epsilon)$ for all $k \ge N$. If a sequence does not converge (has *no limit*), we say that the sequence *diverges*.

Proposition 1.1.14. *If a sequence converges, then its limit is unique.*

Proof. Suppose a convergent sequence $\{a_k\}$ had two limits, say $a_k \to a$ and $a_k \to b$ with $a \ne b$. Then for $\epsilon = \frac{|a-b|}{2} > 0$ there exist positive integers N_1 and N_2 such that $|a_k - a| < \frac{\epsilon}{2}$ for all $k \ge N_1$ and $|a_k - b| < \frac{\epsilon}{2}$ for all $k \ge N_2$. But, then for all $k \ge N_3 = \max\{N_1, N_2\}$ we have $|a - b| = |a - a_k + a_k - b| \le |a_k - a| + |a_k - b| < \frac{\epsilon}{2} + \frac{\epsilon}{2} = \epsilon = \frac{|a-b|}{2}$, that is, $1 < \frac{1}{2}$ which is impossible. $\square$

Example 1.1.15.   1. Let $a_k = \frac{1}{k^p}$ with $p > 0$. Then $\lim_{k \to \infty} \frac{1}{k^p} = 0$.

2. Let $x \in \mathbb{R}$ with $|x| < 1$. Then $\lim_{k \to \infty} x^k = 0$.

Solution.

1. Let $\epsilon > 0$ be any positive number. Then $|\frac{1}{k^p} - 0| = \frac{1}{k^p} < \epsilon$ if and
   only if $k^p > \frac{1}{\epsilon}$ or $k > (\frac{1}{\epsilon})^{\frac{1}{p}}$. Hence, taking $N > (\frac{1}{\epsilon})^{\frac{1}{p}}$, we have
   $|\frac{1}{k^p} - 0| < \epsilon$ for all $k \geq N$.

2. If $x = 0$ this is trivial. Assume $x \neq 0$ and let $\epsilon > 0$. Then

$$|x^k| = |x|^k < \epsilon \quad \text{if and only if} \quad k \log |x| < \log(\epsilon) \quad \text{or} \quad k > \frac{\log(\epsilon)}{\log |x|}.$$

   If $0 < \epsilon < 1$, take $N > \frac{\log(\epsilon)}{\log |x|}$. Then $|x^k| < \epsilon$ for all $k \geq N$.

Proposition 1.1.16. *Let $\{a_k\}_{k=1}^{\infty}$ be a sequence. If $a_k \to a$, then $|a_k| \to |a|$. In particular, $a_k \to 0$ if and only if $|a_k| \to 0$.*

Proof. The first assertion follows from the inequality $||a_k| - |a|| \leq |a_k - a|$. In the particular case where $a = 0$, the converse is also true, for if $|a_k| \to |0| = 0$, then for any $\epsilon > 0$, there is N such that $||a_k| - 0| < \epsilon$ for all $k \geq N$, that is, $|a_k| < \epsilon$ for all $k \geq N$, or $a_k \to 0$.   □

Definition 1.1.17. Let $\{a_k\}_{k=1}^{\infty}$ be a sequence of real numbers.

1. The sequence $\{a_k\}_{k=1}^{\infty}$ is called *increasing* (resp. *decreasing*) if $a_k \leq a_{k+1}$ (resp. $a_k \geq a_{k+1}$) for all $k = 1, 2, ...$. When these inequalities are strict, we say the sequence $\{a_k\}_{k=1}^{\infty}$ is *strictly increasing* (resp. *strictly decreasing*). A sequence that is either (strictly) increasing or (strictly) decreasing is called (strictly) *monotone*.

2. The sequence $\{a_k\}_{k=1}^{\infty}$ is called *bounded* if there exists $M > 0$ with $|a_k| \leq M$ for all $k = 1, 2, ...$. A sequence which is not bounded is called *unbounded*.

Proposition 1.1.18. *Every convergent sequence is bounded.*

Proof. Suppose $a_k \to a$. Then for $\epsilon = 1$ there exists positive integer N such that $|a_k - x| < 1$ for all $k \geq N$. So, for all $k \geq N$ we have

$$|a_k| = |(a_k - a) + a| \leq |a_k - a| + |a| < 1 + |a|.$$

Taking $M = \max\{|a_1|, ..., |a_N|, 1 + |a|\}$, we have $|a_k| \leq M$ for all $k = 1, 2, ...$; That is, $\{a_k\}_{k=1}^{\infty}$ is bounded. $\square$

Corollary 1.1.19. *An unbounded sequence diverges.*

The following result is fundamental.

Theorem 1.1.20. *Every bounded monotone sequence in $\mathbb{R}$ is convergent. More precisely, the limit of an increasing (resp. decreasing) sequence is the supremum (resp., infimum) of its set of values.*

Proof. Suppose $\{a_k\}_{k=1}^{\infty}$ is increasing and bounded. Then the set $\{a_k : k = 1, 2, ...\}$ (the range) of the sequence is bounded above. It follows from the completeness axiom that this set has a supremum, call it L. Since L is an upper bound, we have $a_k \leq L$ for all k. Moreover, since L is the *least* upper bound, for any $\epsilon > 0$ there is some N for which $a_N > L - \epsilon$. Since $\{a_k\}_{k=1}^{\infty}$ is increasing, for all $k \geq N$ we also have $a_k \geq a_N > L - \epsilon$. Therefore, $L - \epsilon < a_k \leq L < L + \epsilon$ for all $k \geq N$ and this shows that $a_k \to L$.

Similarly, a decreasing sequence bounded from below converges to the infimum of the set of its values. $\square$

Example 1.1.21. Let $a_k = \frac{x^k}{k!}$ for $x \in \mathbb{R}$, where $k! = 1 \cdot 2 \cdots (k-1) \cdot k$. Then $a_k \to 0$ as $k \to \infty$ (that is, $k!$ grows faster than exponentially as $k \to \infty$).

Solution.

To see this we consider three cases; If $|x| \leq 1$, then

$$0 \leq \frac{|x|^k}{k!} \leq \frac{1}{k!} \leq \frac{1}{k} \to 0.$$

For $x > 1$, we will argue using Theorem 1.1.20. Note that $\frac{a_{k+1}}{a_k} = \frac{x}{k+1} \to 0$ as $k \to \infty$. Therefore, for $\epsilon = 1$, we can find N such that $a_{k+1} < a_k$ for all $k \geq N$. Hence, $\{a_k\}$ is (eventually) decreasing. Moreover, since $0 < a_k < \max\{a_1, ..., a_N\}$, the sequence is also bounded.

It follows, from Theorem 1.1.20, that $\{a_k\}$ converges to a number, say $a \geq 0$. Now, $a = \lim_{k\to\infty} a_{k+1} = \lim_{k\to\infty}(a_k \cdot \frac{x}{k+1}) = a \cdot 0 = 0$. Thus, $\lim_{k\to\infty} \frac{x^k}{k!} = 0$ for all $x \geq 0$. Finally, if $x < 0$, then by the above argument $|a_k| = \frac{|x|^k}{k!} \to 0$ as $k \to \infty$. But then also $a_k \to 0$ as $k \to \infty$.

The following consequence of Theorem 1.1.20 is also an important property of $\mathbb{R}$.

Theorem 1.1.22. *(Nested intervals property). Let $I_n = [a_n, b_n]$ be a sequence of closed bounded intervals in $\mathbb{R}$ such that $I_{n+1} \subseteq I_n$ for all $n = 1, 2, \ldots$ and the length $b_n - a_n$ of I_n tends to 0 as $n \to \infty$. Then there is exactly one point contained in all of the intervals I_n. That is, there exists $\xi \in \mathbb{R}$ such that $\bigcap_{n=1}^{\infty} I_n = \{\xi\}$.*

Proof. The condition $I_1 \supseteq I_2 \supseteq I_3 \supseteq \ldots$ means that $a_1 \leq a_2 \leq a_3 \leq \ldots$ and $b_1 \geq b_2 \geq b_3 \geq \ldots$. So the sequence $\{a_n\}$ is increasing and the sequence $\{b_n\}$ is decreasing. Since $a_1 < a_n \leq b_n < b_1$, both are bounded. Therefore, by Theorem 1.1.20, both converge, say $a_n \to \xi$ and $b_n \to \eta$. Moreover, since $b_n - a_n \to 0$, it follows that $\xi = \eta$. Obviously ξ is the supremum of $\{a_n : n = 1, 2, \ldots\}$ and the infimum of $\{b_n : n = 1, 2, \ldots\}$ and so $a_n \leq \xi \leq b_n$ for all n. Thus, ξ belongs to all the intervals. Suppose, that $\xi_1 \neq \xi$ were another point common to all the intervals I_n. Suppose, for example, $a_n \leq \xi_1 < \xi \leq b_n$ for all n. Then $b_n - a_n \geq \xi - \xi_1 \neq 0$. Letting $n \to \infty$, this gives $\xi_1 \geq \xi$ which contradicts the fact $\xi_1 < \xi$. Similarly, if $a_n \leq \xi < \xi_1 \leq b_n$ we also get a contradiction. $\qquad\square$

Definition 1.1.23. A number $\xi \in \mathbb{R}$ is called a *limit point* (or *accumulation point*) of a sequence $\{a_k\}_{k=1}^{\infty}$, if every open interval around ξ contains an infinite number of terms of the sequence.

We remark that a limit point ξ of the range $\{a_k : k = 1, 2, \ldots\}$ of the sequence is also a limit point of the sequence. The converse is not true; for instance, the points 1 and -1 are limit points of the sequence $a_k = (-1)^k$. However, the range of the sequence is $\{-1, 1\}$ and has no limit point.

Theorem 1.1.24. *(Bolzano-Weierstrass for sequences) Every bounded sequence in $\mathbb{R}$ has a limit point.*

Proof. Suppose $\{a_k\}_{k=1}^{\infty}$ is bounded. Then $S = \{a_k : k = 1, 2, ...\}$ is a bounded set of real numbers. There are two possiblities. Either S is finite or S is infinite. If S is finite, then there is some $\xi \in \mathbb{R}$ for which $a_k = \xi$ for an infinite number of values of k. Hence ξ is a limit point of $\{a_k\}_{k=1}^{\infty}$. On the other hand, if S is infinite, the Bolzano-Weierstrass theorem, implies that S has a limit point, say ξ. Therefore, for every $\epsilon > 0$ we have $a_k \in (\xi - \epsilon, \xi + \epsilon)$ for infinitely many k and thus ξ is a limit point of $\{a_k\}_{k=1}^{\infty}$. □

Definition 1.1.25. Let $\{a_k\}_{k=1}^{\infty}$ be a sequence and $\{m_k\}_{k=1}^{\infty}$ be an increasing sequence of positive integers, that is, $m_1 < m_2 < ... < m_k <$ .... The sequence $\{a_{m_k}\}_{k=1}^{\infty}$ is called a *subsequence* of $\{a_k\}_{k=1}^{\infty}$.

Exercise 1.1.26. Let $\{a_k\}_{k=1}^{\infty}$ be a sequence. Consider $\{a_p\}$, where p is a prime number. Is this a subsequence?

Proposition 1.1.27. *If a sequence converges to a, then every subsequence converges to a.*

Proof. Suppose $\lim_{k \to \infty} a_k = a$. Given any $\epsilon > 0$ there exists N such that $|a_k - a| < \epsilon$ for all $k \geq N$. Since $m_1 < m_2 < ... < m_k < ...$, it follows by induction on k, that $m_k \geq k$. Hence, for all $m_k \geq k \geq N$, we have $|a_{m_k} - a| < \epsilon$. Thus, $a_{m_k} \to a$. □

The converse is not true; For the sequence $a_k = (-1)^k$, the subsequences $a_{2k} = 1$ and $a_{2k-1} = -1$ converge to 1 and -1 respectively, but the sequence itself does not converge. However we ask the reader to prove in Exercise 2 below, that if $a_{2k} \to a$ and $a_{2k-1} \to a$, then $a_k \to a$.

The Bolzano-Weiertsrass Theorem has the following important corollary.

Corollary 1.1.28. *Every bounded sequence in $\mathbb{R}$ has a convergent subsequence.*

Proof. Suppose $\{a_k\}_{k=1}^{\infty}$ is bounded. By the Bolzano-Weierstrass theorem, $\{a_k\}_{k=1}^{\infty}$ has a limit point, say ξ. Then, given any $\epsilon > 0$ the open interval $(\xi - \epsilon, \xi + \epsilon)$ contains infinitely many terms of the sequence. Let $a_{m_1}, a_{m_2}, ..., a_{m_k}, ...$ be the terms of the seuqence in $(\xi - \epsilon, \xi + \epsilon)$ that correspond to the indices $m_1 < m_2 < ... < m_k < ...$. Then $\{a_{m_k}\}_{k=1}^{\infty}$ is a subsequence of $\{a_k\}_{k=1}^{\infty}$ and $a_{m_k} \to \xi$. □

Definition 1.1.29. A sequence $\{a_k\}_{k=1}^{\infty}$ is called a *Cauchy*[4] *sequence* if $|a_k - a_j| \to 0$ as $k, j \to \infty$, that is, for every $\epsilon > 0$ there exists positive integer N such that $|a_k - a_j| < \epsilon$ for all $k, j \geq N$.

Note that every Cauchy sequence is bounded. In fact, for $\epsilon = 1$ there exists N such that for $k, j \geq N$ we have $|a_k - a_j| < 1$. Hence, for $k > N$ we have $|a_k| = |a_k - a_{N+1} + a_{N+1}| \leq |a_k - a_{N+1}| + |a_{N+1}| < 1 + |a_{N+1}|$. Taking as before $M = \max\{|a_1|, ..., |a_N|, 1 + |a_{N+1}|\}$, we have $|a_k| \leq M$ for all $k = 1, 2, ...$.

Proposition 1.1.30. *1. Every convergent sequence is a Cauchy sequence.*

2. A Cauchy sequence converges, if it has a convergent subsequence.

Proof. 1. Suppose $a_k \to a$. Then $|a_k - a_j| \leq |a_k - a| + |a_j - a| \to 0$ when $k, j \to \infty$. Thus, $\{a_k\}_{k=1}^{\infty}$ is Cauchy.

2. Let $\epsilon > 0$ be given. Since $\{a_k\}$ is Cauchy there exists N such that $|a_k - a_j| < \frac{\epsilon}{2}$ for all $k, j \geq N$. On the other hand suppose that the convergent subsequence $\{a_{m_k}\}$ converges to a. Then we can find a positive integer k large enough so that $m_k > N$ and $|a_{m_k} - a| < \frac{\epsilon}{2}$. So, for $k > N$ we have $|a_k - a| \leq |a_k - a_{m_k}| + |a_{m_k} - a| < \frac{\epsilon}{2} + \frac{\epsilon}{2} = \epsilon$. Thus, $a_k \to a$.

$\square$

The following theorem is also fundamental. This property of $\mathbb{R}$ is also refered to as the *completeness of* $\mathbb{R}$.

Theorem 1.1.31. *Every Cauchy sequence in $\mathbb{R}$ converges.*

Proof. Suppose $\{a_k\}_{k=1}^{\infty}$ is Cauchy. Then it is bounded. By Corollary 1.1.28 $\{a_k\}_{k=1}^{\infty}$ has a convergent subsequence. Therefore from Proposition 1.1.30 (2) $\{a_k\}_{k=1}^{\infty}$ converges. $\square$

This concludes our preliminary work on $\mathbb{R}$. In the next sections we will extend these results to $\mathbb{R}^n$.

---

[4] A. Cauchy (1789-1857). Mathematician who made many important contributions in real and complex analysis.

EXERCISES

1. Let $\{a_k\}$ be a sequence. Show that if $a_k \to a$, then $a_{k+m} \to a$ for any $m \in \mathbb{Z}_+$.

2. Let $\{a_k\}$ be a sequence. Show that

$$\text{if } a_{2k} \to a \text{ and } a_{2k-1} \to a, \text{ then } a_k \to a.$$

3. Let $c_k \le a_k \le b_k$ for all $k > N$. Show that, if $c_k \to a$ and $b_k \to a$, then $a_k \to a$. (This is known as the *squeezing theorem*).

4. Let $\{a_k\}$ be a sequence. Show that

$$\text{if } \lim_{k \to \infty} \frac{a_{k+1}}{a_k} = l \text{ with } -1 < l < 1, \text{ then } \lim_{k \to \infty} a_k = 0.$$

5. Let $a_k = \sqrt[k]{k}$. Show that $\lim_{k \to \infty} a_k = 1$.

6. Let $a_k = \frac{k!}{k^k}$. Show that $\lim_{k \to \infty} a_k = 0$.

7. Let $0 < \alpha < 1$ and $a_k = (\alpha + \frac{1}{k})^k$. Show that $\lim_{k \to \infty} a_k = 0$.

8. Let $a_k = (1 + \frac{1}{k})^k$. Show that $\lim_{k \to \infty} a_k = e$ (this is the definition of the number $e \approx 2.718$ called *Euler's number*). *Hint.* Apply Theorem 1.1.20 to the sequence $b_k = (1 + \frac{1}{k})^{k+1}$ and note that $\lim_{k \to \infty} b_k = \lim_{k \to \infty} a_k$.

9. Show that $\lim_{k \to \infty} (1 - \frac{1}{k})^k = \frac{1}{e}$.

10. Show that $\lim_{k \to \infty} (1 \pm \frac{1}{k^2})^k = 1$.

11. Show that $\lim_{k \to \infty} (1 + \frac{1}{2k})^k = \sqrt{e}$.

12. Show that $\lim_{k \to \infty} (1 + \frac{2}{k})^k = e^2$.

13. Let $a_k = 1 + \frac{k}{k+1} \cos(\frac{k\pi}{2})$. Show that $\{a_k\}$ is bounded. However, the sequence diverges. *Hint.* Use subsequences.

14. Let $x \in \mathbb{R}$ with $|x| < 1$. Show that

$$\lim_{k \to \infty} (1 + x + x^2 + \dots + x^k) = \frac{1}{x - 1}.$$

Hint. See Example 1.1.15 (2).

15. Show that

$$\lim_{k \to \infty} \left[\frac{1}{\sqrt{k^2 + 1}} + \frac{1}{\sqrt{k^2 + 2}} + \ldots + \frac{1}{\sqrt{k^2 + k}} \right] = 1.$$

Hint. Use the squeezing theorem.

1.2 $\mathbb{R}^n$ as a vector space

For $n > 1$, $\mathbb{R}^n$ will denote the set of all n-tuples (pairs, triples, quadruples,...) of real numbers, that is,

$$\mathbb{R}^n = \{x = (x_1, x_2, \ldots, x_n) : x_i \in \mathbb{R}, \ i = 1, 2, \ldots, n\}.$$

Each element $x \in \mathbb{R}^n$ is called a *vector* and the numbers $x_1, x_2, \ldots, x_n$ are called the first, second,..., n^{th} component of x respectively. Any real number will be called a *scalar*. In the important case (from a geometrical point of view) when $n = 2$ or $n = 3$, we shall use the familiar notation

$$\mathbb{R}^2 = \{(x, y) : x, y \in \mathbb{R}\} \text{ for the } xy\text{-}plane \text{ and}$$
$$\mathbb{R}^3 = \{(x, y, z) : x, y, z \in \mathbb{R}\} \text{ for the } xyz\text{-}space.$$

Just as the real numbers are represented by points on the number line, similarly any $x = (x_1, \ldots, x_n) \in \mathbb{R}^n$ represents a point (or vector) in $\mathbb{R}^n$. For each $i = 1, \ldots, n$ the real numbers x_i are called the *Cartesian*[5] *coordinates* of x. The point $(0, \ldots, 0)$ corresponds to origin in $\mathbb{R}^n$. If we agree that all vectors have their initial point at the origin, then a point in $\mathbb{R}^n$ determines a vector in $\mathbb{R}^n$ and vice versa. Thus we can identify (and we do so) a point in $\mathbb{R}^n$ with the vector having initial point the origin and terminal point the given point in $\mathbb{R}^n$.

The set $\mathbb{R}^n$ has both an algebraic (vector space) structure and a metric space structure. To give $\mathbb{R}^n$ its algebraic structure, we define on $\mathbb{R}^n$ two operations, addition and scalar multiplication that make it a vector space over the field of real numbers (or a real vector space). Formally, let $x, y \in \mathbb{R}^n$ with $x = (x_1, \ldots, x_n)$ and $y = (y_1, \ldots, y_n)$. The *sum* of x and y is defined by

---

[5]R. Descartes (1596-1650). Philosopher, mathematician and physicist who made fundamental contributions to scientific thought and knowledge.

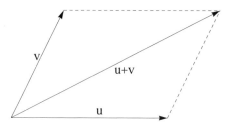

Figure 1.1: Sum of vectors

$$x + y = (x_1 + y_1, ..., x_n + y_n),$$

that is, by adding corresponding components. Since addition in $\mathbb{R}$ is associative and commutative with identity element the number 0, it is evident that addition in $\mathbb{R}^n$ is also associative and commutative and the *zero vector* $0 = (0, ..., 0)$ is the identity element for vector addition. Figure 1.1 in the plane shows that this addition is exactly the parallelogram law for adding vectors in Physics. Note that even in $\mathbb{R}^n$ the space determined by x and y is a plane, so this picture is valid in general.

The second operation is multiplication of a vector $x \in \mathbb{R}^n$ by a scalar $c \in \mathbb{R}$ called *scalar multiplication* and is defined as follows;

$$cx = (cx_1, ...cx_n).$$

This gives a vector pointing in the same direction as x, but of a different magnitude. Here as is easily checked, for $x, y \in \mathbb{R}^n$ and $c, d \in \mathbb{R}$ we have $c(x + y) = cx + cy$, $(c + d)x = cx + dx$, $c(dx) = d(cx) = (cd)x$ and $1x = x$. When all these properties are satisfied we say $\mathbb{R}^n$ is a real vector space. In general, if $\mathbb{R}^n$ above is replaced by any set V on which an addition and a scalar multiplication can be defined so that all the above properties hold V is called a *real vector space*.

Definition 1.2.1. Let V be a real vector space. Given $v_1, v_2, ..., v_k$ in V and scalars $c_1, c_2, ...c_k \in \mathbb{R}$ the sum $\sum_{j=1}^{k} c_j v_j = c_1 v_1 + c_2 v_2 + ... + c_k v_k$ is called a *linear combination* of the vectors $v_1, v_2, ..., v_k$. Moreover, the set of vectors which are linear combinations of $v_1, v_2, ..., v_k$ is called the *linear span* of $\{v_1, v_2, ..., v_k\}$ and we denote it by span $\{v_1, v_2, ..., v_k\}$.

Definition 1.2.2. A set of nonzero vectors $\{v_1, v_2, ..., v_m\}$ in V is said to be *linearly independent* if the relation $\sum_{j=1}^{m} c_j v_j = 0$ implies *all* $c_j = 0$. In general a set S of vectors in V is said to be linearly independent if each finite subset of S is linearly independent.

Clearly, any subset of a linearly independent set is linearly independent. A set of vectors $\{w_1, w_2, ..., w_m\}$ which is not linearly independent is called *linearly dependent*, that is, $\{w_1, w_2, ..., w_m\}$ is linearly dependent if there exists scalars $c_1, ..., c_m$ with at least one $c_j \neq 0$ with $\sum_{j=1}^{m} c_j w_j = 0$.

Definition 1.2.3. A set of vectors B in a vector space V is called a *basis* for V if B is linearly independent and span$(B) = V$. The space V is said to be finite dimensional if it has a finite basis. The number of vectors in the basis is called the *dimension* of V and is denoted by $\dim(V)$.

We remark that any finite dimensional vector space has a finite basis and any two bases have the same number of elements. This number is the dimension of the vector space. Since we won't need this level of detail we leave this fact as an exercise (Exercise 1.8.20).[6]

Example 1.2.4. In $\mathbb{R}^n$ the set of vectors

$$B = \{e_1 = (1, 0, 0, ..., 0), e_2 = (0, 1, 0, ..., 0), ..., e_n = (0, 0, 0, ..., 0, 1)\}$$

is a basis for $\mathbb{R}^n$, called the *standard basis* of $\mathbb{R}^n$. Indeed, suppose $c_1 e_1 + c_2 e_2 + ... + c_n e_n = 0$. Then $(c_1, 0, 0, ..., 0) + (0, c_2, 0, ..., 0) + ... + (0, 0, ..., 0, c_n) = (0, 0, ..., 0)$, that is, $(c_1, c_2, ..., c_n) = (0, 0, ..., 0)$. Hence $c_1 = c_2 = ... = c_n = 0$ and so B is linearly independent. At the same time for any $x = (x_1, x_2, ..., x_n) \in \mathbb{R}^n$ we have $x = x_1 e_1 + x_2 e_2 + ... + x_n e_n$. That is, span$(B) = \mathbb{R}^n$. Thus, B is a basis and $\dim(\mathbb{R}^n) = n$.

Proposition 1.2.5. *A set of vectors* $B = \{v_1, ..., v_n\}$ *is a basis of V if and only if every $v \in V$ can be expressed as a unique linear combination of the v_j $(j = 1, ...n)$.*

---

[6] For a detailed account of vector spaces, see [14].

Proof. Suppose $\{v_1, ..., v_n\}$ is a basis of V. Then any $v \in V$ is a linear combination of the v_j. If this can be done in more than one way, say $\sum_{j=1}^{n} c_j v_j = v = \sum_{j=1}^{n} d_j v_j$. Then $\sum_{j=1}^{n}(c_j - d_j)v_j = 0$. Since the v_j are linearly independent, it follows that $c_j - d_j = 0$ or $c_j = d_j$. Conversely, suppose that every $v \in V$ can be expressed as a unique linear combination of the v_j. This is in particular true for the zero vector. Hence the v_j are linearly independent and so form a basis. $\square$

The unique scalars $c_1, c_2, ..., c_n \in \mathbb{R}$ used to express $v \in V$ as a linear combination of the basis elements $B = \{v_1, ..., v_n\}$ are called the *coordinates* of v relative to the (ordered) basis B and the vector $[v] = (c_1, c_2, ..., c_n) \in \mathbb{R}^n$ is called the *coordinate vector* of $v \in V$. Note that when $V = \mathbb{R}^n$, the coordinates of $v \in V$ relative to the standard basis are just the components of v, that is $v = [v]$.

Definition 1.2.6. Let V be a vector space. A *linear subspace* of V is a subset W of V which is itself a vector space with the operations of addition and scalar multiplication on V.

Equivalently, a non-empty subset W of V is a linear subspace of V if and only if for each $u, v \in W$ and each scalar $c \in \mathbb{R}$ the vector $u + cv \in W$. That is, W is closed under the operations of addition and scalar multiplication.

In any vector space V the subset $\{0\}$ is clearly a linear subspace called the *trivial subspace* of V. The largest linear subspace of V is V itself and the smallest is $\{0\}$. Note that linear subspaces are closed under taking linear combinations. In a vector space V the span(S), of any subset $S \subseteq V$ is a linear subspace called the *subspace generated by* S.

From Example 1.2.4, the dimension of $\mathbb{R}^n$ is n. The dimension of the trivial subspace is taken to be 0 and the dimension of any other proper linear subspace of $\mathbb{R}^n$ is an integer strictly between 0 and n. In particular, in $\mathbb{R}^2$ and $\mathbb{R}^3$ the one-dimensional subspaces are lines passing from the origin. In $\mathbb{R}^3$ the two-dimensional subspaces are planes passing from the origin. In fact, any two linearly independent vectors $u, v \in \mathbb{R}^3$ generate the plane $\{su + tv : s, t \in \mathbb{R}\}$.

1.3 $\mathbb{R}^n$ as an inner product space

1.3.1 The inner product and norm in $\mathbb{R}^n$

Definition 1.3.1. Let V be a real vector space. An *inner product* on V is an operation denoted by $\langle \, , \rangle$ which associates to each pair of vectors $x, y \in V$ a real number $\langle x, y \rangle$ and for all $x, y, z \in V$ and $c \in \mathbb{R}$ satisfies the following properties;

1. $\langle x, x \rangle \geq 0$ and $\langle x, x \rangle = 0$ if and only if $x = 0$

2. $\langle x, y \rangle = \langle y, x \rangle$

3. $\langle x + y, z \rangle = \langle x, z \rangle + \langle y, z \rangle$

4. $\langle cx, y \rangle = c\langle x, y \rangle$

In this case, V is called an *inner product space*.

In $\mathbb{R}^n$ the *usual inner product* of two vectors $x = (x_1, ..., x_n)$ and $y = (y_1, ..., y_n)$ is defined by

$$\langle x, y \rangle = \sum_{i=1}^{n} x_i y_i.$$

That this is an inner product can be readily verified by the reader. It is often called the *dot product* of the vectors x, y and is denoted by $x \cdot y = \langle x, y \rangle$.

Definition 1.3.2. Let V be a vector space. A *norm* on V is a function $||.|| : V \to \mathbb{R}$ satisfying for all $x, y \in V$ and $c \in \mathbb{R}$ the following properties

1. $||x|| \geq 0$ and $||x|| = 0$ if and only if $x = 0$

2. $||cx|| = |c| \cdot ||x||$

3. $||x + y|| \leq ||x|| + ||y||$ (*triangle inequality*).

The space V equipped with a norm is called a *normed linear space*.

Definition 1.3.3. The *Euclidean norm* on $\mathbb{R}^n$ is

$$||x|| = \sqrt{\langle x, x \rangle} = \left(\sum_{i=1}^{n} x_i^2 \right)^{\frac{1}{2}}.$$

The Euclidean norm on $\mathbb{R}^n$ clearly satisfies the conditions (1) and (2) of a norm. The triangle inequality will follow from the important *Cauchy-Schwarz-Bunyakovsky[7] inequality* which is valid in any inner product space and we prove it below. The reader should be aware that there are other norms in $\mathbb{R}^n$ which do not come from an inner product. In Theorem 1.3.7 we characterize the norms on a normed linear space that come from an inner product. The importance of the Euclidean norm resides exactly in the fact that it comes from the usual inner product.

Theorem 1.3.4. *(Cauchy-Schwarz inequality) Let $(V, \langle\,,\rangle)$ be an inner product space. Then for all $x, y \in V$,*

$$|\langle x, y \rangle| \leq ||x||\, ||y||.$$

Moreover, equality occurs if and only if one of these vectors is a scalar multiple of the other.

Proof. If $x = 0$, then both sides are zero and the inequality is true as equality. For $x \neq 0$, let $t \in \mathbb{R}$. Then

$$0 \leq ||tx + y||^2 = \langle tx + y, tx + y \rangle = \langle x, x \rangle t^2 + 2\langle x, y \rangle t + \langle y, y \rangle$$

$$= ||x||^2 t^2 + 2\langle x, y \rangle t + ||y||^2.$$

This expression is a quadratic polynomial in t of the form $\phi(t) = at^2 + bt + c$, with $a = ||x||^2$, $b = 2\langle x, y \rangle$ and $c = ||y||^2$. Since $\phi(t) \geq 0$ for all $t \in \mathbb{R}$, its discriminant $b^2 - 4ac$ must be non positive, that is, $\langle x, y \rangle^2 \leq ||x||^2\,||y||^2$. Therefore $|\langle x, y \rangle| \leq ||x||\,||y||$. Finally, note that equality occurs if and only if $b^2 - 4ac = 0$. But this means that the quadratic polynomial has a double (real) root $t_0 = -\frac{b}{2a}$. That is, $0 = \phi(t_0) = ||x + t_0 y||^2$. It follows that $x + t_0 y = 0$ and so $x = cy$ (with $c = -t_0$). □

Now it is easy to prove the triangle inequality. In fact, we have

$$||x + y||^2 = ||x||^2 + 2\langle x, y \rangle + ||y||^2$$

---

[7]H. Schwarz (1843-1921). A student of Gauss and Kummer. Professor at the University of Göttingen, he is known for his work in complex analysis. V. Bunyakovsky (1804-1889). Mathematician, member of the Petersburg Academy of Science. He is credited of discovering the Cauchy-Schwarz inequality in 1859.

$$\leq ||x||^2 + 2|\langle x,y\rangle| + ||y||^2 \leq ||x||^2 + 2||x||\,||y|| + ||y||^2$$

$$= (||x|| + ||y||)^2 \text{ using the Cauchy-Schwarz inequality.}$$

Therefore,

$$||x + y|| \leq ||x|| + ||y||.$$

Thus we see that $\mathbb{R}^n$ with the Euclidean norm is a normed linear space.

Corollary 1.3.5. *Let V be any normed linear space. Then for $x,y \in V$*

$$|||x|| - ||y||| \leq ||x - y||.$$

Proof. We have $||x|| = ||(x - y) + y|| \leq ||x - y|| + ||y||$. Therefore

$$||x|| - ||y|| \leq ||x - y||.$$

Similarly

$$||y|| - ||x|| \leq ||x - y||$$

and thus the conclusion. $\square$

Proposition 1.3.6. *(The parallelogram law). Let V be an inner product space and $x,y \in V$. Then*

$$||x + y||^2 + ||x - y||^2 = 2||x||^2 + 2||y||^2.$$

Proof. A direct computation gives

$$||x+y||^2 + ||x-y||^2 = \langle x+y, x+y\rangle + \langle x-y, x-y\rangle = ||x||^2 + 2\langle x,y\rangle + ||y||^2$$

$$+ ||x||^2 - 2\langle x,y\rangle + ||y||^2 = 2||x||^2 + 2||y||^2.$$

$\square$

Even for $V = \mathbb{R}^n$ everything takes place in a plane! In fact, Figure 1.1 in $\mathbb{R}^2$ tells us that *in a parallelogram the sum of the squares of the lengths of the diagonals is the sum of the squares of the sides.*

The interest of the parallelogram law is that it characterizes the norms that are derived from an inner product.

Theorem 1.3.7. *Let V be a normed linear space with norm $||\,.||$ satisfying the parallelogram law. Then there is a unique inner product $\langle\,,\rangle$ on V such that $\langle\,,\rangle^{\frac{1}{2}}$ coincides with the given norm on V.*

Proof. For $x, y \in V$, define $\langle x, y \rangle = \frac{1}{4}\left[||x+y||^2 - ||x-y||^2\right]$. Then $\langle x, x \rangle = \frac{1}{4}||2x||^2 = ||x||^2$. So $\langle x, x \rangle^{\frac{1}{2}} = ||x||$. To show that $\langle\,,\rangle$ is an inner product on V, we need show that $\langle\,,\rangle$ satisfies the properties (1), (2), (3), (4) of the Definition 1.3.1 Properties (1) and (2) are obvious. To show (3) we have

$$\langle x, z \rangle + \langle y, z \rangle = \frac{1}{4}\left[||x+z||^2 + ||y+z||^2 - (||x-z||^2 + ||y-z||^2)\right].$$

Setting $x = u + w$ and $y = u - w$, we get

$$\frac{1}{4}\left[||(u+z)+w||^2 + ||(u+z)-w||^2 - (||(u-z)+w||^2 + ||(u-z)-w||^2)\right].$$

By the parallelogram law (applied twice) this is equal to

$$\left[2||u+z||^2 + 2||w||^2 - 2||w-z||^2 - 2||w||^2\right] = \frac{1}{2}\left[||w+z||^2 - ||u-z||^2\right]$$

$$= 2\langle u, z \rangle = 2\langle \frac{x+y}{2}, z \rangle = \langle x+y, z \rangle.$$

Finally, for $c \in \mathbb{R}$ we have

$$\langle cx, y \rangle = \frac{1}{4}\left[||cx+y||^2 - ||cx-y||^2\right] = \frac{1}{4}\left[4c\langle x, y \rangle\right] = c\langle x, y \rangle.$$

To see the uniqueness of this inner product, suppose that $\langle\,,\rangle_1$ is another inner product such that $\langle x, x \rangle_1 = ||x||^2$. Then $\langle x, x \rangle = \langle x, x \rangle_1$ for all $x \in V$. In particular, for all $x, y \in V$ we have $\langle x+y, x+y \rangle = \langle x+y, x+y \rangle_1$ and so $\langle x, y \rangle = \langle x, y \rangle_1$ $\square$

Besides the Euclidean norm on $\mathbb{R}^n$, there are other norms on $\mathbb{R}^n$ of which the following two are the most common.

Definition 1.3.8. For $x = (x_1, x_2, ..., x_n) \in \mathbb{R}^n$, we define

$$||x||_\infty = \max\{|x_1|, |x_2|, ..., |x_n|\} \quad \text{and} \quad ||x||_1 = \sum_{i=1}^{n}|x_i|.$$

It is easy to see that both $||\cdot||_\infty$ and $||\cdot||_1$ are norms in $\mathbb{R}^n$. We ask the reader to show this in Exercise 4 below.

Example 1.3.9. For any $x \in \mathbb{R}^n$, $\quad ||x||_\infty \le ||x|| \le \sqrt{n}||x||_\infty$.

Solution. For each $i = 1, ..., n$ we have

$$|x_i| = \sqrt{x_i^2} \le \sqrt{x_1^2 + ... + x_n^2} = ||x||.$$

Hence, $||x||_\infty \le ||x||$. On the other hand, $|x_i| \le ||x||_\infty$. Therefore,

$$||x||^2 = \sum_{i=1}^{n} x_i^2 \le n||x||_\infty^2.$$

EXERCISES

1. Let $x, y \in \mathbb{R}^3$, where $x = (1, 0, 2)$ and $y = (3, -1, 1)$.

   (a) Find $x + y$, $x - y$, $||x||$, $||y||$, $||x + y||$, $||x - y||$, $\langle x, y \rangle$.

   (b) Verify, the Cauchy-Schwarz inequality, the triangle inequality and the parallelogram law.

   (c) Find $||x||_\infty$, and verify $||x||_\infty \le ||x|| \le \sqrt{3}||x||_\infty$.

   (d) Find $||x||_1$, and verify $||x||_1 \le \sqrt{3}||x|| \le \sqrt{3}||x||_1$.

   (e) Is the paralellogram law valid using the norms $||\cdot||_\infty$ and $||\cdot||_1$?

2. For $x, y \in \mathbb{R}^n$ show that $||x+y|| = ||x|| + ||y||$ if and only if $x = cy$ for some $c > 0$.

3. For $x, y \in \mathbb{R}^n$ show that

$$||x + y||\,||x - y|| \le ||x||^2 + ||y||^2$$

   with equality if and only if $\langle x, y \rangle = 0$.

4. Show that

   (a) $||\cdot||_\infty$ and $||\cdot||_1$ are norms on $\mathbb{R}^n$. Do they satisfy the parallelogram law?

(b) For $x \in \mathbb{R}^n$, show that $||x||_1 \leq \sqrt{n}||x|| \leq \sqrt{n}||x||_1$.

(c) Let $v = (x, y) \in \mathbb{R}^2$. Sketch the sets $\{v \in \mathbb{R}^2 : ||v|| = 1\}$, $\{v \in \mathbb{R}^2 : ||v||_\infty = 1\}$ and $\{v \in \mathbb{R}^2 : ||v||_1 = 1\}$ in $\mathbb{R}^2$ and extend to $\mathbb{R}^n$.

5. Let V be a vector space. Show that any intersection of linear subspaces is again a linear subspace. Is the union of linear subspaces a subspace?

6. Let $S \subseteq V$. Show that span(S) is the smallest linear subspace containing the set S.

1.3.2   Orthogonality

Let V be an inner product space. In view of the Cauchy-Schwarz inequality since $-1 \leq \frac{\langle x, y \rangle}{||x|| \, ||y||} \leq 1$, we can define the *angle* between any two nonzero vectors $x, y \in V$ by

$$\cos\theta = \frac{\langle x, y \rangle}{||x|| \, ||y||} \text{ or } \theta = \cos^{-1}\left(\frac{\langle x, y \rangle}{||x|| \, ||y||}\right), \theta \in [0, \pi].$$

Hence the following definition is a natural consequence.

Definition 1.3.10. Let V be an inner product space and $x, y \in V$. The vectors x and y are called *orthogonal* (or *perpendicular*), denoted by $x \perp y$, when $\langle x, y \rangle = 0$.

The zero vector is orthogonal to *every* vector $x \in V$, since $\langle 0, x \rangle = \langle 0 \cdot 0, x \rangle = 0\langle 0, x \rangle = 0$. Moreover, 0 is the only vector with this property. For if there were another vector $v \in V$ with $\langle v, x \rangle = 0$ for all $x \in V$, then in particular $||v||^2 = \langle v, v \rangle = 0$, and so $v = 0$.

More generally, we say that two linear subspaces U and W of V are orthogonal if every vector in U is orthogonal to every vector in W. We write $U \perp W$. For a subset S of V, the *orthogonal complement* $S^\perp$ of S is defined to be

$$S^\perp = \{y \in V : y \perp x, \, for \, all \, x \in S\}.$$

Clearly $V^\perp = \{0\}$ and $\{0\}^\perp = V$. Note that $S^\perp = \cap_{x \in S}\{x\}^\perp$. Also if $x \perp S$, then $x \perp$ span(S). We denote by $S^{\perp\perp} = (S^\perp)^\perp$.

Proposition 1.3.11. *Let S be a subset of V. Then*

(a) $S^\perp$ *is a linear subspace of V.*

(b) $S \subseteq S^{\perp\perp}$

(c) *If $S_1 \subseteq S_2 \subseteq V$, then $S_2^\perp \subseteq S_1^\perp$*

(d) $S^\perp = (\text{span}(S))^\perp$

Proof.

(a) Let $x, y \in V$. If x and y are orthogonal to every vector of S, then so is any linear combination of x and y. Therefore $S^\perp$ is a linear subspace of V.

(b) If $x \in S$, then for every $y \in S^\perp$ we have $x \perp y$. Hence $x \in (S^\perp)^\perp$

(c) $S_1 \subseteq S_2$ and $x \perp S_2$, then also $x \perp S_1$.

(d) Since $S \subseteq \text{span}(S)$, it follows from (3) that $(\text{span}(S))^\perp \subseteq S^\perp$. The reverse inclusion is trivial.

$\square$

Definition 1.3.12. We say that a set of nonzero vectors $\{v_1, ..., v_k)$ of V is *orthogonal* if $\langle v_i, v_j \rangle = 0$ for $i \neq j$. If in addition, $||v_j|| = 1$ for all $j = 1, ..., k$, we say $\{v_1, ..., v_k)$ is an *orthonormal* set of vectors. Thus, orthonormality means $\langle v_i, v_j \rangle = \delta_{ij} = \begin{cases} 1 \text{ when } i = j, \\ 0 \text{ when } i \neq j \end{cases}$, where the symbol δ_{ij} is the so-called *Kronecker's delta*.

Proposition 1.3.13. *(The law of cosines).* Let x, y be two nonzero vectors in $\mathbb{R}^n$. Then

$$||x - y||^2 = ||x||^2 + ||y||^2 - 2||x|| \, ||y|| \cos(\theta),$$

where θ is the angle between x, y.

Proof. We have

$$||x - y||^2 = \langle x - y, x - y \rangle = ||x||^2 + ||y||^2 - 2\langle x, y \rangle$$
$$= ||x||^2 + ||y||^2 - 2||x|| \, ||y|| \cos(\theta).$$

$\square$

Corollary 1.3.14. *(Pythagorean*[8] *theorem and converse). Let x, y be two nonzero vectors in $\mathbb{R}^n$. Then x is orthogonal to y if and only if*

$$||x + y||^2 = ||x||^2 + ||y||^2.$$

Proof. From the law of cosines we have

$$||x + y||^2 = ||x - (-y)||^2 = ||x||^2 + ||y||^2 - 2||x|| \, ||y|| \cos(\pi - \theta)$$
$$= ||x||^2 + ||y||^2 + 2||x|| \, ||y|| \cos(\theta).$$

Hence, $||x + y||^2 = ||x||^2 + ||y||^2$ iff $\cos(\theta) = 0$ iff $\theta = \frac{\pi}{2}$ iff $\langle x, y \rangle = 0$. $\square$

By induction we get,

Corollary 1.3.15. *If $\{v_1, ..., v_k\}$ is an orthogonal set of vectors of V, then*

$$||\sum_{j=1}^{k} v_j||^2 = \sum_{j=1}^{k} ||v_j||^2.$$

Geometry and algebra cooperate admirably in the following proposition.

Proposition 1.3.16. *An orthogonal set of nonzero vectors is linearly independent.*

Proof. Suppose $\sum_{j=1}^{k} c_j v_j = 0$. Then

$$0 = \langle \sum_{j=1}^{k} c_j v_j, v_i \rangle = \sum_{j=1}^{k} c_j \langle v_j, v_i \rangle = c_i ||v_i||^2.$$

Since $||v_i||^2 > 0$ for all i, it follows that $c_i = 0$ for all $i = 1, ...k$. $\square$

Corollary 1.3.17. *An orthonormal set of vectors $B = \{v_1, ..., v_n\}$ of V which spans V is a basis.*

---

[8]Pythagoras, a famous mathematician and philoshoper of the 6th century B.C. He established the Pythagorean school first in the Greek island of Samos and then in south Italy.

Observe that the *standard basis* $B = \{e_1, ..., e_n\}$ of $\mathbb{R}^n$ is an orthonormal basis of $\mathbb{R}^n$ with the usual inner product.

We now introduce the *Gram-Schmidt orthogonalization process*[9] which assures us that any subspace of V has an orthonormal basis, in fact many.

Theorem 1.3.18. *(Gram-Schmidt) Let $\{v_1, ..., v_m\}$ be a linearly independent set in V. Then there is an orthonormal set $\{w_1, ..., w_m\}$ such that*

$$\text{span}\,\{v_1, v_2, ..., v_j\} = \text{span}\,\{w_1, w_2, ..., w_j\}, \text{ for each } j = 1, ..., m.$$

Proof. The proof is by direct construction of the vectors w_i and the procedure is useful enough to have a name: The Gram-Schmidt process. To begin the construction, we observe that $v_1 \neq 0$ (since the v_i are linearly independent) and we define $w_1 = \frac{v_1}{\|v_1\|}$. Clearly span $\{w_1\} = \text{span}\,\{v_1\}$. The other vectors are then given inductively as follows: Suppose $w_1, w_2, ..., w_k$ have been constructed so that they form an orthonormal set and for $1 \leq j \leq k$, span $\{w_1, ..., w_j\} = \text{span}\,\{v_1, ..., v_j\}$. To construct the next vector w_{k+1}, let

$$u_{k+1} = v_{k+1} - \sum_{j=1}^{k} \langle v_{k+1}, w_j \rangle w_j.$$

Then $u_{k+1} \neq 0$, for otherwise $v_{k+1} \in \text{span}\,\{w_1, ..., w_k\} = \text{span}\,\{v_1, ..., v_k\}$ which contradicts the linear independence of $\{v_1, ..., v_{k+1}\}$. At the same time, for $1 \leq i \leq k$ we have

$$\langle u_{k+1}, w_i \rangle = \langle v_{k+1}, w_i \rangle - \sum_{j=1}^{k} \langle v_{k+1}, w_j \rangle \langle w_j, w_i \rangle = \langle v_{k+1}, w_i \rangle - \langle v_{k+1}, w_i \rangle = 0.$$

Therefore, $u_{k+1} \perp w_i$ for all $i = 1, ..., k$. Moreover,

$$\text{span}\,\{v_1, ..., v_{k+1}\} = \text{span}\,\{w_1, ..., w_k, u_{k+1}\}.$$

---

[9]J. Gram (1850-1916). Actuary and mathematician. He is also know for the *Gram matrix* or *Gramian*. E. Schmidt (1876-1959). A student of Hilbert at the University of Göttingen. Together with Hilbert he made important contributions to functional analysis. The *Gram-Schmidt process* was first published in 1883.

To see this, set $z_{k+1} = \sum_{j=1}^{k} \langle v_{k+1}, w_j \rangle w_j \in \text{span}\{w_1, ..., w_k\}$, so that $v_{k+1} = u_{k+1} + z_{k+1} \in \text{span}\{w_1, ..., w_k, u_{k+1}\}$. But also $z_{k+1} \in \text{span}\{v_1, ..., v_k\}$. Hence $u_{k+1} = v_{k+1} - z_{k+1} \in \text{span}\{v_1, ..., v_{k+1}\}$. Finally, choose $w_{k+1} = \frac{u_{k+1}}{\|u_{k+1}\|}$. Then $\|w_{k+1}\| = 1$ and $\text{span}\{w_1, ..., w_k, u_{k+1}\} = \text{span}\{w_1, ..., w_k, w_{k+1}\}$. This completes the induction and the proof.   □

Since every subspace of $\mathbb{R}^n$ has a basis we see that,

Corollary 1.3.19. *Any subspace of* $\mathbb{R}^n$ *has an orthonormal basis.*

Orthonormal sets have substantial computational advantages. For if $\{v_1, ..., v_n\}$ is a basis for V, then any vector $v \in V$ is a unique linear combination of the vectors $v_1, ..., v_m$. But actually finding this linear combination can be laborius since it involves solving simultaneous linear equations i.e., a linear system. Working with an orthonormal basis makes the task much easier. Some of these advantages are illustrated in the next theorem.

Theorem 1.3.20. *Let* $B = \{v_1, v_2, ..., v_n\}$ *be an orthonormal set in* V. *The following are equivalent.*

1. *Let* $x \in V$. *If* $x \perp v_j$ *for all* $j = 1, ..., n$, *then* $x = 0$.

2. $B = \{v_1, v_2, ..., v_n\}$ *is an orthonormal basis of* V.

3. *Each* $x \in V$ *can be written as* $x = \sum_{j=1}^{n} c_j v_j$, *where* $c_j = \langle x, v_j \rangle$. *The numbers* $c_j = \langle x, v_j \rangle$ *are called the Fourier[10] coefficients of* x *with respect to the orthonormal basis* B

4. *For all* $x, y \in V$, $\langle x, y \rangle = \sum_{j=1}^{n} \langle x, v_j \rangle \langle y, v_j \rangle$.

5. *For* $x \in V$, $\|x\|^2 = \sum_{j=1}^{n} |\langle x, v_j \rangle|^2 = \sum_{j=1}^{n} |c_j|^2$ *(Parseval's identity)[11]*

Proof. We will show that $(1) \Rightarrow (2) \Rightarrow (3) \Rightarrow (4) \Rightarrow (5) \Rightarrow (1)$.

$(1) \Rightarrow (2)$: If $B = \{v_1, ..., v_n\}$ were not an orthonormal basis, then there is an orthonormal set A containing B. Let $x \in A \cap B^c$. Then

---

[10] J. Fourier (1768-1830). Mathematician and physicist best known for initiating the study of *Fourier series* and their applications to problems of *heat transfer*.

[11] M. Parseval (1755-1836), most famous for what is known as *Parseval's equality*.

$x \neq 0$ and $x \perp v_j$ for all $j = 1, ..., n$, contradicting the assumption.

(2) $\Rightarrow$ (3): Since B is a basis of V, each $x \in V$ can be uniquely written as $x = \sum_{j=1}^{n} c_j v_j$. Then

$$\langle x, v_i \rangle = \langle \sum_{j=1}^{n} c_j v_j, v_i \rangle = \sum_{j=1}^{n} c_j \langle v_j, v_i \rangle = c_i.$$

(3) $\Rightarrow$ (4): Let $y \in V$. So that $y = \sum_{j=1}^{n} d_j v_j$, where $d_j = \langle y, v_j \rangle$. Then

$$\langle x, y \rangle = \langle x, \sum_{j=1}^{n} d_j v_j \rangle = \sum_{j=1}^{n} d_j \langle x, v_j \rangle = \sum_{j=1}^{n} \langle y, v_j \rangle \langle x, v_j \rangle.$$

(4) $\Rightarrow$ (5): Taking $x = y$ in (4) we get

$$||x||^2 = \langle x, x \rangle = \sum_{j=1}^{n} (\langle x, v_j \rangle)^2 = \sum_{j=1}^{n} c_j^2.$$

(5) $\Rightarrow$ (1): Clearly, if $x \perp v_j$, that is, if $\langle x, v_j \rangle = 0$ for all $j = 1, ..., n$, then $||x||^2 = 0$ and so $x = 0$. $\qquad\square$

Remark 1.3.21. Note that Parseval's identity in the form

$$|| \sum_{j=1}^{k} c_j v_j ||^2 = \sum_{j=1}^{k} c_j^2$$

is a generalization of the Pythagorean theorem.

The following theorem is of considerable geometric interest.

Theorem 1.3.22. *(Projection Theorem). Let V be an inner product space and W a linear subspace of V. Then $V = W \oplus W^\perp$, that is, any $x \in V$, can be written uniquely as $x = y + z$ with $y \in W$ and $z \in W^\perp$. Moreover, $W^{\perp\perp} = W$.*

Proof. Let $\dim(V) = n$. First observe $W \cap W^{\perp} = \{0\}$. Suppose $(W) = k$ with $0 \le k \le n$. Choose an orthonormal basis $\{w_1, ..., w_k\}$ of W and extend it to a basis for V. Use the Gram-Schmidt process to obtain an orthonormal basis $\{w_1, ..., w_k, w_{k+1}, ..., w_n\}$ of V. Note that $w_{k+1}, ..., w_n \in W^{\perp}$.

Given $x \in V$,

$$x = \sum_{j=1}^{n} \langle x, w_j \rangle w_j = \sum_{j=1}^{k} \langle x, w_j \rangle w_j + \sum_{j=k+1}^{n} \langle x, w_j \rangle w_j = y + z,$$

with $y \in W$ and $z \in W^{\perp}$. If $x \in W^{\perp}$, then $\langle x, w_j \rangle = 0$ for all $j = 1, ..., k$, and so $x = \sum_{j=k+1}^{n} \langle x, w_j \rangle w_j$. Hence $\{w_{k+1}, ..., w_n\}$ is an orthonormal basis of $W^{\perp}$ and $\dim(W) + \dim(W^{\perp}) = \dim(V)$.

To show that this decomposition is unique, suppose $x = y_1 + z_1$, with $y_1 \in W$ and $z_1 \in W^{\perp}$. Then $y - y_1 \in W$ and $z - z_1 \in W^{\perp}$. It follows that $z_1 - z = x - y_1 - (x - y) = y - y_1$, that is, $y - y_1 = z - z_1 \in W \cap W^{\perp} = \{0\}$. Therefore, $y = y_1$ and $z = z_1$.

Finally we show that $W^{\perp\perp} = W$. We know that $W \subseteq W^{\perp\perp}$. To get the reverse inclusion, let $x \in W^{\perp\perp}$. Then $x = y + z$ with $y \in W \subseteq W^{\perp\perp}$ and $z \in W^{\perp}$. So $z = x - y \in W^{\perp\perp} \cap W^{\perp} = \{0\}$. Hence, $z = 0$ and so $x = y \in W$. □

Corollary 1.3.23. *Let S be a subset of V. Then $S^{\perp\perp} = \text{span}(S)$.*

Proof. From Proposition 1.3.11, $S^{\perp} = (\text{span}(S))^{\perp}$. Therefore $S^{\perp\perp} = (\text{span}(S))^{\perp\perp}$. Since $\text{span}(S)$ is a subspace, by the Projection theorem $(\text{span}(S))^{\perp\perp} = \text{span}(S)$. Hence, $S^{\perp\perp} = \text{span}(S)$. □

Corollary 1.3.24. *If W is a proper subspace of V, then there exists $z_0 \ne 0$ such that $z_0 \perp W$.*

Proof. Let $x_0 \in V \backslash W$. Then $x_0 = y_0 + z_0$ with $y_0 \in W$ and $z_0 \in W^{\perp}$ and $z_0 \ne 0$, for if $z_0 = 0$, then $x_0 = y_0 \in W$ which is impossible. □

Definition 1.3.25. Let $V = W \oplus W^{\perp}$ and $v = w + u$ with $w \in W$ and $u \in W^{\perp}$ as in the Projection theorem. The map $P_W : V \to V$ given by $P_W(v) = w$ is called the *orthogonal projection* onto W. Similarly, the map $P_{W^{\perp}} : V \to V$ given by $P_{W^{\perp}}(v) = u$ is the orthogonal projection onto $W^{\perp}$. Observe that $P_W + P_{W^{\perp}} = I$.

Clearly,

$$\mathbb{R}^3 = \mathbb{R}^2 \oplus \mathbb{R}.$$

and the corresponding projections are $P(x,y,z) = (x,y,0)$, $P'(x,y,z) = (0,0,z)$. More generally for $k < n$

$$\mathbb{R}^n = \mathbb{R}^k \oplus \mathbb{R}^{n-k}.$$

Note that the proof of the Projection theorem tells us a way to compute the projection $P = P_W$. In fact, when $\{w_1, w_2, ..., w_k\}$ is an orthonormal basis for W, then $P(v) = w = \sum_{j=1}^{k} \langle v, w_j \rangle w_j$. Note also that for any $v \in V$, we have $P^2(v) = P(P(v)) = P(w) = w = P(v)$. That is, $P^2 = P$.

Example 1.3.26. Let $V = \mathbb{R}^3$ and $W = \text{span}\{v_1 = (1,-1,0), v_2 = (0,-1,1)\}$. Find $W^\perp$, the projections P_W onto W and the projection $P_{W^\perp}$ onto $W^\perp$.

Solution. Note that W is the plane generated by the vectors v_1, v_2. By the linearity of the inner product $W^\perp = \{u \in \mathbb{R}^3 : u \perp v_1 \text{ and } u \perp v_2\}$. Let $u = (x,y,z)$. We have, $u \perp v_1$ if and only if $\langle u, v_1 \rangle = 0$, that is, $x - y = 0$. Similarly, $u \perp v_2$, implies $-y + z = 0$. Solving the system of these two equations we get $u = t(1,1,1)$ where, $t \in \mathbb{R}$. Thus,

$$W^\perp = \text{span}\{u_1 = (1,1,1)\}$$

which is a line through the origin. To apply the Projection theorem and find the projection P_W, we need to construct an orthonormal basis for W. Note that $\{v_1, v_2\}$ is a linear independent set, and so a basis for W. Let us apply Gram-Schmidt orthogonalization process to obtain an orthonormal basis $\{w_1, w_2\}$ of W. Take $w_1 = \frac{v_i}{||v_1||}$, where $||v_1|| = \sqrt{2}$. That is, $w_1 = (\frac{1}{\sqrt{2}}, -\frac{1}{\sqrt{2}}, 0)$. To construct the next vector, let

$$u_2 = v_2 - \langle v_2, w_1 \rangle w_1 = v_2 - \frac{1}{\sqrt{2}} w_1 = (-\frac{1}{2}, -\frac{1}{2}, 0).$$

Now, $||u_2|| = \frac{\sqrt{6}}{2}$ and normalizing u_2 we get $w_2 = (-\frac{1}{\sqrt{6}}, -\frac{1}{\sqrt{6}}, \frac{2}{\sqrt{6}})$. Since orthogonal vectors are linearly independent, taking $w_3 = \frac{u_1}{||u_1||} =$

$(\frac{1}{\sqrt{3}}, \frac{1}{\sqrt{3}}, \frac{1}{\sqrt{3}})$, we obtain an orthonormal basis $\{w_1, w_2, w_3\}$ of $V = \mathbb{R}^3$. Now, let $v \in \mathbb{R}^3$. By the projection theorem, we have $v = w + u$ with $w \in W$ and $u \in W^\perp$. Hence

$$P_W(v) = w = \sum_{j=1}^{2} \langle v, w_j \rangle w_j = \langle v, w_1 \rangle w_1 + \langle v, w_2 \rangle w_2.$$

Therefore for $v = (x, y, z)$, we have

$$P_W(x, y, z) = \frac{x-y}{\sqrt{2}}(\frac{1}{\sqrt{2}}, -\frac{1}{\sqrt{2}}, 0) + \frac{-x-y+2z}{\sqrt{6}}(-\frac{1}{\sqrt{6}}, -\frac{1}{\sqrt{6}}, \frac{2}{\sqrt{6}})$$

$$= \frac{1}{3}(2x - y - z, -x + 2y - z, -x - y + 2z),$$

and $P_{W^\perp}(x, y, z) = \frac{1}{3}(x+y+z, x+y+z, x+y+z)$, since $P_W + P_{W^\perp} = I$.

1.3.3 The cross product in $\mathbb{R}^3$

An important concept that distinguishes $\mathbb{R}^3$ from Euclidean spaces of other dimensions is that of the *cross product* or *vector product*. In Chapter 6, we will make extensive use of the cross product in the study of surface integrals. Here, the usual notation for the elements of the standard basis is $\mathbf{i} = e_1 = (1,0,0)$, $\mathbf{j} = e_2 = (0,1,0)$, $\mathbf{k} = e_3 = (0,0,1)$. Thus, any vector $a = (a_1, a_2, a_3) \in \mathbb{R}^3$ can be written as $a = a_1\mathbf{i} + a_2\mathbf{j} + a_3\mathbf{k}$

Definition 1.3.27. Let $a, b \in \mathbb{R}^3$. The *cross product* of $a = (a_1, a_2, a_3)$ and $b = (b_1, b_2, b_3)$ is defined by

$$a \times b = \det \begin{bmatrix} \mathbf{i} & \mathbf{j} & \mathbf{k} \\ a_1 & a_2 & a_3 \\ b_1 & b_2 & b_3 \end{bmatrix} = (a_2b_3 - a_3b_2)\mathbf{i} + (a_3b_1 - a_1b_3)\mathbf{j} + (a_1b_2 - a_2b_1)\mathbf{k}.$$

Note that

$$\mathbf{i} \times \mathbf{j} = \mathbf{k}, \ \mathbf{j} \times \mathbf{k} = \mathbf{i}, \ \mathbf{k} \times \mathbf{i} = \mathbf{j} \text{ and}$$

$$\mathbf{j} \times \mathbf{i} = -\mathbf{k}, \ \mathbf{k} \times \mathbf{j} = -\mathbf{i}, \ \mathbf{i} \times \mathbf{k} = -\mathbf{j}.$$

The following proposition gives the basic properties of the cross product.

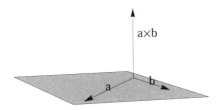

Figure 1.2: Cross product

Proposition 1.3.28. *Let $a, b, u \in \mathbb{R}^3$ and $\lambda, \kappa \in \mathbb{R}$. Then*

1. $a \times a = 0$.

2. $b \times a = -(a \times b)$.

3. $\lambda a \times \kappa b = \lambda \kappa (a \times b)$.

4. $a \times (b + u) = (a \times b) + (a \times u)$ *and* $(a + b) \times u = (a \times u) + (b \times u)$.

5. $(a \times b) \perp a$ *and* $(a \times b) \perp b$.

6. $||a \times b||^2 = ||a||^2 ||b||^2 - \langle a, b \rangle^2$.

Proof. All these properties follow easily from the definition. We prove property (6).

$$||a||^2 ||b||^2 - \langle a, b \rangle^2 = (a_1^2 + a_2^2 + a_3^2)(b_1^2 + b_2^2 + b_3^2) - (a_1 b_1 + a_2 b_2 + a_3 b_3)^2.$$

Multiplying out and rearranging the terms we get

$$(a_2 b_3 - a_3 b_2)^2 + (a_3 b_1 - a_1 b_3)^2 + (a_1 b_2 - a_2 b_1)^2 = ||a \times b||^2.$$

$\square$

Property 5 is illustrated in Figure 1.2.

Example 1.3.29. Find two unit vectors perpendicular to both $a = (2, -1, 3)$ and $b = (-4, 3, -5)$.

Solution. Since $a \times b$ is perpendicular to both a and b, we compute

$$a \times b = \det \begin{bmatrix} \mathbf{i} & \mathbf{j} & \mathbf{k} \\ 2 & -1 & 3 \\ -4 & 3 & -5 \end{bmatrix} = -4\mathbf{i} - 2\mathbf{j} + 2\mathbf{k}.$$

The lenght (norm) of the vector is $||a \times b|| = \sqrt{16 + 4 + 4} = 2\sqrt{6}$. The desired *unit* vector is

$$\mathbf{n} = \frac{1}{\sqrt{6}} \left[-2\mathbf{i} - \mathbf{j} + \mathbf{k} \right].$$

If we had taken $b \times a$ instead, we would have obtained the negative of this vector. Thus the two vectors are

$$\pm\mathbf{n} = \pm\frac{1}{\sqrt{6}} \left[-2\mathbf{i} - \mathbf{j} + \mathbf{k} \right].$$

Corollary 1.3.30. *Let* $a, b \in \mathbb{R}^3$ *and* $\theta \in [0, \pi]$ *the angle between* a *and* b. *Then*

$$||a \times b|| = ||a|| \, ||b|| \sin \theta.$$

In particular, $||a \times b||$ *equals the area of the parallelogram with adjacent sides* a, b.

Proof. We know that $\langle a, b \rangle = ||a|| \, ||b|| \cos \theta$. Hence from (6) we get

$$||a \times b||^2 = ||a||^2 ||b||^2 (1 - \cos^2 \theta).$$

Thus, $||a \times b|| = ||a|| \, ||b|| \sin \theta$. Furthermore, if a, b represent the adjacent sides of a parallelogram and we take b to be the "base", then $||a|| \sin \theta$ is the "height". Hence the area of the parallelogram is equal to $||a \times b||$. $\square$

Note that $a \times b = 0$ if and only if $||a \times b|| = 0$ if and only if $\sin \theta = 0$ if and only if $\theta = 0$ or $\theta = \pi$, viz., a, b are colinear. Equivalently, $a \times b \neq 0$ if and only if a and b are linearly independent.

Example 1.3.31. Find the area of the triangle whose vertices are $A = (2, -1, 3)$ and $B = (1, 2, 4)$, and $C = (3, 1, 1)$.

Solution. Two sides of the triangle are represented by the vectors

$$a = AB = (1 - 2)\mathbf{i} + (2 + 1)\mathbf{j} + (4 - 3)\mathbf{k} = -\mathbf{i} + 3\mathbf{j} + \mathbf{k},$$

$$b = AC = (3-2)\mathbf{i} + (1+1)\mathbf{j}(1-3)\mathbf{k} = \mathbf{i} + 2\mathbf{j} - 2\mathbf{k}.$$

The area of the triangle is clearly half of the erea of the parallelogram with adjacent sides the vectors a and b. Hence the area of the triangle is $\frac{1}{2}||a \times b||$. We compute

$$a \times b = \det \begin{bmatrix} \mathbf{i} & \mathbf{j} & \mathbf{k} \\ -1 & 3 & 1 \\ 1 & 2 & -2 \end{bmatrix} = -8\mathbf{i} - \mathbf{j} - 5\mathbf{k}.$$

Since $||a \times b|| = 3\sqrt{10}$, the area is $\frac{3}{2}\sqrt{10}$.

Example 1.3.32. Find the equation of the plane passing from the point $\mathbf{a} = (a_1, a_2, a_3)$ and perpendicular to $\mathbf{n} = (n_1, n_2, n_3)$. (The vector $\mathbf{n}$ is called the *normal vector* to the plane).

Solution. Let $\mathbf{x} = (x, y, z)$ be any other point on the plane. Then the vector $\mathbf{x} - \mathbf{a}$ lies on the plane and is perpendicular to $\mathbf{n}$. Therefore

$$\langle \mathbf{n}, \mathbf{x} - \mathbf{a} \rangle = 0$$

is the *vector equation* of the plane under consideration. In terms of the coordinates the equation is

$$n_1(x - a_1) + n_2(y - a_2) + n_3(z - a_3) = 0.$$

Example 1.3.33. Find the equation of the plane passing from the points $A = (2, -1, 3)$ and $B = (1, 2, 4)$, and $C = (3, 1, 1)$.

Solution. This is the plane containing the triangle of Example 1.3.31. The normal vector to this plane is $\mathbf{n} = (AB) \times (AC) = -8\mathbf{i} - \mathbf{j} - 5\mathbf{k}$. The problem now is reduced to find the equation of the plane passing from $\mathbf{a} = (2, -1, 3)$ and perpendicular to $\mathbf{n} = (-8, -1, -5)$. So the equation is $-8(x-2) - (y+1) - 5(z-3) = 0$ or $8x + y + 5z = -30$.

Definition 1.3.34. Let $a, b, c \in \mathbb{R}^3$. The *triple scalar product* of a, b and c is defined to be

$$\langle a, (b \times c) \rangle = a_1(b_2c_3 - b_3c_2) + a_2(b_3c_1 - b_1c_3) + a_3(b_1c_2 - b_2c_1)$$

$$= \det \begin{bmatrix} a_1 & a_2 & a_3 \\ b_1 & b_2 & b_3 \\ c_1 & c_2 & c_3 \end{bmatrix}.$$

The following proposition gives a geometric interpretation of the triple scalar product. We invite the reader here to draw a figure in $\mathbb{R}^3$.

Proposition 1.3.35. *Let $a, b, c \in \mathbb{R}^3$. The volume V of the parallelepiped with adjacent sides a, b and c is $|\langle a, (b \times c) \rangle|$. In particular, the vectors a, b and c are coplanar if and only if $\langle a, (b \times c) \rangle = 0$.*

Proof. We know, from Corollary 1.3.30, that $\|b \times c\|$ is the area of the parallelogram with adjacent sides b and c. Moreover $|\langle a, (b \times c) \rangle| = \|a\| \, \|b \times c\| \cos \varphi$, where φ is the acute angle that a makes with $b \times c$ (the normal vector to the plane spanned by b and c). The parallelepiped with adjacent sides a, b and c has height

$$h = \|a\| \cos \varphi = \tfrac{|\langle a, (b \times c) \rangle|}{\|b \times c\|}.$$

From elementary geometry the volume of the parallelepiped is the product of the area of the base times the height. Hence, $V = |\langle a, (b \times c) \rangle|$. □

EXERCISES

1. Let $a = 3\mathbf{i} - \mathbf{j} + 2\mathbf{k}$ and $b = \mathbf{i} + \mathbf{j} + 4\mathbf{k}$. Find $a \times b$, $\|a \times b\|$, $\langle a, b \rangle$ and verify properties (5) and (6).

2. Find the equation of the plane generated by the vectors $a = (3, -1, 1)$ and $b = (1, 2, -1)$.

3. Find the equation of the plane determined by the points $a = (2, 0, 1)$, $b = (1, 1, 3)$ and $c = (4, 7, -2)$.

4. Find a unit vector in the plane generated by the vectors $a = \mathbf{i} + 2\mathbf{j}$ and $b = \mathbf{j} + 2\mathbf{k}$, perpendicular to the vector $v = 2\mathbf{i} + \mathbf{j} + 2\mathbf{k}$

5. Find the equation the line passing through $(1, 1, 1)$ and perpendicular to the plane $3x - y + 2z - 5 = 0$.

6. Find the equation of the plane passing through $(3, 2, -1)$ and $(1, -1, 2)$ and parallel to the line $l(t) = (1, -1, 0) + t(3, 2, -2)$.

7. Find the equation of the plane passing through $(3, 4, -1)$ parallel to the vectors $a = \mathbf{i} - 3\mathbf{k}$ and $b = 2\mathbf{i} + \mathbf{j} + \mathbf{k}$.

8. Find the shortest distance from the point $(3, 4, 5)$ to the line through the origin parallel to the vector $2\mathbf{i} - \mathbf{j} + 2\mathbf{k}$.

9. Find the volume of the tetrahedron determined by the points $P_1 = (2, -1, 4)$, $P_2 = (-1, 0, 3)$, $P_3 = (4, 3, 1)$ and $P_4 = (3, -5, 0)$. *Hint.* The volume of the tetrahedron is one sixth the volume of the parallelepiped with adjacent sides P_1P_2, P_1P_3 and P_1P_4.

10. Let $a, b, v \in \mathbb{R}^3$ with $v \neq 0$. Show that

   (a) If $\langle a, b \rangle = 0$ and $a \times b = 0$, then either $a = 0$ or $b = 0$.
   (b) If $\langle a, v \rangle = \langle b, v \rangle$ and $a \times v = b \times v$, then $a = b$.

11. Let $a, b, c \in \mathbb{R}^3$. Show that the area of the triangle with vertices a, b and c is given by $\frac{1}{2}\|(a - c) \times (b - c)\|$. Find the area of the triangle with vertices the points a, b and c of Exercise 3.

12. Let $a, b, c \in \mathbb{R}^3$. Show that $a \times (b \times c) = (\langle a, c \rangle)b - (\langle a, b \rangle)c$. Deduce that
$$a \times (b \times c) + b \times (c \times a) + c \times (a \times b) = 0$$
(the *Jacobi identity*).

13. Let $a, b \in \mathbb{R}^3$ be nonzero vectors. Show that the vector $v = \|a\|b + \|b\|a$ bisects the angle between a and b.

14. Let $S = \{x = (1, 0, 2), y = (3, -1, 1)\}$

   (a) Find the angle between the vectors x and y.
   (b) Find an orthonormal basis for $W = \text{span}(S)$.
   (c) Find the orthogonal complement, $W^\perp$ of W.
   (d) Find the projections P_W and $P_{W^\perp}$.

Answers to selected Exercises

2. $x - 4y - 7z = 0$.   3. $17x - y + 9z = 43$.   4. $\pm\frac{\sqrt{5}}{25}[5\mathbf{i} + 6\mathbf{j} - 8\mathbf{k}]$.
5. $l(t) = (1, 1, 1) + t(3, -1, 2)$.   6. $y - z + 1 = 0$.   7. $3x - 7y + z + 20 = 0$.
8. $\sqrt{34}$.   9. $\frac{101}{6}$.

1.4 $\mathbb{R}^n$ as a metric space

Definition 1.4.1. A *metric space* is a pair (X, d) where X is a set and d is a function from $X \times X$ into $\mathbb{R}$, called a *distance* or *metric* satisfying the following conditions for x, y and z in X:

1. $d(x, y) \geq 0$ and $d(x, y) = 0$ if and only if $x = y$

2. $d(x, y) = d(y, x)$ (symmetry)

3. $d(x, z) \leq d(x, y) + d(y, z)$ (triangle inequality)

Since we have a norm in $\mathbb{R}^n$ it is natural to consider the *distance* between points (vectors) $x, y \in \mathbb{R}^n$ to be

$$d(x, y) = ||x - y||.$$

The reader should draw a picture to see that this is really the distance between points of $\mathbb{R}^n$. One easily verifies properties (1), (2) and (3) of the distance $d : \mathbb{R}^n \times \mathbb{R}^n \to \mathbb{R}$. For instance, for (3) we have

$$d(x, z) = ||x-z|| = ||(x-y)+(y-z)|| \leq ||x-y||+||y-z|| = d(x,y)+d(y,z).$$

Thus, $\mathbb{R}^n$ equiped with the distance d is an example of a *metric space*. This would apply to any of the norms on $\mathbb{R}^n$ that we considered in Definition 1.3.8. However, we will be chiefly concerned with the Euclidean norm and the Euclidean distance

$$d(x, y) = ||x - y|| = \left(\sum_{i=1}^n (x_i - y_i)^2 \right)^{\frac{1}{2}}.$$

Definition 1.4.2. Let $a \in \mathbb{R}^n$ and $r \in \mathbb{R}$ with $r > 0$. The *open ball* around the point a is defined to be the set

$$B_r(a) = \{x \in \mathbb{R}^n : d(a, x) < r\} = \{x \in \mathbb{R}^n : ||a - x|| < r\}$$

$B_r(a)$ is also called an *r-neighborhood* of a.

For instance, when $n = 2$ and $a = (\alpha, \beta) \in \mathbb{R}^2$ the open ball around a with radius $r > 0$ is the open disk

$$B_r(a) = \left\{ (x, y) \in \mathbb{R}^2 : \sqrt{(\alpha - x)^2 + (\beta - y)^2} < r \right\},$$

and when $n = 1$ and $a \in \mathbb{R}$ it is a (symmetric) open interval around a

$$B_r(a) = \{x \in \mathbb{R} : |a - x| < r\} = (a - r, a + r).$$

We denote by

$$\overline{B}_r(a) = \{x \in \mathbb{R}^n : ||a - x|| \leq r\}.$$

The set $\overline{B}_r(a)$ is called the *closed ball* around a. The reason for the name *closed* will become clear shortly.

Definition 1.4.3. Let $\Omega \subseteq \mathbb{R}^n$. The set Ω is called an *open set* in $\mathbb{R}^n$ if each of its points has a sufficiently small open ball around it completely contained in Ω, that is, for each $x \in \Omega$ there exists $\delta > 0$ (which in general may depend on x) such that $B_\delta(x) \subseteq \Omega$.

The whole space $\mathbb{R}^n$ and the empty set $\emptyset$ are evidently open.

Example 1.4.4. Any open ball $B_r(a)$ in $\mathbb{R}^n$ is an open set. To see this, let $x \in B_r(a)$. We have to find $\delta > 0$ so that $B_\delta(x) \subseteq B_r(a)$.
Take $\delta = r - ||a - x|| > 0$. Then for any $y \in B_\delta(x)$ the triangle inequality implies

$$||a - y|| = ||a - x + x - y|| \leq ||a - x|| + ||x - y|| \leq ||a - x|| + \delta = r,$$

that is, $||a - y|| \leq r$. Hence, $y \in B_r(a)$.

Definition 1.4.5. The *boundary* of a set $\Omega \subseteq \mathbb{R}^n$, denoted by $\partial\Omega$, is the set of points $x \in \mathbb{R}^n$ such that every open ball around x intersects Ω and its complement Ω^c. That is, for every $\epsilon > 0$

$$\partial\Omega = \{x \in \mathbb{R}^n : B_\epsilon(x) \cap \Omega \neq \emptyset \,, \ B_\epsilon(x) \cap \Omega^c \neq \emptyset\}.$$

Clearly $\partial\Omega = \partial(\Omega^c)$.

Example 1.4.6. The boundary of $B_r(a)$ is the sphere $S_r(a)$ centered at $a = (a_1, ..., a_n)$ with radius $r > 0$,

$$S_r(a) = \partial(B_r(a)) = \{x \in \mathbb{R}^n : ||a - x|| = r\}$$

$$= \left\{(x_1, .., x_n) \in \mathbb{R}^n : (a_1 - x_1)^2 + ... + (a_n - x_n)^2 = r^2\right\}.$$

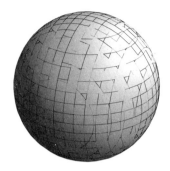

Figure 1.3: Unit sphere in $\mathbb{R}^3$: $x^2 + y^2 + z^2 = 1$

Exercise 1.4.7. Show that an open set can not contain any of its boundary points.

Definition 1.4.8. Let $\Omega \subseteq \mathbb{R}^n$. A point $a \in \Omega$ is called an *interior point* of Ω if there exists $\delta > 0$ such that $B_\delta(a) \subseteq \Omega$. The set of all interior points of Ω is called the *interior* of Ω and is denoted by Ω°.

Note that, $\Omega^\circ \subseteq \Omega$ and $\Omega = \Omega^\circ$ if and only if Ω is open.

Definition 1.4.9. A set S in $\mathbb{R}^n$ is said to be *closed* if and only if its complement S^c in $\mathbb{R}^n$ is open.

The reader should be aware that this is not the same as *not* open! Note that $\mathbb{R}^n$ and $\emptyset$ are closed.

Proposition 1.4.10. *A set S is closed in $\mathbb{R}^n$ if and only if $\partial S \subseteq S$, that is, it contains its boundary.*

Proof. Suppose S is closed. Then S^c is open. So, $S^c \cap \partial S = \emptyset$. Hence $\partial S \subseteq S$. Conversely, let $x \in S^c$. Since $\partial S \subseteq S$, it follows that x is not a boundary point of S. Therefore, from the definition of boundary point, there exists an $r > 0$ such that $B_r(x) \cap S = \emptyset$, that is, $B_r(x) \subseteq S^c$ and so S^c is open. $\qquad\square$

Definition 1.4.11. Let $S \subseteq \mathbb{R}^n$. We define the *closure* of S, denoted by $\overline{S}$, to be the set
$$\overline{S} = S \cup \partial S.$$

Thus, Proposition 1.4.10, tells us that a set S is closed if and only if $\overline{S} = S$. For example, $\overline{B}_r(a) = B_r(a) \cup S_r(a)$, and thus we use the name *closed ball*.

Remark 1.4.12. A set need be neither open nor closed. For example, in $\mathbb{R}$ the set $S = \{x \in \mathbb{R} : 0 \leq x < 1\} = [0, 1)$, whose boundary is $\partial S = \{0, 1\}$, is not open in $\mathbb{R}$ as it contains the boundary point 0 nor is it closed since it does not contain its boundary. Its closure is $\overline{S} = \{x \in \mathbb{R} : 0 \leq x \leq 1\} = [0, 1]$ and its interior $S^\circ = \{x \in \mathbb{R} : 0 < x < 1\} = (0, 1)$.

Similarly, in $\mathbb{R}^2$ the set $\Omega = \{(x, y) \in \mathbb{R}^2 : 0 \leq x \leq 1, 0 < y < 1\}$ a square, is neither open nor closed in $\mathbb{R}^2$. Here

$$\overline{\Omega} = \{(x, y) \in \mathbb{R}^2 : 0 \leq x \leq 1, 0 \leq y \leq 1\} = [0, 1] \times [0, 1]$$

and $\Omega^\circ = (0, 1) \times (0, 1)$.

Example 1.4.13.   1. Consider the set $\mathbb{Z} \subset \mathbb{R}$. We have $\mathbb{Z}^\circ = \emptyset$ and $\partial \mathbb{Z} = \mathbb{Z} = \overline{\mathbb{Z}}$. So $\mathbb{Z}$ is closed.

2. Consider the set $\mathbb{Q} \subset \mathbb{R}$. Theorems 1.1.7, 1.1.9 tell us that $\mathbb{Q}^\circ = \emptyset$ and $\overline{\mathbb{Q}} = \mathbb{R} = \partial \mathbb{Q}$. Notice that $\mathbb{Q}$ is neither open nor closed and thus, $\mathbb{Q}^c$ is also neither open nor closed (why?).

Exactly as in $\mathbb{R}$, replacing $|\cdot|$ by $||\cdot||$, we have

Definition 1.4.14. Let $\Omega \subseteq \mathbb{R}^n$ and $a \in \mathbb{R}^n$ (not necessarily in Ω). We call a an *accumulation point or limit point* of Ω if every open ball around a contains at least one point $x \in \Omega$ distinct from a, that is, for every $r > 0$ we have $(B_r(a) - \{a\}) \cap \Omega \neq \emptyset$.

A point $a \in \Omega$ which is not an accumulation point of Ω is called an *isolated point* of Ω. Hence, a is an isolated point of Ω if there is an open ball $B_{r_0}(a)$ such that $B_{r_0}(a) \cap \Omega = \{a\}$, for some $r_0 > 0$. We remark that as in $\mathbb{R}$, any finite set $\Omega = \{v_1, ..., v_m\}$ in $\mathbb{R}^n$ has no limit points. In fact, each point of Ω is an isolated point. Note also that $\partial(\Omega) = \Omega$ and so any finite set is closed.

The next proposition provides a useful characterization of closedness.

Proposition 1.4.15. *A set S in $\mathbb{R}^n$ is closed if and only if S contains all its limit points.*

Proof. Suppose S is closed and let a be a limit point of S. If $a \notin S$, then $a \in S^c$ which is open. Therefore, there exists $\delta > 0$ such that $B_\delta(a) \subseteq S^c$. It follows that $B_\delta(a) \cap S = \emptyset$. Since a is a limit point of A this is impossible. Conversely, suppose S contains all its limit points. We show that S^c is open (and so S will be closed). Let $x \in S^c$. By hypothesis x is not a limit point of S and consequently there exists $r > 0$ such that $B_r(x) \cap S = \emptyset$. Hence, $B_r(x) \subseteq S^c$ and so S^c is open. $\qquad\square$

EXERCISES

1. Show that if Ω_1 and Ω_2 are open in $\mathbb{R}^n$, then $\Omega_1 \cup \Omega_2$ and $\Omega_1 \cap \Omega_2$ are also open. Generalize to any finite number of open sets.

2. Show that if A_1 and A_2 are closed in $\mathbb{R}^n$, then $A_1 \cup A_2$ and $A_1 \cap A_2$ are also closed. Generalize to any finite number of closed sets.

3. Give an example of a set S in $\mathbb{R}^n$ such that $S^\circ = \emptyset$ and $\overline{S} = \mathbb{R}^n$.

4. Show that $S^\circ = S \setminus \partial S$.

5. Let $x, y \in \mathbb{R}^n$ with $x \neq y$. Show that there exist open sets U and V such that $x \in U$, $y \in V$ and $U \cap V = \emptyset$.[12]

6. Let S_1 and S_2 be sets in $\mathbb{R}^n$. Show that
   $(S_1 \cap S_2)^\circ = S_1^\circ \cap S_2^\circ$ and $S_1^\circ \cup S_2^\circ \subseteq (S_1 \cup S_2)^\circ$

7. Show that $\overline{S_1 \cap S_2} \subseteq \overline{S_1} \cup \overline{S_2}$ and $\overline{S_1 \cup S_2} = \overline{S_1} \cup \overline{S_2}$.

8. Show that $\partial(S_1 \cup S_2) \subseteq \partial S_1 \cup \partial S_2$ and $\partial(S_1 \cap S_2) \subseteq \partial S_1 \cup \partial S_2$.

9. Let $S \subset \mathbb{R}^n$. Show that $\partial S = \overline{S} \cap \overline{S^c}$. Deduce that ∂S is a closed set.

10. Let $S \subset \mathbb{R}^n$. Show that $(S^\circ)^c = \overline{S^c}$. Deduce that $\partial S = \overline{S} \setminus S^\circ$.

11. Let $S \subset \mathbb{R}^n$. Show that

---

[12] A topological space with this property is called a *Hausdorff space*. In particular, $\mathbb{R}^n$ is a Hausdorff space.

(a) $\partial(S^\circ) \subset \partial S$.

(b) $\partial(\partial S) \subset \partial S$.

12. Let $\overline{S} = \mathbb{R}^n$ (such a set is called *dense* in $\mathbb{R}^n$). Show that if Ω is open, then $\overline{S \cap \Omega} = \overline{\Omega}$. Is this true if Ω is not open?

13. Find the limit points of the sets: $A = \left\{ (1 + \frac{1}{m}, \sin \frac{m\pi}{2}) : m \in \mathbb{N} \right\}$ and $B = \left\{ ([1 + \frac{1}{m}] \cos(m\pi), \frac{1}{2^m}) : m \in \mathbb{N} \right\}$

14. Prove that the set of all limit points of any set is always closed.

1.5   Convergence of sequences in $\mathbb{R}^n$

The significance of the metric d is that it enables us to talk about limiting processes. This involves replacing $|\cdot|$ by $||\cdot||$. Here we study the convergence of sequences in $\mathbb{R}^n$.

We reserve the letter n for the dimension and use letters such as i, j, k for the index on a sequence, that is, we denote by $\{x_k\}_{k=1}^\infty$ a sequence of vectors in $\mathbb{R}^n$ and the components of the vector x_k will be $x_k = (x_{k1}, x_{k2}, ..., x_{kn})$. We now define the limit of a sequence in $\mathbb{R}^n$

Definition 1.5.1. Let $\{x_k\}_{k=1}^\infty$ be a sequence of vectors in $\mathbb{R}^n$ and $x \in \mathbb{R}^n$. We say that $\{x_k\}_{k=1}^\infty$ *converges* to a *limit* x if $d(x_k, x) = ||x_k - x|| \to 0$ in $\mathbb{R}$ as $k \to \infty$. We write $x_k \to x$.

Thus $x_k \to x$ if for every $\epsilon > 0$ there exists a positive integer N such that $||x_k - x|| < \epsilon$ whenever $k \geq N$. If a sequence does not converge, we say that the sequence *diverges*. We shall denote a sequence $\{x_k\}_{k=1}^\infty$ simply by $\{x_k\}$ and we shall omit writing $k \to \infty$ whenever this is clear from the context. We remark that $x_k \to x$ if and only if $||x_k - x||^2 \to 0$. As in $\mathbb{R}$, the limit of a convergent sequence is unique.

Example 1.5.2. The sequence $\{x_k\} = \left\{ (3 - \frac{1}{k+1}, \frac{1}{\pi^k}) \right\}$ in $\mathbb{R}^2$ converges to $x = (3, 0)$. Indeed, letting $k \to \infty$ we have

$$||x_k - x||^2 = \left(3 - \frac{1}{k+1} - 3 \right)^2 + \left(\frac{1}{\pi^k} - 0 \right)^2 = \frac{1}{(k+1)^2} + \left(\frac{1}{\pi^2} \right)^k \to 0.$$

Just as in $\mathbb{R}$, a sequence $\{x_k\}$ in $\mathbb{R}^n$ is said to be *bounded* if there exists $M > 0$ with $||x_k|| \leq M$ for all $k = 1, 2, ...$, and a sequence which is not bounded is called *unbounded*. In addition, every convergent sequence is bounded, and any unbounded sequence diverges.

The following theorem lists the basic properties of convergent sequences in $\mathbb{R}^n$.

Theorem 1.5.3. *Let $\{x_k\}$ and $\{y_k\}$ be two sequences in $\mathbb{R}^n$ with $x_k \to x$ and $y_k \to y$, where $x, y \in \mathbb{R}^n$. Then*

1. *For $a, b \in \mathbb{R}$, $ax_k + by_k \to ax + by$*

2. *$||x_k|| \to ||x||$*

3. *$\langle x_k, y_k \rangle \to \langle x, y \rangle$*

4. *$x_k \to x$ if and only if $x_{ki} \to x_i$ for each $i = 1, ...n$*

Proof.   1. By hypothesis and the triangle inequality we have

$$0 \leq ||(ax_k + by_k) - (ax + by)|| = ||a(x_k - x) + b(y_k - y)||$$

$$\leq |a| \cdot ||x_k - x|| + |b| \cdot ||y_k - y|| \to 0.$$

Therefore, $ax_k + by_k \to ax + by$

2. From Corollary 1.3.5, $|\,||x_k|| - ||x||\,| \leq ||x_k - x|| \to 0$.
   Hence, $||x_k|| \to ||x||$

3. First note that

$$|\langle x_k, y_k \rangle - \langle x, y \rangle| = |\langle x_k, y_k \rangle - \langle x, y_k \rangle + \langle x, y_k \rangle - \langle x, y \rangle|$$

$$= |\langle x_k - x, y_k \rangle + \langle x, y_k - y \rangle| \leq |\langle x_k - x, y_k \rangle| + |\langle x, y_k - y \rangle|.$$

Using the Cauchy-Schwarz inequality we obtain

$$|\langle x_k, y_k \rangle - \langle x, y \rangle| \leq ||x_k - x|| \cdot ||y_k|| + ||x||\,||y_k - y||.$$

Since $\{y_k\}$ converges, it is bounded. So there exists $M > 0$ such that $||y_k|| \leq M$. Now, we have

$$|\langle x_k, y_k \rangle - \langle x, y \rangle| \leq ||x_k - x||\,M + ||x||\,||y_k - y|| \to 0.$$

Thus, $\langle x_k, y_k \rangle \to \langle x, y \rangle$.

4. Suppose $x_k \to x$. Then

$$|x_{ki} - x_i| = \sqrt{(x_{ki} - x_i)^2} \leq \sqrt{\sum_{i=1}^{n}(x_{ki} - x_i)^2} = ||x_k - x|| \to 0.$$

Conversely, if $x_{ki} \to x_i$ for all $i = 1, ..., n$, then for every $\epsilon > 0$ there are positive integers N_i such that $|x_{ki} - x_i| < \frac{\epsilon}{\sqrt{n}}$ for all $k \geq N_i$. Letting $N = \max\{N_i : i = 1, 2, ..., n\}$, we have $|x_{ki} - x_i| < \frac{\epsilon}{\sqrt{n}}$ for all $k \geq N$. Now,

$$||x_k - x|| = \sqrt{\sum_{i=1}^{n}(x_{ki} - x_i)^2} < \sqrt{\sum_{i=1}^{n}\frac{\epsilon^2}{n}} = \epsilon.$$

Therefore, $||x_k - x|| < \epsilon$ for all $k \geq N$.   □

By property (4) of the above proposition, we see that to study the covergence of a sequence in $\mathbb{R}^n$ it suffices to study the convergence of its component sequences in $\mathbb{R}$.

Next, we would like to restate the definition of a Cauchy sequence for a sequence in $\mathbb{R}^n$, or more generally in a metric space.

Definition 1.5.4. Let (X, d) be a metric space.

1. A sequence $\{x_k\}$ of points in X is called a *Cauchy sequence* if $d(x_k, x_j) \to 0$ as $k, j \to \infty$, that is, for every $\epsilon > 0$ there exists positive integer N such that $d(x_k, x_j) < \epsilon$ for all $k, j \geq N$.

2. X is called a *complete* metric space if every Cauchy sequence in X converges to a point in X.

As in $\mathbb{R}$, every Cauchy sequence is bounded. Moreover, the definition of *subsequence* of a given sequence in $\mathbb{R}$, as well as, Propositions 1.1.27, 1.1.30 extend with identical proofs for sequences in $\mathbb{R}^n$ (all that is needed is to replace $|\cdot|$ by $||\cdot||$). Thus, for any $\{x_k\}$ in $\mathbb{R}^n$ we have

Proposition 1.5.5.   *1. If $x_k \to x$, then every subsequence $x_{k_m} \to x$.*

2. Every convergent sequence is a Cauchy sequence.

3. *A Cauchy sequence converges, if it has a convergent subsequence.*

Next we prove the completeness of $\mathbb{R}^n$.

Theorem 1.5.6. $\mathbb{R}^n$ *is complete.*

Proof. Since $\{x_k\}$ is Cauchy, for each $i = 1, ..., n$ we have

$$|x_{ki} - x_{ji}| \leq ||x_k - x_j|| \to 0$$

as $k, j \to \infty$. Hence each component sequence $\{x_{k_i}\}$ is Cauchy in $\mathbb{R}$. Since $\mathbb{R}$ is complete, there exist $x_i \in \mathbb{R}$ such that $x_{ki} \to x_i$ for each $i = 1, ..., n$. Taking $x = (x_1, ..., x_n) \in \mathbb{R}^n$, Theorem 1.5.3 yields $x_k \to x$. $\square$

The importance of completeness is that it enables us to prove that a sequence converges without a *priori* knowledge of the limit. It can therefore be used to prove existence of the limit which would enable us to prove important existence theorems in the theory of differential equations and elsewhere.

Sequential convergence can be used to characterize the closure of a set.

Proposition 1.5.7. *Let S be a subset of $\mathbb{R}^n$ and $x \in \mathbb{R}^n$. Then $x \in \overline{S}$ if and only if there exists a sequence of points in S that converges to x.*

Proof. Suppose $x \in \overline{S}$. If x is in S iteslf, let $x_k = x$ for all $k = 1.2....$ If $x \notin S$, then $x \in \partial S$. Therefore, for each k the open ball $B_{\frac{1}{k}}(x)$ contains points of S. So, for each k we can select a point $x_k \in S$ such that $||x_k - x|| < \frac{1}{k}$. Thus, in either case, $x_k \in S$ and $x_k \to x$. Conversely, suppose $\{x_k\}$ is a sequence in S such that $x_k \to x$. Then every open ball around x contains elements of S. In fact, it contains all x_k for sufficiently large k. Hence, $x \in S$ or $x \in \partial S$. That is, $x \in \overline{S}$. $\square$

Of particular interest in $\mathbb{R}$ is the following

Corollary 1.5.8. *Let $\emptyset \neq S \subseteq \mathbb{R}$ bounded from above. Suppose $M = \sup(S)$. Then $M \in \overline{S}$. Hence $M \in S$ if S is closed. Similarly, for $m = \inf(S)$.*

Proof. Since $M = \sup(S)$, for each $k = 1, 2, \ldots$ there exists $x_k \in S$ such that $M - \frac{1}{k} < x_k \leq M$. Hence, $x_k \to M$. Thus, $M \in \overline{S}$.   □

Definition 1.5.9. Let $S \subset \mathbb{R}^n$. The *diameter* $d(S)$ of S is

$$d(S) = \sup \{\|x - y\| : x, y \in S\}.$$

Definition 1.5.10. A set S in $\mathbb{R}^n$ is called *bounded* if there exists $R > 0$ such that $S \subseteq B_R(0)$. Equivalently, if its diameter is finite.

   We conclude this section by generalizing the Bolzano-Weiestrass theorem to $\mathbb{R}^n$.

Theorem 1.5.11. *(Bolzano-Weiertsrass theorem in* $\mathbb{R}^n$*)*

  1. *Every bounded sequence in* $\mathbb{R}^n$ *has a convergent subsequence.*

  2. *Every bounded infinite set in* $\mathbb{R}^n$ *has a limit point.*

Proof.

  1. Suppose $\|x_k\| \leq M$ for all k. Since $|x_{ki}| \leq \|x_k\|$ for all $i = 1, \ldots, n$, it follows that the sequences $\{x_{ki}\}_{k=1}^{\infty}$ of the components are all bounded. Hence, from Corollary 1.1.28, we can extract a convergent subsequence from each component sequence. The trouble is that the indices on these subsequences might all be different, so we can not put them together. (For example, we might have chosen the even-numbered terms for $i = 1$ and the odd-numbered terms for $i = 2$). To construct the desired subsequence, we have to proceed step by step. First we choose a subsequence $\{x_{m_k}\}$ such that the first components converge. This subsequence is bounded. Hence it has a subsequence for which the second components converge. This new subsequence has the property that the first *and* the second components converge. Continuing this way, after a finite number of steps we find a subsequence whose components *all* converge.

  2. Let S be an infinite bounded set in $\mathbb{R}^n$. For $k = 1, 2, \ldots$, we can select $x_k \in S$ with $x_k \neq x_j$ for $k \neq j$. Then $\{x_k\}_{k=1}^{\infty}$ is a bounded sequence in $\mathbb{R}^n$ of distinct elements of S. By part (1), $\{x_k\}$ has a subsequence which converges, say, to a. Hence, every

neighborhood of a contains infinitely many points of S and thus a is a limit point of S.

$\square$

1.6   Compactness

The concept of compactness is of fundamental importance in analysis and in mathematics in general. Although there are several possible definitions of compactness, these all coincide in $\mathbb{R}^n$.

In this section I will denote any (infinite) index set.

Definition 1.6.1. By an *open cover* of a set S in $\mathbb{R}^n$ we mean a collection $\{V_i\}_{i\in I}$ of open set of $\mathbb{R}^n$ such that $S \subset \bigcup_{i\in I} V_i$. A *subcover* is a subcollection which also covers S.

Lemma 1.6.2. *(Lebesgue covering lemma)*[13] *Let S be a closed and bounded subset of $\mathbb{R}^n$ and $\{V_i\}_{i\in I}$ an open cover of S. Then for each $x \in S$, there exists some $\delta > 0$ such that $B_\delta(x) \subseteq V_i$ for at least one $i \in I$. (Such a number $\delta > 0$ is called a Lebesgue number of S for the open cover $\{V_i\}_{i\in I}$.)*

Proof. Suppose that such a $\delta > 0$ does *not* exist. Then for each $k = 1, 2, ...$, there exists some $x_k \in S$ such that $B_{\frac{1}{k}}(x_k) \cap V_i^c \neq \emptyset$ for all $i \in I$. Since S is bounded, $\{x_k\}$ is bounded and by the Bolzano-Weiretsrass theorem it has a convergent subsequence, say, $x_{m_k} \to x$. Then $x \in \overline{S} = S$, by the closedness of S. Since $S \subseteq \bigcup_{i\in I} V_i$, it follows that $x \in V_j$ for some $j \in I$. Moreover, since V_j is open there exists $r > 0$ with $B_r(x) \subset V_j$. Now select some m_k sufficiently large so that $\frac{1}{m_k} < \frac{r}{2}$ and $\| x_{m_k} - x \| < \frac{r}{2}$. It follows that $B_{\frac{1}{m_k}}(x_{m_k}) \subseteq B_r(x) \subset V_j$. That is, $B_{\frac{1}{m_k}}(x_{m_k}) \cap V_j^c = \emptyset$ contradicting the choice of x_k. $\square$

Definition 1.6.3. A subset S of $\mathbb{R}^n$ is said to be *compact* if from every open cover of S we can extract a *finite* subcover.

---

[13]H. Lebesgue (1875-1941). A student of Borel. He made pioneering contributions in analysis and developed what is now called the theory of *Lebesgue integration*.

Definition 1.6.4. A subset S of $\mathbb{R}^n$ is said to be *sequentially compact* if every sequence of points of S has a subsequence which *converges to a point of S.*

The fundamental result is the following theorem and gives a simple characterization of compact sets in $\mathbb{R}^n$: they are the *closed* and *bounded* sets. The implication (1) $\Rightarrow$ (3) is known as the *Heine-Borel theorem*[14]. It tells us, for example that closed and bounded sets are compact.

Theorem 1.6.5. *Let S be a subset of $\mathbb{R}^n$. Then the following are equivalent:*

1. *S is closed and bounded.*

2. *S is sequentially compact.*

3. *S is compact.*

Proof. First we prove (1) $\Leftrightarrow$ (2). Suppose S is closed and bounded. Let $\{x_k\}$ be a sequence of points in S. Since S is bounded, $\{x_k\}$ is bounded. By the Bolzano-Weierstrass theorem $\{x_k\}$ has a convergent subsequence, say $x_{m_k} \to x$. Since S is closed, $x \in S$.

Conversely, suppose S is sequentially compact. Let x be a limit point of S. Then there is a sequence $\{x_k\}$ of points of S with $x_k \to x$. By the sequential compactness of S the sequence $\{x_k\}$ has a subsequence which converges to a point of S. However, since every subsequence of a convergent sequence converges to the same limit as the sequence, it follows that $x \in S$. Hence, from Proposition 1.4.15, S is closed. If S were not bounded, then for each positive integer k there is a point $x_k \in S$ with $\| x_k \| > k$. By the sequential compactness, the sequence $\{x_k\}$ has a convergent subsequence, say $\{x_{m_k}\}$. However, since any convergent sequence is bounded, there is some $M > 0$ with $\| x_{m_k} \| \leq M$ for all m_k. This contradicts $\| x_{m_k} \| > m_k$.

---

[14]E. Heine (1821-1881). He is known for results on special functions (*spherical harmonics, Legendre functions*) and in real analysis. E. Borel (1871-1956). Mathematician and politician in Paris. A student of Darboux, he made important contributions in analysis.

Next we prove (3) $\Leftrightarrow$ (1). Let S be compact. First we show that S is bounded. An open cover of S is given by the open balls $B_k(0)$, $k = 1, 2, ...$ Since S is compact, a finite number of these balls covers S, say, $S \subseteq \bigcup_{k=1}^{m} B_k(0)$. Since $B_1(0) \subset B_2(0) \subset ... \subset B_m(0)$, it follows $S \subseteq B_m(0)$ and so S is bounded. Now we show that S is closed by showing that S^c is open. Let $y \in S^c$. For each $x \in S$, let $r_x = \frac{1}{2} \parallel x - y \parallel > 0$. Then the open balls $\{B_{r_x}(x)\}_{x \in S}$ cover S. Since S is compact, there is a finite number of points $x_1, ..., x_m \in S$ such that $S \subset \bigcup_{j=1}^{m} B_{r_{x_j}}(x_j)$. Take $r = \min \{r_{x_j} : j = 1, ..., m\}$. Then $B_r(y) \cap S = \emptyset$. Hence, $B_r(y) \subset S^c$, so that S^c is open.

Conversely, suppose S is closed and bounded. Let $S \subseteq \bigcup_{i \in I} V_i$ be an open cover of S. We claim that for each $r > 0$, there exist $x_1, ..., x_m \in S$ such that

$$S \subseteq \bigcup_{j=1}^{m} B_r(x_j)$$

For suppose not; let $r > 0$ and fix some $x_1 \in S$. Then choose $x_2 \in (S - B_r(x_1))$. Continuing this way, using induction, choose $x_{k+1} \in (S \backslash \cup_{j=1}^{k} B_r(x_j))$. Then, $\parallel x_k - x_m \parallel \geq r$ for $k \neq m$. This implies that the sequence $\{x_k\}$ of points in S can not have any convergent subsequence. Since S is bounded, $\{x_k\}$ is also bounded and the Bolzano-Weierstrass theorem implies that $\{x_k\}$ has a convergent subsequence. This contradiction proves our claim. Now, by Lemma 1.6.2, let $\delta > 0$ be a Lebesgue number of S for the open cover $\{V_i\}_{i \in I}$ and choose $x_1, ..., x_m \in S$ such that $S \subseteq \bigcup_{j=1}^{m} B_\delta(x_j)$. For each $j = 1, ..., m$ pick some $i_j \in I$ such that $B_\delta(x_j) \subseteq V_{i_j}$. Then $S \subseteq \bigcup_{j=1}^{m} B_r(x_j) \subseteq \bigcup_{j=1}^{m} V_{i_j}$. Hence, S is compact. $\square$

Note that every finite subset of $\mathbb{R}^n$ is obviously bounded and being closed it is therefore compact. As the following exercise shows, both the boundedness and the closedness of the subset S in $\mathbb{R}^n$ are needed to guarantee the compactness of S.

Exercise 1.6.6.   1. Let $S = (0, 1]$ in $\mathbb{R}$. Note that S is bounded but *not* closed. Take $V_k = (\frac{1}{k}, 2)$. Show that $\{V_k\}_{k \in \mathbb{Z}_+}$ is an open cover of S but contains no finite subcover of S.

2. Let $S = [0, \infty)$ in $\mathbb{R}$. Then S is closed but *not* bounded. Take $V_k = (k - 2, k)$. Show that $\{V_k\}_{k \in \mathbb{Z}_+}$ is an open cover of S but contains no finite subcover.

Remark 1.6.7. The proof of (3) $\Rightarrow$ (1) given above is valid in any metric space. Thus, in any metric space, a compact set is closed and bounded. However, the converse, that is, the Heine-Borel theorem does not necessarily hold in any metric space. For example, let X be an infinite set and d the *discrete distance* on X (that is, $d(x, y) = 1$ if $x \neq y$ and $d(x, y) = 0$ if $x = y$.) The set X is bounded since the distance between any two of its points is at most 1. In addition, X being the whole space is closed. Let $r = \frac{1}{2}$. For each $x \in X$, note that $B_r(x) = \{x\}$ and $X = \bigcup_{x \in X} B_r(x)$. But this open cover has no finite subcover. Hence X is not compact. What goes wrong here is the fact that no infinite subset of X (with the discrete distance) has a limit point (in other words, no sequence of distinct points can converge). This means that the Bolzano-Weierstrass theorem is not valid in this case. However, if the Bolzano-Weierstrass theorem were valid in some metric space, then (as in the above proof of (1) $\Rightarrow$ (3)) the Heine-Borel theorem would also be valid. Further nontrivial examples will follow from Exercise 1.8.22.

Proposition 1.6.8. *A closed subset A of a compact set S is itself compact.*

Proof. Let $\{V_i\}_{i \in I}$ be an open cover of A. Then the collection $\{\{V_i\}_{i \in I}, A^c\}$ is an open cover of S. Since S is compact we can extract a finite subcover of S. As $A \cap A^c = \emptyset$, it follows that the open cover $\{V_i\}_{i \in I}$ of A contains a finite subcover of A. $\qquad\square$

Corollary 1.6.9. *If A is closed and S is compact, then $A \cap S$ is compact.*

Next we consider the relationship of completeness to compactness. Notice that in a complete space we don't know anything about the limit of the Cauchy sequence, just that there is one. So completeness is considerably weaker than compactness. For this reason when working with completeness we are not at all restricted to $\mathbb{R}^n$ (although from our arguments it may seem we are).

Proposition 1.6.10. *A closed subspace Y of a complete metric space X is complete.*

Proof. Let $\{y_n\}$ be a Cauchy sequence in Y. Since the metric on Y is the restriction of the one on X, $\{y_n\}$ is a Cauchy sequence in X. By completeness it converges to a point $x \in X$. But since Y is closed, $x \in Y$. □

Definition 1.6.11. A *locally compact* metric space is a space in which each neighborhood of a point contains a compact neighborhood (of that point).

Compact spaces are (obviously and trivially) locally compact. Every infinite set with the discrete metric is locally compact, but as we saw in Remark 1.6.7, it is not compact. Since closed balls are compact in $\mathbb{R}^n$, $\mathbb{R}^n$ *is locally compact.*

We end this section with a proposition which is a special case of the celebrated *Tychonoff Theorem*. This important result states that the product of an arbitrary collection of compact spaces is compact. Here we require only the following,

Proposition 1.6.12. *If X and Y are compact metric spaces, then so is $X \times Y$.*

Proof. A metric on $X \times Y$ is given for instance as

$$d((x,y),(a,b)) = \max(d_X(x,a), d_Y(y,b)).$$

The reader should check that this is a metric and restricted to X or Y gives the original one. The rest is then a simple exercise in sequential compactness and is left to the reader. □

1.7   Equivalent norms (*)

In Definition 1.3.8 we saw that $\|\cdot\|_\infty$ and $\|\cdot\|_1$ are also norms on $\mathbb{R}^n$. It can be shown that for $1 \le p < \infty$

$$\|x\|_p = \left(\sum_{i=1}^n |x_i|^p\right)^{\frac{1}{p}}$$

is a norm on $\mathbb{R}^n$ called the *p-norm* (see Exercise 1.8.21). Of course, for $p = 2$ the 2-norm is just the Euclidean norm. In Example 1.3.9 we saw that for each $\mathbb{R}^n$, $||x||_\infty \leq ||x|| \leq \sqrt{n}||x||_\infty$. When such an estimate holds the norms $|| \cdot ||_\infty$ and $|| \cdot ||$ on $\mathbb{R}^n$ are called *equivalent*. More generally we have

Definition 1.7.1. Given two norms $|| \cdot ||_1$ and $|| \cdot ||_2$ in $\mathbb{R}^n$ (or in any normed linear space), the norms are called *equivalent* if there exist $\alpha > 0$ and $\beta > 0$ such that, for all $x \in \mathbb{R}^n$,

$$\alpha ||x||_1 \leq ||x||_2 \leq \beta ||x||_1.$$

Note that if a sequence converges (or is Cauchy or is bounded) with respect to one norm it also converges (is Cauchy or is bounded) with respect to any other equivalent norm. However, it may happen that a sequence converges with respect to one norm and diverges with respect to another. Fortunaltely, in $\mathbb{R}^n$ we do not have such worries. As we see in the following theorem all norms in $\mathbb{R}^n$ are equivalent. In fact, we prove that *all* norms in a *finite-dimensional* normed linear space are equivalent.

Furthermore note that equivalent norms give equivalent metrics and so a set which is open with respect to one metric is also open with respect to the other and vice versa.[15]

Theorem 1.7.2. *Let V be a finite dimensional normed linear space. Then all norms on V are equivalent. In particular, this holds on $\mathbb{R}^n$.*

Proof. Let $B = \{v_1, ..., v_n\}$ be a basis of V and $x \in V$. Then x can be written uniquely as $x = a_1v_1 + ... + a_nv_n$. Let

$$||x||_1 = \sum_{i=1}^n |a_i|.$$

This is easily seen to be a norm on V. Now let $|| \cdot ||$ be any other norm on V. We will show that $|| \cdot ||$ is equivalent to $|| \cdot ||_1$. We have

$$||x|| = ||a_1v_1 + ... a_nv_n|| \leq |a_1| \cdot ||v_1|| + ... + |a_n| \cdot ||v_n|| \leq \max_{i=1,...n} ||v_i|| \left(\sum_{i=1}^n |a_i| \right).$$

---

[15]In the language of topology this means the norms give rise to the same topology.

That is, $||x|| \leq \beta ||x||_1$ where $\beta = \max_{i=1,...n} ||v_i||$

To find $\alpha > 0$ such that $\alpha ||x||_1 \leq ||x||$, we claim that there exists $c > 0$ such that $||x||_1 \leq c||x||$ or $\sum_{i=1}^{n} |a_i| \leq c|| \sum_{i=1}^{n} a_i v_i ||$. Then with $\alpha = \frac{1}{c}$ it will follow that $\alpha ||x||_1 \leq ||x||$.

Thus, it remains to prove our claim. First note that if $\sum_{i=1}^{n} |a_i| \neq 1$, then taking $b_i = \frac{a_i}{\sum_{i=1}^{n} |a_i|}$ for $i = 1, ...n$, we see, it is enough to show that for $x = \sum_{i=1}^{n} b_i v_i$ with $\sum_{i=1}^{n} |b_i| = 1$ there exists $c > 0$ such that

$$1 \leq c|| \sum_{i=1}^{n} b_i v_i || = c||x||.$$

Suppose that such a $c > 0$ doesn't exist. Then for all $c > 0$ we have $1 > c||x||$ or $||x|| < \frac{1}{c}$. In particular, for $c = k = 1, 2, ...$ we obtain a sequence $\{x_k = \sum_{i=1}^{n} b_{ki} v_i\}$ of vectors in V satisfying $||x_k||_1 = \sum_{i=1}^{n} |b_{ki}| = 1$ and $||x_k|| < \frac{1}{k} \to 0$. Then the sequence $\{[y_k] = (b_{k1}, ..., b_{kn})\}$ of the coordinate vectors of the sequence $\{x_k\}$ relative to the basis B is bounded (with bound 1) in $\mathbb{R}^n$. Therefore from the Bolzano-Weierstass theorem, the sequence $\{(b_{k1}, ..., b_{kn})\}$ in $\mathbb{R}^n$ contains a convergent subsequence, say, $\{d_k = (d_{k1}, ..., d_{kn})\}$ converging to $d = (d_1, ..., d_n)$.

Now, let $x = \sum_{i=1}^{n} d_i v_i$. Then $1 = \sum_{i=1}^{n} |d_{ki}| \to \sum_{i=1}^{n} |d_i|$, that is $\sum_{i=1}^{n} |d_i| = 1$, while

$$||x_k|| = || \sum_{i=1}^{n} d_{ki} v_i || \to || \sum_{i=1}^{n} d_i v_i ||.$$

Since $||x_k|| \to 0$, it follows that $|| \sum_{i=1}^{n} d_i v_i || = 0$ and so $\sum_{i=1}^{n} d_i v_i = 0$ which is impossible since $\{v_1, ..., v_n\}$ are linearly independent. $\qquad \square$

1.8   Solved problems for Chapter 1

Exercise 1.8.1. (*) Let V be a normed linear space and $x, y \in V$ nonzero vectors. Show the following improvement of the triangle inequality.[16]

---

[16]These norm inequalities are due to L. Maligranda, see [22].

$$||x + y|| \le ||x|| + ||y|| - \left(2 - \left\|\frac{x}{||x||} + \frac{y}{||y||}\right\|\right) \min(||x||, ||y||).$$

Solution.
To see this note that without loss of generality we may assume that $||x|| = \min(||x||, ||y||)$. Then by the triangle inequality,

$$||x + y|| = \left\|\left(\frac{||x||}{||x||}x + \frac{||x||}{||y||}y\right) + \left(1 - \frac{||x||}{||y||}\right)y\right\|$$

$$\le ||x|| \cdot \left\|\frac{x}{||x||} + \frac{y}{||y||}\right\| + ||y|| - ||x||$$

$$= ||y|| + \left(\left\|\frac{x}{||x||} + \frac{y}{||y||}\right\| - 1\right)||x||$$

$$= ||x|| + ||y|| + \left(\left\|\frac{x}{||x||} + \frac{y}{||y||}\right\| - 2\right)||x||.$$

A similar computation gives

$$||x + y|| \ge ||x|| + ||y|| - \left(2 - \left\|\frac{x}{||x||} + \frac{y}{||y||}\right\|\right) \max(||x||, ||y||)$$

Furthermore, note that if either $||x|| = ||y|| = 1$ or $x = cy$ with $c > 0$, then equality holds in both inequalities.

Exercise 1.8.2. Let $\{v_1, ..., v_k\}$ be an orthonormal set in V. Then

1. For every $x \in V$, there exists $y \in \text{span}\{v_1, ..., v_k\}$ such that

$$||x - y|| = \min\{||x - z|| : z\} \in \text{span}\{v_1, ..., v_k\}.$$

   In fact, $y = \sum_{j=1}^{k} \langle x, v_j \rangle v_j$

2. For $x \in V$, $\sum_{j=1}^{k} |\langle x, v_j \rangle|^2 \le ||x||^2$ (Bessel's inequality)[17]

Moreover, equality holds if and only if $x \in \text{span}\{v_1, ..., v_k\}$

---

[17]F. Bessel (1784-1846). Mathematician and astronomer. A contemporary of Gauss. He systematized the so-called *Bessel functions* (discovered by D. Bernoulli).

Solution.

1. Let $x \in V$ and $c_j = \langle x, v_j \rangle$. For any $z \in$ span $\{v_1, ..., v_k\}$,
    say $z = \sum_{j=1}^{k} a_j v_j$, we then have

    $$||x - z||^2 = ||x - \sum_{j=1}^{k} a_j v_j||^2 = \langle x - \sum_{j=1}^{k} a_j v_j, x - \sum_{j=1}^{k} a_j v_j \rangle$$

    $$= ||x||^2 - 2 \sum_{j=1}^{k} a_j \langle x, v_j \rangle + \sum_{j=1}^{k} a_j^2 = ||x||^2 - 2 \sum_{j=1}^{k} a_j c_j + \sum_{j=1}^{k} a_j^2$$

    $$= ||x||^2 + \sum_{j=1}^{k} (a_j - c_j)^2 - \sum_{j=1}^{k} c_j^2.$$

    The choice $a_j = c_j$ for all $j = 1, ..., k$, minimizes the quantity $||x - z||$.
    Therefore, the *best approximation* to x by elements of span $\{v_1, ..., v_k\}$
    is $y = \sum_{j=1}^{k} \langle x, v_j \rangle v_j$.

2. This choice $a_j = c_j = \langle x, v_j \rangle$ also gives

    $$0 \le ||x - \sum_{j=1}^{k} \langle x, v_j \rangle v_j||^2 = ||x||^2 - \sum_{j=1}^{k} \langle x, v_j \rangle^2.$$

    Hence,

    $$\sum_{j=1}^{k} |\langle x, v_j \rangle|^2 \le ||x||^2.$$

Exercise 1.8.3. For x and y as in the above exercise, show that $x - y \in$ (span $\{v_1, .., v_k\})^{\perp}$.

Solution.
    This is so because $(x - y) \perp v_i$ for all $i = 1, ..., k$. Indeed,

$$\langle x - y, v_i \rangle = \langle x - \sum_{j=1}^{k} c_j v_j, v_i \rangle = \langle x, v_i \rangle - \sum_{j=1}^{k} c_j \langle v_j, v_i \rangle = c_i - c_i = 0.$$

Exercise 1.8.4. Let X be a metric space. Show that

1. Any union of open sets is open

2. Any intersection of closed sets is closed

Solution.

1. Let $\{V_i\}_{i \in I}$ be any family of open subsets of X and let $x \in \bigcup_{i \in I} V_i$. Then $x \in V_i$ for some $i \in I$. Since V_i is open there exists $r > 0$ such that $B_r(x) \subset V_i \subseteq \bigcup_{i \in I} V_i$, that is, $\bigcup_{i \in I} V_i$ is open.

2. Let $\{S_i\}_{i \in I}$ be any family of closed subsets of X. Let $S = \bigcap_{i \in I} S_i$. Then by the De-Morgan law

$$S^c = \left(\bigcap_{i \in I} S_i \right)^c = \bigcup_{i \in I} S_i^c$$

which is open as a union of open sets. Hence S is closed.

Exercise 1.8.5. Let S be an open subset of a metric space X. Show that $(\partial S)^{\circ} = \emptyset$.

Solution. Suppose $(\partial S)^{\circ} \neq \emptyset$. Let $x \in (\partial S)^{\circ}$. Then there exists an open ball $B_r(x)$ such that

$$B_r(x) \subset (\partial S)^{\circ} \subset \partial S = \overline{S} \cap \overline{S^c} = \overline{S} \cap S^c \subset S^c.$$

Therefore $B_r(x) \cap S = \emptyset$ which is a contradiction.

Remark. A subset A of X is said *nowhere dense* if $(\overline{A})^{\circ} = \emptyset$. Note that since ∂S is closed, we have shown that the boundary of any open set is nowhere dense.

Exercise 1.8.6. (*Nested sets property in $\mathbb{R}^n$*). Let $\{S_k : k = 1, 2, ...\}$ be a sequence of closed sets in $\mathbb{R}^n$ such that $S_{k+1} \subseteq S_k$ for every k and $d(S_k) \to 0$ as $k \to \infty$. Then $\bigcap_{k=1}^{\infty} S_k = \{x\}$.

Solution. For each $k = 1, 2, ...$, let $x_k \in S_k$. We prove that $\{x_k\}$ is a Cauchy sequence. Let $\epsilon > 0$. Since $d(S_k) \to 0$ as $k \to \infty$, there exists N such that $d(S_N) < \epsilon$. At the same time, since $S_{k+1} \subseteq S_k$, $x_k, x_j \in S_N$ for $k, j \geq N$. Hence $\|x_k - x_j\| \leq d(S_N) < \epsilon$, and so $\{x_k\}$ is Cauchy. Since $\mathbb{R}^n$ is complete, it follows that $\{x_k\}$ converges to some point, say x, in $\mathbb{R}^n$. For every m, $x_k \in S_m$ for $k \geq m$, and so $x \in S_m$ since each S_m is closed. That is, $x \in \bigcap_{m=1}^{\infty} S_m$. Finally, if y were any other point such that $y \in S_k$ for all k, then $\|x - y\| \leq d(S_k)$ for all k. Since $d(S_k) \to 0$ as $k \to \infty$, this implies $\|x - y\| = 0$, or $x = y$.

Remark. If we drop the condition that $d(S_k) \to 0$ as $k \to \infty$ and assume instead that S_k are bounded, viz, S_k are compact, the intersection is nonempty (but need no longer be a singleton). Indeed The sequence $\{x_k\}$ with $x_k \in S_k$ is now bounded and by the Bolzano-Weierstrass theorem (Theorem 1.5.11) has a convergent subsequence $\{x_{m_k}\}$. Say $x_{m_k} \to x$ as $k \to \infty$. Since $x_{m_k} \in S_j$ for all $m_k \geq j$, and each S_j is closed, it follows that $x \in S_j$ for all j. Hence $x \in \bigcap_{j=1}^{\infty} S_j$. Of course now, the intersection need no longer be a singleton. Take for example, $S_k = \{x \in \mathbb{R}^n : ||x|| \leq 1 + \frac{1}{k}\}$, then the intersection is $\{x \in \mathbb{R}^n : ||x|| \leq 1\}$, the closed unit ball.

Next we give an alternative proof of the Heine-Borel theorem.

Exercise 1.8.7. (*Heine-Borel theorem*). Let S be a closed and bounded subset of $\mathbb{R}^n$. Then S is compact.

Proof. Since S is bounded it's contained in a cube $C = \prod_{i=1}^{n}[a_i, b_i]$. If we can show that C is compact then so is S since it is closed (see Proposition 1.6.8). Thus we may assume $S = C$. Let U_j be an open covering of C. Suppose it has no finite subcover. By dividing each interval $[a_i, b_i]$ in half we can divide C into 2^n (congruent) parts. Therefore one of these parts say C_1 can't be covered by finitely many of the U_j, for otherwise since there are only finitely many of these and their union is C we could get a finite subcover of C itself. Continue in this way by applying the same reasoning to C_1 we can inductively construct a descending sequence, $\{C_k = \prod_{i_k=1}^{n}[a_{i_k}, b_{i_k}] : k = 1, 2, ...\}$ each of half the size of its predecessor and none of which can be covered by a finite number of the U_j. For x and $y \in C_k$, $||x - y|| \leq \frac{1}{2}d(C_1)$. By Theorem 1.1.22 there is some point $x^* \in \bigcap_{k=1}^{\infty} C_k$. In particular, $x^* \in C_1$ so $x^* \in U_j$ for some j. Since U_j is open there is a ball about x^* of radius $r > 0$ completely contained in U_j. That is, if $||x - x^*|| < r$, then $x \in U_j$. Choose k large enough so that $2^k \geq \frac{d(C_1)}{r}$. Then $\frac{1}{2^k}d(C_1) \leq r$ and so $C_k \subseteq U_j$, a contradiction. □

Exercise 1.8.8. (*Riesz Lemma*[18]). Let W be a closed proper linear subspace of a normed linear space V, and let $\alpha \in \mathbb{R}$ with $0 < \alpha < 1$. Then there exists a unit vector $v_\alpha \in V$ such that

$$||v_\alpha - w|| > \alpha, \quad \text{for all } w \in W.$$

---

[18]F. Riesz (1880-1956). He did some of the fundamental work in developing functional analysis.

Proof. Since W is a proper subspace of V there is some vector v_1 in V not belonging to W. Let

$$d = d(v_1, W) = \inf_{w \in W} ||v_1 - w||.$$

Then $d > 0$, for otherwise, i.e., if $d = 0$, then $v_1 \in \overline{W} = W$, which is contrary to $v_1 \notin W$. For any $0 < \alpha < 1$, since $\frac{d}{\alpha} > d$, the definition of d implies that there exists some $w_0 \in W$ such that $||v_1 - w_0|| < \frac{d}{\alpha}$. Consider now

$$v_\alpha = \frac{v_1 - w_0}{||v_1 - w_0||}.$$

Then $||v_\alpha|| = 1$, and for any vector $w \in W$ we have

$$||v_\alpha - w|| = \left\| \frac{v_1 - w_0}{||v_1 - w_0||} - w \right\| = \frac{1}{||v_1 - w_0||} ||v_1 - w_0 - (||v_1 - w_0||)w|| \geq \frac{d}{||v_1 - w_0||} > \alpha,$$

since $\{w_0 + (||v_1 - w_0||)w\} \in W$. $\qquad\square$

Miscellaneous Exercises

Exercise 1.8.9. Let S be a bounded set in $\mathbb{R}$. Prove that

$$\sup(S) - \inf(S) = \sup\{|x - y| : x, y \in S\}.$$

Exercise 1.8.10. Show that the set $S = \left\{ \frac{m}{10^n} : m \in \mathbb{Z}, n \in \mathbb{Z}_+ \right\}$ is dense in $\mathbb{R}$, that is, show that $\overline{S} = \mathbb{R}$.

Exercise 1.8.11. Show that the set $S = \{n + m\sqrt{2} : n, m \in \mathbb{Z}\}$ is dense in $\mathbb{R}$.

Exercise 1.8.12. Let $\{a_k\}$ be a bounded sequence in $\mathbb{R}$ such that $2a_k \leq a_{k-1} + a_{k+1}$. Show that

$$\lim_{k \to \infty} (a_{k+1} - a_k) = 0.$$

Exercise 1.8.13. Let $\{a_k\}$ be a sequence in $\mathbb{R}$ defined by

$$a_1 = 1 \text{ and } a_{k+1} = \frac{1}{2}\left(a_k + \frac{2}{a_k}\right) \text{ for } k = 1, 2, \dots.$$

Show that $\{a_k\}$ converges and $\lim_{k \to \infty} = \sqrt{2}$.

Exercise 1.8.14. Let $\{x_k\}$ be a sequence in $\mathbb{R}^n$ such that $||x_k - x_j|| < \frac{1}{k} + \frac{1}{j}$. Show that $\{x_k\}$ converges.

Exercise 1.8.15. Let $\{a_k\}$ be a Cauchy sequence. Using the definition of Cauchy sequence, show that the set $\{a_k : k = 1, 2, ...\}$ has at most one limit point.

Exercise 1.8.16. Let Ω be an open subset of $\mathbb{R}^n$ and suppose $a \in \Omega$ and $u \in \mathbb{R}^n$ with $u \neq 0$. Prove that the set $A = \{t \in \mathbb{R} : a + tu \in \Omega\}$ is open in $\mathbb{R}$.

Exercise 1.8.17. Let S_1, S_2 be subsets of $\mathbb{R}^n$. Show that $\partial(S_1 \setminus S_2) \subseteq \partial S_1 \cup \partial S_2$.

Exercise 1.8.18. Let $S \subseteq \mathbb{R}^n$ be bounded. Show that $\overline{S}$ is compact.

Exercise 1.8.19. Call a set *semi-closed* if $(\overline{S})^\circ = S^\circ$. Show that the intersection of two semi-closed sets is semi-closed.

Exercise 1.8.20. (*) Show that every finite dimensional vector space has a basis and any two bases have the same number of elements.

Exercise 1.8.21. (*). Let $1 \leq p < \infty$ and $||x||_p = (\sum_{i=1}^{n} |x_i|^p)^{\frac{1}{p}}$. Show that $||x||_p$ is a norm (called the *p-norm*). The triangle inequality in this case is the so-called *Minkowski's inequality* [19]

$$\left(\sum_{i=1}^{n} |x_i + y_i|^p \right)^{\frac{1}{p}} \leq \left(\sum_{i=1}^{n} |x_i|^p \right)^{\frac{1}{p}} + \left(\sum_{i=1}^{n} |y_i|^p \right)^{\frac{1}{p}}.$$

Hint. Consider the function $f(t) = \frac{t^p}{p} + \frac{t^{-q}}{q}$, where $\frac{1}{p} + \frac{1}{q} = 1$. Use elementary calculus to determine the monotonicity of f on $[1, \infty)$. For an alternative proof of this inequality see Exercise 3.12.27.

Exercise 1.8.22. (*) For a normed linear space X the following are equivalent.

1. The Bolzano-Weierstrass theorem holds.

2. The closed unit ball $\overline{B_1(0)}$ in X is compact.

3. $\dim(X) < \infty$.

Hint. If $\dim(X) = \infty$, use Exercise 1.8.8 (Riesz lemma) to construct a sequence of unit vectors $\{x_n\}$ in X such that $||x_n - x_m|| > \frac{1}{2}$ for all $n \neq m$.

---

[19]H. Minkowski (1864-1909). He used geometrical methods to solve difficult problems in number theory, mathematical physics and the theorey of relativity. He held positions at Göttingen and Zurich and was the thesis advisor of C. Caratheodory.

Chapter 2

Functions on Euclidean Spaces

*In this chapter we introduce functions acting between Euclidean spaces.
In particular, we study continuous functions and their basic properties.*

2.1 Functions from $\mathbb{R}^n$ to $\mathbb{R}^m$

A function f from a subset Ω of $\mathbb{R}^n$, called the *domain* of f, into $\mathbb{R}^m$ is a rule that assigns to each point $x = (x_1, ..., x_n) \in \Omega$ a unique point $y \in \mathbb{R}^m$. We write $f : \Omega \to \mathbb{R}^m$, $y = f(x)$ and we call $f(x)$ the image of x under f. The set of all images is called the *range* of f and is written as

$$\text{range}(f) = \{f(x) : x \in \Omega\} = f(\Omega).$$

Since y is a vector in $\mathbb{R}^m$, $f(x) = y = (y_1, ..., y_m)$. The components y_j are uniquely determined by x and the function f. Hence they define functions $f_j : \Omega \to \mathbb{R}$ by the rules $y_j = f_j(x)$ for all $j = 1, ..., m$. These functions are called the *component functions* of f. We then write

$$f(x) = (f_1(x), ..., f_m(x)),$$

or briefly $f = (f_1, ..., f_m)$. We will see that many properties of a vector-valued function are also shared by its component functions. In the important special case when $m = 1$, f is just real-valued and has no component functions.

The case when $n = 1$ is also important. Here $\Omega \subseteq \mathbb{R}$ and is customary to write $X(t)$ rather than $f(x)$. Thus, $X : \Omega \subseteq \mathbb{R} \to \mathbb{R}^m$ has m component functions X_j which are real-valued functons in one variable $t \in \mathbb{R}$. In this case the function X is identified with its range

$$\{X(t) = (x_1(t), ..., x_n(t)) : t \in \Omega\}$$

and is called a *curve* in $\mathbb{R}^m$. When $\Omega = [a, b]$, the function $\gamma : [a, b] \to \mathbb{R}^m$ is called a *path* in $\mathbb{R}^m$. It is customary for paths $\gamma : [a, b] \to \mathbb{R}^3$ to use (and we shall do so) the notation $\gamma(t) = (x(t), y(t), z(t))$.

Of particular importance are lines and line segments in $\mathbb{R}^m$. Given any $x \in \mathbb{R}^m$ and a nonzero vector $u \in \mathbb{R}^m$, the curve $X(t) = x + tu$ ($t \in \mathbb{R}$) is the line passing from x in the direction of u (that is, parallel to u). In particular, given two distinct points $x, y \in \mathbb{R}^m$, $X(t) = x + t(y - x) = ty + (1 - t)x$ ($t \in \mathbb{R}$) represents the line passing from x and y. Furthermore,

$$\gamma(t) = ty + (1 - t)x$$

for all $t \in [0, 1]$ is the *line segment* joining the points $x = \gamma(0)$ and $y = \gamma(1)$.

Example 2.1.1.   1. The function $f(x, y) = (y, 2xy, e^{x+y})$ is defined everywhere on $\mathbb{R}^2$, it is a vector-valued function $f : \mathbb{R}^2 \to \mathbb{R}^3$ with component functions $f_1(x, y) = y$, $f_2(x, y) = 2xy$, $f_3(x, y) = e^{x+y}$.

2. The function $f(x, y) = \log(1 - x^2 - y^2)$ is real-valued defined on the set $\Omega = \{(x, y) \in \mathbb{R}^2 : x^2 + y^2 < 1\}$. Why is this the domain?

3. For all $x \in \mathbb{R}^n$, $f(x) = ||x|| = \sqrt{x_1^2 + ... + x_n^2}$ defines a function $f : \mathbb{R} \to [0, \infty)$.

4. The function $X(t) = (\cos(\omega t), \sin(\omega t), ct)$ is a vector-valued function in one variable $t \in \mathbb{R}$ with values in $\mathbb{R}^3$, that is, it is a curve in $\mathbb{R}^3$. $X : \mathbb{R} \to \mathbb{R}^3$ with component functions $x(t) = \cos(\omega t)$, $y(t) = \sin(\omega t)$ and $z(t) = ct$.

Associated with any function $f : \Omega \subseteq \mathbb{R}^n \to \mathbb{R}$ are two types of sets which play a special role in the developement of our subject: the *graph* of f and the *level sets* of f.

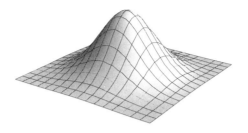

Figure 2.1: Graph of: $z = e^{-(x^2+y^2)}$

Definition 2.1.2. Let $f : \Omega \subseteq \mathbb{R}^n \to \mathbb{R}$. The *graph* of f is the subset of $\mathbb{R}^{n+1}$ (a hypersurface) defined as follows

$$\text{graph}(f) = \{(x, f(x)) : x \in \Omega\}.$$

In terms of components

$$\text{graph}(f) = \left\{(x_1, ..., x_n, f(x_1, ..., x_n)) \in \mathbb{R}^{n+1} : (x_1, ..., x_n) \in \Omega\right\}.$$

Each of these notations has certain advantages. The former is simple and resembles the one variable case while the latter is more detailed and reveals the components. In the important case of a function in two variables $(x, y) \in \Omega \subseteq \mathbb{R}^2$, the graph of f is the surface

$$\text{graph}(f) = \left\{(x, y, z) \in \mathbb{R}^3 : z = f(x, y)\right\}$$
$$= \left\{(x, y, f(x, y)) \in \mathbb{R}^3 : (x, y) \in \Omega\right\}.$$

Definition 2.1.3. Let $f : \Omega \subseteq \mathbb{R}^n \to \mathbb{R}$ and $\alpha \in \mathbb{R}$. The subset S_α of $\mathbb{R}^n$ defined by

$$S_\alpha = \{(x_1, ..., x_n) \in \Omega : f(x_1, ..., x_n) = \alpha\} = \{x \in \Omega : f(x) = \alpha\}$$

is called the *level set* of f of *level* α.

Note that the graph of f can be regarded as the level set S (of level 0) of the function $F : \Omega \times \mathbb{R} :\to \mathbb{R}$ defined by $F(x,y) = f(x) - y$, that is,

$$\operatorname{graph}(f) = S = \{(x_1, ..., x_n, y) : F(x_1, ..., x_n, y) = 0\}$$
$$= \{(x, y) : F(x, y) = 0\}.$$

A particular case of a level set when f is a *linear function* on $\mathbb{R}^n$ given by $f(x) = w_1 x_1 + ... + w_n x_n = \langle w, x \rangle$, where $x = (x_1, ..., x_n)$ and $w = (w_1, ..., w_n)$, is the following:

Definition 2.1.4.   1. Let $w \in \mathbb{R}^n$ be a nonzero vector and $\alpha \in \mathbb{R}$. A set of the form

$$\Pi = \{x \in \mathbb{R}^n : \langle w, x \rangle = \alpha\}$$

is called a *hyperplane* in $\mathbb{R}^n$.

2. An open (resp. closed) *half-space* in $\mathbb{R}^n$ is a set of the form

$$H_w = \{x \in \mathbb{R}^n : \langle w, x \rangle > \alpha\} \text{ (resp. } \overline{H}_w = \{x \in \mathbb{R}^n : \langle w, x \rangle \geq \alpha\}$$

Note that a set of the form

$$H = \{x \in \mathbb{R}^n : \langle w, x \rangle < \alpha\} = \{x \in \mathbb{R}^n : \langle -w, x \rangle > -\alpha\} = H_{-w}$$

is also an open half-space. Similarly, $H = \{x \in \mathbb{R}^n : \langle w, x \rangle \leq \alpha\} = \overline{H}_{-w}$ is a closed half-space.

Definition 2.1.5. Let $f, g : \Omega \subseteq \mathbb{R}^n \to \mathbb{R}^m$. The *sum* of f and g is defined by $(f + g)(x) = f(x) + g(x)$ and the *difference* $(f - g)(x) = f(x) - g(x)$ for all $x \in \Omega$. When $m = 1$ (i.e., when f, g are real-valued), the *product* of f and g is defined by $(fg)(x) = f(x)g(x)$ for all $x \in \Omega$ and the *quotient* by $(\frac{f}{g})(x) = \frac{f(x)}{g(x)}$ for $x \in \Omega$ with $g(x) \neq 0$.

Definition 2.1.6. Given a function $f : \Omega \subseteq \mathbb{R}^n \to \mathbb{R}^m$ and subsets $A \subseteq \mathbb{R}^n$, $B \subseteq \mathbb{R}^m$, we define the *image* of A under f by $f(A) = \{f(x) : x \in A\}$, and the *inverse image* of B under f by $f^{-1}(B) = \{x \in \Omega : f(x) \in B\}$.

Both these function theoretic notions are defined in general for any function $f : X \to Y$ between two sets X and Y. The inverse image behaves well with respect to all the basic set-theoretic operations. The following proposition lists these properties. Their verification is left as an exercise to the reader.

Proposition 2.1.7. *Let $f : X \to Y$ be a function. Then*

1. $B \supseteq f(f^{-1}(B))$ *for any $B \subseteq Y$.*

2. *If $B_1 \subseteq B_2 \subseteq Y$, then $f^{-1}(B_1) \subseteq f^{-1}(B_2) \subseteq X$.*

3. $f^{-1}(B_1 \cup B_2) = f^{-1}(B_1) \cup f^{-1}(B_2)$ *and* $f^{-1}(B_1 \cap B_2) = f^{-1}(B_1) \cap f^{-1}(B_1)$, *for any $B_1 \subseteq Y$ and $B_2 \subseteq Y$.*

4. $f^{-1}(B^c) = [f^{-1}(B)]^c$, *for any $B \subseteq Y$, where B^c is the complement of B in Y.*

However the direct image preserves only inclusions and unions of sets.

Proposition 2.1.8. *Let $f : X \to Y$ be a function. Then*

1. $A \subseteq f^{-1}(f(A))$ *for any $A \subseteq X$.*

2. *If $A_1 \subseteq A_2 \subseteq X$, then $f(A_1) \subseteq f(A_2) \subseteq f(X)$.*

3. $f(A_1 \cup A_2) = f(A_1) \cup f(A_2)$ *and* $f(A_1 \cap A_2) \subseteq f(A_1) \cap f(A_1)$, *for any $A_1, A_2 \subseteq X$.*

Property (3) in 2.1.7 and 2.1.8 generalizes to any (finite or infinite) union or intersection of sets.

Definition 2.1.9. A function $f : \Omega \subseteq X \to Y$ is said to be *one-to-one* (or *injective*), if for all $x_1, x_2 \in \Omega$ with $x_1 \neq x_2$, we have $f(x_1) \neq f(x_2)$. The function f is said to be *onto* (or *surjective*) if range$(f) = f(\Omega) = Y$. Finally f is called a *bijection* if it is both one-to-one and onto.

Definition 2.1.10. Let $f : \Omega \subseteq X \to Y$ be one-to-one. The correspondence $x = f^{-1}(y)$, where x is the unique point of Ω such that $f(x) = y$ defines a function called the *inverse* of f and is denoted by f^{-1}. The inverse function f^{-1} has domain $f(\Omega)$ and range Ω.

Hence $x = f^{-1}(y)$ if and only if $y = f(x)$, for all $x \in \Omega$ and $y \in f(\Omega)$. For example, as is known from one variable calculus $x = \log y$ if and only if $y = e^x$ for $x \in \mathbb{R}$ and $y \in (0, \infty)$.

2.2   Limits of functions

To simplify the notation, the Euclidean norms in $\mathbb{R}^n$ and $\mathbb{R}^m$ will be both denoted by $||\cdot||$. It will be evident from the context when $||\cdot||$ denotes the norm in $\mathbb{R}^n$ and when it denotes the norm in $\mathbb{R}^m$.

Definition 2.2.1. Let $f : \Omega \subseteq \mathbb{R}^n \to \mathbb{R}^m$ and $a \in \overline{\Omega}$. A point $b \in \mathbb{R}^m$ is said to be the *limit* of f as $x \to a$, written as

$$\lim_{x \to a} f(x) = b,$$

if for every $\epsilon > 0$ there exists $\delta > 0$ ($\delta = \delta(\epsilon)$) such that $||f(x) - b|| < \epsilon$ whenever $0 < ||x - a|| < \delta$ and $x \in \Omega$. That is,

$\lim_{x \to a} f(x) = b$ if and only if $\lim_{x \to a} ||f(x) - b|| = 0.$

The idea is of course that $f(x)$ can be made *arbitrarily close* to b by choosing x *sufficiently close*, but not equal to a.

In terms of neighborhoods, the above definition can be written equivalently as follows: $\lim_{x \to a} f(x) = b$ if for every neighborhood $B_\epsilon(b)$ of $b \in \mathbb{R}^m$ there exists a neighborhood $B_\delta(a)$ of $a \in \Omega$ such that whenever $x \in \Omega \cap B_\delta(a)$, this implies $f(x) \in B_\epsilon(b)$, that is,

$$f(\Omega \cap B_\delta(a)) \subseteq B_\epsilon(b).$$

Remark 2.2.2. By a similar argument to that of Proposition 1.1.14 the reader can see that if $\lim_{x \to a} f(x)$ exists, it is unique. We ask the reader to prove this in Exercise 3 below.

Example 2.2.3. Show that $\lim_{(x,y) \to (0,0)} \frac{x^3 + y^3}{x^2 + y^2} = 0$.

Solution. Here $f(x, y) = \frac{x^3 + y^3}{x^2 + y^2}$. Let $\epsilon > 0$ be given. We must find $\delta > 0$ such that $|f(x, y) - 0| < \epsilon$ whenever $0 < ||(x, y) - (0, 0)|| < \delta$. Note that $|x|^3 \le (x^2 + y^2)^{\frac{3}{2}}$ and so by the triangle inequality, we have

$$|f(x, y)| = |\frac{x^3 + y^3}{x^2 + y^2}| \le \frac{2(x^2 + y^2)^{\frac{3}{2}}}{x^2 + y^2} = 2\sqrt{x^2 + y^2}.$$

Whence, choosing $\delta = \frac{\epsilon}{2}$, we get $|f(x, y)| < \epsilon$ provided that $\sqrt{x^2 + y^2} < \delta$. Thus, $\lim_{(x,y) \to (0,0)} f(x, y) = 0$.

Example   2.2.4. Let $f(x,y) = \frac{\sin(x^2+y^2)}{x^2+y^2}$.   Show   that
$\lim_{(x,y)\to(0,0)} f(x,y) = 1$.

Solution. Set $t = x^2 + y^2 = ||(x,y)||^2$. Then the limit becomes the known one variable $\lim_{t\to 0} \frac{\sin(t)}{t} = 1$.

Example   2.2.5. Let $f(x,y) = \frac{\sin^{-1}(xy-2)}{\tan^{-1}(3xy-6)}$.   Show   that
$\lim_{(x,y)\to(2,1)} f(x,y) = \frac{1}{3}$.

Solution. Set $t = xy - 2$. Then the limit becomes

$$\lim_{t\to 0} \frac{\sin^{-1}(t)}{\tan^{-1}(3t)} = \lim_{t\to 0} \frac{\frac{1}{\sqrt{1-t^2}}}{\frac{3}{1+9t^2}} = \frac{1}{3},$$

where we used L'Hopital's Rule.

Clearly, any constant is unchanged under taking limits. The reader can also verify that the usual properties of the limit for real-valued functions of one variable extend to the more general case considered here. The only change that must be made in the "ϵ, δ" proofs of the properties of one variable limits is that expressions of the type $|x-a|$ and $|f(x) - b|$ are replaced by $||x - a||$ and $||f(x) - b||$, respectively. Hence, the limit of a sum or a difference equals the sum or the difference of the limits, respectively. For real-valued functions the limit of a product or a quotient equals the product or the quotient of the limits, respectively. Thus we have:

Proposition 2.2.6. *For $f, g : \Omega \subseteq \mathbb{R}^n \to \mathbb{R}^m$ we have*

$$\lim_{x\to a} (f(x) \pm g(x)) = \lim_{x\to a} f(x) \pm \lim_{x\to a} g(x)$$

and when $m = 1$

$$\lim_{x\to a} (f(x) \cdot g(x)) = \lim_{x\to a} f(x) \cdot \lim_{x\to a} g(x)$$

If in addition $g(x) \neq 0$ everywhere on Ω and $\lim_{x\to a} g(x) \neq 0$, then

$$\lim_{x\to a} \frac{f(x)}{g(x)} = \frac{\lim_{x\to a} f(x)}{\lim_{x\to a} g(x)}.$$

However, the following propostion gives a new featute for the limit of vector-valued functions.

Proposition 2.2.7. *Let* $f : \Omega \subseteq \mathbb{R}^n \to \mathbb{R}^m$ *and* $a \in \overline{\Omega}$. *Then*

$$\lim_{x \to a} f(x) = b \quad \text{if and only if} \quad \lim_{x \to a} f_j(x) = b_j$$

for each $j = 1, ..., m$.

Proof. Let $\epsilon > 0$ and suppose $\lim_{x \to a} f(x) = b$. Then there exist $\delta > 0$ such that for $x \in \Omega$, $0 < ||x - a|| < \delta$ implies $||f(x) - b|| < \epsilon$. Since for each $j = 1, .., m$ we have $|f_j(x) - b_j| \leq ||f(x) = b||$, it follows that $0 < ||x - a|| < \delta$ implies that $|f_j(x) - b_j| < \epsilon$, that is, $\lim_{x \to a} f_j(x) = b_j$. Conversely, suppose $\lim_{x \to a} f_j(x) = b_j$ for all $j = 1, ...m$. Then given any $\epsilon > 0$, for each $j = 1, ...m$, there exists $\delta_j > 0$ such that for $x \in \Omega$ and $0 < ||x - a|| < \delta_j$ we have $|f_j(x) - b_j| < \frac{\epsilon}{\sqrt{m}}$. Take $\delta = \min \{\delta_1, ..., \delta_m\}$. Then $x \in \Omega$ and $0 < ||x - a|| < \delta$ imply

$$||f(x) - b|| = \left(\sum_{j=1}^{m} |f_j(x) - b_j|^2 \right)^{\frac{1}{2}} < \left(m \frac{\epsilon^2}{m} \right)^{\frac{1}{2}} = \epsilon.$$

$\square$

As is known from one variable calculus, whenever $\lim_{x \to a} f(x) = b$, then

$$\lim_{x \to a^+} f(x) = \lim_{x \to a^-} f(x) = b,$$

where $x \to a^+$ means $x > a$ and $x \to a$, while $x \to a^-$ means $x < a$ and $x \to a$. These are called the *one-sided limits* of f at a The following proposition generalizes the idea of one-sided limits from functions in one variable to functions of several variables.

Proposition 2.2.8. *Let* $f : \Omega \subseteq \mathbb{R}^n \to \mathbb{R}^m$ *and* $a \in \overline{\Omega}$. *If* $\lim_{x \to a} f(x) = b$, *then* $\lim_{t \to 0} f(a + tu) = b$ *for any nonzero* $u \in \mathbb{R}^n$.

Proof. Let $\epsilon > 0$ be any positive number. Then there exists $\delta > 0$ such that for all $x \in \Omega$ with $0 < ||x - a|| < \delta$, we have $||f(x) - b|| < \epsilon$. Take $\delta_1 = \frac{\delta}{||u||}$. Now, if $0 < |t| < \delta_1$, then $||(a + tu) - a|| = |t| \cdot ||u|| < \delta$. Therefore, $||f(a + tu) - b|| < \epsilon$. That is, $\lim_{t \to 0} f(a + tu) = b$. $\square$

Proposition 2.2.8 tells us that whenever f has a limit b as x aproaches a, then b is also the limit as x aproaches a along any line passing though the point a. More generally by the uniqueness of the limit of f at a point a, b is also the limit as x approaches a along any curve passing from a. This fact will be used to explain why certain functions fail to have a limit at a point.

Example 2.2.9. Let $f(x,y) = \frac{xy+y^2}{x^2+y^2}$. The domain of f is $\mathbb{R}^2 - \{(0,0)\}$. Does $\lim_{(x,y)\to(0,0)} f(x,y)$ exist?

Solution. If the limit were to exist, it should have the same value no matter how (x,y) approaches $(0,0)$. When $(x,y) \to (0,0)$ along the x-axis $(y=0)$, then $f(x,0) = 0 \to 0$. On the other hand when $(x,y) \to (0,0)$ along the y-axis $(x=0)$, then $f(0,y) = 1 \to 1$. Thus, the limit does not exist.

Example 2.2.10. Let $f(x,y) = \frac{2xy^2}{x^2+y^4}$. Does $\lim_{(x,y)\to(0,0)} f(x,y)$ exist?

Solution. When $(x,y) \to (0,0)$ along any line $y = mx$, then $f(x,mx) = \frac{2m^2x}{1+m^4x^2} \to 0$ as $x \to 0$. However, when $(x,y) \to (0,0)$ along the parabola $x = y^2$, then $f(y^2,y) = 1 \to 1$. Thus, the limit does not exist.

EXERCISES

1. (a) Prove the statements of Proposition 2.1.7.

    (b) Prove the statements of Proposition 2.1.8.

2. Let $f : X \to Y$ be a function. Show that

    (a) If f is onto, then $[f(A)]^c \subseteq f(A^c)$, for any $A \subseteq X$.

    (b) If f is one-to-one and onto, then $[f(A)]^c = f(A^c)$.

    (c) If f is onto, then $B = f(f^{-1}(B))$ for any $B \subseteq Y$.

    (d) If f is one-to-one, then $A = f^{-1}(f(A))$ for any $A \subseteq X$.

    (e) If f is one-to-one, then $f(A_1 \cap A_2) = f(A_1) \cap f(A_1)$, for any $A_1, A_2 \subseteq X$.

3. Let $f : \Omega \subseteq \mathbb{R}^n \to \mathbb{R}^m$. Show that if $\lim_{x \to a} f(x)$ exists, it is unique.

   Hint. See Proposition 1.1.14.

4. Let $f(x,y) = 3x - 2y$. Give an "ϵ, δ" proof that $\lim_{(x,y) \to (4,-1)} f(x,y) = 14$. (Ans. $\delta = \frac{\epsilon}{5}$)

5. Let $f(x,y) = x^2 + xy + y$. Give an "ϵ, δ" proof that $\lim_{(x,y) \to (1,1)} f(x,y) = 3$. (Ans. $\delta = \min\left\{1, \frac{\epsilon}{7}\right\}$.)

6. Determine whether each of the following limits exists, and if does, find it.

   (1) $\lim_{(x,y) \to (0,0)} \frac{x^2 y}{x^2 + y^2}$.   (2) $\lim_{(x,y) \to (0,0)} \frac{xy}{y^4 + x^4}$.

   (3) $\lim_{(x,y) \to (0,0)} \frac{x^3 y^2}{x^6 + y^4}$.   (4) $\lim_{(x,y) \to (0,0)} \frac{(y^2 - x)^2}{y^4 + x^2}$.

   (5) $\lim_{(x,y) \to (0,0)} \frac{x^2 + y}{\sqrt{x^2 + y^2}}$.   (6) $\lim_{(x,y) \to (0,0)} \frac{x^4 + y^4}{x^2 + y^2}$.

7. Show that the following limits do not exist

$$\text{(a) } \lim_{(x,y) \to (1,1)} \frac{x - y^4}{x^3 - y^4} \quad \text{(b) } \lim_{(x,y) \to (0,0)} \frac{1 - \cos(x^2 + y^2)}{(x^2 + y^2) x^2 y^2}$$

8. Show that $\lim_{(x,y) \to (0,0)} \frac{x^2 + y^2}{\sqrt{x^2 + y^2 + 1} - 1} = 2$.

9. Show that $\lim_{(x,y) \to (0,0)} (1 + x^2 y^2)^{-\frac{1}{x^2 + y^2}} = 1$.

10. Show that $\lim_{(x,y) \to (2,-2)} e^{\frac{4(x+y) \log(y^2 x)}{x^2 - y^2}} = 8$.

11. For the following functions f, show that $\lim_{(x,y) \to (0,0)} f(x,y) = 0$

    (a) $f(x,y) = xy \frac{x^2 - y^2}{x^2 + y^2}$.

    (b) $f(x,y) = xy \log(x^2 + y^2)$.

    (c) $f(x,y) = \frac{e^{-\frac{1}{x^2 + y^2}}}{x^4 + y^4}$.

2.3   Continuous functions

Definition 2.3.1. Let $f : \Omega \subseteq \mathbb{R}^n \to \mathbb{R}^m$, where Ω is an open set in $\mathbb{R}^n$ and $a \in \Omega$. We say that f is *continuous at a* if

$$\lim_{x \to a} f(x) = f(a).$$

Moreover, we say that f is *continuous on Ω* if it is continuous at each point of Ω.

Note that by setting $h = x - a \in \mathbb{R}^n$, we have $x \to a$ if and only if $h \to 0$ and continuity of f at a is then equivalent to

$$\lim_{h \to 0} f(a + h) = f(a).$$

We remark that in terms of neighborhoods the function f is continuous at a if for every neighborhood $B_\epsilon(f(a))$ of $f(a)$ there exists a neighborhood $B_\delta(a)$ of a such that

$$f(B_\delta(a)) \subseteq B_\epsilon(f(a)).$$

Example 2.3.2. Let

$$f(x,y) = \begin{cases} \frac{x^2 y}{x^2+y^2} & \text{if } (x,y) \neq (0,0), \\ 0 & \text{if } (x,y) = (0,0). \end{cases}$$

Show that f is continuous at $(0,0)$.

Solution. We must show that

$$\lim_{(x,y) \to (0,0)} f(x,y) = 0 = f(0,0).$$

Let $\epsilon > 0$. To find $\delta > 0$ such that $|\frac{x^2 y}{x^2+y^2} - 0| < \epsilon$ whenever $\sqrt{x^2 + y^2} < \delta$. We have,

$$\left|\frac{x^2 y}{x^2 + y^2}\right| = \frac{|x||xy|}{x^2 + y^2} \leq \frac{|x|\frac{1}{2}(x^2 + y^2)}{x^2 + y^2} = \frac{1}{2}|x| < \frac{1}{2}\sqrt{x^2 + y^2} < \frac{1}{2}\delta = \epsilon.$$

Take $\delta = 2\epsilon$. Thus, $\lim_{(x,y) \to (0,0)} f(x,y) = 0$.

An immediate consequence of the definition of continuity and the properties of limits is the following proposition.

Proposition 2.3.3. *The sum, difference, product and quotient of two continuous functions, whenever defined, is continuous.*

In addition, by Proposition 2.2.7. we get

Corollary 2.3.4. *Let* $f : \Omega \subseteq \mathbb{R}^n \to \mathbb{R}^m$. *Then* $f = (f_1, ..., f_m)$ *is continuous on* Ω *if and only if each* f_j *is continuous on* Ω, *for each* $j = 1, ...m$.

Example 2.3.5. Let $p_i : \mathbb{R}^n \to \mathbb{R}$ be the projection function on the ith coordinate axis, that is, $p_i(x_1, ..., x_n) = x_i$ for $i = 1, ..., n$. Each p_i is continuous on $\mathbb{R}^n$. For any given $\epsilon > 0$, taking $\delta = \epsilon$, for all $x, y \in \mathbb{R}^n$ we have

$$|p_i(x) - p_i(y)| = |x_i - y_i| \leq ||x - y|| < \epsilon.$$

Example 2.3.6. For $x \in \mathbb{R}^n$ the function $f(x) = ||x||$ is continuous on $\mathbb{R}^n$. This follows as above from the inequality

$$| \, ||x|| - ||y|| \, | \leq ||x - y||.$$

Definition 2.3.7. A function $p : \mathbb{R}^n \to \mathbb{R}$ given by

$$p(x_1, ..., x_n) = \sum_{(i_1, ... i_n) \in F} a_{i_1, ..., i_n} x_1^{k_1} \cdots x_n^{k_n},$$

with $a_{i_1, ..., i_n} \in \mathbb{R}$, $k_1, ..., k_n \in \mathbb{Z}^+$ and F a finite set, is called a *polynomial* in n variables. Moreover, a function $r(x_1, ..., x_n) = \frac{p(x_1, ..., x_n)}{q(x_1, ..., x_n)}$ defined when $q(x_1, ...x_n) \neq 0$ is called a *rational* function.

Since the constant function $f(x_1, ..., x_n) = c$ is clearly continuous, a finite number of the operations of addition and multiplication performed on constants and on the functions $p_i(x_1, ..., x_n) = x_i$ leads to a polynomial $p(x_1, ..., x_n)$ in n variables. Thus, polynomials in several variables are (everywhere) continuous. Since rational functions are just quotients of polynomials, it follows that rational functions of several variables are also continuous at every point where are defined (i.e., where the denominator does not vanish).

Example 2.3.8.   1. Identifying $\mathbb{R}^n \times \mathbb{R}^n$ with $\mathbb{R}^{2n}$ the inner producr $\langle\,,\rangle$ on $\mathbb{R}^n$ as a function $\langle\,,\rangle : \mathbb{R}^{2n} \to \mathbb{R}$ is continuous because it is a polynomial

$$\langle x, y \rangle = x_1 y_1 + \ldots + x_n y_n$$

in $2n$ variables.

2. Identifying the set $M_{n \times n}(\mathbb{R})$ of the $n \times n$ real matrices with the Euclidean space $\mathbb{R}^{n^2}$, the determinant can be regarded as a function $\det : \mathbb{R}^{n^2} \to \mathbb{R}$ and as such is continuous as a polynomial in n^2 variables. Similarly, the trace of a $n \times n$ matrix $X = (x_{ij})$ with $i, j = 1, \ldots, n$ is the polynomial

$$\operatorname{tr}(X) = x_{11} + \ldots + x_{nn}.$$

3. Identifying as above $\mathbb{R}^n \times \mathbb{R}^n$ with $\mathbb{R}^{2n}$, the sum of two vectors $(x, y) \to x + y$ regarded as a function $+ : \mathbb{R}^{2n} \to \mathbb{R}^n$ is continuous. Similarly, the scalar multiplication $(c, x) \to cx$ regarded as a function from $\mathbb{R} \times \mathbb{R}^n = \mathbb{R}^{n+1}$ to $\mathbb{R}^n$ is also continuous.

The notion of continuity can be defined equivalently in terms of sequences which often makes it easier to check.

Proposition 2.3.9. *Let* $f : \Omega \subseteq \mathbb{R}^n \to \mathbb{R}^m$. *Then* f *is continuous at* $a \in \Omega$ *if and only if* $f(x_k) \to f(a)$ *whenever* $\{x_k\}$ *is a sequence of points of* Ω *such that* $x_k \to a$.

Proof. Suppose f is continuous at a and let $x_k \to a$. Then, for any $\epsilon > 0$ there exists $\delta > 0$ such that whenever $x \in \Omega$ with $||x - a|| < \delta$, we have $||f(x) - f(a)|| < \epsilon$. Since $x_k \to a$, for this $\delta > 0$ there exists N such that $||x_k - a|| < \delta$ for all $k \geq N$. Thus, for all $k \geq N$ we have $||f(x_k) - f(a)|| < \epsilon$, that is, $f(x_k) \to f(a)$. Conversely, suppose that f is *not* continuous at a. Then there must exist $\epsilon_0 > 0$ such that for every $\delta > 0$ we can pick $x \in \Omega$ with $||x - a|| < \delta$, but $||f(x) - f(a)|| \geq \epsilon_0$. In particular, taking $\delta = \frac{1}{k}$ for $k = 1, 2, \ldots$, we can find $x_k \in \Omega$ with $||x_k - a|| < \frac{1}{k}$, but $||f(x_k) - f(a)|| \geq \epsilon_0$ for all k. We have then obtained a sequence $\{x_k\}$ with $x_k \to a$, but $\{f(x_k)\}$ does *not* converge to $f(a)$. $\square$

Definition 2.3.10. Let $f : \Omega \subseteq \mathbb{R}^n \to \mathbb{R}^m$ and $g : U \subseteq \mathbb{R}^m \to \mathbb{R}^p$ be two functions such that $f(\Omega) \subseteq U$. We define the *composition* of g with

f, denoted by $g \circ f$, to be the (new) function $g \circ f : \Omega \subseteq \mathbb{R}^n \to \mathbb{R}^p$ given by $(g \circ f)(x) = g(f(x))$ for all $x \in \Omega$.

The composition of continuous functions is continuous.

Corollary 2.3.11. *Let $f : \Omega \subseteq \mathbb{R}^n \to \mathbb{R}^m$ and $g : U \subseteq \mathbb{R}^m \to \mathbb{R}^p$ be functions. If f is continuous at $a \in \Omega$ and g is continuous at $b = f(a)$, then $g \circ f$ is continuous at a.*

Proof. Let $\{x_k\}$ be a sequence of points of Ω such that $x_k \to a$. Since f is continuous at a, by Proposition 2.3.9, we have $f(x_k) \to f(a)$. Since g is continuous at $f(a)$, the same Proposition implies $g(f(x_k)) \to g(f(a))$ and so $g \circ f$ is continuous at a. □

Example 2.3.12. For $x \in \mathbb{R}^n$ with $\|x\| < 1$, $f(x) = \log(1 - \|x\|^2)$ is continuous as a composition of continuous functions.

Example 2.3.13. Let

$$f(x, y) = \begin{cases} e^{-\frac{1}{|x-y|}} & \text{if } x \neq y, \\ 0 & \text{if } x = y. \end{cases}$$

Show that f is continuous on $\mathbb{R}^2$.

Solution. Since f is a composition of continuous functions, the only points where f may not be continuous are the points on the line $y = x$, that is, at points (a, a) with $a \in \mathbb{R}$. We investigate the continuity of f at these points. Setting $t = x - y$, the function becomes

$$f(t) = \begin{cases} e^{-\frac{1}{|t|}} & \text{if } t \neq 0, \\ 0 & \text{if } t = 0. \end{cases}$$

Now, since $\lim_{t \to 0} e^{-\frac{1}{|t|}} = 0$, we get

$$\lim_{(x,y) \to (a,a)} f(x, y) = 0 = f(a, a).$$

The next theorem gives a useful characterization of continuity in terms of open sets which is also applicable in more general situations.

Theorem 2.3.14. *A function $f : \mathbb{R}^n \to \mathbb{R}^m$ is continuous if and only if for each open set V in $\mathbb{R}^m$, $f^{-1}(V)$ is open in $\mathbb{R}^n$. In particular, a function $f : \Omega \subseteq \mathbb{R}^n \to \mathbb{R}^m$ is continuous on Ω if and only if for each open set V in $\mathbb{R}^m$, $f^{-1}(V)$ is open relative to Ω; that is, $f^{-1}(V) = G \cap \Omega$ for some open set G in $\mathbb{R}^n$.*

Proof. Suppose that f is continuous on $\mathbb{R}^n$ and let $V \subset \mathbb{R}^m$ be open. For each $x \in f^{-1}(V)$ we have $f(x) \in V$ and since V is open, we can find $\epsilon > 0$ such that $B_\epsilon(f(x)) \subset V$. Now, since f is continuous at x, we can find $\delta > 0$ such that $f(B_\delta(x)) \subseteq B_\epsilon(f(x))$ which implies $B_\delta(x) \subseteq f^{-1}(B_\epsilon(f(x))) \subseteq f^{-1}(V)$. This shows that each point $x \in f^{-1}(V)$ is an interior point and hence $f^{-1}(V)$ is open. Conversely, suppose that $f^{-1}(V)$ is open in $\mathbb{R}^n$ for every open set $V \subseteq \mathbb{R}^m$. Let $x \in \mathbb{R}^n$ and $\epsilon > 0$. The neighborhood $B_\epsilon(f(x)) = V$ is open and hence $f^{-1}(V)$ is also open. Therefore there exists a $\delta > 0$ with $B_\delta(x) \subset f^{-1}(V)$. This implies $f(B_\delta(x)) \subset V$, that is, $f(B_\delta(x)) \subset B_\epsilon(f(x))$ which shows that f is continuous at x. Since x was an arbitrary point of $\mathbb{R}^n$, f is continuous on $\mathbb{R}^n$. $\qquad\square$

Continuity can also be characterized in terms of closed sets.

Corollary 2.3.15. *A function $f : \mathbb{R}^n \to \mathbb{R}^m$ is continuous if and only if for each closed set B in $\mathbb{R}^m$, $f^{-1}(B)$ is closed in $\mathbb{R}^n$. In particular, a function $f : \Omega \subseteq \mathbb{R}^n \to \mathbb{R}^m$ is continuous on Ω if and only if for each closed set B in $\mathbb{R}^m$, $f^{-1}(B)$ is closed relative to Ω; that is, $f^{-1}(B) = S \cap \Omega$ for some closed set S in $\mathbb{R}^n$.*

Proof. Since $f^{-1}(B^c) = [f^{-1}(B)]^c$, and (by definition) B is closed in $\mathbb{R}^m$ if and only if its complement, B^c, is open, the result follows immediately from the theorem. $\qquad\square$

Example 2.3.16. Let $f : \mathbb{R}^n \to \mathbb{R}$ be continuous. Then the sets:

1. $\{x \in \mathbb{R}^n : f(x) = 0\} = f^{-1}(\{0\})$ is closed, since $\{0\}$ is closed in $\mathbb{R}$.

2. $\{x \in \mathbb{R}^n : f(x) \geq 0\} = f^{-1}([0, \infty))$ is closed, since $[0, \infty)$ is closed in $\mathbb{R}$.

3. $\{x \in \mathbb{R}^n : f(x) > 0\} = f^{-1}((0, \infty))$ is open, since $(0, \infty)$ is open in $\mathbb{R}$.

Similarly the sets $\{x \in \mathbb{R}^n : f(x) \leq 0\} = f^{-1}((-\infty, 0])$ and $\{x \in \mathbb{R}^n : f(x) < 0\} = f^{-1}((-\infty, 0))$ are closed and open respectively.

Definition 2.3.17. Let $f : X \to Y$ be a bijection from one metric space (X, d_X) to another (Y, d_Y). If both f and its inverse f^{-1} are continuous functions, then f is called a *homeomorphism* and the spaces X and Y are said to be *homeomorphic*[1].

Clearly if f is a homeomorphism, then so is f^{-1}. Theorem 2.3.14 shows that a homeomorphism maps open subsets of X onto open subsets of Y. By Corollary 2.3.15, it also maps closed subsets of X onto closed subsets of Y. A property of a set which remains invariant under a homeomorphism is called a *topological property*[2]. Thus the properties of being open or closed are topological properties.

Definition 2.3.18. Let (X, d_x) and (Y, d_Y) be metric spaces. A function $f : X \to Y$ is called an *isometry* if

$$d_Y(f(x), f(x')) = d_X(x, x'), \text{ for all } x, x' \in X.$$

An isometry is evidently injective and continuous (in fact uniformly continuous, see Section 2.5) but not necessarily surjective. If however, f is an isometry from X *onto* Y, then the inverse mapping f^{-1} is also an isometry, and hence f is a homeomorphism. If there is an isometry from (X, d_X) onto (Y, d_Y) the two metric spaces are called *isometric*. In the next section we shall characterize the isometries of $\mathbb{R}^n$.

We conclude this section with the *Contraction Mapping Principle*, sometimes known as *Banach's fixed point theorem*.

Definition 2.3.19. Let (X, d) be a metric space. A function $f : X \to X$ is called a *contraction mapping* (or simply a *contraction*) if there exist some $0 < c < 1$ such that

$$d(f(x), f(y)) \leq c\, d(x, y) \text{ for all } x, y \in X.$$

The constant c is called the *contraction constant*.

---

[1]The word *homeomorphic* is derived from the Greek word $o\mu o\iota o' \mu o\rho \phi o\varsigma$, which means of similar structure.

[2]In topology, homeomorphic spaces X and Y may be regarded as indistinguishable. Every topological property which holds in X also holds in Y. For a detailed account to the general theory of metric and topological spaces we refer to [9].

A contraction mapping is automatically (uniformly) continuous. We are now going to prove that a contraction mapping on a *complete metric space* has a unique fixed point. A function $f : X \to X$ has a *fixed point* means that $f(\xi) = \xi$ for some $\xi \in X$. This result is a powerful tool in proving the existence of solutions to differential and integral equations.

Theorem 2.3.20. *(Contraction mapping principle). Let (X, d) be a complete metric spaace. If $f : X \to X$ is a contraction, then f has a unique fixed point.*

Proof. Let $x_0 \in X$ be arbitrary and define the sequence $x_0, x_1, x_2, \ldots$ recursively by $x_{k+1} = f(k_n)$ for $k = 0, 1, 2, \ldots$. We will show that $\{x_k\}$ converges to a point of f. First we show that $\{x_k\}$ is a Cauchy sequence. For $k \geq 1$ we have

$$d(x_{k+1}, x_k) = d(f(x_k), f(x_{k-1})) \leq c\, d(x_k, x_{k-1}) = c\, d(f(x_{k-1}), f(x_{k-2}))$$

$$\leq c^2\, d(x_{k-1}, x_{k-2}) = c^2\, d(f(x_{k-2}), f(x_{k-3})) \leq \ldots \leq c^k\, d(x_1, x_0).$$

Hence, $d(x_{k+1}, x_k) \leq c^k\, d(x_1, x_0)$. Now, if $m \geq k$ then, using the triangle inequality repeatedly,

$$d(x_m, x_k) \leq d(x_m, x_{m-1}) + d(x_{m-1}, x_{m-2}) + \ldots + d(x_{k+1}, x_k)$$

$$\leq (c^{m-1} + c^{m-2} + \ldots + c^k)d(x_1, x_0) = \frac{c^k - c^m}{1 - c}d(x_1, x_0) < \frac{c^k}{1 - c}d(x_1, x_0).$$

Since $0 < c < 1$, it follows that $c^k \to 0$ as $k \to \infty$ (see, Example 1.1.15 (2)). This inequality implies that $\{x_k\}$ is a Cauchy sequence. The completeness of X now implies that there is a point $\xi \in X$ with $x_k \to \xi$. By the continuity of f,

$$f(\xi) = \lim_{k \to \infty} f(x_k) = \lim_{k \to \infty} x_{k+1} = \xi.$$

Hence ξ is a fixed point of f. Finally, if $\eta \in X$ was another fixed point, $f(\eta) = \eta$, then

$$d(\xi, \eta) = d(f(\xi), f(\eta)) \leq c\, d(\xi, \eta).$$

Since $0 < c < 1$, this can only happen when $d(\xi, \eta) = 0$, that is, $\xi = \eta$. $\qquad \square$

The sequence $\{x_k\}$ used in the proof is often called the *sequence of successive approximations*. *The proof shows that this sequence converges no matter what the starting point x_0 in X was.* The method is also called simple iteration. For, $x_1 = f(x_0)$, $x_2 = f(x_1) = f(f(x_0))$, $x_3 = f(f(f(x_0)))$,...., which might be written as $x_k = f^{(k)}(x_0)$, where $f^{(k)}$ are the *iterates* of f.

EXERCISES

1. Let

$$f(x,y) = \begin{cases} \frac{x^3 y^3}{x^2+y^2} & \text{if } (x,y) \neq (0,0), \\ 0 & \text{if } (x,y) = (0,0). \end{cases}$$

Show that $f(x,y)$ is continuous at the origin.

2. Let

$$f(x,y) = \begin{cases} \frac{x^4 - y^4}{x^4+y^4} & \text{if } (x,y) \neq (0,0), \\ 0 & \text{if } (x,y) = (0,0). \end{cases}$$

Show that $f(x,y)$ is not continuous at the origin.

3. Let

$$f(x,y) = \begin{cases} y^2 + x^3 \sin\left(\frac{1}{x}\right) & \text{if } x \neq 0, \\ y^2 & \text{if } x = 0. \end{cases}$$

Show that $f(x,y)$ is continuous at $(0,0)$.

4. Let

$$f(x,y) = \begin{cases} x\cos(\frac{1}{y}) & \text{if } y \neq 0, \\ 0 & \text{if } y = 0. \end{cases}$$

Show that $f(x,y)$ is continuous at any point (x,y) with $y \neq 0$, also at $(0,0)$.

5. Let

$$f(x,y) = \begin{cases} \sqrt{1 - x^2 - 4y^2} & \text{if } x^2 + 4y^2 \leq 1, \\ c & \text{if } x^2 + 4y^2 > 1. \end{cases}$$

For what c will the function be continuous in $\mathbb{R}^2$. (Ans $c = 0$).

6. Is the function

$$f(x, y, z) = \begin{cases} \frac{xy - z^2}{x^2 + y^2 + z^2} & \text{if } (x, y, z) \neq (0, 0, 0) \\ 0 & \text{if } (x, y, z) = (0, 0, 0) \end{cases}$$

continuous at $(0, 0, 0)$?

7. Let

$$f(x, y) = \begin{cases} \frac{\sin^2(x - y)}{|x| + |y|} & \text{if } |x| + |y| > 0, \\ 0 & \text{if } |x| + |y| = 0. \end{cases}$$

Discuss the continuity of f on $\mathbb{R}^2$. (Ans. continuous).

8. Let $f : [0, 1] \to [a, b]$ be $f(x) = a + x(b - a)$ and $a \neq b$. Show that f is a homeomorphism.

9. For what values of r is $f(x) = x^2$ a contraction on $[0, r]$?

10. Let $f : [a, b] \to [a, b]$ be continuously differentiable with $|f'(x)| \leq c < 1$ for all $x \in [a, b]$. Show that f is a contraction. In particular, show that $f : [1, \infty) \to [1, \infty)$ with $f(x) = \sqrt{x}$ is a contraction.

11. Let $f : B \to B$ be the function $f(x, y) = \left(\frac{1-y}{2}, \frac{1+x^2}{3} \right)$, where B is the closed unit disk in $\mathbb{R}^2$. Show that f has a fixed point in B.

12. Show that $f : \mathbb{R}^n \to \mathbb{R}^m$ is continuous[3] if and only if for each subset $A \subseteq \mathbb{R}^n$ we have $f(\overline{A}) \subseteq \overline{f(A)}$.

13. Show that $f : \mathbb{R}^n \to \mathbb{R}^m$ is continuous if and only if for each subset $E \subseteq \mathbb{R}^m$ we have $\overline{f^{-1}(E)} \subseteq f^{-1}(\overline{E})$.

14. Show that $f : \mathbb{R}^n \to \mathbb{R}^m$ is continuous if and only if for each subset $E \subseteq \mathbb{R}^m$ we have $f^{-1}(E^\circ) \subseteq [f^{-1}(E)]^\circ$.

---

[3]This is no longer true if the closure of $A \subseteq \mathbb{R}^n$ be replaced by its interior. For example, consider $f : \mathbb{R} \to \mathbb{R}$ where $f(x) = |x|$. Take $A = \mathbb{R}$. Then $f(A^\circ) = f(\mathbb{R}) = [0, \infty)$. However, $[f(A)]^\circ = [f(\mathbb{R})]^\circ = (0, \infty)$.

2.4   Linear transformations

In this section we study functions between Euclidean spaces that pre-
serve the linear structure of the space. Such functions are called *linear
transformations* or linear operators. Linear tranformations are properly a part of linear algebra. However, we need to know their continuity
properties.

Definition 2.4.1. A function $T : \mathbb{R}^n \to \mathbb{R}^m$ is called a *linear transformation* (or *linear mapping*) if:

1. $T(x + y) = T(x) + T(y)$ for all $x, y \in \mathbb{R}^n$,

2. $T(cx) = cT(x)$ for every $x \in \mathbb{R}^n$ and $c \in \mathbb{R}$.

In particular, when $m = 1$, a linear mapping $\lambda : \mathbb{R}^n \to \mathbb{R}$ is called a
linear functional on $\mathbb{R}^n$.

Exercise 2.4.2. Show that for any linear transformation $T(0) = 0$.

Note that conditions (1) and (2) are equivalent to the single
condition

$$T(cx + dy) = cT(x) + dT(y)$$

for all $x, y \in \mathbb{R}^n$ and scalars $c, d \in \mathbb{R}$. By induction, if T is linear

$$T\left(\sum_{j=1}^{k} c_j v_j\right) = \sum_{j=1}^{k} c_j T(v_j)$$

for every $v_1, ..., v_k \in \mathbb{R}^n$ and scalars $c_1, ..., c_k \in \mathbb{R}$.

Definition 2.4.3. Let $T : \mathbb{R}^n \to \mathbb{R}^m$ be a linear transformation. T is
said to be *bounded* if there exists $M > 0$ such that $||T(x)|| \leq M||x||$ for
all $x \in \mathbb{R}^n$.

Proposition 2.4.4. *Any linear transformation* $T : \mathbb{R}^n \to \mathbb{R}^m$ *is
bounded.*

Proof. Let $\{e_1, ..., e_n\}$ be the standard basis of $\mathbb{R}^n$ and $x = (x_1, ..., x_n) \in \mathbb{R}^n$. Then $x = \sum_{i=1}^{n} x_i e_i$ and since T is linear we have

$$||T(x)|| = ||\sum_{i=1}^{n} x_i T(e_i)|| \leq \sum_{i=1}^{n} |x_i| ||T(e_i)|| \leq ||x||_\infty \sum_{i=1}^{n} ||T(e_i)||.$$

Taking $M = \sum_{i=1}^{n} ||T(e_i)||$, we have $||T(x)|| \leq M||x||_\infty$. Since $||x||_\infty \leq ||x||$ we conclude that $||T(x)|| \leq M||x||$.   □

Note that the constant M depends on the dimension n of $\mathbb{R}^n$ and T. Since in a finite dimensional normed linear space all norms are equivalent (Theorem 1.7.2.), the above proof generalizes to any linear transformation $T : V \to U$ between normed linear spaces with V of finite dimension.

Next we prove that for linear transformations between two normed linear spaces, boundedness and continuity are equivalent.

Theorem 2.4.5. *Let $T : V \to U$ be a linear transformation between the normed linear spaces V and U. Then the following are equivalent:*

1. *T is continous on V.*

2. *T is continuous at 0.*

3. *T is bounded.*

Proof. We prove $(1) \Rightarrow (2) \Rightarrow (3) \Rightarrow (1)$.

$(1) \Rightarrow (2)$: Obvious.

$(2) \Rightarrow (3)$: Suppose T is continuous at 0, but not bounded. Then for each $k = 1, 2, \ldots$ there exists $v_k \in V$ such that $||T(v_k)|| > k||v_k||$. Let $w_k = \frac{v_k}{k||v_k||}$. Then

$$||T(w_k)|| = \frac{1}{k||v_k||}||T(v_k)|| > k\frac{1}{k||v_k||}||v_k|| = 1,$$

that is, $||T(w_k)|| > 1$. Now, $||w_k|| = \frac{1}{k} \to 0$ or $w_k \to 0$, but $T(w_k)$ does not converge to $0 = T(0)$ contradicting the continuity of T at 0.

$(3) \Rightarrow (1)$: Suppose $||T(v)|| \leq M||v||$ for all $v \in V$. Let $v_k \to v$, then

$$||T(v_k) - T(v)|| = ||T(v_k - v)|| \leq M||v_k - v|| \to 0,$$

that is, $T(v_k) \to T(v)$ which proves the continuity of T on V.   □

Corollary 2.4.6. *A linear transformation $T : \mathbb{R}^n \to \mathbb{R}^m$ is continuous.*

Another characterization of bounded linear tranformations is given in Section 2.7. It tells us that a linear transformation is bounded if and only if it maps bounded sets into bounded sets (see Exercise 2.7.4).

As is known from linear algebra the inverse T^{-1} of a linear transformation T, whenever it exists, is itself a linear transformation. The next proposition characterizes the boundedness of the inverse transformation.

Proposition 2.4.7. *Let $T : \mathbb{R}^n \to \mathbb{R}^m$ be a linear transformation. Then the inverse transformation T^{-1} exists and is bounded on $T(\mathbb{R}^n)$ if and only if there exists $\beta > 0$ with $\beta||x|| \leq ||T(x)||$ for all $x \in \mathbb{R}^n$ (that is, T is bounded from below).*

Proof. Suppose T^{-1} exists and is a bounded on $T(\mathbb{R}^n)$. Then there exists $M > 0$ such that for all $x \in \mathbb{R}^n$, we have $||T^{-1}(T(x))|| \leq M||T(x)||$. Hence, $\beta||x|| \leq ||T(x)||$ with $\beta = \frac{1}{M}$. Conversely, suppose T is bounded from below with bound $\beta > 0$. To show that T^{-1} exists on $T(\mathbb{R}^n)$, all we need to show is that T is one-to-one (or equivalently ker$(T) = \{0\}$, see Appendix D). For this, suppose $T(x) = 0$. Then the hypothesis implies $||x|| = 0$, and so $x = 0$. For the boundedness of T^{-1}, let $y \in T(\mathbb{R}^n)$. Then there is $x \in \mathbb{R}^n$ with $x = T^{-1}(y)$. Therefore, $\beta||x|| \leq ||T(x)||$ implies $||T^{-1}(y)|| \leq \frac{1}{\beta}||T(T^{-1}(y))|| = \frac{1}{\beta}||y||$, so that T^{-1} is bounded. □

Definition 2.4.8. A mapping $F : \mathbb{R}^n \to \mathbb{R}^m$ is called *affine* if it is of the form $F(x) = T(x) + w$ for all $x \in \mathbb{R}^n$, where T is linear and $w \in \mathbb{R}^m$.

Note that $F(0) = w$. Hence an affine mapping F is linear if and only if $F(0) = 0$. If $n = m$ and $T = I$ (the identidy map on $\mathbb{R}^n$, $I(x) = x$ for all $x \in \mathbb{R}^n$), then F is said a *translation*, denoted by $\tau_w(x) = x + w$. Evidently, a translation τ_w is continuous with a continuous inverse $\tau_{-w}(x) = x - w$. That is, τ_w is a homeomorphism, and hence any two neighborhoods in $\mathbb{R}^n$ are homeomorphic.

Definition 2.4.9. A linear transformation $O : \mathbb{R}^n \to \mathbb{R}^n$ is called *orthogonal* if

$$\langle O(u), O(v) \rangle = \langle u, v \rangle, \tag{2.1}$$

for all $u, v \in \mathbb{R}^n$. That is, O preserves the usual (Euclidean) inner product.

As is known from linear algebra an orthogonal linear transformation on $\mathbb{R}^n$ can be represented by an orthogonal $n \times n$ matrix O, that is, a matrix satisfying $OO^t = I$, where O^t is the transpose matrix of O. By the properties of determinants (see Appendix D), since for any $n \times n$ matrix A, $\det A^t = \det A$, we see that if O is orthogonal, then

$$1 = \det I = \det(OO^t) = \det O \det O^t = (\det O)^2.$$

Hence $\det O = \pm 1$. If $\det O = 1$, then O is called a *rotation* of $\mathbb{R}^n$. The next theorem characterizes the isometries of $\mathbb{R}^n$.

Theorem 2.4.10. *A mapping $F : \mathbb{R}^n \to \mathbb{R}^n$ is an isometry of $\mathbb{R}^n$ if and only if F is an affine mapping of the form $F(x) = O(x) + w$ for $x \in \mathbb{R}^n$, where O is orthogonal.*

Proof. Suppose O is orthogonal. Then taking $u = v$ in (2.1) we get

$$||O(u)||^2 = \langle O(u), O(u) \rangle = \langle u, u \rangle = ||u||^2.$$

Therefore $||O(u)|| = ||u||$ for all $u \in \mathbb{R}^n$. Hence for $x, y \in \mathbb{R}^n$ we have

$$||O(x) - O(y)|| = ||O(x - y)|| = ||x - y||,$$

and so O is an isometry of $\mathbb{R}^n$. Since $||F(x) - F(y)|| = ||O(x) - O(y)||$, F is also an isometry of $\mathbb{R}^n$.

Convserely, suppose F is an isometry of $\mathbb{R}^n$. Then $||O(x) - O(y)|| = ||x - y||$ for all $x, y \in \mathbb{R}^n$. Since $O(0) = 0$ taking $y = 0$, this gives

$$||O(x)|| = ||x|| \text{ for all } x \in \mathbb{R}^n.$$

Therefore, since $||x - y||^2 = ||x||^2 - 2\langle x, y \rangle + ||y||^2$ for all $x, y \in \mathbb{R}^n$, it follows

$$||O(x)||^2 - 2\langle O(x), O(y) \rangle + ||O(y)||^2 = ||x||^2 - 2\langle x, y \rangle + ||y||^2.$$

Hence $\langle O(x), O(y) \rangle = \langle x, y \rangle$, and O is orthogonal.   □

Theorem 2.4.11. *(Polar Decomposition). Any isomorphism $T : \mathbb{R}^n \to \mathbb{R}^n$ factors as $T = OP$, where O is an orthogonal and P is a symmetric positive transformation.*

Proof. Let T^t be the transpose of T, that is, T^t is the unique isomorphism satisfying

$$\langle Tu, v \rangle = \langle u, T^t v \rangle,$$

for all $u, v \in \mathbb{R}^n$. Consider $T^t T$. Then $(T^t T)^t = T^t (T^t)^t = T^t T$. That is, $T^t T$ is symmetric. By the Spectral theorem (see, Theorem D.0.5), $T^t T$ is similar to a diagonal matrix D with diagonal entries the eigenvalues λ_i ($i = 1, ..., n$) of $T^t T$. Moreover, for all $u \in \mathbb{R}^n$ $\langle T^t T u, u \rangle = \langle T u, T u \rangle = \|Tu\| > 0$, and $T^t T$ is positive. In particular, if $u \in \mathbb{R}^n$ is an eigenvector belonging to eigenvalue λ_i, then

$$0 < \langle T^t T u, u \rangle = \langle \lambda_i u, u \rangle = \lambda_i \|u\|.$$

Hence $\lambda_i > 0$ for all $i = 1, ..., n$. By the spectral theorem $T^t T$ has a unique positive square root $\sqrt{T^t T}$. Take $P^2 = T^t T$, and $O = T P^{-1}$. Then

$$\langle Ou, Ov \rangle = \langle T P^{-1} u, T P^{-1} v \rangle = \langle P^{-1} u, T^t T P^{-1} v \rangle = \langle P^{-1} u, P^2 P^{-1} v \rangle$$

$$= \langle P^{-1} u, P v \rangle = \langle P^t P^{-1} u, v \rangle = \langle P P^{-1} u, v \rangle = \langle u, v \rangle.$$

Thus, O is orthogonal and $T = OP$.

$\square$

We close this section by characterizing the linear functionals on $\mathbb{R}^n$.

Theorem 2.4.12. *A function* $\lambda : \mathbb{R}^n :\to \mathbb{R}$ *is a linear functional if and only if there exists a unique* $w \in \mathbb{R}^n$ *such that* $\lambda(x) = \langle w, x \rangle$ *for all* $x \in \mathbb{R}^n$.

Proof. Let $\lambda(x) = \langle w, x \rangle$, when $w \in \mathbb{R}^n$. Then $\lambda(cx + dy) = \langle w, cx + dy \rangle = c\langle w, x \rangle + d\langle w, y \rangle = c\lambda(x) + d\lambda(y)$ for all $x, y \in \mathbb{R}^n$ and scalars $c, d \in \mathbb{R}$. Hence, $\lambda(x) = \langle w, x \rangle$ defines a linear functional on $\mathbb{R}^n$.
Conversely, suppose $\lambda : \mathbb{R}^n \to \mathbb{R}$ is a linear functional. Let $\{e_1, ..., e_n\}$ be the standard basis of $\mathbb{R}^n$. Then taking $w = (\lambda(e_1), ..., \lambda(e_n)) \in \mathbb{R}^n$, for any $x = (x_1, ..., x_n) \in \mathbb{R}^n$, we have

$$\lambda(x) = \lambda(\sum_{i=1}^{n} x_i e_i) = \sum_{i=1}^{n} x_i \lambda(e_i) = \langle x, w \rangle = \langle w, x \rangle.$$

To see that the vector w is unique, suppose w' is another vector such that $\lambda(x) = \langle w', x \rangle$. Then $\langle w - w', x \rangle = \langle w, x \rangle - \langle w', x \rangle = 0$ for all $x \in \mathbb{R}^n$. In particular for $x = w - w'$, we have $\|w - w'\|^2 = 0$ and so $w = w'$. $\square$

Note that the vector $w = (\lambda(e_1), ..., \lambda(e_n))$ is the $1 \times n$ standard matrix representation of the linear functional $\lambda : \mathbb{R}^n \to \mathbb{R}$, written as vector in $\mathbb{R}^n$.

2.5   Continuous functions on compact sets

In this subsection we look at the relationship between continuity and compactness. An important property of continuous functions is that they map compact sets into compact sets.

Theorem 2.5.1. *If S is a compact subset of $\mathbb{R}^n$ and $f : S \to \mathbb{R}^m$ is continuous on S, then $f(S)$ is a compact subset of $\mathbb{R}^m$.*

Proof. Let $f(S) \subseteq \bigcup_{i \in I} U_i$ be an open cover of $f(S)$. By continuity, each $V_i = f^{-1}(U_i)$ is open in $\mathbb{R}^n$. Moreover,

$$S \subseteq f^{-1}(f(S)) \subseteq \bigcup_{i \in I} f^{-1}(U_i) = \bigcup_{i \in I} V_i.$$

Hence, $\{V_i\}_{i \in I}$ is an open cover of S. Since S is compact, there is a finite subcover $\{V_j\}_{j=1,...,k}$ such that $S \subseteq \bigcup_{j=1}^{k} V_j$. Since $f(f^{-1}(U_j)) \subseteq U_j$ for each j, it follows that

$$f(S) \subseteq \bigcup_{j=1}^{k} f(V_j) \subseteq \bigcup_{j=1}^{k} f(f^{-1}(U_j)) \subseteq \bigcup_{j=1}^{k} U_j.$$

Thus, $f(S)$ is compact. $\square$

As we saw in Theorem 1.6.5, a set in $\mathbb{R}^n$ is compact if and only if it is closed and bounded. We remark that only the *combination* of closedness and boundedness is preserved under continuous functions. The image of a closed set under a continuous function need not be closed nor need the image of a bounded set be bounded. For example, let $f : \mathbb{R} \to \mathbb{R}$ be given by $f(x) = \frac{1}{1+x^2}$. Then $f((-\infty, \infty)) = (0, 1]$ which is not closed.

(This also shows that the image of an open set need not be open). On the other hand, let $g : \mathbb{R} - \{0\} \to \mathbb{R}$ be given by $g(x) = \frac{1}{x}$. Then g maps the bounded set $S = (0, 1]$ onto the unbounded set $f(S) = [1, \infty)$.

Exercise 2.5.2. Prove Theorem 2.5.1, using the sequential compactness of S.

A consequence of Theorem 2.5.1 is the fundamental existence theorem for maxima and minima of real-valued functions.

Corollary 2.5.3. *(Weierstrass' extreme value theorem). Let $S \subset \mathbb{R}^n$ be compact and $f : S \to \mathbb{R}$ be continuous. Then f attains its maximum and minimum values on S. That is, there exist points $a, b \in S$ such that $f(a) \leq f(x) \leq f(b)$ for all $x \in S$.*

Proof. By Theorem 2.5.1, the set $f(S)$ is a compact subset of $\mathbb{R}$. Hence, it is *bounded*, so both $M = \sup f(S)$ and $m = \inf f(S)$ exist, and *closed*, so, by Corollary 1.5.8, both $m, M \in f(S)$. That is, there exist $a, b \in S$ such that $f(a) = m = \min_{x \in S} f(x)$ and $f(b) = M = \max_{x \in S} f(x)$. $\square$

This result may be *not* true if S is not compact.

Example 2.5.4.   1. $f(x) = x^2$, $S = (1, 2)$. The extreme values are attained on the boundary $\partial(S) = \{1, 2\}$.

2. $f(x) = \tan x$, $S = [0, \frac{\pi}{2})$. The minimum is $f(0) = 0$, however f has no maximum.

3. $f(x) = x^3$, $S = \mathbb{R}$. f has neither maximum nor minimum.

4. $f(x) = \tan^{-1} x$, $S = \mathbb{R}$. $f(\mathbb{R}) \subset [-\frac{\pi}{2}, \frac{\pi}{2}]$, however, f approaches but does not achieve the extreme values $\pm\frac{\pi}{2}$.

Example 2.5.5. *(Stereographic projection)*. Let $S = \{(x, y, z) : x^2 + y^2 + z^2 = 1\}$ be the unit sphere in $\mathbb{R}^3$, and $\mathbf{N} = (0, 0, 1)$, (see Figure 2.2). Then for each $(x, y) \in \mathbb{R}^2$ there is a unique line in $\mathbb{R}^3$ joining $(x, y, 0)$ with the point $\mathbf{N}$. Let $f(x, y)$ be the intersection point of this line with S. Show that $f : \mathbb{R}^2 \to S$ is a homeomorphism onto its range $S - \{\mathbf{N}\}$.

Solution. The line in $\mathbb{R}^3$ through $\mathbf{x} = (x, y, 0)$ and $\mathbf{N}$ is given by

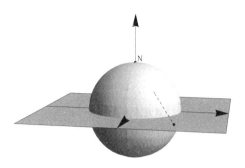

Figure 2.2: Stereographic Projection

$$X(t) = t\mathbf{N} + (1-t)\mathbf{x} = ((1-t)x, (1-t)y, t), \text{ with } t \in \mathbb{R}.$$

$X(t) \in S$ if and only if $(1-t)^2 x^2 + (1-t)^2 y^2 + t^2 = 1$ or $(1-t)^2 ||\mathbf{x}||^2 + t^2 = 1$. For $t \neq 1$ this implies

$$||\mathbf{x}||^2 = \tfrac{1+t}{1-t} \quad \text{or} \quad t = \tfrac{||\mathbf{x}||^2 - 1}{||\mathbf{x}||^2 + 1}.$$

Hence

$$f(x,y) = \left(\frac{2x}{||\mathbf{x}||^2 + 1}, \frac{2y}{||\mathbf{x}||^2 + 1}, \frac{||\mathbf{x}||^2 - 1}{||\mathbf{x}||^2 + 1} \right).$$

Similarly, the line in $\mathbb{R}^3$ from $\mathbf{N}$ to $\mathbf{x} = (x, y, 0)$ intersects the sphere at $\mathbf{p} = (u, v, w)$ with $w \neq 1$. Here $t = \frac{1}{1-w}$ and

$$f^{-1}(u, v, w) = (\frac{u}{1-w}, \frac{v}{1-w}, 0).$$

The inverse of f is called the *stereographic projection*. Since both f and f^{-1} are continuous f is a homeomorphism.[4] It is important remarking that since S is compact, by letting $\mathbf{N}$ correspond to "∞" we obtain $\mathbb{R}^2 \cup \{\infty\}$ the *one-point compactification* of $\mathbb{R}^2$.

Definition 2.5.6. Let $f : X \to Y$ be a map of metric spaces. We shall say that f is *uniformly continuous* on X if given $\epsilon > 0$ there is a $\delta = \delta(\epsilon) > 0$ so that if $d_X(x', x) < \delta$, then $d_Y(f(x'), f(x)) < \epsilon$, for all $x', x \in X$

---

[4]The sphere S is sometimes referred to as the *Riemann sphere*.

The point here is that δ depends *only* on ϵ, that is, $\delta = \delta(\epsilon)$, and not on the point $x \in X$. When f is merely continuous on X, the $\delta > 0$ depends on *both* the $\epsilon > 0$ and the point $x \in X$, that is, $\delta = \delta(x, \epsilon)$.

Example 2.5.7. The function $f(x) = \sin x$ is uniformly continuous on $\mathbb{R}$. This follows from the inequality $|\sin x - \sin y| \leq |x - y|$ and the definition by taking $\delta = \epsilon$.

The reader should check that a uniformly continuous function is continuous, but as the following example shows the converse is not valid.

Example 2.5.8. Let $f(x) = \frac{1}{x}$ on the internval $(0, 1)$ is continuous but not uniformly continuous there. In fact let $\epsilon = \frac{1}{2}$ and δ be any positive number. Choose an integer n so that $\frac{1}{n} < \delta$ (Corollary 1.1.5). Then taking $x = \frac{1}{n}$ and $x' = \frac{1}{n+1}$, as any two points in $(0, 1)$, we have

$$|x - x'| = \frac{1}{n(n+1)} < \frac{1}{n} < \delta, \text{ but } |f(x) - f(x')| = 1 > \epsilon.$$

Hence f is not uniformly continuous on $(0, 1)$.

However, when X is compact we have

Theorem 2.5.9. *Let $f : X \to Y$ be continuous. If X is compact, then f is uniformly continuous.*

Proof. Let $\epsilon > 0$. By continuity for each $x \in X$ choose an open ball of radius $\delta_x = \delta(x, \epsilon) > 0$ so that if $d_X(x', x) < \delta_x$, then $d_Y(f(x'), f(x)) < \epsilon$. Then these balls, $B_{\delta_x}(x)$, cover X. Since X is compact a finite number of these balls $\left\{B_{\delta_{x_j}}(x_j)\right\}$ centered at the points $x_1, ..., x_m \in X$ also cover X. Let $\delta = \delta(\epsilon) = \min\left\{\delta_{x_j} = \delta(x_j, \epsilon) : j = 1, 2, ..., m\right\}$. Then, $\delta > 0$ and $d_X(x', x) < \delta$ implies $d_Y(f(x'), f(x)) < \epsilon$. $\square$

2.6   Connectedness and convexity

2.6.1   Connectedness

In this subsection, we develope the basic properties of connected sets in the general frame of metric spaces, and in particular in $\mathbb{R}^n$. Intuitively, one would call a metric space connected if it is *all in one piece*, that is, if it is not the union of two nonempty subsets that do not touch each other. The precise definition is as follows

Definition 2.6.1.

1. A metric space (X, d) is said to be *connected* if it can *not* be written as the union of two disjoint, non empty open sets; i.e., if it is *not* possible to write $X = A \cup B$, with A, B nonempty open subsets of X and $A \cap B = \emptyset$.

2. A subset $S \subset X$ is said to be *connected* if it is *not* possibble to write $S = (S \cap A) \cup (S \cap B)$, with A, B open subsets of X such that $S \cap A$ and $S \cap B$ both nonempty and $S \cap A \cap B = \emptyset$.

3. A set which is not connected is said to be *disconnected*.

Remark 2.6.2.

1. Note that if a metric space X can be written as the disjoint union $X = A \cup B$ of two nonempty open sets $A, B \subset X$, then the sets A and B are automatically both open and closed in X. Thus, X is connected if and only if the *only* open *and* closed sets in X are the empty set $\emptyset$ and the whole space X itself.

2. Note also that a $S \subset X$ is connected if and only if whenever S can be written as $S = (S \cap A) \cup (S \cap B)$, with A, B open subsets of X and $S \cap A \cap B = \emptyset$, this implies that either $S \cap A = \emptyset$ or $S \cap B = \emptyset$ (equivalently, either $S \subset A$ or $S \subset B$).

Example 2.6.3.

1. Let $X = [a, b]$, an interval in $\mathbb{R}$, with the usual metric and $a < x_0 < b$. Then $X - \{x_0\}$ is not connected. Let $A = \{x \in X : x < x_0\}$ and $B = \{x \in X : x > x_0\}$. Then these are disjoint, non empty open sets whose union is $X - \{x_0\}$. In fact if S were any subset of $\mathbb{R}$ containing $X - \{x_0\}$ then the same argument shows S is not connected.

2. On the other hand the closed interval $X = [a, b]$ itself is connected. For suppose $X = A \cup B$ where A and B are disjoint non empty open sets. Since a and b are in X one must be in A and the other in B. Renaming if necessary we may assume $a \in A$ and

$b \in B$. Since A is non empty and is bounded from above by b let x_0 be the least upper bound of the elements of A. Because A and B are disjoint and open, they are also closed. Since A is closed $x_0 \in A$. On the other hand since A is open, there is a little interval about x_0 completely contained in A. But then x_0 is not even an upper bound of the elements of A. This contradiction proves X is connected.

3. Any discrete space with more than one point is not connected.

4. The set of rational $\mathbb{Q} \subset \mathbb{R}$ is not connected. Take $A = (-\infty, \sqrt{2}) \cap \mathbb{Q}$ and $B = (\sqrt{2}, \infty) \cap \mathbb{Q}$. Then $\mathbb{Q} = A \cup B$ and $A \cap B = \emptyset$.

In Example 2.6.3 (2) we proved that any closed interval $[a, b]$ in $\mathbb{R}$ is connected. As our intuition suggests, we expect all intervals to be connected. Indeed this is the case, *the connected subsets of $\mathbb{R}$ are precisely the intervals*. By an interval we mean a subset $S \subseteq \mathbb{R}$ having the property: for any $a, b \in S$ such that $a < b$, we have $(a, b) \subset S$. The following theorem characterizes the connected sets in $\mathbb{R}$.

Theorem 2.6.4. *A subset S of $\mathbb{R}$ is connected if and only is it is an interval.*

Proof. Suppose S is connected but not an interval. Then there are $a, b \in S$ and $x \notin S$ for some $x \in (a, b)$. Let $A = (-\infty, x)$ and $B = (x, \infty)$. Then A, B are disjoint open sets such that $S = (S \cap A) \cup (S \cap B)$ with $S \cap A \neq \emptyset$ and $S \cap B \neq \emptyset$, contradicting the hypothesis that S is connected. Conversely, suppose S is an interval with endpoints $\alpha, \beta \in \mathbb{R} \cup \{\pm\infty\}$, $\alpha < \beta$. We argue again by contradiction. Suppose S is not connected, that is, suppose $A, B \subset \mathbb{R}$ are open disjoint sets such that $S = (S \cap A) \cup (S \cap B)$ with $S \cap A \neq \emptyset$ and $S \cap B \neq \emptyset$. Pick $a \in S \cap A$, $b \in S \cap B$ and assume (without loss of generality) that $a < b$. Define $x = \sup([a, b] \cap (S \cap A)) = \sup([a, b] \cap A)$.

If $x \in A \cap S$, then $x < b$ and for some $\delta > 0$ we have $[x, x + \delta) \subset [a, b] \cap (A \cap S) = [a, b] \cap A$, contradicting the definition of x. On the other hand, if $x \in B \cap S$, then $x > a$ and for some $\delta > 0$ we have $(x - \delta, x] \subset [a, b] \cap (B \cap S) = [a, b] \cap B$, again contradicting the definition of x. Hence, we must have $x \notin A \cap S$ and $x \notin B \cap S$. But this is impossible since $[a, b] \subset S$. $\square$

Next we prove that continuous images of connected sets are connected.

Proposition 2.6.5. *Let (X, d) and (Y, d') be metric spaces. If $f : X \to Y$ is a continuous function and $S \subseteq X$ is a connected subset of X. Then $f(S)$ is connected.*

Proof. Assume, on the contrary, that $f(S)$ is not connected. If $f(S) = A' \cup B'$, where A' and B' are nonempty disjoint open subsets of $f(S)$, then the sets $A = S \cap f^{-1}(A')$ and $B = S \cap f^{-1}(B')$ are nonempty, disjoint, open sets in S such that $S = A \cup B$, which is impossible since S is connected. □

The above proposition shows the unit circle S^1 in $\mathbb{R}^2$ is connected as it is the continuous image of $[0, 2\pi]$ under $f(t) = (\cos t, \sin t)$. We shall see shortly that the unit sphere S^{n-1} in $\mathbb{R}^n$ with $n > 1$, is also connecetd.

We now come to the Intermediate Value theorem.

Corollary 2.6.6. *(Intermediate value theorem). Let (X, d) be a metric space and $S \subseteq X$. If S is connected and $f : S \to \mathbb{R}$ is continuous, then for any $x_1, x_2 \in S$ and $c \in \mathbb{R}$ with $f(x_1) \leq c \leq f(x_2)$, there exists $x_0 \in S$ with $f(x_0) = c$.*

Proof. Since $f(S)$ is a connected subset (subspace) of $\mathbb{R}$, by Theorem 2.6.4, it is an interval and thus contains all of $[f(x_1), f(x_2)]$. □

2.6.2   Path-connectedness

There is another stronger and quite usefull notion of connecteness. This is the notion of *pathwise connectedness.*

Definition 2.6.7. Let x and y be points of a metric space X. A *path* in X from x to y is a continuous function $\gamma : [a, b] \to X$, where $a, b \in \mathbb{R}$, $a \leq b$ with $\gamma(a) = x$ and $\gamma(b) = y$.

Note that given a path $\gamma : [a, b] \to X$, since the function $\varphi : [0, 1] \to [a, b]$ given by $\varphi(t) = a + t(b - a)$ is continuous, by considering the continuous function $\gamma \circ \varphi$ we can assume that any path is defined on the unit closed interval $[0, 1]$.

Definition 2.6.8. A metric space (X, d) is called *pathwise connected* if any two points in X can be joined by a path in X. Likewise, a set $S \subset X$ is called *pathwise connected* if any two ponts in S can be joined by a path in S

Proposition 2.6.9. *A pathwise connected metric space is connected.*

Proof. Let X be a pathwise connected metric space. If X is not connected, then there are nonempty disjoint open sets A and B of X such that $X = A \cup B$. Let $x \in A$, $y \in B$. Since X is pathwise connected there is a path $\gamma : [a, b] \to X$ in X with $\gamma(a) = x$, $\gamma(b) = y$. At the same time, since γ is continuous, $\gamma^{-1}(A)$ and $\gamma^{-1}(B)$ are open, nonempty disjoint sets in $[a, b]$ and their union is all of $[a, b]$. This contradicts the fact that $[a, b]$ is connected. $\square$

Proposition 2.6.10. *Continuous images of pathwise connected sets are pathwise connected.*

Proof. Let $f : X \to Y$ be a continuous function between the metric spaces X and Y. Suppose $S \subseteq X$ is pathwise connected. To show $f(S)$ is pathwise connecetd. Let $y_1, y_2 \in f(S)$, and let $x_1, x_2 \in S$ be such that $f(x_1) = y_1$ and $f(x_2) = y_2$. Since S is pathwise connected, there is a path in S, $\gamma : [0, 1] \to S$, joining x_1 with x_2. Then $f \circ \gamma$ is a path in $f(S)$ that joins y_1 with y_2. $\square$

Example 2.6.11. For $n > 1$, the *punctured Euclidean space*, $\mathbb{R}^n - \{0\}$, is pathwise connected and hence connected. Given nonzero $x, y \in \mathbb{R}^n$, we can join x and y by the line segment between them if that path does not pass through the origin. If it does, we can choose a third point z not on the line segment joining x and y and take the broken-line path from x to z and then from z to y. Note that when $n = 1$, $\mathbb{R} - \{0\}$ is *not* connected!

Example 2.6.12. For $n > 1$, the *unit sphere*

$$S^{n-1} = \{x \in \mathbb{R}^n : \| x \| = 1\}$$

is pathwise connected (and therefore, S^{n-1} is also connected). To see this consider the function $f : \mathbb{R}^n - \{0\} \to S^{n-1}$ given by $f(x) = \frac{x}{\|x\|}$. Since f is continuous and surjective, S^{n-1} is pathwise connected as the

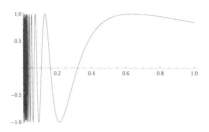

Figure 2.3: $y = \sin\frac{1}{x}$

continuous image of $\mathbb{R}^n - \{0\}$ under f. Here, note also that when $n = 1$, $S^0 = \{-1, 1\}$ is *not* connected!

The following classical example shows, the two notions do not coincide. The same example shows that the closure of a pathwise connected set need not be pathwise connected. The closure of a connected set is connecetd (see Exercise 2.7.5).

Example 2.6.13. Show that the set

$$S = \left\{ (x,y) \in \mathbb{R}^2 : y = \sin(\frac{1}{x}), 0 < x \le 1 \right\}$$

is pathwise connected but its closure $\overline{S}$ is not. See Figure 2.3.

Solution. S is the image of the connected set $(0, 1]$ under the continuous function $\varphi : (0, 1] \to \mathbb{R}^2$ given by $\varphi(x) = (x, \sin(\frac{1}{x}))$ and therefore connected. As $(0, 1]$ is clearly pathwise connected, by Proposition 2.6.10, $S = \varphi((0, 1])$ is also pathwise connected. Since the closure of a connected set is connected

$$\overline{S} = S \cup \{(0, y) : -1 \le y \le 1\}$$

is connected. However, $\overline{S}$ is not pathwise connected. There is no path joining $(0, 0)$ to any point on $\left\{ (x, \sin(\frac{1}{x})) : 0 < x \le 1 \right\}$. In particular, there is no path $\gamma : [0, 1] \to \overline{S}$ joining $(0, 0)$ to the point $(\frac{1}{\pi}, 0)$. One way to see this is to note that, from Corollary 2.3.4, $\gamma(t) = (\gamma_1(t), \gamma_2(t))$ with $\gamma(0) = (0, 0)$ and $\gamma(1) = (\frac{1}{\pi}, 0)$ is continuous if and only if its component functions γ_1 and γ_2 are continuous. If γ_1 is continuous, since $\gamma_1(0) = 0$ and $\gamma_1(1) = \frac{1}{\pi}$, by the Intermediate Value theorem, it takes all the values $\frac{1}{k\pi}$ for $k = 1, 2, \ldots$. Since $\gamma_2(t) = \sin(\frac{1}{\gamma_1(t)})$, it follows that γ_2 must assume the values $-1, 1$ in each neighborhood of 0 in $[0, 1]$. Hence, there is *no* neighborhood $[0, \delta)$ mapped into $(-\frac{1}{2}, \frac{1}{2})$ under γ_2. Thus, γ_2 cannot be continuous.

However, any *open* connected subset of $\mathbb{R}^n$ *is* pathwise connected. Sets which are both *connected* and *open* are of special importance and they are known as *domains*.

Proposition 2.6.14. *Let S be an open connected set in $\mathbb{R}^n$. Then S is pathwise connected.*

Proof. Fix any point $x_0 \in S$ and consider the sets A, B defined by $A = \{x \in S : x$ can be joined to x_0 by a path in $S\}$ and $B = \{z \in S : z \notin A\}$. Then $S = A \cup B$ and $A \cap B = \emptyset$. The result will be established by showing that $B = \emptyset$. For this will show that each of the sets A, B is open in $\mathbb{R}^n$. We first show that A is open. Let $x \in A \subseteq S$. Since S is open there is $\delta > 0$ such that $B_\delta(x) \subset S$. Since balls are convex (Example 2.6.18), any point y in $B_\delta(x)$ can be joined to x_0 by taking a path from x_0 to x and then the line segment from x to y. Hence $B_\delta(x) \subseteq A$ and A is open. Next, we show that B is also open. Let $z \in B \subseteq S$. Then there is $r > 0$ such that $B_r(z) \subseteq S$, since S is open. By the same argument as above, if any point of $B_r(z)$ can be joined by a path to x_0 so can z itself. But since $z \notin A$, no point of $B_r(z)$ can be in A. That is, $B_r(z) \subseteq B$ and so, B is open. Now, since S is connected and $x_0 \in A$, it follows that $B = \emptyset$, so $S = A$.   □

We close our discussion on connectedness, with an improved version of Proposition 2.6.14 which is, in particular, useful in the study of integrals along curves in Chapter 6. By a *polygonal path* joining two points x and y in $\mathbb{R}^n$, we mean a finite collection of line segments $\gamma_1, \ldots, \gamma_k$ such that the initial point of γ_1 is x, the final point of γ_1 is the initial point of γ_2, the final point of γ_2 is the initial point of γ_3 and so on in this manner with the final point of γ_k being the point y.

Proposition 2.6.15. *Let S be an open set in $\mathbb{R}^n$. Then S is connected if and only if any two points of S can be joined by a polygonal path lying in S.*

Proof. Suppose that any two points of S can be joined by a polygonal path. Since a polygonal path is a path (i.e, continuous), it follows that S is pathwise connected and so connected. For the converse, note that the same argument used to prove Proposition 2.6.14 also shows that every open connected set $S \subset \mathbb{R}^n$ is polygonally connected. $\square$

2.6.3   Convex sets

Convexity is an important geometrical property that certain subsets (called convex sets) of $\mathbb{R}^n$ posses.

Definition 2.6.16. A nonempty set C in $\mathbb{R}^n$ is said to be *convex* if the line segment joining any two points x and y of C is contained in C. That is, for $x, y \in C$, $\gamma(t) = tx + (1 - t)y$, where $t \in [0, 1]$ is contained in C.

Clearly, a convex set is pathwise connected and hence connected. In particular, $\mathbb{R}^n$ is connected. Furthermore, any singleton $\{x\}$ and every linear subspace of $\mathbb{R}^n$ are convex. A set S is *a-star shaped* if there is a point $a \in S$ such that for each x in S, the line segment $[a, x]$ joining a with x lies entirely in S. Note that, when $x, y \in S$, we can join x and y by joining first x with a and then a with y. Hence, a star-shaped region is connected.

Example 2.6.17. Any hyperplane and any closed (or open) half-space are convex sets. For instance $H_a = \{x \in \mathbb{R}^n : \langle a, x \rangle \geq c\}$ with $a \neq 0$ is convex. Let $x, y \in H_a$ and $t \in [0, 1]$. Then

$$\langle a, tx + (1 - t)y \rangle = t\langle a, x \rangle + (1 - t)\langle a, y \rangle \geq tc + (1 - t)c = c.$$

Consequently, $\gamma(t) \in H_a$.

Example 2.6.18. Any open or closed ball is convex. Consider for example the open ball $B_r(a)$. Let $x, y \in B_r(a)$ and $t \in [0, 1]$. Then

$$||tx + (1-t)y - a|| = ||tx + (1-t)y - ta - (1-t)a|| = ||t(x-a) + (1-t)(y-a)||$$

$$\leq t||y - a|| + (1 - t)||x - a|| < tr + (1 - t)r = r.$$

Therefore, $\gamma(t) \in B_r(a)$.

More generally in the problem section we prove that an ellipsoid is convex (see, Exercise 2.7.6)

Example 2.6.19. A rectangular parallelepiped $\prod_{i=1}^{n}[a_i, b_i]$ is convex.

Solution. If x and y are in $\prod_{i=1}^{n}[a_i, b_i]$ then $tx + (1 - t)y$ has each of its components of the form $tx_i + (1 - t)y_i$. Now since x_i and y_i are $\geq a_i$ and both t and $1 - t$ can be taken as positive we conclude $tx_i + (1 - t)y_i \geq ta_i + (1 - t)a_i = a_i$. Similarly, $tx_i + (1 - t)y_i \leq b_i$.

Definition 2.6.20. Let $v_1, ..., v_k$ be distinct points in $\mathbb{R}^n$ and $t_1, ..., t_k \in \mathbb{R}$, $t_j \geq 0$ for $j = 1, ..., k$ such that $\sum_{j=1}^{k} t_j = 1$. The sum $\sum_{j=1}^{k} t_j v_j$ is called a *convex combination* of $v_1, ..., v_k$.

Proposition 2.6.21. *A set C is convex if and only if every convex combination of points of C is a point of C.*

Proof. Suppose C is convex. We prove by induction on k that if v is any convex combination of $v_1, ..., v_k \in C$, then $v \in C$. For $k = 2$ this is true by the definition of convexity. Suppose the result is true for $k \geq 2$. Let $v = \sum_{j=1}^{k+1} t_j v_j$ where $t_j \geq 0$ for $j = 1, ..., k + 1$ with $\sum_{j=1}^{k+1} = 1$. If $t_{k+1} = 1$, then $t_1 = ... = t_k = 0$ and $v = v_{k+1} \in C$. If $t_{k+1} < 1$, let $t = 1 - t_{k+1} > 0$ (so that $0 < t \leq 1$) and set $s_j = \frac{t_j}{t}$ for $j = 1, ..., k$. Then note that $\sum_{j=1}^{k} s_j = 1$ and so the induction hypothesis implies $w = \sum_{j=1}^{k} s_j v_j \in C$. Hence, $v = tw + t_{k+1}v_{k+1} = tw + (1 - t)v_{k+1} \in C$.

Conversely, if any convex combination of points of C lies in C, so does any convex combination of two points x, y of C. Hence, C is convex. $\qquad\square$

The convex subsets of $\mathbb{R}^n$ have many remarkable properties which find interesting applications in the theory of *linear programming*. The theory of linear programming is concerned with the problem of maximizing or minimizing a linear function (a functional) subject to a system of linear inequalities. Since a linear inequality in n variables determines a (closed) half space in $\mathbb{R}^n$, a system of such linear inequalities determines an intersection of half-spaces. We saw in Example 2.6.17 that half-spaces are convex subsets of $\mathbb{R}^n$. As the reader can easily see any intersection of convex sets is convex (Exercise 2.7.21), it follows, in particular, an intersection of half-spaces is a convex set. Such a set is called a *convex polytope*. In other words, linear programming is the theory of maximizing or minimizing a linear functional over a convex polytope. Linear programming has several interesting applications in Economics.[5] In Section 3.9 we shall use differential calculus to maximize or minimize a numerical *nonlinear* function in n variables constrained over much more general subsets of $\mathbb{R}^n$.

2.7 Solved problems for Chapter 2

Exercise 2.7.1. Let

$$f(x,y) = \begin{cases} \frac{x^\alpha y^\beta}{x^2+xy+y^2} & \text{if } (x,y) \neq (0,0), \\ 0 & \text{if } (x,y) = (0,0). \end{cases}$$

Under what condition on $\alpha, \beta \in \mathbb{R}$ is the function continuous at $(0,0)$?

Solution. We must have $\lim_{(x,y)\to(0,0)} f(x,y) = 0$. If $(x,y) \to (0,0)$ along a line $y = mx$ passing from $(0,0)$, for $x \neq 0$ we have

$$f(x,y) = f(x,mx) = \frac{m^\beta}{1+m+m^2} x^{\alpha+\beta-2}.$$

If $\alpha + \beta - 2 = 0$, then the limit depends on the slope m, and hence the limit does not exist. If $\alpha + \beta - 2 < 0$, then $x^{\alpha+\beta-2} \to \infty$ as $x \to 0$ and again there is no limit. So the only alternative is $\alpha + \beta - 2 > 0$. Suppose $\alpha + \beta > 2$ and let $x = r\cos\theta$, $y = r\sin\theta$. Then

$$f(x,y) = \frac{r^{\alpha+\beta}\cos^\alpha\theta\sin^\beta\theta}{1+\cos\theta\sin\theta} = r^{\alpha+\beta}g(\theta), \text{ where } g(\theta) = \frac{\cos^\alpha\theta\sin^\beta\theta}{1+\cos\theta\sin\theta}.$$

Since for any θ, $\frac{1}{2} \leq 1 + \cos\theta\sin\theta \leq 2$, we have

$$|f(x,y)| = r^{\alpha+\beta}|g(\theta)| \leq 2r^{\alpha+\beta} \to 0 \text{ as } r \to 0.$$

Exercise 2.7.2. Let X be compact and $C(X)$ be the set of all continuous functions $f : X \to \mathbb{R}$. Then $C(X)$ is a complete metric space under

$$d(f,g) = ||f - g|| = \sup_{x \in X} |f(x) - g(x)|.$$

---

[5]See for example, [8].

Solution. Suppose $\{f_n\}$ is a Cauchy sequence in $C(X)$ and $\epsilon > 0$. Then $||f_n - f_m|| < \epsilon$, if n and m are large enough. Now for fixed $x \in X$, $|f_n(x) - f_m(x)| \le ||f_n - f_m||$. Hence $\{f_n(x)\}$ is a Cauchy sequence of real numbers. By completeness of $\mathbb{R}$, there is an $f(x)$ such that $f_n(x) \to f(x)$. (Notice here the rate of convergence could depend on x. This is called pointwise convergence). Now

$$|f(x) - f_n(x)| \le |f(x) - f_m(x)| + |f_m(x) - f_n(x)| \le |f(x) - f_m(x)| + ||f_n - f_m||.$$

Now take m large enough so that $|f(x) - f_m(x)| < \epsilon$ and n, m large enough so that $||f_n - f_m|| < \epsilon$. Then For all $x \in X$ and n large enough, $|f(x) - f_n(x)| < 2\epsilon$. Taking the sup over X we conclude $||f - f_n|| \le 2\epsilon$. Since $\epsilon > 0$ is arbitrary this means $\{f_n\}$ converges uniformly to f. Since f must be continuous by (see, Theorem C.0.2), this completes the proof.

Exercise 2.7.3. Let $g \in C([0,1])$. Show that there exists a unique $f \in C([0,1])$ satisfying the (integral) equation

$$f(x) - \int_0^x f(x-t)e^{-t^2}dt = h(x).$$

Solution. Let $T : C([0,1]) \to C([0,1])$ be the mapping defined by

$$T(f)(x) = \int_0^x f(x-t)e^{-t^2}dt + h(x).$$

For $f, g \in C([0,1])$, we have

$$||T(f) - T(g)|| = \sup_{x \in [0,1]} \left| \int_0^x (f(x-t) - g(x-t))\, e^{-t^2}dt \right|$$

$$\le \sup_{x \in [0,1]} \int_0^x |f(x-t) - g(x-t)| e^{-t^2}dt$$

$$\le ||f - g|| \sup_{x \in [0,1]} \int_0^x e^{-t^2}dt$$

$$= ||f - g|| \int_0^1 e^{-t^2}dt < c\,||f - g||,$$

where $c = \int_0^1 e^{-t^2}dt < 1$. Hence T is a contraction. Since $C([0,1])$ is complete, the Contraction Mapping Principle tells us that there is a unique f such that $T(f) = f$.

Exercise 2.7.4. Let $T : \mathbb{R}^n \to \mathbb{R}^m$ be a linear transformation. Show that T is bounded if and only if maps bounded sets in $\mathbb{R}^n$ to bounded sets in $\mathbb{R}^m$.

Solution. Suppose T is bounded and let B be a bounded set in $\mathbb{R}^n$. Then there exists $R > 0$ with $\parallel x \parallel \leq R$ for all $x \in B$. Since T is bounded, $\parallel T(x) \parallel \leq M \parallel x \parallel \leq MR$, which means that $T(B)$ is bounded. Conversely, assume T maps bounded sets into bounded sets. Then, in particular it maps the closed unit ball $\overline{B}_1(0)$ in $\mathbb{R}^n$ into a bounded set in $\mathbb{R}^m$. Therefore, there exists $R > 0$ such that $\parallel T(x) \parallel \leq R$ for all $x \in \overline{B}_1(0)$. Now, for any nonzero vector $x \in \mathbb{R}^n$ we have $\frac{x}{\|x\|} \in \overline{B}_1(0)$, and so $\parallel T(\frac{x}{\|x\|}) \parallel \leq R$. Hence, $\parallel T(x) \parallel \leq R \parallel x \parallel$. Since this clearly also holds for $x = 0$, T is a bounded.

Exercise 2.7.5. Let X be a metric space and $S \subseteq X$.

1. Suppose $\overline{S} = X$ and S is connected. Show that X is connected.

2. If S is connected, show that any set E satisfying $S \subset E \subset \overline{S}$ is also connected. In particular, the closure of a connected set is connected.

Solution.

1. Suppose $X = A \cup B$ with A, B open and disjoint subsets of X. We show that either $A = \emptyset$ or $B = \emptyset$. Since $S \subseteq A \cup B$, $S \cap (A \cap B) = \emptyset$ and S is connected, it follows from Remark 2.6.2, that either $S \subset A$ or $S \subset B$. Assume, $S \subset A$. Since $A = B^c$ a closed set, we have $\overline{S} \subset B^c$. That is, $X \subset B^c$. Hence, $B = \emptyset$.

2. The closure of S in E is $E \cap \overline{S}$. However, since $E \subset \overline{S}$, we get $E \cap \overline{S} = E$. Thus, by part (1), E is connected.

Exercise 2.7.6. Show that an ellipsoid is convex.

Solution. An ellipsoid (centered at the origin) with minor axes $a_1, \ldots a_n$ (all positive numbers) is given by $C = \left\{ (x_1, ..., x_n) \in \mathbb{R}^n : \sum_{i=1}^n \frac{x_i^2}{a_i^2} < 1 \right\}$. Notice when all the a_i are equal we get a ball.

Let $0 \leq t \leq 1$ and consider

$$\sum_{i=1}^n \frac{[tx_i + (1-t)y_i]^2}{a_i^2} = \sum_{i=1}^n \frac{t^2 x_i^2 + 2t(1-t)x_i y_i + (1-t)^2 y_i^2}{a_i^2}.$$

This is evidently

$$t^2 \sum_{i=1}^n \frac{x_i^2}{a_i^2} + 2t(1-t) \sum_{i=1}^n \frac{x_i y_i}{a_i^2} + (1-t)^2 \sum_{i=1}^n \frac{y_i^2}{a_i^2}.$$

The first and third coefficients are each < 1. The second can be estimated by Cauchy-Schwarz by letting $u_i = \frac{x_i}{a_i}$ and $v_i = \frac{y_i}{a_i}$. Then $\Sigma_{i=1}^n u_i v_i \leq \sqrt{\sum_{i=1}^n u_i^2 \sum_{i=1}^n v_i^2} < 1 \cdot 1 = 1$. Thus

$$t^2 \Sigma_{i=1}^n \frac{x_i^2}{a_i^2} + 2t(1-t) \sum_{i=1}^n \frac{x_i y_i}{a_i^2} + (1-t)^2 \sum_{i=1}^n \frac{y_i^2}{a_i^2} < t^2 + 2t(1-t) + (1-t)^2 = 1.$$

Miscellaneous Exercise

Exercise 2.7.7. Let $f : \mathbb{R}^3 \to \mathbb{R}$ be the function

$$f(x, y, z) = \begin{cases} \frac{xy|z|^\alpha}{x^{2\alpha} + y^2 + z^2} & \text{if } (x, y, z) \neq (0, 0, 0) \\ 0 & \text{if } (x, y, z) = (0, 0, 0) \end{cases}$$

where $\alpha > 0$. Prove that f is continuous at $(0, 0, 0)$.

Exercise 2.7.8. Let S be a nonempty subset of $\mathbb{R}^n$. The *distance* from a point $x \in \mathbb{R}^n$ to the set S is defined by

$$d(x, S) = \inf \{ ||x - y|| : y \in S \}.$$

Show that

1. $d(x, S) = 0$ if and only if $x \in \overline{S}$.

2. $d(x, \overline{S}) = d(x, S)$ and $\overline{S}$ has the same diameter as S.

3. The function $x \mapsto d(x, S)$ is (uniformly) continuous.

4. If S is compact, then there exist $y_0 \in S$ such that $d(x, S) = ||x - y_0||$.

Exercise 2.7.9. Let $f : B \to \mathbb{R}$ be continuous, where B is the closed unit disk in $\mathbb{R}^2$. Show that f can not be one-to-one.

Exercise 2.7.10. Let S be a compact subset of $\mathbb{R}^n$ and let $f : S \to \mathbb{R}^m$ be continuous and one-to-one on S. Show that the inverse $f^{-1} : f(S) \to S$ is continuous.

Exercise 2.7.11. Let $S \subset \mathbb{R}^n$. Show that if every continuous real-valued function on S attains a maximum value, then S is compact.

Exercise 2.7.12. Let $S \subseteq \mathbb{R}^n$ be compact and $f : S \to S$ an isometry. Show that $f(S) = S$.

Exercise 2.7.13. Show that the open interval $(-1, 1)$ in $\mathbb{R}$ is homeomorphic to $\mathbb{R}$.

Exercise 2.7.14. Is the function $f(x, y) = \frac{1}{x^2+y^2-1}$ uniformly continuous in the open unit disk in $\mathbb{R}^2$?

Exercise 2.7.15. Let $f : \Omega \to \mathbb{R}^m$ be uniformly continuous on $\Omega \subset \mathbb{R}^n$. Show that

1. If $\{x_k\}$ is a Cauchy sequence in Ω, then $\{f(x_k)\}$ is also a Cauchy sequence.

2. If Ω is bounded, then $f(\Omega)$ is bounded.

Exercise 2.7.16. Show that $S \subset \mathbb{R}^n$ is connected if and only if every continuous function $f : S \to \mathbb{R}$ such that $f(S) \subseteq \{0, 1\}$ is constant. *Hint.* Consider the set $\{0, 1\}$ as a metric space with the discrete metric.

Exercise 2.7.17. Let $f : S \to S$ be continuous. Take $S = [0, 1]$. Show that there is a point ξ such that $f(\xi) = \xi$. What happens if $S = [0, 1)$ or if $S = (0, 1)$?

Exercise 2.7.18. Let $S = S^{n-1}$ be the unit sphere in $\mathbb{R}^n$ and let $f : S \to \mathbb{R}$ be a continuous function. Show there must be a pair of diametrically opposite points on S at which f assumes the same value.
Hint. Consider $\varphi(x) = f(x) - f(-x)$.

Exercise 2.7.19. Let $S \subset \mathbb{R}^n$, $x \in S$ and $y \in S^c$. Let $\gamma : [0, 1] \to \mathbb{R}^n$ is a path joining x and y. Show that there is a $t_0 \in [0, 1]$ such that $\gamma(t_0) \in \partial(S)$.

Exercise 2.7.20. Show that the set

$$S = \{(x, y) \in \mathbb{R}^2 : 0 < y \le x^2,\ x \ne 0\} \cup \{(0, 0)\}$$

is connected in $\mathbb{R}^2$, but not polygonally connected.
Hint. There is no polygonal path lying in S joining $(0, 0)$ to other points in the set.

Exercise 2.7.21. Show that any intersection of convex sets is convex.

Exercise 2.7.22.   1. Show that the interior of a convex set is convex.

2. Show that the closure of a convex set is convex.

Exercise 2.7.23.   1. Show that if C is convex then so are the sets λC for $\lambda \in \mathbb{R}$, and $x + C$ for any $x \in \mathbb{R}^n$.

2. Suppose that $\frac{1}{2}(x + y) \in S$ for all $x, y \in S$. Show that if S is closed, then S is convex.

Exercise 2.7.24. Let S be any subset of $\mathbb{R}^n$. The set of all convex combinations of elements of S is called the *convex hull* of S and is denoted by $conv(S)$, that is,

$$conv(S) = \left\{ \sum_j t_j v_j : v_j \in S, t_j \geq 0 \text{ with } \sum_j t_j = 1 \right\}$$

where the sums $\sum_j$ are finite.

1. Show that $conv(S)$ is the smallest convex set containing S. In fact, show that $conv(S)$ is the intersection of all convex sets that contain S.

2. Show that if S is open subset of $\mathbb{R}^n$, then $conv(S)$ is open.

Exercise 2.7.25. Show that if $T : \mathbb{R}^n \to \mathbb{R}^m$ is affine and C convex subset of $\mathbb{R}^n$, then $T(C)$ is convex.

Exercise 2.7.26. (*) Show that there exists a unique continuous function $f : [0, 1] \to \mathbb{R}$ such that

$$f(x) = \sin x + \int_0^1 e^{-(x+y+1)} f(y) dy.$$

Exercise 2.7.27. (*) Let $f : \mathbb{R}^n \to \mathbb{R}$ be continuous. Show that the graph of f is connected in $\mathbb{R}^n \times \mathbb{R}$.

Exercise 2.7.28. (*) Let C be a closed convex set in $\mathbb{R}$. Show that for each $a \in C$ there exists a unique $b \in C$ closest to a, that is,

$$||a - b|| = \min_{x \in C} ||a - x||.$$

Exercise 2.7.29. (*) Let $GL(n, \mathbb{R})$ be the set of all $n \times n$ real invertible matrices. Show that $GL(n, \mathbb{R})$ is not a connected subset of $M_{n \times n}(\mathbb{R}) \cong \mathbb{R}^{n^2}$.

Chapter 3

Differential Calculus in Several Variables

In this chapter we introduce the concept of differentiability for functions of several variables and derive their fundamental properties. Included are the chain rule, Taylor's theorem, maxima - minima, the inverse and implicit function theorems, constraint extrema and the Lagrange multiplier rule, functional dependence, and Morse's lemma.

3.1 Differentiable functions

To motivate the definition of differentiability for functions on $\mathbb{R}^n$, where $n > 1$, we recall the definition of the derivative when $n = 1$. Let $\Omega \subseteq \mathbb{R}$ be an interval and $a \in \Omega$. A function $f : \Omega \to \mathbb{R}$ is said *differentiable at a* if the limit

$$\lim_{x \to a} \frac{f(x) - f(a)}{x - a} = f'(a)$$

exists.

The number $f'(a)$ is called the *derivative* of f at a and geometrically is the slope of the tangent line to the graph of f at the point $(a, f(a))$. As it stands this definition can not be used for functions defined on $\mathbb{R}^n$ with $n > 1$ since division by elements of $\mathbb{R}^n$ makes no sense. However, $f'(a)$ gives information concerning the local behavior of f near the point *a.* In fact, we may write the above formula equivalently as

$$\lim_{x \to a} \frac{|f(x) - f(a) - f'(a)(x - a)|}{|x - a|} = 0,$$

which makes precise the sense in which we approximate the values $f(x)$, for x sufficiently near a, by the values of the linear function $y = f(a) + f'(a)(x - a)$. The graph of this function is the tangent line to the graph of f at the point $(a, f(a))$. In other words, the error $R(x, a)$ made by approximating a point on the graph of f by a point on the tangent line with the same $x - coordinate$ is $f(x) - (f(a) + f'(a)(x - a))$ (for $x \neq a$) and has the property that

$$\lim_{x \to a} \frac{|R(x, a)|}{|x - a|} = 0.$$

Roughly speaking this says that $R(x, a)$ approaches 0 *faster than* $|x - a|$, and $f(a) + f'(a)(x - a)$ is a good approximation to $f(x)$ for $|x - a|$ small. Because $T(x) = f'(a)x$ defines a linear transformation $T : \mathbb{R} \to \mathbb{R}$ and the good approximation to $f(x)$ is $f(a) + T(x - a)$, it is this idea that generalizes to $\mathbb{R}^n$ using the terminology of linear algebra.

Definition 3.1.1. Let $f : \Omega \to \mathbb{R}^m$ be a function defined on an open set Ω in $\mathbb{R}^n$ and $a \in \Omega$. We say that f is *differentiable at a* if there is a linear transformation $T : \mathbb{R}^n \to \mathbb{R}^m$ such that

$$\lim_{x \to a} \frac{||f(x) - f(a) - T(x - a)||_m}{||x - a||_n} = 0.$$

Here in the numerator we use the norm in $\mathbb{R}^m$ and in the denominator that of $\mathbb{R}^n$. In the future we shall surpress the subscripts. As we shall see T is unique and is denoted by $D_f(a)$, the *derivative* of f at the point a. If f is differentiable at every point of Ω, we say that f is *differentiable on Ω*.

Remark 3.1.2. Setting $h = x - a \neq 0$, then $x \to a$ in $\mathbb{R}^n$ is equivalent to $h \to 0$ and so, the limit in Definition 3.1.1 is equivalent to

$$\lim_{h \to 0} \frac{||f(a + h) - f(a) - T(h)||}{||h||} = 0. \tag{3.1}$$

The derivative[1] $T = D_f(a)$ depends on the point a as well as the function f. We are not saying that there exists a T which works for all a, but that for a fixed a such a T exists. Differentiability of f at a says that $f(x)$ is *"well" approximated* by the affine transformation $F(x) = f(a) + T(x - a)$ for x near a. Important particular cases of the derivative are when $m = 1$ or $n = 1$. When $m = 1$, f is real-valued function and the derivative T is a linear functional on $\mathbb{R}^n$, which will turn out to be the so called *gradient* of f. When $n = 1$, for $t \in \Omega \subseteq \mathbb{R}$, $f(t)$ is a curve in $\mathbb{R}^m$ and the derivative T is a vector in $\mathbb{R}^m$, which will turn out to be the *tangent vector* to the curve at $f(a)$.

The following is called the *Linear Approximation theorem*. Since a is fixed it will be convenient in what follows to write $\epsilon(x)$ rather $\epsilon(x, a)$. Here ϵ is a function from $\Omega \subseteq \mathbb{R}^n \to \mathbb{R}^m$.

Theorem 3.1.3. $f : \Omega \subseteq \mathbb{R}^n \to \mathbb{R}^m$ *is differentiable at a if and only if there is a function $\epsilon(x)$ so that for $x \in \Omega$ we have*

$$f(x) = f(a) + T(x - a) + \epsilon(x)||x - a||$$

with $\epsilon(x) \to 0$ as $x \to a$.

Proof. Set

$$\epsilon(x) = \frac{f(x) - f(a) - T(x - a)}{||x - a||}, \quad x \neq a.$$

Now, if f is differentiable at a, then $\lim_{x \to a} \epsilon(x) = 0$. Conversely, if

$$f(x) = f(a) + T(x - a) + \epsilon(x)\,||x - a||$$

holds, then since $x \neq a$, we have

$$\frac{f(x) - f(a) - T(x - a)}{||x - a||} = \epsilon(x) \to 0$$

as $x \to a$ and f is differentiable at a. □

---

[1] In more abstract settings where f is a differentiable mapping between complete normed linear spaces (called *Banach spaces*) the derivative $D_f(a)$ is known as the *Frechet derivative*.

Note that by Remark 3.1 2., differentiability of f at a is equivalent to

$$f(a + h) = f(a) + T(h) + \epsilon(h)||h|| \qquad (3.2)$$

with $\epsilon(h) \to 0$ as $h \to 0$.

Proposition 3.1.4. *Let* $f : \Omega \to \mathbb{R}^m$ *be a function defined on an open set* Ω *in* $\mathbb{R}^n$ *and* $a \in \Omega$. *If* f *is differentiable at* a, *then* $D_f(a)$ *is uniquely determined by* f.

Proof. Suppose T, S are linear transformations satisfying (3.2). We prove $T = S$. By Theorem 3.1.3, we have $f(a + h) = f(a) + T(h) + \epsilon_T(h)||h||$ and $f(a + h) = f(a) + S(h) + \epsilon_S(h)||h||$ with both $\epsilon_T(h)$ and $\epsilon_S(h)$ approaching zero as $h \to 0$. Subtracting we get

$$T(h) - S(h) = (\epsilon_T - \epsilon_S)||h||.$$

Setting $L = T - S$ (again a linear transformation), and dividing by $||h||$, it follows that

$$\frac{||L(h)||}{||h||} = ||\epsilon_T - \epsilon_S|| \to 0$$

as $h \to 0$.

Now, let $x \in \mathbb{R}^n$ be any (but fixed) nonzero vector and for $t \in \mathbb{R}$ take $h = tx$. Then $h \to 0$ is equivalent to $t \to 0$ and since L is linear

$$0 = \lim_{t \to 0} \frac{||L(tx)||}{||tx||} = \frac{|t|\,||L(x)||}{|t|\,||x||} = \frac{||L(x)||}{||x||}.$$

Therefore, $||L(x)|| = 0$. Thus, $Lx = 0$ for all x and so $L = 0$. Hence $T = S$. $\qquad\square$

Example 3.1.5. Let $F : \mathbb{R}^n \to \mathbb{R}^m$ be an affine transformation of the form $F(x) = T(x) + v$, where T is a fixed linear transformation and v is a fixed vector. Then F is everywhere differentiable and $D_F(a) = T$ for all $a \in \mathbb{R}^n$. Indeed,

$$\lim_{h \to 0} \frac{||F(a + h) - F(a) - T(h)||}{||h||} = \lim_{h \to 0} \frac{0}{||h||} = 0.$$

In particular, note that a linear transformation T is its own derivative, that is, $D_T(a) = T$.

One simple, but important consequence of differentiability is continuity.

Proposition 3.1.6. *If f is differentiable at a, then it is continuous at a.*

Proof. Since f is diferentiable at a, we have

$$f(x) = f(a) + D_f(a)(x - a) + \epsilon(x)||x - a||.$$

Let $x \to a$. Then the third term on the right tends to zero "in spades". As for the second term, by Proposition 2.4.3 there is $b > 0$ such that

$$||D_f(a)(x - a)|| \le b||x - a||,$$

and so it also tends to zero. Hence, $\lim_{x \to a} f(x) = f(a)$. $\square$

Next we state the differentiation rules. The proofs of the differentiation rules proceed almost exactly as in the one variable case, with slight modifications in notation, and we leave them to the reader as an exercise.

Theorem 3.1.7. *Let f and g be functions from an open set Ω in $\mathbb{R}^n$ to $\mathbb{R}^m$, differentiable at $a \in \Omega$ and let $c \in \mathbb{R}$. Then*

1. *$f + g$ is differentiable at a and*

$$D_{(f+g)}(a) = D_f(a) + D_g(a).$$

2. *cf is differentiable at a and*

$$D_{cf}(a) = cD_f(a).$$

3. *Suppose $m = 1$, then $(fg)(x) = f(x)g(x)$ is differentiable at a and*

$$D_{fg}(a) = g(a)D_f(a) + f(a)D_g(a).$$

4. *Suppose $m = 1$ and $g(a) \ne 0$, then $(\frac{f}{g})(x) = \frac{f(x)}{g(x)}$ is differentiable at a and*

$$D_{(\frac{f}{g})}(a) = \frac{g(a)D_f(a) - f(a)D_g(a)}{g(a)^2}.$$

We now come to the Chain Rule, which deals with the derivative of a composite function.

Theorem 3.1.8. *(Chain Rule). Let Ω be open in $\mathbb{R}^n$ and $f : \Omega \to \mathbb{R}^m$ and $g : U \to \mathbb{R}^p$, where U is open in $\mathbb{R}^m$ with $f(\Omega) \subseteq U$. If f is differentiable at $a \in \Omega$ and g is differentiable at $f(a)$, then $g \circ f$ is differentiable at a and*

$$D_{(g \circ f)}(a) = D_g(f(a))D_f(a).$$

Proof. We shall apply the Linear Approximation theorem. Let $b = f(a)$ and $y = f(x)$ for $x \in \Omega$. By the differentiability of f and g, we have

$$f(x) = f(a) + D_f(a)(x - a) + \epsilon_1(x)\|x - a\|$$

and

$$g(y) = g(b) + D_g(b)(y - b) + \epsilon_2(y)\|y - b\|.$$

Substituting the first equation into the second yields

$$g(f(x)) = g(f(a)) + D_g(f(a)) \left\{ D_f(a)(x - a) + \epsilon_1(x)\|x - a\| \right\}$$

$$+ \epsilon_2(y) \left\| \left\{ D_f(a)(x - a) + \epsilon_1(x)\|x - a\| \right\} \right\|.$$

Since $D_g(f(a))$ is linear, it follows that

$$g(f(x)) = g(f(a)) + D_g(f(a))D_f(a)(x - a) + \|x - a\|D_g(f(a))\epsilon_1(x)$$

$$+ \epsilon_2(y) \left(\left\| D_f(a)\frac{x - a}{\|x - a\|} + \epsilon_1(x) \right\| \right) \|x - a\|.$$

Factoring $\|x - a\|$ from the last two terms, we get

$$g(f(x)) = g(f(a)) + D_g(f(a))D_f(a)(x - a) + \epsilon(x)\|x - a\|,$$

where $\epsilon(x) = D_g(f(a))\epsilon_1(x) + \epsilon_2(y)\|\frac{D_f(a)(x-a)}{\|x-a\|} + \epsilon_1(x)\|$.

The proof will be complete, if we show that $\epsilon(x) \to 0$ as $x \to a$. Clearly, the first term tends to zero as $x \to a$ (since $\epsilon_1(x) \to 0$ as $x \to a$ and $D_g(f(a))$ is continuous). For the second term, using the triangle inequality and the fact that $D_f(a)$ is bounded (i.e., $||D_f(a)(v)|| \le c||v||$ for some $c > 0$), we have

$$0 \le ||\epsilon_2(y)|| \left\| \frac{1}{||x-a||} D_f(a)(x-a) + \epsilon_1(x) \right\| \le ||\epsilon_2(y)||(c + ||\epsilon_1(x)||).$$

As $x \to a$, the continuity of f at a implies $y \to b$ and so, both $\epsilon_2(y) \to 0$ and $\epsilon_1(x) \to 0$ as $x \to a$. Hence $||\epsilon_2(y)||(c + ||\epsilon_1(x)||) \to 0$ $\qquad\square$

Proposition 3.1.9. *Let Ω be an open set in $\mathbb{R}^n$ and $f : \Omega \to \mathbb{R}^m$, with component functions $f = (f_1, f_2, ..., f_m)$. Then f is differentiable at a if and only if f_j is differentiable at a, for all $j = 1, 2, ..., m$. Moreover,*

$$D_f(a) = (D_{f_1}(a), ..., D_{f_m}(a)).$$

Proof. Let $D_f(a) = T = (\lambda_1, ..., \lambda_m)$. If f is differentiable at a, the vector equality $f(x) = f(a) + T(x - a) + \epsilon(x)||x - a||$ written in terms of the components becomes $f_j(x) = f_j(a) + \lambda_j(x - a) + \epsilon_j(x)||x - a||$, for $j = 1, 2, ..., m$.
Since

$$\max_{j=1,...m} |\epsilon_j(x)| \le ||\epsilon(x)|| \le \sqrt{n} \max_{j=1,...m} |\epsilon_j(x)|,$$

we have $\epsilon(x) \to 0$ as $x \to a$ if and only if $\epsilon_j(x) \to 0$ as $x \to a$ and so the result. At the same time by the uniqueness of the derivative $\lambda_j = D_{f_j}(a)$ for $j = 1, ..., m$ $\qquad\square$

By the above proposition, we see that to study the differentiability of functions $f : \Omega \subseteq \mathbb{R}^n \to \mathbb{R}^m$, it suffices to study the differentiability of its component functions $f_j : \Omega \subseteq \mathbb{R}^n \to \mathbb{R}$, for $j = 1, ..., m$. Hence we turn to real-valued functions of several variables.

3.2 Partial and directional derivatives, tangent space

Here we consider a real-valued function $f : \Omega \subseteq \mathbb{R}^n \to \mathbb{R}$ on an open set Ω which we assume is differentiable at $a \in \Omega$. Thus, there is a linear functional $\lambda : \mathbb{R}^n \to \mathbb{R}$ such that

$$f(x) = f(a) + \lambda(x - a) + \epsilon(x)||x - a||$$

with $\epsilon(x) \to 0$ as $x \to a$. By Theorem 2.4.12, a linear functional $\mathbb{R}^n$ is an inner product by some fixed vector $w \in \mathbb{R}^n$, that is, $\lambda(x) = \langle w, x \rangle$. This vector, which depends on a, is called the *gradient* of f at a and is denoted by $\nabla f(a)$. Thus,

$$f(x) = f(a) + \langle \nabla f(a), (x - a) \rangle + \epsilon(x) ||x - a||$$

We want to get a convenient and explicit form of the gradient. To do so for f as above, we define the notion of directional derivative.

Definition 3.2.1. The *directional derivative of f at a in the direction* of a nonzero vector $u \in \mathbb{R}^n$, denoted by $D_u f(a)$ is defined by

$$D_u f(a) = \lim_{t \to 0} \frac{f(a + tu) - f(a)}{t},$$

whenever the limit exists.

We remark that the function $g : \mathbb{R} \to \mathbb{R}$ given by $g(t) = f(a + tu)$ represents the function f restricted on the line $X(t) = a + tu$ in Ω, passing from a in the direction of u. Since

$$g'(0) = \lim_{t \to 0} \frac{g(t) - g(0)}{t} = \lim_{t \to 0} \frac{f(a + tu) - f(a)}{t} = D_u f(a),$$

the directional derivative is the rate of change of f in the direction u.

Theorem 3.2.2. *If $f : \Omega \subseteq \mathbb{R}^n \to \mathbb{R}$ is differentiable at $a \in \Omega$, then for any direction $u \neq 0$, $u \in \mathbb{R}^n$, $D_u f(a)$ exists and*

$$D_u f(a) = \langle \nabla f(a), u \rangle.$$

Proof. Let $t \in \mathbb{R}$ and normalize $u \neq 0$ so that $||u|| = 1$. Since f is differentiable at a, $f(x) = f(a) + \langle \nabla f(a), (x - a) \rangle + \epsilon(x) ||x - a||$, with $\epsilon(x) \to 0$ as $x \to a$. Setting $x = a + tu$, this yields $f(a + tu) = f(a) + \langle \nabla f(a), tu \rangle + \epsilon(a + tu) ||tu||$ and implies

$$\frac{f(a + tu) - f(a)}{t} - \langle \nabla f(a), u \rangle = \frac{|t|}{t} \epsilon(a + tu).$$

Taking absolute values we get,

$$|\frac{f(a + tu) - f(a)}{t} - \langle \nabla f(a), u \rangle| = |\epsilon(a + tu)|.$$

Now, as $t \to 0$, $x \to a$ and so $\lim_{t \to 0} |\epsilon(a + tu)| = 0$. Hence,

$$\lim_{t \to 0} \frac{f(a + tu) - f(a)}{t} = \langle \nabla f(a), u \rangle.$$

$\square$

Corollary 3.2.3. *For nonzero vectors $u, v \in \mathbb{R}^n$ and $c, d \in \mathbb{R}$, we have*

$$D_{cu+dv}f(a) = cD_u f(a) + dD_v f(a).$$

We now address the question, for which direction u, is $|D_u f(a)|$ the largest?

Corollary 3.2.4. $|D_u f(a)|$ *is maximum in the direction of* $\nabla f(a)$.

Proof. By the Cauchy-Schwarz inequality

$$|D_u f(a)| = |\langle \nabla f(a), u \rangle| \leq ||\nabla f(a)|| \cdot ||u||$$

with equality only if u is a scalar multiple of $\nabla f(a)$. $\square$

Certain directions are special, namely those of the standard basis elements e_i for $i = 1, ..., n$, the directions of the coordinate axes.

Definition 3.2.5. Let $f : \Omega \subseteq \mathbb{R}^n \to \mathbb{R}$. The directional derivative of f at a in the direction of e_i is denoted by

$$\frac{\partial f}{\partial x_i}(a) = D_{e_i} f(a) = \langle \nabla f(a), e_i \rangle$$

is called the *partial derivative* of f at a with respect to x_i.

Writing $f(a_1, ..., a_n)$ in place of $f(a)$, we see that

$$\frac{\partial f}{\partial x_i}(a) = \lim_{t \to 0} \frac{f(a + te_i) - f(a)}{t}$$

is simply the ordinary derivative of f considered as a function of x_i alone, keeping the other components fixed. Since $\lambda(x) = \langle w, x \rangle$ for each $i = 1, ..., n$

$$\nabla f(a) = \left(\frac{\partial f}{\partial x_1}(a), ..., \frac{\partial f}{\partial x_n}(a) \right),$$

and so

$$f(x) = f(a) + \sum_{i=1}^{n} (x_i - a_i) \frac{\partial f}{\partial x_i}(a) + \epsilon(x)||x - a||.$$

Example 3.2.6. Let $f(x, y, z) = 2x^2 + 3y^2 + z^2$. Find the directional derivative of f at $a = (2, 1, 3)$ in the direction $v = (1, 0 - 2)$. What is the largest of the directional derivative of f at a, and in what dirction does it occur?

Solution. $\|v\| = \sqrt{5}$. The unit vector in the given direction is $u = (\frac{1}{\sqrt{5}}, 0, \frac{-2}{\sqrt{5}})$. The gradient is $\nabla f(x, y, z) = (4x, 6y, 2z)$, so that $\nabla f(2, 1, 3) = (8, 6, 6)$. By Theorem 3.2.2

$$D_u f(a) = \langle \nabla f(a), u \rangle = \langle (8, 6, 6), (\frac{1}{\sqrt{5}}, 0, \frac{-2}{\sqrt{5}}) \rangle = -\frac{4}{\sqrt{5}}.$$

The negative sign indicates that f decreases in the given direction. The largest directional derivative at a is $\|\nabla f(a)\| = 2\sqrt{34}$, and it occurs in the direction $\frac{1}{\sqrt{34}}(4, 3, 3)$.

Next we show how matrices arise in connection with derivatives. We shall see that if $f : \Omega \subseteq \mathbb{R}^n \to \mathbb{R}^m$ is differentiable at $a \in \Omega$, then the partial derivatives of its component functions $\frac{\partial f_j}{\partial x_i}(a)$ exist and determine the linear transformation $D_f(a)$ completely.

Theorem 3.2.7. *Let Ω be an open set in $\mathbb{R}^n$ and $f : \Omega \to \mathbb{R}^m$ be differentiable at $a \in \Omega$. Then $\frac{\partial f_j}{\partial x_i}(a)$ exist and the standard matrix representation of $D_f(a)$ is the $m \times n$ matrix whose ji^{th} entry is $\frac{\partial f_j}{\partial x_i}(a)$ for $j = 1, ..., m$ and $i = 1, ..., n$.*

Proof. Let $D_f(a) = T = (\lambda_1, ..., \lambda_m)$ and $\{e_1, ..., e_n\}$ and $\{u_1, ..., u_m\}$ be the standard bases for $\mathbb{R}^n$ and $\mathbb{R}^m$ respectively. By the definition of the matrix of a linear map the ji^{th} entry of the standard matrix of T, say c_{ji}, is given by the jth component of the vector $T(e_i) = \sum_{j=1}^{m} c_{ji} u_j$. Since f is differentiable at a, by Proposition 3.1.9 each f_j is differentiable at a. Hence, by Theorem 3.2.2 each $\frac{\partial f_j}{\partial x_i}(a)$ exists and $\frac{\partial f_j}{\partial x_i}(a) = \langle \nabla f_j(a), e_i \rangle = \lambda_j(e_i)$.

So we have,

$$T(e_i) = (\lambda_1(e_i), ..., \lambda_m(e_i)) = \sum_{j=1}^{m} \lambda_j(e_i) u_j = \sum_{j=1}^{m} \frac{\partial f_j}{\partial x_i}(a) u_j.$$

Therefore,

$$\sum_{j=1}^{m} \left[c_{ji} - \frac{\partial f_j}{\partial x_i}(a) \right] u_j = 0.$$

Since $\{u_1, ..., u_m\}$ is linearly independent, it follows that $c_{ji} = \frac{\partial f_j}{\partial x_i}(a)$ for $j = 1, ..., m$ and $i = 1, ..., n$. $\square$

As the following example shows, the converse of Theorem 3.2.7 is false. Namely, the existence of the partial derivatives of $f : \Omega \to \mathbb{R}$ at $a \in \Omega$, does not imply differentiability of f at the same point a (nor even the continuity at a). However, see Theorem 3.2.16.

Example 3.2.8. Consider the function $f : \mathbb{R}^2 \to \mathbb{R}$ given by

$$f(x, y) = \begin{cases} 0 \text{ if } x = 0 \text{ or } y = 0, \\ 1 \text{ otherwise.} \end{cases}$$

We have

$$\frac{\partial f}{\partial x}(0, 0) = \lim_{t \to 0} \frac{f(t, 0) - f(0, 0)}{t} = \lim_{t \to 0} \frac{0}{t} = 0.$$

Similarly, $\frac{\partial f}{\partial y}(0, 0) = 0$.   But, f is not continuous at a, since $\lim_{(x,y) \to (0,0)} f(x, y)$ does not exist. Hence, f is not differentiable at $(0, 0)$. It is quite simple to understand such behavior. The partial derivatives depend only on what happens in the direction of the coordinate axes, whereas the definition of the derivative D_f involves the combined behavior of f in a whole neighborhood of a given point. Note also, $D_f(0, 0) = \nabla f(0, 0) = (0, 0)$, so that the derivative exists at $(0, 0)$, but still f is *not* differentiable at $(0, 0)$.

Definition 3.2.9. The standard matrix of $D_f(a)$ denoted again by $D_f(a)$ is called the *Jacobian matrix* of f at a. That is, the Jacobian matrix of f is the $m \times n$ matrix

$$D_f(a) = \begin{pmatrix} \frac{\partial f_1}{\partial x_1}(a) & \cdots & \frac{\partial f_1}{\partial x_n}(a) \\ . & \cdots & . \\ . & \cdots & . \\ . & \cdots & . \\ \frac{\partial f_m}{\partial x_1}(a) & \cdots & \frac{\partial f_m}{\partial x_n}(a) \end{pmatrix} = \begin{pmatrix} D_{f_1}(a) \\ . \\ . \\ . \\ D_{f_m}(a) \end{pmatrix}$$

and reduces the problem of computing the derivative of a differentiable function f to that of computing the partial derivatives of its component functions $f_1, ..., f_m$.

When $m = n$, the Jacobian matrix of f is a square $n \times n$ matrix and its determinant is then defined. This determinant is called the *Jacobian*[2] of f at a and is denoted by $J_f(a)$. Thus,

$$J_f(a) = \det\left(D_f(a)\right).$$

Other common notations of the Jacobian are

$$\frac{\partial(f_1, ... f_n)}{\partial(x_1, ..., x_n)}\Big|_{x=a} \text{ or } \frac{\partial(y_1, ... y_n)}{\partial(x_1, ..., x_n)} \text{ for } y = f(x).$$

Note that when $m = 1$, in which case f is a real-valued function of n variables, the derivative $D_f(a)$ is an $1 \times n$ (row) matrix which can be regarded as a vector in $\mathbb{R}^n$. This is the gradient vector

$$\nabla f(a) = \left(\frac{\partial f}{\partial x_1}(a), ..., \frac{\partial f}{\partial x_n}(a)\right).$$

The case (when $n = 1$) of a vector-valued function in one variable is also important. As mentioned earlier, here it is customary to use $X(t)$ rather than $f(x)$. Thus,

$$X(t) = (x_1(t), ..., x_m(t))$$

and X is said a *curve* in $\mathbb{R}^m$. If X is differentiable at $t = a$, we write the derivative $X'(a)$ which is represented by the $m \times 1$ column vector

$$X'(a) = \begin{pmatrix} x_1'(a) \\ \cdot \\ \cdot \\ \cdot \\ x_m'(a) \end{pmatrix} \tag{3.3}$$

---

[2]C. Jacobi (1804-1851). He initiated the theory of *elliptic functions* and made important contributions in differential equations, number theory, the theory of determinats and other fields of mathematics.

Of course, X is differentiable at a if and only if each x_j is differentiable at a, for $j = 1, ..., m$. Since X is differentiable at a if and only if

$$X'(a) = \lim_{h \to 0} \frac{X(a+h) - X(a)}{h}$$

exists, using the fact that $\frac{1}{h}(X(a+h) - X(a))$ is a chord which approximates the tangent line to the curve at $X(a)$, we see that $X'(a)$ represents the tangent vector at $X(a)$. For this reason, we call $X'(a)$ the *tangent vector* to the curve $X(t)$ at $t = a$. If $X'(a) \neq 0$, then the tangent line to $X(t)$ at a is given by $Y(t) = X(a) + tX'(a)$. The physical interpretation here is: if $X(t)$ is the position of a particle moving smoothly on a curve $X(t)$ in space at time t, then $X'(t)$ is the *velocity vector* which is of course tangent to the curve at that time and $||X'(t)||$ is its *speed*.

If $X'(t)$ exists for all $t \in \mathbb{R}$ and is itself differentiable, then a glance at (3.3) shows

$$X''(t) = \begin{pmatrix} x_1''(t) \\ \cdot \\ \cdot \\ \cdot \\ x_m''(t) \end{pmatrix}$$

which physically represents the *accelaration vector* of the moving particle along the curve $X(t)$.

Example 3.2.10. Let $f : \mathbb{R}^3 \to \mathbb{R}^2$ be $f(x, y, z) = (\frac{x^2 - y^4 - 2z}{2}, yz) = (u, v)$. Find $D_f(x, y, z)$. What is $D_f(3, 2, -1)$? Find the Jacobians $\frac{\partial(u,v)}{\partial(x,y)}$, $\frac{\partial(u,v)}{\partial(y,z)}$ and $\frac{\partial(u,v)}{\partial(x,z)}$

Solution. Here, f has two component functions $u = f_1(x, y, z) = \frac{x^2 - y^4 - 2z}{2}$ and $v = f_2(x, y, z) = yz$ and the derivative of f at any point (x, y, z) is

$$D_f(x, y, z) = \begin{pmatrix} \frac{\partial f_1}{\partial x}(x, y, z) & \frac{\partial f_1}{\partial y}(x, y, z) & \frac{\partial f_1}{\partial z}(x, y, z) \\ \frac{\partial f_2}{\partial x}(x, y, z) & \frac{\partial f_2}{\partial y}(x, y, z) & \frac{\partial f_2}{\partial z}(x, y, z) \end{pmatrix}$$

$$= \begin{pmatrix} x & -2y^3 & -1 \\ 0 & z & y \end{pmatrix}.$$

In particular, $D_f(3, 2, -1) = \begin{pmatrix} 3 & -16 & -1 \\ 0 & -1 & 2 \end{pmatrix}$.

Finally, $\frac{\partial(u,v)}{\partial(x,y)} = \det \begin{pmatrix} x & -2y^3 \\ 0 & z \end{pmatrix} = xy$. Similarly $\frac{\partial(u,v)}{\partial(y,z)} = -2y^4 + z$,

and $\frac{\partial(u,v)}{\partial(x,z)} = xy$.

Theorem 3.2.7 enables us to look at the Chain Rule (Theorem 3.1.8) in terms of partial derivatives. Since composition of linear transformations corresponds to matrix multiplication, the Chain Rule $D_{(g \circ f)}(a) = D_g(f(a))D_f(a)$ yields.

Corollary 3.2.11. *(Chain Rule)*

$$
D_{(g \circ f)}(a) = \begin{pmatrix} \frac{\partial g_1}{\partial y_1}(f(a)) & \cdots & \frac{\partial g_1}{\partial y_m}(f(a)) \\ \cdot & \cdots & \cdot \\ \cdot & \cdots & \cdot \\ \cdot & \cdots & \cdot \\ \frac{\partial g_p}{\partial y_1}(f(a)) & \cdots & \frac{\partial g_p}{\partial y_m}(f(a)) \end{pmatrix} \begin{pmatrix} \frac{\partial f_1}{\partial x_1}(a) & \cdots & \frac{\partial f_1}{\partial x_n}(a) \\ \cdot & \cdots & \cdot \\ \cdot & \cdots & \cdot \\ \cdot & \cdots & \cdot \\ \frac{\partial f_m}{\partial x_1}(a) & \cdots & \frac{\partial f_m}{\partial x_n}(a) \end{pmatrix}.
$$

Two important special cases are:

Case 1. $p = 1$. Then

$$
D_{(g \circ f)}(a) = \begin{pmatrix} \frac{\partial g}{\partial y_1}(f(a)) & \cdots & \frac{\partial g}{\partial y_m}(f(a)) \end{pmatrix} \begin{pmatrix} \frac{\partial f_1}{\partial x_1}(a) & \cdots & \frac{\partial f_1}{\partial x_n}(a) \\ \cdot & \cdots & \cdot \\ \cdot & \cdots & \cdot \\ \cdot & \cdots & \cdot \\ \frac{\partial f_m}{\partial x_1}(a) & \cdots & \frac{\partial f_m}{\partial x_n}(a) \end{pmatrix}
$$

Writing this out, we obtain, for $i = 1, ..., n$

$$
\frac{\partial(g \circ f)}{\partial x_i}(a) = \sum_{j=1}^{m} \frac{\partial g}{\partial y_j}(f(a)) \cdot \frac{\partial f_j}{\partial x_i}(a). \tag{3.4}
$$

Case 2. Let $\Omega \subseteq \mathbb{R}$ and $X : \Omega \to \mathbb{R}^n$ and $g : U \to \mathbb{R}$, where U is open in $\mathbb{R}^n$ such that $X(\Omega) \subseteq U$. If X is differentiable at a and g differentiable at $X(a) = b$, then $g \circ X : \Omega \to \mathbb{R}$ is a (one-variable) function differentiable at a and

$$\frac{d(g \circ X)}{dt}(a) = \left(\frac{\partial g}{\partial y_1}(X(a)) \ \dots \ \frac{\partial g}{\partial y_n}(X(a)) \right) \begin{pmatrix} \frac{dx_1}{dt}(a) \\ \cdot \\ \cdot \\ \cdot \\ \frac{dx_n}{dt}(a) \end{pmatrix}.$$

That is,

$$\frac{d(g \circ X)}{dt}(a) = \sum_{i=1}^{n} \frac{\partial g}{\partial x_i}(X(a)) \cdot \frac{dx_i}{dt}(a). \tag{3.5}$$

Alternatively,

$$\frac{d(g \circ X)}{dt}(a) = \langle \nabla g(X(a)), X'(a) \rangle.$$

Example 3.2.12. Let $z = e^{xy^2}$ and suppose $x = t \cos t$, $y = t \sin t$. Compute $\frac{dz}{dt}$ at $t = \frac{\pi}{2}$.

Solution. By (3.5) we have

$$\frac{dz}{dt} = \frac{\partial z}{\partial x}\frac{dx}{dt} + \frac{\partial z}{\partial y}\frac{dy}{dt} = (y^2 e^{xy^2})(\cos t - t \sin t) + (2xy e^{xy^2})(\sin t + t \sin t).$$

At $t = \frac{\pi}{2}$, $x = 0$ and $y = \frac{\pi}{2}$. Hence $\frac{dz}{dt}\big|_{t=\frac{\pi}{2}} = \frac{\pi^2}{4}(\frac{\pi}{2}) = -\frac{\pi^3}{8}$.

Example 3.2.13. Suppose $\varphi : \mathbb{R}^2 \to \mathbb{R}^2$ is the *polar coordinates* mapping defined by

$$(x, y) = \varphi(r, \theta) = (r \cos \theta, r \sin \theta),$$

(see Figure 3.1). Let $f : \mathbb{R}^2 \to \mathbb{R}$ be a differentiable function and $u = f \circ \varphi$. Find $\frac{\partial u}{\partial r}$ and $\frac{\partial u}{\partial \theta}$

Solution. The composition is $u(r, \theta) = f(r \cos \theta, r \sin \theta)$. Then the Chain Rule (3.4) gives

$$\frac{\partial u}{\partial r} = \frac{\partial f}{\partial x}\frac{\partial x}{\partial r} + \frac{\partial f}{\partial y}\frac{\partial y}{\partial r} = \frac{\partial f}{\partial x}\cos \theta + \frac{\partial f}{\partial y}\sin \theta,$$

and

$$\frac{\partial u}{\partial \theta} = \frac{\partial f}{\partial x}\frac{\partial x}{\partial \theta} + \frac{\partial f}{\partial y}\frac{\partial y}{\partial \theta} = -\frac{\partial f}{\partial x}r \sin \theta + \frac{\partial f}{\partial y}r \cos \theta.$$

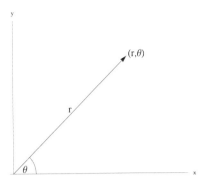

Figure 3.1: Polar coordinates

Example 3.2.14. Let $f(x,y,z) = (x^2y, y^2, e^{-xz})$ and $g(u,v,w) = u^2 - v^2 - w$. Find $g \circ f$ and compute the derivative of $g \circ f$.

1. directly

2. using the Chain Rule.

Solution. Set $F = g \circ f : \mathbb{R}^3 \to \mathbb{R}$. We have

$$F(x,y,z) = (g \circ f)(x,y,z) = g(f(x,y,z)) = g(x^2y, y^2, e^{-xz})$$
$$= (x^2y)^2 - (y^2)^2 - e^{-xz} = x^4y^2 - y^4 - e^{-xz}.$$

1.

$$\nabla F(x,y,z) = (\frac{\partial F}{\partial x}, \frac{\partial F}{\partial y}, \frac{\partial F}{\partial z}) = \left(4x^3y^2 + ze^{-xz}, 2x^4y - 4y^3, xe^{-xz}\right).$$

2. At any point (x,y,z), the Chain Rule gives

$$D_{(g \circ f)} = D_F = \left(\frac{\partial F}{\partial x} \; \frac{\partial F}{\partial y} \; \frac{\partial F}{\partial z}\right) = \left(\frac{\partial g}{\partial u} \; \frac{\partial g}{\partial v} \; \frac{\partial g}{\partial w}\right) \begin{pmatrix} \frac{\partial u}{\partial x} & \frac{\partial u}{\partial y} & \frac{\partial u}{\partial z} \\ \frac{\partial v}{\partial x} & \frac{\partial v}{\partial y} & \frac{\partial v}{\partial z} \\ \frac{\partial w}{\partial x} & \frac{\partial w}{\partial y} & \frac{\partial w}{\partial z} \end{pmatrix}$$

$$= (2u \; -2v \; -1) \begin{pmatrix} 2xy & x^2 & 0 \\ 0 & 2y & 0 \\ -ze^{-xz} & 0 & -xe^{-xz} \end{pmatrix}$$

$$= (2x^2y - 2y^2 - 1) \begin{pmatrix} 2xy & x^2 & 0 \\ 0 & 2y & 0 \\ -ze^{-xz} & 0 & -xe^{-xz} \end{pmatrix}$$

$$= \begin{pmatrix} 4x^3y^2 + ze^{-xz} \\ 2x^4y - 4y^3 \\ xe^{-xz} \end{pmatrix},$$

which is the gradient $\nabla F(x, y, x)$ written as a column vector.

Example 3.2.15. Let $f(x, y) = (e^{x+y}, e^{x-y})$ and $X : \mathbb{R} \to \mathbb{R}^2$ a curve in $\mathbb{R}^2$ with $X(0) = (0, 0)$ and $X'(0) = (1, 1)$. Find the tangent vector to the image of the curve $X(t)$ under f at $t = 0$.

Solution.   Set $Y(t) = f(X(t))$.   By the Chain Rule, $Y'(t) = D_f(X(t))X'(t)$.   At $t = 0$ we get $Y'(0) = D_f(X(0))X'(0) = D_f(0, 0) \begin{pmatrix} 1 \\ 1 \end{pmatrix}$. An easy calculation shows that $D_f(0, 0) = \begin{pmatrix} 1 & 1 \\ 1 & -1 \end{pmatrix}$.
Hence

$$Y'(0) = \begin{pmatrix} 1 & 1 \\ 1 & -1 \end{pmatrix} \begin{pmatrix} 1 \\ 1 \end{pmatrix} = \begin{pmatrix} 2 \\ 0 \end{pmatrix},$$

That is, $Y'(0) = (2, 0)$.

As we saw in Example 3.2.8, if the partial derivatives of the function exist the function need not be differentiable. However, if they are also *continuous*, then the function is differentiable.

Theorem 3.2.16. *(Differentiability Criterion). Let $f : \Omega \to \mathbb{R}^m$ where Ω is open in $\mathbb{R}^n$. If all partial derivatives $\frac{\partial f_j}{\partial x_i}$, $j = 1, ..., m$ and $i = 1, ..., n$ exist in a neighborhood of $a \in \Omega$ and are continuous at a, then f is differentiable at a.*

Proof. In view of Proposition 3.1.9, it suffices to prove the result for a real-valued function f. We must show

$$\lim_{x \to a} \frac{|f(x) - f(a) - \langle \nabla f(a), x - a \rangle|}{||x - a||} = 0.$$

To do this, we write the change $f(x) - f(a)$ as a telescoping sum, by making the change one coordinate at a time. If $x = (x_1, ..., x_n)$ and

$a = (a_1, ..., a_n)$, consider the vectors $v_0 = a$, $v_i = (x_1, ..., x_i, a_{i+1}, ..., a_n)$ for $i = 1, ..., n-1$ and $v_n = x$. Note that if x is in some ball around a, so are all the vectors v_i, for $i = 1, ..., n$. Then

$$f(x) - f(a) = \sum_{i=1}^{n} [f(v_i) - f(v_{i-1})].$$

Set $g_i(t) = f(x_1, ..., x_{i-1}, t, a_{i+1}, ..., a_n)$. Then g_i maps the interval between a_i and x_i into $\mathbb{R}$, with $g_i(a_i) = f(v_{i-1})$, $g_i(x_i) = f(v_i)$ and the derivative of g_i is $\frac{\partial f}{\partial x_i}(x_1, ..., x_{i-1}, t, a_{i+1}, ..., a_n)$. By the one-variable Mean Value theorem (Theorem B.1.2), there are ξ_i strictly between a_i and x_i such that

$$f(v_i) - f(v_{i-1}) = (x_i - a_i)\frac{\partial f}{\partial x_i}(x_1, ..., x_{i-1}, \xi_i, a_{i+1}, ..., a_n)$$

Let $u_i = (x_1, ..., x_{i-1}, \xi_i, a_{i+1}, ..., a_n)$ and note that $||u_i - a|| \leq ||x - a||$, so that as $x \to a$ also $u_i \to a$. Hence

$$\lim_{x \to a} \frac{|f(x) - f(a) - \sum_{i=1}^{n}(x_i - a_i)\frac{\partial f}{\partial x_i}(a)|}{||x - a||}$$

$$= \lim_{x \to a} \frac{|\sum_{i=1}^{n}[\frac{\partial f}{\partial x_i}(u_i) - \frac{\partial f}{\partial x_i}(a)](x_i - a_i)|}{||x - a||}.$$

Since $|x_i - a_i| \leq ||x - a||$, we get

$$\leq \lim_{x \to a} \frac{\sum_{i=1}^{n}|\frac{\partial f}{\partial x_i}(u_i) - \frac{\partial f}{\partial x_i}(a)||x_i - a_i|}{||x - a||} \leq \lim_{x \to a} \sum_{i=1}^{n}|\frac{\partial f}{\partial x_i}(u_i) - \frac{\partial f}{\partial x_i}(a)|.$$

The latter is zero since each $\frac{\partial f}{\partial x_i}$ is continuous at a.   □

Definition 3.2.17. Let $f : \Omega \to \mathbb{R}^m$ where Ω is open in $\mathbb{R}^n$. If all partial derivatives $\frac{\partial f_j}{\partial x_i}(x)$, $j = 1, ..., m$ and $i = 1, ..., n$ exist for every $x \in \Omega$ and are continuous on Ω, we say f is *continuously differentiable* on Ω. We write $f \in C^1(\Omega)$ and say f is of *class C^1* on Ω. We also call a continuous function on Ω of *class C^0* on Ω or $C(\Omega)$.

Theorem 3.2.16 and Proposition 3.1.6 yield the following.

Corollary 3.2.18. *A function of class $C^1(\Omega)$ is of class $C^0(\Omega)$.*

Identifying the set of $m \times n$ real matrices with $\mathbb{R}^{mn}$, we see that $f \in C^1(\Omega)$ if and only if the derivative function $D_f : \Omega \to \mathbb{R}^{mn}$ is continuous on Ω. We have also seen that $f \in C^1(\Omega)$ implies *f is differentiable on Ω* which, in turn, implies that f has a *directional derivative in every direction*, in particular *all partial derivatives exist on Ω*. As the following examples show if the partial derivatives of a function f are *not* continuous at a point, the question on the differentiability of f at the point remains open. One has to use the definition, or even better, the Linear Approximation theorem (Theorem 3.1.2) to decide on the differentiability of f at the point. Hence, the converse of Theorem 3.2.16 is also not true.

Example 3.2.19. Let $f : \mathbb{R}^2 \to \mathbb{R}$ be defined by

$$f(x,y) = \begin{cases} (x^2 + y^2) \sin \dfrac{1}{\sqrt{x^2+y^2}} & \text{if } (x,y) \neq (0,0), \\ 0 & \text{if } (x,y) = (0,0). \end{cases}$$

First we study the function on $\mathbb{R}^2 - \{(0,0)\}$. At any $(x,y) \neq (0,0)$, by partial differentiation of f we see that

$$\frac{\partial f}{\partial x}(x,y) = 2x \sin \frac{1}{\sqrt{x^2 + y^2}} - (x^2 + y^2)\frac{x}{(x^2 + y^2)^{\frac{3}{2}}} \cos \frac{1}{\sqrt{x^2 + y^2}},$$

and by symmetry

$$\frac{\partial f}{\partial y}(x,y) = 2y \sin \frac{1}{\sqrt{x^2 + y^2}} - (x^2 + y^2)\frac{y}{(x^2 + y^2)^{\frac{3}{2}}} \cos \frac{1}{\sqrt{x^2 + y^2}}.$$

Since both functions $\frac{\partial f}{\partial x}(x,y)$ and $\frac{\partial f}{\partial y}(x,y)$ are continuous on $\mathbb{R}^2 - \{(0,0)\}$, f is C^1 on $\mathbb{R}^2 - \{(0,0)\}$ and so f is differentiable at any $(x,y) \neq (0,0)$.
Next we study f at $(0,0)$.

$$\frac{\partial f}{\partial x}(0,0) = \lim_{t \to 0} \frac{f(t,0) - f(0,0)}{t} = \lim_{t \to 0} t \sin \frac{1}{|t|} = 0$$

and similarly $\frac{\partial f}{\partial y}(0,0) = 0$. Since

$$\frac{\partial f}{\partial x}(x,0) = 2x \sin \frac{1}{|x|} - \frac{x}{|x|} \cos \frac{1}{|x|}$$

does not have limit as $x \to 0$, it follows that the function $\frac{\partial f}{\partial x}(x,y)$ is not continuous at $(0,0)$. Similarly $\frac{\partial f}{\partial y}(x,y)$ is not continuous at $(0,0)$. However, f is differentiable at $(0,0)$. To see this we use the Linear Approximation theorem

$$\epsilon(x,y) = \frac{f(x,y) - f(0,0) - \langle \nabla f(0,0), (x,y)\rangle}{||(x,y)||} = \frac{f(x,y)}{\sqrt{x^2+y^2}}$$

For $(x,y) \neq (0,0)$ we have $0 \leq \frac{|f(x,y)|}{\sqrt{x^2+y^2}} \leq \sqrt{x^2+y^2}$. Therefore,

$$\lim_{(x,y)\to(0,0)} \epsilon(x,y) = 0$$

and f is differentiable at $(0,0)$.

Example 3.2.20. Let $f : \mathbb{R}^2 \to \mathbb{R}$ be defined by

$$f(x,y) = \begin{cases} \frac{xy}{\sqrt{x^2+y^2}} & \text{if } (x,y) \neq (0,0), \\ 0 & \text{if } (x,y) = (0,0). \end{cases}$$

It is easy to see as before that $\frac{\partial f}{\partial x}(0,0) = 0$ and $\frac{\partial f}{\partial y}(0,0) = 0$. If $(x,y) \neq (0,0)$ differentiating f with respect x to we get

$$\frac{\partial f}{\partial x}(x,y) = \frac{y}{\sqrt{x^2+y^2}} - \frac{x^2y}{(x^2+y^2)^{\frac{3}{2}}},$$

and by symmetry

$$\frac{\partial f}{\partial x}(x,y) = \frac{x}{\sqrt{x^2+y^2}} - \frac{y^2x}{(x^2+y^2)^{\frac{3}{2}}}.$$

Since both $\lim_{(x,y)\to(0,0)} \frac{\partial f}{\partial x}(0,y)$ and $\lim_{(x,y)\to(0,0)} \frac{\partial f}{\partial y}(x,0)$ do not exist, it follows that $\lim_{(x,y)\to(0,0)} \frac{\partial f}{\partial x}(x,y)$ and $\lim_{(x,y)\to(0,0)} \frac{\partial f}{\partial y}(x,y)$ do not exist and so both partial derivatives $\frac{\partial f}{\partial x}$ and $\frac{\partial f}{\partial y}$ are discontinuous at $(0,0)$. However, the function f itself is continuous at $(0,0)$. This follows from the estimate

$$0 \leq |f(x,y)| = \frac{|xy|}{\sqrt{x^2+y^2}} \leq \frac{1}{2}\sqrt{x^2+y^2}$$

We show that f is not differentiable at $(0,0)$. By the Linear Approximation theorem we must look at

$$\epsilon(x,y) = \frac{f(x,y)}{\sqrt{x^2+y^2}} = \frac{xy}{x^2+y^2}.$$

But,

$$\lim_{(x,x)\to(0,0)} \epsilon(x,x) = \frac{1}{2}.$$

Therefore, f is *not* differentiable at $(0,0)$.

To find the relationship between the gradient of a function f and its level sets, recall that a level set of level $c \in \mathbb{R}$ for $f : \Omega \subseteq \mathbb{R}^n \to \mathbb{R}$ is the set $S_c = \{x \in \Omega : f(x) = c\}$. The set S_c (a hypersurface) in $\mathbb{R}^n$ has dimension $n - 1$.

Proposition 3.2.21. *Let $f : \Omega \subseteq \mathbb{R}^n \to \mathbb{R}$ be a differentiable function and $a \in \Omega$ lie on S_c. Then $\nabla f(a)$ is orthogonal to S_c: If v is the tangent vector at $t = 0$ of a differentiable curve $X(t)$ in S_c with $X(0) = a$, then $\nabla f(a)$ is perpendicular to v.*

Proof. Let $X(t)$ lie in S_c with $X(0) = a$. Then $v = X'(0)$ and $f(X(t)) = c$. So $f(X(t))$ is constant in t, and the Chain Rule tells us

$$0 = \frac{d(f(X(t)))}{dt} = \langle \nabla f(X(t)), X'(t) \rangle$$

For $t = 0$, this gives $\langle \nabla f(a), v \rangle = 0$. $\qquad\square$

Now consider all curves on the level set S_c of f passing through the point $a \in S_c$. As we just saw, the tangent vectors at a of all these curves are perpendicular to $\nabla f(a)$. If $\nabla f(a) \neq 0$, then these tangent vectors determine a hyperplane and $\nabla f(a)$ is the normal vector to it. This plane is called the *tangent hyperplane* to the surface S_c at a, and we denote it by $T_a(S_c)$. We recall from Chapter 1 (Example 1.3.32) that the plane passing through a point $\mathbf{a} \in \mathbb{R}^3$ with normal vector $\mathbf{n}$ consists of all points $\mathbf{x}$ satisfying

$$\langle \mathbf{n}, \mathbf{x} - \mathbf{a} \rangle = 0.$$

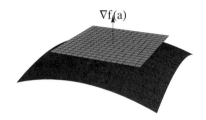

Figure 3.2: Tangent plane and gradient

Hence, we arrive at the following.

Definition 3.2.22. Let $f : \Omega \subseteq \mathbb{R}^n \to \mathbb{R}$ be a differentiable function at $a \in \Omega$. The *tangent hyperplane* to the level set S_c of f at $a \in S_c$ is the set of points $x \in \mathbb{R}^n$ satisfying

$$\langle \nabla f(a), x - a \rangle = 0.$$

The tangent hyperplane translated so that it passes from the origin is called the *tangent space* at a and we denote it again by $T_a(S_c)$.

Example 3.2.23. (*Tangent hyperplane to a graph*). An important special case arises when S_c is the graph of a differentiable function $y = f(x_1, ..., x_n)$. As we saw in Chapter 2 the graph of f may be regarded as the level set $S \subseteq \mathbb{R}^{n+1}$ (of level 0) of the function $F(x, y) = F(x_1, ..., x_n, y) = f(x_1, ..., x_n) - y$. Then $\nabla F(x, y) = (\frac{\partial f}{\partial x_1}(x), ..., \frac{\partial f}{\partial x_n}(x), -1)$. So that,

$$\langle \nabla F(a, f(a)), (x - a, y - f(a)) \rangle = 0$$

implies

$$y = f(a) + \langle \nabla f(a), x - a \rangle.$$

In coordinates the tangent hyperplane at $(a, f(a))$ is written

$$y = f(a) + \sum_{i=1}^{n} \frac{\partial f}{\partial x_i}(a)(x_i - a_i). \tag{3.6}$$

For a differentiable function in two variables $z = f(x, y)$ this yield the equation of the *tangent plane* to the graph of f at $((x_0, y_0), f(x_0, y_0))$

$$z = f(x_0, y_0) + \frac{\partial f}{\partial x}(x_0, y_0))(x - x_0) + \frac{\partial f}{\partial y}(x_0, y_0))(y - y_0).$$

Example 3.2.24. Find the equation of the tangent plane to the graph of $z = f(x, y) = x^2 + y^4 + e^{xy}$ at the point $(1, 0, 2)$.

Solution. The partial derivatives are $\frac{\partial f}{\partial x}(x, y) = 2x + ye^{xy}$, $\frac{\partial f}{\partial y}(x, y) = 4y^3 + xe^{xy}$, so that $\frac{\partial f}{\partial x}(1, 0) = 2$ and $\frac{\partial f}{\partial y}(1, 0) = 1$. Substituting in (3.6) yields

$$z = 2 + 2(x - 1) + 1(y - 0) \quad \text{or} \quad z = 2x + y.$$

Example 3.2.25. Let $y = f(x_1, ..., x_n) = \sqrt{x_1^2 + ... + x_n^2} = ||x||$. Find the tangent hyperplane

1. at $a \in \mathbb{R}^n$ with $a \neq 0$.

2. at $e_j \in \mathbb{R}^n$.

3. at $0 \in \mathbb{R}^n$

Solution.

1. At $a = (a_1, ..., a_n)$ we have $\frac{\partial f}{\partial x_i}(a) = \frac{a_i}{\sqrt{a_1^2 + ... + a_n^2}} = \frac{a_i}{||a||}$. Substitution in (3.6) yields

   $$y = ||a|| + \sum_{i=1}^{n} \frac{a_i}{||a||}(x_i - a_i) = ||a|| + \frac{1}{||a||}\sum_{i=1}^{n}(a_i x_i - a_i^2) = \frac{\langle a, x \rangle}{||a||}.$$

2. At e_j we get $y = x_j$.

3. At 0

   $$\frac{\partial f}{\partial x_i}(0) = \lim_{t \to 0} \frac{f(0 + te_i) - f(0)}{t} = \lim_{t \to 0} \frac{|t|}{t}.$$

   Since this latter limit does not exist, f is not differentiable at 0, and so there is no tangent hyperplane there.

EXERCISES

1. For each of the following functions find the partial derivatives and ∇f.

   (a) $f(x, y) = e^{4x-y^2} + \log(x^2 + y)$.

   (b) $f(x, y) = \cos(x^2 - 3y)$.

   (c) $f(x, y) = \tan^{-1}(xy)$.

   (d) $f(x, y, z) = x^2 e^{\frac{y}{z}}$.

   (e) $f(x, y, z) = zx^y$.

   (f) $f(r, \theta, \phi) = r \cos \theta \sin \phi$.

   (g) $f(x, y, z) = e^{3x+y} \sin(5z)$ at $(0, 0, \frac{\pi}{6})$.

   (h) $f(x, y, z) = \log(z + \sin(y^2 - x))$ at $(1, -1, 1)$.

   (i) $f(x) = e^{-\frac{||x||^2}{2}}$, $x \in \mathbb{R}^n$.

2. Find the directional derivative of f at the given point in the given direction.

   (a) $f(x, y) = 4xy + 3y^2$ at $(1, 1)$ in the direction $(2, -1)$.

   (b) $f(x, y) = \sin(\pi xy) + x^2 y$ at $(1, -2)$ in the direction $(\frac{3}{5}, \frac{4}{5})$.

   (c) $f(x, y) = 4x^2 + 9y^2$ at $(2, 1)$ in the direction of maximum directional derivative.

   (d) $f(x, y, z) = x^2 e^{-yz}$ at $(1, 0, 0)$ in the direction $v = (1, 1, 1)$.

3. Prove Theorem 3.2.2, using the Chain Rule.

4. Find the angles made by the gradient of $f(x, y) = x^{\sqrt{3}} + y$ at the point $(1, 1)$ with the coordinate axes.

5. Find the tangent plane to the surface in $\mathbb{R}^3$ described by the equation

   (a) $z = x^2 - y^3$ at $(2, 1, -5)$.

   (b) $x^2 + 2y^2 + 2z^2 = 6$ at $(1, 1, -1)$.

6. Show that the function $f(x, y) = |x| + |y|$ is continuous, but not differentiable at $(0, 0)$.

7. Let
$$f(x,y) = \begin{cases} xy\frac{x^2-y^2}{x^2+y^2} & \text{if } (x,y) \neq (0,0), \\ 0 & \text{if } (x,y) = (0,0). \end{cases}$$

Show that $\frac{\partial f}{\partial x}(x,0) = 0 = \frac{\partial f}{\partial y}(0,y)$, $\frac{\partial f}{\partial x}(0,y) = -y$ and $\frac{\partial f}{\partial y}(x,0) = x$. Show that f is differentiable at $(0,0)$.

8. Let
$$f(x,y) = \begin{cases} \frac{x^3-y^3}{x^2+y^2} & \text{if } (x,y) \neq (0,0), \\ 0 & \text{if } (x,y) = (0,0). \end{cases}$$

Find $\frac{\partial f}{\partial x}(0,0)$ and $\frac{\partial f}{\partial y}(0,0)$. Show that f is not differentiable at $(0,0)$.

9. Let
$$f(x,y,z) = \begin{cases} \frac{xyz}{(x^2+y^2+z^2)^\alpha} & \text{if } (x,y,z) \neq (0,0,0), \\ 0 & \text{if } (x,y,z) = (0,0,0), \end{cases}$$

where $\alpha \in \mathbb{R}$ a constant. Show that f is differentiable at $(0,0,0)$ if and only if $\alpha < 1$.

10. Show that the function $f(x,y) = \sqrt{|xy|}$ is not differentiable at $(0,0)$.

11. Let $\alpha > \frac{1}{2}$. Show that the function $f(x,y) = |xy|^\alpha$ is differentiable at $(0,0)$.

12. Let
$$\varphi(t) = \begin{cases} \frac{\sin t}{t} & \text{if } t \neq 0, \\ 1 & \text{if } t = 0. \end{cases}$$

Show that φ is differentiable on $\mathbb{R}$.
Let
$$f(x,y) = \begin{cases} \frac{\cos x - \cos y}{x-y} & \text{if } x \neq y, \\ -\sin x & \text{otherwise.} \end{cases}$$

Express f in terms of φ and show that f is differentiable on $\mathbb{R}^2$.

13. For each of the following functions find the derivative D_f and the indicated Jacobians.

(a) $f : \mathbb{R}^2 \to \mathbb{R}^2$, $f(x, y) = (x^2 + y, 2xy - y^2)$. Find the Jacobian $J_f(x, y)$.

(b) $f : \mathbb{R}^2 \to \mathbb{R}^3$, $f(x, y) = (xy, x^2 + xy^2, x^3 y) = (u, v, w)$. Find $\frac{\partial(u,v)}{\partial(x,y)}$, $\frac{\partial(u,w)}{\partial(x,y)}$ and $\frac{\partial(v,w)}{\partial(x,y)}$.

(c) $f : \mathbb{R}^3 \to \mathbb{R}^2$, $f(x, y, z) = (xe^y, x^3 + z^2 \sin x) = (u, v)$. Find $\frac{\partial(u,v)}{\partial(x,y)}$, $\frac{\partial(u,v)}{\partial(x,z)}$ and $\frac{\partial(u,v)}{\partial(y,z)}$.

(d) $f : \mathbb{R}^3 \to \mathbb{R}^3$, $f(r, \theta, z) = (r \cos \theta, r \sin \theta, z)$. Find $J_f(r, \theta, z)$.

(e) $f : \mathbb{R}^3 \to \mathbb{R}^3$, $f(r, \theta, \phi) = (r \cos \theta \sin \phi, r \sin \theta \sin \phi, r \cos \phi)$. Find $J_f(r\theta, \phi)$.

14. Let $u = xyf(\frac{x+y}{xy})$, where $f : \mathbb{R} \to \mathbb{R}$ is differentiable. Show that u satisfies the partial differential equation $x^2 \frac{\partial u}{\partial x} - y^2 \frac{\partial u}{\partial y} = g(x, y)u$ and find $g(x, y)$.

15. Let $x \in \mathbb{R}^n$ and $u = f(r)$, where $r = ||x||$ and f differentiable. Show that

$$\sum_{i=1}^{n} \left(\frac{\partial u}{\partial x_i} \right)^2 = [f'(r)]^2.$$

16. Let $z = e^x \sin y$, $x = \log t$, $y = \tan^{-1}(3t)$. Compute $\frac{dz}{dt}$ in two ways:

(a) Using the Chain Rule.

(b) Finding the composition and differentiating.

17. Let $f(x, y)$ be C^1 and let $x = s \cos \theta - t \sin \theta$, $y = s \sin \theta + t \cos \theta$. Compute

$$\left(\frac{\partial f}{\partial s} \right)^2 + \left(\frac{\partial f}{\partial t} \right)^2.$$

18. Find $\frac{\partial u}{\partial s}$ and $\frac{\partial u}{\partial t}$ in terms of the partial derivatives $\frac{\partial f}{\partial x}$, $\frac{\partial f}{\partial y}$ and $\frac{\partial f}{\partial z}$ for

(a) $u = f(e^{s-3t}, \log(1 + s^2), \sqrt{1 + t^4})$.

(b) $u = \tan^{-1}[f(t^2, 2s - t, -4)]$.

19. Let $f(x,y) = (2x + y, 3x + 2y)$ and $g(u,v) = (2u - v, -3u + v)$. Find D_f, D_g, and $D_{(g \circ f)}$.

20. Let $g : \mathbb{R}^2 \to \mathbb{R}^3$ be $g(x,y) = (x^2 - 5y, ye^2x, 2x - \log(1 + y^2))$. Find $D_g(0,0)$. Let $f : \mathbb{R}^2 \to \mathbb{R}^2$ be of class C^1, $f(1,2) = (0,0)$ and $D_f(1,2) = \begin{pmatrix} 1 & 2 \\ 3 & 4 \end{pmatrix}$. Find $D_{(g \circ f)}(1,2)$.

Answers to selected Exercises

1. (g) $(\frac{3}{2}, \frac{1}{2}, \frac{5\sqrt{3}}{2})$. (h) $(-1, -2, 1))$.

2. (a) $-\frac{2}{\sqrt{5}}$. (b) $-\frac{2}{5}(4 + \pi)$. (c) $2\sqrt{145}$. (d) $\frac{2}{\sqrt{3}}$.

4. x-axis $\alpha = \frac{\pi}{6}$, y-axis $\beta = \frac{\pi}{3}$. 5. (a) $4x - 3y = z = 6$.
   (b) $2x + 4y - 6z = 12$.

14. $g(x,y) = x - y$. 20. $\begin{pmatrix} -15 & -20 \\ 3 & 4 \\ 2 & 4 \end{pmatrix}$.

3.3 Homogeneous functions and Euler's equation

Definition 3.3.1. Let $f : \mathbb{R}^n \to \mathbb{R}$ be a function. We say that f is *homogeneous* of degree $\alpha \in \mathbb{R}$ if $f(tx) = t^\alpha f(x)$ for all $t \in \mathbb{R}$ and $x \in \mathbb{R}^n$.

Observe that for such a function $f(0) = 0$. The simplest homogeneous functions that appear in analysis and its applications are the homogeneous polynomials in several variables, that is, polynomials consisting of monomials all of which have the same degree. For example, $f(x,y) = 2x^3 - 5x^2y$ is homogeneous of degree 3 on $\mathbb{R}^2$. Of course, there are homogeneous functions which are not polynomials. For instance, $f(x,y) = (x^2 + 4xy)^{-\frac{1}{3}}$ is not a polynomial and is a homogeneous function of degree $-\frac{2}{3}$ on $\mathbb{R}^2 - \{(0,0)\}$. In general for $x = (x_1, ..., x_n)$ the function $f(x) = ||x||^\alpha$ with $\alpha \in \mathbb{R}$ is homogeneous of degree α for $t \in \mathbb{R}_+$. For example $f(x,y) = (x^2 + y^2)^{\frac{1}{2}}$ is homogeneous of degree 1 and it is not a polynomial, for if it were, then the function $\varphi(t) = (1+t^2)^{\frac{1}{2}}$ would also be a polynomial. However, since all derivatives of φ are never identically zero, this is impossible.

The following theorem characterizes homogeneous differentiable functions.

Theorem 3.3.2. *(Euler[3]) Let $f : \mathbb{R}^n \to \mathbb{R}$ be a differentiable functon. Then f is homogeneous of degree α if and only if for all $x \in \mathbb{R}^n$ it satisfies Euler's partial differential equation*

$$\sum_{i=1}^{n} x_i \frac{\partial f}{\partial x_i}(x) = \alpha f(x).$$

Proof. Suppose f is homogeneous of degree α. Consider the function $\varphi : \mathbb{R} \to \mathbb{R}$ given by $\varphi(t) = f(tx)$. On the one hand, since $f(tx) - t^\alpha f(x)$, we have $\varphi'(t) = \alpha t^{\alpha-1} f(x)$. On the other, by the Chain Rule we have $\varphi'(t) = \langle \nabla f(tx), \frac{d}{dt}(tx) \rangle = \langle \nabla f(tx), x \rangle$. Taking $t = 1$ we get $\varphi'(1) = \alpha f(x) = \langle \nabla f(x), x \rangle$. That is, $\alpha f(x) = \sum_{i=1}^{n} x_i \frac{\partial f}{\partial x_i}(x)$. Conversely, suppose f satisfies Euler's equation. Then for $t \neq 0$ and $x \in \mathbb{R}^n$, we have

$$\varphi'(t) = \langle \nabla f(tx), x \rangle = \frac{1}{t} \langle \nabla f(tx), tx \rangle = \frac{1}{t} \alpha f(tx) = \frac{\alpha}{t} \varphi(t).$$

Letting $g(t) = t^{-\alpha} \varphi(t)$, it follows that

$$g'(t) = -\alpha t^{-\alpha-1} \varphi(t) + t^{-\alpha} \varphi'(t) = -\alpha t^{-\alpha-1} \varphi(t) + t^{-\alpha} \frac{\alpha}{t} \varphi(t) = 0.$$

That is, $g'(t) = 0$. Therefore, $g(t) = c$ and $\varphi(t) = ct^\alpha$. For $t = 1$ this gives $c = \varphi(1) = f(x)$. Thus, $f(tx) = t^\alpha f(x)$ and f is homogeneous of degree α. □

---

[3]L. Euler (1707-1783), was a pioneering mathematician and physicist. A student of J. Bernoulli and the thesis advisor of J. Lagrange. He made important discoveries in fields as diverse as infinitesimal calculus and graph theory. *Euler's identity $e^{i\pi} + 1 = 0$* was called "the most remarkable formula in mathematics" by R. Feynman. Euler spent most of his academic life between the Academies of Berlin and St. Petersburg. He is also known for his many contributions in mechanics, fluid dynamics, optics and astronomy.

3.4   The mean value theorem

In this section we generalize the Mean Value theorem to functions of several variables.

Theorem 3.4.1. *(Mean value theorem). Suppose $f : \Omega \subseteq \mathbb{R}^n \to \mathbb{R}$ is a differentiable function on the open convex set Ω. Let $a, b \in \Omega$ and $\gamma(t) = a + t(b - a)$ the line segment joining a and b. Then there exists c on $\gamma(t)$ such that*

$$f(b) - f(a) = \langle \nabla f(c), b - a \rangle.$$

Proof. Consider the function $\varphi : [0, 1] \to \mathbb{R}$ defined by $\varphi(t) = f(\gamma(t))$. Then φ is continuous on $[0, 1]$ and by the Chain Rule, φ is differentiable on $(0, 1)$ with derivative $\varphi'(t) = \langle \nabla f(\gamma(t)), \gamma'(t) \rangle = \langle \nabla f(\gamma(t)), b - a \rangle$. By the one-variable Mean Value theorem applied to φ, there is a point $\xi \in (0, 1)$ such that $\varphi(1) - \varphi(0) = \varphi'(\xi)$. Take $c = \gamma(\xi)$. Then

$$f(b) - f(a) = \varphi(1) - \varphi(0) = \varphi'(\xi) = \langle \nabla f(c), b - a \rangle.$$

$\square$

The Mean Value theorem has some important corollaries.

Corollary 3.4.2. *Let $f : \Omega \subseteq \mathbb{R}^n \to \mathbb{R}$ be a differentiable function on a convex subset K of Ω. If $||\nabla f(x)|| \leq M$ for all $x \in K$, then*

$$|f(x) - f(y)| \leq M||x - y||$$

for all $x, y \in K$. This is an example of what is called a Lipschitz condition.

Proof. Let $x, y \in K$. Since K is convex, the line segment joining x with y lies in K. From the Mean Value theorem, we have $f(x) - f(y) = \langle \nabla f(c), x - y \rangle$ for some $c \in K$. The Cauchy-Schwarz inequality tells us

$$|f(x) - f(y)| \leq ||\nabla f(c)|| \cdot ||x - y|| \leq M||x - y||.$$

$\square$

Corollary 3.4.3. *Let f be a differentiable function on an open convex set Ω in $\mathbb{R}^n$. If $\nabla f(x) = 0$ for all $x \in \Omega$, then f is constant on Ω.*

Proof. Let x and y be any two distinct points of Ω. The proof of Corollary 3.4.2 tells us $|f(x) - f(y)| = 0$. That is, $f(x) = f(y)$.   □

Clearly every convex set is pathwise connected (line segments are paths), but most connected sets are not convex. Corollary 3.4.3 can be extended to differentiable functions on an open connected set Ω in $\mathbb{R}^n$. So more generally.

Corollary 3.4.4. *Let f be differentiable on an open connected set $\Omega \in \mathbb{R}^n$. If $\nabla f(x) = 0$ for all $x \in \Omega$, Then f is constant on Ω.*

Proof. Let $a, b \in \Omega$ be any two points in Ω. Since Ω is open and connected it is pathwise connected (see Proposition 2.6.14). Let $\gamma(t)$ be a path in Ω with $\gamma(0) = a$ and $\gamma(1) = b$. Cover the path $\gamma(t)$ by open balls. Since $\gamma([0,1])$ is compact, as a continuous image of the compact interval $[0,1]$, a finite number of these balls to covers $\gamma(t)$. As each ball is convex, by Corollary 3.4.3, we know that f is constant on each ball. The balls intersect nontrivially, so f is the same constant as we move from one ball to the next. After a finite number of steps we conclude that $f(a) = f(b)$. Since $a, b \in \Omega$ were arbitrary, f is constant on Ω.   □

Obviously, the hypothesis of connectness is essential in Corollary 3.4.4., even in one variable. For example, let $\Omega = (-\infty, 0) \cup (1, \infty)$ and $f : \Omega \to \mathbb{R}$ be the function

$$f(x) = \begin{cases} 0 \text{ if } x < 0, \\ 1 \text{ if } x > 1. \end{cases}$$

then $f'(x) = 0$ for all $x \in \Omega$, but $f(x)$ is not constant on Ω.

Exercise 3.4.5. Prove Corollary 3.4.4 using only connectedness. *Hint.* Let $a \in \Omega$ and set $f(a) = c$. Look at the set $A = \{x \in \Omega : f(x) \neq c\}$ and show that $A = \emptyset$.

Corollary 3.4.6. *Let $f : \Omega \to \mathbb{R}^m$ where Ω is open connected set in $\mathbb{R}^n$. If $D_f(x) = 0$ for all $x \in \Omega$, then f is constant on Ω.*

Proof. If $D_f(x) = 0$, then $\nabla f_j(x) = 0$ for all $j = 1, 2, ..., m$. By Corollary 3.4.4, each f_j is constant. Hence so is f.   □

For *vector-valued* functions $f : \Omega \subseteq \mathbb{R}^n \to \mathbb{R}^m$, even if $n = 1$, there can be *no* Mean Value theorem when $m > 1$.

Example 3.4.7. Let $f : \mathbb{R} \to \mathbb{R}^2$ be given by $f(x) = (x^2, x^3)$. Let us try to find a ξ such that $0 < \xi < 1$ and $f(1) - f(0) = f'(\xi)(1 - 0)$. This means $(1,1) - (0,0) = (2\xi, 3\xi^2)$ or $1 = 2\xi$ and $1 = 3\xi^2$, which is impossible.

However, there is a useful inequality which is called the *Lipschitz*[4] *condition*. Before we prove it, we define $f \in C^1(\overline{\Omega})$ to mean that $f \in C^1(\Omega)$ such that f and $\frac{\partial f}{\partial x_i}$ for all $i = 1, 2, ...n$ extend continuously to $\partial(\Omega)$

Corollary 3.4.8. *(Lipschitz condition). Let Ω be an open bounded convex set in $\mathbb{R}^n$ and $f : \overline{\Omega} \to \mathbb{R}^m$ with $f \in C^1(\overline{\Omega})$. Then there exists a constant $M > 0$ such that for all $x, y \in \overline{\Omega}$*

$$||f(x) - f(y)|| \leq M||x - y||.$$

Proof. Let $f = (f_1, f_2, ..., f_m)$. Since Ω is convex, for $x, y \in \Omega$, an application of the Mean Value theorem to each component of f and the Cauchy-Schwarz inequality give,

$$||f(x) - f(y)||^2 = \sum_{j=1}^{m} |f_j(x) - f_j(y)|^2 = \sum_{j=1}^{m} |\langle \nabla f_j(c_j), x - y \rangle|$$

$$\leq \sum_{j=1}^{m} ||\nabla f_j(c_j)||^2 ||x - y||^2.$$

Now $\overline{\Omega}$ is compact and since $\frac{\partial f_j}{\partial x_i}(x)$ are continuous on $\overline{\Omega}$, the mapping $x \mapsto ||\nabla f_j(x)||^2$ being the composition of continuous functions, is itself continuous, and therefore bounded. Let $M_j = \max_{x \in \overline{\Omega}} ||\nabla f_j(x)||^2$ and take $M = \sqrt{\sum_{j=1}^{m} M_j}$, then, for $x, y \in \Omega$, we have $||f(x) - f(y)|| \leq M||x - y||$. Since our inequality is $\leq$, this also applies to $\overline{\Omega}$.   □

---

[4]R. Lipschitz (1832-1903). A student of Dirichlet and professor at the University of Bonn. He worked in a board range of areas including number theory, mathematical analysis, algebras with involution, differential geometry and mechanics.

3.5   Higher order derivatives

For a function $f : \Omega \subseteq \mathbb{R}^n \to \mathbb{R}$ the partial derivatives $\frac{\partial f}{\partial x_i}(x)$ for $i = 1, 2, ..., n$, are functions of $x = (x_1, x_2, ..., x_n)$ and there is no problem defining partial derivatives of higher order, whenever they exist; just iterate the process of partial differentation for $i, j = 1, 2, ..., n$

$$\partial_j \partial_i f = \frac{\partial^2 f}{\partial x_j \partial x_i} = \frac{\partial}{\partial x_j}\left(\frac{\partial f}{\partial x_i}\right).$$

When $i = j$ we write $\frac{\partial^2 f}{\partial x_i^2}$. Repeating the process of partial differentiation we obtain third (and higher) order partial derivatives. For example,

$$\frac{\partial^3 f}{\partial x_j \partial x_i^2} = \frac{\partial}{\partial x_j}\left(\frac{\partial^2 f}{\partial x_i^2}\right).$$

Of course, when $n = 2$, we denote the variables by x, y rather than using subscripts. Thus, for a function $f : \Omega \subseteq \mathbb{R}^2 \to \mathbb{R}$, one obtains the four second partial derivatives

$$\frac{\partial^2 f}{\partial x^2}, \ \frac{\partial^2 f}{\partial y^2}, \ \frac{\partial^2 f}{\partial x \partial y}, \ \frac{\partial^2 f}{\partial y \partial x}.$$

The last two are refered to as the *mixed* second order partial derivatives of f. In certain situations subscript notation for partial derivatives has advantages. We shall also denote by f_x, f_y, and f_{xx}, f_{yy}, f_{xy}, f_{yx} the first and second order partial derivatives of f respectively.

Definition 3.5.1. Let $f : \Omega \subseteq \mathbb{R}^n \to \mathbb{R}$. If the second order partial derivatives $\frac{\partial^2 f}{\partial x_j \partial x_i}$, for $i, j = 1, 2, ..., n$ all exist and are continuous on Ω, then we say that f is of *class* C^2 on Ω and we write $f \in C^2(\Omega)$. Likewise, for each positive integer k, we say f is of *class* $C^k(\Omega)$, when all the kth order partial derivatives of f exist and are continuous on Ω. A function f is said to be *smooth* on Ω or of class $C^\infty(\Omega)$, if f has all its partial derivatives of all orders, that is, $f \in C^\infty(\Omega)$ if $f \in C^k(\Omega)$ for all $k = 1, 2, ...$.

A consequence of Corollary 3.2.18 are the following inclusions

$$C^\infty(\Omega) \subseteq ... \subseteq C^k(\Omega) \subseteq C^{k-1}(\Omega) \subseteq ... \subseteq C^1(\Omega) \subseteq C^0(\Omega).$$

Example 3.5.2. Let $f(x,y) = x^5 - 3x^2y + y^4$. Find $\frac{\partial^2 f}{\partial x^2}$, $\frac{\partial^2 f}{\partial y^2}$, $\frac{\partial^2 f}{\partial x \partial y}$, $\frac{\partial^2 f}{\partial y \partial x}$.

Solution. $\frac{\partial f}{\partial x} = 5x^4 - 6xy$, $\frac{\partial f}{\partial y} = -3x^2 + 4y^3$, $\frac{\partial^2 f}{\partial x^2} = 20x^3 - 6y$, $\frac{\partial^2 f}{\partial y^2} = 12y^3$, and $\frac{\partial^2 f}{\partial x \partial y} = \frac{\partial}{\partial x}(\frac{\partial f}{\partial y}) = -6x$, $\frac{\partial^2 f}{\partial y \partial x} = \frac{\partial}{\partial y}(\frac{\partial f}{\partial x}) = -6x$. Note that the mixed partial derivatives are equal.

In the computation of higher order partial derivatives it turns out that if all derivatives concerned are *continuous* on Ω, then, the order of differentiation is immaterial.

Theorem 3.5.3. *(Equality of mixed partials) Let $f : \Omega \subseteq \mathbb{R}^n \to \mathbb{R}$. If the mixed partials $\frac{\partial^2 f}{\partial x_j \partial x_i}$ and $\frac{\partial^2 f}{\partial x_i \partial x_j}$, $i, j = 1, 2, ..., n$, are continuous on Ω, then on Ω*

$$\frac{\partial^2 f}{\partial x_j \partial x_i} = \frac{\partial^2 f}{\partial x_i \partial x_j}.$$

Proof. Since only two components x_i and x_j are involved, it suffices to consider the case $n = 2$. We show, $\frac{\partial^2 f}{\partial x \partial y} = \frac{\partial^2 f}{\partial y \partial x}$. Let $(a, b) \in \Omega$. Since Ω is open, we can choose h, k small enough so that the rectangle with vertices (a, b), $(a + h, b)$, $(a, b + k)$ and $(a + h, b + k)$, lies in Ω. Consider the "difference of differences",

$$\Delta_{hk} = [f(a + h, b + k) - f(a, b + k)] - [f(a + h, b) - f(a, b)].$$

Let $\varphi(x) = f(x, b + k) - f(x, b)$ for $x \in [a, a + h]$. So that $\Delta_{hk} = \varphi(a + h) - \varphi(a)$. Since φ is differentiable on $(a, a + h)$, the one-variable Mean Value theorem gives $\Delta_{hk} = \varphi'(\xi_1)h$, for some $\xi_1 \in (a, a + h)$, that is,

$$\Delta_{hk} = [\frac{\partial f}{\partial x}(\xi_1, b + k) - \frac{\partial f}{\partial x}(\xi_1, b)]h$$

A similar application of the Mean Value theorem to the second component, on the interval $[b, b + k]$ tells us

$$\Delta_{hk} = \frac{\partial^2 f}{\partial y \partial x}(\xi_1, \eta_1)hk,$$

for some $\eta_1 \in (b, b + k)$. Applying the same procedure to the function $\psi(y) = f(a + h, y) - f(a, y)$, for $y \in [b, b + k]$, we get

$$\Delta_{hk} = \frac{\partial^2 f}{\partial x \partial y}(\xi_2, \eta_2)hk,$$

with $\xi_2 \in (b, b + k)$ and $\eta_2 \in (a, a + h)$. Hence, if $h, k \neq 0$, then

$$\frac{\partial^2 f}{\partial y \partial x}(\xi_1, \eta_1) = \frac{\partial^2 f}{\partial x \partial y}(\xi_2, \eta_2).$$

Now, let $(h, k) \to (0, 0)$. The continuity of $\frac{\partial^2 f}{\partial y \partial x}$ and $\frac{\partial^2 f}{\partial x \partial y}$ at (a, b) implies

$$\frac{\partial^2 f}{\partial y \partial x}(a, b) = \frac{\partial^2 f}{\partial x \partial y}(a, b).$$

Since (a, b) was an arbitrary point of Ω, $\frac{\partial^2 f}{\partial y \partial x} = \frac{\partial^2 f}{\partial x \partial y}$ on Ω. $\qquad\square$

In Exercise 4.6.8 we shall see an alternative proof of Theorem 3.5.3.

Example 3.5.4. Let $f : \mathbb{R}^2 \to \mathbb{R}$ be defined by

$$f(x, y) = \begin{cases} \frac{x^3 y}{x^2 + y^2} & \text{if } (x, y) \neq (0, 0), \\ 0 & \text{if } (x, y) = (0, 0). \end{cases}$$

Then $\frac{\partial^2 f}{\partial x \partial y}(0, 0) = 1 \neq 0 = \frac{\partial^2 f}{\partial y \partial x}(0, 0)$. Note that at $(x, y) \neq (0, 0)$

$$\frac{\partial^2 f}{\partial x \partial y}(x, y) = \frac{\partial^2 f}{\partial y \partial x}(x, y) = \frac{x^6 + 6x^4 y^2 - 3x^2 y^4}{(x^2 + y^2)^3}$$

and $\frac{\partial^2 f}{\partial x \partial y}(x, y)$ is not continuous at $(0, 0)$.

Example 3.5.5. Let $u = f(x, y)$ and suppose $x = x(s, t)$, $y = y(s, t)$. Assume all functions are of class C^2. Compute $\frac{\partial^2 u}{\partial s^2}$, $\frac{\partial^2 u}{\partial t^2}$, and $\frac{\partial^2 u}{\partial s \partial t}$ in terms of the partial derivatives of f.

Solution. By the Chain Rule

$$\frac{\partial u}{\partial s} = f_x \frac{\partial x}{\partial s} + f_y \frac{\partial y}{\partial s},$$

$$\frac{\partial u}{\partial t} = f_x \frac{\partial x}{\partial t} + f_y \frac{\partial y}{\partial t}.$$

Therefore, by the Chain Rule again

$$\frac{\partial^2 u}{\partial s^2} = f_{xx} \left[\frac{\partial x}{\partial s} \right]^2 + 2 f_{xy} \frac{\partial x}{\partial s} \frac{\partial y}{\partial s} + f_{yy} \left[\frac{\partial y}{\partial s} \right]^2 + f_x \frac{\partial^2 x}{\partial s^2} + f_y \frac{\partial^2 y}{\partial s^2}$$

$$\frac{\partial^2 u}{\partial t^2} = f_{xx}\left[\frac{\partial x}{\partial t}\right]^2 + 2f_{xy}\frac{\partial x}{\partial t}\frac{\partial y}{\partial t} + f_{yy}\left[\frac{\partial y}{\partial t}\right]^2 + f_x\frac{\partial^2 x}{\partial t^2} + f_y\frac{\partial^2 y}{\partial t^2}$$

and

$$\frac{\partial^2 u}{\partial s \partial t} = \frac{\partial^2 u}{\partial t \partial s} = f_{xx}\frac{\partial x}{\partial t}\frac{\partial x}{\partial s} + f_{yx}\frac{\partial y}{\partial t}\frac{\partial x}{\partial s} + f_{xy}\frac{\partial x}{\partial t}\frac{\partial y}{\partial s} + f_{yy}\frac{\partial y}{\partial t}\frac{\partial y}{\partial s}$$
$$+ f_x\frac{\partial^2 x}{\partial t \partial s} + f_y\frac{\partial^2 y}{\partial t \partial s}.$$

Definition 3.5.6. Let $f : \Omega \subseteq \mathbb{R}^n \to \mathbb{R}$ be $C^2(\Omega)$. The expression

$$\Delta f = \frac{\partial^2 f}{\partial x_1^2} + \dots + \frac{\partial^2 f}{\partial x_n^2}$$

is caled the *Laplacian* of f. The equation $\Delta f = 0$ is called the *Laplace equation* and a function $f \in C^2(\Omega)$ satisfying the Laplace equation is called a *harmonic function* on Ω.

Example 3.5.7. Let $f(x,y,z) = \frac{1}{\sqrt{x^2+y^2+z^2}}$. Show that f is harmonic in $\mathbb{R}^3 - \{\mathbf{0}\}$.

Solution. $\frac{\partial f}{\partial x} = \frac{x}{(x^2+y^2+z^2)^{\frac{3}{2}}}$ and $\frac{\partial^2 f}{\partial x^2} = \frac{y^2+z^2-2x^2}{(x^2+y^2+z^2)^{\frac{5}{2}}}$. By symmetry

$$\frac{\partial^2 f}{\partial y^2} = \frac{x^2+z^2-2y^2}{(x^2+y^2+z^2)^{\frac{5}{2}}}, \frac{\partial^2 f}{\partial z^2} = \frac{x^2+y^2-2z^2}{(x^2+y^2+z^2)^{\frac{5}{2}}}.$$

Hence

$$\Delta f = \frac{\partial^2 f}{\partial x^2} + \frac{\partial^2 f}{\partial y^2} + \frac{\partial^2 f}{\partial z^2} = 0.$$

Example 3.5.8. (*The Laplacian in polar coordinates*).   Write the Laplacian $\frac{\partial^2 u}{\partial x^2} + \frac{\partial^2 u}{\partial y^2}$ in polar coordinates $x = r\cos\theta$, $y = r\sin\theta$.

Solution. From Example 3.2.13

$$\frac{\partial u}{\partial r} = \frac{\partial u}{\partial x}\cos\theta + \frac{\partial u}{\partial y}\sin\theta,$$

$$\frac{\partial u}{\partial \theta} = -\frac{\partial u}{\partial x}r\sin\theta + \frac{\partial u}{\partial y}r\cos\theta.$$

Solving these for $\frac{\partial u}{\partial x}$ and $\frac{\partial u}{\partial y}$, we get

$$\frac{\partial u}{\partial x} = \frac{\partial u}{\partial r}\cos\theta - \frac{\partial u}{\partial \theta}\frac{\sin\theta}{r},$$

$$\frac{\partial u}{\partial y} = \frac{\partial u}{\partial r}\sin\theta + \frac{\partial u}{\partial \theta}\frac{\cos\theta}{r}.$$

Hence

$$\begin{aligned}
\frac{\partial^2 u}{\partial x^2} &= \frac{\partial}{\partial x}\left[\frac{\partial u}{\partial x}\right] = \frac{\partial}{\partial r}\left[\frac{\partial u}{\partial r}\cos\theta - \frac{\partial u}{\partial \theta}\frac{\sin\theta}{r}\right]\cos\theta \\
&\quad - \frac{\partial}{\partial \theta}\left[\frac{\partial u}{\partial r}\cos\theta - \frac{\partial u}{\partial \theta}\frac{\sin\theta}{r}\right]\frac{\sin\theta}{r} \\
&= \left[\frac{\partial^2 u}{\partial r^2}\cos\theta + \frac{\partial u}{\partial \theta}\frac{\sin\theta}{r^2} - \frac{\partial^2 u}{\partial r\partial \theta}\frac{\sin\theta}{r}\right]\cos\theta \\
&\quad - \left[-\frac{\partial u}{\partial r}\sin\theta + \frac{\partial^2 u}{\partial \theta\partial r}\cos\theta - \frac{\partial u}{\partial \theta}\frac{\cos\theta}{r} - \frac{\partial^2 u}{\partial \theta^2}\frac{\sin\theta}{r}\right]\frac{\sin\theta}{r}.
\end{aligned}$$

Similarly,

$$\begin{aligned}
\frac{\partial^2 u}{\partial y^2} &= \left[\frac{\partial^2 u}{\partial r^2}\sin\theta - \frac{\partial u}{\partial \theta}\frac{\cos\theta}{r^2} + \frac{\partial^2 u}{\partial r\partial \theta}\frac{\cos\theta}{r}\right]\sin\theta \\
&\quad + \left[\frac{\partial u}{\partial r}\cos\theta + \frac{\partial^2 u}{\partial \theta\partial r}\sin\theta - \frac{\partial u}{\partial \theta}\frac{\sin\theta}{r} + \frac{\partial^2 u}{\partial \theta^2}\frac{\cos\theta}{r}\right]\frac{\cos\theta}{r}.
\end{aligned}$$

Adding these expression we get

$$\frac{\partial^2 u}{\partial x^2} + \frac{\partial^2 u}{\partial y^2} = \frac{\partial^2 u}{\partial r^2} + \frac{1}{r^2}\frac{\partial^2 u}{\partial \theta^2} + \frac{1}{r}\frac{\partial u}{\partial r}. \tag{3.7}$$

3.5.1   The second derivative

Let $f : \Omega \subseteq \mathbb{R}^n \to \mathbb{R}$ where Ω is open. If f is differentiable on Ω, then its derivative $D_f(x) = \nabla f(x) = (\frac{\partial f}{\partial x_1}(x), ..., \frac{\partial f}{\partial x_n}(x))$ exists for all $x \in \Omega$ and defines a function $D_f : \Omega \to \mathbb{R}^n$. The second derivative of f is obtained dy differentiating D_f.

Definition 3.5.9. (*Second Derivative*). Let $f : \Omega \subseteq \mathbb{R}^n \to \mathbb{R}$ be differentiable on the open set Ω and $a \in \Omega$. The function f is said to be *twice* *differentiable* at a, if D_f is differentiable at a, i.e., if there is a linear transformation $B : \mathbb{R}^n \to \mathbb{R}^n$ such that

$$\lim_{x \to a} \frac{||D_f(x) - D_f(a) - B(x - a)||}{||x - a||} = 0.$$

The linear transformation B is denoted by $D_f^2(a)$ and is called the *second derivative* of f at a. If f is twice differentiable at every point of Ω, we say f is twice differentiable on Ω.

From Proposition 3.1.4., we see that if f is twice differentiable at $a \in \Omega$, then the second derivative $D_f^2(a)$ is uniquelly determined by f. Just as in the case of the first derivative, we can express the matrix of the second derivative in terms of the (second order) partial derivatives.

Theorem 3.5.10. *Let $f : \Omega \subseteq \mathbb{R}^n \to \mathbb{R}$ be twice differentiable at $a \in \Omega$. Then the second order partial derivatives $\frac{\partial^2 f}{\partial x_i \partial x_j}(a)$ for $i, j = 1, ..., n$ exist and the standard matrix representation of $D_f^2(a)$ is given by*

$$H_f(a) = \begin{pmatrix} \frac{\partial^2 f}{\partial x_1^2}(a) & \cdots & \frac{\partial^2 f}{\partial x_n \partial x_1}(a) \\ . & \cdots & . \\ . & \cdots & . \\ . & \cdots & . \\ \frac{\partial^2 f}{\partial x_1 \partial x_n}(a) & \cdots & \frac{\partial^2 f}{\partial x_n^2}(a) \end{pmatrix}$$

Proof. Since $D_f : \Omega \to \mathbb{R}^n$ is differentiable at $a \in \Omega$ and its component functions are $\frac{\partial f}{\partial x_j}$ for $j = 1, ..., n$, Theorem 3.2.7. implies that $\frac{\partial}{\partial x_i}(\frac{\partial f}{\partial x_j})(a) = \frac{\partial^2 f}{\partial x_i \partial x_j}(a)$ exist and $D_f^2(a)$ has the above standard matrix representation. □

Definition 3.5.11. Let $f : \Omega \subseteq \mathbb{R}^n \to \mathbb{R}$ be twice differentiable at $a \in \Omega$. The $n \times n$ matrix $H_f(a)$ of the second order partial derivarives $\frac{\partial^2 f}{\partial x_i \partial x_j}(a)$ is called the *Hessian*[5] of f at a.

An important consequence of Theorem 3.5.3 is

---

[5]L. Hesse (1811-1874). He worked on algebraic invariants. He held positions at the University of Königsberg, Heidelberg, and the Polytechnic School of Munich.

Corollary 3.5.12. *If f is of class $C^2(\Omega)$, then its Hessian is a symmetric matrix.*

Exercise 3.5.13. Show that the Hessian is the Jacobian matrix of the gradient, and the Laplacian is the trace of the Hessian.

Example 3.5.14. Let $f : \mathbb{R}^2 \to \mathbb{R}$ be the function $f(x, y) = x^3 - x^2y - y^4$. Find its Hessian $H_f(x, y)$.

Solution. The derivative of f at any point (x, y) is

$$D_f(x, y) = \nabla f(x, y) = (3x^2 - 2xy, x^2 - 4y^3).$$

Hence the Hessian of f is

$$H_f(x, y) = \begin{pmatrix} 6x - 2y & 2x \\ 2x & -12y^2 \end{pmatrix}.$$

The second derivative is also defined for *vector-valued* functions $f :$ $\Omega \subseteq \mathbb{R}^n \to \mathbb{R}^m$. Let $L(\mathbb{R}^n, \mathbb{R}^m)$ denote the space of linear tranformations from $\mathbb{R}^n$ to $\mathbb{R}^m$. We recall by choosing bases in $\mathbb{R}^n$ and $\mathbb{R}^m$, $L(\mathbb{R}^n, \mathbb{R}^m)$ can be identified with the space of $m \times n$ matrices $M_{m \times n}(\mathbb{R})$ and hence with $\mathbb{R}^{nm}$, (see Appendix D). Let $f : \Omega \subseteq \mathbb{R}^n \to \mathbb{R}^m$ be differentiable in Ω. Then differentiating the function $D_f : \Omega \subseteq \mathbb{R}^n \to \mathbb{R}^{nm}$ we obtain the second derivative of f. Specifically

Definition 3.5.15. A function $f : \Omega \subseteq \mathbb{R}^n \to \mathbb{R}^m$ is said *twice-differentiable* at $a \in \Omega$, if it is differentiable on Ω and there is a linear transformation $B : \mathbb{R}^n \to \mathbb{R}^{nm}$ such that

$$\lim_{x \to a} \frac{||D_f(x) - D_f(a) - B(x - a)||}{||x - a||} = 0.$$

(Here, the norm in the numerator is the Euclidean norm in $\mathbb{R}^{nm}$). The linear transformation B is called the *second derivative* of f at a and is also denoted by $D_f^2(a)$.

Corollary 3.5.16. *Let $f : \Omega \subseteq \mathbb{R}^n \to \mathbb{R}^m$ be a twice differentiable function at $a \in \Omega$. Then the second order partial derivatives $\frac{\partial^2 f_k}{\partial x_i \partial x_j}(a)$,* *for* $k = 1, ..., m$ *and* $i, j = 1, ..., n$ *exist and the standard block matrix representation of* $D_f^2(a)$ *is*

$$D_f^2(a) = \begin{pmatrix} H_{f_1}(a) & & & \\ & H_{f_2}(a) & & \\ & & \cdot & \\ & & & \cdot \\ & & & & H_{f_m}(a). \end{pmatrix}$$

Proof. Let $f = (f_1, ..., f_m)$ be twice differentiable. By Proposition 3.1.9 each component function $f_1, ..., f_m$ is differentiable at a and $D_f(a) = (D_{f_1}(a), ..., D_{f_m}(a))$. By the same proposition, each f_k is twice differentiable at a and the Jacobian matrix of $D_f(a)$ written in row notation is

$$D_f^2(a) = \begin{pmatrix} H_{f_1}(a) \\ \cdot \\ \cdot \\ \cdot \\ H_{f_m}(a) \end{pmatrix}.$$

$\square$

Now let $u = (u_1, ..., u_n)$ and $v = (v_1, ..., v_n)$ be two vectors in $\mathbb{R}^n$ and consider the mapping $\Phi : \mathbb{R}^n \times \mathbb{R}^n \to \mathbb{R}^m$ given by

$$\Phi(u, v) = \begin{pmatrix} \langle H_{f_1}(a)u, v \rangle \\ \cdot \\ \cdot \\ \cdot \\ \langle H_{f_m}(a)u, v \rangle \end{pmatrix} = \begin{pmatrix} \sum_{i=1}^n \sum_{j=1}^n \frac{\partial^2 f_1}{\partial x_i \partial x_j}(a) u_i v_j \\ \cdot \\ \cdot \\ \cdot \\ \sum_{i=1}^n \sum_{j=1}^n \frac{\partial^2 f_m}{\partial x_i \partial x_j}(a) u_i v_j \end{pmatrix}.$$

Then, Φ is linear in each component and such maps are said to be *bilinear*. In the special case where, $m = 1$

$$\Phi(u, v) = \langle H_f(a)u, v \rangle = \sum_{i=1}^n \sum_{j=1}^n \frac{\partial^2 f}{\partial x_i \partial x_j}(a) u_i v_j.$$

In this case Φ is called the *quadratic form* associated with the (symmetric) matrix $H_f(a)$.

Finally, derivatives of higher order are defined inductively and give rise to *multilinear* maps. So that, the pth-order derivative D_f^p for $p > 2$ is defined by $D_f^p = D(D_f^{p-1})$ and gives a p-linear map.

EXERCISES

1. Show that

   (a) Sums and products of $C^k(\Omega)$ functions are $C^k(\Omega)$ functions (including $k = \infty$).

   (b) Polynomials in n-variables are $C^\infty(\mathbb{R}^n)$.

   (c) Rational functions are C^∞ whenever the denominator does not vanish.

2. Let $u(s,t) = t^k e^{-\frac{s^2}{4t}}$. Find a value of the constant k such that u satisfies

$$\frac{\partial u}{\partial t} = \frac{1}{s^2} \frac{\partial}{\partial s}\left[s^2 \frac{\partial u}{\partial s}\right].$$

3. Let $u = f(x,y)$, $x = r\cos\theta$, $y = r\sin\theta$.

   (a) Show that $\left(\frac{\partial u}{\partial x}\right)^2 + \left(\frac{\partial u}{\partial y}\right)^2 = \left(\frac{\partial u}{\partial r}\right)^2 + \frac{1}{r^2}\left(\frac{\partial u}{\partial \theta}\right)^2$

   (b) Find $\frac{\partial^2 u}{\partial r \partial \theta}$. (See Example 3.5.8).

4. Let $u = f(2s - t^2, s\sin 3t, s^4)$. Compute $\frac{\partial^2 u}{\partial s^2}$ and $\frac{\partial^2 u}{\partial s \partial t}$ in terms of the partial derivatives of f.

5. Let $u = \phi(x + ct) + \psi(x - ct)$. Show that $u_{tt} = c^2 u_{xx}$. Show also that if $\xi = x + ct$, $\eta = x - ct$, the equation becomes $u_{\xi\eta} = 0$.

6. Let $u = f(x, y, g(x,y))$. Compute $\frac{\partial^2 u}{\partial y^2}$ and $\frac{\partial^2 u}{\partial x \partial y}$ in terms of the derivatives of f and g.

7. Let $u = f(x,y)$ and $x = \frac{1}{2}s(e^t + e^{-t})$, $y = \frac{1}{2}s(e^t - e^{-t})$. Show that

$$u_{xx} - u_{yy} = u_{ss} + \frac{1}{s}u_s + \frac{1}{s^2}u_{tt}.$$

8. Let $u = f(x, y)$ and $x = e^s \cos t$, $y = e^s \sin t$. Show that

$$u_{ss} + u_{tt} = e^{2s}(u_{xx} + u_{yy}).$$

9. Let $u = u(x, y)$, $x = r \cos \theta$, $y = r \sin \theta$. Show that the equation

$$xy(u_{xx} - u_{yy}) - (x^2 - y^2)u_{xy} = 0, \text{ becomes } ru_{r\theta} - u_\theta = 0.$$

10. Show that $u(x, y) = e^{(x^2 - y^2)} \sin 2xy$ satisfies $u_{xx} + u_{yy} = 0$.

11. Show that if $u = f(x, y)$ is harmonic, then so is $\varphi(x, y) = f\left(\frac{x}{x^2+y^2}, \frac{y}{x^2+y^2}\right)$.

12. For $x \in \mathbb{R}^n - \{0\}$, let $f(x) = g(r)$ where g is a C^2 function on $(0, \infty)$ and $r = ||x||$. Show that

$$\frac{\partial^2 f}{\partial^2 x_1} + \ldots + \frac{\partial^2 f}{\partial^2 x_n} = g''(r) + \frac{n-1}{r}g'(r).$$

13. Let $f(x) = ||x||^{2-n}$, where $n \geq 3$. Prove that f is harmonic.

14. Let f be the function of Exercise 7 in Section 3.2. Show that

$$f_{xy}(0, 0) = 1 \neq -1 = f_{yx}(0, 0).$$

15. Let $f : \mathbb{R}^2 \to \mathbb{R}$ be defined by

$$f(x, y) = \begin{cases} (x^2 + y^2)\tan^{-1}(\frac{y}{x}) & \text{for } x \neq 0, \\ \frac{\pi y^2}{2} & \text{for } x = 0. \end{cases}$$

Examine the equality of $f_{xy}(0, 0)$ and $f_{yx}(0, 0)$.

16. (a) Assuming all functions are C^2, show that if two functions $u = u(x, y)$ and $v = v(x, y)$ satisfy the *Cauchy-Riemann equations*

$$\frac{\partial u}{\partial x} = \frac{\partial v}{\partial y}, \quad \frac{\partial u}{\partial y} = -\frac{\partial v}{\partial x},$$

then both functions are harmonic.

(b) Show that the Cauchy-Riemann equations in polar coordinates become

$$\frac{\partial u}{\partial r} = \frac{1}{r}\frac{\partial v}{\partial \theta}, \quad \frac{\partial v}{\partial r} = -\frac{1}{r}\frac{\partial u}{\partial \theta}.$$

17. (*Laplace's equation in spherical coordinates*). Let $u = u(x, y, z)$ and let $x = r\cos\theta\sin\phi$, $y = r\sin\theta\sin\phi$, $z = r\cos\phi$ the spherical coordinates in $\mathbb{R}^3$. Show that Laplace's equation $\frac{\partial^2 u}{\partial x^2} + \frac{\partial^2 u}{\partial y^2} + \frac{\partial^2 u}{\partial z^2} = 0$ in the spherical coordinates can be written as

$$\frac{1}{r^2\sin^2\phi}\left[\sin^2\phi\frac{\partial}{\partial r}\left(r^2\frac{\partial u}{\partial r}\right) + \sin\phi\frac{\partial}{\partial\phi}\left(\sin\phi\frac{\partial u}{\partial\phi}\right) + \frac{\partial^2 u}{\partial\theta^2}\right] = 0.$$

Answers to selected Exercises

2. $k - -\frac{3}{2}$.

4. $\frac{\partial^2 u}{\partial s^2} = 4f_{xx} + 4(\sin 3t)f_{xy} + 8s^3 f_{xz} + (\sin^2 3t)f_{yy} + 8s^3(\sin 3t)f_{yz} + 16s^6 f_{zz} + 12s^2 f_z.$

$\frac{\partial^2 u}{\partial s\partial t} = -4tf_{xx} + (3s\cos 3t - 2t\sin 3t)f_{xy} - 8s^3 tf_{xz} + 3s(\sin 3t\cos 3t)f_{yy} + 12s^4(\cos 3t)f_{yz} + 3(\cos 3t)f_y.$

3.6   Taylor's theorem

Since the one variable case may not be familiar to the reader and the situation of several variables depends very much on the one variable situation we shall begin with Taylor's theorem and Taylor's series in one variable.

3.6.1   Taylor's theorem in one variable

To motivate the theorem we consider a polynomial

$$f(x) = \sum_{k=0}^{n} b_k x^k$$

and let $a \in \mathbb{R}$. Replacing x by $(x - a) + a$, we can express $f(x)$ as a sum of powers of $(x - a)$:

$$f(x) = \sum_{k=0}^{n} a_k (x - a)^k$$

Differentiating k times for $0 \le k \le n$, and setting $x = a$, we get $a_k = \frac{1}{k!} f^{(k)}(a)$. Hence,

$$f(x) = \sum_{k=0}^{n} \frac{f^{(k)}(a)}{k!} (x - a)^k.$$

This is known as the Taylor[6] formula for f. As we shall see any function possesing derivatives of high enough order can be approximated by such a polynomial. Taylor's theorem deals with the form that this approximation takes. Let f be an infinitely differentiable function on an interval, I. For example I could be a finite interval, (α, β), or (α, ∞), or $(-\infty, \beta)$, or $(-\infty, \infty) = \mathbb{R}$. Let $a \in I$ be fixed and for each integer $n \ge 0$ we define

$$P_n(x) = \sum_{k=0}^{n} \frac{f^{(k)}(a)}{k!} (x - a)^k$$

the *nth Taylor polynomial* associated with the function f at the point a, which is called the basepoint. The zero derivative being understood to be $f(a)$ itself. P_n is clearly a polynomial of degree $\le n$ in $x - a$. For example, $P_0(x) = f(a)$, the constant function, while $P_1(x) = f(a) + f'(a)(x - a)$, the tangent line to the graph of f at the point $(a, f(a))$. Going further

$$P_2(x) = f(a) + f'(a)(x - a) + \frac{1}{2} f''(a)(x - a)^2.$$

If $f''(a) \ne 0$ this is a parabola passing through the point $(a, f(a))$ and is even slightly "more tangent" to the graph of f than the tangent line.

We want to approximate f by the polynomial P_n on I. For this purpose we consider the remainder $R_{n+1}(x) = f(x) - P_n(x)$. Thus, $f(x) = P_n(x) + R_{n+1}(x)$. To show P_n approximates f on I we need to know $R_{n+1}(x) \to 0$ for all $x \in I$ as $n \to \infty$. Here is Taylor's theorem.

---

[6]B. Taylor (1685-1731), best known for Taylor's theorem and the Taylor series.

Theorem 3.6.1. *Let f be a function possesing derivatives of order up to $n + 1$ and let $x \in I$. Then there is some c between x and a (and, in particular, $c \in I$) so that*

$$f(x) = P_n(x) + \frac{f^{(n+1)}(c)}{(n+1)!}(x - a)^{n+1}.$$

Proof. Recall Cauchy's Mean Value theorem (see, Theorem B.1.3) which is usually dealt with in connection with l'Hopital's rule in first semester calculus. We will now use it here. Let n be fixed and set $\phi(x) = R_{n+1}(x)$ and $\psi(x) = (x - a)^{n+1}$. Notice that $\phi(a)$ and $\psi(a)$ are both zero and that the same is true of each of their derivatives from $i = 1, \ldots, n$. Also for $i = 1, \ldots, n$, the derivative $\psi^{(i)}$ vanishes at no other point. Therefore for $x \in I$

$$\frac{\phi(x)}{\psi(x)} = \frac{\phi(x) - \phi(a)}{\psi(x) - \psi(a)} = \frac{\phi'(c_1)}{\psi'(c_1)},$$

where c_1 is between x and a and therefore is a point of I. Taking this as our new "x" and applying the same reasoning to ϕ' and ψ' we get

$$\frac{\phi\prime(c_1)}{\psi\prime(c_1)} = \frac{\phi'(c_1) - \phi'(a)}{\psi'(c_1) - \psi'(a)} = \frac{\phi''(c_2)}{\psi''(c_2)},$$

where c_2 is between c_1 and a and therefore is between x and a and is a point of I.

Continuing we get

$$\frac{\phi^{(n)}(c_n)}{\psi^{(n)}(c_n)} = \frac{\phi^{(n)}(c_n) - \phi^{(n)}(a)}{\psi^{(n)}(c_n) - \psi^{(n)}(a)} = \frac{\phi^{(n+1)}(c_n)}{\psi^{(n+1)}(c_n)},$$

where c_n is between x and a and is a point of I.

Putting all these equations together gives:

$$\frac{\phi(x)}{\psi(x)} = \frac{\phi^{(n+1)}(c)}{\psi^{(n+1)}(c)},$$

where $c = c_n$ is between x and a.

Now, $\phi(x) = f(x) - P_n(x)$ and $\psi(x) = (x - a)^{n+1}$, we see that $\phi^{(n+1)}(x) = f^{(n+1)}(x)$ and $\psi^{(n+1)}(x) = (n + 1)!$, a constant.

Thus $f(x) - P_n(x) = \frac{f^{(n+1)}(c)}{(n+1)!}(x - a)^{n+1}$ for some c between x and a. $\square$

Thus to see P_n approximates f on I we need only understand why

$$\frac{f^{(n+1)}(c)}{(n+1)!}(x-a)^{n+1} \to 0 \text{ on } I \text{ as } n \to \infty.$$

This we will deal with shortly. Let's first see what this says for $n = 0$. Here our theorem says $f(x) = f(a) + f'(c)(x-a)$. That is, $\frac{f(x)-f(a)}{x-a} = f'(c)$, where c is between x and a. This is the Mean Value theorem. Hence Taylor's theorem is an elaborated version of the Mean Value theorem.

Corollary 3.6.2. *If there are positive constants k and M so that for n sufficiently large $\max_{x\in I} | f^{(n)}(x) |\le kM^n$, then for all $x \in I$*

$$\lim_{n\to\infty} P_n(x) = f(x).$$

Proof. Here we apply Taylor's theorem and get for n sufficiently large

$$|\frac{f^{(n+1)}(c)}{(n+1)!}(x-a)^{n+1}| \le \frac{kM^{n+1}|x-a|^{n+1}}{(n+1)!}.$$

To see that this tends to zero and hence so does $R_{n+1}(x)$ we need only recall (see Example 1.1.21) that for any real number y, $\frac{y^n}{n!}$ tends to 0 as $n \to \infty$. □

Whenever (and for whatever reason) the series below converges in some neighborhood of a to the function we call it the *Taylor series* of f at basepoint a. Thus

$$f(x) = \sum_{k=0}^{\infty} \frac{f^{(k)}}{k!}(x-a)^k.$$

In particular, Corollary 3.6.2 gives a simple condition which guarantees this happens.

We now give some important examples of this.

Example 3.6.3. Let $f(x) = e^x$. Then $f^{(n)}(x) = e^x$ for all n and if I is any interval, $\max_{x\in I} |f^{(n)}(x)| = e^\alpha$, where α is the right end point of I. So we take this value for k and let $M = 1$. Then the conditions of Corollary 3.6.2 are satisfied. Here any interval with a finite right hand end point will work. Thus, for any real x, just take the interval $(x-1, x+1)$. Therefore taking $a=0$ we get:

$$e^x = \sum_{n=0}^{\infty} \frac{x^n}{n!}. \tag{3.8}$$

Example 3.6.4. Similarly, if $f(x) = \sin(x)$, $\max_{x \in \mathbb{R}} |f^{(n)}(x)| \le 1$. Therefore we can take $I = \mathbb{R}$, $k = 1$, $M = 1$ and $a = 0$ and we get

$$\sin(x) = \sum_{n=0}^{\infty} \frac{(-1)^n x^{2n+1}}{(2n+1)!}.$$

and

$$\cos(x) = \sum_{n=0}^{\infty} \frac{(-1)^n x^{2n}}{(2n)!}.$$

Similar results work for many common functions. The statement about cos explains why in calculus of one variable in connection with differentiation of $\sin x$ one has $\lim_{x \to 0} \frac{\cos(x)-1}{x^2} = -\frac{1}{2}$.

The reader is encouraged to sketch the graphs of the first few approximating Taylor polynomials to each of these functions.

Example 3.6.5. Let $f(x) = \log(x)$, which is defined on $I = (0, \infty)$. Here it is easy to check that for each $n \ge 1$,

$$f^{(n)}(x) = \frac{(-1)^{n-1}(n-1)!}{x^n}.$$

Therefore on (α, β), $|f^{(n)}(x)| \le \frac{(n-1)!}{\alpha^n}$, since $\frac{1}{x^n}$ takes its maximum value at the left endpoint, $x = \alpha$. Unfortunately because of the factor $(n-1)!$, there is no k and M for which $\frac{(n-1)!}{\alpha^n} \le kM^n$ for all n sufficiently large. So although our previous reasoning worked well for e^x, $\sin(x)$ and $\cos(x)$, it will not work for $\log(x)$. Nevertheless we shall see that for any a the Taylor series converges in some small neighborhood of a. However, if we fix the basepoint, say at $a = 1$ we don't get a convergent Taylor series at all points of our domain P. To see all these we estimate the remainder on I.

$$\left| \frac{f^{(n+1)}(c)}{n+1!}(x-a)^{n+1} \right| \le \frac{n! \, |x-a|^{n+1}}{n+1! \alpha^{n+1}} \le \frac{1}{n+1} \left(\frac{|x-a|}{\alpha} \right)^{n+1}.$$

Hence if $|x - a| \leq \alpha$, this last term tends to zero as $n \to \infty$. Thus for any a, there is a sufficiently small neighborhood about a where the Taylor series converges. Just take any $\alpha < a$. This will work for all x satisfying

$$| \, x - a \, | \leq \alpha.$$

Now we take $a = 1$. Here $\log(1) = 0$ and $f^{(n)}(1) = (-1)^{n-1}(n-1)!$, for all $n \geq 1$, so the Taylor series is.

$$1 - (x - 1) + \frac{(x-1)^2}{2} - \frac{(x-1)^3}{3} \ldots.$$

The question is, where does this series converge and does it converge to $\log(x)$?

As to where, we can apply the ratio test:

$$\frac{| \, (-1^n)\frac{(x-1)^{n+1}}{n+1} \, |}{| \, (-1^{n-1})\frac{(x-1)^n}{n} \, |} = | \, x - 1 \, | \frac{n}{n+1}.$$

Since $\frac{n}{n+1}$ converges to 1, $| \, x - 1 \, | \frac{n}{n+1}$ converges to $| \, x - 1 \, |$. This means the Taylor series converges if $| \, x - 1 \, | < 1$ and diverges if $| \, x - 1 \, | > 1$. Thus if $x > 2$, the Taylor series diverges! If $x = 2$ or 0 the ratio test gives no information. However notice that at $x = 0$ the series diverges:

$$\sum_{n=1}^{\infty} \frac{(-1)^{n-1}(-1)^n}{n} = \sum_{n=1}^{\infty} \frac{(-1)^{2n-1}}{n} = -\sum_{n=1}^{\infty} \frac{1}{n} = -\infty.$$

This is consistent with the fact that the log is not defined at 0 (or that its value there is $-\infty$).

On the other hand, if $1 \leq x \leq 2$ then $|x - 1| = x - 1 \leq 1$. So here we get convergence

$$\log(x) = \sum_{n=1}^{\infty} \frac{(-1)^{n-1}(x-1)^n}{n}.$$

In particular,

$$\log(2) = 1 - \frac{1}{2} + \frac{1}{3} \ldots.$$

Now consider the case where $0 < x < 1$. Then, since $x < c < 1$, we see $\frac{1}{c} < \frac{1}{x}$ and so

$$\left| \frac{f^{n+1}(c)(x-1)^{n+1}}{n+1!} \right| \le \frac{n!}{n+1!} \frac{|x-1|^{n+1}}{x^{n+1}}$$

$$= \frac{1}{n+1} \left(\frac{x-1}{x} \right)^{n+1} = \frac{1}{n+1} \left(1 - \frac{1}{x} \right)^{n+1}.$$

Since $1 - \frac{1}{x} < 1$, this last term tends to zero.

Thus $\log(x) = \sum_{n=1}^{\infty} \frac{(-1)^{n-1}(x-1)^n}{n}$ for all $0 < x \le 2$.

Exercise 3.6.6. Show that $\sinh(x)$ and $\cosh(x)$ have convergent Taylor series about $x = 0$ and find them.

Next we obtain the integral form of the remainder.

Proposition 3.6.7. *Let f be a function possesing derivatives of order up to $n + 1$. Then*

$$R_{n+1}(x) = \frac{1}{n!} \int_a^x (x-t)^n f^{(n+1)}(t) dt.$$

Proof. From the Fundamental Theorem of Calculus (see, Theorem B.2.1) we know that

$$f(x) = f(a) + \int_a^x f'(t) dt.$$

We now integrate by parts, choosing for the antiderivative of the constant function 1 not t, but rather $-(x - t)$. Thus,

$$\int_a^x f'(t) dt = -f'(t)(x-t) \Big|_a^x + \int_a^x f''(t)(x-t) dt.$$

Integrating by parts again tells us

$$\int_a^x f''(t)(x-t) dt = -f''(t) \frac{(x-t)^2}{2} \Big|_a^x + \int_a^x f^{(3)}(t) \frac{(x-t)^2}{2} dt.$$

Hence,

$$f(x) = f(a) + f'(a)(x-a) + f''(a) \frac{(x-t)^2}{2} + \frac{1}{2!} \int_a^x (x-t)^2 f^{(3)}(t) dt.$$

The pattern is now clear; integrating by parts n times, yields the stated form of the remainder $R_{n+1}(x)$.   □

Corollary 3.6.8. *If f is smooth on an interval I, then*

$$\lim_{x \to a} \frac{R_{n+1}(x)}{(x-a)^n} = 0.$$

Proof. As $x \to a$, from Theorem 3.6.1, we have

$$\frac{R_{n+1}(x)}{(x-a)^n} = \frac{1}{(n+1)!} f^{(n+1)}(c)(x-a) \to 0.$$

□

Note that for $n = 1$ this is just a restatement of the differentiability of f at a.

Exercise 3.6.9. Prove Corollary 3.6.8, using the integal form of the remainder $R_{n+1}(x)$.

An essential fact about the Taylor expansion of a function f about a point a, is that it is the *only* way to write $f(x) = P_n(x) + R_{n+1}(x)$ with

$$\lim_{x \to a} \frac{R_{n+1}(x)}{(x-a)^n} = 0.$$

We prove this in the following corollary, but first we need a lemma.

Lemma 3.6.10. *Let $p(t) = a_0 + a_1 t + \ldots + a_n t^n$ be a polynomial of degree $\leq n$. If*

$$\lim_{t \to 0} \frac{p(t)}{t^n} = 0,$$

then $p(t) = 0$ for all $t \in \mathbb{R}$.

Proof. For $0 \leq k \leq n$, we have $\frac{p(t)}{t^k} = t^{n-k} \cdot \frac{p(t)}{t^n} \to 0$, as $t \to 0$. For $k = 0$, this gives, $0 = \lim_{t \to 0} p(t) = p(0) = a_0$, that is, $a_0 = 0$. Now, for $k = 1$, we have $\frac{p(t)}{t} = a_1 + a_2 t + \ldots + a_n t^{n-1}$ and $0 = \lim_{t \to 0} \frac{p(t)}{t} = a_1$. Hence, $a_1 = 0$. For $k = 2$ we have, $\frac{p(t)}{t^2} = a_2 + a_3 t + \ldots a_n t^{n-2}$ and similarly then, $a_2 = 0$. Continuing this way, we get $a_n = 0$. Thus, $p(t) = 0$ for all $t \in \mathbb{R}$. □

Corollary 3.6.11. *Let f be smooth on an interval I and $a \in I$. If $f(x) = Q_n(x) + E_{n+1}(x)$, where $Q_n(x)$ is a polynomial of degree n and*

$$\lim_{x \to a} \frac{E_{n+1}(x)}{(x-a)^n} = 0,$$

then $Q_n(x)$ is the Taylor polynomial $P_n(x)$.

Proof. By Corollary 3.6.8, $f(x) = P_n(x) + R_{n+1}(x)$ and $\lim_{x \to a} \frac{R_{n+1}(x)}{(x-a)^n} = 0$. Therefore, $Q_n(x) - P_n(x) = R_{n+1}(x) - E_{n+1}(x)$ and so, as $x \to a$, we have

$$\frac{Q_n(x) - P_n(x)}{(x-a)^n} = \frac{R_{n+1}(x) - E_{n+1}(x)}{(x-a)^n} \to 0.$$

Let $p(x) = Q_n(x) - P_n(x)$ and set $h = x - a$. Then, $p(h)$ is a polynomial of degree $\leq n$ satisfying $\lim_{h \to 0} \frac{p(h)}{h^n} = 0$. By Lemma 3.6.10, $p \equiv 0$, and so $Q_n = P_n$. $\qquad \square$

Corollary 3.6.11 gives us another way to compute the Taylor expansion of a smooth function f. If one can find (by any means whatever) a polynomial $Q_n(x)$ of degree n, such that

$$\lim_{x \to a} \frac{f(x) - Q_n(x)}{(x-a)^n} = 0,$$

then $Q_n(x)$ must be the Taylor polynomial of f. This enables one to generate new Taylor expansions from old ones, by substitution or other operations. For instance, if $f(x) = e^x \sin x$, then $P_3(x) = x + x^2 + \frac{1}{3}x^3$. Moreover,

$$\sin(x^2) \approx x^2 - \frac{x^6}{3!} + ... + (-1)^n \frac{x^{4n}}{(2n)!}.$$

Definition 3.6.12. A smooth function f which is represented by the Taylor series

$$f(x) = \sum_{k=0}^{\infty} \frac{f^{(k)}(a)}{k!}(x-a)^k$$

in a neighborhood of a, is called *real analytic* at a. Furthermore, if f is analytic at every point $a \in \Omega$, we say f is (real) analytic on Ω.

Examples of analytic functions on $\mathbb{R}$ are, e^x, $\sin x$, $\cos x$, $\sinh x$, $\cosh x$. But, a C^∞ function may not be analytic, for example, $f(x) = e^{-\frac{1}{x^2}}$ for $x \neq 0$ and $f(0) = 0$ is not analytic at 0. The reader can verify this as an exercise.

Finally, as an application of (3.8) we prove e is irrational. The transcendence of e is much more difficult.

Proposition 3.6.13. *(Hermite). The number e is irrational.*

Proof. Suppose $e = \frac{p}{q}$, where p and q are positive integers. We may assume p and q have no common factors and since $2 < e < 3$ that $q \geq 2$. To see that $2 < e < 3$ examine the power series

$$e^x = 1 + x + \frac{x^2}{2!} + \ldots + \frac{x^n}{n!} + \ldots$$

Since for $x = 1$ all terms are positive $2 < e$. To see the other inequality it's sufficient to see that $\frac{1}{2!} + \ldots + \frac{1}{n!} \ldots < 1$. But since $k! > 2^{k-1}$ for each integer $k > 2$, it follows that

$$\frac{1}{2!} + \ldots + \frac{1}{n!} + \ldots < \frac{1}{2} + \frac{1}{4} + \ldots = 1.$$

Now the derivative of e^x is e^x itself, which is positive. So e^x is monotone increasing and in the finite Taylor expansion, the remainder term $\frac{f^{(n+1)}(t)x^{n+1}}{(n+1)!}$, is of the form $\frac{e^t x^{n+1}}{(n+1)!}$, where $0 \leq t \leq x$. Choose an integer $n > q$ and look at the Taylor expansion of e^x of the n-th order. For any real x,

$$e^x = 1 + x + \frac{x^2}{2!} + \ldots + \frac{x^n}{n!} + \frac{e^t x^{n+1}}{(n+1)!}$$

where $0 \leq t \leq x$. Hence

$$\frac{p}{q} = e^1 = 1 + 1 + \frac{1}{2!} + \ldots + \frac{1}{n!} + \frac{e^t}{(n+1)!}$$

and so $\frac{n!p}{q} =$ an integer $+ \frac{e^t}{n+1}$, where $0 \leq t \leq 1$. But then this means that $\frac{e^t}{n+1}$ is itself an integer. Since e^x is monotone increasing $e^t \leq e < 3$. On the other hand, $n + 1 > n > q$ so that $n + 1 \geq 3$ and so $\frac{e^t}{n+1}$ can't be an integer. This contradiction proves e is irrational. $\qquad\square$

3.6.2   Taylor's theorem in several variables

We now generalize Taylor's theorem to functions of several variables. We first deal with the second order Taylor expansion.

Theorem 3.6.14. *Let* $f : \Omega \subseteq \mathbb{R}^n \to \mathbb{R}$ *be a* C^2 *function on the open convex set* Ω *of* $\mathbb{R}^n$ *and* $a, x \in \Omega$. *Then there exists a point* c *on the line segment joining* a *and* x, *such that*

$$f(x) = f(a) + \langle \nabla f(a), x - a \rangle + \frac{1}{2!} \langle H_f(c)(x - a), x - a \rangle$$

Proof. As in Theorem 3.4.1, let $\gamma(t) = a + t(x - a) \in \Omega$. Define $g(t) = f(\gamma(t)) = f(a + tv)$ where, $v = x - a$. Since g is C^2, we can write the second order Taylor expansion for g about 0 (one variable case), to get

$$g(t) = g(0) + g'(0)t + \frac{1}{2!} g''(\theta t) t^2$$

for $t \in [0, 1]$ and some $\theta \in (0, 1)$. In particular, for $t = 1$, this yields

$$g(1) = g(0) + g'(0) + \frac{1}{2!} g''(\theta) \tag{3.9}$$

Calculating the derivatives of g by the Chain Rule,

$$g'(t) = \frac{d}{dt} f(\gamma(t)) = \langle \nabla f(\gamma(t)), \gamma'(t) \rangle = \langle \nabla f(\gamma(t)), v \rangle.$$

Furthermore,

$$g''(t) = \frac{d}{dt} \langle \nabla f(\gamma(t)), v \rangle = \langle \frac{d}{dt} \nabla f(\gamma(t)), v \rangle$$

$$= \langle D(\nabla f)(\gamma(t) \cdot \gamma'(t), v \rangle = \langle H_f(\gamma(t)) v, v \rangle.$$

But, $g(1) = f(x)$, $g(0) = f(a)$, $g'(0) = \langle \nabla f(a), v \rangle$ and $g''(\theta) = \langle H_f(\gamma(\theta)) v, v \rangle$. Setting $c = \gamma(\theta)$ and substituting in (3.9) we obtain the result.   □

Of course, in coordinates the second order Taylor's expansion takes the form

$$f(x) = f(a) + \sum_{i=1}^{n} \frac{\partial f}{\partial x_i}(a)(x_i - a_i) + \frac{1}{2!} \sum_{i,j=1}^{n} \frac{\partial^2 f}{\partial x_i \partial x_j}(a)(x_i - a_i)(x_j - a_j)$$

We now come to the general finite Taylor expansion of order s.

Theorem 3.6.15. *(Taylor's theorem) Let $f : \Omega \subseteq \mathbb{R}^n \to \mathbb{R}$ be a smooth function on the open convex set Ω of $\mathbb{R}^n$ and $a, x \in \Omega$. Then*

$$f(x) = f(a) + \sum_{i=1}^{n} \frac{\partial f}{\partial x_i}(a)(x_i - a_i) + \frac{1}{2!} \sum_{i,j=1}^{n} \frac{\partial^2 f}{\partial x_i \partial x_j}(a)(x_i - a_i)(x_j - a_j)$$

$$\cdots$$

$$+ \frac{1}{s!} \sum_{i_1,\ldots,i_s=1}^{n} \frac{\partial^s f}{\partial x_{i1}\ldots\partial x_{is}}(a)(x_{i1} - a_{i1})\cdots(x_{is} - a_{is}) + R_{s+1}(x),$$

where $R_{s+1}(x)$ has the same form (involving $s + 1$ order partial derivatives) and is evaluated at some point c on the line segment joining a and x.

Proof. Consider the function $g(t) = f(\gamma(t))$ as above. Write the $s + 1$ Taylor expansion of g about $t = 0$ and set $t = 1$ to get

$$g(1) = g(0) + g'(0) + \frac{1}{2!}g''(0) + \ldots + \frac{1}{s!}g^{(s)}(0) + \frac{1}{(s+1)!}g^{(s+1)}(\theta),$$

with $0 < \theta < 1$. Now, computing the various higher order derivatives of g (just as we have done using the chain rule) evaluating for $t = 0$ and setting $c = \gamma(\theta)$, yields the expansion. $\square$

Definition 3.6.16. $R_{s+1}(x)$ is called the s order remainder. We let $P_s(x)$ stand for the sum of the terms up to and including order s. $P_s(x)$ is called the Taylor polynomial of order s. Thus $f(x) = P_s(x) + R_{s+1}(x)$.

We remark that if f is a polynomial of degree s (and $\Omega = \mathbb{R}^n$), then $f(x)$ equals its s order Taylor expansion for any base point a. This is because all derivatives of order $s + 1$ are identically zero, so the remainder is also zero.

Corollary 3.6.17. *Suppose there are positive constants k and M so that for s sufficiently large*

$$max_{x \in \Omega}\left|\frac{\partial^s f(x)}{\partial x_{i_1} \ldots \partial x_{i_s}}(x)\right| \leq ks!M^s.$$

Then given $a \in \Omega$ there is a sufficiently small ball about a so that for all x in this ball

$$\lim_{s \to \infty} P_s(x) \to f(x).$$

So here we also have an analytic function of several variables.

Proof. For a fixed s (which we can assume is $> n$), how many partial derivatives are there of order s? The number is $\leq$ to the number of ways of choosing n objects or less from s objects. But the number of ways of choosing n or less objects is $\leq$ the number of all subsets of s objects which is 2^s.

For all $x \in \Omega$ and s sufficiently large using the triangle inequality we get

$$|f(x) - P_s(x)| \leq \frac{1}{s!}||x - a||^s k \frac{1}{s!} M^s 2^s = k \frac{1}{s!}(||x - a||2M)^s.$$

Thus if $||x - a|| < \frac{1}{2M}$ this tends to zero as $s \to \infty$. $\qquad\square$

Moreover with a stronger estimate we get a global result.

Corollary 3.6.18. *If there are positive constants k and M so that for s sufficiently large*

$$max_{x \in \Omega}|\frac{\partial^s f(x)}{\partial x_{i_1} \ldots \partial x_{i_s}}(x)| \leq kM^s,$$

then $P_s(x) \to f(x)$ for all $x \in \Omega$.

Proof. Here $|f(x) - P_s(x)| \leq k\frac{1}{s!}(||x - a||2M)^s$. But $\lim_{s \to \infty} \frac{t^s}{s!} = 0$ for any $t \in \mathbb{R}$. $\qquad\square$

Corollary 3.6.19. *With the notation as in Taylor's Theorem*

$$\lim_{x \to a} \frac{|R_{s+1}(x)|}{||x - a||^s} = 0.$$

Proof. Note, as $x \to a$, also $c \to a$. Since $\frac{\partial f^{(s+1)}}{\partial x_{i1} \ldots \partial x_{i(s+1)}}(x)$ are all continuous at a, it follows

$$\frac{\partial f^{(s+1)}}{\partial x_{i1} \ldots \partial x_{i(s+1)}}(c) \to \frac{\partial f^{(s+1)}}{\partial x_{i1} \ldots \partial x_{i(s+1)}}(a).$$

Now,

$$|R_{s+1}(x)| \leq \frac{1}{(s+1)!} \sum_{i_1, \ldots, i_{s+1}=1}^{n} |\frac{\partial^{s+1} f}{\partial x_{i1} \ldots \partial x_{is+1}}(c)||x_{i1} - a_{i1}| \cdots |x_{is+1} - a_{is+1}|$$
$$\leq \frac{1}{(s+1)!} \sum_{i_1, \ldots, i_{s+1}=1}^{n} |\frac{\partial^{s+1} f}{\partial x_{i1} \ldots \partial x_{is+1}}(c)| \, ||x - a||^{s+1}.$$

Therefore, as $x \to a$, we have,

$$\frac{|R_{s+1}(x)|}{||x-a||^s} \le \frac{1}{(s+1)!} \sum_{i_1,\ldots,i_{s+1}=1}^{n} \left|\frac{\partial^{s+1} f}{\partial x_{i1}\ldots\partial x_{is+1}}(c)\right| ||x-a||$$

$$\to \frac{1}{(s+1)!} \sum_{i_1,\ldots,i_{s+1}=1}^{n} \left|\frac{\partial^{s+1} f}{\partial x_{i1}\ldots\partial x_{is+1}}(a)\right| 0 = 0.$$

$\square$

The following integral form of the remainder term $R_{s+1}(x)$ in Taylor's formula for a smooth function of several variables follows directly from the corresponding integral form for the remainder for a function of one variable.

Corollary 3.6.20. *Let $f : \Omega \subseteq \mathbb{R}^n \to \mathbb{R}$ be a smooth function on the open convex set Ω of $\mathbb{R}^n$ and $a, x \in \Omega$. Then*

$$f(x) = P_s(x) + \frac{1}{s!} \int_0^1 (1-u)^s$$

$$\times \sum_{i_1,\ldots,i_{s+1}=1}^{n} \frac{\partial^{s+1} f(a+u(x-a))}{\partial x_{i1}\ldots\partial x_{is+1}} (x_{i1}-a_{i1}) \cdots (x_{is+1}-a_{is+1}) du.$$

(The integral form of the remainder is the last term).

Proof. As before, let $g(t) = f(\gamma(t))$ where $\gamma(t) = a + t(x-a)$. The integral form of the remainder term of order s in the Taylor expansion of the one variable function g of t about 0 is

$$r_{s+1}(t) = \frac{1}{s!} \int_0^t (t-u)^s g^{(s+1)}(u) du$$

Setting $t = 1$ and computing (as before) the derivative $g^{(s+1)}(u)$, we see that

$$f(x) = P_s(x) + \frac{1}{s!} \int_0^1 (1-u)^s$$

$$\times \sum_{i_1,\ldots,i_{s+1}=1}^{n} \frac{\partial^{s+1} f(a+u(x-a))}{\partial x_{i1}\ldots\partial x_{is+1}} (x_{i1}-a_{i1}) \cdots (x_{is+1}-a_{is+1}) du.$$

$\square$

Example 3.6.21. Find the third order Taylor formula of $f(x, y) = e^{x+y}$ about the origin $(0, 0)$.

Solution. Here, $f(0,0) = 1$ and all the first, second and third order partial derivatives at $(0,0)$ are equal to 1. So

$$f(x,y) = 1+x+y+\frac{1}{2!}(x^2+2xy+y^2)+\frac{1}{3!}(x^3+3x^2y+3xy^2+y^3)+R_4(x,y),$$

where $\frac{R_4(x,y)}{||(x,y)||^3} \to 0$ as $(x,y) \to (0,0)$.

Example 3.6.22. Let $f(x,y) = e^{xy}\sin(x+y)$. The Taylor polynomial of f of order 3 about the point $(0,0)$ is

$$P_3(x,y) = x + y + \frac{1}{3!}(-x^3 + 3x^2y + 3xy^2 - y^3).$$

EXERCISES

1. Find the second order Taylor's formula for $f(x,y) = \log(xy)$ about the point $(1,1)$.

2. Find the second order Taylor's formula for $f(x,y) = \frac{1}{xy}$ about the point $(2,-1)$.

3. Verify the quadratic approximation formulas

    (a) $e^x\log(1+y) \approx y + xy - \frac{y^2}{2}$.

    (b) $\cos x\cos y \approx 1 - \frac{1}{2}(x^2 + y^2)$.

4. Find the third order Taylor's formula for $f(x,y) = e^x\log(1+y)$ about $(0,0)$.

5. Show that $\sin x\sin y = xy - \frac{1}{6}(xy^3 + yx^3) + R_4(x,y)$.

Answers to selected Exercises

1. $(x-1) + (y-1) - \frac{(x-1)^2}{2} - \frac{(y-1)^2}{2} + R_3(x,y)$.

2. $-\frac{1}{2}+\frac{1}{4}(x-2)-\frac{1}{2}(y+1)-\frac{1}{8}(x-2)^2+\frac{1}{4}(x-2)(y+1)-\frac{1}{2}(y+1)^2+R_3(x,y)$.

4. $y + \frac{1}{2}(2xy - y^2) + \frac{1}{6}(3x^2y - 3xy^2 + 2y^3) + R_4(x,y)$.

3.7   Maxima and minima in several variables

In this section we shall consider real-valued functions of several variables with a view toward finding those points (if any) at which f has a local extremum, that is, either a local maximum or a local minimum.

We recall briefly the one variable case. Let f be a function defined on an open interval I of $\mathbb{R}$ and $a \in I$. As is known, if f is differentiable at a and has a local extremum at a, then $f'(a) = 0$. If in addition f is twice differentiable, then when $f''(a) > 0$, f has a local minimum at a and when $f''(a) < 0$, f has a local maximum at a (This is the familiar second derivative test of one variable calculus). In case $f''(a) = 0$, no conclusion can be drawn about the behavior of f at a. When $f''(a) = 0$, one can analyze the sign of $f'(x)$ for x left and right of a in a sufficiently small interval around a, and use the local monotonicity of f there to study the behavior of f at a. (i.e., the first derivative test). When Taylor's expansion of f is applicable in a neighborhood of a, we can make use of higher derivatives using the following.

Theorem 3.7.1. *Let I be an open interval of $\mathbb{R}$, $a \in I$ and $f \in C^n(I)$. Suppose $f'(a) = f''(a) = ... = f^{(n-1)}(a) = 0$ and $f^{(n)}(a) \neq 0$. Then*

1. *When n is even. If $f^{(n)}(a) > 0$, then f has a local minimum at a.*

   If $f^{(n)}(a) < 0$, then f has a local maximum at a.

2. *When n is odd, f has neither a local minimum nor a local maximum at a.*

Proof. Since $f^{(n)}(a) \neq 0$ and $f^{(n)}$ is continuous at a, there is $\delta > 0$ so that for $x \in J = (a - \delta, a + \delta)$, the derivative $f^{(n)}(x)$ will have the same sign as $f^{(n)}(a)$. By Taylor's expansion for every $x \in J$ we have

$$f(x) - f(a) = \frac{f^{(n)}(c)}{n!}(x - a)^n$$

for some $c \in J$. Now, when n is even, then $(x - a)^n > 0$ (for all $x \neq a$). Hence, if $f^{(n)}(a) > 0$, then $f^{(n)}(c) > 0$ and so $f(x) > f(a)$ for all $x \in J$, that is, $f(a)$ is a local minimum. If $f^{(n)}(a) < 0$, then $f^{(n)}(c) < 0$ and so $f(x) < f(a)$ for all $x \in J$ and $f(a)$ is a local maximum. When n

is odd, then $(x - a)^n$ changes sign depending whether $x < a$ or $x > a$. Therefore, the difference $f(x) - f(a)$ also changes sign and there can not be extremum at a.   □

Example 3.7.2. Determine the local extrema of the functions

1. $f(x) = x^4 - 4x^3 + 6x^2 - 4x + 1$

2. $f(x) = (x - 2)^3$

Solution.

1. We have $f'(x) = 4x^3 - 12x^2 + 12x - 4 = 4(x - 1)^3$, $f''(x) = 12(x - 1)^2$, $f'''(x) = 24(x - 1)$ and $f^{(4)}(x) = 24$. We see that

$$f'(1) = f''(1) = f'''(1) = 0 \text{ and } f^{(4)}(1) = 24 > 0.$$

Hence by Theorem 3.7.1, f has local minimum at $x = 1$.

2. We have $f'(x) = 3(x - 2)^2$, $f''(x) = 6(x - 2)$, $f'''(x) = 6$. We see that $f'(2) = f''(2) = 0$ and $f'''(2) = 6 > 0$. Hence f has inflection point for $x = 2$.

Remark 3.7.3. Theorem 3.7.1 can not be applied in case a is an endpoint of a closed interval. However, if $f'(x) > 0$ for all x near a or, if $f'(x) < 0$ for all x near a, then f is monotone near a. Hence, in either case, $f(a)$ is a local extremum.

Example 3.7.4. Let $f(x) = e^{-\frac{1}{x^2}}$ for $x \neq 0$ and $f(0) = 0$. The point $x = 0$ is a critical point and $f^{(n)}(0) = 0$ for all n. Theorem 3.7.1 does not apply, since f has no Taylor's expansion near 0. However, analyzing the sign of $f'(x) = \frac{2}{x^3} e^{-\frac{1}{x^2}}$ left and right of zero, we see that $f(0)$ is a global minimum.

3.7.1   Local extrema for functions in several variables

Definition 3.7.5. Let $f : \Omega \subseteq \mathbb{R}^n \to \mathbb{R}$ with Ω open. We say that f has a *local maximum* (resp. *minimum*) at a point $a \in \Omega$, if $f(x) \leq f(a)$ (resp. $f(x) \geq f(a)$) for all x in some neighborhood of a. Local maxima and local minima are referred to as *local extrema*. Moreover, if $f(x) \leq f(a)$ (resp. $f(x) \geq f(a)$) for all $x \in \Omega$, then we say that f has a *global maximum* (resp. *minimum*) on Ω at a.

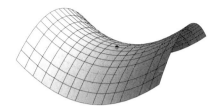

Figure 3.3: Saddle Point

Just as in the one variable case, we have

Theorem 3.7.6. *If f has a local extremum at $a \in \Omega$ and f is differentiable at a, then $\nabla f(a) = 0$ (i.e., the tangent hyperplane at a is flat)*

Proof. Let $v \neq 0$ and $t \in \mathbb{R}$ small enough so that $\gamma(t) = a + tv \in \Omega$. Then $g(t) = f(\gamma(t))$ is a differentiable real-valued function of t, which has a local extremum at $t = 0$. Therefore, $g'(0) = 0$. That is, $D_v f(a) = \langle \nabla f(a)v, v \rangle = 0$ for all $v \neq 0$. Hence $\nabla f(a) = 0$. $\square$

Definition 3.7.7. A point $a \in \Omega$ at which $\nabla f(a) = 0$ is called a *critical point* of f.

Theorem 3.7.6 asserts that local extrema are critical points. However, at a critical point a function could have a local maximum or a local minimum, or neither.

Definition 3.7.8. A critical point a is said to be *non degenerate* if $\det(H_f(a)) \neq 0$. A non degenerate critical point which is not a local extremum is called a *saddle point*. Figure 3.3 illustrates this situation.

We need to be able to determine whether or not a function has a local maximum or a local minimum at a non degenerate critical point. For $f \in C^2(\Omega)$, we shall apply the second order Taylor expansion to obtain sufficient conditions for a local extremum (local maximum, local minimum or saddle point). In this connection, we need to define the notions of positive definite, negative definite and indefinite quadratic form.

Definition 3.7.9. Let $A = (a_{ij})$ be a symmetric matrix. The *quadratic form* associated with A is the real-valued function $Q : \mathbb{R}^n \to \mathbb{R}$ defined by

$$Q(v) = \langle Av, v \rangle = \sum_{i=1}^{n} \sum_{j=1}^{n} a_{ij} v_i v_j,$$

which is a homogeneous quadratic polynomial in the variables $v_1, ..., v_n$. Q is said to be *non degenerate*, if $\det(A) \neq 0$

For instance, when $n = 2$ and $A = \begin{pmatrix} \alpha & \beta \\ \beta & \gamma \end{pmatrix}$ with $v = (x, y)$ the quadratic form is $Q(v) = Q(x, y) = \alpha x^2 + 2\beta xy + \gamma y^2$ and Q is non degenerate if $\alpha\gamma - \beta^2 \neq 0$.

Definition 3.7.10. Let $Q : \mathbb{R}^n \to \mathbb{R}$ be a quadratic form. Then we say that

1. Q is *positive definite*, if $Q(v) > 0$ for all $v \in \mathbb{R}^n$, with $v \neq 0$.

2. Q is *negative definite* if $Q(v) < 0$) for all $v \in \mathbb{R}^n$, with $v \neq 0$.

3. Q is *indefinite*, if $Q(u) > 0 > Q(v)$ for some $v, u \in \mathbb{R}^n$, with $v \neq 0$ and $u \neq 0$.

For example, $Q(x, y) = x^2 + y^2$ is positive definite, $Q(x, y) = -x^2 - y^2$ is negative definite, $Q(x, y) = x^2 - y^2$ and $Q(x, y) = xy$ are indefinite. Of course, Q is negative definite if and only if $-Q$ is positive definite.

Now, let $f : \Omega \subseteq \mathbb{R}^n \to \mathbb{R}$ be a C^2 function on the open set Ω. The behavior of f near a non degenerate critical point a is determined by the quadratic form $Q(v) = \langle H_f(a)v, v \rangle$ associated with the Hessian $H_f(a)$.

Theorem 3.7.11. *(Second derivative test) Let $f : \Omega \subseteq \mathbb{R}^n \to \mathbb{R}$ be $C^2(\Omega)$ and suppose $a \in \Omega$ is a non degenerate critical point of f. Let Q be the quadratic form associated with $H_f(a)$. Then*

1. *If Q is positive definite, f has a (strict) local minimum at a.*

2. *If Q is negative definite, f has a (strict) local maximum at a.*

3. *If Q is indefinite, f has a saddle point at a.*

Proof. Since $\frac{\partial^2 f}{\partial x_i \partial x_j}(x)$ are continuous at a, it follows from Corollary 2.3.4 that the function $x \mapsto H_f(x)$ from Ω to $\mathbb{R}^{n^2}$ is continuous at a. Because the inner product is continuous, for each $v \in \mathbb{R}^n$ and $x \in \Omega$ the function $x \mapsto \langle H_f(x)v, v \rangle$ is also continuous at a. Hence for fixed v there is $\delta > 0$ such that for $x \in B_\delta(a)$ the function $\langle H_f(x)v, v \rangle$ has the same sign as $\langle H_f(a)v, v \rangle$. Now for every $x \in B_\delta(a)$, as $\nabla f(a) = 0$, the second order Taylor expansion of f yields

$$f(x) - f(a) = \frac{1}{2}\langle H_f(c)(x - a), x - a \rangle,$$

for some $c \in B_\delta(a)$, where we used the convexity of $B_\delta(a)$. Hence, for $x \neq a$ we have: (1) If Q is positive definite, then $f(x) > f(a)$, and f has a (strict) local minimum at a. (2) If Q is negative definite, then f has a (strict) local maximum at a. Finally, if Q is indefinite, there exist nonzero vectors $u, v \in \mathbb{R}^n$ with $Q(u) > 0$ and $Q(v) < 0$. Since $Q(u) > 0$, as above there is $\delta_1 > 0$ such that $\langle H_f(x)u, u \rangle > 0$, for all $x \in B_{\delta_1}(a)$. Furthermore, as $u \neq 0$, we can choose $s_1 > 0$ sufficiently small, so small that $a + s_1 u \in B_{\delta_1}(a)$. Again by convexity, $\gamma_1(t) = a + t(s_1 u) \in B_{\delta_1}(a)$, for $t \in [0, 1]$. Now, for $x \in \gamma_1(t)$, by Taylor's expansion there is $c_1 \in B_{\delta_1}(a)$ such that

$$f(x) - f(a) = \frac{1}{2}\langle H_f(c_1)(x - a), x - a \rangle = \frac{1}{2}\langle H_f(c_1)(ts_1 u), ts_1 u \rangle$$

$$= \frac{1}{2}t^2 s_1{}^2 \langle H_f(c_1)u, u \rangle > 0.$$

Hence, $f(x) - f(a) > 0$ for all $x \in B_{\delta_1}(a)$ parallel to u. Similarly, since $Q(v) < 0$, there is $\delta_2 > 0$ such that $f(x) - f(a) < 0$ for all $x \in B_{\delta_2}(a)$ parallel to v. Taking $\delta = \min\{\delta_1, \delta_2\}$, we see that for $x \in B_\delta(a)$, the sign of the difference $f(x) - f(a)$ changes as we change direction. This automatically means that there is no local extremum at a. That is, a is a saddle point. $\qquad\square$

Note that when $n = 1$, v is just a nonzero real number, $H_f(a) = f''(a)$ and $Q(v) = f''(a)v^2$. So (since $v^2 > 0$), the sign of Q is the same as the sign of $f''(a)$, so that Theorem 3.7.11 is a generalization of the familiar *second derivative test* of one variable calculus.

In studying maxima and minima, when $n = 2$, it is easy to sort out the various cases.

Theorem 3.7.12. *(Second derivative test: $n = 2$) Let $\Omega \subseteq \mathbb{R}^2$ be open and $f : \Omega \to \mathbb{R}$ be $C^2(\Omega)$. Suppose $a \in \Omega$ is a non degenerate critical point of f. Let $\alpha = \frac{\partial^2 f}{\partial x^2}(a)$, $\beta = \frac{\partial^2 f}{\partial x \partial y}(a)$ and $\gamma = \frac{\partial^2 f}{\partial y^2}(a)$. Then*

1. *If $\alpha\gamma - \beta^2 > 0$ and $\alpha > 0$, f has a local minimum at a.*

2. *If $\alpha\gamma - \beta^2 > 0$ and $\alpha < 0$, f has a local maximum at a.*

3. *If $\alpha\gamma - \beta^2 < 0$, f has a saddle point at a.*

Proof. The Hessian matrix of f at a is $H_f(a) = \begin{pmatrix} \alpha & \beta \\ \beta & \gamma \end{pmatrix}$ and its associated quadratic form is $Q(x, y) = \alpha x^2 + 2\beta xy + \gamma y^2$. Regarding $Q(x, y)$ as a quadratic polynomial $\varphi(x) = Ax^2 + Bx + C$ in one variable, say x, for *any* but fixed nonzero $y \subset \mathbb{R}$ with coefficients $A = \alpha$, $B = 2\beta y$ and $C = \gamma y^2$, the discriminant of φ is

$$B^2 - 4AC = 4y^2(\beta^2 - \alpha\gamma).$$

Now, $\varphi(x) > 0$ (resp. $\varphi(x) < 0$) for all $x \in \mathbb{R}$ if and only if $\alpha\gamma - \beta^2 > 0$ and $\alpha > 0$ (resp. $\alpha < 0$). Hence if $\alpha\gamma - \beta^2 > 0$ and $\alpha > 0$ (resp. $\alpha < 0$), the quadratic form $Q(x, y)$ is positive definite (resp. negative definite) and so by Theorem 3.7.11 f has a local minimum (resp. local maximum) at a. Furthermore, if $\alpha\gamma - \beta^2 < 0$, the discriminant of φ is positive. Hence, $\varphi(x)$ does not have a fixed sign for all $x \in \mathbb{R}$. Therefore $Q(x, y)$ is indefinite and so f has a saddle point at a. $\square$

Example 3.7.13. Find the critical points of $f(x, y) = x^2y - x^2 - y^2 + y + 5$ and determine their nature.

Solution. We have

$$\frac{\partial f}{\partial x}(x, y) = 2xy - 2x = 0$$

$$\frac{\partial f}{\partial y}(x, y) = x^2 - 2y + 1 = 0$$

From the first equation we get $x = 0$ or $y = 1$ and substituting these into the second equation we obtain $y = \frac{1}{2}$ or $x = \pm 1$ respectively. Hence, there are three critical points $(0, \frac{1}{2})$, $(1,1)$ and $(-1,1)$. To determine whether these are local maxima, local minima, or saddle points we apply Theorem 3.7.12. The second order partial derivatives are

$$\frac{\partial^2 f}{\partial x^2}(x,y) = 2y - 2, \ \frac{\partial^2 f}{\partial x \partial y}(x,y) = 2x \text{ and } \frac{\partial^2 f}{\partial y^2}(x,y) = -2.$$

At $(0, \frac{1}{2})$, we have $\alpha = \frac{\partial^2 f}{\partial x^2}(0, \frac{1}{2}) = -1$, $\beta = \frac{\partial^2 f}{\partial x \partial y}(0, \frac{1}{2}) = 0$ and $\gamma = \frac{\partial^2 f}{\partial y^2}(0, \frac{1}{2}) = -2$. Since $\alpha\gamma - \beta^2 = 2 > 0$ and $\alpha = -1 < 0$, f has a local maximum at $(0, \frac{1}{2})$. $(f(0, \frac{1}{2}) = \frac{21}{4})$. At $(\pm 1, 1)$, $\alpha = \frac{\partial^2 f}{\partial x^2}(\pm 1, 1) = 0$, $\beta = \frac{\partial^2 f}{\partial x \partial y}(\pm 1, 1) = \pm 2$ and $\gamma = \frac{\partial^2 f}{\partial y^2}(\pm 1, 1) = -2$. Since $\alpha\gamma - \beta^2 = -4 < 0$, both points are saddle points.

The question whether a local maximum (resp. minimum) of a differentiable function f on set Ω in $\mathbb{R}^n$ is a *global maximum* (resp. *minimum*) depends, of course, on the set Ω and the function itself. From Weierstrass's Extreme Value theorem (Corollary 2.5.3) we know that if Ω is compact, then f attains its maximum and minimum values on Ω. If an extreme value occurs at an interior point of Ω, that point must be a critical point of f. To find the candidates for extreme values on the boundary of Ω, we can apply the methods for solving extremal problems with constraints discussed in Section 3.9 below (see, extrema over a closed and bounded set). If Ω is unbounded, there is no longer any guarantee that a global maximum or minimum will exist. In this situation, one procceds on a case-by-case basis.

Example 3.7.14. Let $f(x,y) = x^2 e^{-(x^4+y^2)}$. Show that f has a global minimum and maximum on $\mathbb{R}^2$, and find them.

Solution. Clearly $f(x,y) \geq 0$ for all $(x,y) \in \mathbb{R}^2$, and $f(0,y) = 0$, so the minimum is zero, attained at all points on the y-axis. Furthermore, since $e^{-y^2} \leq 1$ and an application of L'Hopital's Rule tells us $\lim_{x \to \infty} x^2 e^{-x^4} = 0$, it follows that $f(x,y) \to 0$ as $||(x,y)|| \to \infty$. Hence any maximum occurs in a bounded region of the plane. To find it, we find the critical points.

$$\frac{\partial f}{\partial x}(x,y) = e^{-y^2}(2xe^{-x^4} - 4x^5 e^{-x^4} e^{-x^4}) = 2xe^{-x^4-y^2}(1 - 2x^4) = 0$$

$$\frac{\partial f}{\partial y}(x,y) = -2x^2 y e^{-x^4-y^2} = 0$$

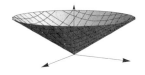

Figure 3.4: Cone: $z = \sqrt{x^2 + y^2}$

Solving for x and y we see that the critical points are: $(0, y)$ and $(\pm\frac{1}{\sqrt[4]{2}}, 0)$. Evaluating we see $f(\pm\frac{1}{\sqrt[4]{2}}, 0) = \frac{1}{\sqrt{2}}e^{-\frac{1}{2}}$, which is the maximum value.

Example 3.7.15. Find the minimum distance from the point $(1, 2, 0)$ to the surface of the cone $z^2 = x^2 + y^2$, $z \geq 0$. (See, Figure 3.4).

Solution. Let (x, y, z) be any point on the surface of the cone. The distance from the point $(1, 2, 0)$ to (x, y, z) is $d = \sqrt{(x-1)^2 + (y-2)^2 + z^2}$. We want to minimize d. To simplify the calculations, since $d > 0$, it suffices to minimize $d^2 = (x-1)^2 + (y-2)^2 + z^2$. Substituting $z^2 = x^2 + y^2$ in d^2, we obtain the function $f(x, y) = 2x^2 + 2y^2 - 2x - 4y + 5$. A short calculation that we leave to the reader shows that the only critical point of f is $(\frac{1}{2}, 1)$ and the Hessian $H_f(x, y) = \begin{pmatrix} 4 & 0 \\ 0 & 4 \end{pmatrix}$. Hence $\alpha\gamma - \beta^2 = 16 > 0$, $\alpha > 4$, and so at $(\frac{1}{2}, 1)$ f has a local minimum. Since $f > 0$, it is a global minimum, and the minimum distance is $d = \frac{\sqrt{10}}{2}$.

A simple and useful application of these techniques in Statistics is to the problem of fitting the "best" straight line to a set of data.

Example 3.7.16. (*Method of least squares*). Given n distinct numbers $x_1, ..., x_n$ and further numbers $y_1, ..., y_n$ (not necessarily distinct), it is generally impossible to find a straight line $f(x) = ax + b$ which passes through all the points (x_i, y_i), that is, $f(x_i) = y_i$ for each $i = 1, ...n$. However, we can seek a linear function which minimizes the "total square error"

$$q = q(a, b) = \sum_{i=1}^{n} [f(x_i) - y_i]^2 = \sum_{i=1}^{n} [(ax_i + b) - y_i]^2.$$

Determine values of a and b which do this.

Solution. The critical points are solutions of the system

$$\frac{\partial q}{\partial a} = \sum_{i=1}^{n} 2(ax_i + b - y_i)x_i = 0$$

and

$$\frac{\partial q}{\partial b} = \sum_{i=1}^{n} 2(ax_i + b - y_i) = 0.$$

Expanding these equations we get

$$a \sum_{i=1}^{n} x_i^2 + b \sum_{i=1}^{n} x_i = \sum_{i=1}^{n} x_i y_i$$

$$a \sum_{i=1}^{n} x_i + nb = \sum_{i=1}^{n} y_i.$$

Since $x_1, ..., x_n$ are distinct, the determinant of the coefficient matrix of this system is nonzero. Hence solving this system by using determinants or the method of elimination, we find

$$a = \frac{n \left(\sum_{i=1}^{n} x_i y_i \right) - \left(\sum_{i=1}^{n} x_i \right) \left(\sum_{i=1}^{n} y_i \right)}{n \left(\sum_{i=1}^{n} x_i^2 \right) - \left(\sum_{i=1}^{n} x_i \right)^2}$$

and

$$b = \frac{\sum_{i=1}^{n} y_i - a \sum_{i=1}^{n} x_i}{n}.$$

Moreover, this *unique* critical point is a local minimum. To see this, we apply the second derivative test. Indeed, $\alpha = \frac{\partial^2 q}{\partial a^2} = 2 \sum_{i=1}^{n} x_i^2 > 0$, $\beta = \frac{\partial^2 q}{\partial a \partial b} = 2 \sum_{i=1}^{n} x_i$, and $\gamma = \frac{\partial^2 q}{\partial b^2} = 2n$. The Cauchy-Schwarz inequality $\left(\sum_{i=1}^{n} x_i \right)^2 \leq \left(\sum_{i=1}^{n} x_i^2 \right) \left(\sum_{i=1}^{n} 1^2 \right) = n \left(\sum_{i=1}^{n} x_i^2 \right)$ tells us $\alpha \gamma - \beta^2 > 0$ (since all the x_i are distinct, equality can not hold in the Cauchy-Schwarz inequality. This is also the reason that the denominator of a is nonzero.)

In general, it is not easy to determine directly from the definition, whether a quadratic form is positive (resp. negative) definite or indefinite. The full story involves a certain amount of linear algebra. In fact, the symmetric matrix A that defines the quadratic form Q can be used to obtain criteria for the quadratic form to be positive (negative) definite or indefinite. One of these criteria involves the eigenvalues of A and assumes the *Spectral theorem,* which states that *a symmetric matrix A is diagonalizable, or equivalently, $\mathbb{R}^n$ has an orthonormal basis of real eigenvectors of A.* (Theorem D.0.5) With this result at hand, we arrive at the following theorem which gives the precise character of a quadratic form in terms of the eigenvectors and eigenvalues.

Theorem 3.7.17. *(Principal axis theorem). Let A be a symmetric matrix and $Q(v) = \langle Av, v \rangle$ its associated quadratic form. Then there is a basis $\{u_1, ..., u_n\}$ of $\mathbb{R}^n$ and real numbers $\lambda_1, ..., \lambda_n$ such that if $v = y_1 u_1 + ... + y_n u_n$, the value of Q is given by the formula*

$$Q(v) = \lambda_1 y_1^2 + ... + \lambda_n y_n^2$$

Proof. By the Spectral theorem, we may find an orthonormal basis $\{u_1, ..., u_n\}$ of $\mathbb{R}^n$ composed of eigenvectors of A. Let $\lambda_1, ..., \lambda_n$ be the corresponding eigenvalues. For $v \in \mathbb{R}^n$, if $v = y_1 u_1 + ... + y_n u_n$, then

$$Q(v) = \langle Av, v \rangle = \langle A(\sum_{i=1}^{n} y_i u_i), \sum_{i=1}^{n} y_i u_i \rangle = \langle \sum_{i=1}^{n} A(y_i u_i), \sum_{i=1}^{n} y_i u_i \rangle$$

$$= \sum_{i=1}^{n} \sum_{j=1}^{n} \langle y_i \lambda_i u_i, y_j u_j \rangle = \sum_{i=1}^{n} \sum_{j=1}^{n} \lambda_i y_i^2 \langle u_i, u_j \rangle = \sum_{i=1}^{n} \lambda_i y_i^2$$

since $u_1, ..., u_n$ are orthonormal.   □

A basis with this property is called a *canonical basis* and the expression

$$Q(v) = \lambda_1 y_1^2 + ... + \lambda_n y_n^2$$

is called a *canonical form* of Q.

Corollary 3.7.18. *Let $\lambda_1, ..., \lambda_n$ be the eigenvalues of A. Then the quadratic form $Q(v) = \langle Av, v \rangle$ is*

1. *positive definite if and only if $\lambda_i > 0$, for all $i = 1, ..., n$.*

2. *negative definite if and only if $\lambda_i < 0$, for all $i = 1, ..., n$.*

3. *indefinite if and only if for some i, j, $\lambda_i > 0 > \lambda_j$.*

Remark 3.7.19. We give an alternative proof of Theorem 3.7.12 on the basis of Corollary 3.7.18. The Hessian in Theorem 3.7.12 is $H_f(a) = \begin{pmatrix} \alpha & \beta \\ \beta & \gamma \end{pmatrix}$. Suppose λ_1, λ_2 are the eigenvalues of $H_f(a)$. The spectral theorem tells us that $H_f(a)$ is similar to the diagonal matrix $\begin{pmatrix} \lambda_1 & 0 \\ 0 & \lambda_2 \end{pmatrix}$. Then

$$\det(H_f(a)) = \alpha\gamma - \beta^2 = \lambda_1\lambda_2 \quad \text{and} \quad \operatorname{tr}(H_f(a)) = \alpha + \gamma = \lambda_1 + \lambda_2.$$

Suppose $\alpha\gamma - \beta^2 > 0$. Then λ_1, λ_2 have the same sign. If $\alpha > 0$, then $\alpha\gamma - \beta^2 > 0$ implies $\gamma > \frac{\beta^2}{\alpha} > 0$. Hence, $\lambda_1 + \lambda_2 = \alpha + \gamma > 0$ and so both eigenvalues λ_1, λ_2 are positive. Therefore, by Corollary 3.7.18, the quadratic form associated to $H_f(a)$ is positive definite and so f has a local minimum at a. If $\alpha < 0$, then $\alpha\gamma - \beta^2 > 0$ implies $\gamma < \frac{\beta^2}{\alpha} < 0$. Hence, $\lambda_1 + \lambda_2 = \alpha + \gamma < 0$ and so both λ_1, λ_2 are negative. Therefore, the quadratic form asociated to $H_f(a)$ is negative definite and so f has a local maximum at a. Finally, suppose $\alpha\gamma - \beta^2 < 0$. Then $\lambda_1\lambda_2 < 0$. Hence, the two eigenvalues have opposite sign. Therefore, the quadratic form asociated to $H_f(a)$ is indefinite and so f has a saddle point at a.

Another more general criterion, to determine the nature of a quadratic form, involving determinants is given in the following theorem. Let $A = (a_{ij})$ be a symmetric matrix with $\det(A) \neq 0$. For $k = 1, ...n$, we denote by A_k the $k \times k$ submatrix of A formed by the first k rows and columns of A, that is,

$$A_k = \begin{pmatrix} a_{11} & ... & a_{1k} \\ . & ... & . \\ . & ... & . \\ . & ... & . \\ a_{k1} & ... & a_{kk} \end{pmatrix}.$$

Thus, $A_1 = a_{11}$, $A_2 = \begin{pmatrix} a_{11} & a_{12} \\ a_{21} & a_{22} \end{pmatrix}$, ..., $A_n = A$. The submatrix A_k is called a *principal submatrix* of A and the determinants $\det(A_k)$, for $k = 1, ..., n$, the *principle minors* of A.

Theorem 3.7.20. *(Sylvester's criterion[7]). Let $A = (a_{ij})$ be a symmetric matrix with $\det(A) \neq 0$ and $Q(v) = \langle Av, v \rangle$ the quadratic form associated with A. Then*

1. *Q is positive definite if and only if $\det(A_k) > 0$, for all $k = 1, ..., n$.*

2. *Q is negative definite if and only if $(-1)^k \det(A_k) > 0$, for all $k = 1, ..., n$.*

3. *Q is indefinite if neither (1) nor (2) hold.*

Proof.   1. Suppose Q is positive definite. Let $Q_k(u) = \langle A_k u, u \rangle$, with $u = (x_1, ..., x_k) \in \mathbb{R}^k$, $u \neq 0$. Consider the vector $v = (x_1, ..., x_k, 0, ..., 0) \in \mathbb{R}^n$. Then

$$Q_k(u) = \langle A_k u, u \rangle = \langle Av, v \rangle = Q(v).$$

So Q_k is positive definite on $\mathbb{R}^k$. Therefore from Corollary 3.7.18, all eigenvalues $\mu_1, ..., \mu_k$ of A_k are positive. Since A_k is symmetric it is diagonalizable with diagonal entries its eigenvalues. Hence $\det(A_k) = \mu_1 \cdots \mu_k > 0$, for all $k = 1, ..., n$.

Conversely, suppose $\det(A_k) > 0$, for all $k = 1, ..., n$. We use induction on n. For $n = 1$, the result is obvious. Assume it is true for $k = n - 1$, that is, assume that all matrices $A_1, ..., A_{n-1}, A_n$ have positive determinant and Q_{n-1} is positive definite. We show that $Q_n = Q$ is positive definite. Let $\{w_1, ..., w_n\}$ be an orthonormal basis of eigenvectors of A belonging to eigenvalues $\lambda_1, ..., \lambda_n$ (such a basis exists by the Spectral theorem). By rearranging

---

[7]J. Sylvester (1814-1897). He made fundamental contributions to matrix theory, invariant theory, number theory, partition theory and combinatorics. He played a leadership role in American mathematics in the later half of the 19th century as a professor at the Johns Hopkins University and as founder of the *American Journal of Mathematics*.

the basis elements we may asuume $\lambda_1 \leq \lambda_2 \leq ... \leq \lambda_n$. Since $\det(A_n) = \det(A) = \lambda_1 \cdots \lambda_n$, the hypothesis implies

$$\lambda_1 \cdots \lambda_n > 0.$$

If all $\lambda_i > 0$, by Corollary 3.7.18 we are done. If not, an *even* number of them are negative, in particular, at least two of them are negative, say, $\lambda_1 < 0$ and $\lambda_2 < 0$. Consider the vector $v = c_1 w_1 + c_2 w_2$, with $c_1, c_2 \in \mathbb{R}$. Then

$$Q(v) = \langle Av, v \rangle = \langle A(c_1 w_1 + c_2 w_2), c_1 w_1 + c_2 w_2 \rangle = c_1^2 \lambda_1 + c_2^2 \lambda_2 < 0.$$

On the other hand suppose $w_1 = (x_1, ..., x_n)$ and $w_2 = (y_1, ... y_n)$. If $x_n \neq 0$ take $c_1 = y_n$ and $c_2 = -x_n$. Then $v = c_1 w_1 + c_2 w_2 = (c_1 x_1 + c_2 y_1, ..., c_1 x_{n-1} + c_2 y_{n-1}, 0)$. If $x_n = 0$, take $c_1 = 1$ and $c_2 = 0$. Then $v = c_1 w_1 + c_2 w_2 = (x_1, ..., x_{n-1}, 0)$. Since w_1, w_2 are linearly independent, it follows that $v \neq 0$ and is of the form $v = (z_1, ..., z_{n-1}, 0)$. Now the vector $v_0 = (z_1, ..., z_{n-1}) \in \mathbb{R}^{n-1}$ is nonzero and since Q_{n-1} is positive definite on $\mathbb{R}^{n-1}$

$$0 < Q_{n-1}(v_0) = \langle A_{n-1} v_0, v_0 \rangle = \langle Av, v \rangle = Q(v),$$

which contadicts $Q(v) < 0$. Thus, Q can not have negative eigenvalues and hence it is positive definite.

2. Q is negative definite if and only if $-Q$ is positive definite. By part (1) this mean $\det(-A_k) > 0$, or equivalently $(-1)^k \det(A_k) > 0$, for all $k = 1, ..., n$.

3. Since $\det(A) = \lambda_1 \cdots \lambda_n \neq 0$ and the eigenvalues can not be all positive nor all negative, some are positive and some are negative. Hence, by Corollary 3.7.18, Q is indefinite.

$\square$

Note that when $n = 2$, Theorem 3.7.20 gives automatically the conditions of Theorem 3.7.12.

Example 3.7.21. Let $f(x, y, z) = x^2 + y^2 + z^2 - xy + x - 2z$. Find the critical points of f and determine their nature.

Solution. We have

$$\frac{\partial f}{\partial x}(x,y,z) = 2x - y + 1 = 0$$

$$\frac{\partial f}{\partial y}(x,y,z) = 2y - x = 0$$

$$\frac{\partial f}{\partial z}(x,y,z) = 2z - 2 = 0$$

Solving the system we obtain $(-\frac{2}{3}, -\frac{1}{3}, 1)$, the only critical point of f. To determine its nature, we apply Theorem 3.7.20. The Hessian matrix of f at any point (x,y,z) is

$$H_f(x,y,z) = \begin{pmatrix} 2 & -1 & 0 \\ -1 & 2 & 0 \\ 0 & 0 & 2 \end{pmatrix}. \text{ In particular, } H_f(-\tfrac{2}{3}, -\tfrac{1}{3}, 1) =$$

$$\begin{pmatrix} 2 & -1 & 0 \\ -1 & 2 & 0 \\ 0 & 0 & 2 \end{pmatrix}.$$

Since $a_{11} = 2 > 0$, $\det \begin{pmatrix} 2 & -1 \\ -1 & 2 \end{pmatrix} = 3 > 0$, and $\det \begin{pmatrix} 2 & -1 & 0 \\ -1 & 2 & 0 \\ 0 & 0 & 2 \end{pmatrix} =$

$6 > 0$, $H_f(-\frac{2}{3}, -\frac{1}{3}, 1)$ is positive definite. Hence f has local minimum at $(-\frac{2}{3}, -\frac{1}{3}, 1)$.

3.7.2   Degenerate critical points

Let $f : \Omega \subseteq \mathbb{R}^n \to \mathbb{R}$ and suppose $a \in \Omega$ is a *degenerate* critical point, i.e., a critical point for f with $\det(H_f(a)) = 0$, or equivalently some eigenvalue of $H_f(a)$ is zero. Then Theorem 3.7.11 does not apply. If one wishes to analyze the behavior of f at a, it would be necessary to look at higher derivatives. On some occassions *ad hoc* methods also apply, for example, by investigating cross sections of the function in a small neighborhood of the critical point. In particular, when $n = 2$ one can use one variable calculus to determine the behavior of the critical point on the cross sections.

Here are some examples of what can happen when a critical point is degenerate.

Example 3.7.22. Let $f(x,y) = x^2 + y^4$ and $g(x,y) = x^2 - y^4$. The point $(0,0)$ is a critical point for both functions. Moreover, $H_f(0,0) =$

$H_g(0,0) = \begin{pmatrix} 2 & 0 \\ 0 & 0 \end{pmatrix}$ and the associated quadratic form $Q(x,y) = 2x^2$

is the same for both f and g and, of course, $\det H_f(0,0) = 0 = \det H_g(0,0)$. Now, $f(0,0) = 0$, but near $(0,0)$ $f(x,y) = x^2 + y^4 > 0$. Thus, f has a local (actually global) minimum at $(0,0)$. On the other hand, $g(0,0) = 0$ but along the x-axis $g(x,0) = x^2 > 0$ and along the y-axis $g(0,y) = -y^4 < 0$. Thus, g has a saddle point at $(0,0)$

Example 3.7.23. Let $f(x,y) = 3x^4 - 4x^2y + y^2$. The origin $(0,0)$ is a degenerate critical point at which f has a local minimum along every line $y = mx$, but f does not have a local minimum in any neighborhood of $(0,0)$.

Solution. Here

$$\frac{\partial f}{\partial x}(x,y) = 12x^3 - 8xy = 0$$
$$\frac{\partial f}{\partial y}(x,y) = -4x^2 = 2y = 0$$

Solving the system we find that $(0,0)$ is a (the only) critical point of f. The Hessian of f at $(0,0)$ is $H_f(0,0) = \begin{pmatrix} 0 & 0 \\ 0 & 2 \end{pmatrix}$. Since its determinant is zero, the origin is a degenerate critical point of f.

The restriction of f on any line $y = mx$ is the (one variable) function $g(x) = f(x,mx) = 3x^4 - 4mx^3 + m^2x^2$. Differentiating we see that $g'(x) = 12x^3 - 12mx^2 + 2m^2x$ and $g''(x) = 36x^2 - 24mx + 2m^2$. Clearly, $x = 0$ is a critical point of g and $g''(0) = 2m^2 > 0$ for $m \neq 0$. When $m = 0$, $g(x) = 3x^4$ which has a minimum at $x = 0$. Thus, at $(0,0)$ f has a local minimum along any straight line through the origin. However, f does not have a local minimum at the origin. To see this, first note that $f(x,y) = (x^2 - y)(3x^2 - y)$ and $f(0,0) = 0$. Taking x arbitrarily small and considering the restriction of f along the parabola $y = 2x^2$ we find $f(x,y) = -x^4 < 0$. On the other hand for the restriction of f along the parabola $y = 4x^2$ we find $f(x,y) = 3x^4 > 0$. Thus there is no local extremum at the origin. In fact, $f(x,y) < 0$ between the parabolas $y = x^2$ and $y = 3x^2$ and $f(x,y) > 0$ elsewhere.

However, when Ω is a bounded domain in $\mathbb{R}^2$, $f : \overline{\Omega} \to \mathbb{R}$ is continous on $\overline{\Omega}$ and $\left(\frac{\partial^2 f}{\partial x^2}\right)\left(\frac{\partial^2 f}{\partial y^2}\right) - \left(\frac{\partial^2 f}{\partial x \partial y}\right)^2 \leq 0$ on Ω, we can describe the over all behavior of f on $\overline{\Omega}$. To simplify the notation, we use subscript notation for the partial derivatives of f.

Theorem 3.7.24. *Let $\Omega \subseteq \mathbb{R}^2$ be open and bounded, $f \in C^2(\Omega)$ and extends continuously on $\partial(\Omega)$. If $f_{xx}f_{yy} - f_{xy}^2 \leq 0$ on Ω, then the maximum and minimum of f in are attained on $\partial(\Omega)$.*

Proof. Since $\overline{\Omega}$ is compact and f is continuous on $\overline{\Omega}$ both the maximum and the minimum of f on $\overline{\Omega}$ exist. We give the proof for the maximum. The result then follows for the minimum by considering $-f$. Let $M = \max_{v \in \overline{\Omega}} f(v)$. Suppose this maximum is not assumed on $\partial(\Omega)$. Then there is a point $a = (x_0, y_0) \in \Omega$ such that $f(a) = M > \max_{v \in \partial(\Omega)} f(v)$. Hence, $f(a) > \max_{v \in \partial(\Omega)} f(v) + \delta$ for some small $\delta > 0$. Let $\epsilon > 0$ be so small that $\epsilon ||v - a||^2 \leq \delta$, for all $v = (x, y) \in \overline{\Omega}$ and consider the function $g(v) = f(v) + \epsilon ||v - a||^2$. Then

$$g(a) = f(a) > \max_{v \in \partial(\Omega)} f(v) + \delta > \max_{v \in \partial(\Omega)} f(v) + \epsilon ||v - a||^2 \geq \max_{v \in \partial(\Omega)} g(v).$$

Therefore the maximum of g on $\overline{\Omega}$ is not attained on $\partial(\Omega)$. So, there is a point $b = (x_1, y_1) \in \Omega$ such that $g(b)$ is the maximum of g on $\overline{\Omega}$. In particular, $g(b)$ is a local maximum along each axis. Therefore, $g_x(b) = g_y(b) = 0$ and $g_{xx}(b) \leq 0$, $g_{yy}(b) \leq 0$. At the same time by direct computation we see $g_{xx}(v) = f_{xx}(v) + 2\epsilon$, $g_{yy}(v) = f_{yy}(v) + 2\epsilon$, and $g_{xy}(v) = f_{xy}(v)$. So at $v = b$, we have

$$g_{xx}g_{yy} - g_{xy}^2 = (f_{xx} + 2\epsilon)(f_{yy} + 2\epsilon) - f_{xy}^2$$

$$= f_{xx}f_{yy} - f_{xy}^2 + 2\epsilon(f_{xx} + f_{yy}) + 4\epsilon^2$$

$$\leq f_{xx}f_{yy} - f_{xy}^2 + 2\epsilon(-2\epsilon - 2\epsilon) + 4\epsilon^2$$

$$= f_{xx}f_{yy} - f_{xy}^2 - 4\epsilon^2 < 0.$$

Thus, $g_{xx}(b)g_{yy}(b) - g_{xy}^2(b) < 0$ and b is a saddle point for g, a contradition.

If $f_{xx}f_{yy} - f_{xy}^2 < 0$ on Ω, then all critical points of f in Ω would be saddle points and so automatically the maximum and minimum of f will occur on $\partial(\Omega)$. $\qquad\qquad\Box$

Theorem 3.7.24 has some important consequences.

Corollary 3.7.25. *(Maximum-minimum principle).* *Let Ω be a bounded domain in $\mathbb{R}^2$, $f \in C^2(\Omega)$ and extends continuously to $\partial(\Omega)$. If f is harmonic on Ω, then f assumes its maximum and minimum values on $\partial(\Omega)$. In particular, if f is harmonic on Ω and $f \equiv 0$ on $\partial(\Omega)$, then $f \equiv 0$ on Ω.*

Proof. Since f is harmonic, $f_{xx} + f_{yy} = 0$ on Ω, $f_{xx}f_{yy} - f_{xy}^2 \leq 0$ on Ω.   □

Corollary 3.7.26. *Let Ω be a bounded domain in $\mathbb{R}^2$. A harmonic function on Ω is uniquely determined by its values on $\partial(\Omega)$.*

Proof. Suppose f_1, f_2 are both harmonic on Ω and $f_1 = f_2$ on $\partial(\Omega)$. Then, $f = f_1 - f_2$ is harmonic on Ω, continuous and $f \equiv 0$ on $\partial(\Omega)$. By Corollary 3.7.25, $f \equiv 0$ on Ω.   □

Corollary 3.7.26 proves the uniqueness of the solution of the Dirichlet problem which we will study in Chapter 7 (see 7.8.3). We remark that the maximum-minimum principle (Corollary 3.7.25) generalizes easily to n-variables.

EXERCISES

In Exercises 1-14, find the critical points of each of following functions and determine their nature.

1. $f(x,y) = x^3 + y^3 - 9xy + 1$.

2. $f(x,y) = x^2 - xy + y^2 - 2x + y$.

3. $f(x,y) = (x-1)(x^2 - y^2)$.

4. $f(x,y) = x^4 + y^4 - 2(x-y)^2$.

5. $f(x,y) = x^2 y^3 (6 - x - y)$.

6. $f(x,y) = xy + \frac{1}{x} + \frac{1}{y}$.

7. $f(x,y) = (x^2 + y^2)e^{-(x^2+y^2)}$.

8. $f(x,y) = (5x + 7y - 25)e^{-x^2-xy-y^2}$.

9. $f(x,y) = \sin x + \sin y + \sin(x+y)$ inside the square $0 \le x \le \frac{\pi}{2}$, $0 \le y \le \frac{\pi}{2}$.

10. $f(x,y,z) = x^2 + y^2 + z^2 + 2xyz$.

11. $f(x,y,z) = 2x^2 + 3y^2 + 4z^2 - 3xy + 8z$.

12. $f(x,y,z) = (x+z)^2 + (y+z)^2 + xyz$.

13. $f(x,y,z) = x^3 - y^3 + z^2 - 3x + 9y$.

14. $f(x,y,z) = x^4 + y^4 + z^4 - 4xyz$.

15. Show that the function $f(x,y) = 2x + 4y - x^2 y^4$ has a critical point but no local extreme points.

16. Show that the function $f(x,y) = x^2 + xy + y^2 + \frac{a^3}{x} + \frac{a^2}{y}$ has a minimum at $\left(\frac{a}{\sqrt[3]{3}}, \frac{a}{\sqrt[3]{3}}\right)$.

17. Show that $f(x,y,z) + xyz(1-x-y-z)$ has maximum at $\left(\frac{1}{4}, \frac{1}{4}, \frac{1}{4}\right)$.

18. Show that the function $f(x,y,z) = (x+y+z)^3 - 3(x+y+z) - 24xyz + 8$ has a maximum at $(-1,-1,-1)$ and a minimum at $(1,1,1)$.

19. Show that the function $f(x,y,z) = \frac{x^2}{2} + xyz - z + y$ has no extrema.

20. Classify the critical point $\left(-1, \frac{\pi}{2}, 0\right)$ of $f(x,y,z) = x\sin z - z\sin y$.

21. Let $x \ge 0$, $y \ge 0$. Show that $\frac{x^2 + y^2}{4} \le e^{x+y-2}$.

22. (*Rolle's theorem in $\mathbb{R}^n$*). Let Ω be an open bounded set in $\mathbb{R}^n$ and let $f : \overline{\Omega} \to \mathbb{R}$ be continuous on $\overline{\Omega}$ and differentiable on Ω. Show that if $f(x) =$ constant on $\partial\Omega$, then there exist $\xi \in \Omega$ such that $\nabla f(\xi) = 0$.

23. Let $f : \mathbb{R}^n \to \mathbb{R}$ be differentiable. Suppose that f has only one critical point where f attains its local minimum. Show that when $n = 1$ this point is necessarily a global minimum. Show that $(0,0)$ is the only critical point of $f(x,y) = x^2 + y^2(1-x)^3$ where f attains its local minimum, but there is no global minimum.

24. Let $f : \mathbb{R}^2 \to \mathbb{R}$ be of class C^3. Suppose $\Delta f = f_{xx} + f_{yy} > 0$ everywhere on $\mathbb{R}^2$. Show that f cannot have a local maximum.

Answers to selected Exercises

1. Local minimum at $(3,3)$, saddle point at $(0,0)$.

2. Global minimum at $(1,0)$.

3. Local minimum at $(\frac{2}{3}, 0$, saddle point at $(0,0)$ and $(1, \pm 1)$.

4. Local minimum at $(\sqrt{2}, -\sqrt{2})$ and $(-\sqrt{2}, \sqrt{2})$, no extremum at $(0,0)$.

5. Local maximum at $(2,3)$, saddle point at $(0,0)$.

6. Local minimum at $(1,1)$.

7. Global maximum at each point of the circle $x^2 + y^2 = 1$, global minimum at $(0,0)$.

8. Global maximum at $(1,3)$, global minimum at $(-\frac{1}{26}, -\frac{3}{26})$.

9. Global maximum $\frac{3\sqrt{3}}{2}$ at $(\frac{\pi}{3}, \frac{\pi}{3})$. Global minimum 0 at $(0,0)$.

10. Local minimum at $(0,0,0)$.

11. Local minimum at $(0, 0 - 1)$.

12. No extrema.

13. Local minimum at $(1,1,1)$.

3.8 The inverse and implicit function theorems

3.8.1 The Inverse Function theorem

In Section 3.1, we saw that for a differentiable function, the derivative at a point provides a good linear approximation to the function near the point. As a result, we may expect properties of the derivative to carry over into local properties of the function. The closely related Inverse Function and Implicit Function theorems are both instances of this phenomenon. Both are of a local character and have numerous applications. Here we will deduce the Implicit Function theorem from the Inverse Function theorem (for the reverse, see Exercise 19).

For a C^1 function f from an open set Ω of $\mathbb{R}^n$ into $\mathbb{R}^n$ and $a \in \Omega$, the Inverse Function theorem states that invertibility of $D_f(a)$ implies local invertibility of f. It is a powerful existence result, showing that the condition $J_f(a) = \det(D_f(a)) \neq 0$ implies the invertibility of $f \in C^1(\Omega)$ in a neighborhood of $a \in \Omega$, even when it may be difficult or even impossible to find the local inverse function f^{-1} explicitly.[8]

It is *local invertibility* that concerns us here. The global invertibility of f is a more delicate matter. However, in the one variable case when Ω is an interval, we do get a global inverse.

Theorem 3.8.1. *(Inverse Function theorem; $n = 1$). Let $\Omega = (a,b)$ be an open interval of $\mathbb{R}$ and $f : \Omega \to \mathbb{R}$ be a $C^1(\Omega)$ with $f'(x) \neq 0$ for all $x \in \Omega$. Then f has a C^1 inverse function $f^{-1} : f(\Omega) \to \Omega$ and*

$$(f^{-1})'(y) = \frac{1}{f'(x)},$$

where $x = f^{-1}(y)$.

Proof. Since $f'(x) \neq 0$, for all $x \in \Omega$ and f' is continuous on Ω, the Intermediate Value theorem applied to f', implies that either $f'(x) > 0$ for all $x \in \Omega$, or $f'(x) < 0$ for all $x \in \Omega$. So, f is either strictly increasing or strictly decreasing on Ω. In either case, f is one-to-one on Ω. Hence, f has a global inverse. Furthermore, since f is continuous, the Intermediate Value theorem applied to f, implies that $f(\Omega)$ is also an interval. In fact, $f(\Omega)$ is an open interval, say (c,d) (or $\mathbb{R}$), for if not $f(\Omega)$ would be of the form $[c,d]$, $[c,d)$, $(c,d]$, $[c,\infty)$ or $(-\infty,d]$. In each of these cases, $f(\Omega)$ has either a largest or smallest element. So, f takes on either a maximum or a minimum at some point $x_0 \in \Omega$. But then $f'(x_0) = 0$ contradicting the hypothesis about f. Thus, $f(\Omega) = (c,d)$. By replacing f by $-f$ if necessary, we may assume that f is strictly

---

[8]The reader may observe from the proof that one can approximate the local inverse near the point $b = f(a)$ by a sequence $\{\varphi_k\}$ of succesive approximations. However we shall not pursue this.

increasing on Ω. Since f is strictly increasing on Ω, it is easy to see that its inverse f^{-1} is also strictly increasing. Moreover, f^{-1} is continuous on (c, d). To see this, let y_0 be any point in (c, d) and $x_0 \in \Omega$ with $f(x_0) = y_0$. For any $\epsilon > 0$, let $f(x_0 - \epsilon) = y_1$ and $f(x_0 + \epsilon) = y_2$. Since f is increasing, we have $y_1 < f(x_0) < y_2$. But since f^{-1} is increasing, for all y such that $y_1 < y < y_2$, we have $f^{-1}(y_1) < f^{-1}(y) < f^{-1}(y_2)$ or $x_0 - \epsilon < f^{-1}(y) < x_0 + \epsilon$, that is,

$$|f^{-1}(y_0) - f^{-1}(y)| < \epsilon.$$

Thus f^{-1} is continuous at y_0. It remains to prove that f^{-1} has a continuous derivative given by $(f^{-1})'(y) = \frac{1}{f'(x)}$ for $x = f^{-1}(y)$. Let $y_0 = f(x_0)$. From the continuity of f and f^{-1} at x_0 and y_0 respectively, for $x \neq x_0$, we have $x \to x_0$ if and only if $y \to y_0$ ($y \neq y_0$). So now,

$$(f^{-1})'(y_0) = \lim_{y \to y_0} \frac{f^{-1}(y) - f^{-1}(y_0)}{y - y_0} = \lim_{x \to x_0} \frac{x - x_0}{f(x) - f(x_0)}$$

$$= \lim_{x \to x_0} \frac{1}{\frac{f(x) - f(x_0)}{x - x_0}} = \frac{1}{f'(x_0)}.$$

Since $f'(x_0) \neq 0$ and f' is continuous at x_0, it follows that $(f^{-1})'$ is also continuous at y_0. $\square$

We now turn to the n variables case.

Theorem 3.8.2. *(Inverse Function theorem). Let Ω be an open set in $\mathbb{R}^n$, $a \in \Omega$ and $f : \Omega \to \mathbb{R}^n$ be in $C^1(\Omega)$. If $\det D_f(a) \neq 0$, then there is a neighborhood U of a such that $f : U \to f(U) = V$ has a continuous inverse f^{-1}. Moreover, $f^{-1} \in C^1(V)$ and*

$$D_{f^{-1}}(y) = [D_f(x)]^{-1},$$

for all $y \in V$. In particular, $f(U)$ is an open set in $\mathbb{R}^n$. Furthermore, if $f \in C^k(\Omega)$, then $f^{-1} \in C^k(V)$, $(k = 1, ..., \infty)$.

Proof. We first make some reductions of the problem. It will suffice to prove the theorem in the case $a = 0$, $f(0) = 0$ and $D_f(0) = I$. To begin with, let $f(a) = b$. There is no loss in taking $a = 0$, $f(0) = 0$, for otherwise, we consider the function $\varphi(x) = f(x + a) - b$, that is, $\varphi = \tau_{-b} \circ f \circ \tau_a$, where the map τ_v is the translation by $v \in \mathbb{R}^n$, given by, $\tau_v(x) = x + v$, for $x \in \mathbb{R}^n$. Now, φ is defined on an open set containing 0 (the set $\tau_{-a}(\Omega) = \{x - a : x \in \Omega\}$), $\varphi(0) = 0$ and $D_\varphi(0) = D_f(a)$. Hence $D_\varphi(0)$ is invertible. If φ had a C^1 inverse φ^{-1} near 0, then, since translations are invertible (with C^1 inverse $\tau_x^{-1} = \tau_{-x}$), we would have $f = \tau_b \circ \varphi \circ \tau_{-a}$ and so $f^{-1} = \tau_a \circ \varphi^{-1} \circ \tau_{-b}$, that is, $f^{-1}(y) = \varphi^{-1}(y - b) + a$ would be the C^1 inverse of f near a. Next, let $D_f(0) = T$. Since $\det(T) = J_f(0) \neq 0$, the linear map T^{-1} exists. Then $D_{T^{-1}} = T^{-1}$ and the Chain Rule gives $D_{(T^{-1} \circ f)}(0) = T^{-1} D_f(0) = I$. As $f = T \circ (T^{-1} \circ f)$, if the theorem were true for $T^{-1} \circ f$, then it would also be true for f. Thus, we may assume $D_f(0) = I$.

We will find a neighborhood U of 0 such that $f : U \to f(U)$ is invertible. Let $g(x) = f(x) - x$. Then, g is also C^1 as f and $D_g(0) = D_f(0) - D_I(0) = I - I = 0$. By the continuity of D_g at 0, there is a small ball $B_r(0)$ of radius $r > 0$ about 0, so that $\|\nabla g_i(x)\| < \frac{1}{2n}$, for all $x \in \overline{B}_r(0)$, where g_i are the component functions of $g = (g_1, ..., g_n)$. Since balls are convex, by Corollary 3.4.2, we have

$$|g_i(x_1) - g_i(x_2)| \le \frac{1}{2n}\|x_1 - x_2\|$$

for $x_1, x_2 \in \overline{B}_r(0)$. Hence

$$\|g(x_1) - g(x_2)\| \le \sum_{i=1}^{n} |g_i(x_1) - g_i(x_2)| \le \frac{1}{2n} n \|x_1 - x_2\| = \frac{1}{2}\|x_1 - x_2\|.$$

Taking $x_2 = 0$ and $x_1 = x \in \overline{B}_r(0)$, we get $\|g(x)\| \le \frac{1}{2}\|x\| \le \frac{r}{2}$. Thus, $g : \overline{B}_r(0) \to \overline{B}_{\frac{r}{2}}(0) \subseteq \overline{B}_r(0)$ and $g : B_r(0) \to B_{\frac{r}{2}}(0)$. We shall show that for each $y \in B_{\frac{r}{2}}(0)$, there is a unique $x \in B_r(0)$, so that $f(x) = y$. Let $y \in B_{\frac{r}{2}}(0)$ and consider the function $\psi_y(x) = y - g(x)$. Then,

$$\|\psi_y(x)\| = \|y - g(x)\| \le \|y\| + \|g(x)\| < r.$$

Hence, $\psi_y : \overline{B}_r(0) \to \overline{B}_r(0)$. Moreover, for $x_1, x_2 \in \overline{B}_r(0)$, we have

$$\|\psi_y(x_1) - \psi_y(x_2)\| = \|g(x_1) - g(x_2)\| \le \frac{1}{2}\|x_1 - x_2\|$$

Therefore, ψ_y is a contraction mapping. Since $\overline{B}_r(0)$ is compact and therefore complete, the Contraction Mapping Principle (Theorem

2.3.20) tells us ψ_y has a unique fixed point, say $x_0 \in \overline{B}_r(0)$. Since, $g(x_0) = f(x_0) - x_0$, this means $x_0 = \psi_y(x_0) = y - (f(x_0) - x_0)$. So, $y = f(x_0)$. Actually, $x_0 \in B_r(0)$, because $||x_0|| = ||\psi_y(x_0)||$ and $||\psi(x)|| < r$ for all $x \in \overline{B}_r(0)$. Thus, $f : B_r(0) \to B_{\frac{r}{2}}(0)$ is bijective. Setting $U = B_r(0)$ and $V = f(U) = B_{\frac{r}{2}}(0)$ (an open set), then $f : U \to V$ is invertible, with inverse $f^{-1} : V \to U$. Next, we show that f^{-1} is continuous on V. If $x_1, x_2 \in U$, then

$$||x_1 - x_2|| = ||f(x_1) - g(x_1) - f(x_2) + g(x_2)||$$

$$\leq ||f(x_1) - f(x_2)|| + ||g(x_1) - g(x_2)|| \leq ||f(x_1) - f(x_2)|| + \frac{1}{2}||x_1 - x_2||.$$

Hence $||x_1 - x_2|| \leq 2||f(x_1) - f(x_2)||$, so for $y_1, y_2 \in V$ with $y_1 = f(x_1)$ and $y_2 = f(x_2)$, we have

$$||f^{-1}(y_1) - f^{-1}(y_2)|| \leq 2||y_1 - y_2||.$$

This means f^{-1} is uniformly continuous on V. We now show that f^{-1} is in $C^1(V)$. Since f is in $C^1(\Omega)$ and the determinant function as a polynomial is continuous, it follows that $J_f = \det(D_f)$ is also continuous on Ω. So, as $J_f(0) \neq 0$, there is $\delta > 0$ so that $J_f(x) \neq 0$ for $x \in B_r(0)$. If $B_\delta(0)$ does not contain $B_r(0)$, we can restrict r further until this is the case. Hence, we may assume that $J_f(x) \neq 0$ for $x \in B_r(0)$, that is, $[D_f(x)]^{-1}$ exists for all $x \in B_r(0)$. Let $x_0 \in B_r(0)$ and set $A = [D_f(x_0)]^{-1}$. Since f is differentiable at x_0, by linear approximation $f(x) - f(x_0) = D_f(x_0)(x - x_0) + \epsilon(x - x_0)||x - x_0||$, where $\epsilon(x, x_0) \to 0$ as $x \to x_0$. Hence, since A is linear,

$$A(f(x) - f(x_0)) = x - x_0 + ||x - x_0||A\epsilon(x - x_0).$$

This implies

$$||f^{-1}(y) - f^{-1}(y_0) - A(y - y_0)|| = ||f^{-1}(y) - f^{-1}(y_0)|| \cdot ||A(\epsilon_*(y, y_0))||$$

where $\epsilon_*(y, y_0) = \epsilon(f^{-1}(y) - f^{-1}(y_0)) = \epsilon(x, x_0)$. Since $||f^{-1}(y) - f^{-1}(y_0)|| \leq 2||y - y_0||$, we get

$$\frac{||f^{-1}(y) - f^{-1}(y_0) - A(y - y_0)||}{||y - y_0||} \leq 2||A\epsilon_*(y, y_0)||.$$

By the continuity of f at x_0 and f^{-1} at y_0, $x \to x_0$ if and only if $y \to y_0$. Since $\epsilon_*(y, y_0) = \epsilon(x, x_0)$ and $\epsilon(x, x_0) \to 0$ as $x \to x_0$, it follows that $\epsilon_*(y, y_0) \to 0$ as $y \to y_0$. Because A is a linear transformation, $\|A\epsilon_*(y, y_0)\| \to 0$ as $y \to y_0$ and so

$$\lim_{y \to y_0} \frac{\|f^{-1}(y) - f^{-1}(y_0) - A(y - y_0)\|}{\|y - y_0\|} = 0$$

which proves that f^{-1} is differentiable at y_0 and $D_{f^{-1}}(y_0) = A = [D_f(x_0)]^{-1}$. Thus, for $y \in V$, $D_{f^{-1}}(y) = [D_f(x)]^{-1}$. Because $D_f(x)$ is a continuous function of x and inversion of matrices is also continuous, we conclude the composition $[D_f(x)]^{-1}$ is itself a continuous function of x. Hence, $D_{f^{-1}}(y) = [D_f(f^{-1}(y))]^{-1}$ is a continuous function of y and f^{-1} is in $C^1(V)$. Finally, by induction, assume that if $f \in C^{k-1}(\Omega)$, then $f^{-1} \in C^{k-1}(V)$. Then $D_f \in C^{k-1}(\Omega)$ and as above, $D_{f^{-1}} \in C^{k-1}(V)$. That is, $f^{-1} \in C^k(V)$. $\qquad\square$

Corollary 3.8.3. *(Open mapping theorem). Let Ω be an open set in $\mathbb{R}^n$ and $f : \Omega \to \mathbb{R}^n$ be in $C^1(\Omega)$. If $J_f(x) \neq 0$ for all $x \in \Omega$, then f is an open mapping (i.e., if $W \subseteq \Omega$ is open, then $f(W)$ is open in $\mathbb{R}^n$.*

Proof. Let $b \in f(W)$ and suppose $b = f(a)$ with $a \in W$. Since $J_f(a) \neq 0$, it follows from the Inverse Function theorem that there exist open neighborhoods U of a and $V = f(U)$ of b such that $f : U \to V$ is a bijection. Moreover, since W is open, we can restrict U, by intersecting it with W, if necessary, so that $U \subseteq W$. Hence, $V \subseteq f(W)$ and so $f(W)$ is open.[9] $\qquad\square$

Definition 3.8.4. Let U, V be open subsets of $\mathbb{R}^n$. A mapping $f : U \to V$ is called a *diffeomorphism* if

1. f is of class $C^1(U)$.

2. f is a bijection and

3. f^{-1} is of class $C^1(V)$.

---

[9]If the set $W \subseteq \Omega$ in Corollary 3.8.3 is closed, then $f(W)$ need not be closed. For example, the function $f(x, y) = (e^x \cos y, e^x \sin y)$ has $J_f(x, y) \neq 0$ for all $(x, y) \in \mathbb{R}^2$ and $f(\mathbb{R}^2) = \mathbb{R}^2 - \{(0, 0)\}$, which is not closed, see Example 3.8.5.

By Corollary 3.8.3, every diffeomorphism is a homeomorphism.

The Inverse Function theorem tells us when we can solve the vector equation $y = f(x)$ with $x, y \in \mathbb{R}^n$ in a sufficiently small neighborhood of a point. Equivalently, displaying the components of $x = (x_1, ..., x_n)$, $y = (y_1, ..., y_n)$ and $f = (f_1, ..., f_n)$, it tells when we can solve a system of equations of the form

$$f_1(x_1, ..., x_n) = y_1$$

$$\cdots$$

$$f_n(x_1, ..., x_n) = y_n$$

for $x_1, ..., x_n$.
Specifically, the system can be locally solved for $x_1, ..., x_n$, in a sufficiently small neighborhood of any point $a = (a_1, ...a_n)$ for which $J_f(a) \neq 0$. In this sense, this is a non-linear local version of Cramer's Rule, which is a global theorem about an $n \times n$ system of linear equations. In fact, if $f : \mathbb{R}^n \to \mathbb{R}^n$ is a linear function, the vector equation $f(x) = y$ becomes $Ax = y$, where $A = (a_{ij})$ is the standard matrix of f. In this case, $J_f(x) = \det(A)$ for all $x \in \mathbb{R}^n$ and the corresponding system is the system of linear equation

$$a_{11}x_1 + ... + a_{1n}x_n = y_1$$

$$\cdots$$

$$a_{n1}x_1 + ... + a_{nn}x_n = y_n.$$

From linear algebra, Cramer's Rule tells us how to solve (uniquely) this system if $\det(A) \neq 0$.

Example 3.8.5. Let $f : \mathbb{R}^2 \to \mathbb{R}^2$ given by $f(x, y) = (e^x \cos y, e^x \sin y)$. Show that f is locally invertible, but is not invertible.

Solution. The derivative is $D_f(x, y) = \begin{pmatrix} e^x \cos y & -e^x \sin y \\ e^x \sin y & e^x \cos y \end{pmatrix}$, and clearly f is C^1. Its Jacobian is $J_f(x, y) = e^{2x} > 0$ for all (x, y), and the Inverse Function theorem tells us that f is locally invertible. However, it is not one-to-one on $\mathbb{R}^2$, because $f(x, y + 2k\pi) = f(x, y)$ for any $k \in \mathbb{Z}$.

Example 3.8.6. Let $f : \mathbb{R}^2 \to \mathbb{R}^2$ be $f(x, y) = (u, v) = (x^2 + y^2, 2xy)$. Where is f invertible? Find $\frac{\partial x}{\partial u}$, $\frac{\partial x}{\partial v}$, $\frac{\partial y}{\partial u}$ and $\frac{\partial y}{\partial v}$.

Solution. The partial derivatives $\frac{\partial u}{\partial x} = 2x$, $\frac{\partial u}{\partial y} = 2y$, $\frac{\partial v}{\partial x} = 2y$, $\frac{\partial v}{\partial y} = 2x$ all exist and are continuous for all (x, y), so f is C^1. The Jacobian is

$$J_f(x, y) = \det \begin{pmatrix} 2x & 2y \\ 2y & 2x \end{pmatrix} = 4(x^2 - y^2).$$ By the Inverse Function theorem,

at points where $J_f(x, y) \neq 0$, that is, $x^2 \neq y^2$, the function f has a local inverse. The lines $y = \pm x$ divide the xy-plane into four open quadrants. The restriction of f to any of these quadrants has an inverse. For instance, consider the quadrant $U = \{(x, y) : |x| < y, y > 0\}$. For $(x, y) \in U$ solving the equations $x^2 + y^2 = u$, $2xy = v$ for x, y in terms of u and v we find

$$(x, y) = f^{-1}(u, v) = (\frac{\sqrt{u + v} - \sqrt{u - v}}{2}, \frac{\sqrt{u + v} + \sqrt{u - v}}{2}).$$

Similarly, one can find the local inverse of the restrictions of f in the other quadrants. Here we managed to find an explicit formula for the local inverse f^{-1} and, of course, then we can compute its partial derivatives by differentiation. However, in general this is not always possible! In such cases, the Inverse Function theorem tells us that the derivatives $\frac{\partial x}{\partial u}$, $\frac{\partial x}{\partial v}$, $\frac{\partial y}{\partial u}$, $\frac{\partial y}{\partial v}$ are obtained by inverting the Jacobian matrix D_f. In the 2×2 case this gives[10]

$$D_{f^{-1}}(u, v) = \begin{pmatrix} \frac{\partial x}{\partial u} & \frac{\partial x}{\partial v} \\ \frac{\partial y}{\partial u} & \frac{\partial y}{\partial v} \end{pmatrix} = \frac{1}{J_f(x, y)} \begin{pmatrix} \frac{\partial v}{\partial y} & -\frac{\partial u}{\partial y} \\ -\frac{\partial v}{\partial x} & \frac{\partial u}{\partial x} \end{pmatrix}.$$

Writing this out

$$\frac{\partial x}{\partial u} = \frac{1}{J_f(x,y)} \frac{\partial v}{\partial y}, \quad \frac{\partial x}{\partial v} = -\frac{1}{J_f(x,y)} \frac{\partial u}{\partial y}.$$

$$\frac{\partial y}{\partial u} = -\frac{1}{J_f(x,y)} \frac{\partial v}{\partial x}, \quad \frac{\partial y}{\partial v} = \frac{1}{J_f(x,y)} \frac{\partial u}{\partial x}.$$

In this example,

$$\frac{\partial x}{\partial u} = \frac{x}{2(x^2 - y^2)}, \quad \frac{\partial x}{\partial v} = -\frac{y}{2(x^2 - y^2)}, \quad \frac{\partial y}{\partial u} = -\frac{y}{2(x^2 - y^2)}, \quad \frac{\partial y}{\partial v} = \frac{x}{2(x^2 - y^2)}.$$

---

[10]We used the fact that for a 2×2 matrix, $\begin{pmatrix} a & b \\ c & d \end{pmatrix}^{-1} = \frac{1}{ad - bc} \begin{pmatrix} d & -b \\ -c & a \end{pmatrix}$.

3.8.2   The Implicit Function theorem

We introduce this problem with an example. Suppose that for $x, y \in \mathbb{R}$ we are given the equation $x^2 + y^2 = 1$. When (if at all) does this equation define y as a function of x? The set of points x and y related by this equation is just the unit circle. For every x with $-1 \leq x \leq 1$, solving for y, we get $y = \pm\sqrt{1 - x^2}$. By the very definition of a function, y should be single-valued. So, without further restriction, we can not solve for y uniquely in terms of x.

In general, it is often inconvenient or even impossible to solve an equation $F(x, y) = 0$ explicitly for y. In addition, even if one could find a unique function $y = f(x)$ satisfying $F(x, f(x)) = 0$, it is reasonable to require the function to be at least continuous, or even better differentiable.

Coming back to our example, consider $F(x, y) = x^2 + y^2 - 1$ and let us work locally near a point (a, b), where $F(a, b) = 0$. For (x, y) near (a, b), we want to solve $F(x, y) = 0$ for y as a function $y = f(x)$, that is, $F(x, f(x)) = 0$. As we saw above, if $b > 0$, then, for $-1 < x < 1$, $y = f(x) = \sqrt{1 - x^2}$. In particular, the derivative is $y' = f'(x) = -\frac{x}{\sqrt{1-x^2}}$. Similarly, if $b < 0$, then $y = f(x) = -\sqrt{1 - x^2}$ with derivative $y' = f'(x) = \frac{x}{\sqrt{1-x^2}}$. If $b = 0$, i.e., $(a, b) = (\pm 1, 0)$, these points are exceptional, since near $a = \pm 1$, we don't have a function. These exceptional points are exactly the places where $\frac{\partial F}{\partial y}(x, y) = 0$. Thus, it seems reasonable to hope that the condition $\frac{\partial F}{\partial y}(a, b) \neq 0$ will guarantee that, in a neighborhood of the point (a, b) at which $F(a, b) = 0$, there is a unique differentiable function $y = f(x)$ such that $F(x, f(x)) = 0$. It is worthwhile remarking that once it is known that a differentiable function $y = f(x)$ such that $F(x, f(x)) \equiv 0$ exists, even in the case where no explicit formula for $y = f(x)$ can be found, we can still find its derivative $\frac{dy}{dx}$ by implicit differentiation. In our example, implicit differentiation of $x^2 + y^2 = 1$ gives $2x + 2y\frac{dy}{dx} = 0$ or $\frac{dy}{dx} = -\frac{x}{y} = f'(x)$, regardless which explicit formula we choose for $y = f(x)$.

The general Implicit Function theorem involves a system of equations

rather than a single equation. One seeks to solve this system for some
of the unknowns in terms of the others. Specifically, let

$$F_1(x_1, ..., x_n, y_1, ...y_m) = 0$$

$$...$$

$$F_m(x_1, ..., x_n, y_1, ...y_m) = 0$$

be a system of m equations in $n + m$ unknowns. One would hope to be
able to solve for m of these unknowns $y_1, ..., y_m$, from the m equations,
in terms of the others $x_1, ..., x_n$.

Before stating the Implicit Function theorem, we will require some no-
tation: We consider the space $\mathbb{R}^{n+m}$ as the Cartesian product $\mathbb{R}^n \times \mathbb{R}^m$
and write the elements of $\mathbb{R}^{n+m}$ as (x, y), where $x = (x_1, ...x_n)$ and
$y = (y_1, ..., y_m)$. Let Ω_n and Ω_m be open sets of $\mathbb{R}^n$ and $\mathbb{R}^m$, respectively. Then $\Omega_n \times \Omega_m = \Omega$ is an open set of $\mathbb{R}^{n+m}$ (by the definition of
the product topology). Let $F : \Omega \subseteq \mathbb{R}^{n+m} \to \mathbb{R}^m$ be a function whose
component functions are $F_1, ..., F_m$. With this notation, the above sys-
tem can be written as the vector equation

$$F(x, y) = 0, \text{ with } (x, y) \in \Omega.$$

Let $F_y(x, y) = F(0, y)$ and

$$\frac{\partial F}{\partial y} = D_{F_y}(x, y) = \begin{pmatrix} \frac{\partial F_1}{\partial y_1} & \cdots & \frac{\partial F_1}{\partial y_m} \\ . & \cdots & . \\ . & \cdots & . \\ . & \cdots & . \\ \frac{\partial F_m}{\partial y_1} & \cdots & \frac{\partial F_m}{\partial y_m} \end{pmatrix}$$

the $m \times m$ Jacobian matrix of F on $\{0\} \times \Omega_m$. Similarly,

$$\frac{\partial F}{\partial x} = D_{F_x}(x, y) = \begin{pmatrix} \frac{\partial F_1}{\partial x_1} & \cdots & \frac{\partial F_1}{\partial x_n} \\ . & \cdots & . \\ . & \cdots & . \\ . & \cdots & . \\ \frac{\partial F_m}{\partial x_1} & \cdots & \frac{\partial F_m}{\partial x_n} \end{pmatrix}$$

the $m \times n$ Jacobian matrix of the restriction $F_x(x, y) = F(x, 0)$ of F on $\Omega_n \times \{0\}$. So that we may write,

$$D_F(x, y) = \left(\frac{\partial F}{\partial x}, \frac{\partial F}{\partial y}\right)$$

Theorem 3.8.7. *(Implicit Function theorem). Let $F : \Omega \to \mathbb{R}^m$ be a C^1 function on the open set $\Omega = \Omega_n \times \Omega_m$, of $\mathbb{R}^{n+m}$, such that $F(a, b) = 0$ for some point $(a, b) \in \Omega$. If $\det(\frac{\partial F}{\partial y}(a, b)) \neq 0$, then there exists a neighborhood $U_a \times U_b$ of (a, b) in Ω and a C^1 function $f : U_a :\to \mathbb{R}^m$ such that $F(x, f(x)) = 0$ on U_a and $f(a) = b$.*

Proof. We will apply the Inverse Function theorem to the function $G : \Omega \to \mathbb{R}^{n+m}$ defined by $G(x, y) = (x, F(x, y))$. Evidently, $G \in C^1(\Omega)$ since F is. Now with respect to the standard bases on $\mathbb{R}^n$ and $\mathbb{R}^m$ and putting them together to get a basis of $\mathbb{R}^{n+m}$, the $(n + m) \times (n + m)$ matrix of $D_G(x, y)$ is a block triangular matrix of the form

$$D_G(x, y) = \begin{pmatrix} I & 0 \\ \frac{\partial F}{\partial x} & \frac{\partial F}{\partial y} \end{pmatrix}$$

Hence, by our hypothesis

$$J_G(a, b) = \det(\tfrac{\partial F}{\partial y}(a, b)) \neq 0 \text{ so } D_G(a, b) \text{ is invertible.}$$

By the Inverse Function theorem, there is an open set about (a, b), which by going smaller if necessary, we can take to be of the form $U_a \times U_b$ and an open set $W = G(U_a \times U_b)$ about $G(a, b) = (a, 0)$ so that G has a C^1 inverse $G^{-1} : W \to U_a \times U_b$. For $(x, y) \in U_a \times U_b$, we have $G(x, y) = (x, F(x, y))$ if and only if $(x, y) = G^{-1}(x, F(x, y))$ and so G^{-1} preserves the first n components as G does. Hence, G^{-1} is of the form $G^{-1}(x, y) = (x, \Phi(x, y))$ for some C^1 function Φ from W into $\mathbb{R}^m$. Let $\pi : \Omega \to \Omega_m$ denote the projection on the y coordinate. Then $\pi \circ G = F$. So, $F(x, \Phi(x, y)) = \pi \circ G(x, \Phi(x, y)) = \pi \circ G \circ G^{-1}(x, y) = \pi(x, y) = y$. In particular, $F(x, \Phi(x, 0)) = 0$. Letting $f(x) = \Phi(x, 0)$ for $x \in U_a$, f is C^1 and satisfies $F(x, f(x)) = 0$. Moreover, $(a, b) = G^{-1}(a, 0) = (a, \Phi(a, 0)) = (a, f(a))$ so $b = f(a)$. $\qquad\square$

Corollary 3.8.8. *(Implicit differentiation).* *With the notation as in the implicit function theorem, the derivative D_f of $f : U_a :\to \mathbb{R}^m$ is given by*

$$D_f(x) = -[\frac{\partial F}{\partial y}]^{-1}[\frac{\partial F}{\partial x}]$$

where the partial derivatives are evaluated at $(x, f(x))$. When $n = m = 1$, this reduces to the familiar formula, from one variable calculus,

$$\frac{dy}{dx} = -\frac{\frac{\partial F}{\partial x}}{\frac{\partial F}{\partial y}}.$$

Proof. Since f is known to be differentiable, to compute the derivative D_f we apply the Chain Rule to the relation $F(x, f(x)) = 0$. For $x \in U_a$ set $g(x) = (x, f(x))$. So that $F(g(x)) = 0$. Now the Chain Rule gives

$$0 = D_{(F \circ g)}(x) = DF(g(x)) \cdot D_g(x) = \left(\frac{\partial F}{\partial x}(g(x)) \ \ \frac{\partial F}{\partial y}(g(x)) \right) \left(\begin{array}{c} I \\ D_f(x) \end{array} \right)$$

$$= \frac{\partial F}{\partial x}(g(x)) + \frac{\partial F}{\partial y}(g(x))Df(x)$$

Therefore,

$$D_f(x) = -[\frac{\partial F}{\partial y}(g(x))]^{-1}[\frac{\partial F}{\partial x}(g(x))]$$

$\square$

The case of numerical functions ($m = 1$), follows directly. We shall make use of this in the next section on constrained extrema.

Corollary 3.8.9. *(Implicit Function theorem; m=1).* *Let $\Omega_n \times \Omega$ be an open subset of $\mathbb{R}^{n+1}$ and let $F : \Omega_n \times \Omega \to \mathbb{R}$ be a function in $C^1(\Omega_n \times \Omega)$ with $F(a, b) = 0$ at some point $(a, b) \in \Omega_n \times \Omega$. If $\frac{\partial F}{\partial y}(a, b)) \neq 0$, then there exist some neighborhood $U_a \times U_b$ of (a, b) in $\Omega_n \times \Omega$ and a C^1 function $f : U_a :\to \mathbb{R}$ such that $F(x, f(x)) = 0$ and $f(a) = b$.*

Corollary 3.8.10. *Let $F : \mathbb{R}^n \to \mathbb{R}$ be a C^1 function and let $S = \{x \in \mathbb{R}^n : F(x) = 0\}$. For every $a \in S$ with $\nabla F(a) \neq 0$, there exists a neighborhood V of a so that $S \cap V$ is the graph of a C^1 function.*

Proof. Since $\nabla F(a) \neq 0$, we have $\frac{\partial F}{\partial x_j}(a) \neq 0$ for some $j = 1, ..., n$. Set $y = x_j$, $\tilde{x} = (x_1, ..., x_{j-1}, x_{j+1}, ..., x_n) \in \mathbb{R}^{n-1}$ and view F as a function from $\mathbb{R}^{n-1}$ into $\mathbb{R}$. By Corollary 3.8.9 the equation $F(x) = F(\tilde{x}, y) = 0$ can be solved to yield x_j as a C^1 function of the remaining variables in a neighborhood V of a.   □

Example 3.8.11. Does the surface $x^3 + 3y^2 + 8xz^2 - 3z^3y = 1$ represent a graph of a C^1 function $z = f(x, y)$? If so, near what points?

Solution. Consider the function $F(x, y, z) = x^3 + 3y^2 + 8xz^2 - 3z^3y - 1$. We attempt to solve $F(x, y, z) = 0$ for z as a function of (x, y). By the Implicit Function theorem this may be done near any point (x_0, y_0, z_0) with $F(x_0, y_0, z_0) = 0$ where $\frac{\partial F}{\partial z}(x_0, y_0, z_0) \neq 0$ or $16x_0z_0 - 9z_0^2y_0 \neq 0$. Near such points the implicit function $z = f(x, y)$ exists, but there is no simple way to solve for it. Observe that this is consistent with what we get by implicit differentiation $\frac{\partial}{\partial x}(x^3 + 3y^2 + 8xz^2 - 3z^3y) = 0$ or $3x^2 + 8z^2 + 16xz\frac{\partial z}{\partial x} - 9z^2y\frac{\partial z}{\partial x} = 0$. Hence $\frac{\partial z}{\partial x} = \frac{-3x^2+8z^2}{16xz-9z^2y}$. Similarly, $\frac{\partial z}{\partial x} = \frac{3z^2-6z}{16xz-9z^2y}$.

Example 3.8.12. Consider the problem of solving the system of equations

$$xy^2 + xzu + yv^2 = 3$$
$$u^3yz + 2xv - u^2v^2 = 2$$

for u and v in terms of x, y, z near $x = y = z = 1$, $u = v = 1$ and computing the partial derivatives $\frac{\partial u}{\partial y}, \frac{\partial v}{\partial y}$ at the point $(1, 1, 1)$.

Solution. Let $a = (1, 1, 1)$, $b = (1, 1)$. Consider the function $F : \mathbb{R}^3 \times \mathbb{R}^2 \to \mathbb{R}^2$ given by

$$F(x, y, z, u, v) = (xy^2 + xzu + yv^2 - 3, u^3yz + 2xv - u^2v^2 - 2).$$

Then $F(a, b) = (0, 0)$. We need to check that $\det \begin{pmatrix} \frac{\partial F_1}{\partial u} & \frac{\partial F_1}{\partial v} \\ \frac{\partial F_2}{\partial u} & \frac{\partial F_2}{\partial v} \end{pmatrix} \neq 0$ at $(a, b) = (1, 1, 1, 1, 1)$, where F_1, F_2 are the component functions of F. In fact, we have

$$\det \begin{pmatrix} xz & 2yv \\ 3u^2yz - 2uv^2 & 2x - 2u^2v \end{pmatrix} = \det \begin{pmatrix} 1 & 2 \\ 1 & 0 \end{pmatrix} = -2$$

Hence, by the Implicit Function theorem there exists a C^1 function $f(x,y,z) = (f_1(x,y,z), f_2(x,y,z))$ defined in a neighborhood of $(1,1,1)$ such that

$$u = f_1(x,y,z), \ v = f_2(x,y,z) \text{ and } f(1,1,1) = (1,1).$$

To compute $\frac{\partial v}{\partial y}$ we use the Chain Rule and implicit differentiation.

$$\frac{\partial F_1}{\partial y} = 2xy + xz\frac{\partial u}{\partial y} + v^2 + 2yv\frac{\partial v}{\partial y} = 0$$

$$\frac{\partial F_2}{\partial y} = 3u^2\frac{\partial u}{\partial y}yz + u^3z + 2x\frac{\partial v}{\partial y} - 2u\frac{\partial u}{\partial y}v^2 - 2v\frac{\partial u}{\partial y}u^2 = 0.$$

Evaluating at $(1,1,1)$ and simplifying we get

$$\frac{\partial u}{\partial y} + 2\frac{\partial v}{\partial y} = -3$$

$$\frac{\partial u}{\partial y} = -1.$$

Thus, $\frac{\partial u}{\partial y}(1,1,1) = -1$ and $\frac{\partial v}{\partial y}(1,1,1) = -1$.

Similarly, differentiating with respect to x and z, one can compute $\frac{\partial u}{\partial x}, \frac{\partial v}{\partial x}$ and $\frac{\partial u}{\partial z}, \frac{\partial v}{\partial z}$ at $(1,1,1)$, respectively.

EXERCISES

1. Show that $f : \mathbb{R}^2 \to \mathbb{R}^2$ defined by $(u,v) = f(x,y) = (x - 2y, 2x - y)$ is globally invertible and find its inverse $f^{-1}(u,v) = (x,y)$. Find the region in the xy-plane that is mapped to the triangle with vertices $(0,0)$, $(2,1)$ and $(-1,2)$ in the uv-plane.

2. Let $f : \mathbb{R}^2 \to \mathbb{R}^2$ be $(u,v) = f(x,y) = (x^2 - y^2, 2xy)$.

   (a) Show that f is one-to-one on the set $U = \{(x,y) : x > 0\}$.

   (b) What is the set $V = f(U)$?

   (c) Find $D_{f^{-1}}(0,1)$.

3. Let $f : \mathbb{R}^2 \to \mathbb{R}^2$ be $f(x,y) = (\cos x + \cos y, \sin x + \sin y)$. Show that f is locally invertible near all points (a,b) such that $a - b \neq k\pi$ with $k \in \mathbb{Z}$ and at all other points no local inverse exists.

4. Let $f : \mathbb{R}^2 \to \mathbb{R}^2$ be $f(x, y) = (x^3 + 2xy + y^2, x^2 + y)$. Show that f is locally invertible near $(1, 1)$. Find $D_{f^{-1}}(4, 2)$ and obtain the affine approximation to $f^{-1}(u, v)$ near $(4, 2)$.

5. Let $f : \mathbb{R}^2 \to \mathbb{R}^2$ be $(u, v) = f(x, y) = (x - y, xy)$.

   (a) Draw some curves $x - y =$ constant and $xy =$ constant in the xy-plane. Which regions in xy-plane map onto the rectangle $[0, 1] \times [1, 4]$ in the uv-plane? Draw a picture; there are two of them.

   (b) Find the Jacobian $J_f(x, y) = \det D_f(x, y)$. Show that $J_f(a, b) = 0$ precisely when the gradients $\nabla u(a, b)$ and $\nabla v(a, b)$ are linearly independent.

   (c) Compute explicitly the local inverse of f near $(2, -3)$ and find $D_{f^{-1}}(5, -6)$.

   (d) Show by matrix multiplication that $D_f(2, -3)D_{f^{-1}}(5, -6) = I$, where I is the 2×2 identity matrix.

6. Let $f : \mathbb{R}^3 \to \mathbb{R}^3$ be $(u, v, w) = f(x, y, z) = (x - xy, xy - xyz, xyz)$ and $U = \{(x, y, z) : xy \neq 0\}$. Show that f is invertible on U and determine a formula for the inverse $f^{-1} : f(U) \to U$. Find the Jacobian of f^{-1} at a point (a, b, c) of $f(U)$.

7. (*Spherical coordinates*). Solve the equations $x = r \cos\theta \sin\phi$, $y = r \sin\theta \sin\phi$, $z = r \cos\phi$ for r, θ, ϕ in terms of x, y, z. When is this inverse function differentiable? Compare with the conclusions of the Inverse Function theorem.

8. Let $f : \mathbb{R}^2 \to \mathbb{R}^2$ be $f(x, y) = (2ye^{2x}, xe^y)$ and $g : \mathbb{R}^2 \to \mathbb{R}^3$ be defined by $g(x, y) = (3x - y^2, 2x + y, xy + y^3)$. Show that there exists a neighborhood of $(0, 1)$ that f carries in a one-to-one fashion onto a neighborhood of $(2, 0)$. Find $D_{g \circ f^{-1}}(2, 0)$.

9. Show that the function $F(x, y) = x^3 y + y^3 x - 2 = 0$ defines y as a function f of x near $(1, 1)$. Find $f'(1)$.

10. Can the equation $(x + 1)^2 y - xy^2 = 4$ be solved uniquely for y in terms of x in a neighborhood of the following points: (a) $(1, 2)$; (b) $(2, 1)$; (c) $(-1, 2)$?

11. Show that $F(x, y) = 0$ defines implicitly a function $y = f(x)$ at the given point (a, b), and find $f'(a)$.

    (a) $F(x, y) = xe^y - y + 1$ at $(-1, 0)$.
    (b) $F(x, y) = x \cos xy$ at $(1, \frac{\pi}{2})$.

12. Show that $F(x, y, z) = 0$ defines implicitly a function $z = f(x, y)$ at the given point (a, b, c), and find $\frac{\partial f}{\partial x}(a, b)$ and $\frac{\partial f}{\partial y}(a, b)$.

    (a) $F(x, y, z) = z^3 - z - xy \sin z$ at $(0, 0, 0)$.
    (b) $F(x, y, z) = x + y + z - e^{xyz}$ at $(0, \frac{1}{2}, \frac{1}{2})$.

13. Does the equation

$$F(x, y, z) = \sqrt{x^2 + y^2 + 2z^2} - \cos z = 0$$

define z as a function of x and y near $(0, 1, 0)$? Does it define y as a function of x and z there?

14. Let $F : \mathbb{R}^3 \to \mathbb{R}^2$ be a C^1 function and write F in the form $F(x, y_1, y_2)$. Assume that $F(3, -1, 2) = 0$ and $D_F(3, -1, 2) = \begin{pmatrix} 1 & 2 & 1 \\ 1 & -1 & 1 \end{pmatrix}$.

    (a) Show that there is a C^1 function $f : U \to \mathbb{R}^2$ defined on an open set U in $\mathbb{R}$ such that $F(x, f_1(x), f_2(x)) = 0$ for $x \in U$, and $f(3) = (-1, 2)$. Find $D_f(3)$.

    (b) Can the equation $F(x, y_1, y_2) = 0$ be solved for an arbitrary pair of the unknowns in terms of the third, near the point $(3, -1, 2)$?

15. Solve the equations

$$x^3 - yu = 0$$

$$xy^2 + uv = 0$$

explicitly for u and v as functions of x, y and compare the results with the Implicit Function theorem.

16. Show that the equations

$$xy^5 + yu^5 + zv^5 = 1$$

$$x^5y + y^5u + z^5v = 1$$

have a unique solution $(u, v) = f(x, y, z) = (f_1(x, y, z), f_2(x, y, z))$ near the point $(0, 1, 1, 1, 0)$ and find the derivative $Df(0, 1, 1)$.

17. Show that there are points which satisfy the equations

$$x - e^u \cos v = 0$$

$$v - e^y \sin x = 0.$$

Let $a = (x_0, y_0, u_0, v_0)$ be such a point. Show that in a neighborhood of a there exists a unique solution $(u, v) = f(x, y)$. Prove that $J_f(x, y) = \frac{v}{x}$.

18. Let $f : \mathbb{R}^n \to \mathbb{R}^n$ be differentiable. Suppose that f is invertible on a neighborhood of a point a and that $J_f(a) = 0$. Show that f^{-1} is not differentiable at $f(a)$.

19. Assume the Implicit Function theorem and use it to prove the Inverse Function theorem. *Hint.* Consider the function $F(x, y) = x - f(y)$.

Answers to selected Exercises

11. (a) $f'(-1) = \frac{1}{2}$. (b) $f'(1) = -\frac{\pi}{2}$.

12. (a) $\frac{\partial f}{\partial x}(0, 0) = 0 = \frac{\partial f}{\partial y}(0, 0)$. (b) $\frac{\partial f}{\partial x}(0, \frac{1}{2}) = -\frac{3}{4}$ and $\frac{\partial f}{\partial y}(0, \frac{1}{2}) = -1$.

13. No. Yes $y = f(x, z)$.   16. $\begin{pmatrix} -\frac{1}{5} & -\frac{1}{5} & 0 \\ \frac{1}{5} & -\frac{24}{5} & 0 \end{pmatrix}$.

3.9   Constrained extrema, Lagrange multipliers

Under certain circumstances we may want to maximize or minimize a function subject to constraints or side conditions. When there is only one constraint, the problem is as follows: Let Ω be open in $\mathbb{R}^n$, $f, g : \Omega \to \mathbb{R}$ be $C^1(\Omega)$ and $S = \{x \in \Omega : g(x) = 0\}$. We want to find necessary (and sufficient) conditions for f to be extremized on S (the constraint set).

Theorem 3.9.1. *(Lagrange multiplier rule)*[11]. *Let $f, g : \Omega \subseteq \mathbb{R}^n \to \mathbb{R}$ be C^1 functions in the open set Ω. Let $S = \{x \in \Omega : g(x) = 0\}$ and $s \in S$ with $\nabla g(s) \neq 0$. If the restriction $f|_S$ of f to S takes on an extreme value at s, then there exist $\lambda \in \mathbb{R}$ (called a* Lagrange multiplier*) such that*

$$\nabla f(s) = \lambda \nabla g(s).$$

Proof. Since $\nabla g(s) \neq 0$, it follows that $\frac{\partial g}{\partial x_i}(s) \neq 0$ for some $i = 1, ..., n$. If necessary, rename the variables so that $\frac{\partial g}{\partial x_n}(s) \neq 0$. By the Implicit Function theorem there is a C^1 function φ defined in a neighborhood V of s so that $g(x_1, ..., x_{n-1}, \varphi(x_1, ..., x_{n-1})) = 0$. Hence, the constraint hypersurface has dimension $n - 1$ in a neighborhood V of s. Let $T_s(S)$ be the tangent hyperplane to S at the point s. For each vector $v \in T_s(S)$ choose a smooth curve $X(t)$ passing through s at $t = 0$ with tangent vector v; that is, $X(0) = s$ and $X'(0) = v$. Since s is an extreme value for $f|_S$, the function $\psi(t) = f(X(t))$ has an extremum at $t = 0$. Therefore,

$$0 = \psi'(0) = \langle \nabla f(X(0)), X'(0) \rangle = \langle \nabla f(s), v \rangle$$

Hence, $\nabla f(s)$ is orthogonal to v. Since $v \in T_s(S)$ was arbitrary, we see that $\nabla f(s)$ is orthogonal to the entire hyperplane. Moreover, since this has dimension exactly $n - 1$, its orthocomplement in $\mathbb{R}^n$ has dimension one and therefore, by Proposition 3.2.21, it is generated by the nonzero vector $\nabla g(s)$. Thus, $\nabla f(s) = \lambda \nabla g(s)$ for some $\lambda \in \mathbb{R}$. $\square$

---

[11] J. Lagrange ((1736-1813) a student of Euler, he succeeded him as the director of mathematics at the Prussian Academy of Sciences in Berlin. Lagrange's treatise *Mecanique Analytique* (1788) offered the most comprehensive treatment of classical mechanics since Newton. He held also a professorship at Ecole Polytechnique upon its opening in 1794. Among his students there were J. Fourier and S. Poisson.

This theorem tells us that to find the constrained extrema of f, we must look for all points $x \in S$ and a constant $\lambda \in \mathbb{R}$ that satisfy $\frac{\partial f}{\partial x_i}(x) = \lambda \frac{\partial g}{\partial x_i}(x)$ for $i = 1, ..., n$ together with the constraint equation $g(x) = 0$. We have a system of $n + 1$ equations in the $n + 1$ unknowns $x_1, ..., x_n, \lambda$

Example 3.9.2. Find the points on the surface $x^2 + 2y^2 - z^2 = 1$ that are nearest to the origin $(0, 0, 0)$.

Solution. The square of the distance from the origin to any point (x, y, z) on the given surface S is $f(x, y, z) = x^2 + y^2 + z^2$. We want to minimize f subject to the constraint $g(x, y, z) = x^2 + 2y^2 - z^2 - 1 = 0$ that defines the surface S. Since for all $(x, y, z) \in S$ the gradient $\nabla g(x, y, z) = (2x, 4y, -2z) \neq (0, 0, 0)$, Theorem 3.9.1 tells us $\nabla f(x, y, z) = \lambda \nabla g(x, y, z)$. Hence we must solve the system of equations

$$2x = \lambda 2x \qquad (1)$$
$$2y = \lambda 4y \qquad (2)$$
$$2z = -\lambda 2z \qquad (3)$$
$$x^2 + 2y^2 - z^2 - 1 = 0 \quad (4).$$

From (3) we see that either $z = 0$ or $\lambda = -1$. If $z \neq 0$ then $\lambda = -1$ and equations (1) and (2) give $x = y = 0$. In that case, (4) gives $z^2 = -1$ which is impossible. Hence any solution must have $z = 0$. If $y \neq 0$, then from (2) we see that $\lambda = 1$ and then (1) and (3) yield $x = z = 0$. Substituting in (4) we get $y = \pm\sqrt{\frac{1}{2}}$. Hence the solutions

$(0, -\sqrt{\frac{1}{2}}, 0)$ and $(0, \sqrt{\frac{1}{2}}, 0)$. At these points $f(0, \pm\sqrt{\frac{1}{2}}, 0) = \frac{1}{2}$.

Similarly, if $x \neq 0$, we find two more solutions $(-1, 0, 0)$ and $(1, 0, 0)$. At these points $f(\pm 1, 0, 0) = 1$. Thus, the minimum is attained at the points $(0, \pm\sqrt{\frac{1}{2}}, 0)$.

Example 3.9.3. Among all triangles with a fixed perimeter p find the one of maximum area.

Solution. Let x, y and z denote the lengths of the sides of the triangle. The area A is given by Heron's formula

$$A = \sqrt{s(s-x)(s-y)(s-z)},$$

where $s = \frac{p}{2}$. Instead of maximizing A it is equivalent to maximize $A^2 = f$. Here $g(x,y,z) = x + y + z - p$. Hence $\nabla g(x,y,z) = (1,1,1)$, which is nonzero everywhere on Ω the first octant in $\mathbb{R}^3$. A direct calculation tells us

$$\nabla f(x,y,z) = -\frac{p}{2}\left((\tfrac{p}{2}-y)(\tfrac{p}{2}-z), (\tfrac{p}{2}-x)(\tfrac{p}{2}-z), (\tfrac{p}{2}-x)(\tfrac{p}{2}-y) \right).$$

Therefore,

$$-\lambda = \tfrac{p}{2}(\tfrac{p}{2}-y)(\tfrac{p}{2}-z).$$
$$-\lambda = \tfrac{p}{2}(\tfrac{p}{2}-x)(\tfrac{p}{2}-z).$$
$$-\lambda = \tfrac{p}{2}(\tfrac{p}{2}-x)(\tfrac{p}{2}-y).$$

Now dividing the first of these by the second yields $x = y$ while dividing the second of these by the third yields $y = z$. This maximal area comes from an equalateral triangle.

Example 3.9.4. Show that the volume of the greatest rectangular parallelepiped that can be inscribed in the ellipsoid $\frac{x^2}{a^2} + \frac{y^2}{b^2} + \frac{z^2}{c^2} = 1$, is $\frac{8abc}{3\sqrt{3}}$.

Solution. We have to find the maximum value of $f(x,y,z) = 8xyz$ subject to the conditions $g(x,y,z) = \frac{x^2}{a^2} + \frac{y^2}{b^2} + \frac{z^2}{c^2} - 1 = 0$, $x > 0$, $y > 0$, $z > 0$ (by symmetry). Again here $\nabla g(x,y,z) \neq (0,0,0)$. Then $\nabla f = \lambda g$ and the constraint $g = 0$ give

$$8yz = \lambda\tfrac{2x}{a^2} \quad (1)$$
$$8xz = \lambda\tfrac{2y}{b^2} \quad (2)$$
$$8xy = \lambda\tfrac{2z}{c^2} \quad (3)$$
$$\tfrac{x^2}{a^2} + \tfrac{y^2}{b^2} + \tfrac{z^2}{c^2} - 1 = 0 \quad (4).$$

Multiplying (1),(2), (3) by x, y, z respectively, adding and using (4) we get

$$24xyz + 2\lambda = 0, \text{ that is, } \lambda = -12xyz.$$

Hence, from (1),(2), (3) we see

$$x = \frac{a}{\sqrt{3}}, \, y = \frac{b}{\sqrt{3}}, \, z = \frac{c}{\sqrt{3}}, \text{ and so the maximum volume is}$$
$$f(\frac{a}{\sqrt{3}}, \frac{b}{\sqrt{3}}, \frac{c}{\sqrt{3}}) = \frac{8abc}{3\sqrt{3}}.$$

Our next result gives an idea, in an important case, of the significance of the λ themselves.

Theorem 3.9.5. *Let $A = (a_{ij})$ be an $n \times n$ real symmetric matrix and $f(x) = \langle Ax, x \rangle$, for $x \in \mathbb{R}^n$. Then the maximum and minimum of f on the unit sphere $S = \{x \in \mathbb{R}^n : ||x|| = 1\}$ are respectively the largest and smallest eigenvalues of A.*

Proof. This is a constrained extremum problem for $f(x) = \sum_{i,j=1}^n a_{ij}x_ix_j$ subject to the constraint $g(x) = 0$, with $g(x) = ||x||^2 - 1 = x_1^2 + ... + x_n^2 - 1$. As $\nabla g(x) = 2(x_1, ..., x_n) \neq 0$ for $x \in S$, Theorem 3.9.1 tells us that there exist $\lambda \in \mathbb{R}$ such that $\nabla f(s) = \lambda \nabla g(s)$. Since $\frac{\partial f}{\partial x_i}(x) = 2 \sum_{j=1}^n a_{ij}x_j$, the system of the equations

$$\nabla f(s) = \lambda \nabla g(s) \text{ and } g(x) = 0$$

is equivalent to,

$$(A - \lambda I)x = 0 \text{ and } ||x||^2 = 1.$$

As $x \neq 0$, we must have $\det(A - \lambda I) = 0$. Hence, λ is an eigenvalue of A and x a corresponding eigenvector. Therefore,

$$f(x) = \langle Ax, x \rangle = \langle \lambda x, x \rangle = \lambda ||x||^2 = \lambda.$$

Thus, the greatest and smallest eigenvalues of A are the extreme values of f on S. $\qquad \square$

Note that, we have also proved along the way that *every real symmetric matrix $A = (a_{ij})$ has at least one real eigenvalue*. This is a result well-known from linear algebra and is fundamental in the proof of the Spectral theorem for symmetric matrices. (Actually once we know it has one real eigenvalue, it has all real eigenvalues by induction).

We now turn to the problem of finding extrema of a numerical function f subject to several constraints $g_1(x) = 0, ..., g_m(x) = 0$.

Theorem 3.9.6. *(Lagrange multiplier rule, several constraints). Let Ω be an open set in $\mathbb{R}^n$, $f \in C^1(\Omega)$ and $g_j \in C^1(\Omega)$, for $j = 1, ..., m < n$. Let $S = \{x \in \Omega : g_1(x) = 0, ..., g_m(x) = 0\}$ and $s \in S$ with $\nabla g_j(s) \neq 0$ for all $j = 1, ..., m$. If the restriction $f|_S$ of f on S takes on an extreme value at $s \in S$, then there exist real numbers $\lambda_1, ..., \lambda_m$ (called Lagrange multipliers) such that*

$$\nabla f(s) = \sum_{j=1}^{m} \lambda_j \nabla g_j(s).$$

Proof. Fix any j and set $S_j = \{x \in \Omega : g_j(x) = 0\}$. Arguing as in Theorem 3.9.1, $\nabla f(s)$ is orthogonal to the tangent hyperplane $T_s(S_j)$ of S_j at s.

Now, since $S = \cap_{j=1}^{m} S_j$, it follows that $T_s(S) = \cap_{j=1}^{m} T_s(S_j)$ and so $\nabla f(s)$ is orthogonal to the whole tangent space $T_s(S)$. Hence, $\nabla f(s) \in (T_s(S))^{\perp}$, the orthocomplement of $T_s(S)$. But $(T_s(S))^{\perp} = \left(\cap_{j=1}^{m} T_s(S_j)\right)^{\perp} = \sum_{j=1}^{m} (T_s(S_j))^{\perp}$ and as we saw above each $(T_s(S_j))^{\perp}$ is generated by $\nabla g_j(s) \neq 0$, it follows that $\nabla f(s)$ is a linear combination of the vectors $\nabla g_1(s), ..., \nabla g_m(s)$. Thus, there exist numbers $\lambda_1, ..., \lambda_m$ such that,

$$\nabla f(s) = \lambda_1 \nabla g_1(s) + ... + \lambda_m \nabla g_m(s)$$

$\square$

Therefore, in order to find the extrema of f on S, we must solve the system of the n equations

$$\frac{\partial f}{\partial x_i}(x) = \lambda_1 \frac{\partial g_1}{\partial x_i}(x) + ... + \lambda_m \frac{\partial g_m}{\partial x_i}(x),$$

$(i = 1, ..., n)$ together with the m constraint equations

$$g_j(x) = 0,$$

$(j = 1, ..., m)$ in the $n + m$ unknowns $x_1, ..., x_n, \lambda_1, ..., \lambda_m$. In practice these equations are non linear and may be difficult to solve. Our objective is to find the $x_1, ..., x_n$, not necessarily the $\lambda_1, ..., \lambda_m$, but sometimes it may be necessary to find the $\lambda_1, ..., \lambda_m$ in order to find $x_1, ..., x_n$.

Example 3.9.7. Find the extreme values of $f(x, y, z) = x + y + z$ subject to the conditions $x^2 + y^2 = 2$ and $x + z = 1$.

Solution. Here the constraint set S is

$$\{(x, y, z) : g_1(x, y, z) = x^2 + y^2 - 2 = 0, \quad g_2(x, y, z) = x + z - 1 = 0\}$$

and represents a curve in $\mathbb{R}^3$. Note that $\nabla g_2(x, y, z) = (1, 0, 1) \neq \mathbf{0}$ on S and since $\mathbf{0} = (0, 0, 0) \notin S$ also $\nabla g_1(x, y, z) = (2x, 2y, 0) \neq \mathbf{0}$ on S. Since S is a compact set and f is continuous, f takes both a maximum and a minimum value on S. To find these extreme values we use the Lagrange multiplier rule (Theorem 3.9.6). The extreme values will occur at ponts (x, y, z) satisfying

$$\nabla f(x, y, z) = \lambda_1 \nabla g_1(x, y, z) + \lambda_2 \nabla g_2(x, y, z)$$
$$g_1(x, y, z) = 0$$
$$g_2(x, y, z) = 0.$$

Thus computing the gradients and equating components, we must solve the system

$$1 = \lambda_1 \cdot 2x + \lambda_2 \cdot 1$$
$$1 = \lambda_1 \cdot 2y + \lambda_2 \cdot 0$$
$$1 = \lambda_1 \cdot 0 + \lambda_2 \cdot 1$$
$$x^2 + y^2 = 2$$
$$x + z = 1.$$

for x, y, z, λ_1 and λ_2. From the third equation $\lambda_2 = 1$ and so $2x\lambda_1 = 0$ and $2y\lambda_1 = 1$. Since the second implies $\lambda_1 \neq 0$, we have $x = 0$. Therefore $y = \pm\sqrt{2}$ and $z = 1$. Hence the points of exterema of f on S are $(0, \pm\sqrt{2}, 1)$. Thus, $f(0, \sqrt{2}, 1) = \sqrt{2} + 1$ and $f(0, -\sqrt{2}, 1) = -\sqrt{2} + 1$ are the maximum and minimum values of f respectively.

Observe that just as in ordinary extrema problems, the Lagrange multiplier rule gives necessary conditions for an extremum. It locates all possible candidates. Just as in Section 3.7, not all such points need be extrema. Here we prove a result which is useful in sufficiency conditions. Before we do so we need the following:

Definition 3.9.8. Let Ω be an open set in $\mathbb{R}^n$, $f \in C^1(\Omega)$, $g = (g_1, ..., g_m) \in C^1(\Omega)$ and $S = \{x \in \Omega : g(x) = 0\} = \cap_{j=1}^m \{x \in \Omega : g_j(x) = 0\}$. A point $s \in S$ at which $\nabla f(s)$ is orthogonal to the tangent space $T_s(S)$ is called a *critical point of f on S*.

In Theorem 3.9.6, we saw that if the restriction $f|_S$ of f on S has an extremum at $s \in S$, then s is a critical point of f on S. Moreover, there exist real numbers $\lambda_1, ..., \lambda_m$ such that $\nabla f(s) = \sum_{j=1}^m \lambda_j \nabla g_j(s)$.

Lagrange proposed the use of the auxiliary (*Lagrangian*) function $L : \mathbb{R}^n \to \mathbb{R}$ given by

$$L(x) = f(x) - \sum_{j=1}^m \lambda_j g_j(x),$$

$x \in \mathbb{R}^n$ with parameters $\lambda_1, ..., \lambda_m$. Note that

$$\nabla L(s) = \nabla f(s) - \sum_{j-1}^m \lambda_j \nabla g_i(s) = 0,$$

so s is an (ordinary) critical point of L.

Suppose $f, g_j \in C^2(\Omega)$ for $j = 1, ..., m$ and let $H_L(x) = \left(\frac{\partial^2 L}{\partial x_i \partial x_j}(x) \right)$ be the Hessian of L at x. As in the case of unconstrained (free) extrema, we are, in particular, interested in the quadratic form $Q(v) = \langle H_L(x)v, v \rangle$ associated with the Hessian matrix of L.

Theorem 3.9.9. *(Second derivative test for constrained extrema). Let Ω be an open set in $\mathbb{R}^n$ and $f \in C^2(\Omega)$. Suppose s is a critical point for f on the constraint set $S = \{x \in \Omega : g(x) = 0\}$, where $g = (g_1, ..., g_m)$ with $g_j \in C^2(\Omega)$ for all $j = 1, .., m$. Let $Q(v) = \langle H_L(s)v, v \rangle$, where L is the Lagrangian function. Then*

1. *If Q is positive definite on $T_s(S)$, f has a (strict) local minimum on S at s.*

2. *If Q is negative definite on $T_s(S)$, f has a (strict) local maximum on S at s.*

3. *If Q is indefinite on $T_s(S)$, f has no extremum on S at s.*

Proof. As $L(x) = f(x)$ for $x \in S$, if we prove that $s \in S$ is a local extremum of the restriction $L|_S$ of L on S, then automatically s will be a local extremum of $f|_S$. By hypothesis for the point s, we have $\nabla L(s) = 0$. Moreover, since $\frac{\partial^2 L}{\partial x_i \partial x_j}(x)$ are continuous at s, it follows as in Theorem 3.7.11, that the function $x \mapsto \langle H_L(x)v, v \rangle$ is also continuous at s.

Now, choose a smooth curve $X(t)$ on S such that $X(0) = s$ and $X'(0) = v$ with $v \in T_s(S)$. The second order Taylor expansion of L gives

$$L(X(t)) - L(X(0)) = \frac{1}{2}\langle H_L(c(t))(X(t) - X(0)), X(t) - X(0) \rangle,$$

for some $c(t)$ in the line segment between $X(0)$ and $X(t)$. Therefore,

$$\frac{L(X(t)) - L(X(0))}{t^2} = \frac{1}{2}\langle H_L(c(t))\frac{X(t) - X(0)}{t}, \frac{X(t) - X(0)}{t} \rangle.$$

Letting $t \to 0$, we have $c(t) \to X(0)$ and $\frac{X(t)-X(0)}{t} \to X'(0) = v$. Hence,

$$\lim_{t \to 0} \frac{L(X(t)) - L(X(0))}{t^2} = \frac{1}{2}\langle H_L(s)v, v \rangle = Q(v).$$

The proof now proceeds just as in the free extrema case.   □

Example 3.9.10. Find the dimensions of a rectangular box of minimal total surface area given that the volume is 1000. Find the minimum surface area.

Solution. Let x, y, z be the dimensions of the box. We wish to minimize $f(x, y, z) = 2(xy + xz + yz)$ subject to the constraint

$$S = \{(x, y, z) : g(x, y, z) = xyz - 1000 = 0\}.$$

Since x, y and z are all positive we have $\nabla g(x, y, z) = (yz, xz, yz) \neq (0, 0, 0)$. Applying the Lagrange multiplier rule, we must solve the system

$$2(y + z) = \lambda yz$$
$$2(x + z) = \lambda xz$$
$$2(x + y) = \lambda xy$$
$$xyz = 1000.$$

Eliminating λ from the first two equations we get $\frac{y+z}{yz} = \frac{x+z}{xz}$, which gives $x = y$. Similarly, the second and third equations give $y = z$.

Hence $x = y = z$, and substituting in the last equation, we obtain $x^3 = 1000$ or $x = 10$. Thus, the only critical point is $\mathbf{s} = (10, 10, 10)$, and $\lambda = \frac{2}{5}$.

We now wish to show that f takes its *minimum* value at the point $\mathbf{s}$. We shall use Theorem 3.9.9. The Lagrangian function is

$$L(x, y, z) = f(x, y, z) - \lambda g(x, y, z) = 2(xy + xz + zy) - \frac{2}{5}xyz + 400.$$

The Hessian of L at $\mathbf{s} = (10, 10, 10)$ is

$$H_L(\mathbf{s}) = H_L(10, 10, 10) = \begin{pmatrix} \frac{\partial^2 L}{\partial x^2} & \frac{\partial^2 L}{\partial y \partial x} & \frac{\partial^2 L}{\partial z \partial x} \\ \frac{\partial^2 L}{\partial x \partial y} & \frac{\partial^2 L}{\partial y^2} & \frac{\partial^2 L}{\partial z \partial y} \\ \frac{\partial^2 L}{\partial x \partial z} & \frac{\partial^2 L}{\partial y \partial z} & \frac{\partial^2 L}{\partial z^2} \end{pmatrix} = \begin{pmatrix} 0 & -2 & -2 \\ -2 & 0 & -2 \\ -2 & -2 & 0 \end{pmatrix}$$

The associated quadratic form is

$$Q(v) = Q(x, y, z) = \langle H_L(\mathbf{s})v, v \rangle = -2xy - 2xz - 2yz.$$

Note that Q is indefinite on $\mathbb{R}^3$. However, it is the behavior of Q on the *tangent space* $T_{\mathbf{s}}(S)$ that interest us. From Proposition 3.2.21, the tangent plane to S at $\mathbf{s}$ has normal $\nabla g(\mathbf{s}) = (100, 100, 100)$, and Example 1.3.32 tells us that its equation is $x + y + z = 30$. Hence the tangent space is

$$T_{\mathbf{s}}(S) = \left\{ (x, y, z) \in \mathbb{R}^3 : x + y + z = 0 \right\},$$

in other words, $T_{\mathbf{s}}(S)$ is the two dimensional subspace generated by the vectors $v_1 = (1, 0, -1)$ and $v_2 = (0, 1, -1)$. Now for any nonzero vector $v \in T_{\mathbf{s}}(S)$, we have $v = xv_1 + yv_2 = (x, y, -x - y)$, and so $Q(v) = 2(x^2 + xy + y^2)$. Since the quadratic form $q(x, y) = x^2 + yx + y^2$ is positive definite on $\mathbb{R}^2$ (by Sylvester's criterion), it follows that Q is positive definite on $T_{\mathbf{s}}(S)$. Thus, the minimum surface area of the box is $f(10, 10, 10) = 600$.

The second derivative test for constrained extrema is more of theoretical importance than practical. In practice, we are usually interested in finding the maximum or the minimum value of a function f on the constraint set S. When we apply the Lagrange method, we usually know beforehand that an extremum of f on S exists. In particular, if S is a compact set, then f will take on both a maximum and a minimum value on S. Thus, to find the extreme values of f on S, it is only necessary to evaluate f at the critical points (found by Lagrange's multiplier rule) and choose the greatest and least values.

Extrema over compact sets

To find the maximum and minimum values of a smooth function f on a compact set B, which we know are certainly attained, we employ a procedure similar to that in one variable calculus. If the interior of B is not empty, we find the critical points of f in the interior of B and use the second derivative test to determine their nature. Then we consider points on ∂B, the boundary of B. Frequently, the boundary of B can be expressed in the form $\partial B = \{x : g(x) = 0\}$ for a C^1 function g. This is a problem with constraints and can be studied by the Lagrange multipliers rule. Finally, we compare all these values and we select the largest and the smallest.

Example 3.9.11. Find the extreme values of $f(x, y) = 3x^2 - 2y^2 + 2y$ on the disk $B = \{(x, y) : x^2 + y^2 \leq 1\}$.

Solution. For interior critical points we solve

$$f_x = 6x = 0 \text{ and } f_y = -4y + 2 = 0.$$

Hence the only critical point is $(0, \frac{1}{2})$ (which is inside the disk). The Hessian is $H_f(x, y) = \begin{pmatrix} 6 & 0 \\ 0 & -4 \end{pmatrix}$. So the point $(0, \frac{1}{2})$ is a saddle point.

Boundary extrema are extrema of f subject to $g(x, y) = x^2 + y^2 - 1 = 0$. Using Lagrange's multiplier rule, we have to solve

$$6x = 2\lambda x, \quad -4y + 2 = 2\lambda y, \quad x^2 + y^2 = 1.$$

The first equation implies that either $x = 0$ or $\lambda = 3$. If $x = 0$, then the third equation gives $y = \pm 1$. If $x \neq 0$, then $\lambda = 3$ and the second equation gives $y = \frac{1}{5}$. Substituting this in the third equation yields $x = \pm \frac{\sqrt{24}}{5}$. Therefore the boundary critical points are

$$(0, \pm 1), \quad (\pm \tfrac{\sqrt{24}}{5}, \tfrac{1}{5}).$$

Evaluating we see that $f(0,1) = 0$, $f(0,-1) = -4$, $f(\pm\frac{\sqrt{24}}{5}, \frac{1}{5}) = \frac{16}{5}$. Thus the minimum value is -4 and the maximum $\frac{16}{5}$.

Remark 3.9.12. Suppose s is a critical point for f on $S = \{x \in \Omega : g(x) = 0\}$, where $g = (g_1, ..., g_m)$. Note that the Sylvester criterion can not be applied directly to the quadratic form $Q(v) = \langle H_L(s)v, v \rangle$ for $v \in T_s(S)$. However, assuming that $\nabla g_1(s), ..., \nabla g_m(s)$ are *linearly independent* (in case $m > 1$, of course, this is a *stronger hypothesis* than the one in Theorem 3.9.6, and therefore a weaker theorem), one can use the Implicit Function theorem to work differently and even prove Theorem 3.9.6 alternatively as follows:

Alternative proof of Theorem 3.9.6. Regard $\mathbb{R}^n$ as $\mathbb{R}^{n-m} \times \mathbb{R}^m$ and write $x \in \mathbb{R}^n$ as $x = (z, y)$ where, $z = (x_1, ..., x_{n-m})$ and $y = (x_{n-m+1}, ..., x_n)$. So that, $g(x) = 0$, becomes $g(z, y) = 0$ and $D_g(x) = D_g(z, y) = (\frac{\partial g}{\partial z}, \frac{\partial g}{\partial y})$, where $\frac{\partial g}{\partial y}$ is the $m \times m$ matrix given by

$$\frac{\partial g}{\partial y} = \begin{pmatrix} \frac{\partial g_1}{\partial x_{(n-m)+1}} & \cdots & \frac{\partial g_1}{\partial x_n} \\ \cdot & \cdots & \cdot \\ \cdot & \cdots & \cdot \\ \cdot & \cdots & \cdot \\ \frac{\partial g_m}{\partial x_{(n-m)+1}} & \cdots & \frac{\partial g_m}{\partial x_n} \end{pmatrix}$$

and similarly for the $(m \times (n-m))$ matrix $\frac{\partial g}{\partial z}$.

Since $\nabla g_1(s), ..., \nabla g_m(s)$ are linearly independent, we know from linear algebra, that the matrix $D_g(s)$ has maximal rank m, or equivalently,

$$\det\left(\frac{\partial g}{\partial y}(s)\right) \neq 0.$$

By the Implicit Function theorem, with $s = (a, b)$, there is a C^1 function φ defined in a neighborhood of a such that $g(z, \varphi(z)) = 0$. That is,

$$g(x_1, ..., x_{n-m}, \varphi(x_{n-m+1}, ..., x_n)) = 0.$$

Now, the function $F(z) = f(z, \varphi(z))$ is a smooth function $F : \mathbb{R}^{n-m} \to \mathbb{R}$ defined in a neighborhood of a. Furthermore, if the restriction of f on S has a local extremum at s, the function F has an (unconditional) local extremum at a and we can use the usual second derivative test for F to determine the nature of the local extremum. This of course requires the computation of the derivatives of F, up to order two. We can do this by implicit differentiation. In fact, implicit differentiation gives $D_F(a) = 0$ if and only if $\nabla f(s) = \sum_{j=1}^{m} \lambda_j \nabla g_j(s)$.

The calculation of $H_F(a)$ is rather involved, even in the case where there is only one constraint, but as the reader can check it leads to an $(n - m) \times (n - m)$ symmetric matrix whose positive (or negative) definiteness can be checked using the Sylvester criterion (or by computing its eigenvalues).

Remark 3.9.13.

1. Remark 3.9.12 describes another method for solving constrained extrema problems: We solve the equation $g(x) = 0$ for m of the variables either implicitly or explicitly (if possible) and thus reduce the problem to finding the critical points of a function of the remaining $n - m$ variables and test them using the second derivative test (Theorem 3.7.11), for (free) extrema. This was what we did in Example 3.7.15. In fact, this is what one *always* does in one variable calculus.

2. Yet, another possible way to solve such problems would be to describe the constraint set S parametrically in terms of parameters $t_1, , ..., t_{n-m}$ and find the critical points of f as a function of these new variables. This is particularly effective when S is a closed curve or surface such as a circle or a sphere that can not be described in its entirety as the graph of a function.

3. Calculus is of no use in finding *extrema of linear functionals*. Indeed, as we saw in Theorem 2.4.12 any linear functional λ on $\mathbb{R}^n$ is of the form $\lambda(x) = \langle w, x \rangle = x_1 w_1 + ... + x_n w_n$. If $a \in \mathbb{R}^n$ is a critical point of λ, then $\nabla \lambda(a) = 0$. Since $\nabla \lambda(x) = w$ for all $x \in \mathbb{R}^n$, this tells us $w = 0$. Thus, $\lambda \equiv 0$. As we have mentioned in the last paragraph of Subsection 2.6.3 in this case one uses methods of *linear programming*.

3.9.1 Applications to economics

In constrained optimization problems in Economics, the second derivative test for constrained extrema is stated in terms of the so called bordered Hessian of the Lagrangian L. The *bordered Hessian* is the Hessian of the Lagrangian function L regarded as a function in the $m + n$ variables $\lambda_1, ..., \lambda_m, x_1, ..., x_n$. In other words, the bordered Hessian is the Hessian of the function

$$\widetilde{L}(\lambda, x) = \widetilde{L}(\lambda_1, ..., \lambda_m, x_1, ..., x_n) = f(x_1, ..., x_n) - \sum_{j=1}^{m} \lambda_j g_j(x_1, ..., x_n).$$

When only one constraint $g(x_1, ..., x_n) = 0$ is present, calculating the bordered Hessian (an $(n + 1) \times (n + 1)$ symmetric matrix), we see it is of the form

$$H_{\widetilde{L}}(\lambda, x) = \begin{pmatrix} 0 & -\frac{\partial g}{\partial x_1} & \cdots & -\frac{\partial g}{\partial x_n} \\ -\frac{\partial g}{\partial x_1} & \frac{\partial^2 L}{\partial x_1^2} & \cdots & \frac{\partial^2 L}{\partial x_n \partial x_1} \\ . & \cdots & . & \cdots \\ . & \cdots & . & \cdots \\ -\frac{\partial g}{\partial x_n} & \frac{\partial^2 L}{\partial x_1 \partial x_n} & \cdots & \frac{\partial^2 L}{\partial x_n^2} \end{pmatrix}$$

When m constraints $g_j(x_1, ..., x_n) = 0$ $(j = 1, ..., m)$ are present the upper left corner of the bordered Hessian is an $m \times m$ block of zeros and there are m border rows at the top and m border columns at the left with entries the partial derivatives $-\frac{\partial g_j}{\partial x_i}(x)$ for $i = 1, ..., n$ and $j = 1, ..., m$. The quadratic form associated to this matrix $Q(v) = \langle H_{\widetilde{L}}(\lambda, x)v, v \rangle$, even in the case where $m = 1$, can not be definite on $\mathbb{R}^{n+1}$; for any $c \neq 0$ the vector $v = ce_1 \neq 0$ but $Q(v) = 0$. The rules concerning positive and negative definite forms do not apply directly to the quadratic form associated with $H_{\widetilde{L}}(\lambda, x)$. However, we have the following

Theorem 3.9.14. *Let Ω be an open set in $\mathbb{R}^n$ and $f, g \in C^2(\Omega)$. Suppose s is a critical point for f on the constraint set $S = \{x \in \Omega : g(x) = 0\}$. Let $H_{\widetilde{L}}(\lambda, x)$ be the bordered Hessian and*

$$d_k = \det(H_{\widetilde{L}, k}(\lambda, s)) = \det \begin{pmatrix} 0 & -\frac{\partial g}{\partial x_1} & \cdots & -\frac{\partial g}{\partial x_k} \\ -\frac{\partial g}{\partial x_1} & \frac{\partial^2 L}{\partial x_1^2} & \cdots & \frac{\partial^2 L}{\partial x_k \partial x_1} \\ . & \cdots & . & \cdots \\ . & \cdots & . & \cdots \\ -\frac{\partial g}{\partial x_k} & \frac{\partial^2 L}{\partial x_1 \partial x_k} & \cdots & \frac{\partial^2 L}{\partial x_k^2} \end{pmatrix}$$

$(k = 2, ..., n)$ *be the principal minors of the bordered Hessian* $H_{\tilde{L}}(\lambda, x)$ *of order* ≥ 3 *evaluated at s. Then*

1. *If* $d_k < 0$ *for all* $k = 2, ..., n$, f *has a local minimum on* S *at* s.

2. *If* $(-1)^{k+1}d_k < 0$ *for all* $k = 2, ..., n$, f *has a local maximum on* S *at* s.

3. *If* d_k *are not all zero and neither (1) nor (2) hold,* f *has no extremum at* s.

Example 3.9.15. *(Cobb-Douglas production function[12]).*
The function $Q(L, K) = AK^\alpha L^{1-\alpha}$ is sometimes used as a simple model of the national economy. $Q(L, K)$ is then the aggregate output of the economy for a given amount of capital K invested and a given amount of labor L used. A and α are positive constants with $0 < \alpha < 1$ (the constant A captures the stages of the division of labor and the technology level during the production). Let r be the price of capital (or the rate of return of capital), w the price of labor (or the wage rate) and B the amount of dollars that can be spent in the production (the production budget). What amount of capital and labor maximizes the output Q?

Solution. We wish to maximize $Q(L, K) = AK^\alpha L^{1-\alpha}$ subject to the constraint $G(L, K) = wL + rK - B = 0$. By the Lagrange multiplier rule, $\nabla Q = \lambda \nabla G$. We compute the partial derivatives of Q and G.

$$\frac{\partial Q}{\partial L} = A(1 - \alpha)L^{-\alpha}K^\alpha = (1 - \alpha)A\left(\frac{K}{L}\right)^\alpha,$$
$$\frac{\partial Q}{\partial K} = A\alpha K^{\alpha-1}L^{1-\alpha} = \alpha A\left(\frac{L}{K}\right)^{1-\alpha},$$

$$\frac{\partial G}{\partial L} = w, \quad \frac{\partial G}{\partial K} = r.$$

Hence we have to solve the system

$$(1 - \alpha)A\left(\frac{K}{L}\right)^\alpha = \lambda w \quad (1)$$

---

[12]It was proposed by Knut Wicksell (1851-1926), and tested against statistical evidence by Charles Cobb and Paul Douglas in 1900-1928.

$$\alpha A \left(\frac{L}{K}\right)^{1-\alpha} = \lambda r \qquad (2)$$

$$wL + rK = B \qquad (3)$$

Dividing equation (2) by (1) we get

$$(\frac{\alpha}{1-\alpha})\frac{L}{K} = \frac{r}{w}.$$

Solving this for K yields $K = (\frac{\alpha}{1-\alpha})L\frac{w}{r}$ (4). Substitution of (4) into (3) gives $L = \frac{(1-\alpha)B}{w}$. Finally, substituting this back into (4) we obtain $K = \frac{\alpha B}{r}$. Thus the output is maximized at the point

$$(L, K) = \left(\frac{(1-\alpha)B}{w}, \frac{\alpha B}{r}\right).$$

EXERCISES

1. Find the extreme values of $f(x, y, z) = xy + yz + zx$ subject to the constraint $x^2 + y^2 + z^2 = 3$.

2. Show that the maximum of $f(x, y, z) = -x^2 + y - 2z^2$ subject to the constraint $x^4 + y^4 - z^2 = $ is $\frac{3}{8}$.

3. Find the extreme values of $f(x, y, z) = x^2 + y^2 + z^2 - xy - yz - zx$ subject to the constraint $x^2 + y^2 + z^2 - 2x + 2y + 6z + 9 = 0$.

4. Find the maximum of $x^2 y^2 z^2$ on the sphere $x^2 + y^2 + z^2 = a^2$.

5. Find the maximum and minimum of $f(x, y, z) = x^2 + y^2 + z^2$ subject to the constraints $\frac{x^2}{16} + y^2 + z^2 = 1$ and $x + y + z = 0$.

6. Find the maximum and minimum of $f(x, y, z) = x^3 + y^3 + z^3$ subject to the constraints $x^2 + y^2 + z^2 = 1$ and $x + y + z = 1$.

7. Let $f(x, y) = x^2 + 2x + y^2$. Find the minimum value of f on $|x| + |y| = 4$ by Lagrange multipliers. Why can't this method be used to find the maximum value of f?

8. Let $f(x,y) = x^2 + 2x + y^2$. Find the maximum and minimum values of f on the set $\{(x,y) : x^2 + y^2 \leq 1\}$.

9. Find the maximum and minimum values of $f(x,y,z) = 6x+3y+2z$ on the set $4x^2 + 2y^2 + z^2 \leq 70$.

10. Let $f(x,y) = (1 - x^2)\sin y$. Find the maximum and minimum values of f on the set $\{(x,y) : |x| \leq 1, |y| \leq \pi\}$.

11. Let $f(x,y,z) = 2(x + y + z) - xyz$. Find the maximum and minimum values of f on the set $\{(x,y,z) : x^2 + y^2 + z^2 \leq 9\}$.

12. Find the minimum of the function

$$f(x,y,z) = (x - 1)^2 + (\tfrac{y}{x} - 1)^2 + (\tfrac{z}{y} - 1)^2 + (\tfrac{4}{z} - 1)^2$$

on the set $\{(x,y,z) : 1 \leq x \leq y \leq z \leq 4\}$.

13. Let $P = (p,q,r)$ be a point in $\mathbb{R}^3$ that doesn't lie in the plane $z = Ax + By + D$. Find the point on this plane nearest to P. Find the shortest distance. (This problem can be done by the usual method, or by Lagrange multipliers. In the latter case one may as well assume the plane to be given by $Ax+By+Cz+D = 0$).

14. Use Lagrange multipliers to find the triangle of largest area with a given perimeter. Can this maximum problem be solved by the methods of calculus of one variable? *Hint.* Let x, y and θ be, respectively, two sides of the triangle and the adjacent angle. Take these as your variables and express the third side z and the area A as functions of these.

15. Find the largest volume closed rectangular box with a total surface area of 10 square feet.

16. Find the shortest distance from a point on the ellipse $x^2 + 4y^2 = 4$ to the line $x + y = 4$.

17. $\Omega = \{(x,y,z) \in \mathbb{R}^3 : 4x^2 + y^2 \leq 16, x + z = 1\}$. Find the points of Ω which are, respectively, nearest to and furthest from the origin.

18. Let $a, x \in \mathbb{R}^n$ with $a \neq 0$.

(a) Show that $\frac{1}{||a||^2} = \min\left\{||x||^2 : \langle a, x \rangle = 1\right\}$.

(b) Show that $||a|| = \max\left\{\langle a, x \rangle : ||x|| = 1\right\}$.

19. Let $A : \mathbb{R}^n \to \mathbb{R}^m$ be a linear map.

(a) Show that the function $f(x) = ||Ax||$ has a maximum value on the unit sphere $S = \{x \in \mathbb{R}^n : ||x|| = 1\}$.

(b) Let M be the maximum in part (1). Show that $||Ax|| \le M||x||$ for all $x \in \mathbb{R}^n$, with equality for at least one unit vector x. Deduce that $M = ||A|| = \sup\{||Ax|| : ||x|| = 1\}$. The number $||A||$ is called the *norm* of A.

20. Show that the maximum of $f(x_1, ..., x_n) = (x_1 \cdots x_n)^2$ on the unit sphere $S = \{x \in \mathbb{R}^n : ||x|| = 1\}$ is equal to n^{-n}. Use this to obtain the inequality

$$|x_1 \cdots x_n| \le ||x||^n n^{-\frac{n}{2}}, \text{ for } x \in \mathbb{R}^n.$$

Now let $a_1, \ldots, a_n$ be any set of n positive real numbers. Use the above to prove the inequality

$$\sqrt[n]{a_1 \cdots a_n} \le \frac{\sum_{i=1}^n a_i}{n} \tag{3.10}$$

This says that the *geometric mean* of any finite set of positive numbers $a_1, \ldots, a_n$ is less than or equal to the *arithmetic mean* of these numbers.

21. Denote by $S = \{x = (x_1, \ldots, x_n) \in \mathbb{R}^n : x_i > 0, \sum_{i=1}^n x_i = 1\}$. Observe that on the set S the function $f(x) = \log(x_1 \cdots x_n)$ is always less than zero. Use Lagrange multipliers to maximize f on S. Exponentiate the estimate you get to show that $x_1 \cdots x_n \le n^{-n}$ for $x \in S$. Use this to deduce the inequality (3.10). *Hint.* For each $i = 1, \ldots, n$, let $x_i = \frac{a_i}{\sum a_i}$.

22. Prove Theorem 3.9.14.
Hint. Let λ_0 be the value of λ determined by the Lagrange multiplier rule. Consider the function

$$F(x_1, ..., x_n, \lambda) = f(x_1, ..., x_n) - \lambda g(x_1, ...x_n) \pm (\lambda - \lambda_0)^2$$

and study F using the unconstrained second derivative test.

Answers to selected Exercises

3. Maximum at $(2, -1, -4)$, minimum at $(0, -1, -2)$.   4. $\left(\frac{a^2}{3}\right)^3$.

5. max $= \frac{8}{3}$, min $= 1$.   6. max $= 1$, min $= \frac{5}{9}$.   11. max $= 10$, min $= -10$.

12. $f(\sqrt{2}, 2, \sqrt{8}) = 12 - 8\sqrt{2}$.   16. $\frac{4-\sqrt{5}}{\sqrt{2}}$.   17. Furthest points $(-\frac{1}{2}, \pm\sqrt{15}, \frac{3}{2})$, nearest $(\frac{1}{2}, 0, \frac{1}{2})$.

3.10   Functional dependence

Functional dependence generalizes the familiar notion, from linear algebra, of linear dependence of a collection of numerical functions $f_1, ... f_m$.

Definition 3.10.1. Let Ω be an open set in $\mathbb{R}^n$ and let $f_j : \Omega \to \mathbb{R}$ be $C^1(\Omega)$ for $j = 1, ..., m$. The functions $f_1, ..., f_m$ are said *functionally dependent* at $a \in \Omega$, if there is a neighborhood V of $b = (f_1(a), ... f_m(a))$ in $\mathbb{R}^m$ and a C^1 function $F : V \to \mathbb{R}$ with $\nabla F \neq 0$ on V and $F(f_1(x), ..., f_m(x)) = 0$ for all x in some neighborhood U of a. Moreover, if $f_1, ..., f_m$ are functionally dependent at every point of Ω, we say that they are *functionally dependent* on Ω. But notice that this is a local condition.

Set $y = (y_1, ..., y_m)$ with $y_j = f_j(x)$ for $j = 1, ..., m$. The nonvanishing of $\nabla F(y)$ on a neighborhood of b, guarantees, via the Implicit Function theorem, that the equation

$$F(y_1, ..., y_m) = 0$$

can be solved locally for (at least) one of the variables $y_1, ..., y_m$ in terms of the others, that is, at least one of the functions, say f_m, is expressed in terms of the other functions by means of an equality

$$f_m(x) = \Phi(f_1(x), ..., f_m(x))$$

$x \in U$, where Φ is a C^1 function of $y_1, ... y_{m-1}$.

Observe that the notion of linear dependence studied in linear algebra, is dependence with respect to linear relations

$$F(y_1, ..., y_m) = c_1 y_1 + ... + c_m y_m$$

Since $\nabla F(y) = (c_1, ..., c_m)$, $\nabla F(y) \neq 0$ means that $c_j \neq 0$ for at least one $j = 1, ..., m$ and so, at least one function is a linear combination of the others. However, the two notions, should not be confused. For instance, for the functions $f_1(x) = \sin x$, $f_2(x) = \cos x$, $\sin^2 x + \cos^2 x = 1$. So $f_1^2(x) + f_2^2(x) - 1 = 0$. Hence f_1, f_2 are functionally dependent on $\mathbb{R}$ (Here $F(y_1, y_2) = y_1^2 + y_2^2 - 1$). On the other hand, if $c_1 \sin x + c_2 \cos x = 0$ for all $x \in \mathbb{R}$, then setting $x = 0$ and $x = \frac{\pi}{2}$, this implies $c_1 = c_2 = 0$. So f_1, f_2 are *not* linearly dependent.

Next, we want to find necessary and sufficient conditions for a collection of C^1 functions $f_1, ..., f_m$ in n variables to be functionally dependent.

To motivate our discussion, we look first at the linear case.

The linear case

When $n = m$, let $\lambda_1, ..., \lambda_n$ be linear functions (functionals) from $\mathbb{R}^n \to \mathbb{R}$ and consider the linear map $\Lambda : \mathbb{R}^n \to \mathbb{R}^n$ given by $\Lambda(x) = (\lambda_1(x), ..., \lambda_n(x))$. Let A be the $n \times n$ standard matrix of Λ. If $\det(A) = 0$, then $\ker(\Lambda) \neq (0)$ and so, the range of Λ (or equivalently, the column space of A) is a proper subspace of $\mathbb{R}^n$. Suppose, $\dim(\text{range}(\Lambda)) = k$ (namely, $\text{rank}(A) = k$), where $k < n$. Then, only k rows of A are linearly independent, say the first k rows, $A_1, ..., A_k$, while the other $n - k$ rows depend on $A_1, ..., A_k$. Since $\lambda_j(x) = A_j x$, for $j = 1, ..., k$, it follows that the functions $\lambda_1, ..., \lambda_k, \lambda_{k+1}, ..., \lambda_n$ are linearly dependent.

Similarly, when $n \neq m$, one can consider linear maps $\Lambda : \mathbb{R}^n \to \mathbb{R}^m$ defined by $m \times n$ matrices A. The range of such a map is a linear subspace of $\mathbb{R}^m$ whose dimension is the rank of A. If $m > n$, then the range of Λ is always a proper subspace of $\mathbb{R}^m$ and so, as before, the component functions $\lambda_1, ..., \lambda_m$ of Λ are (always) linearly dependent. On the other hand, if $m < n$, the range of Λ may be either a proper subspace of $\mathbb{R}^m$, in which case, $\lambda_1, ..., \lambda_m$ are linearly dependent, or it is the whole space $\mathbb{R}^n$, in which case $\lambda_1, ..., \lambda_m$ are linearly independent.

Thus, we see that the question of linear dependence for a collection of linear functions $\lambda_1, ..., \lambda_m$ is determined by examining the range of Λ, that is, the rank of A. Since the Jacobian matrix of a linear map Λ is just its standard matrix A, it is natural to turn our attention at the Jacobian matrix D_f of the (nonlinear) map $f(x) = (f_1(x), ..., f_m(x))$ in the study of functional dependence of more general functions $f_1, ..., f_m$.

The nonlinear case

It should be noted, as the linear case suggests, the question of functional dependence, is interesting only when $m \leq n$, that is, the number of functions does not exceed the number of independent variables; when $m > n$ functional dependence normally holds. For example, *any* two C^1 functions f_1, f_2 of one variable, such that either $f_1' \neq 0$ or $f_2' \neq 0$ on any (open) interval I of $\mathbb{R}$, are automatically functionally dependent. Indeed, say $f_1' \neq 0$ on I. Then f_1 is one-to-one on I. Hence, it has an inverse function f_1^{-1} and $y_1 = f_1(x)$ if and only if $x = f_1^{-1}(y_1)$ for all $x \in I$. Therefore, for $y_2 = f_2(x)$ the C^1 function $F(y_1, y_2) = f_2(f_1^{-1}(y_1)) - y_2$ satisfies $F(f_1(x), f_2(x)) = 0$ for all $x \in I$, and the functions f_1, f_2 are functionally dependent. Thus, we shall concern ourselves, only with the case $m \leq n$.

We now state and prove the main results on functional depedence.

Theorem 3.10.2. *Let $f = (f_1, ..., f_m)$ be a C^1 function from an open set Ω of $\mathbb{R}^n$ into $\mathbb{R}^m$, with $m \leq n$. If $f_1, ..., f_m$ are functionally dependent on Ω, then all the minors of the mth order of the Jacobian matrix $D_f(x)$ are identically zero on Ω. In particular, when $m = n$, $\det(D_f(x)) \equiv 0$ on Ω*

Proof. Let a be any point of Ω. By the functional dependence of $f_1, ..., f_m$ at a, there is a C^1 function F in m variables such that $\nabla F(y) \neq 0$ for $y = f(x)$ in a neighborhood of $f(a)$ and $F(f(x)) = 0$ for x in a neighborhood U of a. Differentiating $F(f(x)) = 0$, we get

$$\nabla F(y) D_f(x) = 0. \tag{3.11}$$

If the conclusion of the Theorem were false, there would be some $m \times m$ submatrix of $D_f(x)$ which would be nonsingular. By relabeling the functions and the variables, we can assume it is the one in the upper left corner. That is, $\det(\frac{\partial f_j}{\partial x_j}(x)) \neq 0$, for $j = 1, ..., m$.

Now, by (3.11)

$$\left(\frac{\partial F}{\partial y_1} \cdots \frac{\partial F}{\partial y_m} \right) \begin{pmatrix} \frac{\partial f_1}{\partial x_1} & \cdots & \frac{\partial f_1}{\partial x_m} \\ \cdot & \cdots & \cdot \\ \cdot & \cdots & \cdot \\ \cdot & \cdots & \cdot \\ \frac{\partial f_m}{\partial x_1} & \cdots & \frac{\partial f_m}{\partial x_m} \end{pmatrix} = 0$$

Since the matrix $(\frac{\partial f_j}{\partial x_j})$, $j = 1, ..., m$ is invertible at $x \in U$, we conclude that $\nabla F(y) = 0$ which contradicts our assumption on F. $\qquad \square$

Theorem 3.10.2 has the following interesting converse.

Theorem 3.10.3. *Let* $f = (f_1, ..., f_m)$ *be a* C^1 *function from an open set* Ω *of* $\mathbb{R}^n$ *into* $\mathbb{R}^m$. *Suppose that all the* m^{th}-*order minors of the Jacobian matrix* $D_f(x)$ *are zero on* Ω *and let* k $(k < m)$ *be the greatest of the indices for which at least one of the* k^{th}-*order minors of the matrix* $D_f(x)$ *is not zero at a point* $a \in \Omega$. *Then there is a neighborhood* U *of* a, *such that* $f_1, ..., f_k$ *are functionally dependent and the other* $m - k$ *functions* $f_{k+1}, ..., f_m$ *depend on* $f_1, ..., f_k$ *in the neighborhood* U. *In other words, there exist* C^1 *functions* Φ_j, $j = k + 1, ..., m$, *such that for* $x \in U$*

$$f_j(x) = \Phi_j(f_1(x), ..., f_k(x)).$$

Proof. By relabeling, we may assume that

$$J_{k,f}(a) = \det \begin{pmatrix} \frac{\partial f_1}{\partial x_1}(a) & \cdots & \frac{\partial f_1}{\partial x_k}(a) \\ \cdot & \cdots & \cdot \\ \cdot & \cdots & \cdot \\ \cdot & \cdots & \cdot \\ \frac{\partial f_k}{\partial x_1}(a) & \cdots & \frac{\partial f_k}{\partial x_k}(a) \end{pmatrix} \neq 0.$$

Since f is C^1 and the determinant function is continuous, it follows that $J_{k,f}(x)) \neq 0$ in a neighborhood U_1 of a. Therefore, by Theorem 3.10.2, the functions $f_1, ..., f_k$ are functionally independent on U_1 (and therefore on any neighborhood of a contained in U_1). By the Implicit Function theorem (applied to the function $\tilde{f} = (f_1, ..., f_k)$ from $\mathbb{R}^k \times \mathbb{R}^{n-k}$ to $\mathbb{R}^k$), there exist C^1 functions ψ_i defined in a neighborhood U_2 contained in U_1 (U_2 is a neighborhood of the point $(0, ..., 0, a_{k+1}, ..., a_n)$) such that

$$x_i = \psi_i(x_{k+1}, ..., x_n), \text{ for } i = 1, ..., k \text{ and}$$

$$f_j(\psi_1, ..., \psi_k, x_{k+1}, ..., x_n) = 0, \ (j = 1, ..., k). \tag{3.12}$$

Let p be an integer with $k < p \le n$. Differentiating (3.12) with respect to x_p we get

$$\sum_{i=1}^{k} \frac{\partial f_j}{\partial x_i} \frac{\partial \psi_i}{\partial x_p} + \frac{\partial f_j}{\partial x_p} = 0, \ (j = 1, ..., k). \tag{3.13}$$

For $j = k+1, ..., n$, let Φ_j be the functions

$$\Phi_j(y_1, ..., y_k, x_{k+1}, ..., x_n) = f_j(\psi_1, ..., \psi_k, x_{k+1}, ..., x_n),$$

where $y_1 = f_1(x), ..., y_k = f_k(x)$, for $x \in U_2$.
Then

$$\frac{\partial \Phi_j}{\partial x_p} = \sum_{i=1}^{k} \frac{\partial f_j}{\partial x_i} \frac{\partial \psi_i}{\partial x_p} + \frac{\partial f_j}{\partial x_p} = 0, (j = k+1, ..., n). \tag{3.14}$$

By hypothesis we have

$$\det \begin{pmatrix} \frac{\partial f_1}{\partial x_1} & \cdots & \frac{\partial f_1}{\partial x_k} & \frac{\partial f_1}{\partial x_p} \\ \cdot & \cdots & \cdot \\ \cdot & \cdots & \cdot \\ \frac{\partial f_k}{\partial x_1} & \cdots & \frac{\partial f_k}{\partial x_k} & \frac{\partial f_k}{\partial x_p} \\ \frac{\partial f_j}{\partial x_1} & \cdots & \frac{\partial f_j}{\partial x_k} & \frac{\partial f_j}{\partial x_p} \end{pmatrix} = 0,$$

on Ω, for $j = k+1, ..., n$. Multiplying the first k columns of this matrix by, respectively, $\frac{\partial \psi_i}{\partial x_p}$ for $i = 1, ..., k$ and adding to the last column, using

(3.13), (3.14), we get

$$\det \begin{pmatrix} \frac{\partial f_1}{\partial x_1} & \cdots & \frac{\partial f_1}{\partial x_k} & 0 \\ . & \cdots & \cdots & \\ . & \cdots & \cdots & \\ \frac{\partial f_k}{\partial x_1} & \cdots & \frac{\partial f_k}{\partial x_k} & 0 \\ \frac{\partial f_j}{\partial x_1} & \cdots & \frac{\partial f_j}{\partial x_k} & \frac{\partial \Phi_j}{\partial x_p} \end{pmatrix} = 0.$$

This implies

$$J_{k,f}(x) \cdot \frac{\partial \Phi_j}{\partial x_p}(x) = 0.$$

Since $J_{k,f}(x) \neq 0$, it follows that $\frac{\partial \Phi_j}{\partial x_p}(x) = 0$ in some neighborhood V of the point $(f_1(a), ..., f_k(a), a_{k+1}, ..., a_n)$. Hence for $j = k+1, ..., n$ the functions

$$\Phi_j(y_1, ..., y_k, x_{k+1}, ..., x_n) = f_j(\psi_1, ..., \psi_k, x_{k+1}, ..., x_n)$$

do not depend on the variables $x_{k+1}, ..., x_n$. Therefore,

$$f_j(\psi_1, ..., \psi_k, x_{k+1}, ..., x_n) = \Phi_j(y_1, ..., y_k).$$

Finally, by the continuity of f_i, $(i = 1, ..., k)$ at a, for every neighborhood of $f_i(a)$, there is a neighborhood W of a such that $f_i(W)$ is contained in any neighborhood of $f_i(a)$. So we can choose the neighborhood W so that $f_i(W) \subseteq V$. Thus, taking $U = U_2 \cap W$, for $x \in U$, we have

$$f_j(x) = f_j(x_1, ..., x_k, x_{k+1}, ..., x_n) = f_j(\psi_1, ..., \psi_k, x_{k+1}, ..., x_n)$$

$$= \Phi_j(y_1, ..., y_k) = \Phi_j(f_1(x), ..., f_k(x)).$$

$$\square$$

Example 3.10.4. Let $f_1(x, y) = e^{x+y}$, $f_2(x, y) = x^2 + y^2 + 2(x + y + xy)$. Show that f_1, f_2 are functionally dependent and find the functional relation.

Solution. We compute the Jacobian

$$\frac{\partial(f_1, f_2)}{\partial(x, y)} = \det \begin{pmatrix} e^{x+y} & e^{x+y} \\ 2x + 2(1+y) & 2y + 2(1+x) \end{pmatrix} = 0.$$

By Theorem 3.10.3 the functions f_1, f_2 are functionally dependent. To find the functional relationship, note that $f_2 = (x + y)^2 + 2(x + y)$ and $\log(f_1(x, y)) = x + y$. Hence $f_2 = [\log(f_1)]^2 + 2\log(f_1)$.

EXERCISES

1. Determine whether the following functions are functionally dependent on some open set $U \subseteq \mathbb{R}^2$. If they are, find the functional relation.

   (a) $f_1(x, y) = \frac{x^2 - y^2}{x^2 + y^2}$, $f_2(x, y) = \frac{2xy}{x^2 + y^2}$.

   (b) $f_1(x, y) = xy + e^{2xy}$, $f_2(x, y) = xy - e^{2xy}$.

2. Determine whether the following functions are functionally dependent on some open set $U \subseteq \mathbb{R}^3$. If they are, find the functional relation.

   (a) $f_1(x, y, z) = \frac{xe^z}{y^2} + \frac{y^2 e^{-z}}{x}$, $f_2(x, y, z) = \frac{xe^z}{y^2} - \frac{y^2 e^{-z}}{x}$.

   (b) $f_1(x, y, z) = xe^y \sin z$, $f_2(x, y, z) = xe^y \cos z$, $f_3(x, y, z) = x^2 e^{2y}$.

3. Let $f = (f_1, ..., f_m)$ be a C^1 function on a subset Ω of $\mathbb{R}^n$ to $\mathbb{R}^m$. Show that the functions $f_1, ..., f_m$ are functionally dependent in an open subset U of Ω if and only if $f(U)$ has no interior point. *Hint.* If $f(U)$ has no interior point, define $\psi : \mathbb{R}^m \to \mathbb{R}$ by $\psi(y) = 0$ for $y \in f(U)$, $\psi(y) = 1$ for $y \notin f(U)$. On the other hand, if $f(U)$ has an interior point, then, whenever $\psi(f(x)) = 0$ for $x \in U$, $\psi(y) = 0$ in an open set containing that point.

Answers to selected Exercises

1. (a) $f_1^2 + f_2^2 = 1$. (b) $2e^{f_1 + f_2} - f_1 + f_2$.
2. (a) $f_1^2 - f_2^2 = 4$. (b) $f_1^2 + f_2^2 - f_3^2 = 0$.

3.11 Morse's lemma (*)

In Section 3.7, we saw that the Hessian of a smooth function $f : \Omega \subseteq \mathbb{R}^n \to \mathbb{R}$, at a nondegenerate critical point $a \in \Omega$, determined the local behavior of f near this point. The Morse[13] lemma carries this step further and gives a complete description of the function near the point a. It is an important tool in more advanced work in topology and analysis.

As we observed in Theorem 3.7.17 a quadratic form can be brought into its canonical form by a linear change of coordinates. Morse's lemma asserts that at a nondegenerate critical point one can make a local change of cordinates $x = g(y)$ so that the function f will have the form

$$(f \circ g)(y) = f(a) - y_1^2 - \ldots - y_k^2 + y_{k+1}^2 + \ldots + y_n^2$$

when expresed in the new y-coordinates. The proof uses the idea of the proof of the corresponding algebraic theorem and relies on the Inverse Function theorem and the following lemma.

Lemma 3.11.1. *Let $f : U \to \mathbb{R}$ be a smooth function defined in a convex neighborhood U of 0 in $\mathbb{R}^n$ with $f(0) = 0$. Then there exist smooth functions g_i ($i = 1, \ldots, n$) defined in U with $g_i(0) = \frac{\partial f}{\partial x_i}(0)$ such that*

$$f(x) = \sum_{i=1}^n x_i g_i(x), \text{ for all } x = (x_1, \ldots x_n) \in U.$$

Proof. Using the Fundamental Theorem of Calculus and the Chain Rule, we can write

$$f(x_1, \ldots, x_n) = \int_0^1 \frac{df(tx_1, \ldots, tx_n)}{dt} dt = \int_0^1 \sum_{i=1}^n \frac{\partial f}{\partial x_i}(tx_1, \ldots tx_n) dt.$$

Hence we set

$$g_i(x_1, \ldots x_n) = \int_0^1 \frac{\partial f}{\partial x_i}(tx_1, \ldots tx_n) dt \text{ for } i = 1, \ldots, n.$$

Now $g_i(0) = \int_0^1 \frac{\partial f}{\partial x_i}(0) dt = \frac{\partial f}{\partial x_i}(0)$. To see that $g_i \in C^\infty(U)$, one only needs to justify differentiation under the integral sign. The reader may accept this for the moment, since in Chapter 5 (Theorem 5.3.2) we shall prove a general rule on differentiation of integrals depending on a parameter of which this is a special case. $\square$

[13]M. Morse (1892-1977), was a professor of mathematics at the Institute for Advanced Study. He is best knwon for his work on Calculus of Variations in the large (meaning its applications to topology).

Theorem 3.11.2. *(Morse Lemma). Let Ω be open in $\mathbb{R}^n$ and $f : \Omega \to \mathbb{R}$ be a smooth function. Suppose $a \in \Omega$ is a non degenerate critical point of f. Then there is a diffeomorphism $g : V \to U$ of some neighborhood V of $f(a)$ in $\mathbb{R}^n$ onto a neighborhood U of a, such that in new coordinates $y \in V$*

$$(f \circ g)(y) = f(a) - y_1^2 - ... - y_k^2 + y_{k+1}^2 + ... + y_n^2, \text{ for all } y \in V,$$

where $0 \le k \le n$.

The non negative integer k is called the index of f at a and equals the number of negative eigenvalues of the Hessian $H_f(a)$.

Proof. By linear change of variables, we can assume that $a = 0$ and $f(a) = 0$. Since these linear changes of variable leave the first partial derivatives unaffected and therefore also the second partial derivatives, we see that the new function has a non degenerate point at 0. From Lemma 3.11.1, for $x = (x_1, ..., x_n)$ in some neighborhood of 0, we write

$$f(x) = \sum_{i=1}^{n} x_i g_i(x).$$

Since 0 is a critical point of f, $g_i(0) = \frac{\partial f}{\partial x_i}(0) = 0$. Applying Lemma 3.11.1 now to the g_i we have

$$g_i(x_1, ...x_n) = \sum_{j=1}^{n} x_j h_{ij}(x_1, ..., x_n)$$

for certain smooth functions h_{ij} in a neighborhood V of 0 (V is the intersection of the finite number of neighborhoods V_i of 0 associated with each g_i to which we apply Lemma 3.11.1). Consequently

$$f(x_1, ..., x_n) = \sum_{i,j=1}^{n} x_i x_j h_{ij}(x_1, ..., x_n). \tag{3.15}$$

Next we symmetrize by making the substitution $\widetilde{h}_{ij} = \frac{1}{2}(h_{ij} + h_{ji})$ if necessary, and we can assume that $h_{ij} \equiv h_{ji}$. We remark also that, by the uniqueness of the Taylor expansion, the continuity of the functions h_{ij} implies that $h_{ij}(0) = \frac{1}{2}\frac{\partial^2 f}{\partial x_i \partial x_j}(0)$. Since 0 is a non degenerate critical point of f, it follows that the matrix $(h_{ij}(0))$ is non singular. The function f has now been written in a way analogous to a quadratic form and we wish, so to speak, to reduce it to its canonical form. Just as in the classical case we proceed by induction.

Suppose that there exist coordinates $u_1, ..., u_n$ in a neighborhood U_1 of $0 \in \mathbb{R}^n$, that is, a diffeomorphism $x = \varphi(u)$, such that

$$(f \circ \varphi)(u) = \pm(u_1)^2 \pm ... \pm (u_{r-1})^2 + \sum_{i,j=1}^{n} u_i u_j H_{ij}(u_1, ...u_n) \quad (3.16)$$

on U_1, where $r \geq 1$ and the matrices H_{ij} are symmetric. Note that from relation (3.15) for $H_{ij} = h_{ij}$ it follows that relation (3.16) holds for $r = 1$.

Since $\det(h_{ij}(0)) \neq 0$, the quadratic form $\sum_{i,j=1}^{n} x_i x_j h_{ij}(0)$ is non degenerate. At the same time since $x = \varphi(u)$ is a diffeomorphism at 0, $\det(D_\varphi(0)) \neq 0$. Hence the matrix of the quadratic form $\pm(u_1)^2 \pm ... \pm (u_{r-1})^2 + \sum_{i,j=1}^{n} u_i u_j H_{ij}(0)$ obtained from the matix $(h_{ij}(0))$ through right multiplication by $D_\varphi(0)$ and left multiplication by the transpose of $D_\varphi(0)$ which is also non singular. By a linear change of variable we can bring the form $\sum_{i,j=1}^{n} u_i u_j H_{ij}(0)$ to diagonal form, and so we may assume that $H_{rr}(0) \neq 0$. Because the functions $H_{ij}(u)$ are continuous at 0, it follows that $H_{rr}(u) \neq 0$ for u in a smaller neighborhood $U_2 \subseteq U_1$ of 0.

Let $\psi(u_1, ..., u_n) = \sqrt{H_{rr}(u)}$. Hence, ψ is a smooth nonzero function throughout U_2. Now introduce new coordinates $v_1, ..., v_n$ by

$$v_i = u_i \text{ for } i \neq j$$

$$v_r = \psi(u_1, ..., u_n) \left(u_r + \sum_{i>r} \frac{u_i H_{ir}(u_1, ..., u_n)}{H_{rr}(u_1, ..., u_n)} \right). \quad (3.17)$$

The Jacobian of the mapping $(u_1, ..., u_n) \mapsto (v_1, ...v_n)$ at 0 is

$$\frac{\partial(v_1, ..., v_n)}{\partial(u_1, ...u_n)} = \det \begin{pmatrix} I_r & & 0 \\ \frac{\partial v_r}{\partial u_1} & ...\psi(0) & ...\frac{\partial v_r}{\partial u_n} \\ 0 & & I_{n-r-1} \end{pmatrix} = \psi(0) \neq 0,$$

where I_r is the $r \times r$ identity matrix and I_{n-r-1} is the $(n-r-1) \times (n-r-1)$ identity matrix. Therefore, by the Inverse Function theorem this mapping has a smooth inverse in some neighborhood $U_3 \subseteq U_2$ of 0 and so the variables $v_1, ..., v_n$ can serve as coordinates of points in U_3. Now, using the symmetry of H_{ij} we can write (3.16)

$$u_r u_r H_{rr}(u_1, ..., u_n) + 2 \sum_{j=r+1}^{n} u_r u_j H_{rj}(u_1, ..., u_n) \quad (3.18)$$

Comparing (3.15) with (3.17), we see that (3.18) can be written

$$\pm v_r v_r - \frac{1}{H_{rr}} \left(\sum_{i>r} u_i H_{ir}(u_1, ..., u_n) \right)^2$$

the sign $\pm$ appears in front of $v_r v_r$ because $H_{rr} = \pm(\psi)^2$, where we use the plus if H_{rr} is positive and the minus if H_{rr} is negative.

Thus, after the substitution $v = \psi(u)$, the expression (3.16) becomes

$$(f \circ \varphi \circ \psi^{-1})(v) = \sum_{i=1}^{r} \pm(v_i)^2 + \sum_{i,j>r} v_i v_j \widetilde{H}_{ij}(v_1, ..., v_n)$$

for new symmetric $\widetilde{H}_{ij}$. The mapping $g = \varphi \circ \psi^{-1}$ is a diffeomorphism. This completes the induction, and proves Morse's lemma. $\qquad \square$

Corollary 3.11.3. *Nondegenerate critical points are isolated.*

3.12   Solved problems for Chapter 3

Exercise 3.12.1. Let $f : \mathbb{R}^2 \to \mathbb{R}$ be defined by

$$f(x,y) = \begin{cases} \frac{x^2 y}{x^4+y^2} & \text{if } (x,y) \neq (0,0), \\ 0 & \text{if } (x,y) = (0,0). \end{cases}$$

Show that the directional derivative of f at $(0,0)$ exist along any direction, but f is not differentiable at the origin.

Solution. Clearly

$$D_{e_1}(0,0) = \frac{\partial f}{\partial x}(0,0) = \lim_{t \to 0} \frac{f(t,0) - f(0,0)}{t} = \lim_{t \to 0} \frac{0}{t} = 0.$$

Similarly $D_{e_2}(0,0) = \frac{\partial f}{\partial y}(0,0) = 0$. Now we compute the directional derivative along any other direction $u = (u_1, u_2)$.

$$D_u f(0,0) = \lim_{t \to 0} \frac{f(0 + tu_1, 0 + tu_2)) - f(0,0)}{t} = \lim_{t \to 0} \frac{f(tu_1, tu_2)}{t} =$$

$$\lim_{t \to 0} \frac{u_1^2 u_2}{t^2 u_1^4 + u_2^2} = \frac{u_1^2}{u_2}.$$

Therefore $D_u f(0,0)$ exists along any direction. However, f is not continuous at $(0,0)$ because $\lim_{(x,y)\to(0,0)} f(x,y)$ does not exist; as $(x,y) \to (0,0)$ along any line $y = mx$ through the origin the limiting value of f is 0, but as $(x,y) \to (0,0)$ along points on the parabola $y = x^2$ the function has the value $\frac{1}{2}$.

Exercise 3.12.2. Find the best affine approximation at $(0,0)$ of the mapping $f : \Omega \subseteq \mathbb{R}^2 \to \mathbb{R}^3$ defined by $f(\theta, \phi) = ((b + a\cos\phi)\cos\theta, (b + a\cos\phi)\sin\theta, a\sin\phi)$, where $\Omega = \{(\theta, \phi) : 0 \leq \theta \leq 2\pi, 0 \leq \phi \leq 2\pi\}$. (The image of f is the *torus*, see Examples 4.3.26 and 6.3.11 below).

Solution. We compute the derivative of f at $(0,0)$

$$D_f(0,0) = \begin{pmatrix} 0 & 0 \\ a+b & 0 \\ 0 & b \end{pmatrix}$$

Hence the best affine approximation is

$$F(\theta, \phi) = f(0,0) + D_f(0,0)(\theta, \phi) = \begin{pmatrix} a+b \\ 0 \\ 0 \end{pmatrix} + \begin{pmatrix} 0 & 0 \\ a+b & 0 \\ 0 & b \end{pmatrix} \begin{pmatrix} \theta \\ \phi \end{pmatrix}$$

$$= \begin{pmatrix} a+b \\ (a+b)\theta \\ b\phi \end{pmatrix}.$$

That is,

$$F(\theta, \phi) = (a+b, (a+b)\theta, b\phi).$$

Exercise 3.12.3. Let f be a C^2 function in n variables. Prove that the Laplacian $\Delta f = \sum_{i=1}^{n} f_{x_i x_i}$ is invariant under translations and rotations of $\mathbb{R}^n$. That is, if T is a translation or a rotation on $\mathbb{R}^n$, then $\Delta(f \circ T) = \Delta f$.

Solution. Let $T : \mathbb{R}^n \to \mathbb{R}^n$ be a translation $y = T(x) = x + a$. That is, $(y_1, ..., y_n) = T(x_1, ..., x_n) = (x_1 + a_1, ..., x_n + a_n)$. Set $F = f \circ T$.

Since $D_T = I$, viz,

$$\frac{\partial y_j}{\partial x_i}(x) = \delta_{ij} = \begin{cases} 1 \text{ when } i = j, \\ 0 \text{ when } i \neq j, \end{cases}$$

the Chain Rule yields

$$F_{x_i}(x) = \sum_{j=1}^{n} f_{y_j}(y) \frac{\partial y_j}{\partial x_i}(x) = \sum_{j=1}^{n} f_{y_j}(y)\delta_{ij} = f_{y_i}(y).$$

Hence

$$\Delta F(x) = \sum_{i=1}^{n} F_{x_i x_i} = \sum_{i=1}^{n} f_{y_i y_i} = \Delta f(y).$$

Next let $y = T(x)$ be a rotation in $\mathbb{R}^n$ represented by the matrix $T = (t_{ji})$. Since $D_T = T$, viz, $\frac{\partial y_j}{\partial x_i}(x) = t_{ji}$, the Chain Rule gives

$$F_{x_i}(x) = \sum_{j=1}^{n} f_{y_j}(y) t_{ji}.$$

and

$$F_{x_i x_i}(x) = \sum_{k=1}^{n} \sum_{j=1}^{n} f_{y_j y_k}(y) t_{ji} t_{ki}.$$

Since T is orthogonal, $TT^t = I$, and so $\sum_{i=1}^{n} t_{ji} t_{ki} = \delta_{kj}$. Hence

$$\Delta F(x) = \sum_{i=1}^{n} F_{x_i x_i} = \sum_{i=1}^{n} \left(\sum_{k=1}^{n} \sum_{j=1}^{n} f_{y_j y_k}(y) t_{ji} t_{ki} \right)$$

$$= \sum_{k=1}^{n} \sum_{j=1}^{n} f_{y_j y_k}(y) \left(\sum_{i=1}^{n} t_{ji} t_{ki} \right) = \sum_{k=1}^{n} \sum_{j=1}^{n} f_{y_j y_k}(y) \delta_{kj} = \sum_{j=1}^{n} f_{y_j y_j}(y) = \Delta f(y).$$

Exercise 3.12.4. For $\mathbf{x} \in \mathbb{R}^3 - \{0\}$, $r = ||\mathbf{x}|| = \sqrt{x^2 + y^2 + z^2}$, and $t \in \mathbb{R}$ let $f(\mathbf{x}, t) = \frac{1}{r} g(ct - r)$, where c is a constant and g is a C^2 function in one variable. Show that f satisfies the *wave equation* $c^2 \Delta f = \frac{\partial^2 f}{\partial t^2}$.

Solution. Differentiating $\frac{\partial f}{\partial t} = \frac{c}{r} g'(ct - r)$ and

$$\frac{\partial^2 f}{\partial t^2} = \frac{c^2}{r} g''(ct - r) = c^2 \frac{1}{r} g''.$$

On the other hand

$$\frac{\partial f}{\partial x} = -\frac{x}{r^3} g - \frac{x}{r^2} g' = -\frac{x}{r^2}(r^{-1} g + g')$$

and

$$\frac{\partial^2 f}{\partial x^2} = -\frac{1}{r^2} \left[(r^{-1} g + g')(1 - 3x^2 r^{-2}) - x^2 r^{-1} g'' \right].$$

By symmetry similar formulas hold for $\frac{\partial^2 f}{\partial y^2}$ and $\frac{\partial^2 f}{\partial z^2}$. Hence

$$\Delta f = -\frac{1}{r^2} \left[(r^{-1} g + g')(3 - 3(x^2 + y^2 + z^2) r^{-2}) - (x^2 + y^2 + z^2) r^{-1} g'' \right] = \frac{1}{r} g''.$$

Exercise 3.12.5. (*). Let $f : \mathbb{R}^2 \to \mathbb{R}$ be harmonic, not identically 0, and homogeneous of degree α, where $\alpha > 0$. Prove that α is an integer.

Solution. Since f is homogeneous of degree α we have $f(tx, ty) = t^\alpha f(x, y)$. Passing to polar coordinates $x = r\cos\theta$, $y = r\sin\theta$, we may write f as $f(r, \theta) = r^\alpha g(\theta)$. From Example 3.5.8, the Laplace equation in polar coordinates is

$$\Delta f(r, \theta) = \frac{\partial^2 f}{\partial r^2} + \frac{1}{r^2}\frac{\partial^2 f}{\partial \theta^2} + \frac{1}{r}\frac{\partial f}{\partial r} = 0.$$

That is,

$$\Delta f = \alpha(\alpha - 1)r^{\alpha-2}g(\theta) + r^{\alpha-2}g''(\theta) + \alpha r^{\alpha-2}g(\theta) = r^{\alpha-2}[\alpha^2 g(\theta) + g''(\theta)] = 0.$$

Hence g must satisfy the ordinary differential equation $g'' + \alpha^2 g = 0$. From Example 7.4.8 of Chapter 7, this equation has the general solution

$$g(\theta) = A\cos(\alpha\theta) + B\sin(\alpha\theta).$$

Since g must be 2π-periodic, α is an integer.

Exercise 3.12.6. Let $f : \mathbb{R}^2 \to \mathbb{R}^2$ be given by $(u, v) = f(x, y) = (xe^y, xe^{-y})$. Show that f is locally invertible near any point (x, y) such that $x \neq 0$, and find its local inverse.

Solution. The Jacobian is $J_f(x, y) = \det\begin{pmatrix} e^y & xe^y \\ e^{-y} & -xe^{-y} \end{pmatrix} = -2x$. By the Inverse Function theorem f has a local inverse $(x, y) = f^{-1}(u, v)$ near any point (x, y) with $x \neq 0$. Moreover,

$$D_{f^{-1}}(u, v) = \begin{pmatrix} \frac{\partial x}{\partial u} & \frac{\partial x}{\partial v} \\ \frac{\partial y}{\partial u} & \frac{\partial y}{\partial v} \end{pmatrix} = \frac{1}{J_f(x, y)}\begin{pmatrix} \frac{\partial v}{\partial y} & -\frac{\partial u}{\partial y} \\ -\frac{\partial v}{\partial x} & \frac{\partial u}{\partial x} \end{pmatrix} = \frac{1}{2x}\begin{pmatrix} xe^{-y} & xe^y \\ e^{-y} & -e^{-y} \end{pmatrix}.$$

The range of f is the set $\{(u, v) : uv \geq 0\}$. Solving the equations $u = xe^y$ and $v = xe^{-y}$ for x, y we see

$$x = \pm\sqrt{uv} \text{ and } y = \log\sqrt{\tfrac{u}{v}}.$$

Hence the local inverse is $f^{-1}(u, v) = (\sqrt{uv}, \log\sqrt{\tfrac{u}{v}})$ if $x > 0$ and $f^{-1}(u, v) = (-\sqrt{uv}, \log\sqrt{\tfrac{u}{v}})$ if $x < 0$. When $x = 0$, there is no incverse because then $u = v = 0$ and so $f(0, y) = (0, 0)$ for any y. Finally the derivative of the local inverse is

$$D_{f^{-1}}(u, v) = \begin{pmatrix} \pm\frac{1}{2}\sqrt{\frac{v}{u}} & \frac{1}{2}\sqrt{\frac{u}{v}} \\ \frac{1}{2u} & -\frac{1}{2v} \end{pmatrix},$$

according as $x > 0$ or $x < 0$.

Next we give a proof of Corollary 3.11.3 independent of Morse' lemma.

Exercise 3.12.7. Let $f : \Omega \to \mathbb{R}$ be twice differentiable on the open set $\Omega \subseteq \mathbb{R}^n$ and let A be the set of critical points of f, i.e., $A = \{x \in \Omega : \nabla f(x) = 0 \text{ and } H_f(x) \text{ is nondegenerate}\}$. Prove that the points of A are isolated.

Solution. As f is twice differentiable, for $x \in A$ we have

$$\nabla f(x + h) = \nabla f(x) + H_f(x)h + \epsilon(h)||h|| = H_f(x)h + ||h||\epsilon(h),$$

with $\epsilon(h) \to 0$ as $h \to 0$. That is,

$\frac{\nabla f(x+h)}{||h||} = H_f(x)\frac{h}{||h||} + \epsilon(h)$, with $\lim_{h \to 0} \epsilon(h) = 0$.

Since $\det(H_f(x)) \neq 0$, by Proposition 2.4.7 there exists $\beta > 0$ with $\beta||v|| \leq ||H_f(x)v||$ for all $v \in \mathbb{R}^n$. In particular, for all vectors u with $||u|| = 1$,

$$0 < \beta \leq ||H_f(x)u||.$$

Now $\lim_{h \to 0} \epsilon(h) = 0$ tells us that there is $\delta > 0$ such that $||\epsilon(h)|| < \frac{\beta}{2}$ whenever $||h|| < \delta$

Then, if $0 < ||h|| < \delta$, we have

$$\left\|\frac{\nabla f(x + h)}{||h||}\right\| \geq \left\|H_f(x)\frac{h}{||h||}\right\| - ||\epsilon(h)|| \geq \beta - \frac{\beta}{2} = \frac{\beta}{2}.$$

Hence for $||h|| \in (0, \delta)$, $\nabla f(x + h) \neq 0$ and so the points of A are isolated.

Miscellaneous Exercises

Exercise 3.12.8. Let $f : \Omega \subseteq \mathbb{R}^n \to \mathbb{R}$. Show that if the partial derivatives $\frac{\partial f}{\partial x_i}$ exist and are bounded in a neighborhood of $a \in \Omega$, then f is continuous at a.

Exercise 3.12.9. Let $f : \Omega \to \mathbb{R}^m$ be differentiable, where Ω is an open connected subset of $\mathbb{R}^n$. Suppose $D_f(x)$ is constant on Ω, that is, $D_f(x) = T$ for all $x \in \Omega$. Show that f is the restriction to Ω of an affine transformation.

Exercise 3.12.10. Let $f : \mathbb{R} \to \mathbb{R}$ be of class C^1. Define

$$g(x, y) = \begin{cases} \frac{f(x) - f(y)}{x - y} & \text{for } x \neq y, \\ f'(x) & \text{for } x = y. \end{cases}$$

Prove that g is continuous and if in addition f is twice differentiable at x_0, then g is differentiable at (x_0, x_0).

Exercise 3.12.11. $f : \mathbb{R}^2 \to \mathbb{R}$ be defined by

$$f(x,y) = \begin{cases} \frac{|y|}{x^2} e^{-\frac{|y|}{x^2}} & \text{for } x \neq 0, \\ 0 & \text{for } x = 0. \end{cases}$$

Study the continuity and differentiability of f at $(0,0)$.

Exercise 3.12.12. Study the continuity and differentiability of the function defined by

$$f(x,y) = \begin{cases} \frac{|x|^\alpha y}{x^2+y^4} & \text{for } x \neq 0, \\ 0 & \text{for } x = 0, \end{cases}$$

where $\alpha > 0$ is a constant.

Exercise 3.12.13. Let $f : \mathbb{R}^n \to \mathbb{R}$ be such that f is differentiable on $\mathbb{R}^n - \{0\}$, f is continuous at 0, and

$$\lim_{x \to 0} \nabla f(x) = 0.$$

Prove that f is differentiable at 0

Exercise 3.12.14. Let $f : \mathbb{R}^2 \to \mathbb{R}$ be defined by

$$f(x,y) = \begin{cases} xy \sin[\frac{\pi}{2}(\frac{x+y}{x-y})] & \text{for } x \neq y, \\ 0 & \text{for } x = y. \end{cases}$$

1. Show that f is differentiable at $(0,0)$.

2. Compute $\frac{\partial^2 f}{\partial x \partial y}(0,0)$ and $\frac{\partial^2 f}{\partial y \partial x}(0,0)$ and show that at least one of derivatives $\frac{\partial^2 f}{\partial x \partial y}$, and $\frac{\partial^2 f}{\partial y \partial x}$ is not continuous at $(0,0)$.

Exercise 3.12.15. Let f be the function

$$f(x,y) = \begin{cases} xy\sqrt{-\log(x^2+y^2)} & \text{for } x^2+y^2 \leq 1, \\ 0 & \text{for } (x,y) = (0,0). \end{cases}$$

Show that $\frac{\partial^2 f}{\partial x^2}$ and $\frac{\partial^2 f}{\partial y^2}$ are both continuous at $(0,0)$. However, show that $\frac{\partial^2 f}{\partial x \partial y}$ is not continuous at $(0,0)$.

Exercise 3.12.16. Let $f : \mathbb{R}^3 \to \mathbb{R}^2$ be of class C^1 satisfying $f(0,0) = (1,2)$ and $D_f(0,0) = \begin{pmatrix} 1 & 2 & 3 \\ 0 & 0 & 1 \end{pmatrix}$. Let $g : \mathbb{R}^2 \to \mathbb{R}^2$ be $g(x,y) = (x+2y+1, 3xy)$. Find $D_{(f \circ g)}(0,0)$

Exercise 3.12.17. Let $u = f(x,y)e^{ax+by}$ and suppose $\frac{\partial^2 f}{\partial x \partial y} = 0$. Find values of the constants a, b such that

$$\frac{\partial^2 u}{\partial x \partial y} - \frac{\partial u}{\partial x} - \frac{\partial u}{\partial y} + u = 0.$$

(Ans. $a = b = 1$).

Exercise 3.12.18. Let $u = u(x,y,z)$, where $x = r \cos\theta \sin\phi$, $y = r\sin\theta \sin\phi$, $z = r\cos\phi$. Show that

$$\left(\frac{\partial u}{\partial x}\right)^2 + \left(\frac{\partial u}{\partial y}\right)^2 + \left(\frac{\partial u}{\partial z}\right)^2 = \left(\frac{\partial u}{\partial r}\right)^2 + \left(\frac{1}{r}\frac{\partial u}{\partial \phi}\right)^2 + \left(\frac{1}{r\sin\phi}\frac{\partial u}{\partial \theta}\right)^2.$$

Exercise 3.12.19. Suppose that f is a homogeneous function of degree α on $\mathbb{R}^n$. Show that

$$\sum_{j,k}^n x_j x_k \frac{\partial^2 f}{\partial x_i \partial x_k} = \alpha(\alpha - 1)f.$$

Exercise 3.12.20. Let $f : \mathbb{R} \to \mathbb{R}$ be such that $f''(t) \neq 0$ for all $t \in \mathbb{R}$ and let $g : \mathbb{R}^n \to \mathbb{R}$ be harmonic. Prove that $\varphi = f \circ g$ is harmonic if and only if g is constant.

Exercise 3.12.21. Show that the function

$$f(x_1, ..., x_n, t) = \frac{1}{(2c\sqrt{\pi t})^n} e^{-\frac{||x||^2}{4c^2 t}},$$

defined for $t > 0$ and $x = (x_1, ... x_n) \in \mathbb{R}^n$, satisfies the *heat equation* $c^2 \Delta f = \frac{\partial f}{\partial t}$.

Exercise 3.12.22. Let $f : \mathbb{R} \to \mathbb{R}$ be a C^1 function such that there is $k \in (0,1)$ with $|f'(t)| \leq k < 1$ for all $t \in \mathbb{R}$. Prove that the function $g : \mathbb{R}^2 \to \mathbb{R}^2$ given by $g(x,y) = (x + f(y), y + f(x))$ is a diffeomorphism.

Exercise 3.12.23. Show that if the point (a,b) satisfies the equation

$$F(x,y) = xy - \log\tfrac{x}{y} = 0, \ (x,y > 0),$$

then there exists a C^1 function $y = f(x)$ which passes through this point and is unique in its neighborhood. Furthermore, show that there is one point (a,b) such that the corresponding function $y = f(x)$ has a maximum at a.

Exercise 3.12.24. Find a point (a,b,c) in a neighborhood of which the equation

$$F(x, y, z) = \sin xy + \sin yz + \sin xz = 0$$

has a unique solution $z = f(x, y)$ (Ans. $(a, 0, 0)$ with $a \neq 0$, $(0, b, 0)$ with $b \neq 0$).

Exercise 3.12.25. Let $g : \mathbb{R}^2 \to \mathbb{R}$ be of class C^2 with $g(0, 0) = 0$ and $D_g(0, 0) = (2\ 3)$. Let $F : \mathbb{R}^3 \to \mathbb{R}$ be defined by

$$F(x, y, z) = g(x + 2y + 3z - 1, x^3 + y^2 - z^2).$$

Note that $F(-2, 3, -1) = g(0, 0) = 0$.

1. Show that the equation $F(x, y, z) = 0$ defines a function $z = f(x, y)$ in a neighborhood of $(-2, 3)$, such that $f(-2, 3) = -1$. Find $D_f(-2, 3)$

2. If the Hessian of g at $(0, 0)$ is $H_g(0, 0) = \begin{pmatrix} 3 & -1 \\ -1 & 5 \end{pmatrix}$. Find $\frac{\partial^2 f}{\partial y \partial x}(-2, 3)$

Exercise 3.12.26. Let $a, b, c, d \in \mathbb{R}^n$. Show that

$$||a - c|| \cdot ||b - d|| \leq ||a - b|| \cdot ||c - d|| + ||a - d|| \cdot ||b - c||.$$

Hint. Reduce the problem to the case $a = 0$ and consider the mapping $\varphi(x) = \frac{x}{||x||^2}$ for $x \neq 0$. When does equality hold?

Exercise 3.12.27.　　1. For $x > 0$, $y > 0$ let $f(x, y) = \frac{x^p}{p} + \frac{y^q}{q}$, where $p > 1$, $q > 1$ with $\frac{1}{p} + \frac{1}{q} = 1$. Show that the

$$\min \{f(x, y) : xy = 1\} = 1.$$

2. Use part (1) to show that if $a > 0$, $b > 0$, then

$$ab \leq \frac{a^p}{p} + \frac{b^q}{q}.$$

3. Let $a_1, ..., a_n$　$b_1, ..., b_n$ be positive real numbers. Prove *Holder's* inequality

$$\sum_{i=1}^{n} a_i b_i \leq \left(\sum_{i=1}^{n} a_i^p \right)^{\frac{1}{p}} \left(\sum_{i=1}^{n} b_i^q \right)^{\frac{1}{q}}.$$

4. Use Holder's inequality in (3) to obtain *Minkowski's inequality*

$$\left(\sum_{i=1}^{n}|a_i+b_i|^p\right)^{\frac{1}{p}} \leq \left(\sum_{i=1}^{n}|a_i|^p\right)^{\frac{1}{p}} + \left(\sum_{i=1}^{n}|b_i|^q\right)^{\frac{1}{q}}.$$

Hint. For (3), set $A = (\sum_{i=1}^{n} a_i^p)^{\frac{1}{p}}$, $B = (\sum_{i=1}^{n} b_i^q)^{\frac{1}{q}}$ and apply (2) with $a = \frac{a_i}{A}$ and $b = \frac{b_i}{B}$. For (4) note $|a+b|^p = |a+b||a+b|^{\frac{p}{q}} \leq |a||a+b|^{\frac{p}{q}} + |b||a+b|^{\frac{p}{q}}$

Exercise 3.12.28. (*) Let $f : \mathbb{R}^n \to \mathbb{R}$ be given by

$$f(x) = \frac{1}{2}\langle Ax, x\rangle - \langle u, x\rangle,$$

where A is a symmetric positive definite $n \times n$ martix and $u \in \mathbb{R}^n$ a fixed vector. Show that f is C^∞. Find ∇f and determine the extrema of f.

Exercise 3.12.29. (*). Maximize the function $f(X) = (\det X)^2$ over the set of matrices $X \in M_{n\times n}(\mathbb{R}) \cong \mathbb{R}^{n^2}$ that meet the constraint $\operatorname{tr}(XX^T) = 1$.

Exercise 3.12.30. (*). Find the maximum value of the function $f(X) = \operatorname{tr}(XX^T)$ over all matrices $X \in M_{n\times n}(\mathbb{R}) \cong \mathbb{R}^{n^2}$ with $\det X = 1$

Exercise 3.12.31. (*). Let the entropy $E(x_1,...,x_n) = -\sum_{i=1}^{n} x_i \log x_i$, where x_i is the probability of the occurence of the ith character of an information source. Thus for each i, $0 \leq x_i \leq 1$. What choice of probabilities x_i will maximize the entropy?

This page intentionally left blank

Chapter 4

Integral Calculus in Several Variables

In this chapter we present the Riemann integration of real-valued function on domains in $\mathbb{R}^n$ where $n > 1$. Among the topics included are the basic properties of multiple integrals, and in particular, Fubini's theorem.

4.1 The integral in $\mathbb{R}^n$

The theory of the Riemann integral in $\mathbb{R}^n$, for $n > 1$, is a direct extension of that of the integral of functions of one variable in $\mathbb{R}$. Nevertheless, certain serious complications arise from the greater variety of regions in $\mathbb{R}^n$ over which integration is to be performed. The two-dimensional case ($n = 2$) has the advantage of simpler notation and clearer geometric intuition and it will be helpful to the reader to turn often to this case as he or she goes over this section. However, we will develop the theory of n-dimensional integrals. We start by defining the integral of functions of n variables over particularly simple subsets of $\mathbb{R}^n$, the analogues of closed intervals in $\mathbb{R}$.

Definition 4.1.1. The set

$$R = \prod_{i=1}^{n} [a_i, b_i] = [a_1, b_1] \times [a_2, b_2] \times ... \times [a_n, b_n]$$

$$= \{(x_1, ..., x_n) \in \mathbb{R}^n : a_i \leq x_i \leq b_i, \ i = 1, 2, ..., n\}$$

231

is called an *n-dimensional rectangle* or simply a (closed) *rectangle* in $\mathbb{R}^n$. Each interval $[a_i, b_i]$ is called a component interval. When $b_1 - a_1 = b_2 - a_2 = \ldots = b_n - a_n$ R is called a *cube* in $\mathbb{R}^n$.

Definition 4.1.2. Let $R = \prod_{i=1}^{n} [a_i, b_i]$ be a closed rectangle in $\mathbb{R}^n$. We define the *n-dimensional volume* or simply the *volume* of R to be the number

$$vol(R) \equiv \nu(R) = (b_1 - a_1)(b_2 - a_2) \cdots (b_n - a_n).$$

For example, when $n = 1$, $R = [a, b]$ is a closed interval in $\mathbb{R}$,

$$\nu(R) = b - a$$

is the *length* of $[a, b]$. When $n = 2$, $R = [a_1, b_1] \times [a_2, b_2]$ is a rectangle in $\mathbb{R}^2$ in the usual sense,

$$\nu(R) = (b_1 - a_1)(b_2 - a_2)$$

is the *area* of R. Finally, when $n = 3$, $R = [a_1, b_1] \times [a_2, b_2] \times [a_3, b_3]$ is rectangular parallelepiped in $\mathbb{R}^3$,

$$\nu(R) = (b_1 - a_1)(b_2 - a_2)(b_3 - a_3)$$

is the *volume* of R in the usual sense.

Given a closed interval $[a, b]$ in $\mathbb{R}$, we recall from one variable calculus that a *partition* $\mathcal{P}$ of $[a, b]$ is a set of points $\mathcal{P} = \{t_0, t_1, \ldots t_m\}$ such that $a = t_0 < t_1 < t_2 < \ldots < t_m = b$. For each $j = 1, \ldots, m$, the interval $I_j = [t_{j-1}, t_j]$ is called a *subinterval* determined by the partition P of $[a, b]$. Setting $\Delta t_j = t_j - t_{j-1}$ the number $|\mathcal{P}| = \max\{\Delta t_j : j = 1, 2, \ldots, m\}$ is called the *mesh* of the partition $\mathcal{P}$. Alternatively, we may write the partition of $[a, b]$ as $\mathcal{P} = \{I_1, \ldots, I_m\}$. Clearly then $[a, b] = \bigcup_{j=1}^{m} I_j$ and $b - a = \sum_{j=1}^{m} \Delta t_j$. The analogue of this in $\mathbb{R}^n$ is the following:

Definition 4.1.3. Let R be a closed rectangle in $\mathbb{R}^n$. A *partition* of R is a collection $\mathcal{P} = \{R_1, \ldots, R_m\}$ of rectangles R_j with mutually disjoint interiors such that $R = \bigcup_{j=1}^{m} R_j$. Each R_j is called a *subrectangle* determined by the partition $\mathcal{P}$ of the rectangle R. Note that

$$\nu(R) = \sum_{j=1}^{m} \nu(R_j) \text{ and } \nu(R) > 0.$$

Definition 4.1.4. The number $||\mathcal{P}|| = \max\{d(R_j) : 1 \leq j \leq m\}$ is called the *mesh* of the partition $\mathcal{P}$, where $d(R_j)$ is the *diameter* of R_j, that is,

$$d(R_j) = \max\{||x - y|| : x, y \in R_j\}.$$

The reader should draw a picture of these in $\mathbb{R}^2$.   Let $\mathcal{P}$ be a partition of R. A partition $\mathcal{Q}$ of R is called a *refinement* of $\mathcal{P}$, if each subrectangle of $\mathcal{Q}$ is contained in a subrectangle of $\mathcal{P}$. In this case $||\mathcal{Q}|| \leq ||\mathcal{P}||$. A refinement of $\mathcal{P}$ is obtained by spliting one or more of its subrectangles into additional subrectangles. Furthermore, given two partitions $\mathcal{P} = \{R_j\}$ and $\mathcal{P}' = \{R_k'\}$ of R, one sees that the partition $\mathcal{Q} = \bigcup_{j,k}\{R_j \cap R_k'\}$ is a refinement of both $\mathcal{P}$ and $\mathcal{P}'$ and it is called their *common refinement*.

Definition 4.1.5. Let $R \subset \mathbb{R}^n$ be a rectangle, $f : R \to \mathbb{R}$ be a *bounded* function and $\mathcal{P}$ a partition of R. We choose an arbitrary point ξ_j in each of the subrectangles R_j, for $j = 1, .., m$, and form the sum

$$S_\mathcal{P}(f) = \sum_{j=1}^{m} f(\xi_j)\nu(R_j). \tag{4.1}$$

Such a sum is called a *Riemann sum* of f corresponding to the partition $\mathcal{P}$ of R. Note that there are many Riemann sums corresponding to the same partition $\mathcal{P}$ of R depending on which ξ_j's are choosen.

Definition 4.1.6. (*Riemann[1] integral*).   A function f defined on a rectangle R in $\mathbb{R}^n$ is said to be *Riemann integrable* over R if there exists a number s such that for every $\epsilon > 0$ there is a partition $\mathcal{P}_\epsilon$ such that $|S_\mathcal{P}(f) - s| < \epsilon$ for any partition $\mathcal{P}$ finer than $\mathcal{P}_\epsilon$ and for any choice of points $\xi_j \in R_j$. The number s is called the *Riemann integral* of f over the rectangle R.

The next proposition explains the reason of assuming that f is bounded to begin with.

Proposition 4.1.7. *If $f : R \to \mathbb{R}$ is integrable on R it must be bounded.*

---

[1]B. Riemann (1826-1866). Besides his contributions in integration theory and complex analysis, he introduced the *Riemannian manifolds* in n dimensions and investigated the zeros of the *Riemann zeta function*.

Proof. Since f is integrable it follows that for $\epsilon = 1$ there is a partition $\mathcal{P}_1$ such that $|S_{\mathcal{P}}(f) - s| < 1$ for any partition $\mathcal{P}$ finer than $\mathcal{P}_1$. In particular, $|S_{\mathcal{P}}(f)| < |s| + 1$. Let $\mathcal{P}$ be a fixed partition finer than $\mathcal{P}_1$. If f is *unbounded* on R, it must be unbounded in at least one subrectangle, say R_k of the partition $\mathcal{P}$. Hence there is $\xi_k \in R_k$ such that $|f(\xi_k)|$ can be arbitrarily large. Form the Riemann sum

$$S_{\mathcal{P}}(f) = \sum_{j \neq k} f(\xi_j)\nu(R_j) + f(\xi_k)\nu(R_k) = \sigma + f(\xi_k) \cdot \nu(R_k),$$

where $\sigma = \sum_{j \neq k} f(\xi_j) \cdot \nu(R_j)$. Now since $|f(\xi_k)|$ can be as large as we wish, $|f(\xi_k) \cdot \nu(R_k)| > N + |\sigma|$ for any integer $N > 0$. Therefore

$$|S_{\mathcal{P}}(f)| = |\sigma + f(\xi_k) \cdot \nu(R_k)| \geq |f(\xi_k) \cdot \nu(R_k)| - |\sigma| > N.$$

In particular, for $N > |s| + 1$ this contradicts $|S_{\mathcal{P}}(f)| < |s| + 1$. $\qquad\square$

4.1.1 Darboux sums. Integrability condition

Let $R \subset \mathbb{R}^n$ be a rectangle, $f : R \to \mathbb{R}$ be a bounded function (i.e., $|f(x)| \leq M$ for some $M > 0$ for all $x \in R$), and $\mathcal{P}$ a partition of R. Since f is bounded on R, let

$$m = \inf\{f(x) : x \in R\} \quad \text{and} \quad M = \sup\{f(x) : x \in R\}.$$

For each of the subrectangles R_j $(j = 1, ..., m)$ determined by $\mathcal{P}$, we set

$$m_j = \inf\{f(x) : x \in R_j\}$$

and

$$M_j = \sup\{f(x) : x \in R_j\}.$$

So that for each j we have

$$m \leq m_j \leq M_j \leq M.$$

Definition 4.1.8. The *lower* and *upper Darboux*[2] *sums* of f are defined respectively to be

$$L_{\mathcal{P}}(f) = \sum_{j=1}^{m} m_j \cdot \nu(R_j) \quad \text{and} \quad U_{\mathcal{P}}(f) = \sum_{j=1}^{m} M_j \cdot \nu(R_j).$$

---

[2]G. Darboux (1842-1917). He made several important contributions to geometry and analysis. A member of Academie des Sciences. Among his students were E. Borel, E. Cartan, E. Goursat, E. Picard, T. Stieltjes and S. Zaremba.

Clearly, for every partition $\mathcal{P}$,

$$L_{\mathcal{P}}(f) \le S_{\mathcal{P}}(f) \le U_{\mathcal{P}}(f).$$

Lemma 4.1.9. *Let $\mathcal{P}$ be a partition of the rectangle R and $f : R \to \mathbb{R}$ be a bounded function. If $\mathcal{Q}$ is a refinement of $\mathcal{P}$, then*

$$L_{\mathcal{P}}(f) \le L_{\mathcal{Q}}(f) \ and \ U_{\mathcal{Q}}(f) \le U_{\mathcal{P}}(f).$$

Proof. It suffices to prove the lemma when $\mathcal{Q}$ is obtained from $\mathcal{P}$ by splitting a single subrectangle of the subrectangles determined by $\mathcal{P}$. Say the R_k for some $k = 1, ..., m$, is split into two subrectangles R_k', R_k'' and $\nu(R_k) = \nu(R_k') + \nu(R_k'')$ (all the other subrectangles R_j for $j \ne k$ remain uneffected). Now since $m_k \le f(x)$ for each $x \in R_k'$ and for each $x \in R_k''$, it follows that

$$m_k \le m_k' \ and \ m_k \le m_k''.$$

and so

$$m_k \cdot \nu(R_k) \le m_k' \cdot \nu(R_k') + m_k'' \cdot \nu(R_k'').$$

Now forming the corresponding lower sums we get $L_{\mathcal{P}}(f) \le L_{\mathcal{Q}}(f)$. Similarly, for the upper sums and upon noticing that $M_k' \le M_k$ and $M_k'' \le M_k$, we obtain $U_{\mathcal{Q}}(f) \le U_{\mathcal{P}}(f)$. $\square$

Corollary 4.1.10. *For any two partitions $\mathcal{P}$ and $\mathcal{P}'$ of the rectangle R we have*

$$L_{\mathcal{P}}(f) \le U_{\mathcal{P}'}(f)$$

Proof. Let $\mathcal{Q} = \mathcal{P} \cup \mathcal{P}'$ be the common refinement of $\mathcal{P}$ and $\mathcal{P}'$. By Lemma 4.1.9,

$$L_{\mathcal{P}}(f) \le L_{\mathcal{Q}}(f) \le U_{\mathcal{Q}}(f) \le U_{\mathcal{P}'}(f).$$

$\square$

Let **P** denote the set of *all* partitions of the rectangle R. Note that by Corollary 4.1.10 the set $\{L_{\mathcal{P}}(f) : \mathcal{P} \in \mathbf{P}\}$ of all lower sums for f is bounded above by $U_{\mathcal{P}'}(f)$ where $\mathcal{P}'$ is any fixed partition of R, while the set $\{U_{\mathcal{P}}(f) : \mathcal{P} \in \mathbf{P}\}$ of all upper sums for f is bounded below by $L_{\mathcal{P}'}(f)$. Now, the following definition is a natural consequence.

Definition 4.1.11. The *lower* and *upper Darboux integrals* of the bounded function $f : R \to \mathbb{R}$ over the rectangle R are defined respectively to be

$$\boldsymbol{J}_* = \sup \{L_{\mathcal{P}}(f) : \mathcal{P} \in \mathbf{P}\}$$

and

$$\boldsymbol{J}^* = \inf \{U_{\mathcal{P}}(f) : \mathcal{P} \in \mathbf{P}\}.$$

If $\boldsymbol{J}_* = \boldsymbol{J}^*$, we say that f is *Darboux integrable* and we define the *Darboux integral* of f over R to be their common value $\boldsymbol{J}_* = \boldsymbol{J} = \boldsymbol{J}^*$.

The integrals $\boldsymbol{J}_*$ and $\boldsymbol{J}^*$ are denoted by $\int_{*R} f$ and $\int_R^* f$ respectively.

As we noticed above, if $\mathcal{P}'$ is any fixed partition of R, then $L_{\mathcal{P}}(f) \leq U_{\mathcal{P}'}(f)$ for every partition $\mathcal{P}$ of R. It follows that $\boldsymbol{J}_* \leq U_{\mathcal{P}'}(f)$. Now since $\mathcal{P}'$ is arbitrary we get $\boldsymbol{J}_* \leq \boldsymbol{J}^*$.

A condition that characterizes integrable functions is the following:

Theorem 4.1.12. *(The Riemann condition). Let R be a rectangle and $f : R \to \mathbb{R}$ be a bounded function. Then $\boldsymbol{J}_* = \boldsymbol{J}^*$ if and only if for every $\epsilon > 0$, there exists a corresponding partition $\mathcal{P}_\epsilon$ of R such that*

$$U_{\mathcal{P}_\epsilon}(f) - L_{\mathcal{P}_\epsilon}(f) < \epsilon.$$

Proof. Suppose $\boldsymbol{J}_* = \boldsymbol{J} = \boldsymbol{J}^*$. Given $\epsilon > 0$, there is a partition $\mathcal{P}'$ of R such that

$$U_{\mathcal{P}'}(f) < \boldsymbol{J}^* + \frac{\epsilon}{2} = \boldsymbol{J} + \frac{\epsilon}{2},$$

and a partition $\mathcal{P}''$ such that

$$L_{\mathcal{P}''}(f) > \boldsymbol{J}_* - \frac{\epsilon}{2} = \boldsymbol{J} - \frac{\epsilon}{2}.$$

Let $\mathcal{P}_\epsilon$ be the common refinement of $\mathcal{P}'$ and $\mathcal{P}''$. Then we have

$$\boldsymbol{J} - \frac{\epsilon}{2} < L_{\mathcal{P}''}(f) \leq L_{\mathcal{P}_\epsilon}(f) \leq U_{\mathcal{P}_\epsilon}(f) \leq U_{\mathcal{P}'}(f) < \boldsymbol{J} + \frac{\epsilon}{2}.$$

Hence

$$0 \leq U_{\mathcal{P}_\epsilon}(f) - L_{\mathcal{P}_\epsilon}(f) < \epsilon.$$

Conversely, suppose that for every $\epsilon > 0$ there is a partition $\mathcal{P}_\epsilon$ such that $0 \le U_{\mathcal{P}_\epsilon}(f) - L_{\mathcal{P}_\epsilon}(f) < \epsilon$. For every partition $\mathcal{P}$ we have

$$L_{\mathcal{P}}(f) \le \boldsymbol{J}_* \le \boldsymbol{J}^* \le U_{\mathcal{P}}(f).$$

In particular, for $\mathcal{P} = \mathcal{P}_\epsilon$, this implies that

$$\boldsymbol{J}^* - \boldsymbol{J}_* < \epsilon.$$

As we have seen several times earlier this means $\boldsymbol{J}^* \le \boldsymbol{J}_*$ and hence $\boldsymbol{J}^* = \boldsymbol{J}_*$. □

Remark 4.1.13. The Riemann condition not only gives an efficient criterion for integrability but also gives us a way for approximating (or even computing) the integral of an integrable function f. Indeed, since $L_{\mathcal{P}} \le \boldsymbol{J} \le U_{\mathcal{P}}(f)$, if $U_{\mathcal{P}}(f) - L_{\mathcal{P}}(f) < \epsilon$ for $\mathcal{P} = \mathcal{P}_\epsilon$, then $U_{\mathcal{P}}(f)$ and $L_{\mathcal{P}}(f)$ are both within ϵ of $\boldsymbol{J}$. Therefore, the integral is the limit of the sums $U_{\mathcal{P}}(f)$ or $L_{\mathcal{P}}(f)$ as $\mathcal{P}$ runs through any sequence of partitions such that $U_{\mathcal{P}}(f) - L_{\mathcal{P}}(f) \to 0$.

Next we prove the equivalence of the two definitions of the integral.

Theorem 4.1.14. *Let R be a rectangle and $f : R \to \mathbb{R}$ be a bounded function. Then f is Riemann integrable over R if and only if f is Darboux integrable over R.*

Proof. Suppose f is Riemann integrable over R. Then for every $\epsilon > 0$ there is a partition $\mathcal{P}_\epsilon$ such that $|S_{\mathcal{P}}(f) - s| < \frac{\epsilon}{4}$ for any partition $\mathcal{P}$ finer than $\mathcal{P}_\epsilon$ and for any choice of points $\xi_j \in R_j$. Fix such a refinement $\mathcal{P}$ of $\mathcal{P}_\epsilon$. Since $M_j = \sup\{f(x) : x \in R_j\}$ there is $\xi_j \in R_j$ such that $f(\xi_j) > M_j - \frac{\epsilon}{4\nu(R)}$. The corresponding Riemann sum then satisfies

$$S_1 = S_{\mathcal{P}}(f) > \sum_{j=1}^{m} \left[M_j - \frac{\epsilon}{4\nu(R)} \right] \nu(R_j) = U_{\mathcal{P}}(f) - \frac{\epsilon}{4}.$$

Similarly there is a Riemann sum $S_2 = S_{\mathcal{P}}(f) < L_{\mathcal{P}}(f) + \frac{\epsilon}{4}$. Since $|S_1 - s| < \frac{\epsilon}{4}$ and $|S_2 - s| < \frac{\epsilon}{4}$, it follows that $|S_1 - S_2| < \frac{\epsilon}{2}$. Hence

$$U_{\mathcal{P}}(f) - L_{\mathcal{P}}(f) < S_1 + \frac{\epsilon}{4} - S_2 + \frac{\epsilon}{4} < \epsilon.$$

Conversely, suppose f is Darboux integrable over R. Then by the Riemann condition for every $\epsilon > 0$ there is a partition $\mathcal{P}_\epsilon$ such that $U_{\mathcal{P}_\epsilon}(f) - L_{\mathcal{P}_\epsilon}(f) < \epsilon$. Since for any partition $\mathcal{P}$,

$$L_{\mathcal{P}}(f) \leq S_{\mathcal{P}}(f) \leq U_{\mathcal{P}}(f)$$

and

$$L_{\mathcal{P}}(f) \leq \boldsymbol{J} \leq U_{\mathcal{P}}(f),$$

it follows, in particular, that for any refinement $\mathcal{P}$ of $\mathcal{P}_\epsilon$ we have

$$|S_{\mathcal{P}}(f) - \boldsymbol{J}| \leq U_{\mathcal{P}}(f) - L_{\mathcal{P}}(f) \leq U_{\mathcal{P}_\epsilon}(f) - L_{\mathcal{P}_\epsilon}(f) < \epsilon.$$

Thus $s = \boldsymbol{J}$.   □

Since the Darboux integral and the Riemann integral are equivalent we write

$$\boldsymbol{J} = s.$$

Another equivalent definition for a bounded function to be integrable over a rectangle R reads as follows: $f : R \to \mathbb{R}$ is Riemann integrable over R if and only if there exist a real number s such that for every $\epsilon > 0$ there exists a $\delta = \delta(\epsilon) > 0$ such that for *any* partition $\mathcal{P}$ with $||\mathcal{P}|| < \delta$ we have $|S_{\mathcal{P}}(f) - s| < \epsilon$ for *any* choice of points $\xi_i \in R_j$. Briefly written:

$$s = \lim_{||\mathcal{P}|| \to 0} S_{\mathcal{P}}(f).$$

We ask the reader to prove this in Exercise 1 at the end of this section.

The Riemann integral s of f over R is also called the *n-fold integral* or the *multiple integral* of f over R and is denoted by

$$s = \int_R f = \int_R f(x)dx$$

or by

$$s = \int \int \cdots \int_R f(x_1, x_2, ..., x_n)dx_1 dx_2 ... dx_n.$$

Thus,

$$\int_R f = \lim_{||\mathcal{P}|| \to 0} \sum_{j=1}^{m} f(\xi_j)\nu(R_j).$$

When $n = 2$ the integral

$$\int\int_R f(x, y)dxdy$$

is called the *double integral* of f over R. For $n = 3$ the integral

$$\int\int\int_R f(x, y, z)dxdydz$$

is called the *triple integral* of f over R.

We now turn to the *geometrical meaning* of the integral. Since the geometry is simplest when $n = 2$, we look at the two-dimensional case. Let $R \subset \mathbb{R}^2$ be a rectangle, $f : R \to \mathbb{R}$ be bounded and $z = f(x, y)$ the graph of f, where $(x, y) \in R$. Given a partition $\mathcal{P}$ of R and assuming $f(x, y)$ is positive, a Riemann sum $S_{\mathcal{P}}(f)$ can be regarded as an approximation of the volume V of the solid in $\mathbb{R}^3$ bounded above by the given surface $z = f(x, y)$ and below by the rectangle R lying in the plane $z = 0$. We can naturally suppose that the smaller the diameters $d(R_j)$ of the subrectangles R_j are, the better the accuracy of the approximation $V \approx S_{\mathcal{P}}(f)$. Hence, the volume V of the solid under consideration is

$$V = \lim_{||\mathcal{P}|| \to 0} S_{\mathcal{P}}(f) = \int\int_R f.$$

In the n-dimensional case the multiple integral $\int_R f$ of a positive function f represents the hypervolume V of the region in $\mathbb{R}^{n+1}$ under the hypersurface of the graph of f and above the rectangle R lying in $\mathbb{R}^n$, that is, $V = \int\int \ldots \int_R f$.

An easy consequence of Theorem 4.1.14 is the following:

Corollary 4.1.15. *The constant function $f(x) \equiv c$ on R is integrable and*

$$\int_R c = c \cdot \nu(R).$$

In particular, if $c = 0$, then $\int_R 0 = 0$, and for $c = 1$

$$\int_R 1 = \nu(R).$$

Proof. Let $\mathcal{P}$ be any partition of R. Then $m_j = c = M_j$ for all $j = 1, ..., m$. Therefore, $L_{\mathcal{P}}(f) = c \sum_{j=1}^{m} \nu(R_j) = U_{\mathcal{P}}(f)$, and the Riemann condition holds trivially. Hence, f is integrable. Morever, since $L_{\mathcal{P}} \leq \int_R f \leq U_{\mathcal{P}}(f)$, it follows that

$$\int_R c = c \sum_{j=1}^{m} \nu(R_j) = c \cdot \nu(R).$$

$\square$

Example 4.1.16. Let $R = [0,1] \times [0,1]$. The function $f : R \to \mathbb{R}$ defined by

$$f(x,y) = \begin{cases} 1 \text{ when } (x,y) \in [0,1] \cap \mathbb{Q} \times [0,1] \cap \mathbb{Q}, \\ 0 \text{ otherwise.} \end{cases}$$

is not Riemann integrable.

Solution. For any partition $\mathcal{P} = \{R_j : j = 1, ..., m\}$ of R, we have $m_j = 0$ and $M_j = 1$. Therefore $L_{\mathcal{P}}(f) = 0$ and $U_{\mathcal{P}}(f) = 1$. It follows that

$$\boldsymbol{J}_* = 0 < 1 = \boldsymbol{J}^*.$$

Thus f is not integrable.

It is natural to expect that one obtains a large class of functions integrable on R. In the sequel we shall characterize the functions which are integrable. The next theorem tells us that continuous functions on R are integrable.

Theorem 4.1.17. *Let $R \subset \mathbb{R}^n$ be a closed rectangle. If $f : R \to \mathbb{R}$ is continuous on R, then f is integrable on R.*

Proof. Since f is continuous on R and R is compact (as a closed and bounded set in $\mathbb{R}^n$), by Corollary 2.5.3 f is bounded, and from Theorem 2.5.9, f is *uniformly continuous* on R. So for any $\epsilon > 0$ there is a $\delta = \delta(\epsilon) > 0$ such that $|f(x) - f(x')| < \frac{\epsilon}{\nu(R)}$ whenever $||x - x'|| < \delta$ for all $x, x' \in R$. Let $\mathcal{P}$ be an arbitrary partition of R with $||\mathcal{P}|| < \delta$. Then since $||x - x'|| < \delta$ for all $x, x' \in R_j$, we have

$$M_j - m_j = \sup_{x,x' \in R_j} |f(x) - f(x')| \leq \frac{\epsilon}{\nu(R)}.$$

Hence,

$$U_{\mathcal{P}}(f) - L_{\mathcal{P}}(f) = \sum_{j=1}^{m}(M_j - m_j)\nu(R_j) \leq \frac{\epsilon}{\nu(R)}\sum_{j=1}^{m}\nu(R_j) = \epsilon.$$

Thus, f is integrable on R.   □

Definition 4.1.18. A subset S of $\mathbb{R}^n$ *has volume zero*, written $\nu(S) = 0$, if for every $\epsilon > 0$ there is a *finite* number of (closed or open or half open) rectangles $\{R_1, ..., R_N\}$ such that $S \subset \bigcup_{i=1}^{N} R_i$ with $\sum_{i=1}^{N}\nu(R_i) < \epsilon$.

Any finite set $S = \{x_1, ..., x_N\}$ of points in $\mathbb{R}^n$ has volume zero. Indeed, clearly S can be covered by rectangles R_i such that $\nu(R_i) < \frac{\epsilon}{N}$ for each $i = 1, ..., N$, and $\sum_{i=1}^{N}\nu(R_i) < \epsilon$.

Definition 4.1.19. A subset S of $\mathbb{R}^n$ *has measure zero*, $\mu(S) = 0$, if for every $\epsilon > 0$ there is a *sequence* of (closed or open or half open) rectangles $\{R_1, R_2, R_3, ...\}$ such that $S \subset \bigcup_{i=1}^{\infty} R_i$ with $\sum_{i=1}^{\infty}\nu(R_i) < \epsilon$.

Clearly, if a set S has volume zero it also has measure zero. As the following example shows the converse is not necessarily true.

Example 4.1.20. Let $S = \mathbb{Q} \cap (0, 1)$ in $\mathbb{R}$. The set S has measure zero but does not have volume (length) zero.

Solution. Let $\epsilon > 0$. Since the rational numbers in $(0, 1)$ can be enumerated (in a single sequence) $\{q_1, q_2, ...\}$, the sequence of closed interval

$$\left\{ I_i = [q_i - \frac{\epsilon}{2^{i+1}}, q_i + \frac{\epsilon}{2^{i+1}}] : i = 1, 2, ... \right\}$$

covers S and, because of the geometric series, $\sum_{i=1}^{\infty}\nu(I_i) = \sum_{i=1}^{\infty}\frac{\epsilon}{2^i} = \epsilon$. So, S has measure zero. However, S has not volume (length) zero. To see this, let $\{I_1, ..., I_N\}$ be *any* finite collection of closed intervals such that $S \subset \bigcup_{i=1}^{N} I_i$. Since $K = \bigcup_{i=1}^{N} I_i$ is a closed set, it follows that $(0, 1) \subset K$, for otherwise, if $x \in (0, 1)$ and $x \notin K$ there would be a neighborhood of x not in K, but from Theorem 1.1.7, this neighborhood would contain rationals in $(0, 1)$. This contradicts the fact that $S \subset K$. Therefore $(0, 1) \subset K$ and hence $\sum_{i=1}^{N}\nu(I_i) \geq 1$.

However when $S \subset \mathbb{R}^n$ is *compact*, since any cover of S has a finite subcover, the concepts of volume zero and measure zero coincide. It is also clear that any subset of a set of volume zero (measure zero) has itself volume zero (measure zero). Moreover, it is easy to see that if S_1 and S_2 have volume zero so does their union $S_1 \cup S_2$. An important generalization of this latter statement is the following proposition.

Proposition 4.1.21. *If* $\{S_k : k = 1, 2, ...\}$ *is a sequence of sets each of measure zero, then* $S = \bigcup_{k=1}^{\infty} S_k$ *also has measure zero.*

Proof. Let $\epsilon > 0$. Since each S_k has measure zero, there is a sequence of rectangles $\{R_{1k}, R_{2k}, R_{3k}, ...\}$ such that they cover S_k and $\sum_{i=1}^{\infty} \nu(R_{ik}) < \frac{\epsilon}{2^k}$.

Now the collection of rectangles $\{R_{ik} : i, k = 1, 2, ...\}$ is countable[3], covers S and has total volume $\sum_{k=1}^{\infty} \sum_{i=1}^{\infty} \nu(R_{ik}) < \sum_{k=1}^{\infty} \frac{\epsilon}{2^k} = \epsilon$. $\square$

Consequently, any countable set in $\mathbb{R}^n$ has measure zero.

Exercise 4.1.22. (*) An interesting and important example of a non countable set of measure zero is the *Cantor set*. In the closed unit interval $[0, 1]$, let U_1 denote the open middle third $(\frac{1}{3}, \frac{2}{3})$. Similarly, $U_2 = (\frac{1}{9}, \frac{2}{9})$ and $U_3 = (\frac{7}{9}, \frac{8}{9})$ etc. Evidently, we have a countable number of U_i. The sum of all their lengths is $\sum_{i=1}^{\infty} \mu(U_i) = \frac{1}{3} + \frac{2}{9} + \frac{4}{27} \ldots$, which is a geometric series with ratio $\frac{2}{3}$. Hence it is convergent and sums to 1 (check!). The Cantor set C is defined by $C = [0, 1] \setminus \cup_{i=1}^{\infty} U_i$. Show $\mu(C) = 0$. Show C is uncountable.

Proposition 4.1.23. *Let* $f : R \subset \mathbb{R}^n \to \mathbb{R}$ *be a continuous function defined on a rectangle* R *in* $\mathbb{R}^n$. *Then the graph of* f, $S = \text{graph}(f)$, *a set in* $\mathbb{R}^{n+1}$ *has* $(n + 1)$*-dimensional volume zero.*

Proof. Since f is uniformly continuous on R, for $\epsilon > 0$ there is $\delta > 0$ such that $|f(x) - f(x')| < \epsilon$ whenever $||x - x'|| < \delta$ for all $x, x' \in R$. If we now take a partition $\mathcal{P}$ of R with $||\mathcal{P}|| < \delta$, then on each subrectangle R_j we have $M_j - m_j < \epsilon$. Therefore, for an arbitrary fixed $x_j \in R_j$, the $(n + 1)$-dimensional rectangle $R'_j = R_j \times [f(x_j) - \epsilon, f(x_j) + \epsilon]$ contains the portion of the graph of f lying over the subrectangle R_j. Hence

---

[3]See Appendix A.

the union $\bigcup_{j=1}^{m} R'_j$ covers the whole graph S of f over R. At the same time,

$$\sum_{j=1}^{m} \nu(R'_j) = \sum_{j=1}^{m} \nu(R_j) \cdot 2\epsilon = 2\epsilon\nu(R)$$

(here $\nu(R'_j)$ is the volume of R'_j in $\mathbb{R}^{n+1}$).   Since ϵ is arbitrary $\nu(S) = 0$.   □

Corollary 4.1.24. *Let $f : \mathbb{R}^n \to \mathbb{R}$ be a continuous. Then the graph of f has $(n+1)$-dimensional measure zero.*

Proof. Cover $\mathbb{R}^n$ with a countable union of rectangles and apply Proposition 4.1.21.   □

Corollary 4.1.25. *Any n-dimensional subspace or any n-dimensional hyperplane Π of $\mathbb{R}^{n+1}$ has $(n+1)$-dimensional volume zero. In particular, any lower dimensional subspace of $\mathbb{R}^{n+1}$ has $(n+1)$-dimensional measure zero.*

Proof. Let $f(x) = \lambda(x) + x_0$, where $\lambda : \mathbb{R}^n \to \mathbb{R}$ is a linear functional and x_0 is a fixed vector in $\mathbb{R}^n$. Then f is (uniformly) continuous and $\Pi = \text{graph}(f)$.   □

The assumption of the continuity of f in Theorem 4.1.17 can be relaxed by allowing the function to be discontinuous on a set of measure zero. In fact, a *necessary* and *sufficient* condition for a bounded function f to be integrable on R is that the set D of points of discontinuity of f has measure zero. This important result is known as the *Lebesgue criterion*. The proof of Lebesgue's criterion is rather technical, however the idea of the proof is quite simple. It is based in splitting the sum of the difference between the upper and lower Darboux sums $U_{\mathcal{P}}(f) - L_{\mathcal{P}}(f) = \sum_{j=1}^{m}(M_j - m_j)\nu(R_j)$ into two sums; the first sum $\sum_1$ corresponding to the subrectangles on which f is *continuous* and the second sum $\sum_2$ corresponding to the subrectangles *meeting the set of discontinuities* D of f. For the first sum the continuity of f tells us that the *oscillations* $M_j - m_j$ can be made (uniformly) small. For the second sum, if D has measure zero, then the total volume $\sum_2 \nu(R_j)$ can be made arbitrarily small. We give the details below.

244   Chapter 4   Integral Calculus in Several Variables

Before we prove Lebesgue's criterion, we define the *oscillation* of a bounded function.

Definition 4.1.26. Let $f : \Omega \subseteq \mathbb{R}^n \to \mathbb{R}$ be a bounded function and let $a \in \Omega$. The *oscillation* $\omega_f(a)$ of f at a is

$$\omega_f(a) = \lim_{r \to 0} \omega_f(B_r(a)) = \lim_{r \to 0} \left[\sup_{x_1, x_2 \in B_r(a)} |f(x_1) - f(x_2)| \right].$$

Proposition 4.1.27. *A bounded function $f : \Omega \subseteq \mathbb{R}^n \to \mathbb{R}$ is continuous at $a \in \Omega$ if and only if $\omega_f(a) = 0$.*

Proof. If $\omega_f(a) = 0$, then for every $\epsilon > 0$ there exists $\delta > 0$ ($\delta = \delta(\epsilon, a)$) such that whenever $r < \delta$ we have $\sup_{x_1, x_2 \in B_r(a)} |f(x_1) - f(x_2)| < \epsilon$. Hence, in particular, $|f(x) - f(a)| < \epsilon$ for all $x \in B_r(a) \cap \Omega$, and f is continuous at a. Conversely, suppose f is continuous at $a \in \Omega$ and let $\epsilon > 0$. Then there is $\delta > 0$ such that $|f(x) - f(a)| < \frac{\epsilon}{2}$ whenever $||x - a|| < \delta$ and $x \in \Omega$. Therefore for $x_1, x_2 \in \Omega \cap B_\delta(a)$ we have $|f(x_1) - f(x_2)| \le |f(x_1) - f(a)| + |f(a) - f(x_2)| < \epsilon$. Hence $\omega_f(a) < \epsilon$, that is, $\omega_f(a) = 0$. $\square$

Proposition 4.1.27 tells us that the oscillation of a bounded function at its points of discontinuity is positive. Hence $D = \{x \in \Omega : \omega_f(x) > 0\}$ is the set of discontinuities of f on Ω.

Lemma 4.1.28. *For $\epsilon > 0$, let $D_\epsilon = \{x \in \Omega : \omega_f(x) \ge \epsilon\}$. Then D_ϵ is relatively closed in Ω.*

Proof. We show that the complement of D_ϵ is relatively open in Ω. Indeed, let $x \in \Omega$ and suppose for $\omega_f(x) < \epsilon$. Then there exist $\delta > 0$ such that if $r < \delta$ we have $|f(x_1) - f(x_2)| < \epsilon$ for all $x_1, x_2 \in B_r(x) \cap \Omega$. Therefore, for $r < \delta$ and any $y \in \Omega$ with $||x - y|| < \frac{r}{2}$ we have $B_{\frac{r}{2}}(y) \subset B_r(x)$. Hence $|f(x_1) - f(x_2)| < \epsilon$ for all $x_1, x_2 \in B_{\frac{r}{2}}(y) \cap \Omega$ and so $\omega_f(y) < \epsilon$. $\square$

When $\Omega = R$ a closed rectangle in $\mathbb{R}^n$, it is important to note that since R is closed the set D_ϵ is also closed. Hence D_ϵ being closed and bounded is compact.

Lemma 4.1.29. *Let R be a closed rectangle in $\mathbb{R}^n$ and $f : R \to \mathbb{R}$ a bounded function. Suppose $\omega_f(x) < \epsilon$ for all $x \in R$. Then there exist a partition $\mathcal{P} = \{R_j : j = 1, ..., m\}$ such that $\omega_f(R_j) < \epsilon$ for all $j = 1, ...m$. Hence*

$$U_{\mathcal{P}}(f) - L_{\mathcal{P}}(f) < \epsilon \cdot \nu(R).$$

Proof. For each $x \in R$ there is $\delta_x > 0$ such that $\omega_f(B_{\delta_x}(x) \cap R) < \epsilon$. Choose now an open rectangle (cube) U_x with $x \in U_x \subset \overline{U_x} \subset B_{\delta_x}(x)$ (this is possible since $||x||_\infty \leq ||x|| \leq \sqrt{n}||x||_\infty$ form Example 1.3.9). The collection $\{U_x\}_{x \in R}$ is an open cover of the compact set R, and so a finite number $U_{x_1}, ..., U_{x_k}$ of them cover R. Let $\mathcal{P} = \{R_j : j = 1, ..., m\}$ be a partition for R such that each subrectangle R_j is contained in some $\overline{U_{x_i}}$. Then $\omega_f(R_j) < \epsilon$. Hence $U_{\mathcal{P}}(f) - L_{\mathcal{P}}(f) = \sum_{j=1}^m (M_j - m_j)\nu(R_j) < \epsilon \sum_{j=1}^m \nu(R_j) = \epsilon\nu(R)$.   □

Theorem 4.1.30. *(Lebesgue's criterion). Let $f : R \subset \mathbb{R}^n \to \mathbb{R}$ be a bounded function. Then f is (Riemann) integrable on R if and only if the set of discontinuities D of f has measure zero.*

Proof. Suppose that D has measure zero. Let $\epsilon > 0$. Then, since $D_\epsilon \subseteq D$ the compact set D_ϵ has measure zero and hence volume zero. Therefore there is a finite number of (open) rectangles $E_1, .., E_N$ which cover D_ϵ such that $\sum_{i=1}^N \nu(E_i) < \epsilon$. Pick a partition $\mathcal{P}$ of R and separate this partition into two classes $\mathcal{P}_1 = \{R_j : j = 1, 2, ..., p_1\}$ and $\mathcal{P}_2 = \{R_k'' : k = 1, ..., p_2\}$, where $\mathcal{P}_1$ consists of subrectangles R_j such that $R_j \cap D_\epsilon = \emptyset$, and $\mathcal{P}_2$ consists of subrectangles R_k'' such that $R_k'' \subset E_i$ for some i $(i = 1, ..., N)$.

Now in any subrectangle R_j of $\mathcal{P}_1$ we have $\omega_f(x) < \epsilon$ for all $x \in R_j$. By Lemma 4.1.29, R_j can be further partitioned into subrectangles R_j' such that $\omega_f(R_j') < \epsilon$. Once this has been done for each such j, they can be incorporated into a refinement $\mathcal{Q}$ of $\mathcal{P}$. Then

$$U_{\mathcal{Q}}(f) - L_{\mathcal{Q}}(f) = \sum_{j=1}^{p_1'}(M_j - m_j)\nu(R_j') + \sum_{k=1}^{p_2}(M_k - m_k)\nu(R_k'') = \sum_1 + \sum_2$$

Since $M_j - m_j = \omega_f(R'_j) < \epsilon$, the first sum

$$\sum_1 < \epsilon \sum_{j=1}^{p'_1} \nu(R'_j) \leq \epsilon\nu(R).$$

For the second sum note that since f is bounded on R, there is $M > 0$ with $|f(x)| < M$ for all $x \in R$. Therefore, in particular,

$$\sum_2 = \sum_{k=1}^{p_2} (M_k - m_k)\nu(R''_k) \leq 2M \sum_{k=1}^{p_2} \nu(R''_k) \leq 2M \sum_{i=1}^{N} \nu(E_i) \leq 2M\epsilon.$$

Hence

$$U_{\mathcal{Q}}(f) - L_{\mathcal{Q}}(f) < (\nu(R) + 2M)\epsilon.$$

Since M and $\nu(R)$ are fixed and $\epsilon > 0$ arbitrary, the function is integrable.

Conversely, suppose that f is integrable. First note that $D = \bigcup_{k=1}^{\infty} D_{\frac{1}{k}}$, for if $\omega_f(x) > 0$, then there is some $k = 1, 2, \dots$ such that $\omega_f(x) \geq \frac{1}{k}$. Since each $D_{\frac{1}{k}}$ is compact, by Proposition 4.1.21 it suffices to show that each $D_{\frac{1}{k}}$ has volume zero. In fact we will show that for every $\epsilon > 0$ each D_ϵ has volume zero. We argue by contradiction. Suppose that for some $\epsilon_0 > 0$, the set D_{ϵ_0} does *not* have volume zero. Then there is $\delta > 0$ such that any finite cover of D_{ϵ_0} has total volume $\geq \delta$. Let $\mathcal{P} = \{R_j : j = 1, \dots, m\}$ be *any* partition of R. Then

$$U_{\mathcal{P}}(f) - L_{\mathcal{P}}(f) = \sum_{j=1}^{m} (M_j - m_j)\nu(R_j) \geq \sum_l (M_l - m_l)\nu(R_l) \geq \sum_l \epsilon_0\nu(R_l),$$

where $\sum_l$ denotes the sum over all subrectangles R_l which contain points of D_{ϵ_0}. Since these subrectangles cover D_{ϵ_0}, their total volume $\sum_l \nu(R_l) \geq \delta$. Hence

$$U_{\mathcal{P}}(f) - L_{\mathcal{P}}(f) \geq \epsilon_0\delta,$$

and the Riemann condition tells us that f is *not* integrable. $\square$

Remark 4.1.31. With this criterion, which is a restatement of Riemann's own condition, Lebesgue[4] in 1904 created new advances in integration theory by stressing the concept of *measure* (a generalization of the concept of volume). Here we have only discussed the very special case when the Lebesgue measure of a set is zero. For a detailed account see [28].

4.1.2   The integral over a bounded set

Thus far we have defined the integral for a bounded function f of several variables over a rectangle R in $\mathbb{R}^n$. It is necessary to extend the definition of the integral so as to be able to integrate over more general sets in $\mathbb{R}^n$ that are not too complicated (or nasty).

Definition 4.1.32. A bounded set $\Omega \subset \mathbb{R}^n$ is said to be *simple* if its boundary, $\partial(\Omega)$ has n-dimensional volume zero.

Remark 4.1.33.   1. By its very definition the boundary of any set Ω in $\mathbb{R}^n$ is a closed set in $\mathbb{R}^n$. Furthermore, if $\Omega \subset \mathbb{R}^n$ is bounded, then the boundary $\partial(\Omega)$ is not only closed, but also bounded in $\mathbb{R}^n$, that is, it is a compact subset of $\mathbb{R}^n$.

2. Since for any sets S_1, S_2 in $\mathbb{R}^n$ (in fact, in any metric space), the following assertions hold: $\partial(S_1 \cup S_2) \subset \partial(S_1) \cup \partial(S_2)$ and $\partial(S_1 \cap S_2) \subset \partial(S_1) \cup \partial(S_2)$, it follows then that finite unions or finite intersections of simple sets are simple sets. In addition, since $\partial(S_1 \setminus S_2) \subset \partial(S_1) \cup \partial(S_2)$, the difference of simple sets is also a simple set.

A rectangle, a cube, and a ball in $\mathbb{R}^n$ are simple sets. Moreover, a bounded set Ω in $\mathbb{R}^3$ whose boundary is finite union of smooth surfaces (or curves if $\Omega \subset \mathbb{R}^2$, or hypersurfaces if $\Omega \subset \mathbb{R}^n$) is simple.

Definition 4.1.34. Let $\Omega \subset \mathbb{R}^n$. The *characteristic function* χ_Ω of Ω is defined by

$$\chi_\Omega(x) = \begin{cases} 1 \text{ when } x \in \Omega, \\ 0 \text{ when } x \notin \Omega. \end{cases}$$

---

[4]However the Riemann integral and the Lebesgue integral coincide for continuous functions.

A consequence of Lebesgue's criterion is the following.

Corollary 4.1.35. *The characteristic function χ_Ω of a bounded set $\Omega \subset \mathbb{R}^n$ is integrable if and only if $\partial(\Omega)$ has volume zero. That is, if and only if Ω is simple.*

Proof. Since the set of points of discontinuity of χ_Ω is just the boundary, $\partial(\Omega)$, and since the concepts of measure zero and volume zero coincide for compact sets, the assertion follows from Lebesgue's criterion. $\square$

Example 4.1.36. In general, the characteristic function of a bounded set might not be integrable. Let $\Omega = [0,1] \cap \mathbb{Q} \times [0,1] \cap \mathbb{Q}$. From Example 1.4.13 (2), we see that $\partial(\Omega) = [0,1] \times [0,1] = R$. Hence, $\nu(\partial(\Omega)) = 1$. So χ_Ω can not be integrable. In fact as we saw in Example 4.1.16, χ_Ω is not integrable.

Definition 4.1.37. Let Ω be a bounded simple set in $\mathbb{R}^n$ and $f : R \to \mathbb{R}$ a bounded function defined on some rectangle R with $\Omega \subseteq R$. We say that f is *integrable on Ω* if $f \cdot \chi_\Omega$ is integrable on R, in which case we define the integral of f over Ω by

$$\int_\Omega f = \int_R f \cdot \chi_\Omega.$$

We now show the integral $\int_R f \cdot \chi_\Omega$ is independent of the choice of the rectangle R that contains Ω.

Proposition 4.1.38. *Let Ω be a bounded simple set in $\mathbb{R}^n$ and R_1, R_2 be two closed rectangles containing Ω. Suppose $f : \mathbb{R}^n \to \mathbb{R}$ is a bounded function defined on both R_1 and R_2. Then $f \cdot \chi_\Omega$ is integrable on R_1 if and only if $f \cdot \chi_\Omega$ is integrable on R_2 and*

$$\int_{R_1} f \cdot \chi_\Omega = \int_{R_2} f \cdot \chi_\Omega.$$

Proof. Consider the rectangle $R = R_1 \cap R_2$. Clearly $\Omega \subset R$. The points of discontinuity of $f \cdot \chi_\Omega$ are either points of discontinuity of f in Ω or points of discontinuity of χ_Ω, in which case they lie on $\partial(\Omega)$. In any case, all these points lie in $R = R_1 \cap R_2$. The Lebesgue criterion then implies that the integrals of $f \cdot \chi_\Omega$ over the rectangles R, R_1 and R_2

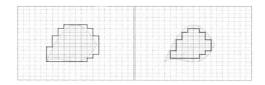

Figure 4.1: Approximations of the outer and inner areas of a region

either all exist or all fail to exist. When they do exist, their value will be independent of the choice of the partitions of R, R_1 and R_2. Therefore, we choose only those partitions of R_1 and R_2 which are obtained as extensions of partitions of the rectangle $R = R_1 \cap R_2$. Since the function is zero outside R, the Riemann sums corresponding to these partitions of R_1 and R_2 reduce to Riemann sums for the corresponding partition $\mathcal{P}$ of R. Letting $\|\mathcal{P}\| \to 0$, it follows that the integrals of f over R_1 and R_2 are equal to the integral of f over R.   □

A natural consequence of the geometrical meaning of the integral is the following:

Definition 4.1.39. Let Ω be a bounded subset of $\mathbb{R}^n$. We say that Ω *has volume* if χ_Ω is integrable over Ω, and the *volume* of Ω is the number

$$\int_\Omega 1 = \nu(\Omega).$$

This extends the concept of volume from rectangles to bounded simple sets (see Corollary 4.1.15).

Corollary 4.1.40. *A bounded set* $\Omega \subset \mathbb{R}^n$ *is simple if and only if it has volume.*

Remark 4.1.41. Note that using upper and lower Darboux integrals we can express the volume of Ω as

$$\nu(\Omega) = \int_\Omega 1 = \int_{R \supset \Omega} \chi_\Omega = \int_{R \supset \Omega}^* \chi_\Omega = \int_{*R \supset \Omega} \chi_\Omega.$$

For any partition $\mathcal{P}$ of the rectangle R, by the definition of χ_Ω the lower Darboux sum $L_{\mathcal{P}}(\chi_\Omega)$ is the sum of the volumes of the subrectangles of the partition $\mathcal{P}$ that are entirely contained in Ω (the volume of the polyhedron *inscribed* in Ω), while the upper Darboux sum $U_{\mathcal{P}}(\chi_{\Omega})$ is the sum of the volumes of the subrectangles each of which contains at least one point of Ω (the volume of the *circumscribed* polyhedron). The volume of Ω is then equal to $\nu(\Omega) = \sup\{L_{\mathcal{P}}(\chi_{\Omega})\} = \inf\{U_{\mathcal{P}}(\chi_{\Omega})\}$ as $\mathcal{P}$ run over all partitions of R. The supremum of the volumes of the inscribed polyhedra in Ω is called the *inner Jordan measure* of Ω, denoted by $\nu_*(\Omega)$, while the infimum of the volumes of the circumscribed polyhedra is called the *outer Jordan measure* of Ω, denoted by $\nu^*(\Omega)$. Thus, Ω has volume if $\nu_*(\Omega) = \nu^*(\Omega)$. Sets for which $\nu_*(\Omega) = \nu^*(\Omega)$ are known as *Jordan*[5] *measurable* sets. See Figure 4.1 for a picture of this in $\mathbb{R}^2$.

Finally we characterize the functions which are Riemann integrable on a bounded simple set Ω in $\mathbb{R}^n$.

Theorem 4.1.42. *Let $f : \Omega \to \mathbb{R}$ be bounded, where $\Omega \subset \mathbb{R}^n$ is a bounded simple set. Then f is integrable on Ω if and only if the set D of discontinuities of f in Ω has measure zero.*

Proof. Let E be the set of discontinuities of $f\chi_{\Omega}$ in some rectangle R containing Ω. Then $E \subseteq D \cup \partial(\Omega)$. Since D and $\partial(\Omega)$ have measure zero so does E. By Theorem 4.1.30, E has measure zero if and only if $f\chi_{\Omega}$ is integrable on R, that is, if and only if f is integrable on Ω. $\square$

4.2   Properties of multiple integrals

Here we develop the basic properties of multiple integrals.

Theorem 4.2.1. *Let f, g be integrable functions on a bounded simple set $\Omega \subset \mathbb{R}^n$ and $c \in \mathbb{R}$ a constant. Then*

---

[5]C. Jordan (1838-1922) is known for his reluts in group theory, linear algebra, analysis and for his influential book *Cour's d' Analyse*. Sometimes the phrase Ω has *content* is used to mean the same as Ω has *volume* or Ω is *Jordan measurable*. Since one really studies measurable sets in *measure theory*, we prefer to say Ω is a *simple* set rather Ω is a *Jordan measurable* set. For a detailed discussion the interested reader can consult for example, [28].

1. *(Linearity). $cf + g$ is integrable and*

$$\int_\Omega (cf + g) = c \int_\Omega f + \int_\Omega g.$$

2. *(Monotonicity). If $f(x) \geq 0$ for all $x \in \Omega$, then $\int_\Omega f \geq 0$.*
   More generally, if $f(x) \geq g(x)$ for all $x \in \Omega$, then

$$\int_\Omega f \geq \int_\Omega g.$$

3. *If $m \leq f(x) \leq M$ for all $x \in \Omega$, then*

$$m \cdot \nu(\Omega) \leq \int_\Omega f \leq M \cdot \nu(\Omega).$$

   In particular, if $\nu(\Omega) = 0$, then $\int_\Omega f = 0$.

4. *$|f|$ is integrable on Ω and*

$$\left| \int_\Omega f \right| \leq \int_\Omega |f|.$$

5. *Let $S \subset \Omega$. If $f(x) \geq 0$ for all $x \in \Omega$ and f is integrable over S, then*

$$\int_S f \leq \int_\Omega f.$$

6. *(Additivity). Let Ω_1, Ω_2 be bounded simple sets and $\Omega = \Omega_1 \cup \Omega_2$. If f is integrable on Ω, then f is integrable on each Ω_1, Ω_2 and $\Omega_1 \cap \Omega_2$ and*

$$\int_\Omega f = \int_{\Omega_1} f + \int_{\Omega_2} f - \int_{\Omega_1 \cap \Omega_2} f.$$

   If in addition $\nu(\Omega_1 \cap \Omega_2) = 0$, then

$$\int_\Omega f = \int_{\Omega_1} f + \int_{\Omega_2} f.$$

Proof.   1. Since $(cf + g)\chi_\Omega = cf\chi_\Omega + g\chi_\Omega$, it suffices to prove the result for a rectangle R containing Ω. We consider first the case where $c = 1$. For *any* partition $\mathcal{P}$ of R, note that since for any set S

$$\inf_{x \in S} \{f(x)\} + \inf_{x \in S} \{g(x)\} \leq \inf_{x \in S} \{f(x) + g(x)\},$$

it follows

$$L_\mathcal{P}(f) + L_\mathcal{P}(g) \leq L_\mathcal{P}(f + g). \tag{4.2}$$

On the other hand, since

$$\sup_{x \in S} \{f(x) + g(x)\} \leq \sup_{x \in S} \{f(x)\} + \sup_{x \in S} \{g(x)\},$$

we also have

$$U_\mathcal{P}(f + g) \leq U_\mathcal{P}(f) + U_\mathcal{P}(g). \tag{4.3}$$

Now, since f, g are integrable, for every $\epsilon > 0$ there exist partitions $\mathcal{P}_1$ and $\mathcal{P}_2$ of R such that $U_{\mathcal{P}_1}(f) - L_{\mathcal{P}_1}(f) < \frac{\epsilon}{2}$ and $U_{\mathcal{P}_2}(g) - L_{\mathcal{P}_2}(g) < \frac{\epsilon}{2}$. Let $\mathcal{P}$ be their common refinement.

By Lemma 4.1.9 and inequalities (4.2) and (4.3) we have

$$L_{\mathcal{P}_1}(f) + L_{\mathcal{P}_2}(g) \leq L_\mathcal{P}(f) + L_\mathcal{P}(g) \leq L_\mathcal{P}(f + g), \text{ and}$$

$$U_\mathcal{P}(f + g) \leq U_\mathcal{P}(f) + U_\mathcal{P}(g) \leq U_{\mathcal{P}_1}(f) + U_{\mathcal{P}_2}(g).$$

Hence

$$U_\mathcal{P}(f + g) - L_\mathcal{P}(f + g) \leq U_{\mathcal{P}_1}(f) + U_{\mathcal{P}_2}(g) - [L_{\mathcal{P}_1}(f) + L_{\mathcal{P}_2}(g)] < \epsilon.$$

Therefore, $f + g$ is integrable. In addition, (4.2) and (4.3) imply

$$\int_R (f + g) = \sup_\mathcal{P} \{L_\mathcal{P}(f + g)\} \geq \sup_\mathcal{P} \{L_\mathcal{P}(f)\} + \sup_\mathcal{P} \{L_\mathcal{P}(g)\} = \int_R f + \int_R g$$

$$\int_R (f + g) = \inf_\mathcal{P} \{U_\mathcal{P}(f + g)\} \leq \inf_\mathcal{P} \{U_\mathcal{P}(f)\} + \inf_\mathcal{P} \{U_\mathcal{P}(g)\} = \int_R f + \int_R g.$$

Hence, $\int_R (f + g) = \int_R f + \int_R g$.

Now let $c > 0$. Note that

$$U_\mathcal{P}(cf) - L_\mathcal{P}(cf) = c[U_\mathcal{P}(f) - L_\mathcal{P}(f)] < c\frac{\epsilon}{2},$$

and since $\epsilon > 0$ is arbitrary cf is integrable and one readily sees
that $\int_R cf = c \int_R f$. To complete the proof for any $c \in \mathbb{R}$, it is
enough to show that $-f$ is integrable and $\int_R(-f) = -\int_R f$. Since
$U_{\mathcal{P}}(-f) = -L_{\mathcal{P}}(f)$ and $L_{\mathcal{P}}(-f) = -U_{\mathcal{P}}(f)$ for any partition $\mathcal{P}$ of
R, it follows that

$$U_{\mathcal{P}}(-f) - L_{\mathcal{P}}(-f) = -L_{\mathcal{P}}(f) + U_{\mathcal{P}}(f).$$

Hence, when f is integrable so is $-f$. Now set $g = -f$ and note
that by the above we have $\int_R f + \int_R g = \int_R(f + g) = \int_R 0 = 0$.

2. Since $\chi_\Omega(x) \geq 0$ for all $x \in \Omega$, it suffices to prove the property for
   the integral over a rectangle $R \supset \Omega$. If $f(x) \geq 0$ for all $x \in R$ then
   $L_{\mathcal{P}}(f) \geq 0$ for every partition $\mathcal{P}$ of R. Hence $\int_R f \geq 0$. For the
   second part, $(f(x) - g(x)) \geq 0$ and by what we just proved and
   the linearity of the intgral we have $\int_R f - \int_R g = \int_R(f - g) \geq 0$.

3. If $m \leq f(x) \leq M$ for all $x \in \Omega$, then $m\chi_\Omega \leq f(x)\chi_\Omega \leq M\chi_\Omega$.
   Therefore

$$m \cdot \nu(\Omega) = \int_R m\chi_\Omega \leq \int_R f\chi_\Omega = \int_\Omega f \leq \int_R M\chi_\Omega = M \cdot \nu(\Omega).$$

Clearly when $\nu(\Omega) = 0$, then $\int_\Omega f = 0$.

4. Again since $|f\chi_\Omega| = |f|\chi_\Omega$, it suffices to prove the property for f
   integrable over a rectangle $R \supset \Omega$. Let $\mathcal{P}$ be a partition of R and
   let us denote by m_j, m_j' and M_j, M_j' the infima and suprema of
   f and $|f|$ in each subrectangle R_j, respectively. For all $x, x' \in R_j$,
   we have

$$\big||f(x)| - |f(x')|\big| \leq |f(x) - f(x')| \leq \sup_{x,x' \in R_j} |f(x) - f(x')| = M_j - m_j.$$

Therefore,

$$M_j' - m_j' = \sup_{x,x' \in R_j} \big||f(x)| - |f(x')|\big| \leq M_j - m_j.$$

Hence,

$$U_{\mathcal{P}}(|f|) - L_{\mathcal{P}}(|f|) \leq U_{\mathcal{P}}(f) - L_{\mathcal{P}}(f).$$

Thus, if f is integrable over R, so is $|f|$. Furthermore, since
$-|f(x)| \leq f(x) \leq |f(x)|$ by property (2) we get $\left|\int_R f\right| \leq \int_R |f|$.

5. Since $S \subseteq \Omega$ it follows that $\chi_S \leq \chi_\Omega$. Therefore $f\chi_S \leq f\chi_\Omega$ and the result follows from property (2).

6. Let $R \supset \Omega = \Omega_1 \cup \Omega_2$. Since $\chi_{\Omega_1 \cup \Omega_2} = \chi_{\Omega_1} + \chi_{\Omega_2} - \chi_{\Omega_1 \cap \Omega_2}$, it follows that

$$f\chi_{\Omega_1 \cup \Omega_2} = f\chi_{\Omega_1} + f\chi_{\Omega_2} - f\chi_{\Omega_1 \cap \Omega_2}.$$

Furthermore, since f is integrable over Ω, the function $f\chi_{\Omega_1}$ is integrable over R because the set of points of its discontinuities are those of f together with part of $\partial(\Omega_1)$, which has volume zero. Similarly the other terms are integrable. Now, by the linearity of the integral

$$\int_\Omega f = \int_{\Omega_1} f + \int_{\Omega_2} f - \int_{\Omega_1 \cap \Omega_2} f.$$

If $\nu(\Omega_1 \cap \Omega_2) = 0$, then, by property (3), $\int_{\Omega_1 \cap \Omega_2} f = 0$ and so

$$\int_{\Omega_1 \cup \Omega_2} f = \int_{\Omega_1} f + \int_{\Omega_2} f.$$

$\square$

Remark 4.2.2. The converse of property 4 in Theorem 4.2.1. is not necessarily true. It is possible, $\int_\Omega |f|$ to exist and $\int_\Omega f$ not. For example the function $f : [0,1] \to \mathbb{R}$ given by $f(x) = 2\chi_{\mathbb{Q} \cap [0,1]}(x) - 1$ is not (Riemann) integrable (the easiest way to see this by looking upper and lower Darboux sums), while $|f(x)| = 1$ and $\int_0^1 |f| = 1$.

Corollary 4.2.3. *Let* Ω_1, Ω_2 *be bounded simple sets in* $\mathbb{R}^n$. *Then*

$$\nu(\Omega_1 \cup \Omega_2) = \nu(\Omega_1) + \nu(\Omega_2) - \nu(\Omega_1 \cap \Omega_2).$$

Proof. Take $f \equiv 1$ in the additivity property of the integral. $\square$

Definition 4.2.4. A relation involving the elements of $\mathbb{R}^n$ is said to hold *essentially* or *almost everywhere* if the set E of all points for which the relation fails to hold has *measure zero*.

Corollary 4.2.5. *(Finite additivity). Suppose* $\Omega = \bigcup_{j=1}^{N} \Omega_j$, *where* $\{\Omega_j : j = 1, ..., N\}$ *is a finite collection of bounded simple sets which are essentially disjoint (i.e., the intersection of any pair* Ω_i, Ω_j *has measure zero). If* f *is integrable on* Ω, *then* f *is integrable on each* Ω_j *and*

$$\int_\Omega f = \int_{\bigcup_{j=1}^N \Omega_j} f = \sum_{j=1}^N \int_{\Omega_j} f.$$

Proof. The case $N = 2$ follows from additivity. The general case follows by induction. □

Corollary 4.2.6. *Let* Ω *be a bounded simple set in* $\mathbb{R}^n$. *If* $f : \overline{\Omega} \to \mathbb{R}$ *is integrable on* $\overline{\Omega}$, *then* f *is integrable on both* Ω, Ω° *and*

$$\int_{\overline{\Omega}} f = \int_\Omega f = \int_{\Omega^\circ} f.$$

Proof. From Section 1.4, $\overline{\Omega} = \Omega \cup \partial(\Omega)$ and $\Omega = \Omega^\circ \cup (\partial(\Omega) \cap \Omega)$. Since Ω is simple, $\nu(\partial(\Omega)) = 0$, and the result follows from property (6). □

Proposition 4.2.7. *Let* $f : \Omega \to \mathbb{R}$ *be integrable on a bounded simple set* $\Omega \subset \mathbb{R}^n$.

1. *If* $f \equiv 0$ *except on a set* E *of volume zero, then* $\int_\Omega f = 0$.

2. *If* $f \geq 0$ *and* $\int_\Omega f = 0$, *then* f *is essentially zero.*

Proof.   1. Since Ω is the disjoint union $\Omega = E \cup (\Omega \setminus E)$ and $\nu(E) = 0$, it follows from properties (3),(6) of Theorem 4.2.1 that

$$\int_\Omega f = \int_E f + \int_{(\Omega \setminus E)} f = 0 + 0 = 0.$$

2. Let $E = \{x \in \Omega : f(x) > 0\}$. All we need show is that E has measure zero. Note that $E = \bigcup_{k=1}^\infty E_k$, where $E_k = \{x \in \Omega : f(x) > \frac{1}{k}\}$. For all $x \in E_k$ we have $0 < \frac{1}{k} < f(x)$. The monotonicity of the integral and property (5) (Theorem 4.2.1) yield

$$0 < \frac{1}{k} \int_{E_k} 1 \leq \int_{E_k} f \leq \int_\Omega f = 0.$$

Therefore, $\frac{1}{k} \int_{E_k} 1 = 0$ or $\nu(E_k) = 0$. Hence each E_k has measure zero, and by Proposition 4.1.21 so does E.

$\square$

Corollary 4.2.8. *Let $f, g : \Omega \to \mathbb{R}$ be bounded, where Ω is a bounded simple set in $\mathbb{R}^n$. Suppose f is integrable and $f = g$ on Ω except on a set E of volume zero. Then g is integrable on Ω and $\int_\Omega g = \int_\Omega f$.*

Proof. Let $\varphi(x) = g(x) - f(x)$ for $x \in \Omega$. Then $\varphi(x) = 0$ for all $x \in (\Omega \backslash E)$. Therefore the set of points of discontinuity of φ is contained in E and so it has volume zero and hence measure zero. Theorem 4.1.42, implies that φ is integrable on Ω. At the same time, by Proposition 4.2.7(1) we have $\int_\Omega \varphi = 0$. Then Theorem 4.2.1 (1) tells us that $g = \varphi + f$ is integrable on Ω and

$$\int_\Omega g = \int_\Omega \varphi + \int_\Omega f = \int_\Omega f.$$

$\square$

Proposition 4.2.9. *Let Ω be a bounded simple set in $\mathbb{R}^n$.*

1. *If f is integrable on Ω, then so is f^2.*

2. *If f and g are integrable on Ω, then so is fg.*

Proof.   1. Since $(f\chi_\Omega)^2 = f^2\chi_\Omega$, it suffices to prove the result for f integrable over a rectangle $R \supset \Omega$. As f is bounded there esists $M > 0$ such that $|f(x)| \leq M$ for all $x \in R$. Let $\mathcal{P}$ be a partition of R and let us denote by m_j, m_j' and M_j, M_j' the infima and suprema of f and f^2 in each subrectangle R_j, respectively. For all $x, x' \in R_j$, we have

$$\left|f(x)^2 - f(x')^2\right| \leq 2M|f(x) - f(x')| \leq 2M \sup_{x,x' \in R_j} |f(x) - f(x')|$$
$$= 2M(M_j - m_j).$$

Therefore,

$$M_j' - m_j' = \sup_{x,x' \in R_j} \left|f(x)^2 - f(x')^2\right| \leq 2M(M_j - m_j).$$

Hence,
$$U_{\mathcal{P}}(f^2) - L_{\mathcal{P}}(f^2) \le 2M[U_{\mathcal{P}}(f) - L_{\mathcal{P}}(f)].$$

Thus, if f is integrable over R, so is f^2.

2. Since $fg = \frac{1}{2}\left[(f+g)^2 - (f^2 + g^2)\right]$, the hypothesis, Theorem 4.2.1 (property 1) and part (1) above tell us that fg is integrable.
$\square$

Corollary 4.2.10. *(Cauchy-Schwarz inequality). Let f, g be integrable on a bounded simple set Ω in $\mathbb{R}^n$. Then*

$$\left| \int_{\Omega} fg \right| \le \left(\int_{\Omega} |f|^2 \right)^{\frac{1}{2}} \cdot \left(\int_{\Omega} |g|^2 \right)^{\frac{1}{2}}. \tag{4.4}$$

Proof. Consider the function $f + \lambda g$, where $\lambda \in \mathbb{R}$. Then

$$0 \le \int_{\Omega} (f + \lambda g)^2 = \int_{\Omega} |f|^2 + 2\lambda \int_{\Omega} fg + \lambda^2 \int_{\Omega} |g|^2.$$

Setting $a = \int_{\Omega} |g|^2$, $b = \int_{\Omega} fg$, $c = \int_{\Omega} |f|^2$, this tells us that the quadratic polynomial $a\lambda^2 + 2b\lambda + c \ge 0$ for all $\lambda \in \mathbb{R}$. Consequently its discriminant must be nonpositive. That is, $b^2 - ac \le 0$ or $|b| \le (ac)^{\frac{1}{2}}$ which is (4.4)
$\square$

Theorem 4.2.11. *(Mean-value theorem for integrals). Suppose f is integrable on a bounded simple set $\Omega \subset \mathbb{R}^n$ and let $m = \inf\{f(\Omega)\}$, $M = \sup\{f(\Omega)\}$. Then there exist λ with $m \le \lambda \le M$ such that $\int_{\Omega} f = \lambda \nu(\Omega)$. If in addition, Ω is connected and f is continuous on Ω, then*

$$\int_{\Omega} f = f(x_0)\nu(\Omega),$$

for some $x_0 \in \Omega$.

Proof. By property (3) we have

$$m \cdot \nu(\Omega) \le \int_{\Omega} f \le M \cdot \nu(\Omega).$$

If $\nu(\Omega) = 0$, any λ will do. If $\nu(\Omega) > 0$, we let $\lambda = \frac{1}{\nu(\Omega)} \int_{\Omega} f$.

Next we prove the second part of the theorem. If $\lambda = M$, there must be some $x_0 \in \Omega$ such that $f(x_0) = M$, for otherwise, we would have $f(x) < M$ for all $x \in \Omega$. Since $\nu(\Omega) > 0$ and $\nu(\partial(\Omega)) = 0$, it follows that the interior of Ω is not empty (for if the interior of Ω is empty, then $\Omega = \partial(\Omega)$). Hence Ω contains an open ball of radius (say $\delta > 0$) and so it also contains a closed rectangle K with $\nu(K) > 0$ (for example any rectangle K with $d(K) < 2\delta$). Since f is continuous on the compact set K, there is $\epsilon > 0$ such that $f(x) \leq M - \epsilon$ for all $x \in K$. Now, since $\Omega = K \cup (\Omega \setminus K)$, the additivity of the integral implies

$$M\nu(\Omega) = \int_\Omega f = \int_K f + \int_{\Omega \setminus K} f \leq (M - \epsilon)\nu(K) + M\nu(\Omega \setminus K) < M\nu(\Omega),$$

a contradiction.

If $\lambda = m$, then a similar argument applies. If $m < \lambda < M$, then there are $x_1, x_2 \in \Omega$ such that $f(x_1) < \lambda < f(x_2)$. Since Ω is connected and f continuous, the Intermediate Value theorem (Corollary 2.6.6) tells us that $\lambda = f(x_0)$ for some $x_0 \in \Omega$. $\square$

Definition 4.2.12. Let f be integrable on a bounded simple set Ω in $\mathbb{R}^n$, with $\nu(\Omega) \neq 0$. The ratio

$$\frac{\int_\Omega f}{\nu(\Omega)}$$

is called the *mean value* of f on Ω.

Corollary 4.2.13. *Let B_r be a ball centered at $a \in \mathbb{R}^n$ and f be a continuous function on B_r. Then*

$$f(a) = \lim_{r \to 0} \frac{\int_{B_r} f}{\nu(B_r)}.$$

Proof. Since the ball is connected, Theorem 4.2.11 tells us that there is $x \in B_r$ such that $f(x) = \frac{\int_{B_r} f}{\nu(B_r)}$. Since $x \to a$ as $r \to 0$, the continuity of f gives

$$f(a) = \lim_{r \to 0} f(x) = \lim_{r \to 0} \frac{\int_{B_r} f}{\nu(B_r)}.$$

$\square$

EXERCISES

1. Show that f is Riemann integrable over a rectangle R with $\int_R f = s$ if and only if for every $\epsilon > 0$ there exists a $\delta = \delta(\epsilon) > 0$ such that for *any* partition $\mathcal{P}$ with

$$||\mathcal{P}|| < \delta \text{ we have } |S_{\mathcal{P}}(f) - s| < \epsilon \text{ for } any \text{ choice of points } \xi_i \in R_j.$$

Moreover, show that s is unique.

2. Prove that if the projection of a bounded set $\Omega \subset \mathbb{R}^n$ onto a hyperplane $\mathbb{R}^{n-1}$ has $(n-1)$-dimensional volume zero, then the set Ω itself has n-dimensional volume zero.

3. Show that a bounded simple set $\Omega \subset \mathbb{R}^n$ whose interior is empty has n-dimensional volume zero.

4. Show that any bounded set $\Omega \subset \mathbb{R}^n$ with only finitely many limit points has n-dimensional volume zero.
   Hint. If a set has no limit points must be finite by the Bolzano-Weierstrass theorem.

5. Let $R = [0, 1] \times [0, 1]$ and $f : R \to \mathbb{R}$ be defined by

$$f(x, y) = \begin{cases} 0 \text{ when } x \notin \mathbb{Q}, \\ 0 \text{ when } x \in \mathbb{Q}, y \notin \mathbb{Q}, \\ \frac{1}{q} \text{ when } x \in \mathbb{Q}, y = \frac{p}{q} \text{in lowest terms.} \end{cases}$$

Prove that f is integrable and $\int_R f = 0$.

6. Show that if f and g are integrable on a bounded simple set $\Omega \subset \mathbb{R}^n$, then the functions $\max\{f, g\}$ and $\min\{f, g\}$ are integrable on Ω.

7. Let Ω_1 and Ω_2 be bounded simple sets in $\mathbb{R}^n$ and $f : \Omega_1 \cup \Omega_2 \to \mathbb{R}$ be a bonded function. Show that if f is integrable over Ω_1 and Ω_2, then f is integrable over $\Omega_1 \backslash \Omega_2$, and

$$\int_{\Omega_1 \backslash \Omega_2} f = \int_{\Omega_1} - \int_{\Omega_1 \cap \Omega_2} f.$$

8. (Holder's inequality). Let $p \geq 1$, $q \geq 1$ such that $\frac{1}{p} + \frac{1}{q} = 1$. Suppose f and g are integrable on a bounded simple set $\Omega \subset \mathbb{R}^n$. Show that

$$\left| \int_\Omega fg \right| \leq \left(\int_\Omega |f|^p \right)^{\frac{1}{p}} \cdot \left(\int_\Omega |f|^q \right)^{\frac{1}{q}}.$$

9. (Minkowski's inequality). Let f and g be integrable on a bounded simple set $\Omega \subset \mathbb{R}^n$. Prove that

(a) If $p \geq 1$

$$\left(\int_\Omega |f+g|^p \right)^{\frac{1}{p}} \leq \left(\int_\Omega |f|^p \right)^{\frac{1}{p}} + \left(\int_\Omega |g|^p \right)^{\frac{1}{p}}.$$

(b) If $0 < p < 1$

$$\left(\int_\Omega |f+g|^p \right)^{\frac{1}{p}} \geq \left(\int_\Omega |f|^p \right)^{\frac{1}{p}} + \left(\int_\Omega |g|^p \right)^{\frac{1}{p}}.$$

4.3   Fubini's theorem

In this section we address the problem of the evaluation of multiple integrals. The idea is to reduce them to successive single integrals (see Examples 4.3.5, 4.3.6 below). The main result in this connection is *Fubini's theorem*[6]. We shall prove Fubini's theorem for integrals over rectangles and then we extend it to integrals over certain bounded simple sets in $\mathbb{R}^n$. Fubini's theorem together with the Change of Variable formula (which we shall study in Chapter 5) are the main tools for computing multiple integrals.

Before we turn to Fubini's theorem we introduce some notation and we make some preliminary remarks. We consider the space $\mathbb{R}^{n+k}$ as the cartesian product $\mathbb{R}^n \times \mathbb{R}^k$ and we write the elements of $\mathbb{R}^{n+k}$ as (x, y) where $x = (x_1, ..., x_n)$ and $y = (y_1, ..., y_k)$. Now given closed rectangles

---

[6]G. Fubini (1870-1943). His main work was in the theory of functions and geometry.

$X \subset \mathbb{R}^n$ and $Y \subset \mathbb{R}^k$ and $f : X \times Y \to \mathbb{R}$ a bounded function on $X \times Y \subset \mathbb{R}^n \times \mathbb{R}^k$, we write f as $f(x, y)$ for $x \in X$ and $y \in Y$. For each fixed $x \in X$ we denote the lower and upper Darboux integrals of the function $f_x(y) = f(x, y)$ for $y \in Y$, over the rectangle Y by

$$g_*(x) = \int_{*Y} f(x, y) dy \quad \text{and} \quad g^*(x) = \int_Y^* f(x, y) dy$$

respectively. Since f is assumed to be bounded both of these Darboux integrals exist and define functions g_* and g^* on X. Similarly, for each fixed $y \in Y$, the lower and upper Darboux integrals of the function $f_y(x) = f(x, y)$ for $x \in X$ over X are denoted by

$$h_*(y) = \int_{*X} f(x, y) dx \quad \text{and} \quad h^*(y) = \int_X^* f(x, y) dx$$

respectively and these define functions h_* and h^* on Y.

It is worthwhile remarking here that if $f(x, y)$ is integrable over $X \times Y$, it does not necessarily follows that either $f_x(y)$ or $f_y(x)$ are integrable over Y or X, respectively. For instance, let $X = [a, b]$, $Y = [c, d]$ and $f : [a, b] \times [c, d] \to \mathbb{R}$ be continuous except on the horizontal line segment $S = \{(x, c) : x \in [a, b]\} \subset \mathbb{R}^2$. Then S has area zero and so $\int_{X \times Y} f(x, y)$ exists. However, the integral $\int_X f(x, c) dx$ does not exist since the function $f(x, c)$ is discontinuous on S which has a positive 1-dimensional volume (length).

Of course, this can not happen when f is *continuous* on $X \times Y$. In this case both $f_x(y)$ and $f_y(x)$ are continuous on Y and X respectively and therefore integrable over Y and X respectively. Moreover, $g_*(x) = g^*(x) = \int_Y f(x, y) dy$ and $h_*(y) = h^*(y) = \int_X f(x, y) dx$. As we shall see in the sequel, Fubini's theorem in this case asserts that the functions

$$g(x) = \int_Y f(x, y) dy \quad \text{and} \quad h(y) = \int_X f(x, y) dx$$

are integrable over X and Y respectively and

$$\int_{X \times Y} f(x, y) = \int_X \left(\int_Y f(x, y) dy \right) dx = \int_Y \left(\int_X f(x, y) dx \right) dy.$$

We shall prove Fubini's theorem under the assumption that f is *integrable* over $X \times Y$ *and* each function $f_x(y)$ (and $(f_y(x))$ is *integrable* over Y (respectively X).

Theorem 4.3.1. *(Fubini's theorem)*[7] *Let $X \subset \mathbb{R}^n$ and $Y \subset \mathbb{R}^k$ be closed rectangles, and let $f : X \times Y \to \mathbb{R}$ be integrable. Suppose that for each $x \in X$ the function $f_x(y) = f(x,y)$ is integrable over Y. Then the function $g(x) = \int_Y f(x,y)dy$ is integrable over X and*

$$\int_{X \times Y} f(x,y) = \int_X g(x)dx = \int_X \left(\int_Y f(x,y)dy \right) dx. \qquad (4.5)$$

Similarly, if for each $y \in Y$ the function $f_y(x) = f(x,y)$ is integrable over X, then the function $h(y) = \int_X f(x,y)dx$ is integrable over Y and

$$\int_{X \times Y} f(x,y) = \int_Y h(y)dy = \int_Y \left(\int_X f(x,y)dx \right) dy. \qquad (4.6)$$

The integrals on the far right in (4.5) and (4.6) are called *iterated integrals*.

Proof. Let $\mathcal{P}_X$ be a partition of X and $\mathcal{P}_Y$ a partition of Y. Together they give a partition $\mathcal{P} = (\mathcal{P}_X, \mathcal{P}_Y)$ of $X \times Y$ for which any subrectangle R is of the form $R = R_X \times R_Y$, where R_X is a subrectangle in the partition $\mathcal{P}_X$, and R_Y is a subrectangle in the partition $\mathcal{P}_Y$. Furthermore, by the definition of the volume of a rectangle, $\nu(R_X \times R_Y) = \nu(R_X) \cdot \nu(R_Y)$, where each of these volumes is computed respectively in the space $\mathbb{R}^{n+k}$, $\mathbb{R}^n$ and $\mathbb{R}^k$ in which that rectangle is situated.

Consider the general subrectangle $R = R_X \times R_Y$ determined by the partition $\mathcal{P}$. Since $Y = \bigcup_{R_Y \in \mathcal{P}_Y} R_Y$, for each $x \in X$ the finite additivity of the integral gives

$$g(x) = \int_Y f(x,y)dy = \sum_{R_Y \in \mathcal{P}_Y} \int_{R_Y} f_x(y)dy.$$

For each fixed $x \in X$, denote $m_{R_Y} = \inf \{f_x(y) : y \in R_Y\}$ and $M_{R_Y} = \sup \{f_x(y) : y \in R_Y\}$. Then since $m_{R_Y} \leq f_x(y) \leq M_{R_Y}$ for all $y \in R_Y$, property (3) of Theorem 4.2.1 implies that

$$m_{R_Y} \cdot \nu(R_Y) \leq \int_{R_Y} f_x(y)dy \leq M_{R_Y} \cdot \nu(R_Y)$$

---

[7]This theorem is a special case of a more general theorem in which the integrals involed are Lebesgue integrals and is known as the *Fubini-Tonelli theorem*.

Let $m_R = \inf\{f(x,y) : x \in R_X, y \in R_Y\}$. For any $x \in R_X$ we have $m_R \leq f(x,y)$ for all $y \in R_Y$. Therefore, $m_R \leq m_{R_Y}$. A similar argument for upper bounds gives that $M_{R_Y} \leq M_R$, where $M_R = \sup\{f(x,y) : x \in R_X, y \in R_Y\}$. That is,

$$m_R \leq m_{R_Y} \leq M_{R_Y} \leq M_R.$$

Hence for any $x \in X$,

$$m_R \cdot \nu(R_Y) \leq \int_{R_Y} f_x(y)dy \leq M_R \cdot \nu(R_Y). \tag{4.7}$$

In particular (4.7) holds for any $x = \xi_X \in R_X$.

Now we have

$$L_{\mathcal{P}}(f) = \sum_{R \in \mathcal{P}} m_R \cdot \nu(R_X \times R_Y) = \sum_{R_X \in \mathcal{P}_X} \left(\sum_{R_Y \in \mathcal{P}_y} m_R \cdot \nu(R_Y) \right) \nu(R_X)$$

$$\leq \sum_{R_X \in \mathcal{P}_X} \left(\sum_{R_Y \in \mathcal{P}_y} \int_{R_Y} f_{\xi_X}(y)dy \right) \nu(R_X) = \sum_{R_X \in \mathcal{P}_X} g(\xi_X)\nu(R_X)$$

$$\leq \sum_{R_X \in \mathcal{P}_X} \left(\sum_{R_Y \in \mathcal{P}_y} M_R \cdot \nu(R_Y) \right) \nu(R_X) = \sum_{R \in \mathcal{P}} M_R \cdot \nu(R_X \times R_Y) = U_{\mathcal{P}}(f).$$

Thus, any Riemann sum of g determined by the partition $\mathcal{P}_X$ of X satisfies

$$L_{\mathcal{P}}(f) \leq \sum_{R_X \in \mathcal{P}_X} g(\xi_X) \cdot \nu(R_X) \leq U_{\mathcal{P}}(f). \tag{4.8}$$

Finally let $\epsilon > 0$. Since f is integrable over $X \times Y$ we can choose the partition $\mathcal{P}$ so fine that $U_{\mathcal{P}}(f) - L_{\mathcal{P}}(f) < \frac{\epsilon}{2}$. Then (4.8) implies that g is integrable over X and

$$\left| \int_{X \times Y} f(x,y) - \int_X g(x)dx \right| < \epsilon.$$

That is, since $\epsilon > 0$ is arbitrary

$$\int_{X \times Y} f(x,y) = \int_X g(x)dx = \int_X \left(\int_Y f(x,y)dy \right) dx.$$

Evidently a similar argument proves the second part of the theorem where the integration is performed in the reverse order.   □

The following example shows that the existence of the iterated integrals is not enough to ensure the function is integrable over $X \times Y$.

Example 4.3.2. Let $X = Y = [0, 1]$ and on the square $[0, 1] \times [0, 1]$, and consider the function $f(x, y) = \frac{x^2 - y^2}{(x^2 + y^2)^2}$ if $(x, y) \neq (0, 0)$ and $f(0, 0) = 0$. Note that

$$f(x, y) = \frac{\partial}{\partial y} \left(\frac{y}{x^2 + y^2} \right) = -\frac{\partial}{\partial x} \left(\frac{x}{x^2 + y^2} \right).$$

Therefore,

$$\int_0^1 \left(\int_0^1 f(x, y) dy \right) dx = \int_0^1 \frac{1}{x^2 + 1} dx = \frac{\pi}{4},$$

while

$$\int_0^1 \left(\int_0^1 f(x, y) dx \right) dy = -\int_0^1 \frac{1}{1 + y^2} dy = -\frac{\pi}{4}.$$

Fubini's theorem shows, of course, that f is not integrable over $[0, 1] \times [0, 1]$.

The reader should also be aware that *even* the existence *and* the equality of the iterated integrals are not enough to ensure the integrability of f over $X \times Y$ (see problem section, Exercise 4.6.6).

The following corollaries are immediate consequences of Fubini's theorem and Theorem 4.1.17.

Corollary 4.3.3. *Let $X \subset \mathbb{R}^n$ and $Y \subset \mathbb{R}^k$ be closed rectangles. If $f : X \times Y \to \mathbb{R}$ is continuous on $X \times Y$, then*

$$\int_{X \times Y} f(x, y) = \int_X \left(\int_Y f(x, y) dy \right) dx = \int_Y \left(\int_X f(x, y) dx \right) dy.$$

Corollary 4.3.4. *Let $R = [a_1, b_1] \times [a_2, b_2] \times ... \times [a_n, b_n]$ be a rectangle in $\mathbb{R}^n$. If $f : R \to \mathbb{R}$ is continuous, then*

$$\int_R f = \int_{a_n}^{b_n} \left(... \left(\int_{a_1}^{b_1} f(x_1, ..., x_n) dx_1 \right) ... \right) dx_n,$$

and the order of integration does not affect the result. In particular, when $R = [a, b] \times [c, d]$ *is a rectangle in* $\mathbb{R}^2$ *the following equalities hold:*

$$\int\int_R f(x, y)dxdy = \int_a^b \left(\int_c^d f(x, y)dy \right) dx = \int_c^d \left(\int_a^b f(x, y)dx \right) dy.$$

Example 4.3.5. Let R be the rectangle $R = [-1, 1] \times [0, \frac{\pi}{2}]$ in $\mathbb{R}^2$ and $f : R \to \mathbb{R}$ be the function $f(x, y) = x \sin y - ye^x$. Compute $\int\int_R f(x, y)dxdy$.

Solution. Clearly f is continuous on R and by Corollary 4.3.4 we have

$$\int\int_R (x \sin y - ye^x)dxdy = \int_{-1}^1 \left(\int_0^{\frac{\pi}{2}} (x \sin y - ye^x dy \right) dx$$

$$= \int_{-1}^1 \left(-x \cos y - \frac{1}{2}y^2 e^x \Big|_{y=0}^{y=\frac{\pi}{2}} \right) dx = \int_{-1}^1 \left(-\frac{\pi^2 e^x}{8} + x \right) dx = \frac{(\frac{1}{e} - e)\pi^2}{8}$$

Performing the calculation by integrating first with respect to x, of course, we expect to yield the same result. Indeed

$$\int\int_R (x \sin y - ye^x)dxdy = \int_0^{\frac{\pi}{2}} \left(\int_{-1}^1 (x \sin y - ye^x)dx \right) dy$$

$$= \int_0^{\frac{\pi}{2}} \left(\frac{x^2}{2} \sin y - ye^x \Big|_{x=-1}^{x=1} \right) dy = \int_0^{\frac{\pi}{2}} \left(-ey + \frac{y}{e} \right) dy = \frac{(\frac{1}{e} - e)\pi^2}{8}.$$

Example 4.3.6. Let the rectangle $R = [0, \pi] \times [-\frac{\pi}{2}, \frac{\pi}{2}] \times [0, 1]$ in $\mathbb{R}^3$ and $f : R \to \mathbb{R}$ the function $f(x, y, z) = z \sin(x + y)$. Compute $\int\int\int_R f(x, y, z)dxdydz$.

Solution. From Corollary 4.3.4

$$\int\int\int_R z \sin(x + y)dxdydz = \int_0^1 \left(\int_{-\frac{\pi}{2}}^{\frac{\pi}{2}} \left(\int_0^{\pi} z \sin(x + y)dx \right) dy \right) dz$$

$$= \int_0^1 \left(\int_{-\frac{\pi}{2}}^{\frac{\pi}{2}} (-z \cos(x + y)|_{x=0}^{x=\pi}) dy \right) dz = \int_0^1 \left(\int_{-\frac{\pi}{2}}^{\frac{\pi}{2}} (2z \cos y) dy \right) dz$$

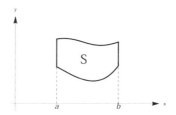

Figure 4.2: Elementary region in $\mathbb{R}^2$

$$= \int_0^1 \left(2z \sin y \Big|_{y=-\frac{\pi}{2}}^{y=\frac{\pi}{2}} \right) dz = \int_0^1 4z\, dz = 2.$$

We now extend Fubini's theorem over more general sets which are especially useful. They are often called *"elementary regions"*.

Definition 4.3.7. Let D be a compact simple set in $\mathbb{R}^{n-1}$ and $\varphi, \psi :$ $D \to \mathbb{R}$ be continuous functions such that $\varphi(x) \leq \psi(x)$ for every $x = (x_1, ..., x_{n-1}) \in D$. The subset S of $\mathbb{R}^n$ defined by

$$S = \{(x,y) : x \in D \ \ and \ \ \varphi(x) \leq y \leq \psi(x)\}$$

is called an *elementary region* in $\mathbb{R}^n$.

Lemma 4.3.8. *An elementary region S is a compact simple set, that is, S is compact and $\partial(S)$ has n-dimensional volume zero.*

Proof. To see that S is compact, we show that S is closed and bounded (see Heine-Borel theorem). Since D is bounded there is $M_1 > 0$ so that $||x|| < M_1$ for all $x \in \Omega$, where the norm is that of $\mathbb{R}^{n-1}$. At the same time, since D is compact and both φ and ψ are continuous, there is $M > 0$ such that $-M \leq \varphi(x) \leq \psi(x) \leq M$ for all $x \in D$. Therefore,

$$||(x,y)|| = \sqrt{x_1^2 + ... + x_{n-1}^2 + y^2} \leq ||x|| + |y| < M_1 + M,$$

and hence S is bounded.

The boundary of S is the union of the following three subsets of $\mathbb{R}^n$. The graph of φ, $B_1 = \{(x,y) : x \in D \, , \, y = \varphi(x)\}$, the graph of ψ,

$B_2 = \{(x, y) : x \in D , y = \psi(x)\}$, and (the $(n-1)$-dimensional cylindrical surface)

$$B_3 = \{(x, y) : x \in \partial(D) \ and \ \varphi(x) \le y \le \psi(x)\}$$

Clearly, each of this sets is contained in S and $\partial(S) = B_1 \cup B_2 \cup B_3$. Hence $\partial(S) \subset S$ and therefore S is closed. To complete the proof of the lemma, note that from Proposition 4.1.23, $\nu(B_1) = \nu(B_2) = 0$. Moreover, since D is a simple set, given any $\epsilon > 0$ we can cover $\partial(D)$ by rectangles $\{R_1, ..., R_N\}$ in $\mathbb{R}^{n-1}$ such that $\sum_{i=1}^{N} \nu(R_i) < \frac{\epsilon}{2M}$. Then the rectangles $R_i \times [-M, M]$ in $\mathbb{R}^n$ cover B_3 and have total volume less than ϵ. That is, $\nu(B_3) = 0$. Thus, $\nu(\partial(S)) = 0$.   □

With the notation as in Lemma 4.3.8 we now prove:

Theorem 4.3.9. *(Fubini's theorem for elementary regions). Let*

$$S = \{(x, y) \in \mathbb{R}^n : x \in D \ and \ \varphi(x) \le y \le \psi(x)\}$$

be an elementary region in $\mathbb{R}^n$ and $f : S \to \mathbb{R}$. If f is integrable over S and for each $x \in D$ the function $f_x(y) = f(x, y)$ is integrable over $[-M, M]$, then

$$\int_S f(x, y) = \int_D \left(\int_{\varphi(x)}^{\psi(x)} f(x, y) dy \right) dx. \qquad (4.9)$$

Proof. Let $R \times [-M, M]$ be a rectangle in $\mathbb{R}^n$ containing S, where R is a rectangle in $\mathbb{R}^{n-1}$ containing D. Set $S_x = \{y \in \mathbb{R} : \varphi(x) \le y \le \psi(x)\}$ if $x \in D$ and $S_x = \emptyset$ if $x \notin D$. Note that $\chi_S(x, y) = \chi_D(x) \cdot \chi_{S_x}(y)$. Now by the definition of the integral over a simple set and Theorem 4.3.1 we have

$$\int_S f(x, y) = \int_{R \times [-M, M]} f \cdot \chi_S(x, y) = \int_R \left(\int_{[-M, M]} f(x, y) \chi_S(x, y) dy \right) dx$$

$$= \int_R \left(\int_{[-M, M]} f(x, y) \chi_{S_x}(y) dy \right) \chi_D(x) dx$$

$$= \int_R \left(\int_{\varphi(x)}^{\psi(x)} f(x, y) dy \right) \chi_D(x) dx = \int_D \left(\int_{\varphi(x)}^{\psi(x)} f(x, y) dy \right) dx.$$

□

Corollary 4.3.10. *Let S be an elementary region in $\mathbb{R}^n$ as above. If $f : S \to \mathbb{R}$ is continuous, then*

$$\int_S f(x,y) = \int_D \left(\int_{\varphi(x)}^{\psi(x)} f(x,y) dy \right) dx.$$

Corollary 4.3.10 gives us an effective method for computing multiple integrals by reducing the n-dimensional integral $\int \int ... \int_S f$ to lower-dimensional integrals, at least when the integrand f is continuous and the set S in $\mathbb{R}^n$ is an elementary region. If the set S is *not* an elementary region, one can in practice express S as a union of essentially disjoint elementary regions and use the finite additivity of the integral to carry out the computation of the $\int \int ... \int_S f$.

Example 4.3.11. When

$$S = \left\{ (x,y,z) \in \mathbb{R}^3 : (x,y) \in D, \ \varphi(x,y) \le z \le \psi(x,y) \right\},$$

where $D = \left\{ (x,y) \in \mathbb{R}^2 : a \le x \le b, \ \alpha(x) \le y \le \beta(x) \right\}$ with both functions $\alpha, \beta : [a,b] \to \mathbb{R}$ continuous, then

$$\int \int \int_S f = \int \int_D \left(\int_{\varphi(x,y)}^{\psi(x,y)} f(x,y,z) dz \right) dxdy$$

$$= \int_a^b \left(\int_{\alpha(x)}^{\beta(x)} \left(\int_{\varphi(x,y)}^{\psi(x,y)} f(x,y,z) dz \right) dy \right) dx.$$

Example 4.3.12. Let $B_r(0) = \left\{ (x,y) \in \mathbb{R}^2 : x^2 + y^2 \le r \right\}$ be the disk in $\mathbb{R}^2$. Then $\nu(B_r(0)) = \pi r^2$.

Solution. Set $S = B_r(0)$ and note that S is the elementary region in $\mathbb{R}^2$ defined by

$$S = \left\{ (x,y) : -r \le x \le r, \ -\sqrt{r^2 - x^2} \le y \le \sqrt{r^2 - x^2} \right\}.$$

Therefore, Corollary 4.3.10 yields

$$\nu(S) = \int \int_S dxdy = \int_{-r}^r \left(\int_{-\sqrt{r^2-x^2}}^{\sqrt{r^2-x^2}} dy \right) dx$$

$$= \int_{-r}^{r} \left(\sqrt{r^2 - x^2} - (-\sqrt{r^2 - x^2}) \right) dx = 4 \int_{0}^{r} \sqrt{r^2 - x^2} dx.$$

Setting $x = r \sin t$, so that $dx = r \cos t$, we get

$$= 4r^2 \int_{0}^{\frac{\pi}{2}} \cos^2 t \, dt = \pi r^2.$$

Example 4.3.13. Evaluate the integral $I = \int_{1}^{2} \int_{0}^{\log y} e^{-x} dx dy$.

Solution. Here the region of integration is

$$S = \left\{ (x, y) \in \mathbb{R}^2 : 1 \le y \le 2, \ 0 \le x \le \log y \right\},$$

such a region is often called *y-simple* elementary region. So integrating first with respect to x we have

$$I = \int_{1}^{2} \left(\int_{0}^{\log y} e^{-x} dx \right) dy = \int_{1}^{2} \left(-e^{-x} \Big|_{x=0}^{x=\log y} \right) dy$$

$$= \int_{1}^{2} \left(-e^{-\log y} + 1 \right) dy = - \int_{1}^{2} \frac{1}{y} dy + 1 = 1 - \log 2.$$

Writing the region S as an *x-simple* elementary region, that is,

$$S = \left\{ (x, y) \in \mathbb{R}^2 : 0 \le x \le \log 2, \ e^x \le y \le 2 \right\},$$

and integrating in the reverse order, we also get

$$I = \int_{0}^{\log 2} \left(\int_{e^x}^{2} e^{-x} dy \right) dx = \int_{0}^{\log 2} \left(e^{-x} \int_{e^x}^{2} dy \right) dx$$

$$= \int_{0}^{\log 2} \left(2e^{-x} - 1 \right) dx = 1 - \log 2.$$

Example 4.3.14. Evaluate the integral $I = \int_{0}^{1} \int_{\sqrt{x}}^{1} \sin(\frac{y^3+1}{2}) dy dx$.

Solution. The region of integration is

$$S = \left\{ (x, y) \in \mathbb{R}^2 : 0 \le x \le 1, \ \sqrt{x} \le y \le 1 \right\},$$

Figure 4.3: Paraboloid: $z = x^2 + y^2$

which is x-simple. At the same time S can be written as

$$S = \left\{ (x,y) \in \mathbb{R}^2 : 0 \leq y \leq 1, \ 0 \leq x \leq y^2 \right\},$$

which is y-simple elementary region. Hence, we can perform the integration in any order. Note that if we choose to do the y integration first the integral cannot be evaluated as it stands, since $\sin(\frac{y^3+1}{2})$ has no elementary antiderivative. However, changing the order of integration we have

$$I = \int_0^1 \left(\int_0^{y^2} \sin(\frac{y^3+1}{2}) dx \right) dy = \int_0^1 \left(\sin(\frac{y^3+1}{2}) \int_0^{y^2} dx \right) dy$$

$$= \int_0^1 y^2 \sin(\frac{y^3+1}{2}) dy = \frac{2}{3} \int_{\frac{1}{2}}^1 \sin u \, du = \frac{2}{3}(\cos(\frac{1}{2}) - \cos 1),$$

where we made the substitution $u = \frac{y^3+1}{2}$.

Example 4.3.15. Let Ω be the region in $\mathbb{R}^3$ bounded by the surface of the paraboloid $z = x^2 + y^2$, and the planes $x + y = 1$, $x = 0$, $y = 0$, and $z = 0$. Write the triple integral $\int \int \int_\Omega f(x,y,z) dx dy dz$ as an iterated integral and compute the volume of Ω by setting $f(x,y,z) = 1$.

Solution. The region Ω lies in the first octant of xyz-space, while the plane $x + y = 1$ is parallel to the z-axis. The curve of intersection is the line $x + y = 1$. The projection of Ω onto the xy-plane is the triangular region $D = \left\{ (x,y) \in \mathbb{R}^2 : 0 \leq x \leq 1, \ 0 \leq y \leq 1-x \right\}$. Now, Ω is the elementary region

$$\Omega = \left\{ (x,y,z) \in \mathbb{R}^3 : (x,y) \in D, \ 0 \leq z \leq x^2 + y^2 \right\}.$$

Therefore by Corollary 4.3.10

$$\int\int\int_\Omega f(x,y,z)dxdydz = \int_0^1 \left(\int_0^{1-x} \left(\int_0^{x^2+y^2} f(x,y,z)dz \right) dy \right) dx.$$

If $f(x,y,z) = 1$, then

$$\nu(\Omega) = \int_0^1 \left(\int_0^{1-x} (x^2+y^2)dy \right) dx$$

$$= \int_0^1 \left(x^2 \int_0^{1-x} dy + \int_0^{1-x} y^2 dy \right) dx = \frac{1}{3} \int_0^1 (1-3x+6x^2-4x^3)dx = \frac{1}{6}.$$

Example 4.3.16. Find the volume of the region Ω in $\mathbb{R}^3$ bounded by the surfaces of the paraboloids $z = x^2 + y^2$, $z = 2(x^2 + y^2)$ and the cylindrical surfaces $y = \sqrt{x}$, $y = x^2$.

Solution. The region Ω represents a "parabolic shoe" cut by the cylindrical surfaces $y = \sqrt{x}$ and $y = x^2$ between the two paraboloids. Ω is bounded by a piece of the surface $z = x^2 + y^2$ from below and by a piece of the surface $z = 2(x^2 + y^2)$ from above. The projection of Ω on the xy-plane is $D = \{(x,y) \in \mathbb{R}^2 : 0 \le x \le 1,\ x^2 \le y \le \sqrt{x}\}$. Hence,

$$\Omega = \{(x,y,z) \in \mathbb{R}^3 : (x,y) \in D,\ x^2 + y^2 \le z \le 2(x^2 + y^2)\}.$$

Thus,

$$\nu(\Omega) = \int\int\int_\Omega dxdydz = \int_0^1 \left(\int_{x^2}^{\sqrt{x}} \left(\int_{x^2+y^2}^{2(x^2+y^2)} dz \right) dy \right) dx$$

$$= \int_0^1 \left(\int_{x^2}^{\sqrt{x}} (x^2+y^2))dy \right) dx = \int_0^1 \left[x^2(\sqrt{x}-x^2) + \frac{1}{3}(x^{\frac{3}{2}} - x^6) \right] dx = \frac{6}{35}.$$

An important consequence of Fubini's theorem is *Cavalieri's principle* which gives a method for calculating volumes "by cross-sections".

Corollary 4.3.17. *(Cavalieri's principle).*[8]   *Let* $\Omega \subset R \times [a,b]$ *be a simple set in* $\mathbb{R}^n$, *where* R *is a closed rectangle in* $\mathbb{R}^{n-1}$ *and* $[a,b]$ *an interval in* $\mathbb{R}$. *For each fixed* $t \in [a,b]$ *and* $x = (x_1, ..., x_{n-1}) \in R$ *let*

$$\Omega_t = \{(x,y) \in \Omega : y = t\} \subset \mathbb{R}^n,$$

the cross-section of Ω *corresponding to the hyperplane* $y = t$. *If* $v(\Omega_t)$ *is the* $(n-1)$-*dimensional volume of* Ω_t, *then*

$$\nu(\Omega) = \int_a^b v(\Omega_t)dt.$$

Proof. Taking $\varphi(x) = t = \psi(x)$ Lemma 4.3.8 tells us Ω_t is a simple set in $\mathbb{R}^n$ for each $t \in [a,b]$. Furthermore $\chi_\Omega(x,t) = \chi_{\Omega_t}(x)$. By the definition of the volume of a bounded subset of $\mathbb{R}^n$ and Fubini's theorem, the n-dimensional volume of Ω is given by

$$\nu(\Omega) = \int_\Omega 1 = \int_{R \times [a,b]} \chi_\Omega = \int_{[a,b]} \left(\int_R \chi_{\Omega_t}(x)dx \right) dt$$

$$= \int_a^b \left(\int_{\Omega_t} 1 dx \right) dt = \int_a^b v(\Omega_t)dt.$$

$\square$

Remark 4.3.18. Let R be a closed rectangle in $\mathbb{R}^{n-1}$ and suppose $\Omega \subset R \times \mathbb{R}$ is a set in $\mathbb{R}^n$ such that the portion Ω_k of Ω contained in $R \times [-k,k]$ is simple for each $k = 1, 2, ...$ If for each $t \in \mathbb{R}$ the cross section Ω_t has $(n-1)$-dimensional volume $v(\Omega_t) = 0$, then Ω has measure zero. Indeed, by Cavalieri's principle

$$\nu(\Omega_k) = \int_{-k}^k v(\Omega_t)dt = 0.$$

Since $\mathbb{R} = \bigcup_{k=1}^\infty [-k,k]$ we have $\Omega = \bigcup_{k=1}^\infty \Omega_k$ and Proposition 4.1.21 tells us that Ω has measure zero.

---

[8]B. Cavalieri (1598-1647). A student of Galileo and a professor at Bolonga. His investigations into area and volume were important for the foundations of calculus.

A typical application of Cavalieri's principle is the computation of *volumes of revolution* in $\mathbb{R}^3$ which we illustrate it with the following example.

Example 4.3.19. Find the volume of the 3-dimensional ball

$$B_r = B_r(0) = \left\{ (x, y, z) \in \mathbb{R}^3 : \sqrt{x^2 + y^2 + z^2} \le r \right\}.$$

Solution. Let $\Omega = B_r$. Consider the function $f(x) = \sqrt{r^2 - x^2}$ on the interval $[-r, r]$. The ball B_r in $\mathbb{R}^3$ is obtained by revolving the region between the graph of f and the interval $[-r, r]$ about the x-axis. The area of each cross section Ω_x for $x \in [-r, r]$ is clearly $v(\Omega_x) = \pi[f(x)]^2$. By Cavalieri's principle

$$\nu(B_r) = \int_{-r}^{r} v(\Omega_x) dx = \pi \int_{-r}^{r} (r^2 - x^2) dx = \frac{4}{3} \pi r^3.$$

We have already seen that the *area* of a plane region Ω in $\mathbb{R}^2$ is equal to $\int \int_\Omega dx dy$ and the *volume* of a region Ω in $\mathbb{R}^3$ is $\nu(\Omega) = \int \int \int_\Omega dx dy dz$. Many other concepts and physical quantities such as *mass, electric charge, center of mass* and *moment of inertia* are defined and computed with the aid of double and triple integrals. These concepts are of special importance in physics and engineering. We now give a brief discussion of some of these topics.

4.3.1 Center of mass, centroid, moment of inertia

Suppose that a quantity of some substance (such as mass, electic charge, or a certain chemical compound, etc) is distributed through-out a region Ω in $\mathbb{R}^3$. Frequently the distribution of the substance is described by a *density* function ρ. The total amount of substance in the set Ω is equal to

$$\int \int \int_\Omega \rho(x, y, x) dx dy dz.$$

This also works in lower dimensions, for example, to describe the distribution of a substance in a plane region or a line.

Definition 4.3.20. (*Center of gravity*). The *center of gravity* or *center of mass* of an object that occupies the region $\Omega \subset \mathbb{R}^3$ (with mass density $\rho(x, y, z)$) is the point $(\overline{x}, \overline{y}, \overline{z})$ whose coordinates are

$$\overline{x} = \frac{1}{m} \int \int \int_\Omega x\rho(x,y,z)dxdydz,$$
$$\overline{y} = \frac{1}{m} \int \int \int_\Omega y\rho(x,y,z)dxdydz,$$
$$\overline{z} = \frac{1}{m} \int \int \int_\Omega z\rho(x,y,z)dxdydz,$$

where m is the total mass $m = \int \int \int_\Omega \rho(x,y,x)dxdydz$.

In the special case where $\rho \equiv 1$, the point $(\overline{x}, \overline{y}, \overline{z})$ is called the *centroid* of Ω, and is the point whose coordinates are the average values of the coordinate functions on Ω. That is,

$$\overline{x} = \frac{1}{\nu(\Omega)} \int \int \int_\Omega xdxdydz,$$
$$\overline{y} = \frac{1}{\nu(\Omega)} \int \int \int_\Omega ydxdydz,$$
$$\overline{z} = \frac{1}{\nu(\Omega)} \int \int \int_\Omega zdxdydz.$$

Definition 4.3.21. (*Moment of inertia*). Given a body with mass density $\rho(x,y,z)$ occupying the region Ω in $\mathbb{R}^3$ and a line L in $\mathbb{R}^3$ the *moment of inertia* of the body about the line L is defined by

$$I_L = \int \int \int_\Omega [d(x,y,z)]^2 \rho(x,y,z)dxdydz,$$

where $d(x,y,z)$ denotes the distance from (x,y,z) to the line L. In particular, when L is the z-axis, then $[d(x,y,z)]^2 = x^2 + y^2$ and

$$I_z = \int \int \int_\Omega (x^2 + y^2)\rho(x,y,z)dxdydz,$$

and similarly for I_x and I_y. The moment of inertia of the body about the xy-plane is defined by

$$I_{xy} = \int \int \int_\Omega z^2 \rho(x,y,z)dxdydz,$$

and similarly for I_{yz} and I_{zx}.

Note that the moments of inertia satisfy such relations as

$$I_{xy} + I_{xz} = I_x, \; I_{xy} + I_{xz} + I_{yz} = I_0 \text{ and } I_x + I_y + I_z = 2I_0,$$

where

$$I_0 = \int \int \int_\Omega (x^2 + y^2 + z^2)\rho(x,y,z)dxdydz$$

is the *moment of inertia about the origin*.

Example 4.3.22. A thin plate of material of variable density occupies the square Ω in $\mathbb{R}^2$ whose vertices are $(0,0)$, $(a,0)$, (a,a) and $(0,a)$. Let the density at a point $(x,y) \in \Omega$ be $\rho(x,y) = xy$. Find the mass of the plate, its center of mass, and its moment of inertia about the x-axis.

Solution. The total mass of the plate is

$$m = \int\int_\Omega \rho(x,y)dxdy = \int_0^a \int_0^a xydxdy$$

$$= \int_0^a \left(y \int_0^a xdx \right) dy = \frac{a^2}{2} \int_0^a ydy = \frac{1}{4}a^4.$$

The x-coordinate of the center of mass is

$$\bar{x} = \frac{1}{m} \int\int_\Omega x\rho(x,y)dxdy = \frac{4}{a^4} \int_0^a \int_0^a x^2ydxdy$$

$$= \frac{4}{a^4} \int_0^a \left(y \int_0^a x^2dx \right) dy = \frac{4}{a^4}\frac{a^3}{3} \int_0^a ydy = \frac{2}{3}a.$$

By symmetry $\bar{x} = \bar{y} = \frac{2}{3}a$, and its center of mass is at $\frac{2}{3}a(1,1)$. The moment of inertial is

$$I_x = \int\int_\Omega y^2\rho(x,y)dxdy = \int_0^a \int_0^a xy^3dxdy = \frac{1}{8}a^6.$$

It is customary to express the moment of inertia of a body about an axis in terms of its total mass m, so here $I_x = \frac{1}{2}ma^2$.

Example 4.3.23. Find the centroid of the prismatic body Ω in $\mathbb{R}^3$ bounded by the planes $z = 0$, $x = 0$, $y = 1$, $y = 3$, and $x + 2z - 3 = 0$.

Solution. First we find the volume of the body

$$\nu(\Omega) = \int\int\int_\Omega dxdydz = \int_0^3 \left(\int_1^3 \left(\int_0^{\frac{3-x}{2}} dz \right) dy \right) dx$$

$$= \int_0^3 (3-x)dx = \frac{9}{2}.$$

Figure 4.4: Cylinder: $x^2 + y^2 = 1$

Now we have

$$\bar{x} = \frac{2}{9} \int \int \int_\Omega x \, dx \, dy \, dz = \int_0^3 \left(x \int_1^3 \left(\int_0^{\frac{3-x}{2}} dz \right) dy \right) dx$$

$$= \frac{2}{9} \int_0^3 x(3-x) \, dx = 1.$$

Similarly,

$$\bar{y} = \frac{2}{9} \int \int \int_\Omega y \, dx \, dy \, dz = 2 \text{ and } \bar{z} = \frac{2}{9} \int \int \int_\Omega z \, dx \, dy \, dz = \frac{1}{2}.$$

Thus the centroid of the body is the point $(1, 2, \frac{1}{2})$.

Example 4.3.24. Let Ω be a circular cylinder of radius a and height h of constant density $\rho = 1$. Find the moment of inertia of the cylinder about the diameter of its base.

Solution. Place the base of the cylinder on the plane $z = 0$ with the center of the base at the origin and let the z-axis be directed along the axis of the cylinder. We shall find the moment of inertia about the y-axis, in which case $[d(x, y, z)]^2 = x^2 + z^2$. We have

$$I_y = \int \int \int_\Omega (x^2 + z^2) \, dx \, dy \, dz = \int_{-a}^a \int_{-\sqrt{a^2-x^2}}^{\sqrt{a^2-x^2}} \left(\int_0^h (x^2 + z^2) \, dz \right) dy \, dx$$

$$= 4h \int_0^a \sqrt{a^2 - x^2} (x^2 + \frac{h^2}{3}) \, dx.$$

Setting $x = a \sin t$ and integrating we find

$$I_y = 4a^2 h \int_0^{\frac{\pi}{2}} \cos^2 t (a^2 \sin^2 t + \frac{h^2}{3}) dt = \frac{a^2 h \pi}{12}(3a^2 + 4h^2).$$

Theorem 4.3.25. *(Pappus' theorem)*[9]. *The volume* $\nu(\Omega)$ *of the solid* Ω *obtained by revolving a plane simple region S of area A about an axis in the plane of the region is*

$$\nu(\Omega) = 2\pi A h,$$

where h is the distance from the centroid of the region to the axis of revolution.

Proof. Suppose the region $S = \{(y, z) : a \le z \le b, \alpha(z) \le y \le \beta(z)\}$ lies in the yz-plane and is revolved aboout the z-axis. Then the distance of the centroid of S from the z-axis is

$$\overline{y} = \frac{1}{A} \int \int_S y \, dy \, dz.$$

Let $\Omega_t = \{(x, y, z) \in \Omega : z = t\}$ be the cross-section of Ω corresponding to the plane $z = t$, where $t \in [a, b]$. Then by Cavalieri's principle we have

$$2\pi A \overline{y} = \int \int_S 2\pi y \, dy \, dz = \int_a^b \left(\int_{\alpha(z)}^{\beta(z)} 2\pi y \, dy \right) dz = \int_a^b \left(\pi y^2 \Big|_{y=\alpha(z)}^{y=\beta(z)} \right) dz$$

$$= \int_a^b [\pi(\beta(z))^2 - \pi(\alpha(z))^2] dz = \int_a^b V(\Omega_t) dt = \nu(\Omega).$$

$\square$

Example 4.3.26. (*The solid torus*). The torus (doughnut) is the solid Ω in $\mathbb{R}^3$ obtained by revolving a circle (disk) of radius a about a line in its plane at distance b from its center, where $b > a > 0$. Use Pappus' theorem to find its volume.

---

[9]Pappus of Alexandria. A geometer of the second half of the 3rd cendury A.D. in the Greek school of Alexandria.

Figure 4.5: Torus

Solution. The centroid of the circle is its center (check!). The center travels a distance $2\pi b$ around the axis of revolution. The area of the circle is $\pi^2 a$. Therefore Pappus' theorem yields

$$\nu(\Omega) = \pi^2 a \cdot 2\pi b = 2\pi^2 a^2 b.$$

In the next chapter we will find the volume of the torus using the Change of Variable formula for multiple integrals (see Example 5.1.21).

EXERCISES

1. Let $f : [0, 1] \times [0, 1] \to \mathbb{R}$ defined by

$$f(x, y) = \begin{cases} x + y \text{ if } x^2 \leq y \leq 2x^2, \\ 0 \quad \text{ elsewhere.} \end{cases}$$

Compute the integral $\int \int f(x, y) dx dy$.

2. Evaluate $\int \int_{\Omega} |x+y| dx dy$, where $\Omega = \{(x, y) \in \mathbb{R}^2 : |x| < 1, |y| < 1\}$.

3. Evaluate $\int \int_{\Omega} y \, dx dy$, where Ω is the triangle bounded by the lines, $y + x = 0$, $y = 0$ and $3x - 5y = 24$.

4. Evaluate $\int \int_{\Omega} \frac{x^2}{y^2} dx dy$, where Ω is the set in $\mathbb{R}^2$ bounded by the curves $x = 2$, $y - x = 0$ and $xy = 1$.

5. Evaluate $\int \int_{\Omega} \frac{1}{\sqrt{x^2 + y^2}} dx dy$, where $\Omega = \{(x, y) \in \mathbb{R}^2 : |x| \leq 1, |y| \leq 1\}$.

6. Compute the integral by changing the order of integration

$$\int_0^1 \int_x^{\frac{1}{x}} \frac{y^2}{(x+y)^2\sqrt{1+y^2}} dy dx.$$

7. Evaluate

$$\int_0^{\frac{\pi}{2}} \int_{\frac{2y}{\pi}}^1 \cos\left(\frac{y}{x}\right) dx dy.$$

8. Evaluate

$$\int_1^2 \int_1^z \int_{\frac{1}{y}}^1 z^2 y \, dx dy dz.$$

9. Evaluate $\int \int \int_\Omega z dx dy dz$, where Ω is the set in $\mathbb{R}^3$ bounded by the surfaces $x + y = 1$, $z = 1 - y^2$, $x = 0$, $y = 0$ and $z = 0$.

10. Evaluate $\int \int \int_\Omega (2x - y - z) dx dy dz$, where

$$\Omega = \left\{ (x, y, z) \in \mathbb{R}^3 : z \leq x + y, \, y \leq x^2, \, 0 \leq x \leq 1, \, y \geq 0, \, z \geq 0 \right\}.$$

11. Find the volume of the tetrahedron bounded by the plane $\frac{x}{a} + \frac{y}{b} + \frac{z}{c} = 1$ and the coordinate planes.

12. Find the volume of the region in $\mathbb{R}^3$ bounded by the surfaces $z = x$, $y - x = 2$ $x = 0$, $y = 0$, $z = 0$ and the parobolic cylinder $y = x^2$.

13. Find the volume of the region in $\mathbb{R}^3$ bounded by the surfaces $z = x^2 + y^2$, $x + y = 1$ $x = 0$, $y = 0$, $z = 0$.

14. Find the volume of the solid in $\mathbb{R}^3$ bounded by the cylinder $x^2 + y^2 = a^2$, the plane $z = x + y$ in the first octan.

15. Find the centroid of the tetrahedron bounded by the plane $\frac{x}{a} + \frac{y}{b} + \frac{z}{c} = 1$ and the coordinate planes.

16. Find the center of mass of a cube of side a if its density at each point is proportional to the square of the distance of this point from one corner of the base. (Take base in the xy-plane and this corner at the origin.)

17. Find the moment of inertia of a cube $[0,a] \times [0,a] \times [0,a]$ about the edge of the cube.

18. Prove that moment of inertia of the torus in Example 4.3.26 with respect to its axis of revolution is $\frac{1}{4}(3a^2 + 4b^2)m$, where m is its mass.

Answers to selected Execises

1. $\frac{1}{40}(21 - 8\sqrt{2})$.   2. $\frac{8}{3}$.   3. -12.   4. $\frac{9}{4}$.   5. $4\log(1 + \sqrt{2})$.

6. $\frac{2\sqrt{2}-1}{2}$.   9. $\frac{3}{8}$.   10. $\frac{8}{35}$.   11. $\frac{abc}{6}$.   12. $\frac{8}{3}$.   13. $\frac{1}{6}$.

14. $\frac{2a^3}{3}$.   15. $\frac{1}{4}(a,b,c)$.   16. $(\overline{x}, \overline{y}, \overline{z}) = \frac{7a}{12}(1,1,1)$.   17. $\frac{2a^5}{3}$.

4.4   Smooth Urysohn's lemma and partition of unity (*)

Definition 4.4.1. Let $f : \Omega \to \mathbb{R}$ defined in a domain $\Omega \subseteq \mathbb{R}^n$. The *support* of f, denoted $\mathrm{supp}(f)$, is the closure in Ω of the set of points $x \in \Omega$ at which $f(x) \neq 0$. That is,

$$\mathrm{supp}(f) = \overline{\{x \in \Omega : f(x) \neq 0\}}.$$

A function f is to be of *compact support* if its support (in Ω) is a compact set. The set of functions f of class $C^m(\Omega)$ having compact support is usually denoted by $C_c^m(\Omega)$, and $C_c^m(\mathbb{R}^n)$ when $\Omega = \mathbb{R}^n$ ($0 \leq m \leq \infty$).

We remark that the class $C_c^\infty(\Omega)$ of *infinitely differentiable functions with compact support* in Ω is of great importance in many applications in analysis, in particular in the theory of function spaces and the theory of distributions (generalized functions).

In this section we first sketch a proof of the following important result called *Urysohn's*[10] *lemma* concerning smooth functions in Euclidean

---

[10]P. Uryshon (1898-1924), best known for his contributions in the theory of dimension, and for developing Urysohn's Metrization Theorem and Urysohn's Lemma, both of which are fundamental results in topology.

space, $\mathbb{R}^n$ (the case of continuous functions being much simpler). Then we show smooth *partitions of unity* can be found in Euclidean space.

We first require a lemma whose proof we leave to the reader.

Lemma 4.4.2. *Let* $0 < a < b$ *and* $f(x) = e^{\frac{1}{x-b} - \frac{1}{x-a}}$ *if* $a < x < b$ *and* 0 *otherwise. Then* f *is* $C^\infty(\mathbb{R})$. *Similarly,* $F(x) = \frac{\int_x^b f(t)dt}{\int_a^b f(t)dt}$ *is smooth and* $0 \le F(x) \le 1$.

Theorem 4.4.3. *(C^∞-Uryshon' lemma). Let* $U \supseteq K$ *be subsets of* $\mathbb{R}^n$, *where* U *is open and* K *is compact. Then there exists a* C^∞ *function* $f : \mathbb{R}^n \to \mathbb{R}$ *which is identically* 1 *on* K *and vanishes outside* U *and* $0 \le f \le 1$.

Proof. Note that F takes the value 1 for $x \le a$ and 0 for $x \ge b$. Define $\psi(x_1, \ldots, x_n) = F(x_1^2 + \ldots + x_n^2)$. Then ψ is a smooth radial function $\mathbb{R}^n \to \mathbb{R}$ and takes the value 1 on the closed ball of radius a and 0 off the closed ball of radius b, and $0 \le \psi \le 1$. Now suppose S_1 and S_2 are any two concentric balls in $\mathbb{R}^n$ centered at the origin. Composing ψ with an appropriate linear transformation of the form $x \mapsto rx$, where $r > 0$ we can always find a smooth function, which we again call ψ, taking the value 1 on S_1 and 0 off S_2 and values between 0 and 1.

Let $B = \mathbb{R}^n \setminus U$. Then B is closed and K and B are disjoint. Since K is compact it can be covered by a finite number of open balls, S_i, $i = 1, ..., p$ in such a way that the $\overline{S_i}$ are each disjoint from B (Prove!). Each of these balls can be shrunk slightly so that the smaller spheres $S_{i,2}$ also cover K. Then, as above, let ψ_i be a smooth function on $\mathbb{R}^n$ which is 1 on $\overline{S_{i,2}}$ and vanishes off S_i. Then $f = 1 - (1 - \psi_1) \cdots (1 - \psi_p)$ is a smooth function on $\mathbb{R}^n$ which is identically 1 on A and identically 0 on B and $0 \le f \le 1$. $\square$

Thus, in Euclidean space one can separate compact sets K from closed ones $\mathbb{R}^n \setminus U$ using C^∞ functions. Since $\mathbb{R}^n$ is locally compact one can cover K by a finite number of compact neighborhoods. Hence the f will have compact support.

We now formulate the concept of a partition of unity which is often used to extend local results to global ones.

Corollary 4.4.4. *(Partition of unity).* *Let A be a compact set and $\{U_j\}$ be an open covering of A. Then there exist a finite number ψ_i, $i = 1, ..., p$ of smooth numerical functions such that*

1. *$0 \leq \psi_i \leq 1$.*

2. *ψ_i has support in some U_j.*

3. *$\sum_{i=1}^{p} \psi_i \equiv 1$ on A.*

We say the partition of unity is subordinate to the cover U_j. It enables us to take any continuous function f on $\mathbb{R}^n$ whose support lies in A and write $f = \sum_{i=1}^{p} \psi_i f$, where each $\psi_i f$ has "small" support, that is, its support lies in some U_j.

Proof. Evidently by compactness of A we may assume we have a finite cover U_i, $i = 1 \ldots p$. Let $x \in A$. Then $x \in U_i$ for some $i = 1, ..., p$. Choose a smooth ϕ_i which is 1 on x and vanishes off U_i with $0 \leq \phi_i \leq 1$. On A define for each $i = 1, ..., p$,

$$\psi_i = \frac{\phi_i}{\sum_{j=1}^{p} \psi_j}.$$

Then the first two conditions are clearly satisfied and

$$\sum_{i=1}^{p} \psi_i = \sum_{i=1}^{p} \frac{\phi_i}{\sum_{j=1}^{p} \psi_j} \equiv 1.$$

$\square$

4.5   Sard's theorem (*)

Theorem 4.5.1. *(Sard[11]).* *Let $f : U \to \mathbb{R}^n$ be a C^∞ map defined on an open set U in $\mathbb{R}^m$ and $A = \{u \in U : \operatorname{rank} D_f(u) < n\}$. Then $f(A)$ has measure zero.*

---

[11] A. Sard (1909-1980) taught for many years at Queens College, NY. He is known for his work in the field of differential topology and is especially famous for *The measure of the critical values of differentiable maps*, Bulletin of the American Mathematical Society 48 (12), 883-890 (1942).

In particular, when $m = n$, $f(\{u \in U : \det D_f(u) = 0\})$ has measure 0.

Definition 4.5.2. $\mathbb{R}^n \setminus f(A)$ is called the *regular values* of f.

Proposition 4.5.3. *If a set $S \subseteq \mathbb{R}^n$ has measure 0, then its complement is dense.*

Proof. If U is any non empty open set in $\mathbb{R}^n$ it must intersect the complement of S. Otherwise it would lie completely in S. But then U would also have measure zero as would any cube C in U. But if C has side δ, its measure δ^n is positive.   □

As a corollary of Sard's theorem we have,

Corollary 4.5.4. *The set of regular values of f is dense in $\mathbb{R}^n$.*

Proof of Sard's theorem.

Proof. Let A_1 denote the points in $u \in U$ whose derivative $D_f(u) = 0$. More generally let A_i be those points $u \in U$ such that all derivatives $\frac{\partial^s(f_j)}{\partial u_{k_1} \ldots \partial u_{k_s}} \equiv 0$, where $j = 1, ..., n$, $k_i = 1, ..., m$ and $s \leq i$. Evidently, $A \supseteq A_1 \supseteq A_2 \supseteq \ldots$ and since all these functions are continuous the A_i, $i \geq 1$ are all closed.

We now show that A is also closed. Given a point $u \in U$, by the Implicit Function theorem 3.8.7, locally we can write the derivative matrix for all points in some neighborhood of u. Hence $u \mapsto \operatorname{rank} D_f(u)$ is continuous. Let $A_c = \{u \in U : \operatorname{rank} D_f(u) = c\}$. Since this function is continuous and A_c is the inverse image of the closed set $\{c\}$, A_c is closed. Hence $A = \bigcup_{c=0}^{n-1} A_c$, the finite union of closed sets is also closed.

Since A and all the A_i are closed they are each locally compact and second countable. Note that Euclidean space, $\mathbb{R}^m$, is second countable[12] and hence so are these closed subsets.

The proof will be divided into three steps.

1. $f(A \setminus A_1)$ has measure 0.
2. $f(A_i \setminus A_{i+1})$ has measure 0 for each $i \geq 1$.
3. $f(A_i)$ has measure 0 for some i sufficiently large.

---

[12]A Hausdorff topological space is called *second countable* if it has a countable basis. See [9], Chapter VIII.

Once we have proved each of the three statements, the conclusion will follow from Proposition 4.1.21 since $f(A)$ would then be a union of a countable number of sets of measure zero. We shall prove the first two of these statements by induction on m. The last statement will be proved directly.

Proof of step 1. Here we may assume $n \geq 2$ since when $n = 1$, $A = A_1$. For each $u^- \in A \setminus A_1$ we will find an open neighborhood V in $\mathbb{R}^m$ so that $f(V \cap A)$ has measure 0. Since A is locally compact and second countable and A_1 is closed, $A \setminus A_1$ can be covered by countably many such neighborhoods. Using Proposition 4.1.21 this will prove $f(A \setminus A_1)$ has measure 0. Since u^- is not in A_1 some partial derivative, say $\frac{\partial f_1(u^-)}{\partial u_1} \neq 0$. Let $h(u) = (f_1(u), u_2, \ldots, u_m)$, where $u \in U$. Since $D_h(u^-)$ is non singular, h maps some open neighborhood V of u^- diffeomorphically onto an open set V'. The composition $g = f \circ h^{-1}$ maps V into $\mathbb{R}^n$. Now the set C of critical points of g is exactly $h(V \cap A)$. Hence the set of critical values $g(C) = f(V \cap A)$. For each $(t, u_2, \ldots, u_m) \in V'$, $g(t, u_2, \ldots, u_m)$ lies in the hyperplane $t \times \mathbb{R}^{n-1} \subseteq \mathbb{R}^n$. Thus g carries hyperplanes to hyperplanes. Let g^t denote the restriction of g to the hyperplane $t \times \mathbb{R}^{m-1}$. Then $g^t : (t \times \mathbb{R}^{m-1}) \cap V' \to t \times \mathbb{R}^{n-1}$. Now the first derivative is given by the matrix

$$\frac{\partial g_i}{\partial u_j} = \begin{bmatrix} 1 & 0 \\ * & \frac{\partial g_i^t}{\partial u_j} \end{bmatrix}$$

Hence a point of $t \times \mathbb{R}^{m-1}$ is a critical point of g^t if and only if it is a critical point of g. By inductive hypothesis the set of critical values of g^t has measure 0 in $t \times \mathbb{R}^{n-1}$. Hence the critical values of g intersect each hyperplane $t \times \mathbb{R}^{n-1}$ in a set of measure 0. Since $g(C) = f(V \cap A)$ is a countable union of compact sets it's measurable (see, [28]) and by Cavalieri's principle (see Remark 4.3.18) has measure 0.

Proof of step 2. For each $u^- \in A_i \setminus A_{i+1}$ there is some $i + 1^{st}$ derivative, $\frac{\partial^{i+1}(f_j)}{\partial u_{s_1} \ldots \partial u_{s_{i+1}}}$ which is not zero at u^-, but all derivatives of order i vanish at u^-. Thus if $\omega(u) = \frac{\partial^i f_j(u)}{\partial u_{s_2} \ldots \partial u_{s_{i+1}}}$, then $\omega(u^-) = 0$, but $\frac{\partial \omega(u^-)}{\partial u_{s_1}} \neq 0$. By changing the names of the variables we may assume $s_1 = 1$. Therefore we have a function ω with $\omega(u^-) = 0$, but $\frac{\partial \omega(u^-)}{\partial u_1} \neq 0$. Let

$h : U \to \mathbb{R}^m$ be defined by, $h(u) = (\omega(u), u_2, \ldots, u_m)$. Then h is smooth and by the Inverse Function theorem 3.8.2 carries some neighborhood V of u^- diffeomorphically onto an open set V'. Let $g = f \circ h^{-1} : V' \to \mathbb{R}^n$ and $g|((0) \times \mathbb{R}^{m-1} \cap V') \to \mathbb{R}^n$ be its restriction to the hyperplane $(0) \times \mathbb{R}^{m-1}$. By induction the set of critical values of $g|$ has measure zero in $\mathbb{R}^n$. But each point of $h(A_i \cap V)$ is certainly a critical point of $g|$ since all derivatives of order $\leq i$ vanish. Therefore $(g|\circ h)(A_i \cap V) = f(A_i \cap V)$ has measure 0. Since by local compactness and second countability of the closed set A_i, $A_i \setminus A_{i+1}$ is covered by countably many subsets, V it follows that $f(A_i \setminus A_{i+1})$ has measure 0 by Proposition 4.1.21, proving step 2.

Proof of step 3. Let B be a cube in U with edge δ and i be sufficiently large ($i > \frac{m}{n} - 1$). We will show that $f(A_i \cap B)$ has measure 0. Since A_i can be covered by countably many such cubes an application of Proposition 4.1.21 would show that $f(A_i)$ itself has measure 0. Let $u \in A_i \cap B$. From Taylor's theorem and the compactness of B we see that if $u + h \in B$, there is a constant M so that

$$||f(u+h) - f(u)|| \leq M||h||^{i+1}. \tag{4.10}$$

Now subdivide B into r^m cubes of edge $\frac{\delta}{r}$. Let I be the cube of the subdivision containing u. Then any point of I can be written $u + h$, where $\| h \| \leq \sqrt{m}\frac{\delta}{r}$. From (4.10) it follows that $f(B)$ lies in a cube of edge $\frac{\alpha}{r^{i+1}}$, where $\alpha = 2M(\sqrt{m}\delta)^{i+1}$. Hence $f(A_i \cap B)$ is contained in a union of at most r^m cubes of having a total measure of $r^m(\frac{\alpha}{r^{i+1}})^n = r^{m-(i+1)n}\alpha^n$. Thus the total measure of $f(A_i \cap B) \leq \alpha^n r^{m-(i+1)n}$ so $i + 1 > \frac{m}{n}$, $\mu(f(A_i \cap B)) \to 0$ as $r \to \infty$ and this means $\mu(f(A_i \cap B)) = 0$. $\qquad\square$

4.6   Solved problems for Chapter 4

Exercise 4.6.1. Evaluate the integral $\int \int_\Omega xy dx dy$, where Ω is the interior of the circle $x^2 + y^2 \leq 1$ bounded by the straight line $2x + y = 1$.

Solution. The points of the intersection of the circle $x^2 + y^2 = 1$ and the line $2x + y = 1$ are $(0, 1)$ and $(\frac{4}{5}, -\frac{3}{5})$. The region of integration is

$$\Omega = \left\{ (x, y) \in \mathbb{R}^2 : -\frac{3}{5} \leq y \leq 1, \ \frac{1-y}{2} \leq x \leq \sqrt{1-y^2} \right\},$$

which is y-simple elementary region. Note that Ω cannot be written as x-simple elementary region. Now we have

$$\iint_\Omega xy\,dx\,dy = \int_{-\frac{3}{5}}^{1} \left(y \int_{\frac{1-y}{2}}^{\sqrt{1-y^2}} x\,dx \right) dy$$

$$= \frac{1}{2} \int_{-\frac{3}{5}}^{1} \left(y \left[1 - y^2 - (\frac{1-y}{2})^2 \right] \right) dy = \frac{32}{125}.$$

Exercise 4.6.2. Let $\Omega = [-1,1] \times [0,2]$. Show that

$$\iint_\Omega \sqrt{|y - x^2|}\,dx\,dy = \frac{8 + 3\pi}{6}.$$

Solution. The integrand is

$$f(x,y) = \sqrt{|y - x^2|} = \begin{cases} \sqrt{y - x^2} & \text{if } y \geq x^2, \\ \sqrt{x^2 - y} & \text{if } y < x^2. \end{cases}$$

We divide the region of integration Ω into two essentially disjoint pieces. That is, we write $\Omega = \Omega_1 \cup \Omega_2$, where $\Omega_1 = \{(x,y) : -1 \leq x \leq 1,\, 0 \leq y \leq x^2\}$ and $\Omega_2 = \{(x,y) : -1 \leq x \leq 1,\, x^2 \leq y \leq 2\}$. Now

$$\iint_\Omega f = \iint_{\Omega_1} f + \iint_{\Omega_2} f$$

$$= \int_{-1}^{1} \left(\int_{0}^{x^2} \sqrt{x^2 - y}\,dy \right) dx + \int_{-1}^{1} \left(\int_{x^2}^{2} \sqrt{y - x^2}\,dy \right) dx$$

$$= \frac{2}{3} \int_{-1}^{1} x^3\,dx + \frac{2}{3} \int_{-1}^{1} (2 - x^2)^{\frac{3}{2}}\,dx = 0 + \frac{4}{3} \int_{0}^{1} (2 - x^2)^{\frac{3}{2}}\,dx$$

$$= \frac{16}{3} \int_{0}^{\frac{\pi}{4}} \cos^4\theta\,d\theta = \frac{4}{3} \int_{0}^{\frac{\pi}{4}} (1 + \cos 2\theta)^2\,d\theta = \frac{4}{3} + \frac{\pi}{2},$$

where we set $x = \sqrt{2}\sin\theta$.

Exercise 4.6.3. Find the volume of the region Ω in $\mathbb{R}^3$ bounded by the surfaces of the paraboloids $z = x^2 + y^2$, and $2z = 12 - x^2 - y^2$.

Solution. The curve of intersection of the two paraboloids is the circle $x^2 + y^2 = 4$, $z = 4$. The projection of Ω on the xy-plane is the set

$$D = \left\{ (x,y) \in \mathbb{R}^2 : -2 \le x \le 2, \ -\sqrt{4 - x^2} \le y \le \sqrt{4 - x^2} \right\}.$$

Hence

$$\Omega = \left\{ (x,y,z) \in \mathbb{R}^3 : (x,y) \in D, \ x^2 + y^2 \le z \le \frac{12 - x^2 - y^2}{2} \right\}.$$

Thus,

$$\nu(\Omega) = \int\int\int_\Omega dx\,dy\,dz = \int_{-2}^{2} \left(\int_{-\sqrt{4-x^2}}^{\sqrt{4-x^2}} \left(\int_{x^2+y^2}^{\frac{12-x^2-y^2}{2}} dz \right) dy \right) dx$$

$$= \frac{3}{2} \int_{-2}^{2} \left(\int_{-\sqrt{4-x^2}}^{\sqrt{4-x^2}} (4 - x^2 - y^2) dy \right) dx$$

$$= 6 \int_{0}^{2} \left(\int_{0}^{\sqrt{4-x^2}} (4 - x^2 - y^2) dy \right) dx = 4 \int_{0}^{2} (4 - x^2)^{\frac{3}{2}} dx = 12\pi.$$

Exercise 4.6.4. Find the volume of the region Ω in $\mathbb{R}^3$ bounded by the surfaces of the cylinders $x^2 + y^2 = a^2$, and $x^2 + z^2 = a^2$.

Solution. By the symmetry of Ω it is enough to compute the volume of the part $\Omega_1 = \Omega \cap \{(x,y,z) \in \mathbb{R}^3 : x \ge 0, \ y \ge 0, \ z \ge 0\}$ of Ω lying in the first octant. Now Ω_1 is bounded below by the plane $z = 0$ and above by the surface $x^2 + z^2 = a^2$. The projection of Ω_1 on the xy-plane is the set $D = \left\{ (x,y) \in \mathbb{R}^2 : 0 \le x \le a, \ 0 \le y \le \sqrt{a^2 - x^2} \right\}$. So Ω_1 is the elementary region $\Omega_1 = \left\{ (x,y,z) \in \mathbb{R}^3 : (x,y) \in D, \ 0 \le z \le \sqrt{a^2 - x^2} \right\}$. Therefore,

$$\nu(\Omega_1) = \int_{0}^{a} \left(\int_{0}^{\sqrt{a^2-x^2}} \left(\int_{0}^{\sqrt{a^2-x^2}} dz \right) dy \right) dx$$

$$= \int_{0}^{a} \left(\int_{0}^{\sqrt{a^2-x^2}} \sqrt{a^2 - x^2} dy \right) dx = \int_{0}^{a} (a^2 - x^2) dx = \frac{2a^3}{3}.$$

Thus, $\nu(\Omega) = 8\nu(\Omega_1) = \frac{16a^3}{3}$.

Exercise 4.6.5. Calculate the volume of a conical type figure Ω in $\mathbb{R}^3$.

Solution. We do this using Cavalieri's principle. Let A be the area of the base and h the vertical height. At height x, $0 \leq x \leq h$, the cross section area $A(x) = V(\Omega_x)$ is proportional to A. Therefore $\frac{A(x)}{A} = \frac{(h-x)^2}{h^2}$. Thus $A(x) = \frac{A}{h^2}(h-x)^2$ and

$$\nu(\Omega) = \frac{A}{h^2} \int_0^h (h-x)^2 dx = \frac{A}{h^2} \left(\frac{-(h-x)^3}{3} \Big|_0^h \right) = \frac{A}{h^2} \left(\frac{h^3}{3} \right) = \frac{Ah}{3}.$$

In particular, if Ω is a circular cone with basis radius r and height h, then $\nu(\Omega) = \frac{\pi r^2 h}{3}$.

Exercise 4.6.6. Let $f : [-1,1] \times [-1,1] \to \mathbb{R}$ defined by

$$f(x,y) = \begin{cases} \frac{xy}{(x^2+y^2)^2} & \text{if } (x,y) \neq (0,0), \\ 0 & \text{if } (x,y) = (0,0). \end{cases}$$

Show that the iterated integals exist and are equal

$$\int_{-1}^1 \left(\int_{-1}^1 f(x,y)dy \right) dx = \int_{-1}^1 \left(\int_{-1}^1 f(x,y)dx \right) dy.$$

However, show that f is *not* integrable on $R = [-1,1] \times [-1,1]$.

Solution. Clearly the function $f_x(y) = f(x,y)$ is continuous in y for every fixed $x \in [-1,1]$, and so it is integrable on $[-1,1]$. Since it is odd, $\int_{-1}^1 f(x,y)dy = 0$. Hence

$$\int_{-1}^1 \left(\int_{-1}^1 f(x,y)dy \right) dx = 0.$$

Similarly,

$$\int_{-1}^1 \left(\int_{-1}^1 f(x,y)dx \right) dy = 0.$$

The function f is not integrable on R. Indeed, if it is integrable on R, then by Theorem 4.2.1, its absolute value $|f|$ is also integrable there. Therefore, again by Theorem 4.2.1, it must also be integrable on the square $[0,1] \times [0,1]$. Fubini's theorem then tells us that the iterated integral

$$\int_0^1 \left(\int_0^1 |f(x,y)|dy \right) dx$$

must be finite. This is not true since

$$\int_0^1 |f(x,y)|dy = \int_0^1 \frac{xy}{(x^2+y^2)^2}dy = \frac{1}{2}\left[\frac{1}{x} - \frac{x}{x^2+1}\right],$$

and the function to the right is not integrable on $[0,1]$.

Exercise 4.6.7. (*) (An alternative proof of the *Fubini theorem*). Let f be a continous real valued function on $X \times Y$, where X and Y are cubes in Euclidean space. Then

$$\int_X \int_Y f(x,y)dxdy = \int_Y \int_X f(x,y)dydx.$$

Solution. The theorem is certainly true of functions of the form $f(x)g(y)$. Hence by linearity it is also true for all variable seperable functions. By the Corollary C.1.3 of the Appendix these functions are uniformly dense in $C(X \times Y)$. Let f be an arbitrary continuous function on $X \times Y$ and $\epsilon > 0$. Choose a variable separable function g so that $||f - g||_{X \times Y} < \epsilon$. Then

$$\int_{X \times Y} |f - g|dxdy \le \epsilon \nu(X \times Y).$$

Since ϵ is arbitrary $\nu(X \times Y)$ is fixed, $\int_{X \times Y} |f - g|dxdy = 0$, from which it follows that $\int_X \int_Y f(x,y)dxdy = \int_Y \int_X f(x,y)dydx$.

Exercise 4.6.8. (*) (An alternative proof of the *equality of mixed partials*). Let $f : \Omega \subseteq \mathbb{R}^n \to \mathbb{R}$ be of class C^2. Then for any $i,j = 1,...n$,

$$\frac{\partial^2 f}{\partial x_j \partial x_i} = \frac{\partial^2 f}{\partial x_i \partial x_j}.$$

Solution. Again since only two components are involved, it suffices to prove this for a function in two variables (x,y). Let $(p,q) \in \Omega$ and suppose

$$\frac{\partial^2 f}{\partial x \partial y}(p,q) - \frac{\partial^2 f}{\partial y \partial x}(p,q) > 0.$$

By the continuity of these mixed partials choose a rectangle $R = [a,b] \times [c,d]$ about (p,q) where this inequality remains true throughout. Then

$$\int \int_R \frac{\partial^2 f}{\partial x \partial y}(x,y)dxdy - \int \int_R \frac{\partial^2 f}{\partial y \partial x}(x,y)dxdy > 0.$$

By Fubini's theorem we get

$$\int_a^b \left(\int_c^d \frac{\partial}{\partial y} \left(\frac{\partial f}{\partial x} \right) dy \right) dx - \int_c^d \left(\int_a^b \frac{\partial}{\partial x} \left(\frac{\partial f}{\partial y} \right) dx \right) dy > 0.$$

The Fundamental Theorem of Calculus implies

$$\int_a^b \left(\frac{\partial f}{\partial x}(x,d) - \frac{\partial f}{\partial x}(x,c) \right) dx - \int_c^d \left(\frac{\partial f}{\partial y}(b,y) - \frac{\partial f}{\partial y}(a,y) \right) dy > 0.$$

Hence again by the Fundamental Theorem of Calculus,

$$f(b,d) - f(a,d) - f(b,c) + f(a,c) - f(b,d) + f(b,c) + f(a,d) - f(a,c) > 0.$$

This contradiction completes the proof.

Miscellaneous Exercises

Exercise 4.6.9. Let f be defined and bounded on $R = [a, b] \times [c, d]$. Suppose that for each fixed $y \in [c, d]$ the function $f_y(x) = f(x, y)$ is an increasing function of x, and that for each fixed $x \in [a, b]$ the function $f_x(y) = f(x, y)$ is an increasing function of y. Prove that f is integrable on R. Prove that the same is true if the functions f_y and f_x are assumed to be decreasing on $[a, b]$ and $[c, d]$ respectively.

Exercise 4.6.10. Let Ω is a bounded simple set in $\mathbb{R}^n$ and $f : \Omega \to \mathbb{R}$ be integrable. Prove that

$$\left(\int_\Omega f(x) \sin(||x||) dx \right)^2 + \left(\int_\Omega f(x) \cos(||x||) dx \right)^2 \leq \nu(\Omega) \int_\Omega (f(x))^2 dx.$$

Exercise 4.6.11. Let $f : [0, \pi] \times [0, 1] \to \mathbb{R}$ defined by

$$f(x, y) = \begin{cases} \cos x & \text{if } x \in \mathbb{Q}, \\ 0 & \text{if } x \notin \mathbb{Q}. \end{cases}$$

Show that $\int_0^1 \left(\int_0^\pi f(x, y) dx \right) dy$ exist, but $\int \int_{[0,\pi] \times [0,1]} f(x, y) dx dy$ and $\int_0^\pi \left(\int_0^1 f(x, y) dy \right) dx$ do not exist.

Exercise 4.6.12. Let $R = [-1, 2] \times [-1, 0] \times [0, 2] \subset \mathbb{R}^3$. Show that

$$\int \int \int_R \left(\frac{z}{1 - |x|y} \right)^2 dx dy dz = 9 \log 6.$$

Exercise 4.6.13. By changing the order of integration show that

$$\int_0^1 \int_0^{\sqrt{1-x^2}} \frac{1}{(1+e^y)\sqrt{1-x^2-y^2}} dy dx = \frac{\pi}{2} \log\left(\frac{2e}{1+e}\right).$$

Exercise 4.6.14. Evaluate

$$\int_0^1 \int_{x^{\frac{1}{3}}}^x e^{\frac{x}{y}} dy dx.$$

Exercise 4.6.15. Let $0 < \alpha < \frac{\pi}{2}$. Compute

$$\int_0^1 \int_0^\alpha \sqrt{1-y\cos^2 x}\, dx dy.$$

Hint. Change the order of integration. (Ans. $\frac{2}{3}(\tan\alpha - \sec\alpha - \cos\alpha + 2)$.)

Exercise 4.6.16. Evaluate

$$\int\int\int_\Omega z\, dx dy dz,$$

where $\Omega = \{(x,y,z) \in \mathbb{R}^3 : x^2 + y^2 \le z^2,\ x^2 + y^2 + z^2 \le 1\}$.
(Ans. $\frac{5}{8}$ if the region of integration is restricted by the condition $z > 0$ otherwise zero).

Exercise 4.6.17. Evaluate

$$\int\int\int_\Omega (1-z^2)\, dx dy dz,$$

where Ω is the pyramid with top vertex at $(0,0,1)$ and base vertices $(0,0,0)$, $(1,0,0)$, $(0,1,0)$ and $(1,1,0)$.

Exercise 4.6.18. Find the volume of the portion of the paraboloid of revolution $\Omega = \{(x,y,z,) \in \mathbb{R}^3 : x^2 + y^2 \le z \le a^2\}$. (Ans. $\frac{1}{2}\pi a^4$).

Exercise 4.6.19. Find the volume of the region in $\mathbb{R}^3$ bounded by the surfaces $x^2 + 4y^2 - 4 = 0$, $z = 0$ and $z = 2x^2 + 3y^2$. (Ans. $\frac{11\pi}{2}$).

Exercise 4.6.20. Find the moment of inertia of a homogeneous right circular cone (base radius a, height h) of constant density k with respect to its axis. (Ans. $\frac{1}{2}\pi a^4 h k = \frac{1}{2}a^2 m$ where m is mass).

Exercise 4.6.21. (*) (*Steiner's theorem*)[13]. Prove that the moment of inertia of a body about some axis equals $md^2 + I_c$, where m is the mass of the body, d the distance between the axis and the center of mass of the body, and I_c the moment of inertia about the axis parallel to the given one and passing through the center of mass of the body.

Hint. Choose a system of coordinates such that its origin coincides with the center of mass of the body and one of the axies is parallel to the axis with respect to which we want to find the moment of inertia.

---

[13] J. Steiner (1796-1863), worked mainly in geometry.

Chapter 5

Change of Variables Formula, Improper Multiple Integrals

In this chapter we discuss the change of variables formula for multiple integrals, improper multiple integrals, functions defined by integrals, Weierstrass' approximation theorem, and the Fourier transform.

5.1 Change of variables formula

The "change of variable" formula, also called the "u-substitution rule" in one variable is known from elementary calculus and tells us that; *if $f : [a, b] \to \mathbb{R}$ is continuous on $[a, b]$ and $\varphi : [a, b] \to \mathbb{R}$ is a one-to-one function of class C^1 with $\varphi'(x) > 0$ for $x \in (a, b)$, then with $u = \varphi(x)$,*

$$\int_a^b f(\varphi(x))\varphi'(x)dx = \int_{\varphi(a)}^{\varphi(b)} f(u)du. \qquad (5.1)$$

To prove this, first we find a differentiable function F such that $F' = f$, that is, an antiderivative F of f (this is possible by the Fundamental Theorem of Calculus). Then

$$\int_{\varphi(a)}^{\varphi(b)} f(u)du = F(\varphi(b)) - F(\varphi(a)).$$

On the other hand, by the Chain Rule $(F \circ \varphi)'(x) = f(\varphi(x))\varphi'(x)$. Hence, again by the Fundamental theorem,

$$\int_a^b f(\varphi(x))\varphi'(x)dx = (F \circ \varphi)(b) - (F \circ \varphi)(a) = F(\varphi(b)) - F(\varphi(a)).$$

However, one has to be careful here, for (5.1) is fine as it stands when φ is increasing. If φ were decreasing, (i.e., $\varphi'(x) < 0$) the right side of (5.1) is

$$\int_{\varphi(b)}^{\varphi(a)} f(u)du = -\int_{\varphi(a)}^{\varphi(b)} f(u)du.$$

This is corrected by rewriting (5.1) as

$$\int_a^b f(\varphi(x))|\varphi'(x)|dx = \int_{\varphi(a)}^{\varphi(b)} f(u)du.$$

Setting $\Omega = [a, b]$, so that $\varphi(\Omega)$ is the interval with endpoints $\varphi(a)$ and $\varphi(b)$ we write

$$\int_{\varphi(\Omega)} f(u)du = \int_\Omega f(\varphi(x))|\varphi'(x)|dx. \tag{5.2}$$

We remark that this result is also valid if f is not continuous, but merely integrable, because sets of measure zero contribute nothing to either integral.

We shall now generalize (5.2) for multiple integrals. First we define what is meant by a "change of variables" in an n-dimensional integral.

Definition 5.1.1. Let U be an open set in $\mathbb{R}^n$. Let $\varphi : U \to \mathbb{R}^n$ be a one-to-one function of class $C^1(U)$, such that the Jacobian $J_\varphi(x) = \det D_\varphi(x) \neq 0$ for all $x \in U$. Then φ is called a *change of variables* in $\mathbb{R}^n$.

Note our hypothesis is slightly redundant, for if $\varphi : U \to \varphi(U)$ is a one-to-one function such that both φ and φ^{-1} are of class C^1, that is, if φ is a diffeomorphism, then the Chain Rule implies that D_φ is nonsingular, so that $\det D_\varphi(x) \neq 0$ and hence φ is a change of variables.

On the other hand, if φ is a change of variables in $\mathbb{R}^n$, then the Inverse Function theorem (Theorem 3.8.2) and Corollary 3.8.3 tell us $\varphi(U)$ is open in $\mathbb{R}^n$ and the function $\varphi^{-1} : \varphi(U) \to U$ is of class C^1. In other words, a change of variables in $\mathbb{R}^n$ is just a *diffeomorphism* in $\mathbb{R}^n$.

The basic issue for changing variables in multiple integrals is as follows: *Let U be open in $\mathbb{R}^n$ and $\varphi : U \to \mathbb{R}^n$ a diffeomorphism. Suppose Ω is a bounded simple set (i.e., its boundary has n-dimensional volume zero) with $\overline{\Omega} \subset U$. Given an integrable function $f : \varphi(\Omega) \to \mathbb{R}$ we want to change the integral* $\int_{\varphi(\Omega)} f(y)dy$ *into an appropriate integral over Ω (which we hope to be easier to compute).* In fact, we shall prove the following n-dimensional analogue of (5.2)

$$\int_{\varphi(\Omega)} f(y)dy = \int_{\Omega} f(\varphi(x))|J_{\varphi}(x)|dx. \tag{5.3}$$

Since the sets over which we integrate are bounded simple sets it is natural to ask, whether a diffeomorphism maps (bounded) simple sets to (bounded) simple sets.

Lemma 5.1.2. *Let U be open in $\mathbb{R}^n$, $\varphi : U \to \mathbb{R}^n$ a diffeomorphism whose resrtiction on Ω° is a diffeomorphism and Ω a bounded set with $\overline{\Omega} \subset U$. Then*

$$\partial(\varphi(\Omega)) = \varphi(\partial(\Omega)).$$

Proof. Since φ is continuous and $\overline{\Omega}$ is compact (by the Heine-Borel theorem), it follows that $\varphi(\overline{\Omega})$ is also compact and so $\varphi(\overline{\Omega})$ is closed and bounded. Hence, $\varphi(\overline{\Omega}) = \overline{\varphi(\Omega)}$. At the same time, since $\varphi \in C^1(\Omega)$ and $J_{\varphi}(x) \neq 0$ for all $x \in \Omega$, by Corollary 3.8.3, φ maps open sets onto open sets. In particular, $\varphi(\Omega^\circ)$ is open. Thus, points of $\partial(\varphi(\Omega))$ can not be images of points of $\Omega^\circ = \Omega \backslash \partial\Omega$, that is, $\partial(\varphi(\Omega)) \subseteq \varphi(\partial(\Omega))$. On the other hand, let $x \in \partial(\Omega)$. Then there are sequences $\{x_k\}$ in Ω and $\{y_k\}$ in $U \backslash \Omega$ such that $x_k \to x$ and $y_k \to x$. The continuity of φ implies $\varphi(x_k) \to \varphi(x)$ and $\varphi(y_k) \to \varphi(x)$. Since φ is one-to-one on U, it follows that $\varphi(y_k) \notin \varphi(\Omega)$ and hence $\varphi(x) \in \partial(\varphi(\Omega))$. Therefore $\varphi(\partial(\Omega)) \subseteq \partial(\varphi(\Omega))$. Thus, $\partial(\varphi(\Omega)) = \varphi(\partial(\Omega))$. $\square$

Remark 5.1.3. In developing the theory of integration one uses n-dimensional rectangles in a number of places. However one could use n-dimensional cubes instead. An n-dimensional open *cube* centered at $a \in \mathbb{R}^n$ of side length $2r$ is the set

$$C_r(a) = \{x \in \mathbb{R}^n : ||a - x||_\infty < r\},$$

where the norm is the "max-norm" (or "box-norm")

$$||x||_\infty = \max\{|x_1|, ..., |x_n|\}.$$

As we have proved in Theorem 1.7.2 all norms in $\mathbb{R}^n$ are equivalent, in particular $||x||_\infty \leq ||x|| \leq \sqrt{n}||x||_\infty$. So that a cube C in $\mathbb{R}^n$ contains a ball B in $\mathbb{R}^n$ and vice versa. Furthermore, knowing beforehand that a function f is integrable over a bounded simple set Ω contained in a cube C, by using a partition $\mathcal{P} = \{C_1, ..., C_k\}$ of C involving subcubes of *equal side length* $2r$, (so that $||\mathcal{P}|| \to 0$ if and only if $r \to 0$ if and only if $k \to \infty$), one obtains approximating Riemann sums of f that have substantial computational advantages. In the sequel we shall use n-dimensional cubes as our basic sets rather than rectangles in $\mathbb{R}^n$.

Proposition 5.1.4. *Let $S \subset \mathbb{R}^n$ be a set of measure zero and suppose that $\varphi : S \to \mathbb{R}^n$ satisfies a Lipschitz condition. Then $\varphi(S)$ has measure zero.*

Proof. Since φ satisfies Lipschitz condition, there is $M > 0$ such that,

$$||\varphi(x) - \varphi(x')|| \leq M||x - x'||,$$

for all $x, x' \in S$. Because S has measure zero, given $\epsilon > 0$, there is a sequence of closed cubes C_j such that $S \subset \bigcup_{j=1}^\infty C_j$ and $\sum_{j=1}^\infty \nu(C_j) < \epsilon$. Let the side length of C_j be r_j and $d_j = \sqrt{n}r_j$ its diameter. Then for $x, x' \in S \cap C_j$ we have

$$||\varphi(x) - \varphi(x')||_\infty \leq ||\varphi(x) - \varphi(x')|| \leq M||x - x'|| \leq M\sqrt{n}||x - x'||_\infty.$$

Therefore $||\varphi(x) - \varphi(x')||_\infty \leq M\sqrt{n}d_j$, and so $\varphi(S \cap C_j)$ is contained in the cube K_j with side length $2M\sqrt{n}d_j$. The cubes K_j cover $\varphi(S)$ and

$$\sum_{j=1}^\infty \nu(K_j) = (Mn)^n \sum_{j=1}^\infty \nu(C_j) \leq (Mn)^n \epsilon.$$

Since $(Mn)^n$ is constant and $\epsilon > 0$ is arbitrary, $\varphi(S)$ has measure zero.

$\square$

Lemma 5.1.5. *Let $U \subseteq \mathbb{R}^n$ be open and $\varphi : U \to \mathbb{R}^n$ be $C^1(U)$. Suppose S is a compact set in U. If S has volume zero, then $\varphi(S)$ also has volume zero.*

Proof. Since U is open, for each $x \in S$ we can choose an open ball $B_x = B_{\delta_x}(x) \subseteq U$. The collection $\{B_x : x \in S\}$ is an open cover for S. Since S is compact a finite number of these balls cover S. That is, $S \subseteq \bigcup_{j=1}^{N} B_{x_j} \subseteq U$.

Let $S_j = S \cap B_{x_j}$. Since $S_j \subset S$ and $\nu(S) = 0$ we have $\nu(S_j) \le \nu(S) = 0$. Since $\varphi \in C^1(U)$, Corollary 3.4.8 tells us that φ satisfies a Lipschitz condition on B_{x_j} ans so on S_j. Proposition 5.1.4 then implies that $\nu(\varphi(S_j) = 0$ for all $j = 1, ..., N$ (here because S is compact the notions of volume zero and measure zero coincide).

Now $S = \bigcup_{j=1}^{N} [S \cap B_{x_j}] = \bigcup_{j=1}^{N} S_j$ and so $\varphi(S) = \bigcup_{j=1}^{N} \varphi(S_j)$. Therefore

$$\nu(\varphi(S)) = \nu \left(\bigcup_{j=1}^{N} \varphi(S_j) \right) \le \sum_{j=1}^{N} \nu(\varphi(S_j)) = 0.$$

$\square$

Proposition 5.1.6. *Let U be open in $\mathbb{R}^n$ and $\varphi : U \to \mathbb{R}^n$ a diffeomorphism. If Ω is a bounded simple set with $\overline{\Omega} \subset U$, then $\varphi(\Omega)$ is also bounded and simple.*

Proof. First note that since $\varphi(\Omega) \subseteq \varphi(\overline{\Omega})$ and $\varphi(\overline{\Omega})$ is compact (i.e., closed and bounded), the set $\varphi(\Omega)$ is also bounded. At the same time, since Ω is simple, $\nu(\partial(\Omega)) = 0$. It follows from Lemma 5.1.5 that $\nu(\varphi(\partial(\Omega))) = 0$. Now since φ is a diffeomorphism, Lemma 5.1.2 tells us $\partial(\varphi(\Omega)) = \varphi(\partial(\Omega))$. Hence $\nu(\partial(\varphi(\Omega))) = \nu(\varphi(\partial(\Omega))) = 0$, and $\varphi(\Omega)$ is simple. $\square$

5.1.1   Change of variables; linear case

In this subsection we prove the linear change of variables formula. Here $\varphi = T$, where T is a linear map on $\mathbb{R}^n$, in which case $J_T = \det T$ is a

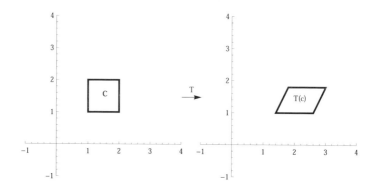

Figure 5.1: Linear Transformation

constant. The proof does not use elementary transformations, rather it uses the polar decomposition.

Proposition 5.1.7. *Let* $T : \mathbb{R}^n \to \mathbb{R}^n$ *be a linear map, and* $\Omega \subset \mathbb{R}^n$ *a bounded simple set. Then*

$$\nu(T(\Omega)) = |\det T| \nu(\Omega). \tag{5.4}$$

Proof. We identify T with its standard matrix representation. If T is singular (i.e., $\det T = 0$), then $T(\mathbb{R}^n)$ is a *proper* subspace of $\mathbb{R}^n$ of dimension $k < n$. By Corollary 4.1.25, $\nu(T(\Omega)) = 0$, and so the statement holds trivially.

Now suppose T is invertible (i.e., $\det T \neq 0$). Then T is an isomorphism on $\mathbb{R}^n$. Proposition 5.1.6 tells us that $T(\Omega)$ is a bounded simple set. By the Polar Decomposition theorem (Theorem 2.4.11), T can be written as $T = OP$, where O is an orthogonal matrix and P a positive definite symmetric matrix. Therefore P is orthogonally conjugate to a diagonal matrix D with positive entries. That is, $P = O_1 D O_1^{-1}$ where O_1 and O_1^{-1} are orthogonal matrices. Since $|\det(AB)| = |\det(A)||\det(B)|$ and if A is orthogonal $|\det A| = 1$, (see, the discussion following Definition 2.4.9)) we see that $|\det T| = |\det((OO_1)DO_1^{-1})| = \det D$.

First we shall prove (5.4) for a cube $C = [-s, s]^n$ in $\mathbb{R}^n$. Since any orthogonal transformation O is an isometry (Theorem 2.4.10), it leaves distances fixed. Hence shapes are fixed and so the geometric effect of the application of O leaves volumes unchanged. Thus it is enough to see how D affects the volume of C. Since D is diagonal with positive diagonal entries $\lambda_1, ..., \lambda_n$ we have

$$D(C) = [-\lambda_1 s, \lambda_1 s] \times ... \times [-\lambda_n s, \lambda_n s].$$

Therefore

$$\nu(D(C)) = (2\lambda_1 s) \cdots (2\lambda_n s) = \lambda_1 \cdots \lambda_n (2s)^n = (\det D)\nu(C).$$

Hence

$$\nu(T(C)) = |\det T|\nu(C).$$

Now, let C be a cube containing Ω and partition C into subcubes of equal side length $2r$. Since Ω is simple its characteristic function χ_Ω is integrable over C, that is, $\nu_*(\Omega) = \nu(\Omega) = \nu^*(\Omega)$ (see Remark 4.1.41). Let C_i be the subcubes for which $C_i \subset \Omega$ with $i = 1, ... p$ and C_j be those subcubes for which $C_j \cap \Omega \neq \emptyset$ with $j = 1, ..., q$. Then $\bigcup_{i=1}^p C_i \subset \Omega \subset \bigcup_{j=1}^q C_j$. Therefore $\bigcup_{i=1}^p T(C_i) \subset T(\Omega) \subset \bigcup_{j=1}^q T(C_j)$. Since the result holds for cubes, it follows that

$$|\det T| \sum_{i=1}^p \nu(C_i) \leq \nu(T(\Omega)) \leq |\det T| \sum_{j=1}^q \nu(C_j).$$

Letting $r \to 0$ we get

$$|\det T|\nu(\Omega) \leq \nu(T(\Omega)) \leq |\det T|\nu(\Omega).$$

Thus again, $\nu(T(\Omega)) = |\det T|\nu(\Omega)$. $\square$

Corollary 5.1.8. *(The volume of a ball in $\mathbb{R}^n$). Let $B^n(r) = \{x \in \mathbb{R}^n : ||x|| \leq r\}$ be the ball of radius $r > 0$ in $\mathbb{R}^n$. Then*

$$\nu(B^n(r)) = c_n r^n,$$

where $c_n = \nu(B^n)$ is the volume of the n-dimensional unit ball B^n.[1]

---

[1] Explicit expressions for c_n are given in Examples 5.1.33 and 5.3.27.

Proof. Since $\|rx\| = r\|x\|$, the ball $B^n(r)$ is the image of B^n under the linear map $T(x) = rx$ on $\mathbb{R}^n$. Furthermore, since $\det T = r^n$, Proposition 5.1.7 gives $\nu(B^n(r)) = \nu(T(B^n)) = r^n c_n$. For example, for $n = 3$, Example 4.3.19 tells us $c_3 = \frac{4}{3}\pi$, and of course $c_2 = \pi$.   □

Theorem 5.1.9. *Let $\Omega \subset \mathbb{R}^n$ be a bounded simple set and let $T : \mathbb{R}^n \to \mathbb{R}^n$ be a nonsingular linear map. If $f : T(\Omega) \to \mathbb{R}$ is an integrable function, then $(f \circ T)|\det T|$ is integrable over Ω and*

$$\int_{T(\Omega)} f(y)dy = \int_{\Omega} f(T(x))|\det T|dx. \tag{5.5}$$

Proof. For $x \in \Omega$ let $F(x) = f(T(x))$ and set $y = T(x)$. Let C be a cube containing Ω and partition C into subcubes $\mathcal{P} = \{C_j : j = 1, ..., m\}$ of equal side length $2r$. Then for each j, the sets $\{T(C_j)\}$ form a partition of $T(\Omega)$. From Proposition 5.1.7, $\nu(T(C_j)) = |\det T|\nu(C_j)$.

Since f is integrable over $T(\Omega)$, by the definition of the integral over a set, $f\chi_{T(\Omega)}$ is integrable over (any) cube C' containing $T(\Omega)$. Moreover, since $\chi_{\Omega} = \chi_{T(\Omega)} \circ T$, for any $y_j = T(x_j) \in T(C_j)$ we can write

$$\sum_{j=1}^{m} f\chi_{T(\Omega)}(y_j)\nu(T(C_j)) = \sum_{j=1}^{m} F\chi_{\Omega}(x_j)|\det T|\nu(C_j) = |\det T|S_{\mathcal{P}}(F\chi_{\Omega}).$$
$$\tag{5.6}$$

Let $d = \max_{j=1,...,m}\{d(T(C_j)\}$. The continuity of T implies that as $r \to 0$ then also $d \to 0$.

Now passing to the limit in (5.6) as $r \to 0$ the first sum converges to $\int_{T(\Omega)} f(y)dy$, which automatically implies the existence of the limit of the second sum (i.e., the integrability of $F = f \circ T$ on Ω) and the equality

$$\int_{T(\Omega)} f(y)dy = |\det T| \int_{\Omega} f(T(x))dx = \int_{\Omega} f(T(x))|\det T|dx.$$

□

The meaning of the determinant

Definition 5.1.10. Let $\{v_1, ..., v_k\}$ be linearly independent vectors in $\mathbb{R}^n$ ($k \leq n$). We define the *k-dimensional parallelepiped* $\Pi = \Pi(v_1, ..., v_k)$ with adjacents sides the vectors $v_1, ..., v_k$ (also called the *edges* of Π) to be the set of all $x \in \mathbb{R}^n$ such that

$$x = c_1 v_1 + ... + c_k v_k,$$

where c_j are scalars with $0 \leq c_j \leq 1$, for $j = 1, ..., k$.

A 2-dimensional parallelepiped is a *parallelogram*, and a higher dimensional analogue is called a *parallelepiped*. The 3-dimensional version of the next result was obtained in Proposition 1.3.35.

Proposition 5.1.11. *Let $v_1, ..., v_n$ be n linearly independent vectors in $\mathbb{R}^n$ and let Π be the parallelepiped $\Pi(v_1, ..., v_n)$. If $A = [v_1...v_n]$ is the $n \times n$ matrix with columns the vectors $v_1, ..., v_n$, then*

$$\nu(\Pi) = |\det A|.$$

Proof. Let T be the linear transformation $T : \mathbb{R}^n \to \mathbb{R}^n$ given by $T(x) = Ax$. Then $T(e_i) = v_i$, where $\{e_i : i = 1, ..n\}$ are the standard (unit) basic vectors in $\mathbb{R}^n$. Therefore, since T is linear it maps the unit cube $C^n = [0, 1]^n$ onto the parallelepiped $\Pi = \Pi(v_1, ..., v_n)$. Proposition 5.1.7 implies

$$\nu(\Pi) = |\det T| \nu(C^n) = |\det T|.$$

$\square$

5.1.2   Change of variables; the general case

Here we shall extend Theorem 5.1.9 for nonlinear change of variables. The principal idea is based on the local approximation of a C^1 mapping $\varphi : U \subset \mathbb{R}^n \to \mathbb{R}^n$ at a point $a \in U$ by its differential $D_\varphi(a)$. *The local replacement of a nonlinear relation by a linear one is a basic idea of mathematics.* Recall from the Linear Approximation theorem (Theorem 3.1.3) that $D_\varphi(a) : \mathbb{R}^n \to \mathbb{R}^n$ is such that

$$D_\varphi(a)(h) \approx \varphi(a + h) - \varphi(a)$$

in a neighborhood of the point $a \in U$, and from Theorem 3.2.7, that the standard matrix of $D_\varphi(a)$ is the $n \times n$ matrix $D_\varphi(a) = (\frac{\partial \varphi_j}{\partial x_i}(a))$, where φ_j are the component functions of φ and x_i the components of x for $i, j = 1, ..., n$. Using the Linear Approximation theorem, we will show that for a sufficiently small cube C centered at $a \in U$

$$\nu(\varphi(C)) \approx |\det D_\varphi(a)|\nu(C).$$

Lemma 5.1.12. *Let $U \subset \mathbb{R}^n$ be open, and let $\varphi : U \to \varphi(U)$ be a diffeomorphism in $\mathbb{R}^n$ and $a \in U$ be fixed. Let $C \subset U$ be a cube centered at a. Then*

$$\lim_{C \downarrow a} \frac{\nu(\varphi(C))}{\nu(C)} = |\det D_\varphi(a)|,$$

where $C \downarrow a$ means that C shrinks to its center $a \in C$.

Here the C can be either a cube or a ball in U centered at a. It is sufficient to consider only balls $B_r(a)$, for replacing the norm $|| \cdot ||$ by the norm $|| \cdot ||_\infty$ the same type of argument works also for cubes. Thus what we are trying to prove is

$$\lim_{r \to 0} \frac{\nu(\varphi(B_r(a)))}{\nu(B_r(a))} = |\det D_\varphi(a)|.$$

Proof. We begin with the Linear Approximation theorem:

$$\varphi(x) = \varphi(a) + D_\varphi(a)(x - a) + \epsilon(x)||x - a||,$$

where $\epsilon(x)$ tends to 0 as $x \to a$. Since φ is a C^1 *change of variable*, $\det D_\varphi(a) \neq 0$ and so $D_\varphi(a)^{-1}$ exists. Applying this linear transformation to the equation above yields,

$$D_\varphi(a)^{-1}\varphi(x) = D_\varphi(a)^{-1}\varphi(a) + (x - a) + ||x - a||D_\varphi(a)^{-1}\epsilon(x),$$

Hence by the triangle inequality[2]

$$||D_\varphi(a)^{-1}\phi(x) - D_\phi(a)^{-1}\varphi(a)|| \leq ||x - a|| + ||x - a|| \, ||D_\varphi(a)^{-1}\epsilon(x)||.$$

---

[2]For the case of the cube, here use the triangle inequality for the box-norm $|| \cdot ||_\infty$ and the estimate $||x - a|| \leq \sqrt{n}||x - a||_\infty$.

Finally, estimating the effect of the linear transformation $D_\varphi(a)^{-1}$ there is a positive constant β so that

$$||D_\varphi(a)^{-1}\epsilon(x)|| \le \beta||\epsilon(x)||.$$

Thus,

$$||D_\varphi(a)^{-1}\varphi(x){-}D_\varphi(a)^{-1}\varphi(a)|| \le ||x{-}a||{+}\beta||\epsilon(x)||\,||x{-}a|| < (1{+}\beta||\epsilon(x)||)r.$$

Since $||x - a|| < r$, $||\epsilon(x)|| \to 0$ as $r \to 0$. Now let $0 < t < 1$ and choose $r > 0$ small enough so that $1 + \beta||\epsilon(x)||$ lies strictly between $1 - t$ and $1 + t$. This means $D_\varphi(a)^{-1}(\varphi(B_r(a)))$ is contained in a ball centered at $D_\varphi(a)^{-1}\varphi(a)$ of radius $(1 + t)r$. If for some x, $D_\varphi(a)^{-1}\varphi(x)$ lies on the boundary of this ball, then

$$||D_\varphi(a)^{-1}\varphi(x) - D_\varphi(a)^{-1}\varphi(a)|| = (1 + t)r > (1 - t)r,$$

so that $D_\varphi(a)^{-1}\varphi(x)$ lies outside this smaller ball of radius $(1 - t)r$. Since, $D_\varphi(a)^{-1}$ is a *homeomorphism* in a neighborhood of a, $D_\varphi(a)^{-1}(\varphi(B_r(a)))$ also contains this smaller ball. In particular, by Corollary 5.1.8 we have

$$c_n(1 - t)^n r^n \le \nu(D_\varphi(a)^{-1}(\varphi(B_r(a)))) \le c_n(1 + t)^n r^n.$$

That is,

$$(1 - t)^n \le \frac{\nu(D_\varphi(a)^{-1}(\varphi(B_r(a))))}{\nu(B_r(a))} \le (1 + t)^n.$$

But since $D_\phi(a)^{-1}$ is linear, Proposition 5.1.7 tells us

$$\nu(D_\varphi(a)^{-1}(\varphi(B_r(a)))) = |\det D_\varphi(a)^{-1}|\nu(\varphi(B_r(a))).$$

Thus,

$$(1 - t)^n \le \frac{\nu(\varphi(B_r(a)))}{|\det D_\varphi(a)|\nu(B_r(a))} \le (1 + t)^n. \tag{5.7}$$

Letting $t \to 0$ (so that $r \to 0$) yields the conclusion. $\qquad\square$

Lemma 5.1.13. *Let $U \subset \mathbb{R}^n$ be open and let Ω be a bounded simple set with $\overline{\Omega} \subset U$. Then there exist a compact set S of the form $S = \bigcup_{i=1}^N C_i$, where $\{C_1, ..., C_N\}$ are essential disjoint closed cubes, such that $\Omega \subset S \subset U$.*

Proof. Since $\overline{\Omega}$ is compact and U is open we may cover $\overline{\Omega}$ by open cubes lying completely inside U. By the compactness of $\overline{\Omega}$ a finite number of these cubes $\{C_{r_j}(x_j) : x_j \in \overline{\Omega}, \ j = 1, ..., p\}$ of sides $2r_j$ respectively, also cover $\overline{\Omega}$. Let $\delta = \min\{r_1, ..., r_p\}$. Then the open cubes $\{C_\delta(x) : x \in \overline{\Omega}\}$ cover $\overline{\Omega}$ and by compactness so does a finite number of them $\{C_\delta(x_k) : x_k \in \overline{\Omega}, \ k = 1, ..., q\}$ and each closed cube $\overline{C_\delta(x_k)} \subset U$. In general these closed cubes overlap. Partitioning each of these to smaller closed subcubes and counting the overlaping subcubes once, we obtained the required finite collection of essential disjoint closed (sub)cubes $\{C_i : i = 1, ..., N\}$ whose union S is contained in U. $\quad\square$

Next we prove the principal result[3] of this section.

Theorem 5.1.14. *(Change of variables formula). Let U be an open set in $\mathbb{R}^n$ and let $\varphi : U \to \mathbb{R}^n$ be a diffeomorphism. Let Ω be a bounded simple set with $\overline{\Omega} \subset U$. If $f : \varphi(\Omega) \to \mathbb{R}$ is integrable on $\varphi(\Omega)$, then $(f \circ \varphi)|\det D_\varphi|$ is integrable over Ω and*

$$\int_{\varphi(\Omega)} f(y)dy = \int_\Omega f(\varphi(x))|\det D_\varphi(x)|dx. \tag{5.8}$$

Proof. Set $J_\varphi = \det D_\varphi$. We show that $(f \circ \varphi)|J_\varphi|$ is integrable over Ω. We shall use Theorem 4.1.42 which characterizes integrable functions in terms of their points of discontinuity. Let D be the set of discontinuities of f in $\varphi(\Omega)$. Then $E = \varphi^{-1}(D)$ is the set of discontinuities of $f \circ \varphi$ in Ω. Since f is integrable on $\varphi(\Omega)$ the set D has measure zero. As $D \subset \varphi(\Omega) \subset \varphi(\overline{\Omega}) \subset \varphi(U)$ and $\varphi(\overline{\Omega})$ is compact contained in the open set $\varphi(U)$, we can find a finite number of closed cubes $\{C_i : i = 1, ..., N_1\}$ such that $D \subset \cup_{j=1}^{N_1} C_i \subset \varphi(U)$ (as in Lemma 5.1.13). Set $K = \cup_{i=1}^{N_1} C_i$, then K is compact and $D \subset K \subset \varphi(U)$. As

---

[3]The Change of Variables formula was first proposed by Euler when he studied double integrals in 1769, and it was generalized to triple integrals by Lagrange in 1773. Although it was used by Legendre, Laplace, Gauss, and first generalized to n variables by Mikhail Ostrogradski in 1836, it resisted a fully rigorous proof for a surprisingly long time. The theorem was first completely proved 125 years later, by Elie Cartan in a series of papers beginning in the mid-1890s. A popular proof adapted by many authors is the one given by J. Schwartz (1954) in [29]. In the proof given here effort has been made in avoiding as many technicalities as possible. A quite different approach to the problem can be found in P. Lax [20] and [21].

in the proof of Theorem 4.1.30 we may write $D = \bigcup_{k=1}^{\infty} D_{\frac{1}{k}}$ with each $D_{\frac{1}{k}}$ compact. Since $\varphi^{-1} \in C^1(\varphi(U))$ and each $D_{\frac{1}{k}}$ has measure zero, Lemma 5.1.5 and Proposition 4.1.21 tell us that $E = \varphi^{-1}(D)$ has also measure zero. Hence $f \circ \varphi$ is integrable. Since $|J_\varphi|$ is continuous on Ω, the set of points of discontinuity of $(f \circ \varphi)|J_\varphi|$ is the set E and so $(f \circ \varphi)|J_\varphi|$ is integrable on Ω.

We first prove (5.8) for a closed cube $C \subset U$. Let $\mathcal{P} = \{C_j : j = 1, ..., m\}$ be a partition of C into subcubes of equal side length $2r$ centered at x_j. Then $\varphi(C_j)$ is a bounded simple set for each j. In addition $\{\varphi(C_j)\}$ are mutually essentially disjoint and $\varphi(C) = \bigcup_{j=1}^{m} \varphi(C_j)$. From (5.7) of Lemma 5.1.12 we have

$$(1 - t)^n |J_\varphi(x_j)|\nu(C_j) \le \nu(\varphi(C_j)) \le |J_\varphi(x_j)|\nu(C_j)(1 + t)^n. \qquad (5.9)$$

Let $y_j = \varphi(x_j)$. We set $F = f \circ \varphi$ and *suppose* $f \ge 0$. From (5.9) we have

$$(1 - t)^n \sum_{j=1}^{m} F(x_j)|J_\varphi(x_j)|\nu(C_j) \le \sum_{j=1}^{m} f(y_j)\nu(\varphi(C_j)) \le$$

$$\le (1 + t)^n \sum_{j=1}^{m} F(x_j)|J_\varphi(x_j)|\nu(C_j).$$

Letting $t \to 0$ yields

$$\sum_{j=1}^{m} F(x_j)|J_\varphi(x_j)|\nu(C_j) \le \sum_{j=1}^{m} f(y_j)\nu(\varphi(C_j)) \le \sum_{j=1}^{m} F(x_j)|J_\varphi(x_j)|\nu(C_j).$$

$$(5.10)$$

The far left and far right are each a Riemann sum $S_\mathcal{P}(F|J_\varphi|)$ for the function $F|J_\varphi|$. Let $d = \max_{j=1,...,m}\{d(\varphi(C_j))\}$. Since φ is (uniformly) continuous on C, if $r \to 0$ then also $d \to 0$. Now since the functions f and $F|J_\varphi|$ are integrable on $\varphi(C)$ and C respectively, letting $r \to 0$ in (5.10) we get

$$\int_{\varphi(C)} f(y)dy = \int_{C} f(\varphi(x))|J_\varphi(x)|dx. \qquad (5.11)$$

To remove the assumption that $f \geq 0$, we write $f = (f + c) - c$ where the constant $c \geq 0$ is sufficiently large that $f + c \geq 0$ on Ω, for example $c = \sup |f(x)|$. The argument just given applies to $f + c$ and to the constant function c. By the linearity of the integral (Theorem 4.2.1) subtracting the results we get (5.11).

Next we prove (5.8) for a finite union $S = \bigcup_{i=1}^{N} C_i$ of essential disjoint closed cubes with $\Omega \subset S \subset U$. That such a set S exists is guaranteed from Lemma 5.1.13. Now the finite additivity of the integral (Corollary 4.2.5) yields

$$\int_{\varphi(S)} f(y)dy = \sum_{i=1}^{N} \int_{\varphi(C_i)} f(y)dy = \sum_{i=1}^{N} \int_{C_i} (f(\varphi(x))|J_\varphi(x)|dx$$

$$= \int_{S} (f(\varphi(x))|J_\varphi(x)|dx.$$

Finally, we obtain (5.8) for Ω. Let $y = \varphi(x)$ with $x \in S$, where $\Omega \subset S \subset U$. Since $\chi_\Omega = \chi_{\varphi(\Omega)} \circ \varphi$, where χ_Ω is the characteristic function of Ω, the definition of the integral over a set gives

$$\int_{\varphi(\Omega)} f(y)dy = \int_{\varphi(S)} (f\chi_{\varphi(\Omega)})(y)dy = \int_{S} [(f\chi_{\varphi(\Omega)}) \circ \varphi |J_\varphi|](x)dx$$

$$= \int_{S} [(f\circ\varphi)|J_\varphi|\chi_\Omega](x)dx = \int_{\Omega} ((f\circ\varphi)|J_\varphi|)(x)dx = \int_{\Omega} f(\varphi(x))|J_\varphi(x)|dx.$$

$\square$

Taking $f \equiv 1$ which is integrable everywhere we get

Corollary 5.1.15. *Let U be open in $\mathbb{R}^n$ and $\varphi : U \to \mathbb{R}^n$ a diffeomorphism. Let Ω be a bounded simple set with $\overline{\Omega} \subset U$. Then*

$$\nu(\varphi(\Omega)) = \int_{\Omega} |\det D_\varphi(x)|dx.$$

In certain circumstances a direct application of the Change of Variables formula (5.8)) is not valid, mainly because φ fails to be one-to-one on the boundary of Ω. The following slight improvement of Theorem 5.1.14 is then useful.

Corollary 5.1.16. *Let U be open in $\mathbb{R}^n$ and $\varphi : U \to \mathbb{R}^n$ be a C^1 map. Let Ω be a bounded simple set with $\overline{\Omega} \subset U$. Suppose $f : \varphi(\Omega) \to \mathbb{R}$ is integrable. Then the change of variables formula (5.8) remains valid if φ is only asssumed to be a diffeomorphism on the interior of Ω.*

Proof. Let C be a cube containing Ω and partition C into subcubes of equal side length $2r$. Let $C_1, ..., C_p$ be the subcubes contained in Ω, and $A = \Omega \backslash K$, where $K = \bigcup_{i=1}^{p} C_i$. Then $K \subset \Omega$ and so $\nu(K) \leq \nu(\Omega)$. Since Ω is simple, the difference $\nu(\Omega) - \nu(K)$ can be made arbitrarily small by taking r sufficiently small. That is, $\nu(A) = 0$ ans so also $\nu(\overline{A}) = 0$. Therefore by Lemma 5.1.5 (with $S = \overline{A}$) it follows that $\nu(\varphi(A)) \leq \nu(\varphi(\overline{A})) = 0$. Theorem 5.1.14 and the finite additvity of the integral yield

$$\int_{\varphi(K)} f(y)dy = \sum_{i=1}^{p} \int_{\varphi(C_i)} f(y)dy$$

$$= \sum_{i=1}^{p} \int_{C_i} (f(\varphi(x))|J_\varphi(x)|dx = \int_{K} (f(\varphi(x))|J_\varphi(x)|dx.$$

Finally, as $\Omega = K \cup A$, $\varphi(\Omega) = \varphi(K) \cup \varphi(A)$ and the integral over sets of volume zero is zero, the additivity of the integral gives $\int_{\varphi(\Omega)} f(y)dy = \int_{\Omega} (f(\varphi(x))|J_\varphi(x)|dx.$ □

More generally a further extension of the Change of Variables formula can be made allowing singularities of φ by using a special case of *Sard's theorem*: Let $\varphi : U \to \mathbb{R}^n$ be of class C^1 on an open set $U \subset \mathbb{R}^n$ and

$$A = \{x \in U : \det D_\varphi(x) = 0\}.$$

Then $\varphi(A)$ has *measure zero*, (see Theorem 4.5.1). Hence the places where $|\det D_\varphi(x)| \equiv 0$ on Ω contribute nothing to the integrals in (5.8).

Next we give an extension of the Change of Variables formula[4] for continuous real valued functions f on $\mathbb{R}^n$ with *compact support* (see, Section 4.4).

---

[4]Although the integral of f is taken over $\mathbb{R}^n$ and such integrals are studied in the next section, since the integrand f has compact support it is an integral over any cube containing the support of f.

Theorem 5.1.17. *Let U be an open set in $\mathbb{R}^n$ and φ be a diffeomorphism of U with its image $\varphi(U)$. Then*

$$\int_{\mathbb{R}^n} f(y)dy = \int_{\mathbb{R}^n} f(\varphi(x))|\det D_\varphi(x)|dx, \qquad (5.12)$$

for any continuous function f with compact support whose support lies in $\varphi(U)$.

Proof. Let $y \in \operatorname{supp}(f)$ and V_y be a neighborhood of y contained in the open set $\varphi(U)$. Since the $\{V_y\}$ cover $\operatorname{supp}(f)$ we can find a finite number of them $\{V_{y_i} : i = 1, ..., k\}$ which also cover. Let ψ_i be a partition of unity subordinate to this cover. By Corollary 4.4.4 $f = \sum_{i=1}^k \psi_i f$, where $\psi_i f$ is a continuous function whose support lies in V_{y_i}. By the Change of Variables formula (5.8) equation (5.12) holds for each $\psi_i f$, $i = 1, ..., k$ and hence by linearity of the integral also for their sum. $\qquad \square$

We now apply the Change of Variables formula to a number of examples.

5.1.3   Applications, polar and spherical coordinates

Let $\Omega \subset \mathbb{R}^n$ be a bounded simple set and $f : \Omega \to \mathbb{R}$ an integrable function. One of the purposes of the Change of Variables formula is to simplify the computation of the multiple integral $\int_\Omega f$ on which either the integrand f or the region Ω is complicated and for which direct computation is difficult. Therefore, a C^1 change of variables $\varphi : \Omega^* \to \Omega$ is chosen, with $\Omega = \varphi(\Omega^*)$, so that the integral is easier to compute with the new integrand $(f \circ \varphi)|J_\varphi|$, or with the new region $\Omega^* = \varphi^{-1}(\Omega)$. In this notation the Change of Variables formula becomes

$$\int_\Omega f = \int_{\varphi(\Omega^*)} f = \int_{\Omega^*} (f \circ \varphi)|J_\varphi|. \qquad (5.13)$$

The Change of Variables formula (5.13) for *double integrals* is

$$\int\int_\Omega f(x,y)dxdy = \int\int_{\Omega^*} f(x(u,v),y(u,v))|J_\varphi(u,v)|dudv,$$

where $\varphi(u,v) = (x(u,v), y(u,v))$ for $(u,v) \in \Omega^* = \varphi^{-1}(\Omega)$ and

$$|J_\varphi(u,v)| = |\det D_\varphi(u,v)| = \left| \frac{\partial(x,y)}{\partial(u,v)} \right| = \left| \det \begin{pmatrix} \frac{\partial x}{\partial u} & \frac{\partial x}{\partial v} \\ \frac{\partial y}{\partial u} & \frac{\partial y}{\partial u} \end{pmatrix} \right|.$$

The Change of Variables formula (5.13) for *triple integrals* is

$$\int \int \int_\Omega f(x,y,z)\,dxdydz$$

$$= \int \int \int_{\Omega^*} f(x(u,v,w), y(u,v,w), z(u,v,w))|J_\varphi(u,v,w)|\,dudvdw,$$

where $\varphi(u,v,w) = (x(u,v,w), y(u,v,w), z(u,v,w))$ for $(u,v,w) \in \Omega^* = \varphi^{-1}(\Omega)$ and

$$|J_\varphi(u,v,w)| = |\det D_\varphi(u,v,w)| = \left| \frac{\partial(x,y,z)}{\partial(u,v,w)} \right| = \left| \det \begin{pmatrix} \frac{\partial x}{\partial u} & \frac{\partial x}{\partial v} & \frac{\partial x}{\partial w} \\ \frac{\partial y}{\partial u} & \frac{\partial y}{\partial v} & \frac{\partial y}{\partial w} \\ \frac{\partial z}{\partial u} & \frac{\partial z}{\partial u} & \frac{\partial z}{\partial w} \end{pmatrix} \right|.$$

Example 5.1.18. Let $f(x,y) = xy$ and let Ω be the parallelogram bounded by the lines $x - y = 0$, $x - y = 1$, $x + 2y = 0$ and $x + 2y = 6$. Using the Change of Variables formula compute $\int \int_\Omega f(x,y)\,dxdy$.

Solution. The equations of the bounding lines of Ω suggest the linear change of variables $u = x - y$, $v = x + 2y$. That is, the linear mapping $L : \mathbb{R}^2 \to \mathbb{R}^2$ given by $L(x,y) = (x - y, x + 2y)$. The standard matrix of L is $L = \begin{pmatrix} 1 & -1 \\ 1 & 2 \end{pmatrix}$, and $\det L = 3$. Hence L is an isomorphism on $\mathbb{R}^2$.

The vertices of the parallelogram Ω are $(0,0)$, $(2,2)$, $(\frac{8}{3}, \frac{5}{3})$, and $(\frac{2}{3}, -\frac{1}{3})$ and these are mapped by L to $(0,0)$, $(0,6)$, $(1,6)$, and $(1,0)$, respectively. Since L is linear, it follows that the image of the parallelogram Ω under L is the rectangle

$$\Omega^* = \{(u,v) : 0 \le u \le 1, 0 \le v \le 6\}.$$

In fact, we are actually interested in the inverse mapping $T = L^{-1}$ from the uv-plane to the xy-plane defined by

$$x = \tfrac{1}{3}(2u + v), \quad y = \tfrac{1}{3}(v - u).$$

That is, the linear transformation $T(u,v) = \frac{1}{3}(2u+v, v-u)$ with Jacobian $J_T(u,v) = \det T = \frac{1}{3}$ and $T(\Omega^*) = \Omega$.

Hence the Change of Variables formula yields

$$\int\int_\Omega xy\,dxdy = \frac{1}{9}\int\int_{\Omega^*}(2u+v)(v-u)|\frac{1}{3}|dudv$$

$$= \frac{1}{27}\int_0^1\int_0^6(2u+v)(v-u)dudv = \int_0^6(-\frac{2}{3}+\frac{1}{2}v+v^2)dv = \frac{77}{27}.$$

Example 5.1.19. Compute the integral

$$\int\int_\Omega e^{\frac{y-x}{y+x}}\,dxdy,$$

where $\Omega = \{(x,y) : x \geq 0, \ y \geq 0, \ x+y \leq 1\}$.

Solution. Here the difficulty in the computation of the integral comes from the integrand. Note that the function $f(x,y) = e^{\frac{y-x}{y+x}}$ is not defined at $(0,0)$ and, in fact, does not even have a limit there. However, no matter how one defines f at $(0,0)$ the function is integrable by Theorem 4.1.42. The set Ω is a simple region in $\mathbb{R}^2$. Actually, Ω is the triangle bounded by the line $x+y = 1$ and the coordinate axes. The integrand suggests the use of the new variables

$$u = y - x, \ v = y + x,$$

i.e., the linear transformation $L(x,y) = (y-x, y+x)$. L is an isomorphism on $\mathbb{R}^2$ and its inverse $T = L^{-1}$ is the linear transformation

$$T(u,v) = (\frac{v-u}{2}, \frac{v+u}{2}).$$

The standard matrix representation is $T = \begin{pmatrix} -\frac{1}{2} & \frac{1}{2} \\ \frac{1}{2} & \frac{1}{2} \end{pmatrix}$, and its Jacobian $J_T(u,v) = \det T = -\frac{1}{2}$. To find the image Ω^* of Ω under L in the uv-plane, note that the lines $x = 0$, $y = 0$ and $x+y = 1$ are mapped onto the lines $u = v$, $u = -v$ and $v = 2$, respectively. The points inside Ω satisfy $0 < x+y < 1$ and these are mapped into points of Ω^* satisfying $0 < v < 1$. Hence Ω^* is the triangle in the uv-plane bounded by the

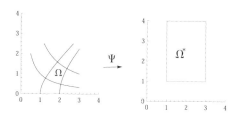

Figure 5.2: $\psi(x,y) = (xy, x^2 - y^2)$

lines $v = u$, $v = -u$ and $v = 2$. Now $T(\Omega^*) = \Omega$ and the Change of Variables formula yields

$$\int\int_\Omega e^{\frac{y-x}{y+x}} dxdy = \frac{1}{2}\int\int_{\Omega^*} e^{\frac{u}{v}} dudv$$

$$= \frac{1}{2}\int_0^1 \left(\int_{-v}^v e^{\frac{u}{v}} du\right) dv = \frac{1}{2}\int_0^1 (e - \frac{1}{e})vdv = \frac{1}{4}(e - e^{-1}) = \frac{1}{2}\sinh(1).$$

Example 5.1.20. Let $f(x,y) = x^2 + y^2$. Evaluate $\int\int_\Omega f(x,y)dxdy$, where Ω is the region in the first quadrant of the xy-plane bounded by the curves $xy = 1$, $xy = 3$, $x^2 - y^2 = 1$, and $x^2 - y^2 = 4$.

Solution. To simplify the region Ω we make the substitution $u = xy$, $v = x^2 - y^2$. This transformation $\psi : \mathbb{R}^2 \to \mathbb{R}^2$

$$\psi(x,y) = (xy, x^2 - y^2)$$

is one-to-one on Ω and maps the region Ω in the xy-plane onto the rectangle Ω^* in the uv-plane bounded by the lines $u = 1$, $u = 3$, $v = 1$, and $v = 4$, ie $\psi(\Omega) = \Omega^* = [1,3] \times [1,4]$. Its Jacobian is

$$J_\psi(x,y) = \frac{\partial(u,v)}{\partial(x,y)} = \det\begin{pmatrix} y & x \\ 2x & -2y \end{pmatrix} = -2(x^2 + y^2) \neq 0$$

for all $(x,y) \in \Omega$. As before, we are interested in its inverse $\varphi = \psi^{-1}$ given by

$$\varphi(u,v) = \left(\left[\frac{(4u^2 + v^2)^{\frac{1}{2}} + v}{2}\right]^{\frac{1}{2}}, \left[\frac{(4u^2 + v^2)^{\frac{1}{2}} - v}{2}\right]^{\frac{1}{2}}\right).$$

The Jacobian of φ is

$$J_\varphi(u,v) = \frac{\partial(x,y)}{\partial(u,v)} = \frac{1}{J_\psi(x,y)} = -\frac{1}{2(x^2+y^2)} = -\frac{1}{2(4u^2+v^2)^{\frac{1}{2}}}.$$

Now $\Omega = \varphi(\Omega^*)$ and the Change of Variables formula gives

$$\int\int_\Omega f(x,y)dxdy = \int\int_{\Omega^*} f(\varphi(u,v))|J_\varphi(u,v)|dudv$$

$$= \int_1^3\int_1^4 (4u^2+v^2)^{\frac{1}{2}} \cdot \frac{1}{2(4u^2+v^2)^{\frac{1}{2}}}dudv = \frac{1}{2}\int_1^3\int_1^4 dudv = 3.$$

Example 5.1.21. (*The solid torus*).   Let $0 < a < b$ be fixed. The solid torus in $\mathbb{R}^3$ is the region Ω obtained by revolving the disk $(y-b)^2 + z^2 \le a^2$ in the yz-plane, about the z-axis. Calculate the volume of Ω. (See, Figure 4.5)

Solution.   Note that the mapping $\varphi : \mathbb{R}^3 \to \mathbb{R}^3$ defined by the equations

$$x = (b + r\cos\phi)\cos\theta$$
$$y = (b + r\cos\phi)\sin\theta$$
$$z = r\sin\phi,$$

maps the rectangle $R = \{(\theta,\phi,r) : \theta, \phi \in [0, 2\pi], \ r \in [0, a]\}$ onto the torus, ie $\varphi(R) = \Omega$. In particular, φ is one-to-one in the interior of R. Its Jacobian is

$$J_\varphi(\theta,\phi,r) = \frac{\partial(x,y,z)}{\partial(\theta,\phi,r)} = r(b + r\cos\phi),$$

which is positive in the interior of R. By Corollary 5.1.16, the Change of Variables formula applies, and Corollary 5.1.15 gives

$$\nu(\Omega) = \nu(\varphi(R)) = \int\int\int_R r(b + r\cos\phi)d\theta d\phi dr = 2\pi^2 a^2 b.$$

Polar coordinates

The map which changes from polar coordinates to rectangular coordinates is

$$\varphi(r, \theta) = (r \cos \theta, r \sin \theta) = (x, y).$$

Its Jacobian is

$$J_\phi(r, \theta) = \frac{\partial(x, y)}{\partial(r, \theta)} = \det \begin{pmatrix} \cos \theta & -r \sin \theta \\ \sin \theta & r \cos \theta \end{pmatrix} = r \cos^2 \theta + r \sin^2 \theta = r,$$

which is zero for $r = 0$. φ maps the set $[0, +\infty) \times [0, 2\pi)$ in the $r\theta$-plane onto $\mathbb{R}^2$. Note that on this set φ is not one-to-one, since for $0 \le \theta < 2\pi$ it sends all points $(0, \theta)$ into $(0, 0)$. However φ restricted to $S = (0, +\infty) \times (0, 2\pi)$ (the interior of $[0, +\infty) \times [0, 2\pi)$) is one-to-one and $J_\phi(r, \theta) \ne 0$ on S. Although $\varphi(S)$ excludes the non-negative x-axis, this *set has 2-dimensional volume zero and therefore contributes nothing to the value of an integral.* Note in particular, that φ maps the rectangle $R_\alpha = [0, \alpha] \times [0, 2\pi]$ onto the disk $B_\alpha = \{(x, y) : x^2 + y^2 \le \alpha^2\}$. Thus by restricting φ in the interior of R_α we can apply the Change of Variables theorem (Corollary 5.1.16) to convert integration over B_α into integration over R_α.

Example 5.1.22. Compute

$$\int \int_{B_\alpha} e^{-(x^2 + y^2)} dx dy,$$

where $B_\alpha = \{(x, y) : x^2 + y^2 \le \alpha^2\}$.

Solution. Changing to polar coordinates and using (5.13) with $\Omega = B_\alpha$ and $\Omega^* = [0, \alpha] \times [0, 2\pi]$ we get

$$\int \int_{B_\alpha} e^{-(x^2 + y^2)} dx dy = \int_0^\alpha \int_0^{2\pi} e^{-r^2} r \, dr \, d\theta$$

$$= \int_0^\alpha \left(e^{-r^2} r \int_0^{2\pi} d\theta \right) dr = \pi \int_{-\alpha^2}^0 e^u du = \pi [1 - e^{-\alpha^2}].$$

In particular, letting $\alpha \to \infty$ we find[5]

$$\int\int_{\mathbb{R}^2} e^{-(x^2+y^2)}dxdy = \pi.$$

Example 5.1.23. Compute

$$\int\int_\Omega \sin\sqrt{x^2+y^2}dxdy,$$

where $\Omega = \{(x,y) : \pi^2 < x^2 + y^2 \le 4\pi^2\}$.

Solution. Changing to polar coordinates we see that $\Omega = \varphi(\Omega^*)$, where $\Omega^* = (\pi, 2\pi] \times [0, 2\pi]$ and we get

$$\int\int_\Omega \sin\sqrt{x^2+y^2}dxdy = \int_\pi^{2\pi}\int_0^{2\pi} \sin r \cdot r\,dr\,d\theta$$

$$= 2\pi\int_\pi^{2\pi} r\sin r\,dr = 2\pi\left[-r\cos r\Big|_\pi^{2\pi} + \int_\pi^{2\pi}\cos r\,dr\right]$$

$$= -6\pi^2 + 2\pi\sin r\Big|_\pi^{2\pi} = -6\pi^2.$$

Example 5.1.24. Compute

$$\int\int_\Omega \sqrt{\frac{a^2b^2 - b^2x^2 - a^2y^2}{a^2b^2 + b^2x^2 + a^2y^2}}dxdy,$$

where $\Omega = \left\{(x,y) : \frac{x^2}{a^2} + \frac{y^2}{b^2} \le 1,\ x > 0,\ y > 0\right\}$.

Solution. Changing to polar coordinates using the transformation

$$\varphi(r,\theta) = (ar\cos\theta, br\sin\theta) = (x,y),$$

we see that $\Omega^* = (0,1] \times [0, \frac{\pi}{2})$ and $J_\varphi(r,\theta) = abr$. Now the integral becomes

---

[5]The integral $\int\int_{\mathbb{R}^2} e^{-(x^2+y^2)}dxdy = \pi$ is improper. See Section 5.2, for a detailed discussion on improper multiple integrals.

$$ab \int_0^1 \int_0^{\frac{\pi}{2}} \sqrt{\frac{1-r^2}{1+r^2}} r\,dr\,d\theta = ab\frac{\pi}{2} \int_0^1 \sqrt{\frac{1-r^2}{1+r^2}} r\,dr$$

$$= ab\frac{\pi}{2} \int_0^{\sqrt{2}} \sqrt{2-\rho^2}\,d\rho,$$

where $\rho^2 = 1 + r^2$. Finally, setting $\rho = \sqrt{2}\sin t$, this latter integral yields $\frac{\pi}{2}(\frac{\pi}{4} - \frac{1}{2})ab$.

Example 5.1.25. Find the area of the region S in $\mathbb{R}^2$ bounded by the circles $r = 1$ and $r = \frac{2}{\sqrt{3}}\cos\theta$ (outside of the circle $r = 1$).

Solution. Solving $1 = \frac{2}{\sqrt{3}}\cos\theta$ we find the (polar) coordinates of the points of intersections of the given circles to be $(1, \frac{\pi}{6})$ and $(1, \frac{11\pi}{6})$ (the reader is recommended to make a drawing; in cartesian coordinates the circle $r = \frac{2}{\sqrt{3}}\cos\theta$ is $(x - \frac{1}{\sqrt{3}})^2 + y^2 = \frac{1}{3}$ and the circle $r = 1$ is, of course, $x^2 + y^2 = 1$). By the symmetry of the region the area is

$$A(S) = \int\int_S r\,dr\,d\theta = 2\int_0^{\frac{\pi}{6}} \left(\int_1^{\frac{2}{\sqrt{3}}\cos\theta} r\,dr \right) d\theta$$

$$= \int_0^{\frac{\pi}{6}} (\frac{4}{3}\cos^2\theta - 1)d\theta = \frac{3\sqrt{3} - \pi}{18}.$$

Cylindrical coordinates

The *cylidrical coordinates* are just the polar coordinates in the xy-plane with the z-coordinate added on, (Figure 5.3),

$$\varphi(r, \theta, z) = (r\cos\theta, r\sin\theta, z).$$

To get a one-to-one mapping we must keep $r > 0$ and restrict θ, say, in $0 < \theta \le 2\pi$. The Jacobian of φ is easily seen to be $J_\varphi(r, \theta, z) = r$, and the Change of Variables formula (5.13) becomes

$$\int\int\int_\Omega f(x, y, z)dx\,dy\,dz = \int\int\int_{\Omega^*} f(r\cos\theta, r\sin\theta, z)r\,dr\,d\theta\,dz.$$

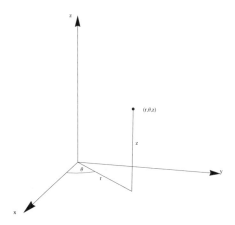

Figure 5.3: Cylindrical coordinates

Example 5.1.26. Find the volume of the region Ω bounded by the surfaces $(\frac{x^2+y^2}{a^2})^2 + \frac{z}{b} = 1$, $z = 0$, where $a, b > 0$.

Solution. As we know from Definition 4.1.39

$$\nu(\Omega) = \int_\Omega 1 = \int\int\int_\Omega dxdydz.$$

Changing to cylidrical coordinates we get

$$\nu(\Omega) = \int\int\int_{\Omega^*} rdrd\theta dz = \int_0^a \int_0^{2\pi} \left(\int_0^{b(1-\frac{r^4}{a^4})} dz \right) drd\theta$$

$$= 2\pi b \int_0^a r\left(1 - \frac{r^4}{a^4}\right) dr = \frac{2}{3}\pi a^2 b.$$

Example 5.1.27. Find the moment of inertia I_L of a cylinder $x^2 + y^2 = a^2$ of height h, if its density at each point is propotional to the distance of this point from the axis of the cylinder, with respect to a line L parallel to the axis of the cylinder and at distance b from it.

Solution. Let Ω be the cylinder. The moment of inertia of Ω about the line L is

$$I_L = \int\int\int_\Omega [d(x, y, z)]^2 \rho(x, y, z)dxdydz,$$

where $d(x, y, z)$ is the distance from (x, y, z) to the line L, and $\rho(x, y, z)$ is the density. Placing the base of the cylinder on the xy-plane with its center at the origin, the line L is described by $x = b$, $y = 0$. The density function is $\rho(x, y, z) = k\sqrt{x^2 + y^2}$ and $[d(x, y, z)]^2 = (x - b)^2 + y^2$. Passing to cylindrical coordinates we get

$$
\begin{aligned}
I_L &= \int_0^h \int_0^{2\pi} \int_0^a [(r\cos\theta - b)^2 + (r\sin\theta)^2](kr) \, dr \, d\theta \, dz \\
&= \int_0^h \int_0^{2\pi} \int_0^a kr^2 [r^2 + b^2 - 2r\cos\theta] \, dr \, d\theta \, dz \\
&= k \int_0^h \int_0^{2\pi} \int_0^a r^2 [r^2 + b^2] \, dr \, d\theta \, dz + 0 \\
&= 2\pi k a^3 h \left(\frac{a^2}{5} + \frac{b^2}{3} \right).
\end{aligned}
$$

Spherical coordinates

Given $(x, y, z) \in \mathbb{R}^3$, the *spherical coordinates* (r, θ, ϕ) are defined by

$$
\Phi(r, \theta, \phi) = (r\cos\theta\sin\phi, r\sin\theta\sin\phi, r\cos\phi) = (x, y, z).
$$

Here $r = \sqrt{x^2 + y^2 + z^2}$, θ is the longitude (the angle from the x-axis to the vector $(x, y, 0)$), and ϕ is the co-latitude (the angle from the positive z-axis to the vector (x, y, z)). (see, Figure 5.4)

It is readily seen that Φ maps the rectangle $R = [0, \alpha] \times [0, 2\pi] \times [0, \pi]$ in the $r\theta\phi$-space into the ball $B_\alpha = \{(x, y, z) : x^2 + y^2 + z^2 \leq \alpha^2\}$.

Again Φ is not globally invertible; we have $\Phi(r, \theta, \phi) = \Phi(r, \theta + 2k\pi, \phi + 2m\pi) = \Phi(r, \theta + (2k + 1)\pi, \phi + (2m + 1)\pi)$ for each $k, m \in \mathbb{Z}$. The map Φ is one-to-one on $\{(r, \theta, \phi) : r > 0, 0 \leq \theta < 2\pi, 0 < \phi < \pi\}$ and maps this set onto $\mathbb{R}^3 - \{(0, 0, z) : z \in \mathbb{R}\}$. Since $\{(0, 0, z) : z \in \mathbb{R}\}$, has volume zero does not affect the value of an integral. Now

$$
J_\Phi(r, \theta, \phi) = \frac{\partial(x, y, z)}{\partial(r, \theta, \phi)}
$$

$$
= \det \begin{pmatrix} \cos\theta\sin\phi & -r\sin\theta\sin\phi & r\cos\theta\cos\phi \\ \sin\theta\sin\phi & r\cos\theta\sin\phi & r\sin\theta\cos\phi \\ \cos\phi & 0 & -r\sin\phi \end{pmatrix} = -r^2\sin\phi.
$$

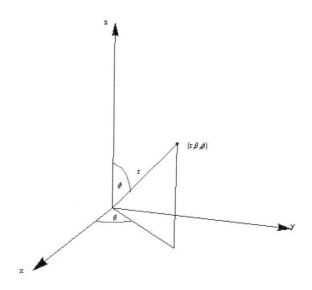

Figure 5.4: Spherical coordinates (r, θ, ϕ)

Hence the formula for integration in spherical coordinates is

$$\int \int \int_{\Omega} f(x, y, z) dx dy dz$$

$$= \int \int \int_{\Omega^*} f(r \cos \theta \sin \phi, r \sin \theta \sin \phi, r \cos \phi) r^2 \sin \phi dr d\theta d\phi.$$

Example 5.1.28. Compute the integral $\int \int \int_{\Omega} e^{(x^2+y^2+z^2)^{\frac{3}{2}}} dx dy dz$, where Ω is the unit ball in $\mathbb{R}^3$.

Solution. Changing to spherical coordinates $\Omega = \Phi(\Omega^*)$, with $\Omega^* = [0, 1] \times [0, 2\pi] \times [0, \pi]$. Now the formula of integration in spherical coordinates gives

$$\int \int \int_{\Omega} e^{(x^2+y^2+z^2)^{\frac{3}{2}}} dx dy dz = \int \int \int_{\Omega^*} e^{r^3} r^2 \sin \phi dr d\theta d\phi$$

$$= \int_0^1 \int_0^{\pi} \left(r^2 e^{r^3} \sin \phi \int_0^{2\pi} d\theta \right) d\phi = 2\pi \int_0^1 \left(r^2 e^{r^3} \int_0^{\pi} \sin \phi d\phi \right) dr$$

$$= 4\pi \int_0^1 r^2 e^{r^3}\, dr = \frac{4}{3}\pi \int_0^1 e^t\, dt = \frac{4\pi(e-1)}{3},$$

where we set $t = r^3$.

Example 5.1.29. Compute $\int\int\int_\Omega \left(\frac{1}{\sqrt{x^2+y^2}} + \frac{1}{z} \right) dxdydz$, where Ω is
the region in $\mathbb{R}^3$ bounded below by the cone $z = \sqrt{x^2 + y^2}$ and above
by the sphere $x^2 + y^2 + z^2 = 1$, and $z > 0$.

Solution. The region of integration Ω represents an "ice cream cone".
Note that Ω is an elementary region[6] in $\mathbb{R}^3$. However, both the region
Ω and the integrand suggest a change to spherical coordinates. Then
$\Omega = \Phi(\Omega^*)$, where

$$\Omega^* = \left\{ (r,\theta,\phi) : 0 < r < 1,\ 0 \le \theta < 2\pi,\ 0 < \phi < \frac{\pi}{4} \right\}.$$

Now the Change of Variables formula yields

$$\int\int\int_\Omega \left(\frac{1}{\sqrt{x^2 + y^2}} + \frac{1}{z} \right) dxdydz$$

$$= \int\int\int_{\Omega^*} \left(\frac{1}{r \sin\phi} + \frac{1}{r \cos\phi} \right) r^2 \sin\phi\, dr d\theta d\phi$$

$$= \int_0^{\frac{\pi}{4}} \left((1 + \tan\phi) \int_0^1 r \left(\int_0^{2\pi} d\theta \right) dr \right) d\phi$$

$$= 2\pi \int_0^{\frac{\pi}{4}} \left((1 + \tan\phi) \int_0^1 r dr \right) d\phi$$

$$= \pi \int_0^{\frac{\pi}{4}} (1 + \tan\phi) d\phi = \pi[(\phi - \log(\cos\phi))|_0^{\frac{\pi}{4}}] = \pi \left[\frac{\pi}{4} - \log\left(\frac{\sqrt{2}}{2} \right) \right].$$

---

[6]$\Omega = \{(x,y,z) : -\frac{1}{\sqrt{2}} \le x \le \frac{1}{\sqrt{2}}, -\sqrt{\frac{1}{2}-x^2} \le y \le \sqrt{\frac{1}{2}-x^2}, \sqrt{x^2+y^2} \le z \le$
$\sqrt{1-x^2-y^2}\}$. Note that working with rectangular coordinates the integral becomes
quite tedious.

Example 5.1.30. Find the volume of the region Ω in $\mathbb{R}^3$ above the cone $z^2 = x^2 + y^2$ and inside the sphere $x^2 + y^2 + z^2 = z$.

Solution. Here $\Omega = \Phi(\Omega^*)$, where

$$\Omega^* = \left\{ (r, \theta, \phi) : 0 \leq r < \cos\phi, \ 0 \leq \theta < 2\pi, \ 0 < \phi < \frac{\pi}{4} \right\}.$$

Hence the volume $\nu(\Omega)$ is equal to

$$\iiint_\Omega 1 dx dy dz = \int_0^{\frac{\pi}{4}} \left(\int_0^{\cos\phi} \left(\int_0^{2\pi} d\theta \right) r^2 \sin\phi dr \right) d\phi$$

$$= 2\pi \int_0^{\frac{\pi}{4}} \left(\sin\phi \int_0^{\cos\phi} r^2 dr \right) d\phi = \frac{2\pi}{3} \int_0^{\frac{\pi}{4}} \cos^3\phi \sin\phi d\phi$$

$$= \frac{2\pi}{3} \int_{\frac{\sqrt{2}}{2}}^1 t^3 dt = \frac{2\pi}{3} \cdot \frac{3}{16} = \frac{\pi}{8},$$

where we set $t = \cos\phi$.

Example 5.1.31. Find the centroid of a ball of radius a. Find the moment of inertia of a ball of constant density with respect to a diameter.

Solution. Place the ball B with its center at the origin, so that B is described by $x^2 + y^2 + z^2 \leq a^2$. From Example 4.3.19 the volume of the ball is $\nu(B) = \frac{4}{3}\pi a^3$. Using spherical coordinates

$$\overline{x} = \frac{1}{\nu(B)} \iiint_B x dx dy dz = \frac{1}{\nu(B)} \int_0^a \int_0^{2\pi} \int_0^\pi (r\cos\phi) r^2 \sin\phi d\phi d\theta dr = 0.$$

By symmetry $\overline{x} = \overline{y} = \overline{z}$. Hence its centroid is its center $(0, 0, 0)$.

Let the density of the ball be $\rho(x, y, z) = k$. By symmetry $I_x = I_y = I_z$. Since $I_x + I_y + I_z = 2I_0$, where I_0 is the moment of inertia about the origin, $I_0 = \int \int \int_B (x^2 + y^2 + z^2) k dx dy dz$, we have

$$I_x = I_y = I_z = \frac{2}{3} I_0 = \frac{2}{3} \int_0^a \int_0^\pi \int_0^{2\pi} k r^4 \sin\phi d\theta d\phi dr = \frac{8\pi k a^5}{15}.$$

Since the mass of the ball is $m = \frac{4\pi k a^3}{3}$, the answer can be expressed $\frac{2}{5} a^2 m$.

Exercise 5.1.32. Show that the centroid of the portion of the ball of radius a in the first octant is the point $\frac{3a}{2}(1,1,1)$.

Spherical coordinates in $\mathbb{R}^n$

This is the higher dimensional analogue of the spherical coordinates in $\mathbb{R}^3$. Given $(x_1, ..., x_n) \in \mathbb{R}^n$ we let $r = ||x|| = \sqrt{x_1^2 + ... + x_n^2}$, and $\Theta = (\theta, \phi_1, ..., \phi_{n-2})$, a generic point on the unit sphere, S^{n-1}. For $n \geq 2$ the n-dimensional spherical mapping $\Phi_n : \mathbb{R}^n \to \mathbb{R}^n$ written as

$$\Phi_n(r, \Theta) = \Phi_n(r, \theta, \phi_1, ..., \phi_{n-2}) = (x_1, ..., x_n)$$

is defined by

$$x_1 = r \cos \phi_1,$$

$$x_2 = r \sin \phi_1 \cos \phi_2,$$

$$x_3 = r \sin \phi_1 \sin \phi_2 \cos \phi_3,$$

$$\cdots\cdots\cdots\cdots\cdots\cdots\cdots\cdots\cdots\cdots$$

$$x_{n-1} = r \sin \phi_1 \cdots \sin \phi_{n-2} \cos \theta,$$

$$x_n = r \sin \phi_1 \cdots \sin \phi_{n-2} \sin \theta.$$

Φ_n maps the rectangle

$$R_\alpha = [0, \alpha] \times [0, 2\pi] \times [0, \pi] \times \cdots \times [0, \pi]$$

$$= \{(r, \theta, \phi_1, ..., \phi_{n-2}) \in \mathbb{R}^n : r \in [0, \alpha], \ \theta \in [0, 2\pi], \ \phi_i \in [0, \pi], i = 1, ..., n-2\}$$

onto the ball $B_\alpha = \{x \in \mathbb{R}^n : ||x|| \leq \alpha\}$.

Thus for $n = 2$

$$\Phi_2(r, \Theta) = \Phi_2(r, \theta) = (r \cos \theta, r \sin \theta),$$

which are just the polar coordinates in $\mathbb{R}^2$, while for $n = 3$

$$\Phi_3(r, \Theta) = \Phi_3(r, \theta, \phi_1) = (r \cos \phi_1, r \sin \phi_1 \cos \theta, r \sin \phi_1 \sin \theta),$$

which are the spherical coordinates in $\mathbb{R}^3$ (viewing the x_1-axis as the z-axis and ϕ_1 as ϕ).

The Jacobian of Φ_n can be calculated by induction and is

$$J_{\Phi_n}(r,\Theta) = J_{\Phi_n}(r,\theta,\phi_1,...,\phi_{n-2}) = \frac{\partial(x_1,x_2,x_3,...,x_n)}{\partial(r,\theta,\phi_1,...,\phi_{n-2})}$$

$$= r^{n-1}\sin^{n-2}\phi_1\sin^{n-3}\phi_2\cdots\sin^2\phi_{n-3}\sin\phi_{n-2} = r^{n-1}\prod_{k=1}^{n-2}\sin^{n-k-1}\phi_k.$$

Clearly, Φ_n is not one-to-one on R_α, however Φ_n restricted to the interior of R_α it is one-to-one with $J_{\Phi_n}(r,\Theta) > 0$ and the Change of Variables formula applies.

Example 5.1.33. (*The volume of the unit ball in $\mathbb{R}^n$*). Find the volume of the n-dimensional unit ball,

$$B^n = \{x \in \mathbb{R}^n : ||x|| \le 1\}.$$

Solution. In Corollary 5.1.8 we denoted the volume $\nu(B^n) = c_n$. Now

$$c_n = \int_{B^n} 1 = \int_{R_1} J_{\Phi_n}(r,\Theta)$$

$$= \frac{2\pi}{n}\left[\int_0^\pi\cdots\int_0^\pi\sin^{n-2}\phi_1\sin^{n-3}\phi_2\cdots\sin\phi_{n-2}d\phi_1\cdots d\phi_{n-2}\right]$$

$$= \frac{2\pi}{n}\prod_{k=1}^{n-2}\left[\int_0^\pi\sin^k\phi d\phi\right] = \frac{2\pi}{n}\prod_{k=1}^{n-2}I_k = \frac{2\pi}{n}[I_1 I_2\cdots I_{n-2}],$$

where $I_k = \int_0^\pi\sin^k\phi d\phi$. From elementary calculus we know that for $k = 1,2,...$

$$I_{2k} = \int_0^\pi\sin^{2k}\phi d\phi = \frac{(2k)!\pi}{2^{2k}(k!)^2},$$

$$I_{2k-1} = \int_0^\pi\sin^{2k-1}\phi d\phi = \frac{2^{2k-1}((k-1)!)^2}{(2k-1)!}.$$

Figure 5.5: Ellipsoid: $x^2 + 2y^2 + 3z^2 = 1$

Now, since $I_{k-1}I_k = \frac{2\pi}{k}$, when $n = 2m$ $(m = 1, 2, ...)$, we have

$$c_{2m} = \frac{2\pi}{2m}\left[\frac{\pi^{m-1}}{(m-1)!}\right] = \frac{\pi^m}{m!},$$

while when $n = 2m + 1$ $(m = 0, 1, ...)$,

$$c_{2m+1} = \frac{2\pi}{2m+1}\left[\frac{\pi^{m-1}}{(m-1)!} \cdot I_{2m-1}\right] = \frac{2^{m+1}\pi^m}{1 \cdot 3 \cdot 5 \cdots (2m+1)}.$$

Thus,

$$c_{2m} = \frac{\pi^m}{m!}, \quad c_{2m+1} = \frac{2^{m+1}\pi^m}{1 \cdot 3 \cdot 5 \cdots (2m+1)}. \tag{5.14}$$

Hence $c_1 = 2$, $c_2 = \pi$, $c_3 = \frac{4\pi}{3}$, $c_4 = \frac{\pi^2}{2}$, $c_5 = \frac{8\pi^2}{15}$, $c_6 = \frac{\pi^3}{6}$, $c_7 = \frac{16\pi^3}{105}$, and so on.

Example 5.1.34. Find the volume of the n-dimensional solid ellipsoid

$$E^n = \left\{x \in \mathbb{R}^n : \frac{x_1^2}{a_1^2} + ... + \frac{x_n^2}{a_n^2} \le 1\right\}.$$

Solution. Consider the linear transformation $T : \mathbb{R}^n \to \mathbb{R}^n$ given by

$$T(x_1, ..., x_n) = (a_1 x_1, ..., a_n x_n).$$

The solid ellipsoid E^n is the image of the unit ball B^n under T. By Proposition 5.1.7

$$\nu(E^n) = \nu(T(B^n)) = |\det T|\nu(B^b) = a_1 a_2 \cdots a_n \nu(B^n).$$

In particular, the volume of the 3-dimensional ellipsoid $\frac{x^2}{a^2} + \frac{y^2}{b^2} + \frac{z^2}{c^2} \le 1$ is $\frac{4}{3}\pi abc$.

Example 5.1.35. (*Invariant integrals on groups*)(*).
We now give an application of the Change of Variables formula to calculating *Invariant integrals on groups* . Rather than deal with generalities, we shall do this with an example which has enough complexity to illustrate the general situation well.

Consider the set G of all matrices of the following form.

$$G = \left\{ x = \begin{pmatrix} u & v \\ 0 & 1 \end{pmatrix} : u \ne 0, v \in \mathbb{R} \right\}.$$

which evidently can be identified with the (open) left and right half planes. We write $x = (u, v)$. As the reader can easily check, G is a group under matrix multiplication (in particular G is closed under multiplication). G is called the *affine group of the real line* $\mathbb{R}$. We define left and right translation (L_g and R_g respectively) in G as follows: $L_g(x) = gx$, $R_g(x) = xg$. The reader can easily check that each L_g and R_g is a diffeomorphism of G. Now as such they also operate on functions. Namely if $f \in C_c(G)$, the continuous functions on G with compact support, we write $L_g(f)(x) = f(gx)$ and $R_g(f)(x) = f(xg)$. Notice that $L_{gh} = L_g L_h$ and $R_{gh} = R_g R_h$ for each g and $h \in G$.

We say an integral over G is *left invariant* if

$$\int_G L_g(f) = \int_G f$$

for every $g \in G$ and $f \in C_c(G)$. Similarly, an integral over G is called *right invariant* if

$$\int_G R_g(f) = \int_G f$$

for every $g \in G$ and $f \in C_c(G)$. We now calculate such left or right invariant integrals with the help of the Change of Variables formula. To do this we first find the derivatives and Jacobians of the L_g's and R_g's. A direct calculation, which we leave to the reader, shows that if $g = (u, v)$, then $D(L_g) = uI$ and hence $|\det D(L_g)| = u^2$. Similarly $|\det D(R_g)| = |u|$. The first observation we make is that these Jacobians are independant of x (as well as v) and only depend on g. They are also nowhere zero. (This latter point is obvious in our case, but is true in general since L_g and R_g are diffeomorphisms).

Now consider $f \in C_c(G)$. Then $\frac{f}{|\det D(L_g)|}$ is also $\in C_c(G)$ and in fact for each g, as f varies over $C_c(G)$ so does $\frac{f}{|\det D(L_g)|}$. Similarly for each g, as f varies over $C_c(G)$ so does $\frac{f}{|\det D(R_g)|}$. Now let $\phi \in C_c(G)$ and L_g be a change of variables. The Change of Variables formula (5.8) tells us

$$\int_G \phi(gx)|\det D(L_g)|dudv = \int_G \phi(x)dudv.$$

Since this holds for all ϕ and we can take $\phi = \frac{f}{|\det D(L_g)|}$, we get

$$\int_G \frac{f(gx)}{|\det D(L_g)(gx)|}|\det D(L_g)|dudv = \int_G \frac{f(x)}{|\det D(L_g)(x)|}dudv.$$

After taking into account that $L_{gh} = L_g L_h$, the Chain Rule and the fact that $|\det(AB)| = |\det(A)||\det(B)|$ we get

$$\int_G f(gx)\frac{dudv}{|\det D(L_g)|} = \int_G f(x)\frac{dudv}{|\det D(L_g)|},$$

for every $f \in C_c(G)$. Thus

$$\frac{dudv}{|\det D(L_g)|} = \frac{dudv}{u^2}$$

is a left invariant integral for G. Similaly, $\frac{dudv}{|u|}$ is a right invariant integral for G. Notice that here left invariant and right invariant integrals are not the same.

EXERCISES

1. Evaluate the integral

$$\int\int_\Omega \sqrt{x+y}dxdy,$$

where Ω is the parallelogram bounded by the lines $x + y = 0$, $x + y = 1$, $2x - 3y = 0$ and $2x - 3y = 4$.

2. Let $n > 0$. Compute the integral

$$\int\int_\Omega (x + y)^n dx dy,$$

where $\Omega = \{(x, y) : x \geq 0, \ y \geq 0, \ x + y \leq 1\}$.

3. Evaluate the integral

$$\int_0^1 \int_0^x \sqrt{x^2 + y^2} dy dx,$$

using the transformation $x = u, \ y = uv$.

4. Evaluate the integral

$$\int\int_\Omega \frac{3y}{\sqrt{1 + (x + y)^3}} dx dy,$$

where $\Omega = \{(x, y) : x + y < a, x > 0, y > 0\}$.
Hint. Set $u = x + y, \ v = x - y$.

5. Evaluate

$$\int\int_\Omega \sin(\frac{x - y}{x + y}) dx dy,$$

where $\Omega = \{(x, y) : x \geq 0, \ y \geq 0, \ x + y \leq 1\}$.

6. Evaluate

$$\int\int_\Omega \sin x \sin y \sin(x + y) dx dy,$$

where $\Omega = \{(x, y) : x \geq 0, \ y \geq 0, \ x + y \leq \frac{\pi}{2}\}$.

7. Evaluate

$$\int\int_\Omega \sqrt{x^2 + y^2} dx dy,$$

where Ω is the region by the circles $x^2 + y^2 = 4, \ x^2 + y^2 = 9$.

8. Evaluate

$$\int\int_\Omega \tan^{-1}\left(\frac{y}{x}\right) dx dy,$$

where $\Omega = \left\{(x, y) : x^2 + y^2 \geq 1, \ x^2 + y^2 \leq 9, \ y \geq \frac{x}{\sqrt{3}}, y \leq x\sqrt{3}\right\}$.
Hint. Change to polar coordinates.

9. Evaluate
$$\int\int_{\Omega}(x-y)^4 e^{x+y}dxdy,$$
where Ω is the square with vertices $(1,0)$, $(2,1)$, $(1,2)$, and $(0,1)$.

10. Evaluate
$$\int_0^1\int_0^{1-x} y\log(1-x-y)dydx,$$
using the transformation $x=u-uv$, $y=uv$.

11. Show that if $\Omega=\{(x,y):y\geq 0,\ x^2+y^2\leq 1\}$,
$$\int\int_{\Omega}\frac{(x+y)^2}{\sqrt{1+x^2+y^2}}dxdy=\frac{2-\sqrt{2}}{3}\pi.$$

12. Find the volume of the region in $\mathbb{R}^3$ which is above the xy-plane, under the paraboloid $z=x^2+y^2$, and inside the elliptic cylinder $\frac{x^2}{9}+\frac{y^2}{4}=1$. *Hint.* Use elliptical coordinates $x=3r\cos\theta$, $y=2r\sin\theta$.

13. Use the transformation in $\mathbb{R}^2$ given by $x=r\cos^3\theta$, $y=r\sin^3\theta$ to prove that the volume of the set $\Omega=\{(x,y,z):x^{\frac{2}{3}}+y^{\frac{2}{3}}+z^{\frac{2}{3}}\leq 1\}$ is $\frac{4\pi}{35}$.

14. Let Ω be the region in the octant with $x,y,z\geq 0$, which is bounded by the plane $x=y+z=1$. Use the change of variables $x=u(1-v)$, $y=uv(1-w)$ $z=uvw$ to compute the integral
$$\int\int\int_{\Omega}\frac{1}{y+z}dxdydz.$$

15. Evaluate the integral
$$\int\int\int_{\Omega}\frac{z}{1+x^2+y^2}dxdydz,$$
where $\Omega=\{(x,y,z):1\leq x^2+y^2\leq 3, x\geq 0, x\leq y, 1\leq z\leq 5\}$. *Hint.* Use cylindrical coordinates.

16. Integrate the function $f(x, y, z) = z^4$ over the ball in $\mathbb{R}^3$ centered at the origin with radius $a > 0$.

17. Evaluate

$$\int\int\int_\Omega \frac{1}{\sqrt{x^2 + y^2 + (z-2)^2}} dxdydz,$$

where Ω is the unit ball $x^2 + y^2 + z^2 \le 1$.

18. Evaluate

$$\int\int\int_\Omega \frac{yz}{1+x} dxdydz,$$

where Ω is the portion of the closed unit ball in $\mathbb{R}^3$ which lies in the positive octant $x, y, z \ge 0$.

19. Evaluate

$$\int\int\int_\Omega \frac{1}{(x^2 + y^2 + z^2)^{\frac{3}{2}}} dxdydz,$$

where Ω is the solid bounded by the two spheres $x^2 + y^2 + z^2 = a^2$ and $x^2 + y^2 + z^2 = b^2$, where $0 < b < a$.

Answers to selected Exercises

1. $\frac{8}{15}$.   2. $\frac{1}{6(n+4)}$.   3. $\frac{1}{6}[\sqrt{2} + \log(1 + \sqrt{2})])$.

5. 0.   6. $\frac{\pi}{16}$.   7. $\frac{38\pi}{3}$.   8. $\frac{\pi^2}{6}$.   9. $\frac{1}{5}(e^3 - e)$.   10. $-\frac{11}{36}$.

14. $\frac{1}{2}$.   15. $\frac{3}{2}\pi \log 2$.   16. $\frac{4\pi a^7}{35}$.   17. $\frac{2\pi}{3}$.   18. $\frac{19 - 24\log 2}{36}$.

19. $4\pi \log(\frac{a}{b})$.

5.2   Improper multiple integrals

The situation we encounter here is as follows: we are given a function f defined on a set $\Omega \subseteq \mathbb{R}^n$ where f may not be integrable on Ω according to Definition 4.1.37, either because Ω is *unbounded* or because f is *unbounded* on Ω. Integrals over unbounded regions or integrals of unbounded functions are refered to as *improper integrals*. It is useful and often necessary to be able to integrate over unbounded regions in $\mathbb{R}^n$ or to integrate unbounded functions. As might be expected, by analogy with improper integrals of functions of a single variable, this involves a process of taking a limit of a (proper) multiple integral. However the process of defining improper integrals in dimension $n > 1$ is trickier than in dimension $n = 1$, (this is due to the *great variety* of ways in which a limit can be formed in $\mathbb{R}^n$). In this connection we have the following.

Definition 5.2.1. Let Ω be a set in $\mathbb{R}^n$ (possibly unbounded). An *exhaustion* of Ω is a sequence of bounded simple sets $\{\Omega_k\}$ such that $\Omega_1 \subset \Omega_2 \subset \Omega_3 \subset ... \subset \Omega$ with $\Omega = \bigcup_{k=1}^\infty \Omega_k$.

Lemma 5.2.2. *Let $\Omega \subset \mathbb{R}^n$ be a bounded simple set and let ν be n-dimensional volume. If $\{\Omega_k\}$ is an exhaustion of Ω, then*

$$\lim_{k \to \infty} \nu(\Omega_k) = \nu(\Omega).$$

Proof. Clearly $\Omega_k \subset \Omega_{k+1} \subset \Omega$ implies that $\nu(\Omega_k) \subset \nu(\Omega_{k+1}) \subset \nu(\Omega)$ and $\lim_{k\to\infty} \nu(\Omega_k) \le \nu(\Omega)$. To get equality, let $\epsilon > 0$. Note that since $\nu(\partial(\Omega)) = 0$, we can cover $\partial(\Omega)$ by a finite number of open rectangles $R_1, ..., R_N$ of total volume less than ϵ. Let $E = \bigcup_{j=1}^N R_j$. Then the set $\Omega \cup E$ is open in $\mathbb{R}^n$ and by construction it contains $\overline{\Omega}$. At the same time, $\nu(\Omega \cup E) \le \nu(\Omega) + \nu(E) < \nu(\Omega) + \epsilon$.

For each $k = 1, 2, ...$, applying this construction to each set Ω_k of the exhaustion $\{\Omega_k\}$ with $\epsilon_k = \frac{\epsilon}{2^k}$, we obtain a sequence of open sets $\{\Omega_k \cup E_k\}$ such that $\Omega = \bigcup_{k=1}^\infty \Omega_k \subset \bigcup_{k=1}^\infty (\Omega_k \cup E_k)$, and $\nu(\Omega_k \cup E_k) \le \nu(\Omega_k) + \nu(E_k) < \nu(\Omega_k) + \epsilon_k$.

Now the collection of open sets $\{E, \Omega_k \cup E_k : k = 1, 2, ...\}$ is an open cover of the compact set $\overline{\Omega}$. So that there is a finite number of these open sets, say, $E, (\Omega_1 \cup E_1), ..., (\Omega_m \cup E_m)$ covering $\overline{\Omega}$. Since $\Omega_1 \subset \Omega_2 \subset \cdots \subset \Omega_m$, the sets $E, E_1, E_2, ..., E_m, \Omega_m$ also form an open cover for $\overline{\Omega}$. Hence,

$$\nu(\Omega) \le \nu(\overline{\Omega}) \le \nu(\Omega_m) + \nu(E) + \sum_{i=1}^m \nu(E_i) < \nu(\Omega_m) + 2\epsilon.$$

Therefore, $\nu(\Omega) \le \lim_{k\to\infty} \nu(\Omega_k)$. $\square$

We begin by defining the improper integral of a *non-negative* function. Let Ω be a (possibly unbounded) set in $\mathbb{R}^n$ and $\{\Omega_k\}$ an exhaustion of Ω. Suppose $f : \Omega \to \mathbb{R}$ is a non-negative (possibly unbounded) function such that f is integrable over the sets $\Omega_k \in \{\Omega_k\}$. Then the integrals $s_k = \int_{\Omega_k} f$ exist for all $k \in \mathbb{N}$ and by the monotonicity of the integral they form an increasing sequence of real numbers $\{s_k\}$. If this sequence is bounded by Theorem 1.1.20 we get a finite limit

$$\lim_{k \to \infty} s_k = \lim_{k \to \infty} \int_{\Omega_k} f < \infty,$$

otherwise it diverges to ∞. Hence the following definition is a natural consequence.

Definition 5.2.3. Let $\{\Omega_k\}$ be an exhaustion of the set $\Omega \subseteq \mathbb{R}^n$ and suppose $f : \Omega \to [0, \infty)$ is integrable over each Ω_k. If the limit

$$\int_\Omega f = \lim_{k \to \infty} \int_{\Omega_k} f \tag{5.15}$$

exists and has a value independent of the choice of the sets in the exhaustion of Ω, this limt is called the *improper integral of f over Ω*. When this limit is *finite*, we say that the integral *converges* and f is *integrable* over Ω. If there is no common limit for all exhaustions of Ω or its value is $+\infty$, we say the integral *diverges* and f is *not integrable* over Ω.

Definition 5.2.3 extends the concept of multiple integral to the case of an unbounded region of integration or an unbounded integrand. This calls for the following remark.

Remark 5.2.4. If Ω is a bounded simple set in $\mathbb{R}^n$ and f is integrable over Ω, then the integral over Ω in the sense of Definition 5.2.3 converges and has the same value as the (proper) integral $\int_\Omega f$ of Definition 4.1.37.

Indeed, that $f|_{\Omega_k}$ is integrable over Ω_k follows from the Lebesgue criterion (as in Theorem 4.2.1 (6)). Since f is integrable over Ω, it is bounded. Hence $|f(x)| \leq M$ for all $x \in \Omega$, for some $M > 0$. From the additivity of the integral we have

$$\left| \int_\Omega f - \int_{\Omega_k} f \right| = \left| \int_{\Omega \setminus \Omega_k} f \right| \leq M \nu(\Omega \setminus \Omega_k). \tag{5.16}$$

Now, since $\nu(\Omega \backslash \Omega_k) = \nu(\Omega) - \nu(\Omega_k)$, letting $k \to \infty$ in (5.16) and using Lemma 5.2.2 we get

$$\int_\Omega f = \lim_{k \to \infty} \int_{\Omega_k} f,$$

and the two definitions agree.

The verification that an improper integral converges is simplified by the following proposition. We prove that for a non-negative function the existence of the limit (5.15) is independent of the choice of the exhaustion $\{\Omega_k\}$ of Ω.

Proposition 5.2.5. *Let Ω be a set in $\mathbb{R}^n$ and $f : \Omega \to [0, \infty)$. Let $\{\Omega_k\}$ and $\{\Omega'_m\}$ be exhaustions of Ω. Suppose f is integrable over each Ω_k and each Ω'_m. Then*

$$\lim_{k \to \infty} \int_{\Omega_k} f = \lim_{m \to \infty} \int_{\Omega'_m} f,$$

where the limit may be finite or $+\infty$.

Proof. Let $s = \lim_{k \to \infty} \int_{\Omega_k} f$ be finite. For each fixed $m = 1, 2, ...$, the sets $G_{km} = \Omega_k \cap \Omega'_m$, $k = 1, 2, ...$ form an exhaustion of the set Ω'_m. Since each Ω'_m is a bounded simple set and f is integrable over Ω'_m it follows from Remark 5.2.4 that

$$\int_{\Omega'_m} f = \lim_{k \to \infty} \int_{G_{km}} f \leq \lim_{k \to \infty} \int_{\Omega_k} f = s.$$

Since $f \geq 0$ and $\Omega'_m \subset \Omega'_{m+1} \subset \Omega$, it follows that $\lim_{m \to \infty} \int_{\Omega'_m} f$ exists and

$$\lim_{m \to \infty} \int_{\Omega'_m} f = s' \leq s.$$

Reversing the roles of the exhaustions shows that $s \leq s'$ also. Thus $s = s'$. If $s = +\infty$, then arguing as in the above paragraph, we have $s \leq s'$ and so $s' = +\infty$ also. $\qquad \square$

Example 5.2.6.   1. If $\Omega = \mathbb{R}^n$ and $f : \mathbb{R}^n \to [0, \infty)$ is continuous (or essentially continuous), we can take for the exhaustion to be

$$\Omega_k = B_k = \{x \in \mathbb{R}^n : ||x|| \leq k\},$$

the ball of radius k centered at the origin. Another choice can be

$$\Omega_k = C_k = [-k, k] \times [-k, k] \times \cdots \times [-k, k],$$

the cube of side length $2k$ centered at the origin.

2. If $\Omega = B^n$ the unit ball in $\mathbb{R}^n$ and $f : B^n \to [0, \infty)$ is continuous (or essentially continuous) but $f(x) \to \infty$ as $x \to 0$, we can take Ω_k to be the spherical shells

$$\Omega_k = \left\{ x \in B^n : \frac{1}{k} \leq ||x|| \leq 1 \right\}.$$

Note here that the union of the Ω_k's is $B^n \setminus \{0\}$, but this is immaterial, since, as we know, omission of a single point or even a set of volume zero from a region of integration does not effect the value of an integral.

Our first example of improper double integrals is a classic calculation that leads to the following important integral.

Example 5.2.7. (*Euler-Poisson integral*).

$$\int_{-\infty}^{\infty} e^{-x^2} dx = \sqrt{\pi}. \tag{5.17}$$

Solution. Exhausting the plane $\mathbb{R}^2$ by a sequence of disks

$$B_k = \left\{ (x, y) \in \mathbb{R}^2 : x^2 + y^2 < k \right\}$$

and recalling Example 5.1.22 we have

$$\int\int_{\mathbb{R}^2} e^{-(x^2+y^2)} dx dy = \pi.$$

If we now consider the exhaustion of the plane by the squares $C_k = [-k, k] \times [-k, k]$, then by Fubini's theorem and the fact that $e^{a+b} = e^a e^b$ we have

$$\int\int_{C_k} e^{-(x^2+y^2)} dx dy = \int_{-k}^{k} \left(e^{-x^2} \int_{-k}^{k} e^{-y^2} dy \right) dx$$

$$= \left(\int_{-k}^{k} e^{-y^2} dy \right) \left(\int_{-k}^{k} e^{-x^2} dx \right) = \left(\int_{-k}^{k} e^{-x^2} dx \right)^2.$$

Letting $k \to \infty$, Proposition 5.2.5 tells us that

$$\left(\int_{-\infty}^{\infty} e^{-x^2} dx \right)^2 = \pi.$$

Since any exponential is positive, taking the positive square root of both sides yields the desired integral.

In particular,

$$\int \cdots \int_{\mathbb{R}^n} e^{-(x_1^2 + \ldots + x_n^2)} dx_1 \cdots dx_n = \int_{\mathbb{R}^n} e^{-||x||^2} dx = (\pi)^{\frac{n}{2}}.$$

The Euler-Poisson integral is inaccessible by one variable calculus (the antiderivative of e^{-x^2} is not an elementary function). The function e^{-x^2} is known as the *Gaussian*[7] function and comes up in many contexts. Its graph is the "bell-shaped curve". Rescaling it by the factor $\frac{1}{\sqrt{\pi}}$, so that the total area under its graph is 1, gives the *normal distribution* in probability and statistics.

Corollary 5.2.8. *(Comparison test). Let f and g be functions defined on the set $\Omega \subseteq \mathbb{R}^n$ and integrable over exactly the same bounded simple subsets of Ω. Suppose $0 \leq f(x) \leq g(x)$ for all $x \in \Omega$. If the improper integral $\int_\Omega g$ converges, then the integral $\int_\Omega f$ also converges.*

Proof. Let $\{\Omega_k\}$ be an exhaustion of Ω on whose elements both f and g are integrable. By the monotonicity of the integral (Theorem 4.2.1 (2)) we have $\int_{\Omega_k} f \leq \int_{\Omega_k} g$ for all $k = 1, 2, \ldots$. Since the functions are non-negative, letting $k \to \infty$

$$\int_\Omega f \leq \int_\Omega g < \infty.$$

$\square$

---

[7]C. Gauss (1777-1855). One of the great mathematicians with numerous pioneering contributions in several fields in mathematics, physics and astronomy.

The basic properties of multiple integrals stated in Theorem 4.2.1 are readily extended to improper integrals of non-negative functions using Definition 5.2.3. Thus far we worked with non-negative functions. We shall now define the improper integral for a function $f : \Omega \subseteq \mathbb{R}^n \to (-\infty, \infty)$. The essential point is that the precceding theory can be applied to $|f|$, so that it makes sense to say $\int_\Omega |f|$ converges. Before we go on we need to define the so-called positive and negative parts of f.

Definition 5.2.9. (The functions f^+, f^-).
Let $f : \Omega \subseteq \mathbb{R}^n \to \mathbb{R}$ be a function. For $x \in \Omega$ we define the *positive part* f^+ of f by

$$f^+(x) = \max\{f(x), 0\} = \frac{|f(x)| + f(x)}{2}$$

and the *negative part* f^- of f by

$$f^-(x) = \max\{-f(x), 0\} = \frac{|f(x)| - f(x)}{2}.$$

Lemma 5.2.10. *Let* $f : \Omega \subseteq \mathbb{R}^n \to \mathbb{R}$. *Then* $f^+ \geq 0$, $f^- \geq 0$, $f^- = (-f)^+$ *and*

$$f = f^+ - f^-.$$

Furthermore, $|f| = \max\{f, -f\}$ *and*

$$|f| = f^+ + f^-.$$

The proof of the lemma is left to the reader as an exercise.

Definition 5.2.11. Let Ω be a set in $\mathbb{R}^n$ and $f : \Omega \to \mathbb{R}$. We say f is *absolutely integrable* if the integral $\int_\Omega |f|$ converges.

Proposition 5.2.12. *Let* $f : \Omega \subseteq \mathbb{R}^n \to \mathbb{R}$. *If* $|f|$ *is integrable over* Ω, *then* f *is integrable.*

Proof. Since $0 \leq f^+ \leq |f|$ and $0 \leq f^- \leq |f|$, the comparison test tells us that both $\int_\Omega f^+$ and $\int_\Omega f^-$ converge. Thus $\int_\Omega f = \int_\Omega f^+ - \int_\Omega f^-$ also converges and f is integrable. $\square$

Next we prove the change of variable formula for improper integrals. We begin with a preliminary result, which gives a useful exhaustion of any open set in $\mathbb{R}^n$ in terms of compact simple sets.

Proposition 5.2.13. *Let* Ω *be an open set in* $\mathbb{R}^n$. *Then there exists a sequence* $\{\Omega_k : k = 1, 2, ...\}$ *of compact simple subsets of* Ω *such that* $\Omega_k \subset (\Omega_{k+1})^\circ$ *and* $\Omega = \bigcup_{k=1}^\infty \Omega_k$.

Proof. For $x, y \in \mathbb{R}^n$ let $d(x, y) = ||x - y||_\infty$ be the box-norm. First note that since $\Omega^c = \mathbb{R}^n \backslash \Omega$ is closed, $d(x, \Omega^c) > 0$ if $x \in \Omega$. Let

$$S_k = \left\{ x \in \Omega : ||x||_\infty \leq k \text{ and } d(x, \Omega^c) \geq \frac{1}{k} \right\}.$$

Clearly S_k is bounded (it is contained in a cube centered at the origin with side lenght $2k$). At the same time, since $||\cdot||$ and $d(\cdot, \Omega^c)$ are continuous (see, Exercise 2.7.8), it follows that S_k is closed as an intersection of two closed sets (Section 1.4, Exercise 2). Therefore S_k is compact. Moreover, S_k, cannot meet Ω^c, because $d(x, \Omega^c) \geq \frac{1}{k}$.

The interior of S_k is $(S_k)^\circ = \left\{ x \in \Omega : ||x||_\infty < k \text{ and } d(x, \Omega^c) > \frac{1}{k} \right\}$ and $S_k \subset (S_{k+1})^\circ$, for if $x \in S_k$ then $||x||_\infty \leq k < k + 1$ and $d(x, \Omega^c) \geq \frac{1}{k} > \frac{1}{k+1}$.

Clearly $\bigcup_{k=1}^\infty S_k \subseteq \Omega$. To get equality, let $x \in \Omega$. Since Ω is open $d(x, \Omega^c) > 0$. Hence there is a k such that $d(x, \Omega^c) \geq \frac{1}{k}$ and $||x||_\infty \leq k$. Thus,

$$\Omega = \bigcup_{k=1}^\infty S_k.$$

The sets S_k so constructed have the required properties stated in the proposition, except they might not be simple. To get the required exhaustion of Ω into compact simple sets, we proceed as follows: for each $x \in S_k$, we choose a closed cube C_{k_x} centered at x such that $C_{k_x} \subset (S_{k+1})^\circ$. Then $S_k \subset \bigcup_{x \in S_k} (C_{k_x})^\circ$. Since S_k is compact we can choose a finite number $C_{k_1}, ... C_{k_m}$ of these cubes whose interiors cover S_k.

Set

$$\Omega_k = \bigcup_{j=1}^m C_{k_j}.$$

Then Ω_k as the union of a finite number of closed cubes is compact and simple. Furthermore $S_k \subset (\Omega_k)^\circ \subset \Omega_k \subset (S_{k+1})^\circ$ for each k. Thus $\Omega = \bigcup_{k=1}^\infty \Omega_k$, $\Omega_k \subset (\Omega_{k+1})^\circ$ and the Ω_k are simple. $\square$

Theorem 5.2.14. *(Change of Variables formula for improper integrals). Let $\varphi : \Omega \to \mathbb{R}^n$ a diffeomorphism of the open $\Omega \subseteq \mathbb{R}^n$, and let the function $f : \varphi(\Omega) \to \mathbb{R}$ be integrable over all simple compact subsets of Ω. If the improper integral $\int_{\varphi(\Omega)} f(y)dy$ converges, then $\int_\Omega f(\varphi(x))|\det D_\varphi(x)|dx$ also converges and*

$$\int_{\varphi(\Omega)} f(y)dy = \int_\Omega f(\varphi(x))|\det D_\varphi(x)|dx. \qquad (5.18)$$

Proof. By Proposition 5.2.13, the open set Ω can be exhausted by a sequence of compact simple sets $\{\Omega_k\}$. Since $\varphi : \Omega \to \varphi(\Omega)$ is a diffeomorphism the sequence of compact simple sets $\{\varphi(\Omega_k)\}$ is an exhaustion of $\varphi(\Omega)$ (see Proposition 5.1.6). Assume first that $f \geq 0$. By the Change of Variables formula for proper integrals (Theorem 5.1.14) we have

$$\int_{\varphi(\Omega_k)} f(y)dy = \int_{\Omega_k} f(\varphi(x))|\det D_\varphi(x)|dx.$$

Letting $k \to \infty$, the left-hand side of this equality has a (finite) limit by hypothesis. Therefore the right-hand side also has the same limit. The general case now follows by writting $f = f^+ - f^-$. □

We close this section with several examples. In order to make effective use of the comparison test of Corollary 5.2.8 in studing the convergence of improper integrals, it is natural and useful to have a store of standard functions for comparison. As for improper integrals in one variable, the basic comparison functions $f(x) = ||x||^\alpha$ are powers of $||x|| = \sqrt{x_1^2 + ... + x_n^2}$, but the critical exponent $\alpha \in \mathbb{R}$ depends on the dimension n.

EXAMPLES

a. Improper multiple integrals over unbounded regions

Example 5.2.15. Let $\Omega = (1, \infty) \times (1, \infty)$ and $f(x, y) = \frac{1}{x^2 y^2}$. Does the improper integral $\int \int_\Omega f(x, y)dxdy$ converge?

Solution. The function is bounded (and nonnegative) on Ω, but Ω is unbounded. Using the exhaustion $\Omega_k = (1, k) \times (1, k)$ of Ω we see that f is integrable on each Ω_k and by Fubini's theorem, we have

$$\iint_\Omega f(x, y) dx dy = \lim_{k \to \infty} \iint_{\Omega_k} \frac{1}{x^2 y^2} dx dy$$

$$= \lim_{k \to \infty} \int_1^k \left(\frac{1}{x^2} \int_1^k \frac{1}{y^2} dy \right) dx = \lim_{k \to \infty} [1 - \frac{1}{k}]^2 = 1.$$

Example 5.2.16. Let $\Omega = \{(x, y, z) \in \mathbb{R}^3 : x^2 + y^2 + z^2 \geq 1\}$. For $\alpha \in \mathbb{R}$, the improper integral

$$\iiint_\Omega \frac{1}{(\sqrt{x^2 + y^2 + z^2})^\alpha} dx dy dz$$

converges for $\alpha > 3$ and diverges for $\alpha \leq 3$.

Solution. We construct an exhaustion of Ω by the annular regions $\Omega_k = \{(x, y, z) \in \mathbb{R}^3 : 1 < \sqrt{x^2 + y^2 + z^2} \leq k\}$. Passing to spherical coordinates, we get $\iiint_\Omega \frac{1}{(\sqrt{x^2+y^2+z^2})^\alpha} dx dy dz$

$$= \lim_{k \to \infty} \iiint_{\Omega_k} \frac{1}{(\sqrt{x^2 + y^2 + z^2})^\alpha} dx dy dz$$

$$= \lim_{k \to \infty} 2\pi \int_1^k \left(r^{-\alpha+2} \int_0^\pi \sin \phi d\phi \right) dr$$

$$= \lim_{k \to \infty} \frac{4\pi}{3 - \alpha} (k^{3-\alpha} - 1) = \begin{cases} \frac{4\pi}{\alpha-3} & \text{when } \alpha > 3, \\ \infty & \text{when } \alpha < 3. \end{cases}$$

For $\alpha = 3$, the integral becomes $\iiint_\Omega \frac{1}{(\sqrt{x^2+y^2+z^2})^3} dx dy dz$

$$= \lim_{k \to \infty} [4\pi \int_1^k r^{-1} dr] = 4\pi \lim_{k \to \infty} \log k = \infty.$$

Example 5.2.17. Let $\Omega = \{(x, y, z) \in \mathbb{R}^3 : x^2 + y^2 + z^2 \geq 1\}$. For $\alpha > \frac{5}{2}$ the improper integral

$$\iiint_\Omega \frac{(x^2 + y^2) \log(x^2 + y^2 + z^2)}{(x^2 + y^2 + z^2)^\alpha} dx dy dz$$

converges. In fact it value is $\frac{16\pi}{3(2\alpha-5)^2}$.

Solution. Taking the same exhaustion $\{\Omega_k\}$ of Ω as in the above example and passing to spherical coordinates, the integral becomes

$$2\pi \int_1^k \int_0^\pi \frac{r^2 \sin^2 \phi \log(r^2)}{r^{2\alpha}} r^2 \sin^2 \phi \, d\phi \, dr = \frac{16\pi}{3} \int_1^k r^{4-2\alpha} \log r \, dr.$$

Now integrating by parts and letting $k \to \infty$ yields $\frac{16\pi}{3(2\alpha-5)^2}$.

The following example generalizes Example 5.2.16 to n-dimensions.

Example 5.2.18. Let $\Omega = \{x : ||x|| > 1\}$ the outside of the unit ball B^n in $\mathbb{R}^n$. For $\alpha \in \mathbb{R}$, the integral

$$\int_\Omega ||x||^{-\alpha} dx$$

converges for $\alpha > n$ and diverges for $\alpha \le n$.

Solution. As in Example 5.2.16 we exhaust Ω by the sequence of spherical shells $\Omega_k = \{x \in \mathbb{R}^n : 1 < ||x|| \le k\}$. Passing to spherical coordinates in $\mathbb{R}^n$, we get

$$\int_{\Omega_k} ||x||^{-\alpha} dx = \int_1^k \left(r^{-\alpha} r^{n-1} \int_{S^{n-1}} d\Theta \right) dr = \sigma_n \int_1^k r^{n-\alpha-1} dr$$

$$= \frac{\sigma_n}{n-\alpha}(k^{n-\alpha} - 1),$$

where

$$\sigma_n = \int_{S^{n-1}} d\Theta$$

$$= \int_0^{2\pi} \int_0^\pi \cdots \int_0^\pi \sin^{n-2} \phi_1 \sin^{n-3} \phi_2 \cdots \sin \phi_{n-2} d\theta d\phi_1 \cdots d\phi_{n-2}.$$

Now let $k \to \infty$. Then for $n < \alpha$, we get

$$\int_\Omega ||x||^{-\alpha} dx = \frac{\sigma_n}{\alpha - n}$$

and for $\alpha \le n$,

$$\int_\Omega ||x||^{-\alpha} dx = \infty.$$

The calculation above suggests the following general result.

Theorem 5.2.19. *Let $f : \mathbb{R}^n \to [0, \infty)$ be an integrable function such that $f(x) = g(||x||)$ for some function g on $(0, \infty)$, and let (r, Θ) be the spherical coordinates in $\mathbb{R}^n$. Then*

$$\int_{\mathbb{R}^n} f(x)dx = \sigma_n \int_0^\infty g(r)r^{n-1}dr. \tag{5.19}$$

Proof. Exhausting $\mathbb{R}^n$ with a sequence of ball $\{x \in \mathbb{R}^n : ||x|| \leq k\}$ and passing to spherical coordinates, we have

$$\int_{\mathbb{R}^n} f(x)dx = \int_{\mathbb{R}^n - \{0\}} f(x)dx = \lim_{k \to \infty} \int_{||x|| \leq k} f(x)dx$$

$$= \lim_{k \to \infty} \int_0^k \left(g(r)r^{n-1} \int_{S^{n-1}} d\Theta \right) dr$$

$$= \lim_{k \to \infty} \left[\sigma_n \int_0^k g(r)r^{n-1}dr \right] = \sigma_n \int_0^\infty g(r)r^{n-1}dr.$$

$\square$

Example 5.2.20. For $s > 0$, the integral

$$\int_{\mathbb{R}^n} \frac{dx}{(1 + ||x||^2)^s}$$

converges if $s > \frac{n}{2}$.

Solution. Take $g(r) = \frac{1}{(1+r^2)^s}$ in Theorem 5.2.19. Then $\int_{\mathbb{R}^n} \frac{dx}{(1+||x||^2)^s}$

$$= \sigma_n \int_0^\infty \frac{1}{(1 + r^2)^s} r^{n-1}dr \leq \sigma_n \left[\int_0^1 \frac{r^{n-1}}{(1 + r^2)^s} dr + \int_1^\infty r^{n-1-2s}dr \right].$$

The first integral is a (proper) integral of a continuous function and so finite. The second improper (single) integral converges if $s > \frac{n}{2}$ since $\int_1^\infty \frac{1}{r^p}dr < \infty$ if $p > 1$.

Example 5.2.21. For $\alpha > 0$,

$$\int_{\mathbb{R}^n} e^{-\alpha||x||^2} dx = (\frac{\pi}{\alpha})^{\frac{n}{2}}. \tag{5.20}$$

Solution. Apply the linear transformation $x \mapsto \frac{x}{\sqrt[n]{\alpha}}$, Theorem 5.2.14 and $\int_{\mathbb{R}^n} e^{-||x||^2} dx = (\pi)^{\frac{n}{2}}$ from Example 5.2.7.

More generally we have

Example 5.2.22. (*Euler-Poisson integral*). Let A be a positive definite symmetric $n \times n$ matrix. Then

$$\int_{\mathbb{R}^n} e^{-\langle Ax, x \rangle} dx = \frac{\pi^{\frac{n}{2}}}{(\det A)^{\frac{1}{2}}}.$$

Solution. Let $Q(x) = \langle Ax, x \rangle$ be the associated quadratic form. By the Principal Axis theorem (Theorem 3.7.17) there exists an orthonormal basis with respect to which $Q(x) = \lambda_1 y_1^2 + ... + \lambda_n y_n^2$, with $\lambda_i > 0$. Then $\det A = \lambda_1 \cdots \lambda_n$. At the same time, as we know from the Spectral theorem, the linear mapping that changes variables $O(x) = y$ is orthogonal and so $|\det O| = 1$. Now, the Change of Variables formula and Fubini's theorem give

$$\int_{\mathbb{R}^n} e^{-\langle Ax, x \rangle} dx = \int_{\mathbb{R}^n} e^{-Q(x)} dx = \int_{\mathbb{R}^n} e^{-(\lambda_1 y_1^2 + \cdots + \lambda_n y_n^2)} dy$$

$$= \prod_{i=1}^{n} \int_{-\infty}^{\infty} e^{-\lambda_i y_i^2} dy_i = \prod_{i=1}^{n} \sqrt{\frac{\pi}{\lambda_i}} = \frac{\pi^{\frac{n}{2}}}{\sqrt{\det A}}.$$

b. Improper multiple integrals over bounded regions

Example 5.2.23. Let $\Omega = [0,1] \times [0,1]$ and $f(x,y) = \frac{1}{x^2 y^2}$. Does the improper integral $\int \int_\Omega f(x,y) dx dy$ converge?

Solution. Here Ω is bounded, but the function is unbounded on Ω.

In fact, $f(x,y) \to \infty$ as either $x \to 0$ or $y \to 0$. Using the exhaustion $\Omega_k = [\frac{1}{k}, 1] \times [\frac{1}{k}, 1]$ of Ω, we have

$$\int\int_{\Omega} \frac{1}{x^2 y^2} dxdy = \lim_{k\to\infty} \int_{\frac{1}{k}}^{1} \left(\frac{1}{x^2} \int_{\frac{1}{k}}^{1} \frac{1}{y^2} dy \right) dx = \lim_{k\to\infty} [1 - k]^2 = \infty.$$

Example 5.2.24. Does the integral

$$\int\int_{x^2+y^2<4} \log \sqrt{x^2 + y^2} dxdy$$

converge?

Solution. The integrand is not only unbounded but it also *changes sign* in the region of integration. So we will work with its absolute value. Using the exhaustion $\Omega_k = \{(x,y) \in \mathbb{R}^2 : \frac{1}{k} \le x^2 + y^2 < 4\}$ and passing to polar coordiantes, we have

$$\int\int_{\frac{1}{k}\le x^2+y^2<4} |\log \sqrt{x^2 + y^2}| dxdy = \int_{\frac{1}{k}}^{2} \left(r|\log r| \int_0^{2\pi} d\theta \right) dr$$

$$= 2\pi \int_{\frac{1}{k}}^{2} r|\log r| dr = 2\pi \left(\int_{\frac{1}{k}}^{1} r(-\log r) dr + \int_1^2 r \log r \, dr \right)$$

$$= -2\pi \left(\left[\frac{1}{2}r^2 \log r - \frac{1}{4}r^2 \right]_{\frac{1}{k}}^{1} - \left[\frac{1}{2}r^2 \log r - \frac{1}{4}r^2 \right]_1^2 \right).$$

Evaluating and letting $k \to \infty$ yields $\pi(\log 16 - 1)$. Therefore by Proposition 5.2.12 the integral converges.

Example 5.2.25. Determine the values of $\alpha \in \mathbb{R}$ for which the integral

$$\int\int_{x^2+y^2<1} \frac{1}{(x^2 + y^2)^{\frac{\alpha}{2}}} dxdy$$

converges.

Solution. When $\alpha > 0$ the function $f(x,y) = \frac{1}{(x^2+y^2)^{\frac{\alpha}{2}}}$ is clearly unbounded at the origin. Letting $\Omega_k = \left\{ (x,y) : \frac{1}{k} \leq x^2 + y^2 < 1 \right\}$ and changing to polar coordinates

$$\int\int_{\Omega_k} (x^2+y^2)^{-\frac{\alpha}{2}} dxdy = \int_{\frac{1}{k}}^1 \left(r^{-\alpha+1} \int_0^{2\pi} d\theta \right) dr = \frac{2\pi}{2-\alpha} \left[1 - k^{\alpha-2} \right],$$

if $\alpha \neq 2$, (if $\alpha = 2$, the integral is equal to $2\pi \log k \to \infty$ as $k \to \infty$).

Letting $k \to \infty$, it follows that the integral converges if and only if $\alpha < 2$ and its value in this case is $\frac{2\pi}{2-\alpha}$

Example 5.2.26. Determine the values of $\alpha \in \mathbb{R}$ for which the integral

$$\int\int_{x^2+y^2<1} \frac{1}{(1-x^2-y^2)^\alpha} dxdy$$

converges.

Solution. When $\alpha > 0$ the integrand is clearly unbounded in a neighborhood of the circle $x^2 + y^2 = 1$. Letting $\Omega_k = \left\{ (x,y) : x^2 + y^2 < 1 - \frac{1}{k} \right\}$ and changing to polar coordinates

$$\int\int_{\Omega_k} \frac{1}{(1-x^2-y^2)^\alpha} dxdy = \int_0^{1-\frac{1}{k}} \left(\frac{r}{(1-r^2)^\alpha} \int_0^{2\pi} d\theta \right) dr$$

$$= 2\pi \int_0^{1-\frac{1}{k}} \frac{r}{(1-r^2)^\alpha} dr = \frac{\pi}{1-\alpha} \left[1 - \left(\frac{2}{k} - \frac{1}{k^2} \right)^{1-\alpha} \right].$$

Letting $k \to \infty$ we see that if $\alpha < 1$ the integral is equal to $\frac{\pi}{1-\alpha}$, while if $\alpha \geq 1$ the integral diverges.

We now generalize Example 5.2.25 to $\mathbb{R}^n$.

Example 5.2.27. Let $\Omega = \{ x : ||x|| < 1 \}$ the interior of the unit ball B^n in $\mathbb{R}^n$. For $\alpha \in \mathbb{R}$, the integral

$$\int_\Omega ||x||^{-\alpha} dx$$

converges for $\alpha < n$ and diverges for $\alpha \geq n$.

Solution. The integral is improper when $\alpha > 0$, since the integrand $f(x) = ||x||^{-\alpha}$ is unbounded in Ω. Exhausting Ω by the sequence of spherical shells

$$\Omega_k = \left\{ x \in B^n : \frac{1}{k} \leq ||x|| \leq 1 \right\}$$

and as before passing to spherical coordinates, we get

$$\int_\Omega f = \lim_{k \to \infty} \int_{\Omega_k} f = \lim_{k \to \infty} \sigma_n \int_{\frac{1}{k}}^{1} r^{n-\alpha-1} dr$$

$$= \frac{\sigma_n}{n-\alpha} \lim_{k \to \infty} [1 - (\frac{1}{k})^{n-\alpha}] = \begin{cases} \frac{\sigma_n}{n-\alpha} & \text{when } \alpha < n, \\ \infty & \text{when } \alpha \geq n. \end{cases}$$

EXERCISES

1. Compute the following improper integrals (if they converge).

   (a)
   $$\int\int_{x,y>0} \frac{1}{(1+x^2+y^2)^2} dx dy.$$

   (b)
   $$\int\int_{x^2+y^2<1} \frac{x^2}{(x^2+y^2)^{\frac{3}{2}}} dx dy.$$

   (c)
   $$\int\int_{x^2+y^2\geq 1} \frac{\log(x^2+y^2)}{x^2+y^2} dx dy.$$

   (d)
   $$\int\int_{x^2+y<1} \frac{\log(x^2+y^2)}{\sqrt{x^2+y^2}} dx dy.$$

2. Compute the following improper integrals (if they converge).

(a)
$$\int\int_{\mathbb{R}^2} \frac{1}{(1+x^2+y^2)^{\frac{3}{2}}} dxdy.$$

(b)
$$\int\int\int_{\mathbb{R}^3} \frac{1}{(1+x^2+y^2+z^2)^{\frac{3}{2}}} dxdydz.$$

(c)
$$\int\int\int_{x,y,z>0} \frac{xy}{(1+x^2+y^2+z^2)^3} dxdydz.$$

(d)
$$\int\int\int_{x,y,z>0} \frac{1}{\sqrt{(1+x+y+z)^7}} dxdydz.$$

3. Evaluate
$$\int_0^1 \int_0^1 \frac{1}{\sqrt{|x-y|}} dxdy.$$

   Hint. Divide the region of integration into two pieces.

4. Evaluate
$$\int\int_{x>0} xe^{-(x^2+y^2)} dxdy.$$

5. Evaluate
$$\int\int_{x,y>0} ye^{-y^2(1+x^2)} dxdy.$$

6. Evaluate
$$\int\int_{\mathbb{R}^2} e^{-|x|-|y|} dxdy.$$

7. Prove that the integral
$$\int\int_{x^2+y^2\leq 1} \frac{\sin^2(x-y)}{\sqrt{1-x^2-y^2}} dxdy$$

   converges.
   Hint. Comparison test.

8. Prove that the integral

$$\int\int_{x^2+y^2\le1}\frac{e^{-x^2-y^2}}{x^2+y^2}dxdy$$

diverges.

9. Let $\Omega = \{(x,y) : 0 \le x < \infty,\ 0 \le y \le x\}$. Prove that the integral

$$\int\int_\Omega x^{-\frac{3}{2}}e^{y-x}dydx$$

converges.

10. Let Ω be a closed and bounded region in $\mathbb{R}^2$ containing the origin. For what values of p is the integral

$$\int\int_\Omega \frac{xy}{(x^2+y^2)^p}dxdy$$

absolutely convergent?

11. Let $f(x,y) = \frac{x^2-y^2}{(x^2+y^2)^2}$, and let $\Omega = [0,1] \times [0,1]$.

(a) Show that $\int\int_\Omega |f(x,y)|dxdy = \infty$

(b) Show by direct calculation that both iterated integrals exist but

$$\int_0^1\int_0^1 f(x,y)dxdy \ne \int_0^1\int_0^1 f(x,y)dydx.$$

12. Let $f(x)$ and $g(y)$ be non-negative functions on $\mathbb{R}$, and suppose that the improper integrals $\int_{-\infty}^{\infty} f(x)dx$ and $\int_{-\infty}^{\infty} g(y)dy$ are convergent. Show that the double integral $\int\int_{\mathbb{R}^2} f(x)g(y)dxdy$ is convergent and it is equal to the product

$$\left(\int_{-\infty}^{\infty} f(x)dx\right)\left(\int_{-\infty}^{\infty} g(y)dy\right).$$

Answers to selected Exercises

1. (a) $\frac{\pi}{4}$.   (b) π.   (c) -4π.   (d) diverges.

2. (a) 2π.   (b) diverges.   (c) $\frac{\pi}{16}$.   (d) $\frac{8}{15}$.   3. $\frac{8}{3}$.

4. $\frac{\sqrt{\pi}}{4}$,   5. $\frac{\pi}{4}$.   6. 4.

5.3   Functions defined by integrals

Let $X \subset \mathbb{R}^n$ and $Y \subset \mathbb{R}^k$ be closed rectangles. As in Section 4.3, we consider $X \times Y$ as a subset of $\mathbb{R}^{n+k} = \mathbb{R}^n \times \mathbb{R}^k$, and write the elements of $\mathbb{R}^{n+k}$ as (x, y), with $x \in \mathbb{R}^n$ and $y \in \mathbb{R}^k$. If the function $f : X \times Y \to \mathbb{R}$ is integrable over Y as a function of y for each fixed $x \in X$, we consider the function[8] F on X defined by

$$F(x) = \int_Y f(x, y) dy. \qquad (5.21)$$

A natural question which arises here is: *what conditions on f will ensure the properties of integrability, continuity or differentiability for F?* Actually the integrability property has already been addressed in Fubini's theorem (Theorem 4.3.1). Indeed, Fubini's theorem shows that, if f is integrable over $X \times Y$, then F is integrable over X and

$$\int_X F(x) dx = \int_{X \times Y} f(x, y) = \int_X \left(\int_Y f(x, y) dy \right) dx.$$

The following theorem deals with the continuity of F.

Theorem 5.3.1. *Let $X \subset \mathbb{R}^n$ and $Y \subset \mathbb{R}^k$ be closed rectangles. If the function $f(x, y)$ is continuous on $X \times Y$, then the function $F : X \to \mathbb{R}$ defined by (5.21) is continuous on X.*

Proof. The continuity of $f(x, y)$ implies the continuity of $f(x, .)$, and hence its integrability over Y. Let $\nu(Y) < \infty$ be the k-dimensional volume of Y. By Proposition 1.6.12, $X \times Y$ is compact, and so by Theorem 2.5.9, f is uniformly continuous on $X \times Y$. Therefore, given $\epsilon > 0$, there is a $\delta > 0$ such that $|f(x, y) - f(x', y)| < \frac{\epsilon}{\nu(Y)}$ whenever $y \in Y$, $x, x' \in X$, and $||x - x'|| < \delta$. Now, for $x, x' \in X$, with $||x - x'|| < \delta$, we have

$$|F(x) - F(x')| \leq \int_Y |f(x, y) - f(x', y)| dy < \frac{\epsilon}{\nu(Y)} \int_Y dy = \epsilon.$$

Thus F is uniformly continuous on X. $\qquad \square$

---

[8]Functions defined by integrals are called *integrals depending on a parameter*, where we view x as a parameter t belonging to a subset of $\mathbb{R}$ or $\mathbb{R}^n$.

Next we address the differentiability of F.

Theorem 5.3.2. *(Differentiation under the integral sign). Let $X \subset \mathbb{R}^n$ and $Y \subset \mathbb{R}^k$ be closed rectangles. If $f(x, y)$ is continuous and has continuous partial derivatives with respect to x, then the function $F(x) = \int_Y f(x, y) dy$ has continuous partial derivatives in X, and*

$$\frac{\partial F}{\partial x_i}(x) = \int_Y \frac{\partial f}{\partial x_i}(x, y) dy \qquad (i = 1, ..., n). \tag{5.22}$$

Proof. First note that for computing the $\frac{\partial F}{\partial x_i}(x)$, the other variables $x_j \neq x_i$ play no role, so we may assume $n = 1$ and take X to be an interval $[a, b]$ in $\mathbb{R}$ (in fact, this is the most important case). Formula (5.22) becomes

$$F'(x) = \frac{d}{dx} \int_Y f(x, y) dy = \int_Y \frac{\partial f}{\partial x}(x, y) dy. \tag{5.23}$$

Let $x \in [a, b]$. We shall prove

$$\lim_{h \to 0} \left| \frac{F(x + h) - F(x)}{h} - \int_Y \frac{\partial f}{\partial x}(x, y) dy \right| = 0.$$

By the Mean Value theorem, we have $f(x + h, y) - f(x, y) = h \frac{\partial f}{\partial x}(\xi, y)$, for some ξ strictly between x and $x + h$. Hence

$$\left| \frac{F(x + h) - F(x)}{h} - \int_Y \frac{\partial f}{\partial x}(x, y) dy \right|$$

$$= \left| \int_Y \frac{f(x + h) - f(x)}{h} - \int_Y \frac{\partial f}{\partial x}(x, y) dy \right|$$

$$= \left| \int_Y \left(\frac{\partial f}{\partial x}(\xi, y) - \frac{\partial f}{\partial x}(x, y) \right) dy \right| \leq \int_Y \left| \frac{\partial f}{\partial x}(\xi, y) - \frac{\partial f}{\partial x}(x, y) \right| dx.$$

Since $\frac{\partial f}{\partial x}$ is continuous on $[a, b] \times Y$, it is uniformly continuous there. Therefore, for any $\epsilon > 0$, there is $\delta > 0$ such that $\left| \frac{\partial f}{\partial x}(\xi, y) - \frac{\partial f}{\partial x}(x, y) \right| < \epsilon$ if $|x - \xi| < \delta$. Thus for $|h| < \delta$, we have $\int_Y \left| \frac{\partial f}{\partial x}(\xi, y) - \frac{\partial f}{\partial x}(x, y) \right| dx \leq \epsilon \nu(Y)$. $\qquad \square$

Remark 5.3.3. Frequently $F(x)$ is defined on some *open* set Ω rather a closed rectangle X in $\mathbb{R}^n$. However, formula (5.22) is still valid in this case. To see this, we may take Ω to be an open (finite or infinite) interval in $\mathbb{R}$ and modify the proof of (5.23) as follows: for any $x_0 \in \Omega$, choose an $r > 0$ such that $B_r(x_0) \subset \Omega$. Then for $x \in B_r(x_0)$, replacing the set $X \times Y$ by $\overline{B_r(x_0)} \times Y$, which is also compact, the proof of (5.23) is identical with the one given above. Finally, since x_0 is an arbitrary point in Ω, (5.23) is valid for all $x \in \Omega$.

Corollary 5.3.4. *Suppose $f : [a, b] \times [c, d] \to \mathbb{R}$ is continuous and has continous partial derivative $\frac{\partial f}{\partial x}$ there. Suppose also $\phi(x)$ and $\psi(x)$ are continuously differentiable functions on $[a, b]$ with values in $[c, d]$. Then the function*

$$F(x) = \int_{\phi(x)}^{\psi(x)} f(x, y) dy \tag{5.24}$$

is continuously differentiable on $[a, b]$ and

$$F'(x) = f(x, \psi(x)) \cdot \psi'(x) - f(x, \phi(x)) \cdot \phi'(x) + \int_{\phi(x)}^{\psi(x)} \frac{\partial f}{\partial x}(x, y) dy. \tag{5.25}$$

Proof. For $u, v \in [c, d]$ and $x \in [a, b]$ consider the function

$$\Phi(u, v, x) = \int_u^v f(x, y) dy.$$

By the Fundamental Theorem of Calculus and (5.23), its partial derivatives are

$$\tfrac{\partial \Phi}{\partial u} = f(x, u), \quad \tfrac{\partial \Phi}{\partial v} = -f(x, v), \text{ and } \tfrac{\partial \Phi}{\partial x} = \int_u^v \tfrac{\partial f}{\partial x}(x, y) dy.$$

The hypothesis and Theorem 5.3.1 tell us that these partials are continuous and so Φ is continuously differentiable. Setting $u = \phi(x)$ and $v = \psi(x)$ function (5.24) is the composition $F(x) = \Phi(\phi(x), \psi(x), x)$ and the Chain Rule yields (5.25). □

Example 5.3.5. Let $F(x) = \int_0^1 \log(x^2 + y^2) dy$. Find $F'(x)$.

Solution. Applying Theorem 5.3.2 with $Y = [0, 1]$ and x in any closed interval not containg 0, we have $F'(x) = \int_0^1 \frac{2x}{x^2 + y^2} dy$. Integrating, we find $F'(x) = 2 \tan^{-1}(\frac{1}{x})$.

Formula (5.23) sometimes makes it possible to compute the integral.

Example 5.3.6. Let $F(x) = \int_0^{\frac{\pi}{2}} \log(x^2 - \sin^2 \theta)d\theta$, $x > 1$. Find $F(x)$.

Solution. Differentiating under the integral sign

$$F'(x) = \int_0^{\frac{\pi}{2}} \frac{x}{x^2 - \sin^2 \theta} d\theta = \frac{\pi}{\sqrt{x^2 - 1}}.$$

Therefore $F(x) = \pi \log(x + \sqrt{x^2 - 1}) + c$. To find the constant c, note that

$$F(x) - \pi \log x \to \pi \log 2 + c$$

as $x \to \infty$. On the other hand, since $\log(\frac{x^2 - \sin^2 \theta}{x^2}) \to 0$ as $x \to \infty$ uniformly in θ, from the definition of $F(x)$, we have

$$F(x) - \pi \log x = \int_0^{\frac{\pi}{2}} \log(\frac{x^2 - \sin^2 \theta}{x^2})d\theta \to 0, \quad (x \to \infty).$$

So $c = -\pi \log 2$. Hence $F(x) = \pi \log(\frac{x + \sqrt{x^2 - 1}}{2})$.

Allowing *complex-valued* functions in *two real variables* and using the differentiation under the integral sign, one can obtain the following simple proof of the *Fundamental Theorem of Algebra*.[9]

Lemma 5.3.7. *Let* $p(z) = z^n + a_{n-1}z^{n-1} + ... + a_1 z + a_0$ *be a polynomial with complex coefficients of degree* $n \geq 1$. *Then* $\lim_{|z| \to \infty} |p(z)| = \infty$.

Proof. Note $|p(z)| = |z|^n \left|1 + \frac{a_{n-1}}{z} + ... + \frac{a_0}{z^n}\right|$ and $1 + \frac{a_{n-1}}{z} + ... + \frac{a_0}{z^n}$ is bounded as $|z| \to \infty$. Hence $\lim_{|z| \to \infty} |p(z)| = \infty$.   □

Theorem 5.3.8. *(Fundamental Theorem of Algebra)(*)*. *Every polynomial of degree* $n \geq 1$ *with complex coefficients has a zero in* $\mathbb{C}$.

Proof. Let $p(z) = z^n + a_{n-1}z^{n-1} + ... + a_1 z + a_0$ be a polynomial of degree $n \geq 1$. Suppose $p(z) \neq 0$ for all $z \in \mathbb{C}$. We write z in polar form $z = re^{i\theta}$, where $r = |z|$ and $\theta \in [0, 2\pi]$. Consider the function $f : [0, \infty) \times [0, 2\pi] \to \mathbb{C}$ given by $f(r, \theta) = \frac{1}{p(re^{i\theta})}$. Then f is continuous

---

[9]This proof is due to A. Schep [30].

$[0, \infty) \times [0, 2\pi]$ and has continuous partial derivatives on $(0, \infty) \times (0, 2\pi)$ satisfying $\frac{\partial f}{\partial \theta} = ir\frac{\partial f}{\partial r}$. Define $F : [0, \infty) \to \mathbb{C}$ by

$$F(r) = \int_0^{2\pi} f(r, \theta)d\theta.$$

Now, for $r > 0$, differentiation under the integral sign and the Fundamental Theorem of Calculus yield

$$irF'(r) = ir \int_0^{2\pi} \frac{\partial f}{\partial r}(r, \theta)d\theta = \int_0^{2\pi} \frac{\partial f}{\partial \theta}(r, \theta)d\theta = f(r, 2\pi) - f(r, 0) = 0.$$

Hence $F'(r) = 0$ for $r > 0$. So for all $r > 0$, $F(r) = c$, for some constant c. The continuity of F implies $F(0) = c$. Thus $c = F(0) = \frac{2\pi}{p(0)} \neq 0$. On the other hand, since $|p(z)| \to \infty$ as $|z| \to \infty$, it follows $f(r, \theta) \to 0$ as $r \to \infty$ uniformly in θ. Therefore $F(r) \to 0$ as $r \to \infty$, that is, $c = 0$ a contradiction. $\square$

EXERCISES

1. (a) Let $F(x) = \int_0^1 \frac{x}{\sqrt{1-x^2y^2}}dy$. Find $F'(x)$ by two different methods.

   (b) If $F(x) = \int_1^{3x} \frac{e^{xy}}{y}dy$, find $F'(x)$.

   (c) If $F(x) = \int_1^{x^2} \frac{\cos(xy^2)}{y}dy$, find $F'(x)$.
      Your answers in all these exercises should not contain integrals.

2. For $x \neq \pm 1$, let

$$F(x) = \int_0^{\pi} \log(1 - 2x \cos y + y^2)dy.$$

   Prove that $F'(x) = 0$ and hence evaluate $F(x)$ for $|x| < 1$ and for $|x| > 1$.

3. Let

$$F(x) = \int_0^{\frac{\pi}{2}} \cos(x \sin \theta)d\theta.$$

   Prove that $xF'' + F' + xF = 0$.

4. Let
$$F(x) = \int_0^x (x - y)e^{x-y}g(y)dy,$$

where g is a continuous function on $\mathbb{R}$. Prove that $F''-2F'+F=g$.

5. Let
$$u(x,t) = \int_{x-ct}^{x+ct} f(y)dy,$$

where f has a continuous derivative. Prove that $u_{tt} = c^2 u_{xx}$.

6. Let $R = [a,b] \times [c,d]$ be a closed rectangle in $\mathbb{R}^2$ and $f : R \to \mathbb{R}$ be continuous. For each interior point $(s,t) \in R$ define

$$F(s,t) = \int_a^s \left(\int_c^t f(x,y)dy \right) dx.$$

Prove that
$$\frac{\partial^2 F}{\partial s \partial t} = \frac{\partial^2 F}{\partial t \partial s} = f(s,t).$$

5.3.1 Functions defined by improper integrals

The results we have so far obtained in this section have analogues for functions defined by improper integrals. Here we shall prove results for functions F of the form

$$F(x) = \int_c^\infty f(x,y)dy, \tag{5.26}$$

where f is bounded on $[c, \infty)$. Other improper integrals are treated in the same way with the obvious modifications. An important example of a function defined via an integral of the form (5.26) is Euler's *Gamma function*, which we discuss below in Example 5.3.26. The notion of uniform convergence plays a key role in all these cases.

Definition 5.3.9. (*Uniform convergence*). Let $f(x,y)$ be a continuous function on $[a,b] \times [c,\infty)$. Suppose $\int_c^\infty f(x,y)dy$ converges for each $x \in [a,b]$. We say that the integral $F(x) = \int_c^\infty f(x,y)dy$ *converges* *uniformly* on $[a, b]$ if for any $\epsilon > 0$ there is a number N depending *only* on ϵ such that

$$\left| F(x) - \int_c^d f(x, y)dy \right| < \epsilon,$$

whenever $d \geq N$ for all $x \in [a, b]$, that is,

$$\sup_{x \in [a,b]} \left| \int_d^\infty f(x, y)dy \right| \to 0, \qquad (5.27)$$

as $d \to \infty$.

Example 5.3.10. The integral $\int_0^\infty e^{-xy}dy$, obviously converges only when $x > 0$. Furthermore it converges uniformly on every set $\{x \in \mathbb{R} : x \geq x_0 > 0\}$. Indeed, if $x \geq x_0 > 0$, then

$$0 \leq \int_d^\infty e^{-xy}dy = \frac{1}{x}e^{-dx} \leq \frac{1}{x_0}e^{-dx_0} \to 0$$

as $d \to \infty$.

The following *Cauchy criterion for uniform convergence* holds:

Theorem 5.3.11. *(Cauchy's Criterion).* The integral $\int_c^\infty f(x, y)dy$ *converges uniformly on* $[a, b]$ *if and only if for every* $\epsilon > 0$ *there is a positive number* $N = N(\epsilon)$ *such that*

$$\left| \int_{d_1}^{d_2} f(x, y)dy \right| < \epsilon \qquad (5.28)$$

for all $d_2 > d_1 \geq N$.

Proof. Assume the integral converges uniformly. Given $\epsilon > 0$, there exist a number N such that, for every $x \in [a, b]$, $d > N$ implies

$$\left| \int_d^\infty f(x, y)dy \right| < \frac{\epsilon}{2}.$$

Therefore for $d_2 > d_1 \geq N$ we have

$$\left| \int_{d_1}^{d_2} f(x, y)dy \right| = \left| \int_{d_1}^\infty f(x, y)dy - \int_{d_2}^\infty f(x, y)dy \right|$$

$$\leq \left| \int_{d_1}^{\infty} f(x,y)dy \right| + \left| \int_{d_2}^{\infty} f(x,y)dy \right| < \frac{\epsilon}{2} + \frac{\epsilon}{2} = \epsilon.$$

Conversely, assume Cauchy's criterion (5.28) hold. By the Cauchy criterion on simple convergence, the integral $\int_c^{\infty} f(x,y)dy$ converges for every $x \in [a,b]$. Suppose that this convergence is *not* uniform on $[a,b]$. Then by the negation of the definition of uniform convergence, there exist $2\epsilon_0 > 0$ such that for any $N > 0$ there exist $x_0 \in [a,b]$ and $d_1 > N$ such that $\left| \int_{d_1}^{\infty} f(x_0,y)dy \right| \geq 2\epsilon_0$. Since $2\epsilon_0 > \epsilon_0$ and

$$\lim_{k \to \infty} \left| \int_{d_1}^{k} f(x_0,y)dy \right| = \left| \int_{d_1}^{\infty} f(x_0,y)dy \right|,$$

there must exist $d_2 > d_1$ such that $\left| \int_{d_1}^{d_2} f(x_0,y)dy \right| > \epsilon_0$. This contradicts (5.28). □

A very useful test for uniform convergence is the following *Weierstrass M-test* for integrals.

Theorem 5.3.12. *(Weierstrass M-test). If $|f(x,y)| \leq M(y)$ on $[c,\infty)$ for all $x \in [a,b]$, and the improper integral $\int_c^{\infty} M(y)dy < \infty$, then $\int_c^{\infty} f(x,y)dy$ is (absolutely) and uniformly convergent for $x \in [a,b]$.*

Proof. The convergence of $\int_c^{\infty} |f(x,y)|dy$ for $x \in [a,b]$ is an immediate consequence of the Comparison test (Corollary 5.2.8) and the hypothesis. So the integral is absolutely convergent. For the uniform convergence, let $\epsilon > 0$. Since $\int_c^{\infty} M(y)dy$ converges, there exists $N = N(\epsilon)$ such that $\int_{d_1}^{d_2} M(y)dy < \epsilon$ for all $d_2 > d_1 > N$. This implies that for all $d_2 > d_1 > N$ and all $x \in [a,b]$ we have

$$\left| \int_{d_1}^{d_2} f(x,y)dy \right| \leq \int_{d_1}^{d_2} |f(x,y)|dy \leq \int_{d_1}^{d_2} M(y)dy < \epsilon.$$

Therefore, by Theorem 5.3.11, the integral $\int_c^{\infty} f(x,y)dy$ converges uniformly for $x \in [a,b]$. □

Example 5.3.13. The integral $\int_0^{\infty} \frac{\cos xy}{1+y^2} dy$ converges uniformly for $x \in \mathbb{R}$.

Solution. Here $|f(x,y)| = \left|\frac{\cos xy}{1+y^2}\right| \leq \frac{1}{1+y^2} = M(y)$. Since $\int_0^\infty \frac{1}{1+y^2}dy =$

$tan^{-1}y\Big|_0^\infty = \frac{\pi}{2}$, it follows from the Weierstrass M-test that the integral converges uniformly on $\mathbb{R}$.

Theorem 5.3.14. *If the function $f(x,y)$ is continuous and the integral $\int_c^\infty f(x,y)dy$ is uniformly convergent on $[a,b]$, then $F(x)$ is continuous in $[a,b]$.*

Proof. From Theorem 5.3.1, the functions

$$F_k(x) = \int_c^k f(x,y)dy \quad (k=1,2,...)$$

are continuous in $[a,b]$. The uniform convergence of the integral $\int_c^\infty f(x,y)dy$ implies the uniform convergence of the sequence $\{F_k(x)\}$ to $F(x)$. It follows (see Theorem C.0.2) that F is also continuous in $[a,b]$.   □

Theorem 5.3.15. *If the integral $\int_c^\infty f(x,y)dy$ converges uniformly on $[a,b]$, then*

$$\int_a^b \left(\int_c^\infty f(x,y)dy\right)dx = \int_c^\infty \left(\int_a^b f(x,y)dx\right)dy. \qquad (5.29)$$

Proof. Let $F_k(x) = \int_c^k f(x,y)dy$ as above. Since $F_k \to F$ uniformly on $[a,b]$ as $k \to \infty$, taking also into account Fubini's theorem, we have

$$\int_a^b \left(\int_c^\infty f(x,y)dy\right)dx = \int_a^b F(x)dx = \int_a^b \lim_{k\to\infty} F_k(x)dx$$

$$= \lim_{k\to\infty} \int_a^b F_k(x)dx = \lim_{k\to\infty} \int_a^b \left(\int_c^k f(x,y)dy\right)dx$$

$$= \lim_{k\to\infty} \int_c^k \left(\int_a^b f(x,y)dx\right)dy = \int_c^\infty \left(\int_a^b f(x,y)dx\right)dy.$$

□

Theorem 5.3.16. *(Changing the order of integration).* *Let $f(x,y)$ be continuous on $[a,\infty) \times [c,\infty)$. Suppose that the integrals*

$$\int_c^\infty |f(x,y)|dy \text{ and } \int_a^\infty |f(x,y)|dx$$

converge uniformly for x *in every finite subinterval of* $[a, \infty)$, *and for* y *in every finite subinterval of* $[c, \infty)$, *respectively. If one of the integrals*

$$\int_a^\infty \left(\int_c^\infty |f(x,y)|dy\right) dx \ \text{ or } \ \int_c^\infty \left(\int_a^\infty |f(x,y)|dx\right) dy$$

converges, then the other converges also, and

$$\int_a^\infty \left(\int_c^\infty |f(x,y)|dy\right) dx = \int_c^\infty \left(\int_a^\infty |f(x,y)|dx\right) dy. \qquad (5.30)$$

Proof. Suppose that $\int_c^\infty \left(\int_a^\infty |f(x,y)|dx\right) dy$ converges. It is enough to prove (5.30) for non-negative functions. For an arbitrary f, we may write $f = f^+ - f^-$ and the linearity of the integral yields (5.30) for f. Hence we assume that $f \geq 0$. From Theorem 5.3.15 we have

$$\int_a^b \left(\int_c^\infty f(x,y)dy\right) dx = \int_c^\infty \left(\int_a^b f(x,y)dx\right) dy$$
$$\leq \int_c^\infty \left(\int_a^\infty f(x,y)dx\right) dy.$$

Therefore

$$\lim_{b\to\infty} \int_a^b \left(\int_c^\infty f(x,y)dy\right) dx \leq \int_c^\infty \left(\int_a^\infty f(x,y)dx\right) dy.$$

Hence the integral $\int_a^\infty \left(\int_c^\infty f(x,y)dy\right) dx$ exists and

$$\int_a^\infty \left(\int_c^\infty f(x,y)dy\right) dx \leq \int_c^\infty \left(\int_a^\infty f(x,y)dx\right) dy.$$

We can now reverse the roles of the integrals in the above argument to conclude their equality (5.30).   □

Theorem 5.3.17. *(Differentiation under the integral sign). Suppose both* $f(x,y)$ *and* $\frac{\partial f}{\partial x}(x,y)$ *are continuous in* $[a,b] \times [c,\infty)$. *Suppose also that the integral (5.26) converges for* $x \in [a,b]$ *and the integral* $\int_c^\infty \frac{\partial f}{\partial x}(x,y)dy$ *converges uniformly on* $[a,b]$. *Then* F *is differentiable on* $[a,b]$ *and*

$$F'(x) = \int_c^\infty \frac{\partial f}{\partial x}(x,y)dy,$$

that is,

$$\frac{d}{dx}\int_c^\infty f(x,y)dy = \int_c^\infty \frac{\partial f}{\partial x}(x,y)dy. \tag{5.31}$$

Proof. Define

$$G(x) = \int_c^\infty \frac{\partial f}{\partial x}(x,y)dy.$$

By Theorem 5.3.15, for $a < \xi \leq b$ we have,

$$\int_a^\xi G(x)dx = \int_c^\infty \left(\int_a^\xi \frac{\partial f}{\partial x}(x,y)dx\right)dy$$

$$= \int_c^\infty [f(\xi,y) - f(a,y)]\,dy = F(\xi) - F(a).$$

Now, since $G(x)$ is continuous (by Theorem 5.3.14), the Fundamental Theorem of Calculus tells us

$$G(\xi) = \frac{d}{d\xi}\int_a^\xi G(x)dx = \frac{d}{d\xi}[F(\xi) - F(a)] = F'(\xi).$$

Thus, $F'(x) = G(x)$ for all $x \in [a,b]$. $\square$

As the following examples show, the hypothesis of *uniform convergence* is essential to each of the preceding theorems.

Example 5.3.18. Let $f(x,y) = 2xye^{-xy^2}$ on $[0,\infty) \times [0,\infty)$.
Then

$$F(x) = \int_0^\infty f(x,y)dy = \lim_{k\to\infty}\int_0^k 2xye^{-xy^2}\,dy = \begin{cases} 1 \text{ if } x > 0, \\ 0 \text{ if } x = 0. \end{cases}$$

Hence $\lim_{x\to 0+} F(x) = 1 \neq 0 = F(0)$, and F is *not* continuous. This is connected with the fact that the integral converges *non-uniformly* on $\{x \in \mathbb{R} : x > 0\}$. Indeed, the remainder of the integral is $\int_d^\infty 2xye^{-xy^2}\,dy = e^{-xd^2}$. Since $\sup_{x\in(0,\infty)} e^{-xd^2} = 1$, of course, this does not tend to zero as $d \to \infty$.

Example 5.3.19. Consider the function $f(x,y) = (2 - xy)xye^{-xy}$ on the set $[0,\infty) \times [0,1]$. Noting that $\frac{d}{dt}(t^2e^{-t}) = (2-t)te^{-t}$, it is easy to see

$$\int_0^\infty \left(\int_0^1 f(x,y)dy\right)dx = 0 \neq 1 = \int_0^1 \left(\int_0^\infty f(x,y)dx\right)dy.$$

Thus, *the value of the iterated integral (5.29) may depend on the order of integration.*

Example 5.3.20. Consider the function

$$F(x) = \int_0^\infty x^3 e^{-x^2 y} dy.$$

Clearly $F(x) = x$ for all $x \in \mathbb{R}$, even for $x = 0$. Therefore $F'(x) = 1$ for all $x \in \mathbb{R}$. On the other hand, differentiation under the integral sign gives

$$F'(x) = \int_0^\infty (3x^2 - 2x^4 y) e^{-x^2 y} dy.$$

For $x \neq 0$ this gives $F'(x) = 1$. However, for $x = 0$, we get $F'(0) = 0$. Hence at $x = 0$ *differentiation under the integral sign is not valid.*

Remark 5.3.21. Note that Example 5.3.19, in particular, tells us

$$1 = \lim_{x \to 0^+} \int_0^\infty f(x,y) dy \neq \int_0^\infty \left(\lim_{x \to 0^+} f(x,y) \right) dy = 0.$$

Thus, *the limit of the integral need not equal the integral of the limit.* The careful reader will notice that the basic question underlying the question which arose at the begining of this section is actually the following: *can we interchange the operations of integrating with respect to y and taking a limit with respect to x?* Putting it more generally, *is the limit of the integral equal to the integral of the limit?* As was shown above the answer in general is no. The following theorem tells us when this is legitimate.[10]

Theorem 5.3.22. *Suppose that the integral $\int_c^\infty f(x,y) dy$ converges uniformly on $[a,b]$. If*

$$\lim_{x \to x_0} f(x,y) = g(y),$$

uniformly in y on every interval $[c,d]$, then the integral $\int_c^\infty g(y) dy$ converges and

$$\lim_{x \to x_0} \int_c^\infty f(x,y) dy = \int_c^\infty \lim_{x \to x_0} f(x,y) dy = \int_c^\infty g(y) dy. \qquad (5.32)$$

---

[10]See the note on the Lebesgue integral at the end of the chapter.

Proof. Since $f(x, y)$ converges to $g(y)$ as $x \to x_0$ uniformly on an interval $[c, d]$,

$$\lim_{x \to x_0} \int_c^d f(x, y) dy = \int_c^d g(y) dy. \tag{5.33}$$

Let $\epsilon > 0$. The uniform convergence of $\int_c^\infty f(x, y) dy$ implies that there exist $N > c$ such that for $d, d' > N$

$$\left| \int_c^d f(x, y) dy - \int_c^{d'} f(x, y) dy \right| < \epsilon$$

for all $x \in [a, b]$. Letting $x \to x_0$, we get by (5.33)

$$\left| \int_c^d g(y) dy - \int_c^{d'} g(y) dy \right| < \epsilon.$$

This tells us that the integral $\int_c^\infty g(y) dy$ converges. Now we prove (5.32). For any $\epsilon > 0$, there is $N > c$ such that for $d > N$, we have

$$\left| \int_d^\infty f(x, y) dy \right| < \epsilon,$$

for all $x \in [a, b]$ and

$$\left| \int_d^\infty g(y) dy \right| < \epsilon.$$

Furthermore from (5.33) there exist $\delta = \delta(d, \epsilon) > 0$ such that

$$\left| \int_c^d f(x, y) dy - \int_c^d g(y) dy \right| < \epsilon,$$

whenever $|x - x_0| < \delta$. Consequently, for $|x - x_0| < \delta$, we see that

$$\left| \int_c^\infty f(x, y) dy - \int_c^\infty g(y) dy \right| < 3\epsilon.$$

$\square$

Remark 5.3.23. Instead of the function defined by the integral $\int_c^\infty f(x, y) dy$), we can consider integrals of the form

$$F(x) = \int_{-\infty}^{\beta} f(x,y)dy \quad \text{or} \quad F(x) = \int_{-\infty}^{\infty} f(x,y)dy,$$

with the condition (5.27) of uniform convergence replaced by the condition

$$\sup_{x \in [a,b]} \left| \int_{-\infty}^{-d} f(x,y)dy \right| \to 0,$$

as $d \to \infty$ for the first integral, and by the two conditions $\sup_{x \in [a,b]} \left| \int_{-\infty}^{-d} f(x,y)dy \right| \to 0$ and $\sup_{x \in [a,b]} \left| \int_{d}^{\infty} f(x,y)dy \right| \to 0$, as $d \to \infty$, for the second integral. The theorems just proved above remain valid for integrals of this form. In fact, the proofs reduce to the case already considered if we replace y by $-y$ in the first integral, and write the second integral as

$$\int_{-\infty}^{\infty} f(x,y)dy = \int_{-\infty}^{0} f(x,y)dy + \int_{0}^{\infty} f(x,y)dy.$$

Next we illustrate the results obtained in this section with a number of important examples.

Example 5.3.24. For $x \geq 0$,

$$F(x) = \int_{0}^{\infty} e^{-xy} \frac{\sin y}{y} dy = \frac{\pi}{2} - \tan^{-1} x. \tag{5.34}$$

Solution. We consider the cases $x > 0$ and $x = 0$ separately.

1. For $x > 0$. Comparing the integrand $f(x,y) = e^{-xy} \frac{\sin y}{y}$ with e^{-xy} we see immediately that the integral exists. We have $\frac{\partial f}{\partial x}(x,y) = -e^{-xy} \sin y$. Clearly $\frac{\partial f}{\partial x}$ is continuous on $\mathbb{R}^2$, and by the Weierstrass M-test, the integral

$$\int_{0}^{\infty} (-e^{-xy} \sin y) dy$$

   is uniformly convergent on any interval $[c,d]$ with $c > 0$. Applying Theorem 5.3.17, differentiation under the integral sign yields

$$F'(x) = -\int_{0}^{\infty} e^{-xy} \sin y \, dy = -\frac{1}{1+x^2},$$

   for $x \in [c,d]$. Therefore there is constant C such that

$$F(x) = C - \tan^{-1} x, \tag{5.35}$$

for $x \in [c, d]$. Since $[c, d]$ is an arbitrary interval with $c > 0$, (5.35) holds for all $x \in (0, \infty)$. To determine C, note that $|F(x)| \leq \int_0^\infty e^{-xy} dy = \frac{1}{x}$. Therefore, letting $x \to \infty$ in (5.35), we get $C = \frac{\pi}{2}$.

2. For $x = 0$, the integral becomes

$$F(0) = \int_0^\infty \frac{\sin y}{y} dy.$$

All we need to do is to prove that the integral (5.34) is uniformly convergent for $x \in [0, \infty)$. For $y > 0$, the function $\frac{e^{-1xy}}{y}$ is positive and decreasing in y. By the Second Mean Value theorem for integrals (see, Appendix B) we have

$$\left| \int_{y_1}^{y_2} e^{-xy} \frac{\sin y}{y} dy \right| \leq \frac{e^{-y_1 x}}{y_1} \left| \int_{y_1}^{\xi} \sin y \, dy \right| \leq \frac{2}{y_1},$$

when $y_2 > y_1 > 0$ and $x \geq 0$, for some $\xi \in [y_1, y_2]$. Since $\frac{2}{y_1} \to 0$ as $y_1 \to \infty$, Cauchy's criterion for uniform convergence shows that the integral (5.34) converges uniformly for $x \in [0, \infty)$. Now Theorem 5.3.14 tells us that $F(x)$ is continuous on $[0, \infty)$, and by part (1) above,

$$F(0) = \lim_{x \to 0^+} F(x) = \lim_{x \to 0^+} \left(\frac{\pi}{2} - \tan^{-1} x \right) = \frac{\pi}{2}.$$

Thus,

$$\int_0^\infty \frac{\sin y}{y} dy = \frac{\pi}{2}. \tag{5.36}$$

Integral (5.36) is known as the *Dirichlet integral*. The integral is *not* absolutely convergent. Indeed, for $k = 1, 2, \ldots$

$$\int_{(k-1)\pi}^{k\pi} \frac{|\sin x|}{x} dx \geq \frac{1}{k\pi} \int_{(k-1)\pi}^{k\pi} |\sin x| dx = \frac{2}{k\pi}.$$

Therefore

$$\int_0^{n\pi} \frac{|\sin x|}{x} dx = \sum_{k=1}^n \int_{(k-1)\pi}^{k\pi} \frac{|\sin x|}{x} dx \geq \frac{2}{\pi} \sum_{k=1}^n \frac{1}{k}.$$

Hence, letting $n \to \infty$

$$\int_0^\infty \frac{|\sin x|}{x}\,dx \geq \frac{2}{\pi}\sum_{k=1}^\infty \frac{1}{k} = \infty,$$

since the harmonic series $\sum_{k=1}^\infty \frac{1}{k}$ diverges.

Example 5.3.25. Show that

$$\lim_{x\to 0^+}\int_0^\infty x^2 e^{-x^2 y^2}\sin y\,dy = 0.$$

Solution. Set $u = xy$ so that $du = x\,dy$. Then

$$\int_0^\infty x^2 e^{-x^2 y^2}\sin y\,dy = \int_0^\infty xe^{-u^2}\sin(\frac{u}{x})\,du.$$

By the Weierstrass M-test (Theorem 5.3.12) the last integral converges uniformly on $(0,\infty)$. At the same time, as $x \to 0^+$, the function $f(x,u) = xe^{-u^2}\sin(\frac{u}{x})$ converges to zero uniformly on $[0,\infty)$. Hence from Theorem 5.3.22

$$\lim_{x\to 0^+}\int_0^\infty xe^{-u^2}\sin(\frac{u}{x})\,du = 0.$$

Example 5.3.26. (*The Gamma Function*). The *Gamma function* is defined for $x > 0$ by the improper integral

$$\Gamma(x) = \int_0^\infty t^{x-1}e^{-t}\,dt \qquad\qquad (5.37)$$

The integral may be written

$$\Gamma(x) = \int_0^1 t^{x-1}e^{-t}\,dt + \int_1^\infty t^{x-1}e^{-t}\,dt,$$

and is improper at $t = 0$ and $t = \infty$. If $x > 0$, the integral

$$\int_0^1 t^{x-1}e^{-t}\,dt$$

is convergent. Moreover, for any $\xi > 0$, the integral is uniformly convergent for $\xi \leq x < \infty$. Indeed, since $t^{x-1}e^{-t} < t^{\xi-1}$ and the integral $\int_0^1 t^{\xi-1}\,dt$ converges, the analogue of the Weierstrass M-test (for

improper integrals over bounded intervals) implies the uniform convergence of $\int_0^1 t^{x-1}e^{-t}dt$.

Next the integral

$$\int_1^\infty t^{x-1}e^{-t}dt$$

converges uniformly on any interval $0 \le x \le C$, for any constant $C > 0$,
since $t^{x-1}e^{-t} \le t^{C-1}e^{-t}$ and the integral $\int_1^\infty t^{C-1}e^{-t}dt$ converges.

Therefore for any $\xi > 0$ and $C > 0$ the integral (5.37) defining $\Gamma(x)$
converges uniformly on the interval $\xi \le x \le C$.

It follows (from Theorem 5.3.14) that $\Gamma(x)$ is *continuous* on $(0, \infty)$.
Moreover, Γ is of class C^∞ on $(0, \infty)$, and its *derivatives* can be calculated by differentiating under the integral sign, that is,

$$\Gamma^{(k)}(x) = \int_0^\infty (\log t)^k t^{x-1}e^{-t}dt, \quad (k = 1, 2, 3, ...) \tag{5.38}$$

To see this, first note that, if $0 < a \le x$, then, since $t^{\frac{x}{2}}(\log t)^k \to 0$
as $t \to 0^+$, there exists $C_k > 0$ such that

$$|t^{x-1}(\log t)^k e^{-t}| < t^{\frac{x}{2}-1}$$

for $0 < t \le C_n$. Hence by the Weierstrass M-test, the integral

$$\int_0^{C_k} t^{x-1}(\log t)^k e^{-t}dt$$

converges uniformly with respect to x on the interval $[a, \infty)$.

If $x \le a < \infty$, then for $t \ge 1$,

$$|t^{x-1}(\log t)^k e^{-t}| \le t^{a-1}|\log t|^k e^{-t},$$

and we conclude similarly that the integral

$$\int_{C_k}^\infty t^{x-1}(\log t)^k e^{-t}dt$$

converges uniformly with respect to x on the interval $(0, a]$. Hence the
integral (5.38) converges uniformly on $(0, \infty)$ and differentation under
the integral sign is legitimate.

An important property of Γ is that it satisfies the *reduction formula*

$$\Gamma(x+1) = x\Gamma(x). \tag{5.39}$$

Its proof is a simple integration by parts. Indeed,

$$\Gamma(x+1) = \int_0^\infty t^x e^{-t} dt = -t^x e^{-t}\big|_0^\infty + x \int_0^\infty t^{x-1} e^{-t} dt = 0 + x\Gamma(x).$$

Since $\Gamma(1) = \int_0^\infty e^{-t} dt = 1$, the reduction formula (5.39) tells us

$$\Gamma(n+1) = n! \quad n = 1, 2, \dots \tag{5.40}$$

This formula resembles the recursive formula $(n+1)! = (n+1)n!$ that defines factorials. Thus, the gamma function provides the natural extension of the factorial function to *non-integers*. Some other values of Γ are:

$$\Gamma(\tfrac{1}{2}) = \int_0^\infty t^{-\frac{1}{2}} e^{-t} dt = 2 \int_0^\infty e^{-u^2} du = 2\frac{\sqrt{\pi}}{2} = \sqrt{\pi},$$

where we made the substitution $u = \sqrt{t}$ and used Example 5.2.7. From (5.39) we now deduce:

$$\Gamma\left(\frac{3}{2}\right) = \Gamma\left(\frac{1}{2}+1\right) = \frac{1}{2}\Gamma\left(\frac{1}{2}\right) = \frac{1}{2}\sqrt{\pi},$$

$$\Gamma\left(\frac{5}{2}\right) = \frac{3}{2}\cdot\frac{1}{2}\sqrt{\pi}, \ \dots, \ \Gamma\left(n+\frac{1}{2}\right) = (n-\frac{1}{2})\cdots\frac{3}{2}\cdot\frac{1}{2}\sqrt{\pi}.$$

Thus,

$$\Gamma\left(n+\frac{1}{2}\right) = \frac{1\cdot 3\cdot 5\cdots(2n-1)}{2^n}\sqrt{\pi}. \tag{5.41}$$

A final remark: although Γ is defined by (5.37) only for $x > 0$, the reduction formula (5.39) can be used to extend the domain of definition to all non-integer negative numbers by means of the formula

$$\Gamma(x) = \frac{\Gamma(x+1)}{x}.$$

For example, $\Gamma\left(-\frac{3}{2}\right) = \frac{\Gamma(-\frac{1}{2})}{-\frac{3}{2}} = -\frac{2}{3}\Gamma\left(-\frac{1}{2}\right) = -\frac{2}{3}\frac{\Gamma(\frac{1}{2})}{-\frac{1}{2}} = \frac{4}{3}\sqrt{\pi}.$

Example 5.3.27. (*Volume of the unit ball in* $\mathbb{R}^n$). The n-dimensional volume c_n of the unit ball $B^n = \{x \in \mathbb{R}^n; ||x|| \le 1\}$ is given by

$$c_n = \frac{2\pi^{\frac{n}{2}}}{n\Gamma\left(\frac{n}{2}\right)}. \tag{5.42}$$

Solution. From Example 5.1.33 the volume of B^n is

$$c_{2m} = \frac{\pi^m}{m!}, \qquad c_{2m+1} = \frac{2^{m+1}\pi^m}{1\cdot3\cdot5\cdots(2m+1)}.$$

If n is an even integer, say $n = 2m$ for some $m = 1, 2, ...$, then

$$c_n = c_{2m} = \frac{\pi^m}{m!} = \frac{\pi^m}{\Gamma(m+1)} = \frac{\pi^{\frac{n}{2}}}{\Gamma(\frac{n}{2}+1)} = \frac{\pi^{\frac{n}{2}}}{\frac{n}{2}\Gamma(\frac{n}{2})} = \frac{2\pi^{\frac{n}{2}}}{n\Gamma\left(\frac{n}{2}\right)}.$$

While if n is odd, say $n = 2m + 1$ for some $m = 0, 1, 2, ...$, then using (5.41) we have

$$c_n = c_{2m+1} = \frac{2^{m+1}\pi^m}{1\cdot3\cdot5\cdots(2m+1)} = \frac{2\pi^{m+\frac{1}{2}}}{\frac{1\cdot3\cdot5\cdots(2m-1)}{2^m}\pi^{\frac{1}{2}}(2m+1)}$$

$$= \frac{2\pi^{m+\frac{1}{2}}}{\Gamma(m+\frac{1}{2})(2m+1)} = \frac{2\pi^{m+\frac{1}{2}}}{\Gamma(\frac{2m+1}{2})(2m+1)} = \frac{2\pi^{\frac{n}{2}}}{n\Gamma\left(\frac{n}{2}\right)}.$$

A typical application of the gamma function is to evaluate integrals.

Example 5.3.28. Prove that for any $\alpha > 0$
(a)

$$\int_0^\infty e^{-x^\alpha} dx = \frac{1}{\alpha}\Gamma\left(\frac{1}{\alpha}\right).$$

(b)

$$\int_0^1 \left(\log\frac{1}{x}\right)^{\alpha-1} dx = \Gamma(\alpha).$$

Solution. (a) Set $u = x^\alpha$. Then $x = u^{\frac{1}{\alpha}}$ and $dx = \frac{1}{\alpha}u^{\frac{1}{\alpha}-1}du$, and the integral becomes

$$\frac{1}{\alpha}\int_0^\infty u^{\frac{1}{\alpha}-1}e^{-u}du = \frac{1}{\alpha}\Gamma\left(\frac{1}{\alpha}\right).$$

(b) Set $u = \log \frac{1}{x} = -\log x$. Then $x = e^{-u}$ and $dx = -e^{-u}du$. As x varies from 0 to 1, the variable u varies from ∞ to 0. Substituting in the integral we get

$$\int_0^\infty u^{\alpha-1}e^{-u}du = \Gamma(\alpha).$$

Example 5.3.29. (*The Beta function*). The *Beta function* is a function of two variables, defined on $(0, \infty) \times (0, \infty)$ by

$$B(x,y) = \int_0^1 t^{x-1}(1-t)^{y-1}dt. \tag{5.43}$$

Since the integrand is approximately equal to t^{x-1} for t near 0 and to $(1-t)^{y-1}$ for t near 1, the integral is proper when $x, y \geq 1$. If either $0 < x < 1$ or $0 < y < 1$ or both, the integral is improper, however, it can be shown to converge by methods similar to those used with the Gamma function, and we leave this verification as an exercise to the interested reader. Note also that the Beta function is symmetric in x and y. That is, $B(x,y) = B(y,x)$.

Setting $t = \sin^2 \theta$, with θ varying from 0 to $\frac{\pi}{2}$, then (5.43) takes the form

$$B(x,y) = 2\int_0^{\frac{\pi}{2}} \sin^{2x-1}\theta \cos^{2y-1}\theta d\theta,$$

$$\int_0^{\frac{\pi}{2}} \sin^m \theta \cos^n \theta d\theta = \frac{1}{2}B\left(\frac{m+1}{2}, \frac{n+1}{2}\right) \tag{5.44}$$

with $m, n > 0$.

Many integrals encountered in applications can be reduced to integrals (5.43) and (5.44).

To compute the values of the Beta function, use is made of the following relation between the Beta function and the Gamma function

Theorem 5.3.30. *For $x, y > 0$,*

$$B(x,y) = \frac{\Gamma(x)\Gamma(y)}{\Gamma(x+y)}. \tag{5.45}$$

Proof. Substitute $t = u^2$ into the integral $\Gamma(x) = \int_0^\infty t^{x-1}e^{-t}dt$, to obtain

$$\Gamma(x) = 2\int_0^\infty u^{2x-1}e^{-u^2}du.$$

From this, using Exercise 12 of Section 5.2, we get

$$\Gamma(x)\Gamma(y) = 4\left(\int_0^\infty u^{2x-1}e^{-u^2}du\right)\left(\int_0^\infty v^{2y-1}e^{-v^2}dv\right)$$

$$= 4\int_0^\infty\int_0^\infty u^{2x-1}v^{2y-1}e^{-(u^2+v^2)}dudv$$

$$= 4\lim_{k\to\infty}\int\int_{\Omega_k} u^{2x-1}v^{2y-1}e^{-(u^2+v^2)}dudv,$$

where $\Omega_k = \left\{(x,y) : x > 0, y > 0, \frac{1}{k} < x^2 + y^2 < k\right\}$.

Introducing polar coordinates (r,θ) the domain of integration Ω_k becomes $R_k - \left\{(r,\theta) : \frac{1}{k} < r < k, \frac{1}{k} < \theta < \frac{\pi}{2} - \frac{1}{k}\right\}$ and so

$$\Gamma(x)\Gamma(y) = 4\lim_{k\to\infty}\int\int_{R_k}(r\cos\theta)^{2x-1}(r\sin\theta)^{2y-1}e^{-r^2}rdrd\theta$$

$$= \lim_{k\to\infty}\left(2\int_{\frac{1}{k}}^{k} r^{2(x+y)-1}e^{-r^2}dr\right)\left(2\int_{\frac{1}{k}}^{\frac{\pi}{2}-\frac{1}{k}}\cos^{2x-1}\theta\sin^{2y-1}\theta d\theta\right)$$

$$= \Gamma(x+y)B(x,y).$$

$\square$

Example 5.3.31. Evaluate

$$\int_0^{\frac{\pi}{2}}\sin^6\theta\cos^8\theta d\theta.$$

Solution. By formula (5.44) for $m = 6$, $n = 8$, and (5.45) we have

$$\int_0^{\frac{\pi}{2}}\sin^6\theta\cos^8\theta d\theta = \frac{1}{2}B\left(\frac{7}{2},\frac{9}{2}\right) = \frac{1}{2}\frac{\Gamma(\frac{7}{2})\Gamma(\frac{9}{2})}{\Gamma(8)} = \frac{5}{2}\frac{\pi}{7!}.$$

Example 5.3.32. Evaluate

$$\int_0^\pi \frac{1}{\sqrt{3-\cos x}}dx.$$

Solution. Set $\cos x = 1 - 2\sqrt{t}$. Then $dx = \frac{1}{2\sqrt[4]{t^3}\sqrt{1-\sqrt{t}}}dt$, with t varying from 0 to 1. Substituting

$$\int_0^\pi \frac{1}{\sqrt{3-\cos x}}dx = \frac{1}{2\sqrt{2}}\int_0^1 t^{-\frac{3}{4}}(1-t)^{-\frac{1}{2}}dt = \frac{1}{2\sqrt{2}}B\left(\frac{1}{4},\frac{1}{2}\right)$$

$$= \frac{1}{2\sqrt{2}}\cdot\frac{\Gamma(\frac{1}{4})\Gamma(\frac{1}{2})}{\Gamma(\frac{3}{4})} = \frac{\sqrt{\pi}}{2\sqrt{2}}\cdot\frac{[\Gamma(\frac{1}{4})]^2}{\Gamma(\frac{3}{4})\Gamma(\frac{1}{4})}.$$

We close this section with a remark concerning functions defined by improper multiple integrals.

Remark 5.3.33. (*Functions defined by improper multiple integrals*). Let $X \subset \mathbb{R}^n$ and Ω an open subset of $\mathbb{R}^m$. Consider the function F defined by the integral

$$F(x) = \int_\Omega f(x,y)dy = \int\cdots\int_\Omega f(x_1,...,x_n,y_1,...,y_m)dy_1\cdots dy_m, \quad (5.46)$$

where the function $f(x,y)$ is assumed to be defined and continuous on $X\times\Omega$. If the set Ω or the function $f(x,y)$ in the integral (5.46) is *unbounded*, the integral is an improper multiple integral and is understood as the limit (5.15) of (proper) integrals over sets of a suitable exhaustion $\{\Omega_k\}$ of Ω (see, Definition 5.2.3). The analogue of Definition 5.3.9, on uniform convergence of the integral reads (equivalently) here as follows: Suppose in the integral (5.46) Ω is unbounded and the integral converges for each $x \in X$. Let Ω_k be the portion of the set Ω obtained by removing from Ω a k-neighborhood of infinity from it[11] and let

$$F_k(x) = \int_{\Omega_k} f(x,y)dy, \quad (k=1,2,...).$$

---

[11]Or an $\frac{1}{k}$-neighborhood of the point where f is unbounded, in case the integral (5.46) is improper because the function is unbounded at a point in a bounded set Ω.

Chapter 5 Change of Variables Formula

We say that the integral (5.46) *converges uniformly* on X, if $F_k(x) \to F(x)$ uniformly on X as $k \to \infty$.

Under the assumption of uniformly convergence, Theorems 5.3.14, 5.3.15 and 5.3.17 can be extended for functions defined by integrals of the form (5.46), but we shall not go into particulars here.

5.3.2 Convolution of functions

The *convolution product* (or simply *convolution*) of two functions is a new function defined by an integral. The definition is as folllows.

Definition 5.3.34. The *convolution* of the functions $f : \mathbb{R} \to \mathbb{C}$ and $g : \mathbb{R} \to \mathbb{C}$ is the function $f * g : \mathbb{R} \to \mathbb{C}$ defined by the integral

$$(f * g)(x) = \int_{-\infty}^{\infty} f(y)g(x - y)dy, \qquad (5.47)$$

provided this integral exists for all $x \in \mathbb{R}$.

Various conditions[12] can be imposed on the functions f and g to guarantee that $f * g$ exists as an improper Riemann integral.

Recall from Section 4.4 that $C_c^{\infty}(\mathbb{R}^n)$ denotes the set of continuous function on $\mathbb{R}^n$ with compact support. Although not obvious that there many functions in $C_c^{\infty}(\mathbb{R}^n)$, Uryshon's lemma shows that there are, indeed non-negative ones. Let $\psi \in C_c^{\infty}(\mathbb{R}^n)$ which has been normalized so that $\int_{\mathbb{R}^n} \psi = 1$. Then we can arrange that supp ψ is the closed unit ball. Using this single function ψ we can construct other elements of $C_c^{\infty}(\mathbb{R}^n)$ by letting $\epsilon > 0$ and define

$$\psi_\epsilon(x) \equiv \epsilon^{-n}\psi\left(\frac{x}{\epsilon}\right).$$

Then $\psi_\epsilon \in C_c^{\infty}(\mathbb{R}^n)$ with $supp\,\psi_\epsilon = \{x : ||x|| \leq \epsilon\}$ and $\int_{\mathbb{R}^n} \psi_\epsilon = 1$. These functions are known as *molifiers*.[13]

---

[12] A detailed study of convolutions involves the Lebesgue integral and is discussed in more advanced courses in functions of real variables. See for example, [12].

[13] The family of functions $\{\psi_\epsilon\}_{\epsilon>0}$ is also called an *approximate identity* because

$$||\psi_\epsilon * f - f|| \to 0, \text{ as } \epsilon \to 0,$$

for any integrable f and various norms.

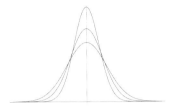

Figure 5.6: Approximate identity based on e^{-x^2}

Definition 5.3.35. A function $f : \Omega \to \mathbb{C}$ is called *absolutely integrable on* Ω if $\int_\Omega |f| < \infty$.

We now give sufficient conditions for which the existence of the convolution (5.47) can be established without difficulty.

Theorem 5.3.36. *Let* $f : \mathbb{R} \to \mathbb{C}$ *and* $g : \mathbb{R} \to \mathbb{C}$ *be integrable on every closed interval* $[a, b]$. *Then each of the following implies the existence of the convolution*

$$(f * g)(x) = \int_{-\infty}^{\infty} f(y)g(x - y)dy.$$

1. *One of the functions f, g is absolutely integrable on $\mathbb{R}$ and the other is bounded on $\mathbb{R}$.*

2. *One of the functions f or g has compact support.*

3. *The functions $|f|^2$ and $|g|^2$ are integrable on $\mathbb{R}$.*

Proof.   1. Suppose f is absolutely integrable on $\mathbb{R}$ and $|g(x)| \leq M$ on $\mathbb{R}$, then

$$|(f * g)(x)| \leq \int_{-\infty}^{\infty} |f(y)g(x - y)|dy \leq M \int_{-\infty}^{\infty} |f(y)|dy < \infty.$$

2. Suppose $\text{supp}(f) \subset [a, b] \subset \mathbb{R}$. Then

$$(f * g)(x) = \int_{-\infty}^{\infty} f(y)g(x - y)dy = \int_{a}^{b} f(y)g(x - y)dy.$$

By the hypothesis, this last integral exists for all $x \in \mathbb{R}$. In case g has compact support, setting $x - y = z$, we get

$$(f * g)(x) = \int_{-\infty}^{\infty} f(x - z)g(z)dz = (g * f)(x)$$

and this case reduces to the case considered above. Note that as a by-product we also proved the commutativity property for the convolution

$$f * g = g * f.$$

3. The Cauchy-Schwarz inequality yields

$$\left(\int_{-\infty}^{\infty} |f(y)g(x - y)|dy \right)^2$$

$$\leq \left(\int_{-\infty}^{\infty} |f(y)|^2 dy \right) \left(\int_{-\infty}^{\infty} |g(x - y)|^2 dy \right) < \infty,$$

since $\int_{-\infty}^{\infty} |g(x - y)|^2 dy = \int_{-\infty}^{\infty} |g(z)|^2 dz$, where $z = x - y$.

$\square$

Example 5.3.37. For $a > 0$, let $f(x) = e^{-a|x|}$ and $g(x) = \cos x$. Find the convolution $f * g$.

Solution. We have

$$(f * g)(x) = \int_{-\infty}^{\infty} f(y)g(x - y)dy = \int_{-\infty}^{\infty} e^{-a|y|} \cos(x - y)dy$$

$$= \int_{-\infty}^{0} e^{ay} \cos(x - y)dy + \int_{0}^{\infty} e^{-ay} \cos(x - y)dy$$

$$= \int_{0}^{\infty} e^{-ay} \cos(x + y)dy + \int_{0}^{\infty} e^{-ay} \cos(x - y)dy$$

$$= 2 \cos x \int_{0}^{\infty} e^{-ay} \cos y \, dy = \frac{2a \cos x}{1 + a^2}.$$

Proposition 5.3.38. *Let f and g be continuous, absolutely integrable and bounded on $\mathbb{R}$. Then $f * g$ is absolutely integrable.*

Proof. We have

$$\int_{-\infty}^{\infty} |(f * g)(x)| dx = \int_{-\infty}^{\infty} \left| \int_{-\infty}^{\infty} f(y)g(x-y)dy \right| dx$$

$$\leq \int_{-\infty}^{\infty} \left(\int_{-\infty}^{\infty} |f(y)g(x-y)| dy \right) dx$$

$$= \int_{-\infty}^{\infty} \left(\int_{-\infty}^{\infty} |f(y)g(x-y)| dx \right) dy = \int_{-\infty}^{\infty} |f(y)| \left(\int_{-\infty}^{\infty} |g(x-y)| dx \right) dy$$

$$= \int_{-\infty}^{\infty} |f(y)| \left(\int_{-\infty}^{\infty} |g(u)| du \right) dy = \int_{-\infty}^{\infty} |g(u)| du \int_{-\infty}^{\infty} |f(y)| dy < \infty.$$

The change of the order of integration is legitimate, since by the Weierstrass M-test both integrals $\int_c^{\infty} |f(y)g(x-y)| dy$ and $\int_a^{\infty} |f(y)g(x-y)| dx$ converge uniformly and Theorem 5.3.16 applies. $\square$

Exercise 5.3.39. Let f be continuous and $g \in C_c^1(\mathbb{R})$. Show that $f * g \in C^1(\mathbb{R})$ and

$$\frac{d}{dx}(f * g) = f * \left(\frac{\partial}{\partial x} g \right).$$

Furthermore, if $g \in C_c^k(\mathbb{R})$, show that $f * g \in C^k(\mathbb{R})$ and

$$\frac{d^k}{dx^k}(f * g) = f * \left(\frac{\partial^k}{\partial x^k} g \right).$$

EXERCISES

1. Compute
$$\int_0^{\infty} \left(\int_x^{\infty} e^{-y^2} dy \right) dx.$$

2. Compute
$$\int_0^{\infty} \left(\int_{2x}^{\infty} xe^{-y} \frac{\sin y}{y^2} dy \right) dx.$$

Hint. Change the order of integration.

3. Let
$$F(x) = \int_0^\infty \frac{\tan^{-1} xy}{y(1 + y^2)} dy.$$
Show that $F(x) = \frac{\pi}{2} \log(1 + x)$ for $x > -1$.

4. Let
$$F(x) = \int_0^\infty \left(\frac{\sin xy}{y} \right)^2 dy.$$
Show that $F'(x) = \frac{\pi}{2}$ for $x > 0$ and hence evaluate $F(x)$ for all x. Use a similar argument and the formula $4\sin^3 \theta = 3\sin \theta - \sin 3\theta$ to evaluate
$$\int_0^\infty \left(\frac{\sin xy}{y} \right)^3 dy.$$

5. For $x \in \mathbb{R}$ show that
$$\int_0^\infty e^{-y^2} \cos(xy) dy = \frac{\sqrt{\pi}}{2} e^{-\frac{x^2}{4}}.$$

   Hint. Call the integral by $F(x)$. Show that $\frac{F'}{F} = -\frac{1}{2x}$ and integrate.

6. For $x > 0$ show that
$$\int_0^\infty e^{-y^2 - \frac{x^2}{y^2}} dy = \frac{\sqrt{\pi}}{2} e^{-2|x|}.$$

   Hint. Call the integral $F(x)$ and differentiate under the integral sign.

7. Prove by induction on n that
$$\int_0^\infty y^n e^{-xy} dy = \frac{n!}{x^{n+1}} \quad (x > 0).$$

8. Verify that $\int_0^\infty \frac{1}{y^2 + x} dy = \frac{\pi}{2\sqrt{x}}$. Show that differentiation under the integral sign is legitimate and deduce
$$\int_0^\infty \frac{1}{(y^2 + x)^n} dy = \frac{1 \cdot 3 \cdots (2n - 3)}{2 \cdot 4 \cdots (2n - 2)} \frac{\pi}{2x^{\frac{2n-1}{2}}}.$$

9. Prove that

$$\lim_{x\to\infty} \int_0^\infty \sin(y^2)\tan^{-1}(xy)dy = \frac{\pi}{4}\sqrt{\frac{\pi}{2}}.$$

Hint. Use Theorem 5.3.22 and Exercise 5.6.10.

10. Evaluate

$$\int_0^\infty x^4 e^{-x^2}dx.$$

11. Prove

$$\int_0^\infty x^m e^{-x^n}dx = \frac{1}{n}\Gamma\left(\frac{m+1}{n}\right).$$

Hint. Use the substitution $u = x^n$.

12. Show that for $\alpha, \beta > 0$

$$\int_0^1 \left(\log\frac{1}{x}\right)^{\alpha-1} x^{\beta-1}dx = \frac{\Gamma(\alpha)}{\beta^\alpha}.$$

Use this result to show that

(a) $\int_0^1 \left(\frac{\log\frac{1}{x}}{x}\right)^{\frac{1}{2}} = \sqrt{2\pi}.$ (b) $\int_0^1 \left(\frac{x}{\log\frac{1}{x}}\right)^{\frac{1}{2}} = \sqrt{\frac{2\pi}{3}}.$

13. The function defined on $(0, \infty)$ by

$$F(s) = \int_0^\infty e^{-st}f(t)dt$$

is called the *Laplace transform* of f, and it is denoted by $L[f](s)$. Prove that if $f(t)$ and $f'(t)$ are continuous and bounded in $0 \leq t < \infty$, then

$$L[f'](s) = sL[f](s) - f(0).$$

Answers to selected Exercises

1. $\frac{1}{2}$.   2. $\frac{1}{16}$.   10. $\frac{3}{8}\sqrt{\pi}$.

5.4 The Weierstrass approximation theorem (*)

This important result tells us that any continuous real function on a closed interval of $\mathbb{R}$ can be uniformly approximated by polynomials. As we shall see in the Appendix generalizes to several variables and beyond. Before proving the *Weierstrass approximation theorem* we need a lemma.

Lemma 5.4.1. *Let $0 < \alpha < 1$ and c be a fixed positive real number, then*

$$\lim_{n \to \infty} \alpha^n n^c \to 0.$$

Proof. Let $y_n = \alpha^n n^c$. Then $\log y_n = n \log(\alpha) + c \log n$. Hence $\frac{\log y_n}{n} = \log(\alpha) + c \frac{\log n}{n}$. Now $\frac{\log n}{n} \to 0$ as $n \to \infty$ because $\frac{\log x}{x} \to 0$ as $x \to \infty$. This follows by L'Hopital's Rule. Hence also $c \frac{\log n}{n} \to 0$ and hence $\frac{\log y_n}{n} \to \log(\alpha)$. Using the continuity of the exponential function this shows $\lim_{n \to \infty} y_n = \lim_{n \to \infty} e^{-n} \alpha = \alpha \lim_{n \to \infty} e^{-n} = 0$. $\square$

Theorem 5.4.2. *Every real valued continuous function f on $[a, b]$ can be uniformly approximated by real polynomials.*

Proof. Since the composition of polynomials is again a polynomial we can assume $[a, b] = [0, 1]$. This is because the two intervals are homeomorphic by a linear function (and therefore polynomial function with polynomial inverse), $\phi : [0, 1] \to [a, b]$. Here $\phi(t) = tb + (1 - t)a$.

Next we can assume $f(0) = 0 = f(1)$. For suppose we could do it for this case. Let $g(x) = f(x) - f(0) - x(f(1) - f(0))$. Then g is a continuous function on $[0, 1]$ and $g(1) = 0 = g(0)$. Approximating g by a polynomial on $[0, 1]$ also approximates f since $g - f$ is itself a polynomial.

Now extend f to a *uniformly* continuous function on all of $\mathbb{R}$ by taking it to be identically zero outside of $[0, 1]$. Let $q_n = c_n(1 - x^2)^n$, where c_n is chosen so that $\int_{-1}^{1} q_n(x) dx = 1$. We need to get some idea as to how fast c_n grows. Since $x \mapsto (1 - x^2)^n$ is even and non-negative on $[0, 1]$ we have

$$\int_{-1}^{1} (1 - x^2)^n dx = 2 \int_{0}^{1} (1 - x^2)^n dx \geq 2 \int_{0}^{\frac{1}{\sqrt{n}}} (1 - x^2)^n dx.$$

On the other hand on $[0, 1]$, $(1 - x^2)^n \geq 1 - nx^2$ (Bernoulli's inequality). This is because the function $\phi(x) = (1 - x^2)^n - 1 + nx^2$ vanishes at 0 and its derivative, $\phi'(x) = 2nx(1 - (1 - x^2)^{n-1}) \geq 0$ on $[0, 1]$. Hence ϕ is increasing and so is always ≥ 0.

Hence,

$$2 \int_0^{\frac{1}{\sqrt{n}}} (1 - x^2)^n dx \geq 2 \int_0^{\frac{1}{\sqrt{n}}} (1 - nx^2) dx = \frac{4}{3\sqrt{n}} > \frac{1}{\sqrt{n}}.$$

This means for all n, $c_n < \sqrt{n}$. Hence $q_n < \sqrt{n}(1 - x^2)^n$. In particular, for $1 \geq \delta > 0$ and $\delta \leq |x| \leq 1$, $q_n(x) \leq \sqrt{n}(1 - \delta^2)^n$. Moreover, since $1 - \delta^2 < 1$ and therfore its powers tend to zero we see $q_n \to 0$ uniformly on $\delta \leq |x| \leq 1$.

Let $p_n(x) = \int_{-1}^1 f(x+t) q_n(t) dt$, for $x \in [0, 1]$. Since f is supported on $[0, 1]$, $f(x+t) = 0$ if either $x+t \geq 1$, or ≤ 0. That is if $t \geq 1 - x$, or $\leq -x$. Thus we may assume $-x \geq t \geq 1 - x$ and so $p_n(x) = \int_{1-x}^{-x} f(x+t) q_n(t) dt$. Now perform the translation $t \mapsto t - x$. Then $t + x \mapsto t$, $1 \mapsto 1 - x$ and $0 \mapsto -x$ so this last integral is $p_n(x) = - \int_0^1 f(t) q_n(t - x) dt$, a real polynomial in x.

Let $\epsilon > 0$. Since f is uniformly continuous there is a $0 < \delta < 1$ so that if $|x - y| < \delta$, then $|f(x) - f(y)| < \epsilon$. Since f is also bounded $|f(x)| \leq M$ for some constant M. Now recall the original definition of $p_n(x) = \int_{-1}^1 f(x + t) q_n(t) dt$. Hence $p_n(x) - f(x) = \int_{-1}^1 (f(x + t) - f(x)) q_n(t) dt$ and since $q_n \geq 0$ and $\int_{-1}^1 q_n(t) dt = 1$ and $q_n(t) \geq 0$ we see

$$|p_n(x) - f(x)| \leq \int_{-1}^1 |f(x + t) - f(x)| q_n(t) dt.$$

But this is

$$\leq 2M \int_{-1}^{-\delta} q_n(t) dt + \epsilon \int_{-\delta}^{+\delta} q_n(t) dt + 2M \int_{\delta}^1 q_n(t) dt.$$

The first and last term together is $4M \int_\delta^1 q_n(t) dt$ which is $\leq 4M(1 - \delta^2)^n \sqrt{n}$ while the middle term is $\leq \epsilon \int_{-1}^1 q_n(t) dt = \epsilon$. Thus

$$|p_n(x) - f(x)| \leq \epsilon + 4M(1 - \delta^2)^n \sqrt{n}.$$

Since the second term tends to zero by Lemma 5.4.1, it is eventually $< 2\epsilon$. $\qquad\square$

5.5   The Fourier transform (*)

The Fourier transform is a very useful and flexible tool with extensive theory and applications. In this section, we introduce the Fourier tranform on $\mathbb{R}$ and study its basic properties, in particular those which we shall use in Chapter 7 to solve certain partial differential equations in two real variables. The discussion of the Fourier transform here is within the context of Riemann integrable functions.[14]

We recall that a function $f : \mathbb{R} \to \mathbb{C}$ is said to be *absolutely integrable* on $\mathbb{R}$ if $\int_{-\infty}^{\infty} |f(x)|dx$ exists as an improper Riemann integral.

Definition 5.5.1. Let $f : \mathbb{R} \to \mathbb{C}$ be an absolutely integrable function on $\mathbb{R}$. The *Fourier transform* $\mathcal{F}(f) \equiv \widehat{f}$ of f is defined by the integral

$$\widehat{f}(\xi) = \frac{1}{\sqrt{2\pi}} \int_{-\infty}^{\infty} f(x)e^{-i\xi x}dx. \tag{5.48}$$

The integral here is understood in the sense of the principal value

$$\int_{-\infty}^{\infty} f(x)e^{-i\xi x}dx = \lim_{N \to \infty} \int_{-N}^{N} f(x)e^{-i\xi x}dx.$$

We first note that the integral (5.48) converges absolutely and uniformly with respect to ξ on the entire line $\mathbb{R}$. Indeed, since

$$|e^{-i\xi x}| = |\cos(\xi x) - i\sin(\xi x)| = 1,$$

we have $|f(x)e^{-i\xi x}| = |f(x)|$ and as $\int_{-\infty}^{\infty} |f(x)|dx < \infty$, the assertion follows from the Weierstrass M-test. At the same time we have

$$|\widehat{f}(\xi)| = \frac{1}{\sqrt{2\pi}} \left| \int_{-\infty}^{\infty} f(x)e^{-i\xi x}dx \right| \leq \frac{1}{\sqrt{2\pi}} \int_{-\infty}^{\infty} |f(x)e^{-i\xi x}|dx$$

$$= \frac{1}{\sqrt{2\pi}} \int_{-\infty}^{\infty} |f(x)|dx < \infty.$$

Hence $\sup_{\xi \in \mathbb{R}} |\widehat{f}(\xi)| < \infty$, and $\widehat{f}$ is *bounded* in $\mathbb{R}$.

---

[14] A completely general treatment of the Fourier transform on either $\mathbb{R}$ or $\mathbb{R}^n$ must necessarily be based on Lebesgue integration and, as such, is beyond the scope of this book. See for example [12].

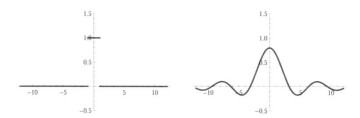

Figure 5.7: $f(x) = \chi_{(-1,1)}(x)$ and its Fourier transform $\widehat{f}(\xi) = \sqrt{\frac{2}{\pi}}\left(\frac{\sin\xi}{\xi}\right)$

Example 5.5.2. Find the Fourier transform of the characteristic function

$$f(x) = \chi_{(-a,a)}(x) = \begin{cases} 1 \text{ if } |x| < a \\ 0 \text{ if } |x| \geq a. \end{cases}$$

Solution. We have

$$\widehat{f}(\xi) = \frac{1}{\sqrt{2\pi}}\int_{-\infty}^{\infty}\chi_{(-a,a)}(x)e^{-i\xi x}dx = \frac{1}{\sqrt{2\pi}}\int_{-a}^{a}e^{-i\xi x}dx$$

$$= \frac{1}{\sqrt{2\pi}}\int_{-a}^{a}\cos\xi x\,dx = \sqrt{\frac{2}{\pi}}\left(\frac{\sin a\xi}{\xi}\right).$$

Note that the function $\widehat{f}(\xi) = \sqrt{\frac{2}{\pi}}\left(\frac{\sin a\xi}{\xi}\right)$, $\widehat{f}(0) = \sqrt{\frac{2}{\pi}}a$ is continuous on $\mathbb{R}$ and $\lim_{\xi\to\infty}|\widehat{f}(\xi)| = 0$, but it is not absolutely integrable.

Example 5.5.3. Find the Fourier transform of $f(x) = e^{-a|x|}$, $a > 0$. Note that $f(x)$ is continuous, decreases rapidly at infinity, but is not differentiable at $x = 0$.

Solution. We have

$$\widehat{f}(\xi) = \frac{1}{\sqrt{2\pi}}\int_{-\infty}^{\infty}e^{-a|x|-i\xi x}dx = \frac{1}{\sqrt{2\pi}}\left[\int_{-\infty}^{0}e^{(a-i\xi)x}dx + \int_{0}^{\infty}e^{-(a+i\xi)x}dx\right]$$

$$= \frac{1}{\sqrt{2\pi}}\left[\frac{1}{a-i\xi} + \frac{1}{a+i\xi}\right] = \sqrt{\frac{2}{\pi}}\frac{a}{a^2+\xi^2}.$$

Here also $\widehat{f}(\xi) = \sqrt{\frac{2}{\pi}} \frac{a}{a^2 + \xi^2}$ is continuous on $\mathbb{R}$ and $\lim_{\xi \to \infty} |\widehat{f}(\xi)| = 0$.

We now derive the basic properties of the Fourier transform.

The *linearity* of the Fourier transform is an immediate consequence of the linearity of the integral. So, if $f(x)$ and $g(x)$ are two functions with Fourier transforms $\widehat{f}(\xi)$ and $\widehat{g}(\xi)$ respectively, and $c_1, c_2 \in \mathbb{R}$, then

$$(\widehat{c_1 f + c_2 g})(\xi) = c_1 \widehat{f}(\xi) + c_2 \widehat{g}(\xi).$$

Theorem 5.5.4. *(The continuity of the Fourier transform). Let f be absolutely integrable on $\mathbb{R}$. Then $\widehat{f}$ is continuous on $\mathbb{R}$.*[15]

Proof. Let $\epsilon > 0$ be given. Since f is absolutely integrable, there exists $R = R(\epsilon) > 0$ such that

$$\int_{|x| > R} |f(x)| dx < \frac{\epsilon}{4}.$$

Let $\delta = \delta(\epsilon) > 0$ be chosen such that

$$R\delta \int_{-R}^{R} |f(x)| dx < \frac{\epsilon}{2}.$$

For $h \in \mathbb{R}$ with $|h| < \delta$ and any $\xi \in \mathbb{R}$, we have

$$|\widehat{f}(\xi+h) - \widehat{f}(\xi)| = \left| \int_{-\infty}^{\infty} f(x) e^{-i\xi x} \left(e^{-ihx} - 1 \right) dx \right| \leq \int_{-\infty}^{\infty} |f(x)| |e^{-ihx} - 1| dx$$

$$= \int_{-\infty}^{\infty} |f(x)| \left| \sin\left(\frac{hx}{2}\right) \right| dx \leq 2 \int_{|x| > R} |f(x)| dx + \int_{-R}^{R} |f(x)| |hx| dx$$

$$< \frac{\epsilon}{2} + R\delta \int_{-R}^{R} |f(x)| dx < \frac{\epsilon}{2} + \frac{\epsilon}{2} = \epsilon.$$

Hence $\widehat{f}$ is continuous. In fact we proved $\widehat{f}$ is *uniformly continuous* on $\mathbb{R}$.   □

---

[15] If f is absolutely integrable *and* continuous on $\mathbb{R}$, then the continuity of $\widehat{f}$ follows automatically from the uniform convergence of the integral $\widehat{f}(\xi) = \frac{1}{\sqrt{2\pi}} \int_{-\infty}^{\infty} f(x) e^{-i\xi x} dx$. See, Remark 5.3.23.

Theorem 5.5.5. *(Riemann-Lebesgue lemma).* *Let* $f : [a, b] \to \mathbb{C}$ *be absolutely integrable and piecewise continuous on* $[a, b]$. *Then*

$$\lim_{\xi \to \infty} \int_a^b f(x)e^{-i\xi x}dx = 0. \tag{5.49}$$

Proof. Let $\epsilon > 0$. Since f is piecewise continuous in $[a, b]$ it is integrable there. So there exists a lower Darboux sum $\sum_{j=1}^m m_j \Delta x_j$, where $m_j = \inf_{x \in [x_{j-1}, x_j]} f(x)$, such that

$$0 < \int_a^b f(x)dx - \sum_{j=1}^m m_j \Delta x_j < \frac{\epsilon}{2}.$$

Consider the piecewise constant function defined in $[a, b]$ by $g(x) = m_j$ for $x \in [x_{j-1}, x_j]$ for $j = 1, ..., m$. Then $g(x) \leq f(x)$ on $[a, b]$ and

$$0 < \int_a^b (f(x) - g(x))dx = \int_a^b f(x)dx - \sum_{j=1}^m m_j \Delta x_j < \frac{\epsilon}{2}.$$

Now

$$\left| \int_a^b f(x)e^{-i\xi x}dx \right| = \left| \int_a^b (f(x) - g(x))e^{-i\xi x}dx + \int_a^b g(x)e^{-i\xi x}dx \right|$$

$$\leq \int_a^b (f(x) - g(x))e^{-i\xi x}dx + \left| \int_a^b g(x)e^{-i\xi x}dx \right|. \tag{5.50}$$

But

$$\int_a^b g(x)e^{-i\xi x}dx = \sum_{j=1}^m m_j \int_{x_{j-1}}^{x_j} e^{-i\xi x}dx = -\frac{1}{i\xi}\sum_{j=1}^m (m_j e^{-i\xi x})\Big|_{x_{j-1}}^{x_j} \to 0,$$

as $\xi \to \infty$. Therefore, for sufficiently large ξ we have $\left| \int_a^b g(x)e^{-i\xi x}dx \right| < \frac{\epsilon}{2}$. Thus, for sufficiently large ξ, (5.50) yields

$$\left| \int_a^b f(x)e^{-i\xi x}dx \right| < \epsilon.$$

$\square$

Separating the real and imaginary parts in (5.49) we obtain

Corollary 5.5.6. *Let $f : [a, b] \to \mathbb{C}$ be absolutely integrable and piecewise continuous on $[a, b]$. Then*

$$\lim_{\xi \to \infty} \int_a^b f(x) \cos(\xi x) dx = 0,$$

$$\lim_{\xi \to \infty} \int_a^b f(x) \sin(\xi x) dx = 0.$$

Corollary 5.5.7. *(Riemann-Lebesgue lemma). Let f be absolutely integrable and piecewise continuous on $\mathbb{R}$. Then*

$$\lim_{\xi \to \infty} |\widehat{f}(\xi)| = 0.$$

Proof. Let $\epsilon > 0$. Since $\left| \int_{-\infty}^{\infty} f(x) e^{-i\xi x} dx \right|$ converges there exist $R = R(\epsilon) > 0$ such that

$$\left| \int_{|x|>R} f(x) e^{i\xi x} dx \right| < \frac{\epsilon}{2}.$$

On the other hand Theorem 5.5.5 shows for sufficiently large ξ

$$\left| \int_{-R}^{R} f(x) e^{i\xi x} dx \right| < \frac{\epsilon}{2}.$$

Thus, for sufficiently large ξ

$$\left| \int_{-\infty}^{\infty} f(x) e^{-i\xi x} dx \right| \leq \left| \int_{|x|>R} f(x) e^{-i\xi x} dx \right| + \left| \int_{-R}^{R} f(x) e^{-i\xi x} dx \right| < \epsilon.$$

$\square$

Theorem 5.5.8. *(The smoothness of the Fourier transform). Let f be continuous on $\mathbb{R}$. If the function $x^k f(x)$ is absolutely integrable on $\mathbb{R}$, then $\widehat{f}$ belongs to $C^k(\mathbb{R})$ $(k = 1, 2, ...)$ and*

$$\frac{d^k}{d\xi^k} (\widehat{f})(\xi) = (-i)^k \widehat{x^k f(x)}(\xi). \tag{5.51}$$

Proof. For $m < k$ and $|x| > 1$ we have $|x^m f(x)| \leq |x^k f(x)|$, and so $x^m f(x)$ is absolutely integrable. Since $|x^m f(x)e^{-i\xi x}| = |x^m f(x)|$, it follows (by the Weierstrass M-test) that the integrals

$$\frac{1}{\sqrt{2\pi}} \int_{-\infty}^{\infty} x^m f(x)e^{-i\xi x} dx$$

converge uniformly for all $\xi \in \mathbb{R}$. Hence differentiation under the integral sign of $\widehat{f}(\xi) = \frac{1}{\sqrt{2\pi}} \int_{-\infty}^{\infty} f(x)e^{-i\xi x} dx$, and of all these integrals, is legitimate. This successively yields

$$\frac{d}{d\xi}(\widehat{f})(\xi) = \frac{-i}{\sqrt{2\pi}} \int_{-\infty}^{\infty} x f(x)e^{-i\xi x} dx = (-i)\widehat{x f(x)}(\xi)$$

$$\cdots$$

$$\cdots$$

$$\frac{d^k}{d\xi^k}(\widehat{f})(\xi) = \frac{(-i)^k}{\sqrt{2\pi}} \int_{-\infty}^{\infty} x^k f(x)e^{-i\xi x} dx = (-i)^k \widehat{x^k f(x)}(\xi).$$

The uniform convergence of the above integrals also tells us that all these derivatives are continuous. Thus, $\widehat{f} \in C^k(\mathbb{R})$ □

An important property of the Fourier transform is that it turns *differentiation* of the original function into *multiplication* of the image (or the transform side) by i times the independent variable. In fact

Theorem 5.5.9. *Let $f \in C^1(\mathbb{R})$. If f and f' are absolutely integrable on $\mathbb{R}$, then*

$$\widehat{f'}(\xi) = (i\xi)\widehat{f}(\xi).$$

Proof. We first note that the assumptions on f and f' imply

$$\lim_{x \to \pm\infty} f(x) = 0.$$

This follows from the Fundamental Theorem of Calculus

$$f(x) = f(0) + \int_0^x f'(t)dt. \tag{5.52}$$

Since f' is absolutely integrable on $\mathbb{R}$ it is integrable on $\mathbb{R}$, and so (5.52) tells us that both limits $\lim_{x \to \pm\infty} f(x)$ exist. Moreover, the integrability of f on $\mathbb{R}$ implies that both these limits must be zero.

Now, we integrate by parts,

$$\widehat{f'}(\xi) = \frac{1}{\sqrt{2\pi}} \int_{-\infty}^{\infty} f'(x) e^{-i\xi x} dx$$

$$= \frac{1}{\sqrt{2\pi}} \left(f(x) e^{-i\xi x} |_{x=-\infty}^{\infty} + (i\xi) \int_{-\infty}^{\infty} f(x) e^{-i\xi x} dx \right) = (i\xi)\widehat{f}(\xi).$$

$\square$

Corollary 5.5.10. *Let* $f \in C^k(\mathbb{R})$ $(k = 1, 2, ...)$. *If all the functions* $f, f', ..., f^{(k)}$ *are absolutely integrable on* $\mathbb{R}$, *then*

$$\widehat{f^{(k)}}(\xi) = (i\xi)^k \widehat{f}(\xi). \tag{5.53}$$

In particular, $|\xi^k \widehat{f}(\xi)| \to 0$ *as* $\xi \to \infty$.

Proof. The hypothesis implies that the functions $f, f', ..., f^{(k-1)}$ tend to zero as $x \to \pm\infty$. Now, induction on k and integration by parts, yield the result. At the same time, by the Riemann-Lebesgue lemma $|\xi^k \widehat{f}(\xi)| = |\widehat{f^{(k)}}(\xi)| \to 0$ as $\xi \to \infty$. $\square$

Formulas (5.51) and (5.53) illustrate a fundamental property of the Fourier transform: *the smoother the function* f *is, the faster its Fourier transform* $\widehat{f}$ *tends to zero at infinity, and vice versa.*

The requirement that the function f is absolutely integrable on $\mathbb{R}$, in the definition of the Fourier transform is a rather restrictive requirement on f. For example, many common functions that occur frequently in applications, such as constant functions, polynomials, the trigonometric functions $\sin(ax)$, $\cos(ax)$, exponential functions are not absolutely integrable on $\mathbb{R}$ and don't have Fourier transforms. In addition, the Fourier transform of an absolutely integrable function may not be absolutely integrable, as is the case of Example 5.5.2. In such a situation, one will encounter serious problems when trying to recover the function from its Fourier transform. These are very unsatisfactory features in the theory of Fourier transforms. However, as we shall see below, the theory of the Fourier transform works very well with a smaller class of functions, the class of *rapidly decreasing functions at infinity*, known also as the *Schwartz space*.

5.5.1   The Schwartz space

Definition 5.5.11. An infinitely differentiable function $f : \mathbb{R} \to \mathbb{C}$ is called *rapidly decreasing at infinity* if the function $x^m f^{(k)}(x)$ is bounded on $\mathbb{R}$ for all non-negative integers m and k, that is, if

$$\sup_{x \in \mathbb{R}} |x^m f^{(k)}(x)| < \infty.$$

This set of functions is called the *Schwartz space*[16] and is denoted by $\mathcal{S} = \mathcal{S}(\mathbb{R})$.

Note, in particular, every $f \in \mathcal{S}$ is bounded (take $m = 0$, $k = 0$). Moreover, since $x^{m+1} f(x)$ is bounded, $|x^{m+1} f(x)| \le M$ for some $M > 0$, and so

$$\lim_{x \to \pm\infty} |x^m f(x)| = 0.$$

In other words, $\mathcal{S}$ is the space of C^∞ functions on $\mathbb{R}$ which decrease to zero as $x \to \pm\infty$ more rapidly than $|x|^{-m}$ for any positive integer m.

For $f \in \mathcal{S}$ and $c \in \mathbb{C}$ it follows immediately from the definition that $cf \in \mathcal{S}$. At the same time, if $f, g \in \mathcal{S}$, then the triangle inequality yields $f + g \in \mathcal{S}$. Hence $\mathcal{S}$ is a complex vector space. The product rule tells us that $fg \in \mathcal{S}$ whenever $f, g \in \mathcal{S}$. Furthermore, applying the product rule repeatedly we see that by multiplication and differentiation one obtains new functions in $\mathcal{S}$. Putting it differently, $\mathcal{S}$ is invariant under multiplication and differentiation. We shall see below that $\mathcal{S}$ has also the important property of being invariant under the Fourier transform, viz, if $f \in \mathcal{S}$, then $\widehat{f} \in \mathcal{S}$.

Does such a "nice" space, as $\mathcal{S}$, have functions in it besides the zero function $f(x) \equiv 0$? Of course it has, and in fact plenty! Clearly all C^∞ functions with compact support belong to $\mathcal{S}$, that is,

$$C_c^\infty(\mathbb{R}) \subset \mathcal{S}(\mathbb{R}).$$

---

[16]L. Schwartz (1915-2002). Best known for his work on the *theory of distributions*, which gives a well-defined meaning to objects such as the Dirac delta function. For a long time he taught at the Ecole Polytechnique.

In addition, the *Gausian* function $f(x) = e^{-ax^2}$, $(a > 0)$ is in $\mathcal{S}$. As the reader can check by repeated application of L'Hopital's rule

$$\lim_{x \to \pm\infty} x^m e^{-ax^2} = 0,$$

for any non-negative integer m, and this tells us that $e^{-ax^2} \in \mathcal{S}$ Note that the function $\frac{1}{1+x^2} \notin \mathcal{S}$, since $\frac{x^m}{1+x^2}$ does not approach zero as $x \to \pm\infty$ if $m \geq 2$. Also the function $e^{-|x|} \notin \mathcal{S}$ but for different reason.

The Gaussian function plays a central role in the theory of the Fourier transform. We will compute its Fourier transform shortly. Before doing so, it is important to note that every function $f \in \mathcal{S}$ is absolutely integrable on $\mathbb{R}$ and hence has Fourier transform. This follows from the estimate $|x| \leq \frac{1}{2}(1 + |x|^2)$ and the fact that the integral $\int_{-\infty}^{\infty} \frac{1}{1+x^2} dx$ converges (to π).

Next we compute the Fourier transform of the Gaussian function.

Proposition 5.5.12. *(The* Gaussian *function). If* $f(x) = e^{-ax^2}$, $a > 0$, *then*

$$\widehat{f}(\xi) = \frac{1}{\sqrt{2a}} e^{-\frac{\xi^2}{4a}}.$$

Proof. We have

$$\widehat{f}(\xi) = \frac{1}{\sqrt{2\pi}} \int_{-\infty}^{\infty} e^{-ax^2} e^{-i\xi x} dx.$$

Differentiating under the integral sign (which is legitimate)

$$\frac{d}{d\xi}\widehat{f}(\xi) = \frac{1}{\sqrt{2\pi}} \int_{-\infty}^{\infty} e^{-ax^2} \frac{\partial}{\partial\xi}(e^{-i\xi x}) dx = \frac{-i}{\sqrt{2\pi}} \int_{-\infty}^{\infty} x e^{-ax^2} e^{-i\xi x} dx$$

$$= \frac{i}{2a\sqrt{2\pi}} \int_{-\infty}^{\infty} \frac{\partial}{\partial x}(e^{-ax^2}) e^{-i\xi x} dx = \frac{-\xi}{2a\sqrt{2\pi}} \int_{-\infty}^{\infty} e^{-ax^2} e^{-i\xi x} dx,$$

where for last equality we integrated by parts and used $\lim_{x \to \pm\infty} e^{-ax^2} = 0$. Hence

$$\frac{d}{d\xi}\widehat{f}(\xi) = -\frac{\xi}{2a}\widehat{f}(\xi). \tag{5.54}$$

Equation (5.54) is a first order ordinary differential equation in $\widehat{f}' = \widehat{f}'(\xi)$. By Theorem 7.2.1, its general solution is

$$\widehat{f}(\xi) = Ce^{-\frac{\xi^2}{4a}}. \tag{5.55}$$

Note $C = \widehat{f}(0) = \frac{1}{\sqrt{2\pi}}\int_{-\infty}^{\infty} e^{-ax^2}\,dx$. Setting $x^2 = \frac{t^2}{a}$ and using Example 5.2.7, we get

$$\widehat{f}(0) = \frac{1}{\sqrt{2a\pi}}\int_{-\infty}^{\infty} e^{-t^2}\,dt = \frac{1}{\sqrt{2a\pi}}\sqrt{\pi} = \frac{1}{\sqrt{2a}}.$$

So $C = \frac{1}{\sqrt{2a}}$, and equation (5.55) becomes

$$\widehat{f}(\xi) = \frac{1}{\sqrt{2a}}e^{-\frac{\xi^2}{4a}}.$$

$\square$

Hence for $a = \frac{1}{2}$ the Gaussian function $f(x) = e^{-\frac{x^2}{2}}$ is its own Fourier transform $\widehat{f}(\xi) = e^{-\frac{\xi^2}{2}}$, that is, $\mathcal{F}\left\{e^{-\frac{x^2}{2}}\right\} = e^{-\frac{\xi^2}{2}}$. In particular, $f = \widehat{f}$ and so forth. We shall generalize this to arbitrary functions in $\mathcal{S}$, and find that $\widehat{\widehat{f}}(x) = f(-x)$. In the special case just considered, the minus sign does not appear because the Gaussian function is even.

Theorem 5.5.13. *If $f \in \mathcal{S}$ then $\widehat{f} \in \mathcal{S}$.*

Proof. Let $f \in \mathcal{S}$. By Theorem 5.5.8 $\widehat{f} \in C^{\infty}(\mathbb{R})$. As we noted above $\mathcal{S}$ is invariant under both multiplication and differentiation, and so the function $(x^m f(x))^{(k)} \in \mathcal{S}$ for any non-negative integers k and m. Using formulas (5.51) and (5.53) we have

$$\mathcal{F}[(x^m f(x))^{(k)}](\xi) = (i\xi)^k (\widehat{x^m f(x)})(\xi) = i^{k+m}\xi^k \widehat{f}^{(k)}(\xi).$$

The Riemann-Lebesgue lemma now tells us that $\xi^k \widehat{f}^{(k)}(\xi) \to 0$ as $\xi \to \infty$ and hence $\widehat{f} \in \mathcal{S}$. $\square$

We saw above that the Fourier transform maps the Schwartz space $\mathcal{S}$ into itself, i.e., $\mathcal{F}(\mathcal{S}) \subseteq \mathcal{S}$. Now we show that it is invertible on $\mathcal{S}$.

Theorem 5.5.14. *(Fourier inversion formula). Let* $f \in \mathcal{S}$. *Then*

$$f(x) = \frac{1}{\sqrt{2\pi}} \int_{-\infty}^{\infty} \widehat{f}(\xi) e^{i\xi x} d\xi. \tag{5.56}$$

Proof. For any $g \in \mathcal{S}$ we have

$$\int_{-\infty}^{\infty} g(\xi)\widehat{f}(\xi) e^{i\xi x} d\xi = \int_{-\infty}^{\infty} g(\xi) \left(\frac{1}{\sqrt{2\pi}} \int_{-\infty}^{\infty} f(y)(\xi) e^{-i\xi y} dy \right) e^{i\xi x} d\xi$$

$$= \frac{1}{\sqrt{2\pi}} \int_{-\infty}^{\infty} \left(\int_{-\infty}^{\infty} g(\xi) e^{-i\xi(y-x)} d\xi \right) f(y) dy$$

$$= \int_{-\infty}^{\infty} \widehat{g}(y-x) f(y) dy = \int_{-\infty}^{\infty} \widehat{g}(u) f(u+x) du,$$

where in the second equality the change of the order of integration is legitimate by Theorem 5.3.16 since f and g are in $\mathcal{S}$. Thus

$$\int_{-\infty}^{\infty} g(\xi)\widehat{f}(\xi) e^{i\xi x} d\zeta = \int_{-\infty}^{\infty} \widehat{g}(y) f(y+x) dy. \tag{5.57}$$

For any $\epsilon > 0$, let $g_\epsilon(\xi) = g(\epsilon\xi)$. Then

$$\widehat{g}_\epsilon(y) = \frac{1}{\sqrt{2\pi}} \int_{-\infty}^{\infty} g(\epsilon\xi) e^{-i\xi y} d\xi = \frac{\epsilon^{-1}}{\sqrt{2\pi}} \int_{-\infty}^{\infty} g(t) e^{-it\frac{y}{\epsilon}} dt = \epsilon^{-1} \widehat{g}\left(\frac{y}{\epsilon}\right).$$

Now (5.57) with g replaced by g_ϵ gives

$$\int_{-\infty}^{\infty} g(\epsilon\xi)\widehat{f}(\xi) e^{i\xi x} d\xi = \int_{-\infty}^{\infty} \epsilon^{-1} \widehat{g}\left(\frac{y}{\epsilon}\right) f(y+x) dy = \int_{-\infty}^{\infty} \widehat{g}(u) f(\epsilon u + x) du.$$

By the Weierstrass M-test the first and the third integral above both converge absolutely and uniformly with respect to ϵ, and so they define continuous functions in ϵ (see, Theorem 5.3.14). Letting $\epsilon \to 0$ we find

$$g(0) \int_{-\infty}^{\infty} \widehat{f}(\xi) e^{i\xi x} d\xi = f(x) \int_{-\infty}^{\infty} \widehat{g}(u) du.$$

Taking $g(x) = e^{-\frac{x^2}{2}}$ and recalling the Poisson-Euler integral $\int_{-\infty}^{\infty} e^{-x^2} dx = \sqrt{\pi}$ we see that $g(0) = 1$ and $\int_{-\infty}^{\infty} \widehat{g}(u) du = \int_{-\infty}^{\infty} e^{-\frac{u^2}{2}} du = \sqrt{2\pi}$. That is, we obtain (5.56). $\square$

Corollary 5.5.15. *(Parseval's equality). If $f, g \in \mathcal{S}$, then*

$$\int_{-\infty}^{\infty} \widehat{f}(\xi) g(\xi) d\xi = \int_{-\infty}^{\infty} f(y) \widehat{g}(y) dy. \tag{5.58}$$

Proof. Set $x = 0$ in (5.57).   □

Definition 5.5.16. Let $f \in \mathcal{S}$. The *inverse Fourier transform* $\mathcal{F}^{-1}(f) = \widetilde{f}$ is defined by

$$\widetilde{f}(x) = \frac{1}{\sqrt{2\pi}} \int_{-\infty}^{\infty} f(\xi) e^{i\xi x} d\xi. \tag{5.59}$$

Note that Theorem 5.5.14 tells us $\widetilde{\widehat{f}} = f = \widehat{\widetilde{f}}$. Thus we have proved that the Fourier transform $\mathcal{F} : \mathcal{S} \to \mathcal{S}$ is a bijection.[17] At the same time a glance at (5.59) tells us $\widehat{f}(x) = \widetilde{f}(-x)$ and so, $\widehat{\widehat{f}}(x) = f(-x)$. Moreover,

$$\overline{\widehat{f}} = \widetilde{\overline{f}} \quad \text{and} \quad \overline{\widetilde{f}} = \widehat{\overline{f}}.$$

The Schwartz space can be equipped with an inner product

$$\langle f, g \rangle = \int_{-\infty}^{\infty} f(x) \overline{g(x)} dx$$

with associated norm

$$||f||^2 = \langle f, f \rangle = \int_{-\infty}^{\infty} |f(x)|^2 dx.$$

Theorem 5.5.17. *(Plancherel). Let $f, g \in \mathcal{S}$. Then*

$$\langle f, g \rangle = \langle \widehat{f}, \widehat{g} \rangle. \tag{5.60}$$

In particular, for $f = g$

$$||\widehat{f}||^2 = ||f||^2.$$

---

[17]In more advanced works on the Fourier transform, a topology is introduced in $\mathcal{S}$ via the family of seminorms $\rho_{m,k}(f) = \sup_{x \in \mathbb{R}} |x^m f^{(k)}(x)|$ and $\mathcal{S}$ is considered as a topological vector space. It turns out then that the Fourier transform and its inverse viewed as linear operators acting $\mathcal{S}$ are both continuous (in this topology). The conclusion then reads *the Fourier transform is a (topological) isomorphism of $\mathcal{S}$ onto itself.*

Proof. Since $\widehat{\widehat{g}} = \overline{g}$, replacing g by $\overline{g}$ in (5.58) we obtain (5.60).   □

Finally we look into the question of how the Fourier transform transforms the convolution of two functions. Since functions in $\mathcal{S}$ are absolutely integrable and bounded, Theorem 5.3.36 tells us that the *convolution*

$$(f * g)(x) = \int_{-\infty}^{\infty} f(y)g(x - y)dy$$

of two functions $f, g \in \mathcal{S}$ always exists (and is bounded). Moreover from Proposition 5.3.38 $f * g$ is absolutely integrable on $\mathbb{R}$.

The Fourier transform converts convolution to pointwise multiplication.

Theorem 5.5.18. *Let $f, g \in \mathcal{S}$. Then $f * g \in \mathcal{S}$ and*

$$\mathcal{F}(f * g)(\xi) = (2\pi)^{\frac{1}{2}} \widehat{f}(\xi)\widehat{g}(\xi). \tag{5.61}$$

Proof. Since $\mathcal{F} : \mathcal{S} \to \mathcal{S}$ is a bijection, all we need to show is (5.61). We have

$$\mathcal{F}(f * g)(\xi) = \frac{1}{\sqrt{2\pi}} \int_{-\infty}^{\infty} (f * g)(x)e^{-i\xi x}dx$$

$$= \frac{1}{\sqrt{2\pi}} \int_{-\infty}^{\infty} \left(\int_{-\infty}^{\infty} f(y)g(x - y)dy \right) e^{-i\xi x}dx$$

$$= \frac{1}{\sqrt{2\pi}} \int_{-\infty}^{\infty} f(y) \left(\int_{-\infty}^{\infty} g(x - y)e^{-i\xi(x-y)}dx \right) e^{-i\xi y}dy$$

$$= \frac{1}{\sqrt{2\pi}} \int_{-\infty}^{\infty} f(y) \left(\int_{-\infty}^{\infty} g(u)e^{-i\xi u}dx \right) e^{-i\xi y}dy$$

$$= \int_{-\infty}^{\infty} f(y)\widehat{g}(\xi)e^{-i\xi y}dy = \sqrt{2\pi}\widehat{f}(\xi)\widehat{g}(\xi).$$

The change of the order of integration in the second equality is legitimate from Theorem 5.3.16 since $f, g \in \mathcal{S}$.   □

Given a function f we denote by f_- the function defined by $f_-(x) = f(-x)$. It is straight forward from the definitions to verify that

$$(f_-)_- = f, \ (fg)_- = f_- g_-, \ (f * g)_- = f_- * g_-.$$

We recall that for any $f \in \mathcal{S}$ (5.59) tells us $\widehat{\widehat{f}} = f_-$ or $\mathcal{F}(\mathcal{F}(f)) = f_-$.

Corollary 5.5.19. *Let $f, g \in \mathcal{S}$. Then*

$$\mathcal{F}(fg) = (2\pi)^{-\frac{1}{2}} \widehat{f} * \widehat{g}. \tag{5.62}$$

Proof. Let $f = \widehat{\varphi}$ and $g = \widehat{\psi}$ with $\varphi, \psi \in \mathcal{S}$ Apllying (5.61) we have

$$\mathcal{F}(fg) = \mathcal{F}(\widehat{\varphi}\widehat{\psi}) = (2\pi)^{-\frac{1}{2}} \mathcal{F}(\mathcal{F}(\phi * \psi)) = (2\pi)^{-\frac{1}{2}} (\varphi * \psi)_-$$

$$= (2\pi)^{-\frac{1}{2}} (\varphi_- * \psi_-)) = (2\pi)^{-\frac{1}{2}} \widehat{\varphi} * \widehat{\psi} = (2\pi)^{-\frac{1}{2}} \widehat{f} * \widehat{g}.$$

$$\square$$

5.5.2   The Fourier transform on $\mathbb{R}^n$

The results obtained in this section generalize essentially without change to functions of several variables.

For each $\xi = (\xi_1, ..., \xi_n)$ and $x = (x_1, ..., x_n)$ in $\mathbb{R}^n$ consider the function $e_\xi : \mathbb{R}^n \to \mathbb{C}$ defined by

$$e_\xi(x) = e^{-i\langle x, \xi \rangle} = e^{-i(x_1\xi_1 + ... + x_n\xi_n)} = \prod_{j=1}^{n} e^{-ix_j\xi_j}.$$

Note that $e_\xi(x+y) = e_\xi(x)e_\xi(y)$, $|e_\xi(x)| = 1$ and e_ξ is continuous. Such a function is called a *character*.

Definition 5.5.20. For $f : \mathbb{R}^n \to \mathbb{C}$ absolutely integrable the *Fourier transform* $\mathcal{F}(f) \equiv \widehat{f}$ of f is defined by

$$\widehat{f}(\xi) = \frac{1}{(2\pi)^{\frac{n}{2}}} \int_{\mathbb{R}^n} f(x)e^{-i\langle x, \xi \rangle} dx, \tag{5.63}$$

where the integral is understood as an improper multiple Riemann integral.

As before we have

$$|\widehat{f}(\xi)| = \frac{1}{(2\pi)^{\frac{n}{2}}} \int_{\mathbb{R}^n} |f(x)| dx < \infty.$$

The Schwartz space is defined similarly replacing the absolute value $|\cdot|$ by the Euclidean norm $||\cdot||$ and ordinary derivatives by partial derivatives. The standard notation here is to introduce multi-index notation. Let $\alpha = (\alpha_1, ..., \alpha_n)$ and $\beta = (\beta_1, ..., \beta_n)$ be multi-indices consisting of non-negative integers α_j, β_j, $j = 1, ..., n$, and denote by ∂^α the differentiation operator

$$\partial^\alpha = \frac{\partial^{|\alpha|}}{\partial x_1^{\alpha_1} \cdots \partial x_n^{\alpha_n}}$$

of order $|\alpha| = \alpha_1 + ... + \alpha_n$ and $x^\beta = x_1^{\beta_1} \cdots x_1^{\beta_1}$.

Definition 5.5.21. The *Schwartz space* $\mathcal{S} = \mathcal{S}(\mathbb{R}^n)$ is the set of all function $f \in C^\infty(\mathbb{R}^n)$ such that

$$\sup_{x \in \mathbb{R}^n} |x^\beta \partial^\alpha f(x)| < \infty$$

for all non-negative multi-indices α and β.

For $x \in \mathbb{R}^n$ and any non-negative integer m, the estimate

$$||x||^m \leq 2^{-m}(1 + ||x||^2)^m \leq C \sum_{|\beta| \leq m} x^{2\beta}$$

tells us that $f \in \mathcal{S}(\mathbb{R}^n)$ if and only if

$$\sup_{x \in \mathbb{R}^n} |(1 + ||x||^2)^m \partial^\alpha f(x)| < \infty,$$

for any multi-index α and m non-negative integer.

Here also any infinite differentiable function with compact support belongs in $\mathcal{S}$,

$$C_c^\infty(\mathbb{R}^n) \subset \mathcal{S}(\mathbb{R}^n).$$

Example 5.5.22. The multidimensional Gaussian function is $f(x) = e^{-\frac{||x||^2}{2}}$ and $\widehat{f} = f$. Indeed, using Theorem 5.3.16 and Proposition 5.5.12, we see that

$$\widehat{f}(\xi) = \frac{1}{(2\pi)^{\frac{n}{2}}} \int_{\mathbb{R}^n} e^{-\frac{||x||^2}{2}} e^{-i\langle x, \xi \rangle} dx$$

$$= \prod_{j=1}^n \int_{-\infty}^\infty e^{-\frac{x_j^2}{2}} e^{-ix_j \xi_j} dx_j = \prod_{j=1}^n e^{-\frac{\xi_j^2}{2}} = e^{-\frac{||\xi||^2}{2}} = f(\xi).$$

Proposition 5.5.23. *Let $f \in \mathcal{S}(\mathbb{R}^n)$. Then f is absolutely integrable on $\mathbb{R}^n$.*

Proof. Since $(1 + ||x||^2)^m f$ is bounded on $\mathbb{R}^n$ for any positive integer m, we can choose $m > \frac{n}{2}$ such that $(1 + ||x||^2)^m |f(x)| \le M$ for all $x \in \mathbb{R}^n$. Now from Example 5.2.20, we have

$$\int_{\mathbb{R}^n} |f(x)| dx = \int_{\mathbb{R}^n} (1 + ||x||^2)^{-m} \left[(1 + ||x||^2)^m |f(x)| \right] dx$$

$$\le M \int_{\mathbb{R}^n} \frac{1}{(1 + ||x||^2)^m} dx < \infty.$$

$\square$

For any $f \in \mathcal{S}$ formulas (5.51) and (5.53) generalize to functions of several variables.

Theorem 5.5.24. *Let $f \in \mathcal{S}$. Then*

$$\partial^\alpha \widehat{f}(\xi) = (-i)^{|\alpha|} (\widehat{x^\alpha f(x)})(\xi). \tag{5.64}$$

$$\widehat{\partial^\alpha f}(\xi) = i^{|\alpha|} \xi^\alpha \widehat{f}(\xi). \tag{5.65}$$

Proof. The first of these is obtained by differentiation under the integral sign in (5.63) with respect to $\xi_1, ..., \xi_n$. The second can be obtained like (5.53), using Theorem 5.3.16 and integration by parts. $\square$

Corollary 5.5.25. *For any $f \in \mathcal{S}$*

$$\lim_{||\xi|| \to \infty} \widehat{f}(\xi) = 0.$$

Proof. Since $|\widehat{\partial^\alpha f}(\xi)| < \infty$, this follows from (5.65) $\square$

As in the case $n = 1$, the Fourier transform maps $\mathcal{S}(\mathbb{R}^n)$ to itself.

Theorem 5.5.26. *If $f \in \mathcal{S}$, then $\widehat{f} \in \mathcal{S}$.*

Proof. Formulas (5.64) and (5.65) imply

$$\mathcal{F}[\partial^\beta (x^\alpha f(x))](\xi) = i^{|\beta|} \xi^\beta (\widehat{x^\alpha f(x)})(\xi) = i^{|\alpha| + |\beta|} \xi^\beta \partial^\alpha \widehat{f}(\xi),$$

from which it follows that

$$\lim_{\|\xi\|\to\infty} \xi^\beta \partial^\alpha \widehat{f}(\xi) = 0.$$

Hence $\xi^\beta \partial^\alpha \widehat{f}(\xi)$ is bounded on $\mathbb{R}^n$ for any multi-indices α and β, i.e., $\widehat{f} \in \mathcal{S}$. $\qquad\qquad\square$

The Fourier inversion formula (5.59) is proved similarly and tells us that the Fourier transform is a bijection on $\mathcal{S}$.

Definition 5.5.27. For $f \in \mathcal{S}$ the *inverse Fourier transform* $\widetilde{f}$ of f is defined by

$$\widetilde{f}(x) = \frac{1}{(2\pi)^{\frac{n}{2}}} \int_{\mathbb{R}^n} f(\xi) e^{i\langle x, \xi \rangle} d\xi.$$

As before we have $\widetilde{\widehat{f}} = f = \widehat{\widetilde{f}}$ and $\widehat{\widetilde{f}}(\xi) = f(-\xi)$.

Finally the convolution property proved similarly generalizes to

$$\mathcal{F}(f * g)(\xi) = (2\pi)^{\frac{n}{2}} \widehat{f}(\xi) \widehat{g}(\xi).$$

A note on the Lebesgue Integral (*)

In Section 5.3 we saw that the *uniform convergence* of an integral is essential in passing a limit inside an integral. We noted there (Remark 5.3.21) that if the convergence is *not* uniform, then this is *no* longer legitimate. This is a serious weakness of the Riemann integral. As we mentioned in Remark 4.1.31 a more general and richer theory of integration with powerfull convergence theorems (the main one being the *Lebesgue dominated convergence theorem*) was developed by Henri Lebesgue in 1904. The so-called *Lebesgue integral* although more complicated than the Riemann integral, it brought with it the theory of measure and revolutionized analysis. The Lebesgue theory of integration enables us to extend the class of functions that can be integrated and can be itself extended to abstract settings.

We state without proof the following results on the relation of the Riemann integral and the Lebesgue integral.

Theorem 5.5.28. *Let R be a closed rectangle in $\mathbb{R}^n$ and $f : R \to \mathbb{R}$ a bounded function. If f is Riemann integrable over R, then it is Lebesgue integrable and its Lebesgue integral*

$$\int_R f(x)dx = (Riemann) \int_R f(x)dx.$$

Theorem 5.5.29. *Let $f : [a,\infty) \to \mathbb{R}$ be Riemann integrable on every closed subinterval of $[a,\infty)$. Then f is Lebesgue integrable over $[a,\infty)$ if and only if $\int_a^\infty |f(x)|dx$ exists, and in this case the Lebesgue integral*

$$\int_{[a,\infty)} f(x)dx = \int_a^\infty f(x)dx.$$

Applications of Lebesgue integration arise in all of modern analysis, in Fourier analysis, Functional analysis (the basic theorems of functional analysis require complete spaces of Lebesgue integrable functions), the theory of Distributions (or generalized functions) and Probability theory. Detailed accounts of the Lebesgue integral can be found, for example, in [28] or [12].

5.6 Solved problems for Chapter 5

Exercise 5.6.1. Evaluate the integral

$$I = \int_0^\pi \int_0^\pi |\cos(x+y)|dxdy.$$

Solution. Set $u = x + y$ and $v = y$, so that $x = u - v$ and $\frac{\partial(x,y)}{\partial(u,v)} = 1$. The Change of Variables formula (5.13) yields

$$\int_0^\pi \int_0^\pi |\cos(x+y)|dxdy = \int_0^\pi \left(\int_v^{v+\pi} |\cos u|du \right) dv.$$

Now

$$\int_v^{v+\pi} |\cos u|du = \int_v^{\frac{\pi}{2}} |\cos u|du + \int_{\frac{\pi}{2}}^\pi |\cos u|du + \int_\pi^{v+\pi} |\cos u|du$$

$$= \sin u \Big|_v^{\pi/2} + (-\sin u)bigg|_{\pi/2}^{\pi} + (-\sin u)\Big|_\pi^{\pi+v} = 1 - \sin v + 1 - \sin(\pi + v) = 2.$$

Hence $I = \int_0^\pi 2dv = 2\pi$.

Example 5.6.2. Evaluate the integral

$$I = \int_0^1 \int_0^x \sqrt{x^2 + y^2}\,dydx$$

by changing to polar coordinates.

Solution. The region of integration Ω is the triangle bounded by the lines $y = 0$, $x = 1$, $y = x$. In polar coordinates these lines are $\theta = 0$, $r\cos\theta = 1$, $\theta = \frac{\pi}{4}$ and the new region of integration is $\left\{(r, \theta) : 0 \le \theta \le \frac{\pi}{4}, 0 \le r \le \sec\theta\right\}$. Therefore

$$\int_0^1 \int_0^x \sqrt{x^2 + y^2}\,dydx = \int_0^{\frac{\pi}{4}} \left(\int_0^{\sec\theta} r^2\,dr \right) d\theta$$

$$= \frac{1}{3} \int_0^{\frac{\pi}{4}} \sec^3\theta\,d\theta = \frac{1}{6}[\sqrt{2} + \log(1 + \sqrt{2})].$$

Exercise 5.6.3. Evaluate the integral

$$I = \int\int\int_\Omega \sqrt{1 - \frac{x^2}{a^2} - \frac{y^2}{b^2} - \frac{z^2}{c^2}}\,dxdydz,$$

where Ω is the *ellipsoid* $\frac{x^2}{a^2} + \frac{y^2}{b^2} + \frac{z^2}{c^2} \le 1$.

Solution. We pass to spherical coordinates, where $\frac{x}{a} = r\cos\theta\sin\phi$, $\frac{y}{b} = r\sin\theta\sin\phi$, $\frac{z}{c} = r\cos\phi$ with $0 \le r \le 1$, $0 \le \theta \le 2\pi$, $0 \le \phi \le \pi$. Then $\frac{\partial(x,y,z)}{\partial(r,\theta,\phi)} = abcr^2\sin\phi$, and the Change of Variables formula yields

$$I = abc \int_0^{2\pi} \left(\int_0^\pi \sin\phi \left(\int_0^1 (\sqrt{1 - r^2})r^2\,dr \right) d\phi \right) d\theta = \frac{\pi^2 abc}{4}.$$

In particular,

$$\int\int\int_{x^2+y^2+y^2\le 1} \sqrt{1 - x^2 - y^2 - z^2}\,dxdydz = \frac{\pi^2}{4}.$$

Exercise 5.6.4. Let Ω be a bounded simple set in $\mathbb{R}^n$ which is symmetric about the origin (i.e., if $x \in \Omega$, then $-x \in \Omega$). Let f be an integrable on Ω such that $f(-x) = -f(x)$. Show that

$$\int_\Omega f = 0.$$

Solution. Set $y = \varphi(x) = -x$. Then by the Change of Variables formula

$$\int_{\varphi(\Omega)} f(y)dy = \int_\Omega f(\varphi(x))|J_\varphi(x)|dx.$$

Since $\varphi(\Omega) = \Omega$, $|J_\varphi(x)| = 1$ and $f(\varphi(x)) = -f(x)$ we get $\int_\Omega f(y)dy = -\int_\Omega f(x)dx$. Hence $\int_\Omega f = 0$.

Exercise 5.6.5. Evaluate the integral

$$\int\int_{\mathbb{R}^2} e^{-(x^2+(y-x)^2+y^2)}dxdy.$$

Solution. Make the linear change of variables $u = x + y$, $v = y - x$. Then $\frac{\partial(x,y)}{\partial(u,v)} = \frac{1}{2}$ and the Change of Variables formula yields

$$\int\int_{\mathbb{R}^2} e^{-(x^2+(y-x)^2+y^2)}dxdy = \frac{1}{2}\int\int_{\mathbb{R}^2} e^{-\frac{1}{2}(u^2+3v^2)}dudv$$

$$= \frac{1}{2}\int\int_{\mathbb{R}^2} e^{-\frac{1}{2}u^2}e^{-\frac{3}{2}v^2}dudv$$

$$= \frac{1}{2}\left(\int_{\mathbb{R}} e^{-\frac{1}{2}u^2}du\right)\left(\int_{\mathbb{R}} e^{-\frac{3}{2}v^2}dv\right) = \frac{1}{2}\left(\sqrt{2\pi}\right)\left(\sqrt{\frac{2\pi}{3}}\right) = \frac{\pi}{\sqrt{3}},$$

where we used

$$\int_{\mathbb{R}} e^{-\alpha x^2}dx = \left(\frac{\pi}{\alpha}\right)^{\frac{1}{2}}$$

from Example 5.2.21.

Exercise 5.6.6. Show that

$$\int\int_{x,y>0} \frac{1}{(1+y)(1+x^2y)}dxdy = \frac{\pi^2}{2}.$$

Use this to show

$$\int_0^\infty \frac{\log x}{x^2-1}dx = \frac{\pi^2}{4}.$$

Solution. As the integrand is positive

$$\int_0^\infty \int_0^\infty \frac{1}{(1+y)(1+x^2y)} dx dy = \int_0^\infty \frac{1}{1+y} \left(\int_0^\infty \frac{1}{1+x^2y} dx \right) dy$$

$$= \frac{\pi}{2} \int_0^\infty \frac{1}{(1+y)\sqrt{y}} dy = \frac{\pi}{2} \frac{\pi}{\sin(\frac{\pi}{2})} = \frac{\pi^2}{2}.$$

Now notice

$$\frac{\pi^2}{2} = \int_0^\infty \left(\int_0^\infty \frac{1}{(1+y)(1+x^2y)} dy \right) dx = \int_0^\infty \left(\frac{1}{x^2-1} \int_0^\infty \left[\frac{x^2}{1+x^2y} - \frac{1}{1+y} \right] dy \right) dx$$

$$= \int_0^\infty \left(\frac{1}{x^2-1} \left[\log \frac{1+x^2y}{1+y} \right]_{y=0}^{y=\infty} \right) dx = 2 \int_0^\infty \frac{\log x}{x^2-1}.$$

Hence $\int_0^\infty \frac{\log x}{x^2-1} dx = \frac{\pi^2}{4}$.

Exercise 5.6.7. Evaluate the integral

$$I = \int \int \int_\Omega \frac{1}{(1+x^2z^2)(1+y^2z^2)} dx dy dz,$$

where $\Omega = \{(x, y, z) : 0 < x < 1, 0 < y < 1, z > 0\}$.
Use the answer to show that

$$\int_0^\infty \left(\frac{\tan^{-1} t}{t} \right)^2 dt = \pi \log 2.$$

Solution.

$$I = \int_0^1 \int_0^1 \left(\int_0^\infty \frac{1}{(1+x^2z^2)(1+y^2z^2)} dz \right) dx dy$$

$$= \int_0^1 \int_0^1 \left(\frac{1}{x^2-y^2} \int_0^\infty \left[\frac{x^2}{1+x^2z^2} - \frac{y^2}{1+y^2z^2} \right] dz \right) dx dy$$

$$= \frac{\pi}{2} \int_0^1 \int_0^1 \frac{1}{x+y} dx dy = \frac{\pi}{2} \int_0^1 \left(\int_0^1 \frac{1}{x+y} dy \right) dx$$

$$= \frac{\pi}{2} \int_0^1 (\log(x+1) - \log x) dx = \pi \log 2.$$

Now we have

$$I = \int_0^\infty \left(\int_0^1 \int_0^1 \frac{1}{(1 + x^2 z^2)(1 + y^2 z^2)} dx dy \right) dz$$

$$= \int_0^\infty \left(\int_0^1 \frac{1}{1 + x^2 z^2} dx \right)^2 dz = \int_0^\infty \left(\frac{\tan^{-1} z}{z} \right)^2 dz.$$

Hence $\int_0^\infty \left(\frac{\tan^{-1} t}{t} \right)^2 dt = \pi \log 2$.

The following extends Example 5.2.21.

Exercise 5.6.8. Let $T : \mathbb{R}^n \to \mathbb{R}^n$ be an invertible linear transformation, and let $B = \{x \in \mathbb{R}^n : ||x|| \le R\}$. Show that

$$\int_{T^{-1}(B)} e^{-||Ty||^2} dy = \frac{1}{|\det T|} \int_B e^{-||x||^2} dx.$$

Let $R \to \infty$ to conclude that

$$\int_{\mathbb{R}^n} e^{-||Ty||^2} dy = \frac{\pi^{\frac{n}{2}}}{|\det T|}.$$

Solution. Taking $\varphi = T^{-1}$ in the Change of Variables formula (Theorem 5.1.14) we get

$$\int_{T^{-1}(B)} e^{-||Ty||^2} dy = \int_B e^{-||TT^{-1}x||^2} |\det T^{-1}| dx = \frac{1}{|\det T|} \int_B e^{-||x||^2} dx.$$

Letting $R \to \infty$ and using Example 5.2.21 (with $\alpha = 1$)

$$\int_{\mathbb{R}^n} e^{-||Ty||^2} dy = \frac{1}{|\det T|} \int_{\mathbb{R}^n} e^{-||x||^2} dx = \frac{\pi^{\frac{n}{2}}}{|\det T|}.$$

Exercise 5.6.9. Give a direct proof of Theorem 5.3.15. *If the integral $F(x) = \int_c^\infty f(x, y) dy$ converges uniformly on $[a, b]$, then*

$$\int_a^b \left(\int_c^\infty f(x, y) dy \right) dx = \int_c^\infty \left(\int_a^b f(x, y) dx \right) dy.$$

Solution. Since the integral is uniformly convergent, Definition 5.3.9 tells us that for any $\epsilon > 0$ there is $N = N(\epsilon)$ such that $\left| F(x) - \int_c^d f(x,y)dy \right| < \frac{\epsilon}{b-a}$ for $d \geq N$. Therefore

$$\left| \int_a^b F(x)dx - \int_a^b \left(\int_c^d f(x,y)dy \right) dx \right| \leq \int_a^b \left| F(x) - \int_c^d f(x,y)dy \right| dx < \epsilon.$$

On the other hand, by Fubini's theorem

$$\int_a^b \left(\int_c^d f(x,y)dy \right) dx = \int_c^d \left(\int_a^b f(x,y)dx \right) dy.$$

Hence for $d \geq N$ we have

$$\left| \int_a^b F(x)dx - \int_c^d \left(\int_a^b f(x,y)dx \right) dy \right| < \epsilon.$$

That is, the integral $\int_c^\infty (\int_a^b f(x,y)dx)dy$ exists, and is equal to $\int_a^b F(x)dx$.

Exercise 5.6.10. (*Fresnel's integrals*). Prove that

$$\int_0^\infty \sin(x^2)dx = \sqrt{\frac{\pi}{8}}$$

and

$$\int_0^\infty \cos(x^2)dx = \sqrt{\frac{\pi}{8}}.$$

Proof. Setting $x = \sqrt{t}$, so that $dx = \frac{1}{2\sqrt{t}}dt$, we get

$$\int_0^\infty \sin(x^2)dx = \frac{1}{2} \int_0^\infty \frac{\sin t}{\sqrt{t}}dt$$

and

$$\int_0^\infty \cos(x^2)dx = \frac{1}{2} \int_0^\infty \frac{\cos t}{\sqrt{t}}dt.$$

We shall compute the first integral. The second can be computed similarly. From Example 5.2.21 (for $n = 1$) we have

$$\frac{1}{\sqrt{t}} = \frac{2}{\sqrt{\pi}} \int_0^\infty e^{-ty^2}dy. \tag{5.66}$$

For $a > 0$ we consider

$$\int_0^\infty \frac{\sin t}{\sqrt{t}} e^{-at} dt.$$

Using (5.66) and changing the order of integration (Theorem 5.3.16) we have

$$\int_0^\infty \frac{\sin t}{\sqrt{t}} e^{-at} dt = \frac{2}{\sqrt{\pi}} \int_0^\infty \sin t e^{-at} \left(\int_0^\infty e^{-ty^2} dy \right) dt$$

$$= \frac{2}{\sqrt{\pi}} \int_0^\infty \left(\int_0^\infty \sin t e^{-(a+y^2)t} dy \right) dt$$

$$= \frac{2}{\sqrt{\pi}} \int_0^\infty \left(\int_0^\infty \sin t e^{-(a+y^2)t} dt \right) dy$$

$$= \frac{2}{\sqrt{\pi}} \int_0^\infty \frac{1}{1 + (a+y^2)^2} dy.$$

Now from Theorem 5.3.22 we get

$$\int_0^\infty \frac{\sin t}{\sqrt{t}} dt = \lim_{a \to 0^+} \int_0^\infty \frac{\sin t}{\sqrt{t}} e^{-at} dt = \lim_{a \to 0^+} \frac{2}{\sqrt{\pi}} \int_0^\infty \frac{1}{1 + (a+y^2)^2} dy$$

$$= \frac{2}{\sqrt{\pi}} \int_0^\infty \frac{1}{1 + y^4} dy = \frac{2}{\sqrt{\pi}} \frac{\pi \sqrt{2}}{4} = \sqrt{\frac{\pi}{2}}.$$

$\square$

Exercise 5.6.11. (*Volume and Surface Area of the unit ball in $\mathbb{R}^n$*). Show that the volume c_n of unit ball B^n in $\mathbb{R}^n$ is

$$c_n = \frac{2\pi^{\frac{n}{2}}}{n\Gamma(\frac{n}{2})},$$

while its surface area $((n-1)$-dimensional volume$)$ σ_n is

$$\sigma_n = \frac{2\pi^{\frac{n}{2}}}{\Gamma(\frac{n}{2})}.$$

Solution. Consider $\mathbb{R}^n$ and the usual integral $\int_{\mathbb{R}^n} f(x)dx$. Passing to spherical coordinates in $\mathbb{R}^n$, for $x \neq 0$ we write $x = r\Theta$, where r is the radial part of $x \in \mathbb{R}^n$ and $\Theta \in S^{n-1}$, the unit sphere of $\mathbb{R}^n$. Then $dx = r^{n-1}drd\Theta$, where

$$d\Theta = \sin^{n-2}\phi_1 \sin^{n-3}\phi_2 \cdots \sin\phi_{n-2}d\theta d\phi_1 \cdots d\phi_{n-2}.$$

For a *radial* function such as $f(x) = e^{-\|x\|^2}$,

$$\int_{\mathbb{R}^n - (0)} e^{-\|x\|^2} dx = \int_0^\infty e^{-r^2} r^{n-1} dr \int_{S^{n-1}} d\Theta = \sigma_n \int_0^\infty e^{-r^2} r^{n-1} dr,$$

where

$$\sigma_n = \int_{S^{n-1}} d\Theta = \int_0^{2\pi} \int_0^\pi \cdots \int_0^\pi \sin^{n-2}\phi_1 \sin^{n-3}\phi_2 \cdots \sin\phi_{n-2} d\theta d\phi_1 \cdots d\phi_{n-2}.$$

On the other hand, $\int_{\mathbb{R}^n - (0)} = \int_{\mathbb{R}^n}$ and as we know from (Example 5.2.21), $\int_{\mathbb{R}^n} e^{-\|x\|^2} dx = \pi^{\frac{n}{2}}$. Hence

$$\sigma_n = \frac{\pi^{\frac{n}{2}}}{\int_0^\infty e^{-r^2} r^{n-1} dr}.$$

If we can calculate the denominator, this last formula gives the total $n - 1$ dimensional volume (surface area) of S^{n-1}. Setting $r^2 = t$, the integral $\int_0^\infty e^{-r^2} r^{n-1} dr = \frac{1}{2} \Gamma(\frac{n}{2})$. Hence

$$\sigma_n = \frac{2\pi^{\frac{n}{2}}}{\Gamma(\frac{n}{2})}.$$

From knowing the surface area σ_n of a sphere one easily finds the volume (n dimensional volume) of the unit ball B^n. To do so let $V(r) = \nu(B^n(r)) = c_n r^n$. Then $\frac{dV(r)}{dr} = nc_n r^{n-1}$. Now consider two concentric balls of radial distance dr from one another. The volume of this shell is $c_n (r + dr)^n - c_n r^n$. Hence $\frac{dV(r)}{dr} = S(r)$. This means $S(r) = nc_n r^{n-1}$ and evaluating at $r = 1$ yields $\sigma_n = nc_n$. Thus the volume of B^n is

$$c_n = \frac{2\pi^{\frac{n}{2}}}{n\Gamma(\frac{n}{2})}$$

In particular, when $n = 4$, $\sigma_4 = \frac{2\pi^2}{\Gamma(2)} = 2\pi^2$ and $c_4 = \frac{\pi^2}{2}$.

Miscellaneous Exercises

Exercise 5.6.12. Let $0 < a < b$ and let Ω be the region in the first quadrant of the xy-plane bounded by the curves $xy = a$, $xy = b$, $x^2 - y^2 = 1$ and $y - x = 0$. Compute

$$\int \int_\Omega (x^2 - y^2)^{xy} (x^2 + y^2) dx dy.$$

Hint. Make the substitution $u = xy$, $v = x^2 - y^2$, and consult Example 5.1.20. (Ans. $\frac{1}{2} \log \frac{1+b}{1+a}$).

Exercise 5.6.13. Evaluate the integral

$$\int\int_\Omega \frac{xy}{(1+x^2+y^2)^2}\,dxdy,$$

where $\Omega = \{(x,y) : y^2 \leq 2x\}$ (Ans. $\frac{\pi\sqrt{2}}{4}$).

Exercise 5.6.14. Evaluate the integral

$$\int\int\int_\Omega \frac{xz}{1+x^2+y^2}\,dxdydz,$$

where $\Omega = \{(x,y,z) : 1 \leq x^2+y^2 \leq 3, 0 \leq z \leq 3\}$.
Hint. Use cylindrical coordinates. (Ans. 0).

Exercise 5.6.15. Evaluate the integral

$$\int\int\int_\Omega (x^2+y^2+z^2)\,dxdydz,$$

where $\Omega = \{(x,y,z) : 1 \leq x^2+y^2+z^2 \leq 5, z \geq 0, x^2+y^2 \leq z^2\}$.
Hint. Use spherical coordinates. (Ans. $\frac{\pi}{5}(2-\sqrt{2})(25\sqrt{5}-1)$).

Exercise 5.6.16. Evaluate the integral

$$\int\int\int_\Omega \frac{xy}{x^2+y^2+z^2}\,dxdydz,$$

where $\Omega = \{(x,y,z) : 1 \leq x^2+y^2+z^2 \leq 5, z \geq 0, x^2+y^2 \leq z^2\}$
Hint. Use spherical coordinates. (Ans. 0).

Exercise 5.6.17. Evaluate the integral

$$\int\int\int_\Omega e^{\sqrt{\frac{x^2}{a^2}+\frac{y^2}{b^2}+\frac{z^2}{c^2}}}\,dxdydz,$$

where $\Omega = \left\{(x,y,z) : \frac{x^2}{a^2}+\frac{y^2}{b^2}+\frac{z^2}{c^2} \leq 1\right\}$ (Ans. $4\pi abc(e-2)$).

Exercise 5.6.18. The linear transformation with standard matrix $T = \begin{pmatrix} 1 & 0 & -1 \\ 2 & 1 & 1 \\ 0 & -1 & 2 \end{pmatrix}$ maps the unit cube in $\mathbb{R}^3$ with vertices $(0,0,0)$, $(1,0,0)$, $(0,1,0)$, $(0,0,1)$, $(1,1,0)$, $(0,1,1)$, $(1,0,1)$, $(1,1,1)$ into a parallelepiped Π. Find the area of each of the faces of Π. Find the volume of Π. (Ans. $\sqrt{6}$, $\sqrt{11}$, $\sqrt{29}$. $\nu(\Pi) = 5$).

Exercise 5.6.19. In $\mathbb{R}^4$ "double" polar coordinates are defined by the equations

$$x = r\cos\theta,\ y = r\sin\theta,\ z = \rho\cos\phi,\ w = \rho\sin\phi.$$

Using this transformation in a four-fold integral, show that the volume of the 4-dimensional ball $x^2 + y^2 + z^2 + w^2 \leq R^2$ is $\frac{\pi^2}{2}R^4$.

Exercise 5.6.20. (*The volume of a "ball" in the norm* $||\cdot||_1$). Find the volume of the region V^n in $\mathbb{R}^n$ defined by

$$V^n = \left\{x : ||x||_1 = |x_1| + \cdots + |x_n| \leq r\right\}.$$

(Ans. $\nu(V^n) = \frac{2^n}{n!}$).

Exercise 5.6.21. Let $c_n = \nu(B^n)$ the volume of the unit ball in $\mathbb{R}^n$. Use Cavalieri's principle and appropriate substitutions to obtain the recursion formula

$$c_n = \frac{2\pi}{n}c_{n-2},\ \text{for } n \geq 2.$$

Furthermore use induction to deduce that

$$c_{2m} = \frac{\pi^m}{m!}\ \text{ and }\ c_{2m+1} = \frac{2^{m+1}\pi^m}{1\cdot3\cdot5\cdots(2m+1)}.$$

Exercise 5.6.22. Let

$$F(x) = \int_0^\pi \frac{\log(1 + x\cos y)}{\cos y}\,dy.$$

Show that $F(x) = \pi\sin^{-1}x$, for $x^2 < 1$.

Exercise 5.6.23. Let $f(x,y) = \frac{x-y}{(x+y)^3}$. Show that

$$\int_1^\infty \left(\int_1^\infty f(x,y)dx\right)dy \neq \int_1^\infty \left(\int_1^\infty f(x,y)dy\right)dx.$$

Exercise 5.6.24. Show that

$$\lim_{x\to\infty}\int_0^\infty \frac{e^{-x^2y}\sqrt{y}}{1+y^2}\,dy = 0.$$

Exercise 5.6.25. Calculate

$$I = \int_0^1 \log(\Gamma(x))dx.$$

Hint. First show $\lim_{x\to0^+} x\Gamma(x) = 1$ and deduce the convergence of the integral. By Theorem 5.3.30 deduce that $\Gamma(x)\Gamma(1-x) = \frac{\pi}{\sin\pi x}$. Then note $2I = \int_0^1 \log(\frac{\pi}{\sin\pi x})dx$.

Exercise 5.6.26. For $x > 0$ prove the *duplication formula*

$$2^{2x} \frac{\Gamma(x)\Gamma(x + \frac{1}{2})}{\Gamma(2x)} = 2\sqrt{\pi}.$$

Hint. First show $B(x, x) = \frac{1}{2^{2x-1}} B(\frac{1}{2}, x)$. Then use Theorem 5.3.30.

Exercise 5.6.27. Prove that

$$\int_0^{\frac{\pi}{2}} \frac{1}{\sqrt{\cos x}} dx = \frac{[\Gamma(\frac{1}{4})]^2}{2\sqrt{2\pi}}.$$

Exercise 5.6.28. (*)

1. Let $f(x) = xe^{-\frac{x^2}{2}}$. Show that $\widehat{f}(\xi) = -if(\xi)$.

2. Let $f(x) = x^2 e^{-\frac{x^2}{2}}$. Show that $\widehat{f}(\xi) = -f(\xi)$.

3. Let $f(x) = \frac{1}{a^2 + x^2}$. Show that $\widehat{f}(\xi) = \frac{\pi}{a} e^{-|a\xi|}$.

Exercise 5.6.29. (*) Let $f : \mathbb{R}^n \to \mathbb{R}$ be absolutely integrable and let $T : \mathbb{R}^n :\to \mathbb{R}^n$ be an invertible linear map. Prove that

$$\widehat{f(Tx)} = \frac{1}{|\det T|} \widehat{f}((T^t)^{-1}\xi).$$

In particular, if T is orthogonal (so that $(T^t)^{-1} = T$, and $|\det T| = 1$),

$$\widehat{f \circ T} = \widehat{f} \circ T.$$

Exercise 5.6.30. (*) Let Δ be the Laplacian in $\mathbb{R}^n$ and $f \in \mathcal{S}(\mathbb{R}^n)$. Show that

$$\widehat{\Delta f}(\xi) = ||\xi||^2 f(\xi).$$

Moreover, show that Δ commutes with translations and rotations, that is, if T is any translation or rotation in $\mathbb{R}^n$, then

$$\Delta(f \circ T) = (\Delta f) \circ T.$$

Exercise 5.6.31. (*) Let A be a positive symmetric $n \times n$ matrix. Prove that

$$\widehat{e^{-\langle Ax, x \rangle}} = \frac{\pi^{\frac{n}{2}}}{(\det A)^{\frac{1}{2}}} e^{-\frac{\langle A^{-1}\xi, \xi \rangle}{4}}.$$

Exercise 5.6.32. (*) Show that

$$\widehat{e^{-||x||}} = 2^n \pi^{\frac{n-1}{2}} \Gamma\left(\frac{n+1}{2}\right) (1 + ||\xi||^2)^{-\frac{n+1}{2}}.$$

This page intentionally left blank

Chapter 6

Line and Surface Integrals

In this chapter we study line and surface integrals and we prove Green's, Stokes', and the Divergence theorems. Include also are differential forms, vector fields on spheres and the Brouwer fixed point theorem.

6.1   Arc-length and Line integrals

6.1.1   Paths and curves

We recall from Section 2.1 that a *path* in $\mathbb{R}^n$ is a map $\gamma : [a, b] \to \mathbb{R}^n$. The range of γ denoted it by C is called a *curve* in $\mathbb{R}^n$ and γ is said a *parametrization* of C. If γ is a continuously diferentiable function, then γ is called a C^1 *path*. A simple C^1 curve $C = \{\gamma(t) : t \in [a, b]\}$ in $\mathbb{R}^n$ with $\gamma'(t) \neq 0$ for all $t \in [a, b]$ is often refered as a *regular* or *smooth* curve. If we need to use coordinates we let $\gamma(t) = (x_1(t), ..., x_n(t))$ when C lies in $\mathbb{R}^n$ and for $n = 3$, $\gamma(t) = (x(t), y(t), z(t))$. It is helpful to think of t as representing time and $\gamma(t)$ as the position of particle moving along the curve C at time t from the initial point $\gamma(a)$ to the terminal point $\gamma(b)$.

Definition 6.1.1. A path $\gamma : [a, b] \to \mathbb{R}^n$ is called *simple* if γ is one-to-one on $[a, b]$. If γ is one-to-one on $[a, b]$ and $\gamma(a) = \gamma(b)$, then γ is called a *simple closed* path.

A simple path corresponds to a curve having no self intersections. Paths have orientations. The *positive orientation* is the one in which the curve is traced out in the direction of increasing t.

Example 6.1.2. Some simple examples are:

1. Given points $\mathbf{p}$ and $\mathbf{q}$ in $\mathbb{R}^3$, to parametrize the line segment joining $\mathbf{p}$ with $\mathbf{q}$, we set

$$\gamma(t) = \mathbf{p} + t(\mathbf{q} - \mathbf{p}), \quad t \in [0, 1].$$

   Note that $\gamma(0) = \mathbf{p}$ and $\gamma(1) = \mathbf{q}$.

2. To parametrize the circle $x^2 + y^2 = r^2$ in the plane $z = 0$, we use the trigonometric functions cos and sin to obtain the very natural parametrization

$$\gamma(t) = (r \cos t, r \sin t), \ 0 \le t \le 2\pi.$$

3. Let $\alpha, \beta > 0$. To parametrize the ellipse $\frac{x^2}{\alpha^2} + \frac{y^2}{\beta^2} = 1$ in $\mathbb{R}^2$, note that the linear map $T : \mathbb{R}^2 \to \mathbb{R}^2$, given by $T(x, y) = (\alpha x, \beta y)$ maps the unit circle $x^2 + y^2 = 1$ to the ellipse $\frac{x^2}{\alpha^2} + \frac{y^2}{\beta^2} = 1$. Hence

$$\gamma(t) = T(\cos t, \sin t) = (\alpha \cos t, \beta \sin t), \quad 0 \le t \le 2\pi$$

   gives a natural parametrization of the ellipse.

4. The graph of a C^1 function $\phi : [a, b] \to \mathbb{R}$ is parametrized by

$$\gamma(t) = (t, \phi(t)), \quad a \le t \le b.$$

5. Let $\alpha, \beta > 0$. The path $\gamma(t) = (\alpha \cos t, \alpha \sin t, \beta t)$, with $0 \le t \le t_0$ is called a *helix*. A threaded screw is a helix.

   Let $\gamma : [a, b] \to \mathbb{R}^n$ be a C^1 path. We recall from Section 3.2 that $\gamma'(t)$ represents the *tangent vector* to the curve C at the point $\gamma(t)$ for $t \in [a, b]$. Thinking of $\gamma'(t)$ as the *velocity* and $||\gamma'(t)||$ the *speed* at time t, since *distance = speed · time*, it is not surpsising that the length of the "infinitesimal" piece of the curve C corresponding to the "infinitesimal" time interval $[t, t + dt]$ is

$$ds = ||\gamma'(t)||dt. \tag{6.1}$$

This gives the global formula.

Proposition 6.1.3. *If* $\gamma : [a, b] \to \mathbb{R}^n$ *is a* C^1 *path, then the total arc length of* C *is*

$$L(C) = \int_a^b ||\gamma'(t)|| dt \tag{6.2}$$

Proof. Since γ is differentiable, we have for all t and $t + h$ in $[a, b]$

$$\gamma(t + h) = \gamma(t) + \gamma'(t)h + \epsilon(h)|h|,$$

where $\epsilon(h) \to 0$ as $h \to 0$. Hence $\gamma(t + h) - \gamma(t) \approx \gamma'(t)h$ for h close to zero. For a partition of $[a, b]$, $a = t_0 < t_1 < ... < t_m = b$, we get a corresponding partition of C and an approximation of the length $L(C)$ of C by the sum $\sum_{j=1}^m ||\gamma(t_j) - \gamma(t_{j-1})||$. Therefore

$$L(C) \approx \sum_{j=1}^m ||\gamma(t_j) - \gamma(t_{j-1})|| \approx \sum_{j=1}^m ||\gamma'(t_j)|| h_j,$$

where $h_j = \Delta t_j = t_j - t_{j-1}$. The latter sum is a Riemann sum for the function $t \mapsto ||\gamma'(t)||$. Since γ' is continuous (and the norm is continuous) this function is also continuous on $[a, b]$ and therefore integrable. Letting the mesh of the partition go to zero we get (6.2).   □

In terms of coordinates we have $\gamma'(t) = (x_1'(t), ..., x_n'(t))$ and

$$L(C) = \int_a^b ||\gamma'(t)|| dt = \int_a^b \sqrt{x_1'(t)^2 + ... + x_n'(t)^2} \, dt.$$

In the important special case in which the curve is the graph of a smooth function $\phi : [a, b] \to \mathbb{R}$ its length is

$$L = \int_a^b \sqrt{1 + (\phi'(t))^2} dt. \tag{6.3}$$

Example 6.1.4. Find the arc length of the circle $x^2 + y^2 = r^2$.

Solution. Parametrizing the circle by $\gamma(t) = (x(t), y(t))$, where $x(t) = r \cos t$, $y(t) = r \sin t$ and $t \in [0, 2\pi]$, we have $\gamma'(t) = (x'(t), y'(t)) = (-r \sin t, r \cos t)$ and $||\gamma'(t)|| = \sqrt{r^2 \sin^2 t + r^2 \cos^2 t} = r$. Hence the arc length of the circle is

$$L = \int_0^{2\pi} r \, dt = 2\pi r.$$

Example 6.1.5. (*The cycloid*). The path $\gamma(t) = (t - \sin t, 1 - \cos t)$, $t \geq 0$ describes the position of a point on a rolling circle of radius 1. The circle lies in the xy-plane and rolls along the x-axis at constant speed, that is, its center is moving to the right along the line $y = 1$ at a constant speed of 1 radian per unit of time. The point $\gamma(t)$ traces a curve C known as the *cycloid*. Find the speed and the arc length of C for $t \in [0, 2\pi]$.

Solution. The velocity vector is $\gamma'(t) = (1 - \cos t, \sin t)$ and the speed of the point $\gamma(t)$ is

$$||\gamma'(t)|| = \sqrt{(1 - \cos t)^2 + \sin^2 t} = \sqrt{2 - 2\cos t}.$$

Note that $\gamma(t)$ moves at variable speed, although the circle rolls at constant speed. Furthermore, note that the speed of $\gamma(t)$ is zero when t is an integral multiple of 2π and at these values of t the point $\gamma(t)$ lies on the x-axis. The arc length of one cycle is

$$L = \int_0^{2\pi} \sqrt{2 - 2\cos t}\, dt = 2 \int_0^{2\pi} \sin\left(\frac{t}{2}\right) dt = 8.$$

Definition 6.1.6. Let C be a curve parametrized by the path $\gamma : [a, b] \to \mathbb{R}^n$ and let $\varphi : [c, d] \to [a, b]$ be a bijective C^1 mapping taking boundary points to boundary points. The composition $\gamma \circ \varphi : [c, d] \to \mathbb{R}^n$ is called a *reparametrization* of C.

The arc-length of a curve C is an intrinsic property of the geometric object C and should not depend on the particular parametrization we use. Indeed we have

Corollary 6.1.7. *The arc-length of a curve C is independent of the parametrization of C.*

Proof. Let $u \in [c, d]$ be a new parameter related to t by $t = \varphi(u)$, where $\gamma \circ \varphi : [c, d] \to \mathbb{R}^n$ is a reparametrization of $C = \{\gamma(t) : t \in [a, b]\}$. Then the curve C described by $\gamma(t)$ is also described by $(\gamma \circ \varphi)(u)$ and by the Change of Variables formula and the Chain Rule we have

$$\int_c^d ||(\gamma \circ \varphi)'(u)||du = \int_c^d ||\gamma'(\varphi(u))||\, |\varphi'(u)|du = \int_a^b ||\gamma'(t)||dt.$$

$\square$

A useful reparametrization of a smooth curve C is given by *arc-length*

$$s(t) = \int_a^t ||\gamma'(u)||du,$$

for $t \in [a, b]$. Note that by the Fundamental Theorem of Calculus, $s'(t) = ||\gamma'(t)|| > 0$. Therefore the function $s : [a, b] \to [0, L]$ is strictly increasing, and hence a bijection. For its inverse $t = s^{-1}(u)$ we have

$$(s^{-1})'(u) = \frac{1}{s'(s^{-1}(u))} = \frac{1}{||\gamma'(s^{-1}(u))||}.$$

Thus we can use $s^{-1} : [0, L] \to [a, b]$ as (an orientation preserving) reparametrization of C and we write $C = \{(\gamma \circ s^{-1})(u) : u \in [0, L]\}$. Moreover, note that $||(\gamma \circ s^{-1})'(u)|| = ||\gamma'(s^{-1}(u))||(s^{-1})'(u) = 1$. Thus, a simple C^1 curve with $\gamma'(t) \neq 0$ for all $t \in [a, b]$ admits a *unit speed parametrization*. Whenever we use $\rho(u) = (\gamma \circ s^{-1})(u)$ we say C is paramatrized by arc length. In this case the unit tangent vector is $T(u) = \rho'(u)$ and the unit normal vector $N(u) = \frac{T'(u)}{||T'(u)||}$ which we can write as $T'(u) = \kappa(u)N(u)$ where $\kappa(u) = ||T'(u)||$ is called the *curvature* of C at the point $\rho(u)$. The curvature of a curve is a measure of the rate at which the direction of motion along the curve is changing. Clearly, a curve with $\kappa = 0$ (zero curvature) is just a straight line.

The notion of arc length in formula (6.2) extends in an obvious way to *piecewise smooth* curves, obtained by joining finitely many smooth curves together end-to-end but allowing corners or cusps at the joining points; we simply compute the lengths of the smooth pieces and add them up.

6.1.2   Line integrals

The functions that we shall work with are of two types. Real-valued functions $f : \mathbb{R}^n \to \mathbb{R}$ called *scalar fields* and vector-valued functions $F : \mathbb{R}^n \to \mathbb{R}^n$ called *vector fields*. A vector field assigns to each point $x = (x_1, ..., x_n) \in \Omega \subseteq \mathbb{R}^n$ a vector

$$F(x) = (F_1(x), ..., F_n(x)).$$

In Physics, vector fields are the natural language of fluid flow, heat flow, gravitational, magnetic, electric fields, etc. Although our main concern is to work in $\mathbb{R}^3$ (so also in $\mathbb{R}^2$ as the subspace $z = 0$), in this section the results are actually valid in Euclidean space of any dimension since we use only the inner product in $\mathbb{R}^n$.

Definition 6.1.8. Let $C = \{\gamma(t) : t \in [a, b]\}$ be a smooth (or piecewise smooth) curve in $\mathbb{R}^n$ and let $f : \mathbb{R}^n \to \mathbb{R}$ be a continuous scalar field whose domain contains C. The *line integral* of f along C is defined by

$$\int_C f ds = \int_a^b f(\gamma(t))||\gamma'(t)||dt. \tag{6.4}$$

If C is piecewise smooth, we define the integral along C as the sum of the integrals over the pieces on which γ is C^1 (the fact that γ may fail to be C^1 at a few points does not effect the integral). That is, if C is the union of the oriented curves $C_1, ..., C_m$, we let

$$\int_C f ds = \sum_{j=1}^m \int_{C_j} f ds.$$

The integral $\int_C f ds$ is independent of the parametrization (and the orientation), by the same argument we gave above for the case $f \equiv 1$. Note also that, when $f \equiv 1$ this gives

$$\int_C 1 ds = \int_a^b ||\gamma'(t)||dt = L.$$

Example 6.1.9. Compute $\int_C xyz ds$, where the curve C is the helix parametrized by $\gamma(t) = (\cos t, \sin t, t)$, with $0 \le t \le 2\pi$. What is the arc-length of this helix?

Solution. In coordinates we have $x(t) = \cos t$, $y(t) = \sin t$, $z(t) = t$, and so $\gamma'(t) = (x'(t), y'(t), z'(t)) = (-\sin t, \cos t, 1)$. Hence

$$||\gamma'(t)|| = \sqrt{\sin^2 t + \cos^2 t + 1} = \sqrt{2}$$

and

$$\int_C xyz\,ds = \int_0^{2\pi} t\sin t\cos t\sqrt{2}\,dt = \frac{\sqrt{2}}{2}\int_0^{2\pi} t\sin 2t\,dt$$

$$= \frac{\sqrt{2}}{2}\left[\frac{1}{4}\sin 2t - \frac{t}{2}\cos 2t\right]_0^{2\pi} = -\frac{\sqrt{2}}{2}\pi.$$

The length of C is $L = \int_C 1\,ds = \int_0^{2\pi}\sqrt{2}\,dt = 2\sqrt{2}\pi$.

Example   6.1.10. Compute   the   line   integral   of $f(x,y) = \sqrt{\left(\frac{bx}{a}\right)^2 + \left(\frac{ay}{b}\right)^2}$ over the ellipse $\frac{x^2}{a^2} + \frac{y^2}{b^2} = 1$.

Solution. The ellipse is parametrized by $x = a\cos t$, $y = b\sin t$ with $t \in [0, 2\pi]$. We have

$$\int_C f(x,y)\,ds = \int_0^{2\pi}\sqrt{b^2\cos^2 t + a^2\sin^2 t}\sqrt{a^2\sin^2 t + b^2\cos^2 t}\;dt$$

$$= \int_0^{2\pi}(a^2\sin^2 t + b^2\cos^2 t)\,dt = \pi(a^2 + b^2).$$

We now turn to the integration of vector fields along curves. We use *dot* and *inner product* interchangeably.

Definition 6.1.11. Let $C = \{\gamma(t) : t \in [a,b]\}$ be a smooth (or piecewise smooth) regular curve in $\mathbb{R}^n$ and let $F : \mathbb{R}^n \to \mathbb{R}^n$ be a continuous vector field whose domain contains C. The *line integral* of F along C is defined by

$$\int_C F \cdot d\mathbf{s} = \int_a^b F(\gamma(t)) \cdot \gamma'(t)\,dt = \int_a^b \langle F(\gamma(t)), \gamma'(t)\rangle\,dt. \qquad (6.5)$$

The dot is meant to denote the dot or inner product of two vectors. If C is piecewise smooth the definition is extended by summing the integrals along the finite number of constituent pieces.

Line integrals are of fundamental importance in both pure and applied mathematics. In physics, they occur in connection with work, potential energy, heat flow, fluid flow, change of entropy, circulation of a fluid, and other physical situations in which the behavior of a vector or scalar field is studied along a curve. For example, if F were a force field, then we would be computing the *work* done against this array of forces in moving a particle along C.

Note that since $\gamma'(t) \neq 0$ the integral $\int_C F \cdot d\mathbf{s}$ can be written as the line integral of the scalar field $\langle F, T \rangle$ along C, where $T(t) = \frac{\gamma'(t)}{||\gamma'(t)||}$ is the unit tangent vector of C at $\gamma(t)$. The scalar field $\langle F, T \rangle$ is refered to as the *tangential component* of the vector field F along the curve C. Indeed

$$\int_C F \cdot d\mathbf{s} = \int_a^b \left(F(\gamma(t)) \cdot \frac{\gamma'(t)}{||\gamma'(t)||} \right) ||\gamma'(t)|| dt$$

$$= \int_a^b (F(\gamma(t)) \cdot T(t)) ||\gamma'(t)|| dt = \int_C \langle F, T \rangle ds.$$

Thus,

$$\int_C F \cdot d\mathbf{s} = \int_C \langle F, T \rangle ds. \tag{6.6}$$

If C is closed, the integral (6.6) is called the *circulation* of the vector field F around the curve C.

We know that the line integral of a scalar field along a curve C is independent of the parametrization and orientation of C. By (6.6) the line integral $\int_C F \cdot d\mathbf{s}$ is independent of the parametrization as long as the orientation is unchanged, but it acquires a factor of -1 if the orientation is reversed. For, if we make a change of parameters, say $t = \varphi(u)$ where $\varphi : [c, d] \to [a, b]$, then

$$\int_C \langle F, T \rangle ds = \int_c^d \langle F(\gamma(\varphi(u))), \gamma'(\varphi(u)) \rangle \varphi'(u) du = \int_{\varphi(c)}^{\varphi(d)} \langle F(\gamma(t)), \gamma'(t) \rangle dt.$$

This latter integral is equal to $- \int_C F \cdot d\mathbf{s}$, if φ is orientation reversing since $\varphi(c) = b$ and $\varphi(d) = a$.

The line integral of a vector field $F = (F_1, F_2, F_3)$ in $\mathbb{R}^3$ written in terms of the components becomes

$$\int_C F \cdot d\mathbf{s} = \int_a^b F(\gamma(t)) \cdot \gamma'(t)dt$$

$$= \int_a^b F(x(t), y(t), z(t)) \cdot (x'(t), y'(t), z'(t))dt$$

$$= \int_a^b \left[F_1 x'(t) + F_2 y'(t) + F_3 z'(t) \right] dt.$$

Writing $dx = x'(t)dt$, $dy = y'(t)dt$ and $dz = z'(t)dt$, this is frequently written as

$$\int_C F \cdot d\mathbf{s} = \int_C F_1 dx + F_2 dy + F_3 dz.$$

Example 6.1.12. Let F be the vector field $F(x, y, z) = (yz, 2y, -x^2)$. Evaluate $\int_C F \cdot d\mathbf{s} = \int_C yz dx + 2y dy - x^2 dz$, where C is:

1. the straight line segment from $(0, 0, 0)$ to $(1, 1, 1)$.

2. the curve $x(t) = t$, $y(t) = t^2$ and $z(t) = t^3$ with $t \in [0, 1]$ joining $(0, 0, 0)$ to $(1, 1, 1)$.

Solution.

1. The parametrization of the line segment is $\gamma(t) = (t, t, t)$ with $t \in [0, 1]$ and $\gamma'(t) = (1, 1, 1)$.

$$\int_C F \cdot d\mathbf{s} = \int_0^1 (t^2, 2t, -t^2) \cdot (1, 1, 1)dt = \int_0^1 2t dt = t^2 \Big|_0^1 = 1.$$

2. Here $\gamma(t) = (t, t^2, t^3)$ with $t \in [0, 1]$ and $\gamma'(t) = (1, 2t, 3t^2)$.

$$\int_C F \cdot d\mathbf{s} = \int_0^1 (t^5, 2t^2, -t^2) \cdot (1, 2t, 3t^2)dt = \int_0^1 (t^5 + 4t^3 - 3t^4)dt = \frac{17}{30}.$$

Example 6.1.13. Evaluate $\int_C y dx + z dy + x dz$, where C is the ellipse formed by the intersection of the cylinder $x^2 + y^2 = 1$ and the plane $z = 2y + 1$, oriented counterclockwise as viewed from above.

Solution. The vector field is $F(x, y, z) = (y, z, x)$ and we parametrize C by $\gamma(t) = (\cos t, \sin t, 2\sin t + 1)$ with $t \in [0, 2\pi]$. Hence $\gamma'(t) = (-\sin t, \cos t, 2\cos t)$ and so

$$\int_C F \cdot ds = \int_0^{2\pi} F(\gamma(t)) \cdot \gamma'(t) dt$$

$$= \int_0^{2\pi} [-\sin^2 t + (2\sin t + 1)\cos t + 2\cos^2 t] dt$$

$$= \int_0^{2\pi} [\cos(2t) + \sin(2t) + \cos t + \cos^2 t] dt = 0 + 0 + 0 + \pi = \pi.$$

As Example 6.1.12 shows the line integral of a vector field generally speaking depends on the path of integration. We shall see in the next section that there are certain vector fields whose line integral along a curve *depend only on the endpoints of the curve and independent of the curve itself.* These are the vector fields that are obtained as the gradients of a scalar field and are called *conservative* vector fields.

EXERCISES

1. Find the arc length of the following curves.

   (a) $y = x^3$, $0 \leq x \leq 2$.

   (b) $\gamma(t) = (a\cos^3 t, a\sin^3 t)$, $0 \leq t \leq 2\pi$.

   (c) $\gamma(t) = (e^t \cos t, e^t \sin t)$, $0 \leq t \leq \frac{\pi}{2}$.

   (d) $\gamma(t) = (\frac{1}{3}t^3 - t, t^2)$, $0 \leq t \leq 2$.

   (e) $\gamma(t) = (\sin 3t, \cos 3t, 2t^{\frac{3}{2}})$, $0 \leq t \leq 1$.

   (f) $\gamma(t) = (\log t, 2t, t^2)$, $1 \leq t \leq e$.

2. Show that the arc length of the ellipse $\frac{x^2}{a^2} + \frac{y^2}{b^2} = 1$, where $a > b$, is

$$4a \int_0^{\frac{\pi}{2}} \sqrt{1 - k^2 \sin^2 t}\, dt,$$

   where $k = \sqrt{1 - \left(\frac{b}{a}\right)^2}$ is the eccentricity of the ellipse. (This integral is one of the so-called *elliptic integrals*).

3. Compute $\int_C (x+y)ds$, where C is the triangle with vertices $(0,0)$, $(1,0)$ and $(0,1)$, oriented counterclockwise.

4. Compute $\int_C y^2 ds$, where C is parametrized by $\gamma(t) = (t-\sin t, 1-\cos t)$, $t \in [0, 2\pi]$.

5. Compute $\int_C e^{\sqrt{z}} ds$, where C is parametrized by $\gamma(t) = (1, 2, t^2)$, $t \in [0, 1]$.

6. Compute $\int_C z ds$, where C is parametrized by $\gamma(t) =$ $(t\cos t, t\sin t, t)$, $t \in [0, a]$.

7. Compute $\int_C \frac{x+y}{y+z} ds$, where C is parametrized by $\gamma(t) = (t, \frac{2}{3}t^{\frac{3}{2}}, t)$, $t \in [0, 1]$.

8. Show that the line integral of $f(x, y)$ along a curve given in polar coordinates by $r = r(\theta)$, $\theta \in [\theta_1, \theta_2]$ is

$$\int_{\theta_1}^{\theta_2} f(r\cos\theta, r\sin\theta)\sqrt{r^2 + \left(\frac{dr}{d\theta}\right)^2}\, d\theta.$$

Use this to compute the arc length of $r = 1 + \cos\theta$, $\theta \in [0, 2\pi]$.

9. Find the mass of a wire whose shape is that of the curve of intersection of the sphere $x^2 + y^2 + z^2 = 1$ and the plane $x + y + z = 0$ if the density of the wire at (x, y, z) is $\rho(x, y, z) = x^2$.

10. Compute $\int_C F \cdot ds$ for the following F and C:

   (a) $F(x, y, z) = (x, y, z)$ and C is the line segment from $(0, 0, 0)$ to $(1, 1, 1)$.

   (b) $F(x, y, z) = (x, y, z)$ and C is parametrized by $\gamma(t) = (\cos t, \sin t, 0)$.

   (c) $F(x, y) = (y^2, -2x)$ and C is the triangle with vertices $(0, 0)$, $(1, 0)$ and $(1, 1)$, oriented counterclockwise.

   (d) $F(x, y) = (x^2 y, x^3 y^2)$ and C is the closed curve formed by portions of the line $y = 4$ and the parabola $y = x^2$, oriented counterclockwise.

(e) $F(x, y, z) = (yz, x^2, xz)$ and C is the portion of the curve $y = x^2$, $z = x^3$ from $(0, 0, 0)$ to $(1, 1, 1)$.

(f) $F(x, y, z) = (y, z, xy)$ and C is parametrized by $\gamma(t) = (\cos t, \sin t, t)$, $t \in [0, 2\pi]$.

11. Evaluate the line integrals

(a) $\int_C y\,dx + x\,dy$, where C is the curve $y = x^2$ joining the points $(0, 0)$ and $(2, 4)$.

(b) $\int_C xy\,dx + x^2\,dy$, where C is the curve $y = x^{\frac{2}{3}}$, $0 \le x \le 1$.

(c) $\int_C y^n\,dx + x^n\,dy$, $n = 0, 1, 2, ...$, where C is the circle $x^2 + y^2 = a^2$.

(d) $\int_C y|y|\,dx + x|x|\,dy$, where C is the boundary of $\{(x, y) : |x| + |y| < 1\}$.

12. Evaluate $\int_C (z + y)\,dx + (x + z)\,dy + (y + x)\,dz$, where C is the polygonal line with vertices $(0, 0, 0)$, $(1, 0, 0)$, $(1, 1, 0)$ and $(1, 1, 1)$.

13. Evaluate $\int_C y\,dx + z\,dy + x\,dz$, where

(a) C is the curve of intersection of the plane $x + y = 2$ and the sphere $x^2 + y^2 + z^2 = 2(x + y)$, oriented clockwise when viewed from the origin.

(b) C is the intersection of the cylinder $x^2 + y^2 = 1$ and the hyperboloid $z = xy$, oriented counterclockwise when viewed from the z-axis.

14. Let $F : \mathbb{R}^n \to \mathbb{R}^n$ be a continuous vector field and C a smooth (or piecewise smooth) curve in $\mathbb{R}^n$ with length L. Show that

$$\left| \int_C F \cdot ds \right| \le \left(\max_{x \in C} \|F(x)\| \right) L.$$

Answers to selected Exercises

1. (a) $\int_0^2 \sqrt{1+9t^4}\,dt$. (b) $6a$. (c) $\sqrt{2}(e^{\frac{\pi}{2}}-1)$. (d) $\frac{14}{3}$. (e) $4\sqrt{2}-2$. (f) e^2.

3. $1 + \sqrt{2}$.   4. $\frac{256}{15}$.   5. 2.   6. $\frac{1}{3}[(2 - a^2)^{\frac{3}{2}} - 2\sqrt{2}]$.   7. $\frac{16}{3} - 2\sqrt{3}$.

9. $\frac{2\pi}{3}$.   10. (a) $\frac{3}{2}$. (b) 0. (c) $-\frac{4}{3}$. (d) -2π. (e) $\frac{23}{21}$. (f) $-\pi$.

11. (a) $\frac{4}{3}$. (b) $\frac{3}{8}$. (c) 0. (d) $-\frac{8}{3}$.   12. 3.   13. (a) $-2\sqrt{2}$. (b) $-\pi$.

6.2   Conservative vector fields and Poincare's lemma

Definition 6.2.1. Let Ω be a domain in $\mathbb{R}^n$ and $f : \Omega \to \mathbb{R}$ be a smooth numerical function (a scalar field). We say that a vector field $F : \Omega \to \mathbb{R}^n$ is a *conservative* or a *gradient vector field* on Ω if

$$F(x) = \nabla f(x) = \left(\frac{\partial f}{\partial x_1}(x), ..., \frac{\partial f}{\partial x_n}(x) \right), \quad x \in \Omega.$$

The function f is called the (scalar) *potential* of F.

Theorem 6.2.2. *(Fundamental Theorem of Line Integrals). Let $F = \nabla f$ be a conservative vector field on a domain Ω in $\mathbb{R}^n$ and C a smooth curve in Ω parametrized by $\gamma(t)$, for $a \leq t \leq b$. Then*

$$\int_C F \cdot ds = f(\gamma(b)) - f(\gamma(a)).$$

Proof.

$$\int_C F \cdot ds = \int_a^b \langle \nabla f(\gamma(t)), \gamma'(t) \rangle dt.$$

But, $\frac{d}{dt}(f(\gamma(t))) = \langle \nabla f(\gamma(t)), \gamma'(t) \rangle$. Therefore, by the Fundamental Theorem of Calculus we have

$$\int_C F \cdot ds = \int_a^b \frac{d}{dt}(f(\gamma(t))dt = f(\gamma(a)) - f(\gamma(b)).$$

$\square$

Corollary 6.2.3. *If F is conservative on Ω and C is a smooth (or piecewise smooth) closed curve in Ω, then $\int_C F \cdot ds = 0$.*

We shall now give an example of a conservative smooth vector field which arises in physics.

Example 6.2.4. This is the situation of *gravitational attraction* between two bodies. According to Newton's Law of Gravitation the force of gravitational attraction between two bodies, of masses m_1 and m_2, is given by

$$F = \frac{gm_1m_2}{r^2},$$

where g is a universal constant and r is the distance between their centers of mass. We shall assume we have point masses, the first of which is located at the origin and the second at the variable point $\mathbf{x} \in \mathbb{R}^3$. Letting $k = gm_1m_2$ we see that $F(\mathbf{x})$, the force exerted at the point $\mathbf{x}$, has magnitude $\frac{k}{||\mathbf{x}||^2}$. Now this force always acts along $\mathbf{x}$, but in the *opposite* direction. Therefore to determine $F(\mathbf{x})$ we write $F(\mathbf{x}) = \phi(\mathbf{x})\frac{\mathbf{x}}{||\mathbf{x}||}$, where ϕ is a positive numerical function. Now since $\frac{\mathbf{x}}{||\mathbf{x}||}$ has length 1 and $F(\mathbf{x})$ has length $\frac{k}{||\mathbf{x}||^2}$ we see that this is also the value of $\phi(\mathbf{x})$. Thus

$$F(\mathbf{x}) = -\frac{k\mathbf{x}}{||\mathbf{x}||^3},$$

F being defined and smooth on the domain consisting of all of $\mathbb{R}^3$, *excluding* the origin.

Why is this smooth vector field conservative? Let $f(\mathbf{x}) = \frac{-k}{||\mathbf{x}||}$. Then $\nabla f(\mathbf{x}) = -\frac{k}{||\mathbf{x}||^3}\mathbf{x} = F(\mathbf{x})$. Hence f is a potential function and F is conservative.

We will explain, presently, why the term conservative vector field is used. Suppose F is a conservative field of forces in physics. Since $\nabla(-f)(x) = -\nabla f(x)$, it will be convenient to assume $F(x) = -\nabla f(x)$, rather than the gradient itself. Then this f is called the *potential energy function* of the vector field F, i.e, $f(x)$ is the potential energy at the point $x \in \Omega$. In contrast the *kinetic energy* of a particle of mass m depends on things other than the position. It is $\frac{1}{2}mv^2$, where v is the velocity. If a particle of mass m travels along a curve C, as we saw the vector velocity at time t is $\gamma'(t)$. Its acceleration is $\gamma''(t)$. Newton's law of motion is then expressed by the vector equation

$$F(\gamma(t)) = m\gamma''(t),$$

$t \in [a, b]$. We define the *total energy* of the particle traveling along the curve C under the action of a conservative field F to be the sum of the *potential energy* and its *kinetic energy*

$$E(t) = f(\gamma(t)) + \frac{1}{2}m \parallel \gamma'(t) \parallel^2 .$$

We now come to the *Principle of Conservation of Energy*.

Theorem 6.2.5. *(Principle of Conservation of Energy). For any smooth curve $C = \{\gamma(t) : t \in [a, b]\}$, the total energy of a particle traveling on such a path under the effect of a conservative field of forces is constant.*

Proof. Let us calculate $\frac{dE}{dt}$ at the point $\gamma(t)$. Using the Chain Rule and a formula for differentiating the scalar product of two vector functions

$$\frac{d}{dt}\langle Y(t), Z(t)\rangle = \langle Y'(t), Z(t)\rangle + \langle Y(t), Z'(t)\rangle,$$

we see that

$$\frac{dE}{dt} = \langle \nabla f(\gamma(t)), \gamma'(t)\rangle + \frac{1}{2}m2\langle \gamma'(t), \gamma''(t)\rangle,$$

that is, $\frac{dE}{dt} = \langle -F(\gamma(t)) + m\gamma''(t), \gamma'(t)\rangle$. Taking account of Newton's law, this is $\langle 0, \gamma'(t)\rangle = 0$. Thus, E is constant along the curve.   □

Next we give an application of the Principle of Conservation of Energy.

Example 6.2.6. (*Escape velocity*). We will use this principle to compute the escape velocity of an object, for example a rocket, leaving the earth's gravitational field. In other words, when a rocket is given an initial impetus and no other forces are exerted on it, the question is how fast must it be going in order to not be pulled back to earth by gravity. Here v_0 is the minimal such speed. It is not obvious that there is any speed for which this would happen. We may assume that the object leaves the earth on some radial line, so we have here a 1-dimensional problem. Also, from the above, when the rocket is a distance r from the center of the earth the gravitational force on it is given by

$$F(r) = \frac{k}{r^2}.$$

In order to evaluate the positive constant k, observe that when $r = R$, the radius of the earth, then the force is its weight, mg. Thus $k = mgR^2$. Let $v(r)$ be the velocity at position r. The Principle of Conservation of Energy tells us that the constant total energy is given by

$$E = \frac{1}{2}mv^2(r) - \int_R^r F(u)du.$$

We can determine this constant by taking $r = R$. This yields $E = \frac{1}{2}mv^2(R) = \frac{1}{2}mv_0^2$. Thus $\frac{1}{2}mv_0^2 = \frac{1}{2}mv^2(r) - \int_R^r F(u)du$. Evaluating the integral and taking into account the value of k, we see that

$$\int_R^r F(u)du = mgR^2 \left(\frac{1}{R} - \frac{1}{r} \right).$$

Substituting this into the above equation and dividing by m yields

$$v_0^2 - v^2(r) = 2gR^2 \left(\frac{1}{R} - \frac{1}{r} \right).$$

The fact that m doesn't appear in this equation shows that, whatever the outcome, it must be *independent of m*. Now the object falls back if and only if at some point $v(r) = 0$. Alternatively, to escape we need only be sure that $v(r) > 0$ for all r, as $r \to \infty$. But, in that case

$$v_0^2 > v_0^2 - v^2(r) = 2gR^2 \left(\frac{1}{R} - \frac{1}{r} \right),$$

and as $r \to \infty$ the right side of this equation approaches $2gR$. Thus $v_0 \geq \sqrt{2gR}$. Conversely, if this is so, or even if $v_0 = \sqrt{2gR}$, then $v_0^2 = 2gR$ so $v_0^2 - v^2(r) = 2gR - v^2(r) = 2gR^2(\frac{1}{R} - \frac{1}{r})$. Thus $v^2(r) = \frac{2gR^2}{r}$ and so $v(r) > 0$. Therefore; *The earth's escape velocity is $\sqrt{2gR}$.*

We now look into what it takes for a vector field to be conservative. We shall see that this is closely related to global properties of the domain $\Omega \subseteq \mathbb{R}^n$.

Theorem 6.2.7. *For a conservative vector field, F, the $n \times n$ matrix, $D_F(x) = (\frac{\partial F_i}{\partial x_j}(x))$, is a symmetric matrix for each $x \in \Omega$.*

Proof. Since F is conservative, for $x \in \Omega$ we have

$$F(x) = (F_1(x), \dots, F_n(x)) = \nabla f(x).$$

So that $F_i = \frac{\partial f}{\partial x_i}$, for each $i = 1, \dots n$. But then taking $\frac{\partial}{\partial x_j}$ of this equation tells us that $\frac{\partial F_i}{\partial x_j}(x) = \frac{\partial^2 f}{\partial x_j \partial x_i}(x)$, for each $i, j = 1, \dots n$. Since f is smooth we also know that $\frac{\partial^2 f}{\partial x_j \partial x_i}(x) = \frac{\partial^2 f}{\partial x_i \partial x_j}(x)$ for $i, j = 1, \dots n$. Thus we conclude for $i, j = 1, \dots n$, and $x \in \Omega$

$$\frac{\partial F_i}{\partial x_j}(x) = \frac{\partial F_j}{\partial x_i}(x).$$

Put another way, the $n \times n$ matrix, $(\frac{\partial F_i}{\partial x_j}(x))$ is symmetric. Of course when $i = j$ this is always satisfied; the real force of these conditions is on the off diagonal terms. □

Now a natural question might be whether these conditions on a vector field are sufficient to guarantee that it is conservative. We shall prove a result, called the *Poincare lemma*,[1] which shows that under certain circumstances they are. It also suggests that whether this is so, in general, may depend on properties of the domain.

For concreteness we will prove the Poincare[2] lemma in dimension 3. Of course, the argument can be modified to work in all dimensions ≥ 2. We leave as an exercise for the reader to show it is also true when the dimension is 1 and Ω is an interval.

Theorem 6.2.8. *(Poincare's lemma).* Let $F = (F_1, F_2, F_3)$ be a smooth vector field whose domain, Ω, is an open ball, or rectangle in $\mathbb{R}^3$. If, for each $i, j = 1, 2, 3$ and $\boldsymbol{x} \in \Omega$, we have

$$\frac{\partial F_i}{\partial x_j}(\boldsymbol{x}) = \frac{\partial F_j}{\partial x_i}(\boldsymbol{x}),$$

then F is conservative.

---

[1] This is a special case of Poincare's lemma which is a general result on differential forms in $\mathbb{R}^n$ that reads; if a differential form ω is closed in a ball, (i.e., $d\omega = 0$), then it is exact there. See Remark 6.8.11

[2] H. Poincare (1854-1912) a student of Hermite, he invented topology, and in particular the fundamental group. He also worked in mathematical physics and ordinary differential equations. He proposed the Poincare conjecture which was only solved in 2006!

Proof. By translation, we can for convenience, assume Ω is centered at the origin. Let $\mathbf{x} = (x, y, z)$ be a point in Ω and define

$$f(x, y, z) = \int_0^x F_1(t, 0, 0)dt + \int_0^y F_2(x, t, 0)dt + \int_0^z F_3(x, y, t)dt.$$

Now these integrals all make sense since the F_i are all defined on Ω and the paths along which the integration takes place all lie in Ω (the latter point having to do with geometric properties of the domain). Thus f is a function on Ω. We must see that for each i, $\frac{\partial f}{\partial x_i} = F_i$. This will also demonstrate smoothness of f, since the F_i are given as smooth. Taking $\frac{\partial}{\partial z}$ and applying the Fundamental Theorem of Calculus we see that

$$\frac{\partial f}{\partial z}(x, y, z) = F_3(x, y, z).$$

Now we take $\frac{\partial}{\partial y}$ of the equation above. Using the Fundamental Theorem of Calculus and the formula for differentiating under the integral sign, this yields

$$\frac{\partial f}{\partial y}(x, y, z) = F_2(x, y, 0) + \int_0^z \frac{\partial F_3}{\partial y}(x, y, t)dt.$$

But, our hypothesis tells us

$$\frac{\partial F_3}{\partial y} = \frac{\partial F_2}{\partial z}.$$

So applying the Fundamental Theorem of Calculus again yields

$$\frac{\partial f}{\partial y}(x, y, z) = F_2(x, y, 0) + F_2(x, y, z) - F_2(x, y, 0) = F_2(x, y, z).$$

Finally,

$$\frac{\partial f}{\partial x}(x, y, z) = F_1(x, 0, 0) + \int_0^y \frac{\partial F_2}{\partial x}(x, t, 0)dt + \int_0^z \frac{\partial F_3}{\partial x}(x, y, t)dt.$$

Using the fact that

$$\frac{\partial F_2}{\partial x} = \frac{\partial F_1}{\partial y}$$

and

$$\frac{\partial F_3}{\partial x} = \frac{\partial F_1}{\partial z}$$

and arguing as above we get

$$\frac{\partial f}{\partial x}(x,y,z) = F_1(x,0,0)+F_1(x,y,0)-F_1(x,0,0)+F_1(x,y,z)-F_1(x,y,0)$$

Since this is just $F_1(x,y,z)$, we are done. $\square$

What we have done here is to prove the *local* existence of a solution of a system of partial differential equations,

$$\frac{\partial f}{\partial x_i} = F_i,$$

when the given data, the F_i, satisfy the necessary conditions

$$\frac{\partial F_i}{\partial x_j}(x) = \frac{\partial F_j}{\partial x_i}(x).$$

The question of global solutions in *simply connected*[3] domains is answered below in Corollary 6.2.15.

Remark 6.2.9. Given a conservative vector field F, to find a function f whose gradient is F, one in practice proceeds as follows: For simplicity we consider the 2-dimensional case where $F = (F_1, F_2)$ is a smooth vector field defined in a rectangle $[a_1, b_1] \times [a_2, b_2]$ in $\mathbb{R}^2$. Since $\frac{\partial F_2}{\partial x} = \frac{\partial F_1}{\partial y}$, we begin by integrating F_1 with respect to x, including a "constant" of integration that can depend on the other variable y:

$$f(x,y) = \int_c^x F_1(t,y)dt + \phi(y),$$

where c is any point in the interval $[a_1, b_1]$. By the Fundamental Theorem of Calculus any such f will satisfy $\frac{\partial f}{\partial x} = F_1$. To obtain $\frac{\partial f}{\partial y} = F_2$,

---

[3]A domain Ω in $\mathbb{R}^n$ is called *simply connected* if any simple closed curve C in Ω can be shrunk to a point (all within Ω). For example, $\mathbb{R}^3 - \{0\}$ is simply connected. If $\Omega \subseteq \mathbb{R}^2$ this simply means Ω has no "holes", so $\mathbb{R}^2 - \{0\}$ is not simply connected.

differentiate the formula for f with respect to y and use Theorem 5.3.2 to differentiate under the integral sign:

$$\frac{\partial f}{\partial y}(x,y) = \int_c^x \frac{\partial F_1}{\partial y}(t,y)dt + \phi'(y) = \int_c^x \frac{\partial F_2}{\partial x}(t,y)dt + \phi'(y)$$

$$= F_2(x,y) - F_2(c,y) + \phi'(y).$$

Hence we obtain f by taking $\phi(y) = \int F_2(c,y)dy + C$. The same idea extends similarly to n variables.

Example 6.2.10. Let $F(x,y) = (y^2 e^{xy}, (xy+1)e^{xy} + \cos y)$. Show that F is a conservative vector field and find a function f satisfying $\nabla f = F$.

Solution. Differentiating we have

$$\frac{\partial F_2}{\partial x} = \frac{\partial F_1}{\partial y} = (2y + xy^2)c^{xy},$$

so by the Poincare lemma F is conservative. To find a function f such that $\nabla f = F$, we set

$$f(x,y) = \int F_1(x,y)dx + \phi(y) = \int y^2 e^{xy}dx + \phi(y) = ye^{xy} + \phi(y).$$

Therefore

$$\frac{\partial f}{\partial y}(x,y) = (1 + xy)e^{xy} + \phi'(y).$$

Since we want $\frac{\partial f}{\partial y}(x,y) = F_2(x,y) = (xy+1)e^{xy} + \cos y$, this tells us $\phi'(y) = \cos y$, and integrating $\phi(y) = \sin y + C$. Thus

$$f(x,y) = ye^{xy} + \sin y + C$$

is a potential of F.

We can now see in the following example that even in the 2-dimensional case the converse of the Poincare lemma is false in general, although as we shall see in Corollary 6.2.15, the converse is true when Ω is simply connected.

Example 6.2.11. Consider the vector field

$$F(x,y) = \left(\frac{y}{x^2 + y^2}, \frac{-x}{x^2 + y^2} \right)$$

defined in $\Omega = \mathbb{R}^2 \setminus \{(0,0)\}$. This is clearly a smooth vector field Ω and one can verify by direct calculation that on Ω, $\frac{\partial F_2}{\partial x} = \frac{\partial F_1}{\partial y}$. However, this field is *not conservative* in Ω.

Solution. If p and q are points in Ω, which have different y coordinates, then they can be joined by a line in Ω. If they have the same y coordinates and it is nonzero, they can be joined by a horizontal line in Ω. If both y coordinates are zero they can be joined by a line and a circle as indicated below. Thus any two points can be joined by a piecewise smooth curve. Consider the unit circle $\gamma(t) = (\cos t, \sin t)$, where $0 \leq t \leq 2\pi$. This is a smooth closed curve C in Ω.
We calculate $\int_C F \cdot d\mathbf{s} = \int_0^{2\pi} \langle F(\gamma(t)), \gamma'(t) \rangle dt$. Now $F(\gamma(t)) = (\sin t, -\cos t)$ and $\gamma'(t) = (-\sin t, \cos t)$. So that

$$\langle F(\gamma(t)), \gamma'(t) \rangle = -(\sin^2 t + \cos^2 t) = -1.$$

Thus

$$\int_C F \cdot d\mathbf{s} = \int_0^{2\pi} (-1)dt = -2\pi.$$

Since $\int_C F \cdot d\mathbf{s} \neq 0$, F is not conservative in Ω. It is worthwhile remarking that in a small neighborhood *not* containing the origin the function $f(x,y) = \arctan \frac{x}{y}$ can serve as a potential of F.

Combining Corollary 6.2.3 with the Poincare lemma we get

Corollary 6.2.12. *In particular, if Ω is a ball or a rectangle and F is a vector field such that the $n \times n$ matrix, $(\frac{\partial F_i}{\partial x_j}(x))$ is symmetric. Then $\int_C F \cdot d\mathbf{s} = 0$ for every piecewise smooth closed curve in Ω.*

We now come to the converse of Corollary 6.2.3. That is, a vector field F *is conservative in Ω if and only if $\int_C F \cdot d\mathbf{s} = 0$, for any closed curve C in Ω.*

Theorem 6.2.13. *Let F be a vector field on a connected domain Ω with the property that for any closed curve C in Ω, we have $\int_C F \cdot d\mathbf{s} = 0$. Then F is conservative.*

Proof. Let $C_1 = \gamma_1(t)$ and $C_2 = \gamma_2(t)$ be curves in Ω both beginning and ending at the same points, with the same parameter domain $a \le t \le b$. Then along the *closed* piecewise smooth curve C_1 followed by $-C_2$ we get

$$0 = \int_{C_1 - C_2} F \cdot ds = \int_{C_1} F \cdot ds + \int_{-C_2} F \cdot ds = \int_{C_1} F \cdot ds - \int_{C_2} F \cdot ds.$$

Thus $\int_{C_1} F \cdot ds = \int_{C_2} F \cdot ds$ so this line integral is independent of the path and depends *only* on the endpoints.

Let p_0 be a fixed point of Ω. For each point $p \in \Omega$ we can find a smooth path $C = \gamma(t)$, for $a \le t \le b$ in Ω joining $p_0 = \gamma(a)$ with $p = \gamma(b)$. Suppose $\int_C F \cdot ds$ is independent of which path we choose and only depends on the endpoints. Then we get a well defined numerical function on Ω

$$f(p) = \int_{p_0}^{p} F \cdot ds.$$

We will show $\nabla f = F$. Thus F would be conservative. Since F is given smooth, this would also show f is smooth. We calculate

$$\frac{\partial f(p)}{\partial x_i} = \lim_{t \to 0} \frac{f(p + te_i) - f(p)}{t}.$$

The numerator of the latter term is $\int_{p_0}^{p+te_i} F \cdot ds - \int_{p_0}^{p} F \cdot ds$, that is, $f(p+te_i) - f(p) = \int_{p}^{p_0} F \cdot ds + \int_{p_0}^{p+te_i} F \cdot ds = \int_{p}^{p+te_i} F \cdot ds$. Since integrals of F in Ω are independent of the path and only depend on endpoints, the integral $\int_{p}^{p+te_i} F \cdot ds$ can be calculated by means of the line segment from p to $p + te_i$, in which the i-th coordinate moves linearly and all others remain fixed, so long as this path lies completely in Ω. But, since Ω is an open set and we are interested in what happens as $t \to 0$, we can take t small enough so these segments do lie in Ω, (i.e. for all small $|t|$ and $i = 1, \ldots, n$, $p + te_i$ is in a small sphere about p contained in Ω). So on this path all coordinates coincide with those of p, except for the i-th one, which varies linearly. This line segment is parametrized by $\sigma(\lambda) = p + \lambda te_i$, where $\lambda \in [0, 1]$. Therefore

$$\int_{p}^{p+te_i} F \cdot ds = \int_{0}^{1} tF_i(p + \lambda te_i)d\lambda.$$

To calculate this last integral let $u = t\lambda$, where λ is regarded as varying. Then $du = td\lambda$ and $\int_0^1 tF_i(p + \lambda te_i)d\lambda = \int_0^t F_i(p + ue_i)du$.

Thus,

$$\frac{\partial f(p)}{\partial x_i} = \lim_{t \to 0} \frac{\int_0^t F_i(p + ue_i)du}{t} = F_i(p),$$

Since this is true for all $i = 1, \ldots, n$ and all $p \in \Omega$, this means $\nabla f = F$. □

As a corollary of the proof we also have

Corollary 6.2.14. *Let F be a vector field on a connected domain Ω with the property that for any curve C joining any two points p and $q \in \Omega$, the line integral, $\int_C F \cdot ds$, is independent of the path and depends only on the endpoints. Then F is conservative.*

By the same methods and the use of Corollary 6.2.12 we get

Corollary 6.2.15. *Let F be a vector field on a simply connected domain Ω. Then F is conservative if and only if*

$$\frac{\partial F_i}{\partial x_j}(x) = \frac{\partial F_j}{\partial x_i}(x),$$

for all $i, j = 1, \ldots, n$.

Example 6.2.11 shows that the simple connectivity of Ω cannot be removed.

EXERCISES

1. Let $F(x, y) = (xy^2 + 3x^2y, x^3 + yx^2)$. Show that there exist a function f such that $\nabla f = F$. Find such an f. Calculate $\int_C F \cdot ds$, where C is the curve consisting of the line segments $(1, 1)$, $(0, 2)$ to $(3, 0)$.

2. Let $F(\mathbf{x}) = -\frac{k\mathbf{x}}{||\mathbf{x}||^j}$, where $j = 1, 2, 3 \ldots$. Here each such F is defined on all of $\mathbb{R}^3$ excluding the origin. For which j is this vector field conservative?

3. Let $g(x,y)$ be a smooth function of two real variables and F be a vector field on $\mathbb{R}^2$ given by $F(x,y) = (\frac{\partial g}{\partial y}(x,y), -\frac{\partial g}{\partial x}(x,y))$. If F were a conservative field what would have to be true of g? For example, suppose $g(x,y) = x^2 + y^2$. Is F conservative?

4. Let $F(x,y) = (\frac{y}{x^2+y^2}, \frac{-x}{x^2+y^2})$ and C be any simple closed curve (one that goes around only once, say in the counterclockwise direction) which has the origin in its interior. Show that $\int_C F \cdot ds = -2\pi$. Also observe that integrating F along a circle that goes around twice gives a result of -4π.

5. Let $F(x,y,z) = xy\mathbf{i} + (xz+y)\mathbf{j} + (xy-z)\mathbf{k}$. Show that there exists a function f such that $\nabla f = F$. Find such an f.

6. Let $F(x,y,z) = (2xy, x^2 + \log z, \frac{y+2}{z})$. Show that F is a conservative vector field and find a potential function f of F.

7. Calculate the work done in bringing a particle of mass m from the earths surface to a point 100 miles above the earths surface. Does this depend on the route? Calculate the escape velocity of this object of mass m from a point 100 miles above the earths surface.

8. Let $F(x,y) = (f(x), g(y))$, where f and g are each smooth functions of one real variable. Is this a smooth vector field in $\mathbb{R}^2$? Is it conservative? If $F(x,y) = (x, -y)$ find a curve perpendicular to F at every point.

9. Calculate $\int_C F \cdot ds$, where $F(x,y) = (x, -y)$ and C is a circle of radius $r > 0$, centered at the origin going around once counterclockwise. Does the result depend on r?

Answers to selected Exercises

5. $f(x,y,z) = xyz + \frac{1}{2}(y^2 - z^2) + c$.   6. $f(x,y,z) = x^2 y + (y+2)\log z + c$.

6.3 Surface area and surface integrals

This is an important application of both differential and integral calculus.

6.3.1 Surface area

Generally speaking an embedded *surface* is the locus of a point moving in $\mathbb{R}^3$ with two degrees of freedom. Another way of thinking of a surface is as a deformation of a subset of $\mathbb{R}^2$. For example a cylinder is obtained by rolling a rectangle and joining the edges; a torus is obtained by bending a cylinder and joining the ends. A useful method of describing surfaces is the parametric representation.

Definition 6.3.1. A *parametrized surface* S in $\mathbb{R}^3$ is the range of a smooth mapping $\Phi : \Omega \subset \mathbb{R}^2 \to \mathbb{R}^3$. The mapping

$$\Phi(u, v) = (x(u, v), y(u, v), z(u, v))$$

where (x, y, z) are the usual coordinates in $\mathbb{R}^3$ and (u, v) runs over the parameter set $\Omega \subseteq \mathbb{R}^2$ is called a parametrization of S. The surface $S = \Phi(\Omega)$ is said *simple* if Φ is a bijection.

When a surface is described parametrically by Φ there are two families of curves that are naturally associated with the parametrization. Each such curve is obtained by setting one parameter equal to a constant (that is, fixing one of the paremeters). Let $(u_0, v_0) \in \Omega$. Fixing $v = v_0$, we get a mapping $\Phi_{v_0}(u) = \Phi(u, v_0) = (x(u, v_0), y(u, v_0), z(u, v_0))$ whose image is a curve lying on the surface S. As we know, the tangent vector to this curve at the point $\Phi(u_0, v_0)$ is given by

$$T_u = \frac{\partial \Phi}{\partial u} = \left(\frac{\partial x}{\partial u}, \frac{\partial y}{\partial u}, \frac{\partial z}{\partial u} \right)$$

evaluated at (u_0, v_0).

Similarly, for u fixed, say $u = u_0$, we get a curve lying on S parametrized by $\Phi_{u_0}(v) = \Phi(u_0, v) = (x(u_0, v), y(u_0, v), z(u_0, v))$ and the tangent vector to this curve at $\Phi(u_0, v_0)$ is given by

$$T_v = \frac{\partial \Phi}{\partial v} = \left(\frac{\partial x}{\partial v}, \frac{\partial y}{\partial v}, \frac{\partial z}{\partial v} \right)$$

evaluated at (u_0, v_0). These curves are known as *parameter curves*. Since the vectors T_u and T_v are tangent to two curves on the surface S at $\Phi(u_0, v_0)$ they ought to determine the plane tangent to the surface at this point. Since the cross product $T_u \times T_v$ is perpendicular to both T_u and T_v it follows that the vector

$$T_u \times T_v = \det \begin{bmatrix} \mathbf{i} & \mathbf{j} & \mathbf{k} \\ \frac{\partial x}{\partial u} & \frac{\partial y}{\partial u} & \frac{\partial z}{\partial u} \\ \frac{\partial x}{\partial v} & \frac{\partial y}{\partial v} & \frac{\partial z}{\partial v} \end{bmatrix} = \frac{\partial(y, z)}{\partial(u, v)}\mathbf{i} + \frac{\partial(z, x)}{\partial(u, v)}\mathbf{j} + \frac{\partial(x, y)}{\partial(u, v)}\mathbf{k}$$

is *normal* to the surface at that point. If $T_u \times T_v \neq 0$ at every point $\Phi(u, v)$ the surface S is said *regular*. For a smooth regular surface we can normalize $T_u \times T_v$ to get the *unit normal* vector

$$\mathbf{n} = \mathbf{n}(u, v) = \frac{T_u \times T_v}{||T_u \times T_v||}.$$

The regularity condition $T_u \times T_v \neq 0$ implies, in particular, that T_u and T_v are both nonzero, in fact that T_u and T_v are linearly independent. Hence they form a basis for the tangent space at $\Phi(u, v)$ and the vectors T_u, T_v and $\mathbf{n}(u, v)$ give local coordinates in a neighborhood of $\Phi(u, v)$.

Example 6.3.2. An important special case arises when the surface is the *graph*[4] of a smooth function $z = f(x, y)$, that is

$$S = \{(x, y, f(x, y)) : (x, y) \in \Omega\}.$$

In this case we can take x and y as the parameters and write

$$\Phi(x, y) = (x, y, f(x, y)).$$

Here $T_x = (1, 0, \frac{\partial f}{\partial x})$ and $T_y = (0, 1, \frac{\partial f}{\partial y})$, so

$$T_x \times T_y = \det \begin{bmatrix} \mathbf{i} & \mathbf{j} & \mathbf{k} \\ 1 & 0 & \frac{\partial f}{\partial x} \\ 0 & 1 & \frac{\partial f}{\partial y} \end{bmatrix} = (-\frac{\partial f}{\partial x}, -\frac{\partial f}{\partial y}, 1).$$

Since $||T_x \times T_y|| \geq 1$, the surface is regular.

---

[4]Note that our surface is a level set of the function $g(x, y, z) = z - f(x, y)$ and $(-\frac{\partial f}{\partial x}, -\frac{\partial f}{\partial y}, 1) = \nabla g$ is normal to the surface as we know from Proposition 3.2.21.

Figure 6.1: Surface of Revolution

Example 6.3.3. Let $S = \left\{(x, y, z) : x^2 + y^2 + z^2 = r^2\right\}$ be the *sphere* centered at the origin in $\mathbb{R}^3$ and of radius $r > 0$. Then S may be parametrized by the spherical coordinates

$$\Phi(\theta, \phi) = (r\cos\theta\sin\phi, r\sin\theta\sin\phi, r\cos\phi),$$

where the parameter set $\Omega = \{(\theta, \phi) : 0 \leq \theta \leq 2\pi,\ 0 \leq \phi \leq \pi\}$. Here

$$T_\theta \times T_\phi = (r^2\cos\theta\sin^2\phi, r^2\sin\theta\sin^2\phi, r^2\sin\phi\cos\phi) = r\sin\phi\Phi(\theta, \phi).$$

This says, as is geometrically evident, that at each point $\Phi(\theta, \phi)$ the normal points from the center through the surface S at that point.

Example 6.3.4. Let $S = \left\{(x, y, z) : x^2 + y^2 = r^2,\ 0 \leq z \leq 1\right\}$ be a *cylinder* in $\mathbb{R}^3$. A parametrization of S is

$$\Phi(\theta, z) = (r\cos\theta, r\sin\theta, z),$$

where the parameter set $\Omega = \{(\theta, z) : 0 \leq \theta \leq 2\pi,\ 0 \leq z \leq 1\}$. Here

$$T_\theta \times T_z = (r\cos\theta, r\sin\theta, 0).$$

Thus here the normal is horizontal pointing from the axis through the surface S. Since $\|T_\theta \times T_z\| = r > 0$, the surface is regular.

Example 6.3.5. (*Surface of revolution*). Let C be a directed simple regular curve in the yz-plane in $\mathbb{R}^3$ parametrized by

$$\gamma(t) = (0, y(t), z(t)) \quad t \in [a, b].$$

The surface S obtained by revolving this curve about the z-axis is called the *surface of revolution* of C about the z-axis. When we revolve a typical point $(0, y(t), z(t))$ on the curve C about the z-axis the third coordinate remains unchanged. The second coordinate will generate a circle in the plane $z = z(t)$ whose projection in the xy-plane is a circle centered at $(0,0)$ with radius $y(t)$. Using the standard parametrization of the circle (Example 6.1.2), we see that a point on this circle has coordinates $(y(t) \cos \theta, y(t) \sin \theta)$. Putting these together we get a parametrization of S by (t, θ),

$$\Phi(t, \theta) = (y(t) \cos \theta, y(t) \sin \theta, z(t)), \tag{6.7}$$

where $t \in [a, b]$ and $\theta \in [0, 2\pi]$. Now

$$T_t = (y'(t) \cos \theta, y'(t) \sin \theta, z'(t))$$

and

$$T_\theta = (-y(t) \sin \theta, y(t) \cos \theta, 0).$$

Therefore

$$T_t \times T_\theta = (z'(t)y(t) \cos \theta, z'(t)y(t) \sin \theta, y'(t)y(t)).$$

Since $\|T_t \times T_\theta\| = |y(t)| \sqrt{y'(t)^2 + z'(t)^2} = |y(t)| \, \|\gamma'(t)\| \neq 0$, we have $T_t \times T_\theta \neq \mathbf{0}$, and the surface is regular.

Many classical surfaces may be obtained as surfaces of revolution. For example the *cone* of base radius $r > 0$ and height $h > 0$ is obtained by revolving about the z-axis the curve $\gamma(t) = (0, rt, ht)$, $t \in [0, 1]$. By (6.7) a parametrization of the cone is

$$\Phi(t, \theta) = (rt \cos \theta, rt \sin \theta, ht),$$

where $(t, \theta) \in [0, 1] \times [0, 2\pi]$. Here

$$T_t = (r \cos \theta, r \sin \theta, h),$$

$$T_\theta = (-rt \sin \theta, rt \cos \theta, 0)$$

and

$$T_t \times T_\theta = (rht \cos \theta, rht \sin \theta, r^2 t).$$

Therefore $||T_t \times T_\theta|| = rt\sqrt{r^2 + h^2}$. Hence if $t \neq 0$ the cone is regular while, if $t = 0$ the cone has a singularity at the origin. When $r = h = 1$ we get the *right circular cone* and $||T_t \times T_\theta|| = t\sqrt{2}$.

Similarly, for the *cylinder* we revolve about the z-axis the vertical segment C in the yz-plane parametrized by $\gamma(z) = (0, r, t)$. Here $y(t) = r$ and $z(t) = t$ for all $t \in [a, b]$ and (6.7) gives

$$\Phi(t, \theta) = (r\cos\theta, r\sin\theta, t),$$

which is the parametrization of Example 6.3.4. The *sphere* centered at the origin of radius $r > 0$ is obtained by revolving the half circle $\gamma(\phi) = (0, r\sin\phi, r\cos\phi)$ with $\phi \in [0], \pi]$ about the z-axis. Here (6.7) gives

$$\Phi(\phi, \theta) = (r\cos\phi\sin\theta, r\sin\phi\sin\theta, r\cos\phi),$$

the parametrization of Example 6.3.3. One, similarly, can parametrize the *torus*. We do this in Example 6.3.11.

Now consider the problem of computing the area of a surface, just as in Proposition 6.1.3 we considered the problem of finding the arc length of a curve.

Proposition 6.3.6. *Let* $S = \Phi(\Omega)$ *be a smooth regular surface in* $\mathbb{R}^3$ *parametrized by* $\Phi : \Omega \subset \mathbb{R}^2 \to \mathbb{R}^3$, *with* $\Phi(u, v) = (x(u, v), y(u, v), z(u, v))$. *Then the area of* S *is given by*

$$A(S) = \int\int_\Omega ||T_u \times T_v|| \, du dv. \tag{6.8}$$

Proof. Let R denote a small rectangle in Ω (in the uv-plane) with vertices (u_0, v_0), $(u_0 + \Delta u, v_0)$, $(u_0, v_0 + \Delta v)$ and $(u_0 + \Delta u, v_0 + \Delta v)$. Its image under the map Φ is a small piece $\Phi(R)$ on the surface (a quadrilateral with curved sides with vertices $\Phi(u_0, v_0)$, $\Phi(u_0 + \Delta u, v_0)$, $\Phi(u_0, v + \Delta v)$ and $\Phi(u_0 + \Delta u, v_0 + \Delta v)$). Since Φ is smooth

$$\Phi(u_0 + \Delta u, v_0) - \Phi(u_0, v_0) \approx T_u(u_0, v_0)\Delta u,$$

and

$$\Phi(u_0, v_0 + \Delta v) - \Phi(u_0, v_0) \approx T_v(u_0, v_0)\Delta v.$$

From Corollary 1.3.30, the area of the parallelogram with adjacent sides the vectors $\Delta u T_u$ and $\Delta v T_v$ is $||\Delta u T_u \times \Delta v T_v|| = ||T_u \times T_v|| \Delta u \Delta v$. For small Δu and Δv this area approximates $\Phi(R)$, that is,

$$A(\Phi(R)) \approx ||T_u \times T_v|| \Delta u \Delta v.$$

Arguing just as in Proposition 6.1.3 we get (6.8).   □

As with the infinitesimal length (6.1), the "infinitesimal" area on the surface is

$$dS = ||T_u \times T_v|| du dv.$$

This takes on other guises. From Proposition 1.3.28 (6) we have

$$\| T_u \times T_v \| = \sqrt{\| T_u \|^2 \| T_v \|^2 - \langle T_u, T_v \rangle^2},$$

and setting $E = E(u, v) = ||T_u||^2$, $F = F(u, v) = \langle T_u, T_v \rangle$, and $G = G(u, v) = ||T_v||^2$ we see $||T_u \times T_v|| = \sqrt{EG - F^2}$. Hence $T_u \times T_v \neq 0$ if and only if $EG - F^2 > 0$. This gives the following useful formula for surface area

$$A(S) = \int \int_\Omega \sqrt{EG - F^2} du dv. \qquad (6.9)$$

In general, a surface may admit several different parametrizations but, as we will see below, they all give the *same value* for surface area. The quantities E, F and G are known as the *Gaussian first fundamental form* of the surface and quantify the deformation or *distortion* of a rectangle by the parametrization Φ. The stretching or contraction of the sides is measured by E and G while F measures the change in angle between T_u and T_v. Thus we see that shape is preserved by Φ if $E = G$ and $F = 0$, in this case Φ is called a *conformal* map.

Now the 2×2 symmetric matrix $\begin{bmatrix} E & F \\ F & G \end{bmatrix}$ is positive definite since $E > 0$ and $\det \begin{bmatrix} E & F \\ F & G \end{bmatrix} = EG - F^2 > 0$ (see, Sylvester's criterion) and of course the infinitesimal area on the surface is

$$dS = \sqrt{EG - F^2} du dv. \qquad (6.10)$$

Example 6.3.7. Let S be the graph of a smooth function $z = f(x,y)$. We have seen in Example 6.3.2 that $T_x = (1,0,\frac{\partial f}{\partial x})$ and $T_y = (0,1,\frac{\partial f}{\partial y})$. So that $E = ||T_x||^2 = 1 + \left(\frac{\partial f}{\partial x}\right)^2$, $F = \langle T_x, T_y\rangle = (\frac{\partial f}{\partial x})(\frac{\partial f}{\partial y})$, $G = ||T_y||^2 = 1 + \left(\frac{\partial f}{\partial y}\right)^2$ and $\sqrt{EG - F^2} = \sqrt{1 + \left(\frac{\partial f}{\partial x}\right)^2 + \left(\frac{\partial f}{\partial y}\right)^2}$. Thus here the surface area of S is given by

$$A(S) = \int\int_\Omega \sqrt{1 + \left(\frac{\partial f}{\partial x}\right)^2 + \left(\frac{\partial f}{\partial y}\right)^2}\, dxdy. \tag{6.11}$$

Remark 6.3.8. (*) We remark that formula (6.10) has an important generalization, (See, 8.2.1). If a Riemannian metric on a "surface" of dimension $n \geq 2$ is given at each point as the $n \times n$ positive definite, symmetric matrix $g_{ij}(x_1, \ldots, x_n)$, which replaces $\begin{bmatrix} E & F \\ F & G \end{bmatrix}$, where $(x_1, \ldots, x_n)$ are local coordinates, then the infinitesimal volume is given by

$$d\mu(x) = \sqrt{\det(g_{ij}(x))}dx_1 \ldots dx_n.$$

Example 6.3.9. Find the surface area of the sphere S of radius $r > 0$.

Solution. As we saw in Example 6.3.3, a parametrization of S is given by $\Phi(\theta, \phi) = (r\cos\theta\sin\phi, r\sin\theta\sin\phi, r\cos\phi)$ with $\Omega = [0, 2\pi) \times [0, \pi]$, and

$$T_\theta \times T_\phi = (r^2\cos\theta\sin^2\phi, r^2\sin\theta\sin^2\phi, r^2\sin\phi\cos\phi) = r\sin\phi\Phi(\theta, \phi).$$

Therefore

$$||T_\theta \times T_\phi|| = r\sin\phi||\Phi(\theta, \phi)||$$
$$= r\sin\phi\sqrt{r^2[\cos^2\theta\sin^2\phi + \sin^2\theta\sin^2\phi + \cos^2\phi]}$$
$$= r^2\sin\phi\sqrt{(\cos^2\theta + \sin^2\theta)\sin^2\phi + \cos^2\phi} = r^2\sin\phi.$$

Hence, by (6.7) the area of the sphere is

$$A(S) = \int\int_\Omega ||T_\theta \times T_\phi||d\theta d\phi = \int_0^{2\pi}\left(\int_0^\pi r^2\sin\phi d\phi\right)d\theta = 2r^2\int_0^{2\pi}d\theta = 4\pi r^2.$$

Example 6.3.10. (*Surface area of a surface of revolution*). We saw in Example 6.3.5 that a surface of revolution S obtained by revolving the curve $\gamma(t) = (0, y(t), z(t))$ about the z-axis is parametrized by

$$\Phi(t, \theta) = (y(t)\cos\theta, y(t)\sin\theta, z(t)),$$

where $t \in [a, b]$ and $\theta \in [0, 2\pi]$. Furthermore, $||T_t \times T_\theta|| = |y(t)|\,||\gamma'(t)||$. So the surface area of S is

$$A(S) = \int_a^b \int_0^{2\pi} |y(t)|\,||\gamma'(t)||dtd\theta = 2\pi \int_a^b |y(t)|\,||\gamma'(t)||dt. \quad (6.12)$$

In particular,

1. For the *cylinder* S of radius $r > 0$ and height $h > 0$ we have $\gamma(t) = (0, r, t)$ so that $\gamma'(t) = (0, 0, 1)$ and $||\gamma'(t)|| = 1$. Hence

$$A(S) = 2\pi \int_0^h r\,dt = 2\pi rh.$$

2. For the *cone* S of height $h > 0$ and (base) radius $r > 0$, we have $\gamma(t) = (0, rt, ht)$, $\gamma'(t) = (0, r, h)$ and $||\gamma'(t)|| = \sqrt{r^2 + h^2}$. Hence the surface area of the cone is

$$A(S) = 2\pi \int_0^1 rt\sqrt{r^2 + h^2}dt = 2\pi r\sqrt{r^2 + h^2}\int_0^1 t\,dt = \pi r\sqrt{r^2 + h^2}.$$

Example 6.3.11. Find the surface area of the *torus*.

Solution. The torus is obtained by revolving about the z-axis a circle of radius a in the yz-plane centered on the y-axis at distance b from the origin, where $0 < a < b$. This circle is parametrized by $\gamma(\phi) = (0, b + a\cos\phi, a\sin\phi)$, $\phi \in [0, 2\pi]$. When we revolve a point of this circle about the z-axis the third coordinate remains unchanged while the first two coordinates describe a circle of radius $b + a\cos\phi$ about the origin. So that (6.7) gives the parametrization

$$\Phi(\theta, \phi) = ((b + a\cos\phi)\cos\theta, (b + a\cos\phi)\sin\theta, a\sin\phi),$$

where the parameter set $\Omega = \{(\theta, \phi) : (\theta, \phi) \in [0, 2\pi) \times [0, 2\pi)\}$.

Now
$$T_\theta = (-(b + a\cos\phi)\sin\theta, (b + a\cos\phi)\cos\theta, 0)$$

and
$$T_\phi = (-a\sin\phi\cos\theta, -a\sin\phi\sin\theta, a\cos\phi).$$

Hence $E = ||T_\theta||^2 = (b + a\cos\phi)^2$, $F = \langle T_\theta, T_\phi \rangle = 0$, $G = ||T_\phi||^2 = a^2$. So that
$$\sqrt{EG - F^2} = a(b + a\cos\phi),$$

and the total area of the torus S is

$$A(S) = \int\int_\Omega \sqrt{EG - F^2}\,d\theta d\phi = \int_0^{2\pi}\int_0^{2\pi} a(b + a\cos\phi)\,d\theta d\phi = 4\pi^2 ab.$$

Note that here $|y(\phi)|\,||\gamma'(\phi)|| = |b + a\cos\phi|a = a(b + a\cos\phi)$ and alternatively one could use (6.12) to get the same result.

Corollary 6.3.12. *(Pappus' theorem). The area $A(S)$ of the surface S obtained by revolving a plane curve C of length L about an axis in the plane of the curve is*
$$A(S) = 2\pi Lh,$$

where h is the distance from the centroid of the curve to the axis of revolution.

Proof. Suppose the curve C lies in the yz-plane ($y > 0$) and is revolved about the z-axis. Then by (6.12), the surface of revolution S so generated has area

$$A(S) = 2\pi \int_a^b y(t)\,||\gamma'(t)||dt,$$

where $\gamma(t) = (0, y(t), z(t))$, $t \in [a, b]$ is a parametrization of C.

Expressing the integral as a line integral we have,

$$\int_a^b |y(t)|\,||\gamma'(t)||dt = \int_C yds = L\left(\frac{1}{L}\int_C yds\right) = L\overline{y}.$$

Thus $A(S) = 2\pi L\overline{y}$. $\square$

Definition 6.3.13. Let $S = \Phi(\Omega)$ be a surface parametrized by the smooth mapping $\Phi : \Omega \subset \mathbb{R}^2 \to \mathbb{R}^3$ and let $\Omega^* \subset \mathbb{R}^2$ be another parameter set in the st-plane. If $\Omega = G(\Omega^*)$ where $G : \Omega^* \to \Omega$ is a change of variables in $\mathbb{R}^2$, then the mapping $\Phi \circ G : \Omega^* \to \mathbb{R}^3$ is called a *reparametrization* of S.

We now show surface area is independent of the parametrization of the surface.

Proposition 6.3.14. *Let S be a surface in $\mathbb{R}^3$. The surface area of $A(S)$ is independent of the parametrization of S.*

Proof. Let $(s, t) \in \Omega^*$ be the new parameters related to $(u, v) \in \Omega$ by

$$(u, v) = G(s, t) = (u(s, t), v(s, t)),$$

where $\Phi \circ G : \Omega^* \to \mathbb{R}^3$ is a reparametrization of

$$S = \{\Phi(u, v) = (x(u, v), y(u, v), z(u, v)) : (u, v) \in \Omega\}.$$

Set $\Psi(s, t) = \Phi(G(s, t))$. By the Chain Rule

$$\frac{\partial \Psi}{\partial s} = \frac{\partial \Phi}{\partial u}\frac{\partial u}{\partial s} + \frac{\partial \Phi}{\partial v}\frac{\partial v}{\partial s} = T_u \frac{\partial u}{\partial s} + T_v \frac{\partial v}{\partial s},$$

$$\frac{\partial \Psi}{\partial t} = \frac{\partial \Phi}{\partial u}\frac{\partial u}{\partial t} + \frac{\partial \Phi}{\partial v}\frac{\partial v}{\partial t} = T_u \frac{\partial u}{\partial t} + T_v \frac{\partial v}{\partial t},$$

where T_u and T_v are evaluated at $(u(s, t), v(s, t))$. Using Proposition 1.3.28 we see that

$$\frac{\partial \Psi}{\partial s} \times \frac{\partial \Psi}{\partial t} = (T_u \times T_v)\left[\frac{\partial u}{\partial s}\frac{\partial v}{\partial t} - \frac{\partial u}{\partial t}\frac{\partial v}{\partial s}\right]$$

$$= (T_u \times T_v)\left(\frac{\partial(u, v)}{\partial(s, t)}\right) = (T_u \times T_v)\, J_G(s, t).$$

Now the Change of Variables formula for double integrals gives

$$\iint_{\Omega^*} \|\frac{\partial \Psi}{\partial s} \times \frac{\partial \Psi}{\partial t}\| ds\, dt = \iint_{\Omega^*} \|T_u \times T_v\|\, |J_G(s, t)| ds\, dt$$

$$= \iint_{\Omega} \|T_u \times T_v\| du\, dv.$$

$\square$

Surfaces, like curves, have *orientations*. Most surfaces we meet in practice have two sides and we distinguish between the sides of a surface by using the *normal* vectors **n** and **-n**. Given an oriented surface S we call the side of S containing the unit *outer* normal **n** the *positive side* and the side which contains the *inner* normal **-n** the *negative side*. In case of a simple oriented surface S a parametrization Φ is said *orientation preserving* or *consistent* with the orientation if $\frac{T_u \times T_v}{\|T_u \times T_v\|} = \mathbf{n}(u,v)$ for all u, v.

There exist, however, surfaces which are *one-sided* and do *not* admit orientation. Integration theory can not be defined over such surfaces. The *Mobius*[5] *strip* is a famous example of this. It can be constructed by pasting together the two ends of a long strip of paper after a half-twist has been given to one of them. It can be parametrized by

$$\Phi(t,\theta) = \left(\left(2 + t\sin\frac{\theta}{2}\right)\cos\theta, \left(2 + t\sin\frac{\theta}{2}\right)\sin\theta, t\cos\frac{\theta}{2}\right),$$

where $t \in [-1, 1]$ and $\theta \in [0, 2\pi]$.

Once a surface S is oriented, we automatically get an orientation of the surface boundary $\partial(S)$ by thinking of S as a piece of a *"curved plane"*. The surface boundary ∂S is either a closed curve or a finite number of closed curves or empty. If $\partial S = \emptyset$, then we call S a *closed surface* (or a surface without boundary). The sphere and the torus are examples of (smooth) closed surfaces. For the upper hemisphere

$$S = \left\{(x,y,z) : x^2 + y^2 + z^2 = 1 \text{ and } z \geq 0\right\}$$

the surface boundary ∂S is the circle

$$\partial S = \left\{(x,y,0) : x^2 + y^2 = 1\right\}.$$

EXERCISES

1. Find the magnitute of the normal vector and the tangent plane to each of the following surfaces.

---

[5]A. Mobius (1790-1868). A student of Gauss and professor of astronomy at Leipzig. He made important contributions to surface theory and projective geometry.

(a) $\Phi(u, v) = (u + v, u - v, 4v^2)$.

(b) $\Phi(u, v) = (u \cos v, u \sin v, \frac{1}{2}u^2 \sin 2v)$.

(c) $\frac{x^2}{a^2} + \frac{y^2}{b^2} + \frac{z^2}{c^2} = 1$ at (x_0, y_0, z_0).

2. Find the area of each of the following surfaces.

   (a) $z = x^2 + y^2$ inside the cylinder $x^2 + y^2 = 1$.

   (b) $z = \sqrt{x^2 + y^2}$ inside the cylinder $x^2 + y^2 = 1$.

   (c) $z = xy$ inside the cylinder $x^2 + y^2 = a^2$.

3. Find the area of the plane $x + y + z = 6$ in the first octant cut off by the cylinder $x^2 + y^2 = 4$.

4. Find the area of the part of the cylinder $x^2 + z^2 = 16$ in the first octant cut off by the cylinder $x^2 + y^2 = 16$.

5. Show that the area of the cylinder $x^2 + y^2 = 6y$ cut off by the sphere $x^2 + y^2 + z^2 = 36$ and the plane $z = 0$ is 72.

6. Show that the area of the part of the cone $z^2 = x^2 + y^2$ lying inside the cylinder $x^2 + y^2 = 2ax$ is $4\sqrt{2}a^2\pi$.

7. Find the area of the part of the cylinder $z^2 = 4x$ cut off by the cylinder $y^2 = 4x$ and the plane $x = 1$.

8. Find the area of the part of the cone $z = \sqrt{x^2 + y^2}$

   (a) cut off by the plane $x + 2z = 3$.

   (b) cut off by the sphere $x^2 + y^2 + z^2 = 2ax$.

9. Find the area of the cone $x^2 = y^2 + z^2$ inside the cylinder $x^2 + y^2 = a^2$.
   Hint. Project on the yz-plane.

10. Show that the area of the cylinder $y^2 + z^2 = 2z$ cut off by the cone $x^2 = y^2 + z^2$ is 16.

11. Show that the area of the cone $x^2 = y^2 + z^2$ between the cylinder $y^2 = z$ and the plane $z = y + 2$ is $9\sqrt{2}$.

12. Find the area of the part of the sphere $x^2 + y^2 + z^2 = c^2$ inside the paraboloid $\frac{x^2}{a^2} + \frac{y^2}{b^2} = 2(z + c)$.

Answers to selected Exercises

1. (a) $\sqrt{128v^2 + 4}$. (b) $\sqrt{2u^4 + u^2}$. (c) $\frac{xx_0}{a^2} + \frac{yy_0}{b^2} + \frac{zz_0}{a^2} = 1$.

2. (a) $\frac{5\sqrt{5}-1}{6}\pi$. (b) $\sqrt{2}\pi$. (c) $\frac{2}{3}[(1 + a^2)^{\frac{3}{2}} - 1]\pi$.   3. $4\sqrt{3}\pi$.

4. 16.   7. $\frac{16(\sqrt{8}-1)}{3}$.   8. (a) $2\pi a^2$. (b) $\frac{\sqrt{2}}{4}a^2\pi$.   9. $2\pi a^2$.   12. $4\pi c\sqrt{ab}$.

6.3.2   Surface integrals

A surface integral can be thought of as the two-dimensional analog of a line integral where the region of integration is a surface rather than a curve.

Definition 6.3.15. Let $S = \Phi(\Omega)$ be a smooth oriented surface in $\mathbb{R}^3$ parametrized by Φ and let $f : \mathbb{R}^3 \to \mathbb{R}$ be a continuous scalar field whose domain contains S. The *surface integral* of f over S is defined by

$$\int\int_S f dS = \int\int_\Omega f(\Phi(u, v))\|T_u \times T_v\|dudv. \qquad (6.13)$$

Note that the continuity of f and the smoothness of Φ guarantee the continuity of the integrand in (6.13) and so the existence of the integral. If a surface S can be decomposed into a finite number of pieces $S_1, S_2, ..., S_m$, the integral over the whole surface S (say, over its positive side) is equal to the sum of the integrals taken over the corresponding (i.e., positive) sides of the surfaces $S_1, S_2, ..., S_m$.

Note also that just as with curves and line integrals if $f \equiv 1$, we recover the formula (6.8) for the surface area. Like the surface area, the surface integral (6.13) is independent of the parametrization of the surface S.

Corollary 6.3.16. *The surface integral $\int\int_S f dS$ is independent of the parametrization of the surface S.*

Proof. With the notation as in Proposition 6.3.14 the Change of Variables formula for double integrals gives

$$\int\int_{\Omega^*} f(\Psi(s,t))\|\frac{\partial\Psi}{\partial s}\times\frac{\partial\Psi}{\partial t}\|dsdt$$

$$=\int\int_{\Omega^*} f(\Phi(G(s,t)))\|T_u\times T_v\|\ |J_G(s,t)|dsdt$$

$$=\int\int_{\Omega} f(\Phi(u,v))\|T_u\times T_v\|dudv.$$

$\square$

Example 6.3.17. Let $f(x,y,z)=x^2z$. Compute $\int\int_S fdS$, where S is the cylindrical surface $S=\{(x,y,z):x^2+y^2=1,\ 0\le z\le 1\}$.

Solution. Form Example 6.3.4 a parametrization of the cylinder S is $\Phi(\theta,z)=(\cos\theta,\sin\theta,z)$ with $(\theta,z)\in\Omega=[0,2\pi]\times[0,1]$. The normal vector to S is $T_\theta\times T_z=(\cos\theta,\sin\theta,0)$ and so $\|T_\theta\times T_z\|=1$. Hence

$$\int\int_S fdS=\int\int_\Omega z\cos^2\theta d\theta dz=\int_0^{2\pi}\cos^2\theta\left(\int_0^1 zdz\right)d\theta$$

$$=\frac{1}{2}\left[\frac{\theta}{2}+\frac{1}{4}\sin 2\theta\right]_0^{2\pi}=\frac{\pi}{2}.$$

Example 6.3.18. Compute $\int\int_S zdS$, where S is the surface defined by $z=x^2+y^2$ and $x^2+y^2\le 1$.

Solution. The equation $z=x^2+y^2$ describes a paraboloid and the inequality $x^2+y^2\le 1$ describes the interior and the boundary of a cylinder. The surface S is the graph of the smooth function $z=g(x,y)=x^2+y^2$ defined on $\Omega=\{(x,y):x^2+y^2\le 1\}$ and can be parametrized by $\Phi(x,y)=(x,y,g(x,y))$. Form Example 6.3.2 we know that $T_x\times T_y=(-\frac{\partial g}{\partial x},-\frac{\partial g}{\partial y},1)=(-2x,-2y,1)$. Hence

$$\|T_x\times T_y\|=\sqrt{4x^2+4y^2+1}.$$

The scalar field we integrate is $f(x, y, z) = z$, so that $f(\Phi(x, y)) = x^2 + y^2$. Now (6.13) gives

$$\int\int_S z\, dS = \int\int_\Omega (x^2 + y^2)\sqrt{4(x^2 + y^2) + 1}\, dx\, dy.$$

Using polar coordinates (r, θ), the double integral becomes

$$\int_0^1 \int_0^{2\pi} r^2 \sqrt{4r^2 + 1}\, r\, dr\, d\theta = 2\pi \int_0^1 r^3 \sqrt{4r^2 + 1}\, dr = \frac{5\sqrt{5}\pi}{12} + \frac{\pi}{60},$$

where we made the substitution $r = \frac{1}{2}\tan\phi$. We remark that S could also be parametrized using the parameters $(r, \theta) \in \Omega = [0, 1] \times [0, 2\pi]$ and $\Phi(r, \theta) = (r\cos\theta, r\sin\theta, r^2)$. We leave this calculation of the surface integral as an exercise to the reader.

Next we define the surface integral of a vector field F on $\mathbb{R}^3$ over a smooth oriented surface $S = \Phi(\Omega)$.

Definition 6.3.19. Let $F : \mathbb{R}^3 \to \mathbb{R}^3$ be a continuous vector field whose domain contains a smooth oriented surface $S = \Phi(\Omega)$. The *surface integral* of F over S denoted by $\int\int_S F \cdot d\mathbf{S}$ is defined by

$$\int\int_S F \cdot d\mathbf{S} = \int\int_\Omega F(\Phi(u, v)) \cdot (T_u \times T_v)\, du\, dv, \qquad (6.14)$$

where the dot means the inner product in $\mathbb{R}^3$.

Since $T_u \times T_v = \mathbf{n}\|T_u \times T_v\|$, we can write

$$\int\int_S F \cdot d\mathbf{S} = \int\int_\Omega F \cdot (T_u \times T_v)\, du\, dv = \int\int_\Omega (F \cdot \mathbf{n})\|T_u \times T_v\|\, du\, dv.$$

Thus,

$$\int\int_S F \cdot d\mathbf{S} = \int\int_S (F \cdot \mathbf{n})\, dS. \qquad (6.15)$$

The scalar field $(F \cdot \mathbf{n})$ is called the *normal component* of the vector field F over the surface S.

From (6.15) and Corollary 6.3.16 we see that the integral is independent of the choice of the parametrization as long as the parametrization induces the specified orientation, but switching to the opposite orientation (using the unit normal **-n**) results in multiplying the integral by -1.

Expressing the normal vector $T_u \times T_v = \frac{\partial(y,z)}{\partial(u,v)}\mathbf{i} + \frac{\partial(z,x)}{\partial(u,v)}\mathbf{j} + \frac{\partial(x,y)}{\partial(u,v)}\mathbf{k}$ in terms of its components, the surface integral (6.14) of a vector field $F = F_1\mathbf{i} + F_2\mathbf{j} + F_3\mathbf{k}$ can be written as

$$\int\int_S F \cdot d\mathbf{S} = \int\int_\Omega F_1(\Phi(u,v)) \left(\frac{\partial(y,z)}{\partial(u,v)}\right) dudv$$

$$+ \int\int_\Omega F_2(\Phi(u,v)) \left(\frac{\partial(y,x)}{\partial(u,v)}\right) dudv + \int\int_\Omega F_3(\Phi(u,v)) \left(\frac{\partial(x,y)}{\partial(u,v)}\right) dudv$$

$$= \int\int_S F_1 dydz + \int\int_S F_2 dzdx + \int\int_S F_3 dxdy.$$

This is frequently written briefly as

$$\int\int_S F \cdot d\mathbf{S} = \int\int_S F_1 dydz + F_2 dzdx + F_3 dxdy. \tag{6.16}$$

The integrals which appear in (6.16) are also referred to as surface integrals. For example, $\int\int_S F_1 dydz = \int\int_S F_1 dydz + 0 dzdx + 0 dxdy$.

Note also that if the unit normal **n** is expressed in terms of its direction cosines, say $\mathbf{n} = \cos\alpha\mathbf{i} + \cos\beta\mathbf{j} + \cos\gamma\mathbf{k}$ and we write F in vector notation $F = F_1\mathbf{i} + F_2\mathbf{j} + F_3\mathbf{k}$, then

$$\int\int_S (F \cdot \mathbf{n})dS = \int\int_S (F_1\cos\alpha + F_2\cos\beta + F_3\cos\gamma)dS,$$

where α, β and γ are the angles that the vector **n** makes with the x-axis, y-axis and z-axis, respectively.

We now try to understand the significance of $\int\int_S(F \cdot \mathbf{n})dS$. We know from Proposition 1.3.35 that the absolute value of the triple scalar product $F \cdot (T_u \times T_v)$ is the volume of the parallelepiped with adjacent sides the vectors F, T_u and T_v. Thinking of F as representing the flow of some substance (for example, a gas, fluid, or heat) the magnitude (norm) of $F(\Phi(u,v))$ is the rate of flow of the substance near the point $(x(u,v), y(u,v), z(u,v))$ and its direction is the direction of the flow.

Thus the integral $\int \int_S (F \cdot \mathbf{n}) dS$ represents the net quantity of a substance to flow across the surface S per unit time and is called the *flux* of F across the surface S.

Example 6.3.20. Let $F(x, y, z) = (x^2, xy, yz)$. Compute $\int \int_S F \cdot d\mathbf{S}$, where S is the surface parametrized by $\Phi(t, \theta) = (t \cos \theta, t \sin \theta, t^2)$ with $(t, \theta) \in \Omega = [0, 1] \times [0, \pi]$

Solution. The surface S is a portion of the paraboloid $z = x^2 + y^2$ obtained by making a half-revolution about the z-axis of the parabola $z = x^2$ lying in the xz-plane. Here $T_t = (\cos \theta, \sin \theta, 2t)$ and $T_\theta = (-t \sin \theta, t \cos \theta, 0)$. Hence $T_t \times T_\theta = (2t^2 \cos \theta, 2t^2 \sin \theta, -t)$.
Thus

$$
\int \int_S F \cdot d\mathbf{S} = \int \int_\Omega F(\Phi(t, \theta) \cdot (T_t \times T_\theta) dt d\theta
$$

$$
= \int \int_\Omega (t^2 \cos^2 \theta, t^2 \cos \theta \sin \theta, t^3 \sin \theta)
$$

$$
\cdot (2t^2 \cos \theta, 2t^2 \sin \theta, -t) dt d\theta
$$

$$
= \int_0^\pi \left(\int_0^1 [2t^4 \cos^3 \theta + 2t^4 \cos \theta \sin^2 \theta - t^4 \sin \theta] dt \right) d\theta
$$

$$
= \int_0^\pi [2 \cos^3 \theta + 2 \cos \theta \sin^2 \theta - \sin \theta] \left(\int_0^1 t^4 dt \right) d\theta
$$

$$
= \frac{1}{5} \left[2 \sin \theta - \frac{2}{3} \sin^3 \theta + \frac{2}{3} \sin^3 \theta + \cos \theta \right]_0^\pi = -\frac{2}{5}.
$$

Example 6.3.21. Let $F(x, y, z) = (2x, -3y, z)$. Compute $\int \int_S F \cdot \mathbf{n} dS$, where S is the entire surface of the solid formed by the cylindrical surface $x^2 + y^2 = 1$, the planes $z = 0$, $z = x + 2$ and $\mathbf{n}$ is the outer unit normal.

Solution. This surface S consists of three pieces; the circular base S_1 in the xy-plane, the lateral surface S_2 of the cylinder $x^2 + y^2 = 1$ and the elliptic plane section S_3, i.e., the part of the plane $z = x + 2$ inside the cylinder. So

$$\iint_S F \cdot \mathbf{n} dS = \iint_{S_1} F \cdot \mathbf{n} dS + \iint_{S_2} F \cdot \mathbf{n} dS + \iint_{S_3} F \cdot \mathbf{n} dS.$$

The surface S_1 is parametrized by

$$\Phi_1(x, y) = (x, y, 0),$$

where the parameter set $\Omega_1 = \{(x, y) : x^2 + y^2 = 1\}$. Here $T_x = (1, 0, 0)$, $T_y = (0, 1, 0)$ and $T_x \times T_y = (0, 0, 1) = \mathbf{k}$. Hence $\mathbf{n} = -\mathbf{k}$, so that

$$F(\Phi_1(x, y)) \cdot \mathbf{n} = 0.$$

Therefore

$$\iint_{S_1} F \cdot \mathbf{n} dS = \iint_{\Omega_1} 0 \, dx dy = 0.$$

The surface $S_2 = \{(x, y, z) : x^2 + y^2 = 1, \ 0 \le z \le x + 2\}$ is parametrized by

$$\Phi_2(\theta, z) = (\cos\theta, \sin\theta, z),$$

where the parameter set $\Omega_2 = \{(\theta, z) : 0 \le \theta \le 2\pi, 0 \le z \le 2 + \cos\theta\}$. Here $\mathbf{n} = T_\theta \times T_z = (\cos\theta, \sin\theta, 0)$. Hence

$$F(\Phi_2(\theta, z)) \cdot \mathbf{n} = 2\cos^2\theta - 3\sin^2\theta.$$

Therefore

$$\iint_{S_2} F \cdot \mathbf{n} dS = \iint_{\Omega_2} (2\cos^2\theta - 3\sin^2\theta) d\theta dz$$

$$= \int_0^{2\pi} (2\cos^2\theta - 3\sin^2\theta) \left(\int_0^{2+\cos\theta} dz\right) d\theta$$

$$= \int_0^{2\pi} (2 - 5\sin^2\theta)(2 + \cos\theta) d\theta = -2\pi.$$

Finally, the surface $S_3 = \{(x, y, z) : x^2 + y^2 = 1, \ z = x + 2\}$ is parametrized by

$$\Phi_3(x, y) = (x, y, x + 2),$$

where the parameter set $\Omega_3 = \Omega_1$. Here $T_x = (1, 0, 1)$, $T_y = (0, 1, 0)$ and $T_x \times T_y = (-1, 0, 1)$. Hence $\mathbf{n} = (-\frac{1}{\sqrt{2}}, 0, \frac{1}{\sqrt{2}})$, so that

$$F(\Phi_3(x, y)) \cdot \mathbf{n} = \frac{1}{\sqrt{2}}(-x + 2).$$

Therefore

$$\int\int_{S_3} F \cdot \mathbf{n} dS = \frac{1}{\sqrt{2}} \int\int_{\Omega_3} (-x + 2)\sqrt{2}dxdy$$

$$= \int_0^{2\pi} \int_0^1 (2 - r\cos\theta)r\,dr\,d\theta = \int_0^{2\pi} (1 - \frac{1}{3}\cos\theta)d\theta = 2\pi.$$

Thus,

$$\int\int_S F \cdot \mathbf{n} dS = 0 + 2\pi + (-2\pi) = 0.$$

EXERCISES

1. Compute $\int\int_S xyz\,dS$, where S is the portion of the plane $x + y + z = 1$, lying in the first octant.

2. Compute $\int\int_S x^2z\,dS$, where S is the part of the surface of cylinder $x^2 + y^2 = 1$, cut off by the planes $z = 0$ and $z = 1$.

3. Compute $\int\int_S z^2\,dS$, where S is the portion of the surface of the cylinder $x^2 + y^2 = 4$ between the planes $z = 0$ and $z = x + 3$.

4. Compute $\int\int_S x^2y^2z\,dS$, where S is the positive side of the lower half of the sphere $x^2 + y^2 + z^2 = a^2$.

5. Compute $\int\int_S \frac{1}{\sqrt{z-y+1}}\,dS$, where S is the surface $2z = x^2 + 2y$, $0 \le x \le 1, 0 \le y \le 1$.

6. Compute $\int\int_S \frac{1}{x^2+y^2+z^2}\,dS$, where S the surface of the cylinder between the planes $z = 0$ and $z = h$.

7. Let $f(x, y, z) = x^4 - y^4 + y^2z^2 - z^2x^2 + 1$. Find the surface integral $\int\int_S f(x, y, z)dS$, where S is the portion of the cone $z^2 = x^2 + y^2$, $z \ge 0$, cut off by the cylinder $x^2 + y^2 = 2x$.

8. The Newtonian potential at $(0, 0 - a)$ due to a mass with constant density ρ on the hemisphere $S : x^2 + y^2 + z^2 = a^2$, $z \geq 0$ is

$$U = \int \int_S \frac{\rho}{\sqrt{x^2 + y^2 + (z + a)^2}} dS.$$

Show that $U = 2\pi\rho a(2 - \sqrt{2})$

9. Find $\int \int_S F \cdot \mathbf{n} dS$, where $F(x, y, z) = (x^2, y^2, z^2)$ and S is parametrized by $\Phi(u, v) = (u + v, u - v, u)$, $0 \leq u \leq 2$ and $1 \leq v \leq 3$.

10. Find the flux of $F(x, y, z) = 3xy^2\mathbf{i} + 3x^2y\mathbf{j} + z^3\mathbf{k}$ out of the unit sphere.

11. Compute $\int \int_S F \cdot \mathbf{n} dS$ for the following F and S, where $\mathbf{n}$ is the outward unit normal of the surface.

   (a) $F(x, y, z) = (x, y^2, z)$ and S is the triangle determined by the plane $x + y + z = 1$ and the coordinate planes.

   (b) $F(x, y, z) = (x^2, z, -y)$ and S is the unit sphere $x^2 + y^2 + z^2 = 1$.

   (c) $F(x, y, z) = (x, 0, 0)$ and S is the part of the unit sphere inside the upper part of the cone $z^2 = x^2 + y^2$.

   (d) $F(x, y, z) = (x^2, y^2, z^2)$ and S is the part of the cone $z^2 = x^2 + y^2$ between the planes $z = 1$ and $z = 2$.

   (e) $F(x, y, z) = y^2, z, -x)$ and S is the part of the cylinder $y^2 = 1 - x$ between the planes $z = 0$ and $z = x$, $x \geq 0$ with $\mathbf{n} \cdot \mathbf{i} > 0$.

12. Find the surface integal of the vector field $F(x, y, z) = (x, y, z^2)$ over the cylinder $x^2 + y^2 = a^2$, cut off by the planes $z = 0$ and $z = 1$,

   (a) including the top and bottom.

   (b) excluding the top and bottom.

13. Compute the surface integral

$$\int \int_S xydydz + yzdzdx + xzdxdy,$$

where S is the outer side of the pyramid formed by the plane
$x + y + z = 1$ and the coordinate planes.

14. Compute the surface integral

$$\iint_S xz\,dydz + x^2y\,dzdx + y^2z\,dxdy,$$

where S is the outer side of the surface in the first octant formed
by the cylinder $x^2+y^2 = 1$, the paraboloid of revolution $z = x^2+y^2$
and the coordinate planes.

15. (*) Let $S = \Phi(\Omega)$ be a parametrized surface in $\mathbb{R}^3$. The *Dirichlet's
functional*[6] is defined by

$$J(\Phi) = \frac{1}{2} \iint_\Omega \left(\left\| \frac{\partial \Phi}{\partial u} \right\|^2 + \left\| \frac{\partial \Phi}{\partial v} \right\|^2 \right) dudv.$$

Show that $A(S) \le J(\Phi)$, and equality holds if

(a) $\left\| \frac{\partial \Phi}{\partial u} \right\|^2 = \left\| \frac{\partial \Phi}{\partial v} \right\|^2$ and (b) $\langle \frac{\partial \Phi}{\partial u}, \frac{\partial \Phi}{\partial v} \rangle = 0$.

A parametrization Φ satisfying (a) and (b) is called *conformal*.
Hint. Use the Cauchy-Schwarz inequality for integrals.

Answers to selected Exercises

1. $\frac{\sqrt{3}}{120}$. 2. $\frac{\pi}{2}$. 3. π. 4. $\frac{2\pi a^7}{105}$. 5. $\sqrt{2}$. 6. $2\pi \tan^{-1}\left(\frac{h}{a}\right)$. 7. $\pi\sqrt{2}$.

9. $\frac{104}{3}$. 10. $\frac{12}{5}\pi$. 11. (a) $\frac{5}{12}$. (b) 0. (c) $\frac{8-5\sqrt{2}}{12}\pi$. (d) $\frac{15}{4}\pi$. (e) $\frac{4}{15}$.

12. (a) $3\pi a^2$. (b) $2\pi a^2$. 13. $\frac{1}{8}$. 14. $\frac{\pi}{8}$.

---

[6]The Dirichlet functional played a major role in the mathematics of the 19th
century. Georg Riemann used it to give a proof of the famous Riemann mapping
theorem. Today it is still used extensively as a tool in the theory of partial differential
equations and the calculus of variations. See also Theorem 8.4.7 in Chapter 8.

6.4   Green's theorem and the divergence theorem in $\mathbb{R}^2$

Green's theorem[7] is a two dimensional analog of the Fundamental Theorem of Calculus. It expresses a double integral over a plane region Ω as a line integral along a closed curve forming the boundary of Ω. In the next section we shall look into the analogous situation in three dimensional space. Before we turn to Green's theorem, we establish some notation and conventions. We shall consider regions Ω in $\mathbb{R}^2$ whose bounding curve, $\partial\Omega$, is closed and piecewise smooth. We shall consider $\partial\Omega$ to have the positive (i.e., *counterclockwise*) orientation. This means that when traversing the boundary, $\partial\Omega$, the region Ω always lies on the left of the boundary. If $F(x, y) = (F_1(x, y), F_2(x, y))$ is a continuous vector field in $\mathbb{R}^2$ and C is a curve in $\mathbb{R}^2$ parametrized by $\gamma(t) = (x(t), y(t))$ with $t \in [a, b]$, we deal with the line integral

$$\int_C F \cdot d\mathbf{s} = \int_a^b F(\gamma(t)) \cdot \gamma'(t)dt = \int_C F_1 dx + F_2 dy.$$

Theorem 6.4.1. *(Green's Theorem). Let $F = (F_1, F_2)$ be a C^1 vector field on an open set containing Ω, with Ω a domain in $\mathbb{R}^2$ whose boundary curve, $C = \partial\Omega$, is as above. Then*

$$\int\int_\Omega \left(\frac{\partial F_2}{\partial x} - \frac{\partial F_1}{\partial y} \right) dxdy = \int_{\partial\Omega} (F_1 dx + F_2 dy). \qquad (6.17)$$

Proof. First we assume Ω is an elementary region which is both x-simple and y-simple, that is, it has the (x-simple) form

$$\Omega = \{(x, y) : a \le x \le b, \phi_1(x) \le y \le \psi_1(x)\}$$

where ϕ_1 and ψ_1 are continuous functions on $[a, b]$, and the (y-simple) form

$$\Omega = \{(x, y) : c \le y \le d, \phi_2(y) \le x \le \psi_2(x)\},$$

where ϕ_2 and ψ_2 are continuous functions on $[c, d]$.

---

[7]G. Green (1793-1841). Mathematician and mathematical physicist. He is also known for *Green's functions* and *Green's identities*.

Since F is C^1 the function F_1 is also C^1 on $\overline{\Omega}$. We first show

$$-\int\int_\Omega \frac{\partial F_1}{\partial y} dx dy = \int_{\partial\Omega} F_1(x,y) dx.$$

If we write Ω as x-simple, then $\partial\Omega$ consists of the curve C_1 followed by the portion of vertical line C_2, the curve C_3 and the vertical line segment C_4. We compute

$$\int_{\partial\Omega} F_1 dx = \int_{C_1} F_1 dx + \int_{C_2} F_1 dx + \int_{C_3} F_1 dx + \int_{C_4} F_1 dx.$$

On the portions of vertical lines $x = a$ and $x = b$ (which may reduce to single points), x is constant, so $x'(t) = 0$ in any parametrization and hence $dx = 0$. Therefore, $\int_{C_2} F_1(x,y)dx = 0$ and $\int_{C_4} F_1(x,y)dx = 0$. The curve C_1 can be parametrized by $\gamma_1(x) = (x, \phi_1(x))$, while the curve C_3 by $\gamma_3(x) = (x, \psi_1(x))$ for $x \in [a,b]$, and C_3 has the negative orientation. Hence $\int_{C_1} F_1(x,y)dx = \int_b^a F_1(x, \phi_1(x))dx$ and $\int_{C_3} F_1(x,y)dx = -\int_a^b F_1(x, \psi_1(x))dx$. Thus,

$$\int_{\partial\Omega} F_1(x,y)dx = \int_a^b (F_1(x, \phi_1(x)) - F_1(x, \psi_1(x)))dx.$$

Turning to the double integral, we know

$$\int\int_\Omega \frac{\partial F_1}{\partial y}(x,y)dx dy = \int_a^b \left(\int_{\phi_1(x)}^{\psi_1(x)} \frac{\partial F_1}{\partial y}(x,y)dy \right) dx.$$

Since, in the inner integral x is held fixed both in taking the partial derivative and in the integration we can apply the Fundamental Theorem of Calculus to it. This tells us that this last integral is

$$\int_a^b (F_1(x, \psi_1(x)) - F_1(x, \phi_1(x)))dx,$$

so that

$$\int\int_\Omega \frac{\partial F_1}{\partial y} dx dy = -\int_{\partial\Omega} F_1(x,y) dx.$$

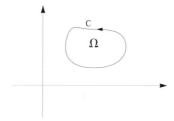

Figure 6.2: The region in Green's theorem

Similarly, expressing Ω as y-simple region, we obtain

$$\int \int_\Omega \frac{\partial F_2}{\partial x} dx dy = \int_{\partial \Omega} F_2(x, y) dy.$$

Adding these last two equalities we obtain (6.17) proving Green's theorem for regions of this type.

To deal with more general regions such as ones with holes we divide them, by introducing cross cuts, into a finite number of regions each of which is of the type we have just considered (see Figure 6.3). All curves are now only piecewise smooth rather than smooth, but, as we have already seen, this causes no trouble. Notice that when we impose a positive orientation on the boundary, although when traversing the outermost curve we go around counterclockwise, when traversing the inner curve we must go around clockwise. Since in computing the line integral along the boundary the integrals along the cross cuts each occur twice and in opposite directions *all cancel*! Since the result does hold for each of the finite number of subregions, the discussion above immediately shows that it also holds here.   □

In a more common notation, by setting $F_1 = P$ and $F_2 = Q$, Green's theorem takes in the form

$$\int \int_\Omega \left(\frac{\partial Q}{\partial x} - \frac{\partial P}{\partial y} \right) dx dy = \int_{\partial \Omega} (P dx + Q dy).$$

Green's theorem is obviously a generalization of our result that in a conservative field the line integral around a piecewise smooth closed curve lying completely in the domain is zero.

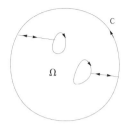

Figure 6.3: A region with holes

Example 6.4.2. Compute the line integral

$$\int_C [x^2 e^x + y - \log(1 + x^2)]dx + [8x - \sin y]dy,$$

where C is the unit circle, oriented counterclockwise.

Solution. A direct calculation of the integral is difficult, however it is easily computed using Green's theorem.

Let $F_1(x, y) = x^2 e^x + y - \log(1 + x^2)$, $F_2(x, y) = 8x - \sin y$ and $C = \partial\Omega$ where Ω is the unit disk in $\mathbb{R}^2$. Then by Green's theorem,

$$\int_C [x^2 e^x + y - \log(1 + x^2)]dx + [8x - \sin y]dy = \int\int_\Omega \left(\frac{\partial F_2}{\partial x} - \frac{\partial F_1}{\partial y} \right) dx dy$$

$$= \int\int_\Omega 7 dx dy = 7 Area(\Omega) = 7\pi.$$

We now get some corollaries of Green's theorem.

Corollary 6.4.3. *Let Ω be a region in $\mathbb{R}^2$ and $\partial\Omega$ be its boundary, with a counter clockwise orientation. Then the area of Ω is*

$$A(\Omega) = \frac{1}{2}\int_{\partial\Omega} (x dy - y dx). \tag{6.18}$$

Proof. Take $F_2(x, y) = x$ and $F_1(x, y) = -y$. Then Green's theorem tells us $A(\Omega) = \int\int_\Omega 1 dx dy = \frac{1}{2}\int_{\partial\Omega}(x dy - y dx)$. $\square$

Example 6.4.4. Compute the area of the region Ω in $\mathbb{R}^2$ bounded by the ellipse $\frac{x^2}{\alpha^2} + \frac{y^2}{\beta^2} = 1$.

Solution. We parametrize the ellipse by $\gamma(t) = (\alpha \cos t, \beta \sin t)$, $t \in [0, 2\pi]$ (see Example 6.1.2). According to (6.18), we have

$$A(\Omega) = \frac{1}{2} \int_{\partial\Omega} (x\,dy - y\,dx) = \frac{1}{2} \int_0^{2\pi} \alpha\beta(\cos^2 t + \sin^2 t)\,dt$$

$$= \frac{1}{2}\alpha\beta \int_0^{2\pi} dt = \pi\alpha\beta.$$

Example 6.4.5. Compute the area inside the loop Ω of the *Descarte folium*
$x^3 + y^3 = 3axy$ (see Figure 6.4 with $a = 1$).

Solution. To obtain a parametrization γ of the boundary of the folium, we set $y = tx$, where $t = \tan\theta = \frac{y}{x}$ with $\theta \in [0, \frac{\pi}{2}]$. Then

$$\gamma(t) = \left(\frac{3at}{1 + t^3}, \frac{3at^2}{1 + t^3} \right),$$

where t varies from 0 to ∞. By (6.18)

$$A(\Omega) = \frac{1}{2} \int_{\partial\Omega} (x\,dy - y\,dx) = \frac{9a^2}{2} \int_0^\infty \frac{t^2}{(1 + t^3)2}\,dt = \frac{3}{2}a^2.$$

For our second corollary, let F be the vector field on $\mathbb{R}^2$ given by $F(x, y) = (\frac{\partial f}{\partial x}, -\frac{\partial f}{\partial y})$, where f is a smooth numerical function. Then we get

Corollary 6.4.6. *For any smooth function f, defined on Ω, we have*

$$\int_{\partial\Omega} \frac{\partial f}{\partial x}\,dy - \frac{\partial f}{\partial y}\,dx = \int\int_\Omega \left(\frac{\partial^2 f}{\partial x^2} + \frac{\partial^2 f}{\partial y^2} \right) dx\,dy.$$

In particular, if f is a harmonic function, then

$$\int_{\partial\Omega} \frac{\partial f}{\partial x}\,dy - \frac{\partial f}{\partial y}\,dx = 0.$$

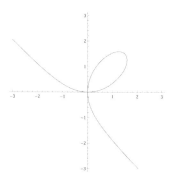

Figure 6.4: $x^3 + y^3 = 3xy$

The following result is very useful in computing line integrals. For example, in the case of the vector field F given in Example 6.2.11, our computation shows the result is 2π.

Corollary 6.4.7. *Let $F = (F_1, F_2)$, be a smooth vector field on $U - \{p_0\}$ satisfying $\frac{\partial F_1}{\partial y} = \frac{\partial F_2}{\partial x}$ also on $U - \{p_0\}$, but having a singularity at a point $p_0 \in U$. Let C any piecewise smooth simple closed curve in U with a counterclockwise orientation containing p_0 in its interior. Then $\int_C F \cdot ds = \int_{C_0} F \cdot ds$, where C_0 is a small circle inside U centered at p_0 going around counterclockwise.*

Proof. Let Ω be the region in $\mathbb{R}^2$ between the piecewise smooth closed curve, C, and the circle C_0. Then F is a smooth vector field on Ω and the boundary of Ω consists of C together with $-C_0$. This has a *counterclockwise* orientation. By Green's theorem

$$\int_C F \cdot ds + \int_{-C_0} F \cdot ds = \int\int_\Omega \left(\frac{\partial F_2}{\partial x} - \frac{\partial F_1}{\partial y} \right) dx\, dy.$$

By hypothesis the double integral is zero. Hence

$$\int_C F \cdot ds = \int_{C_0} F \cdot ds.$$

$\square$

We now use Green's theorem to prove the Change of Variables For-
mula in $\mathbb{R}^2$. It is also possible to prove a similar formula in $\mathbb{R}^3$ using
the Divergence theorem in $\mathbb{R}^3$(see Exercise 6.10.20)

Theorem 6.4.8. *(Change of variables formula for double integrals).*
Let Ω and D be regions in $\mathbb{R}^2$ to which Green's theorem is applicable.
Suppose $\varphi : \Omega \to \mathbb{R}^2$ given by $\varphi(u,v) = (x(u,v), y(u,v))$ is a C^2 change
of variables with $D = \varphi(\Omega)$. Let $f : \varphi(\Omega) \to \mathbb{R}$ be a continuous function.
Then

$$\int\int_{\varphi(\Omega)} f(x,y)dxdy = \int\int_{\Omega} f(x(u,v), y(u,v)) \left| \frac{\partial(x,y)}{\partial(u,v)} \right| dudv.$$

Proof. Since φ is a diffeomorphism Lemma 5.1.2 tells us that the bound-
ary $\partial D = \varphi(\partial\Omega)$. Let $\gamma(t) = (u(t), v(t))$ with $t \in [a,b]$ be a parametrization of $\partial\Omega$. Then

$$\varphi(\gamma(t)) = (x(u(t), v(t)), y(u(t), v(t)))$$

is a parametrization of ∂D.

Consider the function

$$F(x,y) = \int_0^x f(\xi, y)d\xi.$$

Then $\frac{\partial F}{\partial x} = f$ in $\mathbb{R}^2$. By Green's theorem

$$\int\int_D f(x,y)dxdy = \int\int_D \frac{\partial F}{\partial x}(x,y)dxdy = \int_{\partial D} F(x,y)dy$$

$$= \int_a^b F(\varphi(\gamma(t)))\frac{d}{dt}[y(u(t), v(t))]\, dt$$

$$= \int_a^b F(\varphi(\gamma(t)))\left[\frac{\partial y}{\partial u}\frac{\partial u}{\partial t} + \frac{\partial y}{\partial v}\frac{\partial v}{\partial t}\right] dt$$

$$= \pm \int_{\partial\Omega} F\frac{\partial y}{\partial u}du + F\frac{\partial y}{\partial v}dv$$

the sign depending on whether φ preserves or reverses the orientation.

Now we apply Green's theorem in Ω to write the last integral as

$$\pm \int \int_{\Omega} \left[\frac{\partial}{\partial u} \left(F \frac{\partial y}{\partial v} \right) - \frac{\partial}{\partial v} \left(F \frac{\partial y}{\partial u} \right) \right] du\,dv$$

$$= \pm \int \int_{\Omega} \left[\frac{\partial F}{\partial u} \frac{\partial y}{\partial v} - \frac{\partial F}{\partial v} \frac{\partial y}{\partial u} \right] du\,dv,$$

the other two terms cancelling due to the equality of the mixed partials $\frac{\partial^2 y}{\partial u \partial v} = \frac{\partial^2 y}{\partial v \partial u}$ (it is here we use the C^2 hypothesis on φ). The last integrand is equal to

$$\left(\frac{\partial F}{\partial x} \frac{\partial x}{\partial u} + \frac{\partial F}{\partial y} \frac{\partial y}{\partial u} \right) \frac{\partial y}{\partial v} - \left(\frac{\partial F}{\partial x} \frac{\partial x}{\partial v} + \frac{\partial F}{\partial y} \frac{\partial y}{\partial v} \right) \frac{\partial y}{\partial u} = f \frac{\partial(x, y)}{\partial(u, v)}.$$

Therefore

$$\int \int_{D} f(x, y) dx\,dy = \pm \int \int_{\Omega} f(x(u, v), y(u, v)) \frac{\partial(x, y)}{\partial(u, v)} du\,dv.$$

Clearly the $\pm$ is independent of f. Since when $f \equiv 1$ we get the area of D which is positive the ambiguous sign must be that of the Jacobian $\frac{\partial(x,y)}{\partial(u,v)}$ itself. Thus $\int \int_{\varphi(\Omega)} f(x, y) dx\,dy = \int \int_{\Omega} f(x(u, v), y(u, v)) \left| \frac{\partial(x,y)}{\partial(u,v)} \right| du\,dv.$ □

6.4.1   The divergence theorem in $\mathbb{R}^2$

Previously in (6.6), we performed integration of the *tangential* component of vector field along a curve, C parametrized by $\gamma(t) = (x(t), y(t))$ with $t \in [a, b]$. Now we are going to perform integration of the *normal* component of the vector field along a C. Since the tangent vector at t is $\gamma'(t) = (x'(t), y'(t))$, we see that $N(t) = (y'(t), -x'(t))$ is orthogonal to $\gamma'(t)$ for each t. We call it the *right normal* since it points to the right of the curve at every point($-N(t)$ is also orthogonal to the curve at every point, but points to the left of the curve). Another way to say this is if the curve is closed, the normal $N(t)$ points outside the curve.

Clearly, for all t, $||N(t)|| = ||\gamma'(t)||$. If we write $\mathbf{n}(t) = \frac{N(t)}{||N(t)||}$, then $\mathbf{n}(t)$ is the *unit* normal and $N(t) = ||N(t)||\mathbf{n}(t)$.

We now define the *divergence* of a vector field on a domain in $\mathbb{R}^2$. Given a vector field F defined on a domain Ω in $\mathbb{R}^2$ we write

$$divF = \frac{\partial F_1}{\partial x} + \frac{\partial F_2}{\partial y}.$$

The $divF$ is a smooth numerical function on Ω. We shall give a physical interpretation of it shortly.

Our next result is called the Divergence theorem.

Theorem 6.4.9. *(The Divergence theorem in $\mathbb{R}^2$) Let $F = (F_1, F_2)$ be a C^1 vector field on Ω, a region in $\mathbb{R}^2$ (with boundary curve $\partial\Omega$ having a counterclockwise orientation). Then*

$$\int\int_\Omega divF = \int_{\partial\Omega} \langle F, \boldsymbol{n}\rangle ds$$

Proof. Consider the vector field $H = (-F_2, F_1)$. By Green's theorem

$$\int\int_\Omega divF dxdy = \int_{\partial\Omega} (-F_2 dx + F_1 dy) - \int_a^b \langle F(\gamma(t)), N(t)\rangle dt$$

$$= \int_a^b \langle F(\gamma(t)), \mathbf{n}(t)\rangle \parallel N(t) \parallel dt = \int_{\partial\Omega} \langle F, \mathbf{n}\rangle ds$$

□

Corollary 6.4.10. *If F is perpendicular to N at each point of $\partial\Omega$, then*

$$\int\int_\Omega divF = 0.$$

In particular, if F zero at each point of $\partial\Omega$, then $\int\int_\Omega divF = 0$.

We can use the Divergence theorem to get an intuitive sense, in $\mathbb{R}^2$, of what the divergence of a vector field at a point means.

Corollary 6.4.11. *Let D_r be the disk of radius $r > 0$ centered at the point $a \in \mathbb{R}^2$. Let F be a C^1 vector field on D_r. Then*

$$divF(a) = \lim_{r\to 0} \frac{\int_{\partial D_r} \langle F, \boldsymbol{n}\rangle ds}{Area(D_r)}.$$

Thus, $divF(a)$ is the flow of F going out, in the normal direction, of a small circle centered at a, per unit area.

Proof. Corollary 4.2.13 of the Mean-Value theorem for double integrals, with $f = divF$, gives

$$divF(a) = \lim_{r \to 0} \frac{\int\int_{D_r} divF}{Area(D_r)} = \lim_{r \to 0} \frac{\int_{\partial D_r} \langle F, \mathbf{n} \rangle ds}{Area(D_r)}.$$

$\square$

We conclude this section with some applications to harmonic functions. First we prove the mean value property of harmonic functions (in two variables)[8]. This means that the value of f at the center of a disk is the mean value of the function around the circumference. Then we show that this implies that extreme values or even local extreme values of harmonic functions must occur on the boundary of the region and *not in the interior.*

Corollary 6.4.12. *Let f be a harmonic function on a bounded domain $\Omega \subset \mathbb{R}^2$ containing a point $p = (p_1, p_2)$. Then, for any disk, $D(r, p)$, of radius $r > 0$, centered at p contained in Ω we have*

$$f(p) = \frac{1}{2\pi} \int_0^{2\pi} f(p_1 + r\cos t, p_2 + r \sin t)dt.$$

Proof. We can, by translation, assume $p = (0,0)$. Let $r_0 > 0$ be chosen small enough so that the disk of this radius is inside Ω. Then for $r_0 \geq r > 0$ and $0 \leq t \leq 2\pi$, let $\gamma(t) = (r\cos t, r \sin t)$. Then $\gamma(t)$ is the boundary of the disk D_r of radius r centered at the origin. Because the boundary is a circle, the normal at $N(t)$ is also $(r\cos t, r \sin t)$. If we denote

$$\phi(r) = \frac{1}{2\pi} \int_0^{2\pi} f(r\cos t, r \sin t)dt,$$

differentiating under the integral sign gives

$$\phi'(r) = \frac{1}{2\pi r} \int_0^{2\pi} \langle \nabla f(r\cos t, r \sin t), N(t) \rangle dt.$$

---

[8]The analogue of this result in one variable x is the following. The Laplacian $L(f) = \frac{d^2 f}{dx^2}$. Thus a harmonic function is one satisfying $\frac{d^2 f}{dx^2} = 0$. By the mean value theorem for derivatives of functions of 1 variable this means that $f(x)$ is *linear.* If $D(p, r)$ is any "disk" in the domain centered at p and of radius r we see that $D(p, r)$ is actually the interval $(p - r, p + r)$. Thus the "boundary" consists of the two points $p - r$ and $p + r$. Clearly, the linear function f has the property that $f(p) = \frac{1}{2}(f(p - r) + f(p + r))$, the average value of f on the "boundary".

On the other hand,

$$\int\int_{D_r} div(\nabla f) = \int\int_{D_r} \Delta f,$$

and since f is harmonic on Ω, $\int\int_{D_r} \Delta f = 0$. Thus, by the Divergence theorem, $\phi'(r) = 0$ for all $0 < r \le r_0$. Since everything is smooth, this also holds for $r = 0$. Therefore ϕ is a constant. We can find what this constant is by taking $r = 0$. Thus $\phi(0) = \frac{1}{2\pi}\int_0^{2\pi} f(0,0)dt = f(0,0)$. □

An alternative proof of the following result was already given in Corollary 3.7.25.

Corollary 6.4.13. *Let f be a harmonic function on Ω. Then any extreme value of f must occur on $\partial\Omega$.*

Proof. Suppose not, suppose a local extreme value of f occured at a point p which was in the interior of Ω. Choose a sufficiently small disk D about p which is contained in Ω and on which $f(p)$ is larger than (resp. smaller than) any other value of f on this disk and its boundary. Applying the mean value theorem for harmonic functions we see that $f(p)$ equals the mean value of f on the boundary of D. By the mean value theorem for integrals the mean value of f on the boundary of D equals $f(q)$ for some point q on the boundary of D. But this is impossible since $f(p)$ is larger than (resp. smaller than) $f(q)$. □

EXERCISES

1. Evaluate the following line integrals along C, both directly and by using Green's theorem.

   (a) $\int_C (1 - x^2)ydx + (1 + y^2)xdy$, where C is $x^2 + y^2 = a^2$.

   (b) $\int_C xy^2 dy - x^2 ydx$, where C is the boundary of the annulus $1 \le x^2 + y^2 \le 4$.

   (c) $\int_C (y - \sin x)dx + \cos x dy$, where C is the perimeter of the triangle with vertices $(0,0)$, $(\frac{\pi}{2},0)$ and $(\frac{\pi}{2},1)$ traced counter-clockwise.

   (d) $\int_C (xy + x + y)dx + (xy + x - y)dy$, where C is the ellipse $\frac{x^2}{a^2} + \frac{y^2}{b^2} = 1$.

(e) Line integral (d), when C is the circle $x^2 + y^2 = ax$.

2. Show that $\int_C (yx^3 + xe^y)dx + (xy^3 + ye^x - 2y)dy = 0$, where C is any closed curve symmetric with respect to the origin.

3. Compute $\int_C \frac{-y}{x^2+4y^2}dx + \frac{x}{x^2+4y^2}dy$, where C is the circle $x^2+y^2 = 1$ oriented counterclockwise.

4. Using the line integral compute the area of the figures enclosed by the

   (a) Astroid, $x^{\frac{2}{3}} + y^{\frac{2}{3}} = a^{\frac{2}{3}}$, parametrized by $x = a\cos^3 t$, $y = a\sin^3 t$, $t \in [0, 2\pi]$.

   (b) Cardioid, $x = 2a\cos t - a\cos 2t$, $y = 2a\sin t - a\sin 2t$.

   (c) *Bernoulli's lemniscate*, $(x^2 + y^2)^2 = 2a(x^2 - y^2)$. *Hint.* Set $y = x\tan t$.

5. Explain why for a harmonic functions, f, on Ω

$$\int_{\partial\Omega} \langle \operatorname{grad} f, N\rangle dt = 0.$$

   Suppose f is an arbitrary function on $\mathbb{R}^2$ and this holds for any domain. Must f be harmonic?

6. Let f and g be C^2 functions on a domain Ω in $\mathbb{R}^2$ in which Green's theorem is applicable.

   (a) Apply the Divergence theorem to the vector field $g\nabla f - f\nabla g$ to show

$$\int\int_\Omega g\Delta f - f\Delta g = \int_{\partial\Omega} \langle \nabla f - f\nabla g, N\rangle dt.$$

   (*Green's formula*).

   (b) In particular, take $g = 1$ and show for any f

$$\int\int_\Omega \Delta f = \int_{\partial\Omega} \langle \nabla f, N\rangle dt.$$

(c) In particular, (as in problem 5) if f is harmonic then

$$\int_{\partial\Omega} \langle \nabla f, N \rangle dt = 0.$$

Answers to selected Exercises

1. (a) $\frac{1}{2}\pi a^2$. (b) $\frac{15\pi}{2}$. (c) $-\left(\frac{\pi}{4} + \frac{2}{\pi}\right)$. (d) 0. (e) $-\frac{\pi a^3}{8}$.

3. π.   4. (a) $\frac{3\pi a^2}{8}$. (b) $6\pi a^2$. (c) $2a^2$.

6.5   The divergence and curl

Let $\nabla = (\frac{\partial}{\partial x_1}, ..., \frac{\partial}{\partial x_n})$ denote the *"del"-operator*. Given a C^1 function on $\mathbb{R}^n$, we are already familiar with the notation in connection with the *gradient* of f

$$\text{grad } f = \nabla f = \left(\frac{\partial f}{\partial x_1}, ..., \frac{\partial f}{\partial x_n} \right).$$

Definition 6.5.1. Let $F = (F_1, F_2, F_3)$ be a C^1 vector field on a domain $\Omega \subseteq \mathbb{R}^3$. The *divergence* of F is the numerical function

$$divF = \nabla \cdot F = \frac{\partial F_1}{\partial x} + \frac{\partial F_2}{\partial y} + \frac{\partial F_3}{\partial z}.$$

The Divergence Theorem, in the next section, will tell us that the divergence $divF$ measures how much the vector field is spreading out (expanding or compressing) at a point. Thus, if $divF \equiv 0$, the vector field F is said *incompressible*.

The divergence can, of course, be defined for vector fields on $\mathbb{R}^n$ by

$$divF = \sum_{i=1}^{n} \frac{\partial F_i}{\partial x_i}.$$

Note that given a twice differentiable function $f : \Omega \subseteq \mathbb{R}^n \to \mathbb{R}$ the Laplacian, Δf, can be written as

$$\Delta f = \sum_{i=1}^{n} \frac{\partial^2 f}{\partial x_i^2} = \nabla \cdot (\nabla f) = div(\nabla f),$$

that is, the Laplacian is the divergence of the gradient. (This leads some authors to denote the Laplacian by $\nabla^2 f$).

Definition 6.5.2. Let $F = (F_1, F_2, F_3)$ be a C^1 vector field on $\Omega \subseteq \mathbb{R}^3$. The *curl* of F is the vector field

$$\operatorname{curl} F = \nabla \times F = \det \begin{bmatrix} \mathbf{i} & \mathbf{j} & \mathbf{k} \\ \frac{\partial}{\partial x} & \frac{\partial}{\partial y} & \frac{\partial}{\partial z} \\ F_1 & F_2 & F_3 \end{bmatrix}$$

$$= \left(\frac{\partial F_3}{\partial y} - \frac{\partial F_2}{\partial z}, \frac{\partial F_1}{\partial z} - \frac{\partial F_3}{\partial x}, \frac{\partial F_2}{\partial x} - \frac{\partial F_1}{\partial y} \right).$$

Stokes' Theorem, in the next section, will tell us that the curl F measures how much the vector field is curling or twisting or rotating. Thus, if curl $F = \mathbf{0}$, we say F is *irrotational*. (Some authors denote the curl F by rot F; "rot" stands for "rotation".)

We shall use the notations grad $f = \nabla f$, $divF = \nabla \cdot F$ and curl $F = \nabla \times F$ interchangeably.

Example 6.5.3. Let $F(x, y, z) = (x^2, xyz, yz^2)$. Compute the divergence, $divF$, and curl F.

Solution. We have

$$divF = \nabla \cdot F = \frac{\partial}{\partial x}(x^2) + \frac{\partial}{\partial y}(xyz) + \frac{\partial}{\partial z}(yz^2) = 2x + xy + 2yz.$$

For the curl we have

$$\operatorname{curl} F = \nabla \times F = \det \begin{bmatrix} \mathbf{i} & \mathbf{j} & \mathbf{k} \\ \frac{\partial}{\partial x} & \frac{\partial}{\partial y} & \frac{\partial}{\partial z} \\ x^2 & xyz & yz^2 \end{bmatrix} = (z^2 - xy, 0, yz).$$

Proposition 6.5.4. . *For any C^2 function f, $\operatorname{curl}(\nabla f) = \mathbf{0}$, that is, any conservative vector field is irrotational.*

Proof. Since $\nabla f = (\frac{\partial f}{\partial x}, \frac{\partial f}{\partial y}, \frac{\partial f}{\partial z})$, by definition

$$\operatorname{curl}(\nabla f) = \det \begin{bmatrix} \mathbf{i} & \mathbf{j} & \mathbf{k} \\ \frac{\partial}{\partial x} & \frac{\partial}{\partial y} & \frac{\partial}{\partial z} \\ \frac{\partial f}{\partial x} & \frac{\partial f}{\partial y} & \frac{\partial f}{\partial z} \end{bmatrix}$$

$$= \left(\frac{\partial^2 f}{\partial y \partial z} - \frac{\partial^2 f}{\partial z \partial y}, \frac{\partial^2 f}{\partial z \partial x} - \frac{\partial^2 f}{\partial x \partial z}, \frac{\partial^2 f}{\partial x \partial y} - \frac{\partial^2 f}{\partial y \partial x} \right) = \mathbf{0},$$

by the equality of the mixed partials.   □

In the next section with the aid of Stokes' theorem we will see that the condition $\operatorname{curl} F = \mathbf{0}$ is also sufficient for the vector field F to be conservative[9].

We now list some properties of the divergence and the curl, which constitute the basic identities of vector analysis. In the following identities f and g are real-valued functions and $F = (F_1, F_2, F_3)$ and $G = (G_1, G_2, G_3)$ are vector fields in $\mathbb{R}^3$.

Basic identities satisfied by the divergence and curl

1. $div(\operatorname{curl} F) = 0$.

2. $div(F + G) = divF + divG$.

3. $\operatorname{curl}(F + G) = \operatorname{curl} F + \operatorname{curl} G$.

4. $\operatorname{curl}(\operatorname{curl} F) = \operatorname{grad}(divF) - \Delta F$, where $\Delta F = (\Delta F_1, \Delta F_2, \Delta F_3)$.

5. $div(fF) = f divF + (\operatorname{grad} f) \cdot F$.

6. $\operatorname{curl}(fF) = f \operatorname{curl} F + (\operatorname{grad} f) \times F$.

7. $div(F \times G) = G \cdot \operatorname{curl} F - F \cdot \operatorname{curl} G$.

8. $\operatorname{curl}(F \times G) = (divG)F - (divF)G + (G \cdot \nabla)F - (F \cdot \nabla)G$.

9. $\nabla(F \cdot G) = (F \cdot \nabla)G + (G \cdot \nabla)F + F \times (\operatorname{curl} G) + G \times (\operatorname{curl} F)$.

10. $\Delta(fg) = f\Delta g + g\Delta f + 2(\nabla h f \cdot \nabla g)$.

11. $div(f\nabla g - g\nabla f) = f\Delta g - g\Delta f$.

12. $\nabla(\nabla \cdot F) - \nabla \times (\nabla \times F) = \Delta F = (\Delta F_1, \Delta F_2, \Delta F_3)$.

---

[9]This is Theorem 6.6.6 below. In this connection, see also Theorems 6.2.8, 6.2.13.

For properties (8) and (9) in $(F \cdot \nabla)G$ the del operator ∇ is meant to operate to the components G_j of G, that is, $F \cdot \nabla G_j = F_1 \frac{\partial G_j}{\partial x} + F_2 \frac{\partial G_j}{\partial y} + F_3 \frac{\partial G_j}{\partial z}$, for $j = 1, 2, 3$.

Corollary 6.5.5. *Let f and g be any C^2 functions. Then*

$$div(\nabla f \times \nabla g) = 0.$$

That is, the cross product of any two conservative vector fields is incompressible.

Proof. From identity (7) and Proposition 6.5.4, we have

$$div(\nabla f \times \nabla g) = \nabla g \cdot \mathbf{0} - \nabla f \cdot \mathbf{0} = 0.$$

$\square$

EXERCISES

1. Compute the divergence $\nabla \cdot F$ and the curl $\nabla \times F$ of each of the following vector fields.

   (a) $F(x, y, x) = x\mathbf{i} + y\mathbf{j} + z\mathbf{k}$

   (b) $F(x, y, z) = xy^2\mathbf{i} + xy\mathbf{j} + xy\mathbf{k}$

   (c) $F(x, y, z) = (x^2 z, 4xyz, y - 3xz^2)$

   (d) $F(x, y, z) = (\sin xy, e^{xz}, 2x + yz^4)$

2. Let $f(x, y, z) = xyz^2$ and $F(x, y, z) = (xy, yz, zy)$. Compute

   $$div(fF), \quad \nabla \times (fF), \quad div(\nabla \times F), \quad curl(\nabla \times F).$$

   Verify identities (1), (4), (5) and (6).

3. Prove that for *any* vector field F the equation $(\nabla \times F)(x, y, z) = (x, y, z)$ is impossible.

4. Let $\mathbf{r}(x, y, z) = (x, y, z)$ and $r = ||\mathbf{r}||$. Show that for any fixed vector $\mathbf{a}$ in $\mathbb{R}^3$ we have

   (a) $curl(\mathbf{a} \times \mathbf{r}) = 2\mathbf{a}$.

(b) $div[(\mathbf{a} \cdot \mathbf{r})\mathbf{a}] = ||\mathbf{a}||^2$.

(c) $div[(\mathbf{a} \times \mathbf{r}) \times \mathbf{a}] = 2||\mathbf{a}||^2$.

(d) $\nabla \left(\frac{\mathbf{a} \cdot \mathbf{r}}{r^3}\right) = -\operatorname{curl}\left(\frac{\mathbf{a} \times \mathbf{r}}{r^3}\right)$.

(e) $\operatorname{curl}\left[\frac{\mathbf{a} \times \mathbf{r}}{r^m}\right] = \left(\frac{m(\mathbf{a} \cdot \mathbf{r})}{r^{m+2}}\right)\mathbf{r} - \left(\frac{m-2}{r^m}\right)\mathbf{a}$.

5. Prove identities (1), (2) and (3).

6. Prove identities (4), (5) and (6).

7. Prove identities (7), (8) and (9).

8. Prove identities (10), (11) and (12).

6.6   Stokes' theorem

We can now formulate extensions of our results in planar domains to domains in $\mathbb{R}^3$. There are two types of such extensions. The first is *Stokes' theorem*.[10] The second is the *Divergence theorem* which we shall discuss in the next section.

Stokes' theorem is a generalization of Green's theorem in which the planar region is replaced by an oriented surface S with boundary ∂S consisting of a finite number of piecewise smooth oriented curves. In its simplest form, Stokes' theorem relates the line integral of a vector field F around a simple closed curve C in $\mathbb{R}^3$ to a surface integral of the curl F over a surface S for which C is the boundary, i.e., $C = \partial S$. We assume that the *positive side* of S lies on the *left-hand side* as we move along ∂S in the *positive direction*. In practice this consistency between the orientations of the surface and its boundary may be verified by sketching (if you happen to choose the wrong induced orientation for the boundary ∂S in Stokes' theorem you will be off merely by the factor -1).

Both the surface S and its boundary $C = \partial S$ are defined in terms of their respective parameters. In fact, the surface $S = \Phi(\Omega)$ is parametrized by the smooth bijection

---

[10]G. Stokes (1819-1903). Mathematician and physicist who made important contributions in hydrodynamics and electrostatics.

$$\Phi(u,v) = (x(u,v), y(u,v), z(u,v)), \quad (u,v) \in \Omega,$$

where Ω is a region in $\mathbb{R}^2$ to which Green's theorem applies. The boundary of Ω, $\partial\Omega$, is parametrized by $\rho(t) = (u(t), v(t))$ with $t \in [a, b]$. Since Φ is one-to-one, the image of $\partial\Omega$ is a simple closed curve $C = \Phi(\partial\Omega) = \partial S$ in $\mathbb{R}^3$, described by the composite function

$$\gamma(t) = \Phi(\rho(t)) = (x(u(t), v(t)), y(u(t), v(t)), z(u(t), v(t)), \ t \in [a, b].$$

We are now ready to state and prove Stokes' theorem.

Theorem 6.6.1. *(Stokes' Theorem) Let $F = (F_1, F_2, F_3)$ be a C^1 vector field defined in a region in $\mathbb{R}^3$ containing a simple smooth parametrized surface $S = \Phi(\Omega)$ with boundary $C = \partial S$ with S and its boundary properly oriented. Then*

$$\int\int_S (\operatorname{curl} F) \cdot dS = \int_{\partial S} F \cdot ds, \qquad (6.19)$$

Proof. Using (6.6) and (6.15) Stoke's formula (6.19) can also be written

$$\int\int_S \langle \operatorname{curl} F, \mathbf{n} \rangle dS = \int_{\partial S} \langle F, T \rangle ds,$$

where $\mathbf{n}$ is the unit normal to S and T is the induced unit tangent vector to the curve ∂S.

Before we prove (6.19) note that if S is a region in $\mathbb{R}^2$, then $\mathbf{n} = \mathbf{k} = (0, 0, 1)$. So that $\langle \operatorname{curl} F, \mathbf{n} \rangle = \frac{\partial F_2}{\partial x} - \frac{\partial F_1}{\partial y}$ and $\langle F, T \rangle = F_1 dx + F_2 dy$. Hence here Stokes' theorem reduces to Green's theorem.

Now (in $\mathbb{R}^3$) to compute the line integral

$$\int_{\partial S} F \cdot ds = \int_a^b F(\gamma(t)) \cdot \gamma'(t) dt = \int_{\partial S} F_1 dx + F_2 dy + F_3 dz,$$

we consider the components of F separately, just as we did in the proof of Green's theorem. The boundary ∂S of the surface S is parametrized by $\gamma(t) = (x(u(t), v(t)), y(u(t), v(t)), z(u(t), v(t))$ and the chain rule gives

$$\gamma'(t) = \left(\frac{\partial x}{\partial u} u'(t) + \frac{\partial x}{\partial v} v'(t), \frac{\partial y}{\partial u} u'(t) + \frac{\partial y}{\partial v} v'(t), \frac{\partial z}{\partial u} u'(t) + \frac{\partial x}{\partial v} v'(t) \right).$$

We consider

$$\int_{\partial S} F_1 dx = \int_a^b F_1(\gamma(t))(\frac{\partial x}{\partial u}u'(t) + \frac{\partial x}{\partial v}v'(t))dt = \int_{\partial \Omega} F_1 \frac{\partial x}{\partial u}du + F_1 \frac{\partial x}{\partial v}dv,$$

where in the last integral F_1 is evaluated at $\Phi(u,v)$. Applying Green's theorem to this line integral we get

$$\int_{\partial \Omega} F_1 \frac{\partial x}{\partial u}du + F_1 \frac{\partial x}{\partial v}dv = \int \int_{\Omega} \left[\frac{\partial}{\partial u}(F_1 \frac{\partial x}{\partial v}) - \frac{\partial}{\partial v}(F_1 \frac{\partial x}{\partial u}) \right] dudv.$$

Let us simplify the integrand on the right. By the product rule it equals

$$\frac{\partial F_1}{\partial u}\frac{\partial x}{\partial v} + F_1 \frac{\partial^2 x}{\partial u \partial v} - \frac{\partial F_1}{\partial v}\frac{\partial x}{\partial u} - F_1 \frac{\partial^2 x}{\partial v \partial u} = \frac{\partial F_1}{\partial u}\frac{\partial x}{\partial v} - \frac{\partial F_1}{\partial v}\frac{\partial x}{\partial u}. \quad (6.20)$$

By the Chain Rule

$$\frac{\partial F_1}{\partial u} = \frac{\partial F_1}{\partial x}\frac{\partial x}{\partial u} + \frac{\partial F_1}{\partial y}\frac{\partial y}{\partial u} + \frac{\partial F_1}{\partial z}\frac{\partial z}{\partial u}$$

and

$$\frac{\partial F_1}{\partial v} = \frac{\partial F_1}{\partial x}\frac{\partial x}{\partial v} + \frac{\partial F_1}{\partial y}\frac{\partial y}{\partial v} + \frac{\partial F_1}{\partial z}\frac{\partial z}{\partial v}.$$

Substituting these equations in (6.20) we see that

$$\frac{\partial F_1}{\partial u}\frac{\partial x}{\partial v} - \frac{\partial F_1}{\partial v}\frac{\partial x}{\partial u} = -\frac{\partial F_1}{\partial y}\left(\frac{\partial(x,y)}{\partial(u,v)}\right) + \frac{\partial F_1}{\partial z}\left(\frac{\partial(z,x)}{\partial(u,v)}\right)$$

$$= \left(-\frac{\partial F_1}{\partial y}\langle \mathbf{k}, \mathbf{n} \rangle + \frac{\partial F_1}{\partial z}\langle \mathbf{j}, \mathbf{n} \rangle \right) \|T_u \times T_v\|.$$

Therefore,

$$\int_{\partial S} F_1 dx = \int \int_{\Omega} \langle -\frac{\partial F_1}{\partial y}\mathbf{k} + \frac{\partial F_1}{\partial z}\mathbf{j}, \mathbf{n} \rangle \|T_u \times T_v\|dudv,$$

that is,

$$\int_{\partial S} F_1 dx = \int \int_S \langle -\frac{\partial F_1}{\partial y}\mathbf{k} + \frac{\partial F_1}{\partial z}\mathbf{j}, \mathbf{n} \rangle dS.$$

Similarly

$$\int_{\partial S} F_2 dy = \int \int_S \langle -\frac{\partial F_2}{\partial z}\mathbf{i} + \frac{\partial F_2}{\partial x}\mathbf{k}, \mathbf{n} \rangle dS$$

and
$$\int_{\partial S} F_3 dz = \int\int_S \langle -\frac{\partial F_3}{\partial x}\mathbf{j} + \frac{\partial F_3}{\partial y}\mathbf{i}, \mathbf{n}\rangle dS.$$
Adding these equations we get (6.19). □

One can obtain Stokes' theorem more generally for surfaces that can be decomposed into a finite number of pieces that each admit a parametrization by applying (6.19) to the pieces and adding the results.

Corollary 6.6.2. *If* curl $F = 0$ *on* S, *or even if* curl F *is perpendicular to the normal vector* $\mathbf{n}$ *at each point of* S, *then*
$$\int_{\partial S} \langle F, T\rangle ds = 0.$$

Corollary 6.6.3. *If* $F = 0$ *on* ∂S, *or even if* F *is perpendicular to the tangent vector* T *at each point of* ∂S, *then*
$$\int\int_S \langle \text{curl}\, F, \mathbf{n}\rangle dS = 0.$$

Corollary 6.6.4. *If* S *has no boundary then for any smooth vector field* F, *then*
$$\int\int_S \langle \text{curl}\, F, \mathbf{n}\rangle dS = 0.$$

Corollary 6.6.5. *Let* $a \in \mathbb{R}^3$ *and* $\mathbf{n}$ *a unit vector. Suppose* D_r *is a small disk of radius* r, *centered at the point* a *and perpendicular to the vector* $\mathbf{n}$. *Let* F *be a* C^1 *vector field on* D_r. *Then*
$$\text{curl}\, F(a) \cdot \mathbf{n} = \lim_{r\to 0} \frac{\int_{\partial D_r} \langle F, T\rangle ds}{\pi r^2}. \tag{6.21}$$

Thus, the $\mathbf{n}$*-component of* curl $F(a)$ *is the circulation (or rotation) of* F *on a smal circular surface perpendicular to* $\mathbf{n}$, *per unit area.*

Proof. By the definition of the surface integral, it follows that the Mean-Value theorem for (double) integrals (Corollary 4.2.13) is also valid for surface integrals. Together with Stokes' theorem this yields
$$\text{curl}\, F(a) \cdot \mathbf{n} = \lim_{r\to 0} \frac{\int\int_{D_r} \text{curl}\, F(a) \cdot \mathbf{n} dS}{Area(D_r)} = \lim_{r\to 0} \frac{\int_{\partial D_r} \langle F, T\rangle ds}{\pi r^2}.$$
□

It is worthwhile remarking that according to Stokes' theorem, in the computation of $\int \int_S \operatorname{curl} F \cdot d\mathbf{S}$, we can replace the surface S by any other surface having the same boundary $C = \partial S$. Any such surface is called a *capping surface* of C. In most cases, we change to a planar surface (see Example 6.6.8 below.)

The next result gives a simple and useful characterization of conservative vector fields.

Theorem 6.6.6. *Let F be a smooth vector field on a connected domain Ω in $\mathbb{R}^3$. Then F is conservative if and only if $\operatorname{curl} F = \mathbf{0}$.*

Proof. If $F = \nabla f$ is conservative, then Proposition 6.5.4, tells us $\operatorname{curl} F = \mathbf{0}$. Conversely, if $\operatorname{curl} F = \mathbf{0}$, then for any closed curve C choose a capping surface[11] S for C (if F has exceptional points, choose S to avoid them). Now $C = \partial S$ and Stokes' theorem gives

$$\int_C F \cdot d\mathbf{s} = \int \int_S \mathbf{0} \cdot d\mathbf{S} = 0.$$

Hence, from Theorem 6.2.13, F is conservative.   □

Example 6.6.7. Let $F(x, y, z) = (x^2, y^2, -z)$. Compute $\int_C F \cdot d\mathbf{s}$, where C is the perimeter of the triangle with vertices $(0, 0, 0)$, $(0, 2, 0)$ and $(0, 0, 3)$ oriented counterclockwise.

Solution. Let S be the surface of the triangle, so that $C = \partial S$. Then Stokes' theorem gives $\int_C F \cdot d\mathbf{s} = \int \int_S (\operatorname{curl} F) \cdot d\mathbf{S}$.

$$\operatorname{curl} F = \nabla \times F = \det \begin{bmatrix} \mathbf{i} & \mathbf{j} & \mathbf{k} \\ \frac{\partial}{\partial x} & \frac{\partial}{\partial y} & \frac{\partial}{\partial z} \\ x^2 & y^2 & -z \end{bmatrix} = \mathbf{0}. \text{ Hence } \int_C F \cdot d\mathbf{s} = 0.$$

Alternatively, observe that $F = \nabla f$ is a conservative vector field with potential function $f(x, y, z) = \frac{x^3}{3} + \frac{y^3}{3} - \frac{z^2}{2}$. Since C is a closed curve and F is conservative Corollary 6.2.3 implies $\int_C F \cdot d\mathbf{s} = 0$. Finally, note that a direct computation is also possible by evaluating separately and adding up the three line integrals along the sides of the triangle.

---

[11]A rigorous proof of this would require ideas from topology which are beyond the scope of this book.

Example 6.6.8. Let $F(x, y, z) = (z - y)\mathbf{i} + (x + z)\mathbf{j} - (x + y)\mathbf{k}$ and let C be the boundary of the surface of the paraboloid $z = 4 - x^2 - y^2$, with $0 \le z \le 4$, oriented counterclockwise as viewed from above.

1. Compute $\int_C F \cdot d\mathbf{s}$ using Stokes' theorem.

2. Compute $\int_C F \cdot d\mathbf{s}$ directly.

Solution.

1. The curve C is the circle $x^2 + y^2 = 4$. Let Ω be the disk $x^2 + y^2 \le 4$. We choose the capping surface for C to be the plane surface $S = \{(x, y, 0) : (x, y) \in \Omega\}$. By Stokes' theorem we have

$$\int_C F \cdot d\mathbf{s} = \int\int_S (\operatorname{curl} F) \cdot \mathbf{n} dS.$$

A simple calculation gives $\operatorname{curl} F = \nabla \times F = (-2, 2, 2)$. The unit normal to S is $\mathbf{n} = \mathbf{k} = (0, 0, 1)$. Hence

$$\int\int_S (\operatorname{curl} F) \cdot \mathbf{n} dS = 2 \int\int_\Omega dx dy = 8\pi.$$

2. The parametrization of the circle C is $\gamma(t) = (2\cos t, 2\sin t, 0)$ with $t \in [0, 2\pi]$ and $\gamma'(t) = (-2\sin t, 2\cos t, 0)$. Hence

$$\int_C F \cdot d\mathbf{s} = \int_0^{2\pi} (-2\sin t, 2\cos t, -2\cos t - 2\sin t)$$
$$\cdot (-2\sin t, 2\cos t, 0) dt$$
$$= \int_0^{2\pi} [4\sin^2 t + 4\cos^2 t] dt = 4 \int_0^{2\pi} dt = 8\pi.$$

Example 6.6.9. Use Stoke's theorem to compute the line integral

$$\int_C x^2 y^3 dx + dy + z dz,$$

where C is the circle $x^2 + y^2 = a^2$ on the xy-plane.

Solution. Take S to be the disk $x^2 + y^2 \le a^2$, so that $C = \partial S$. By Stoke's theorem

$$\int_C x^2 y^3 dx + dy + z dz = \int \int_S \left(\frac{\partial z}{\partial y} - \frac{\partial 1}{\partial z} \right) dy dz + \left(\frac{\partial x^2 y^3}{\partial z} - \frac{\partial z}{\partial x} \right) dz dx$$

$$+ \left(\frac{\partial 1}{\partial x} - \frac{\partial x^2 y^3}{\partial y} \right) dx dy$$

$$= -3 \int \int_S x^2 y^2 dx dy$$

$$= -3 \int_0^{2\pi} \int_0^a r^5 \cos^2 \theta \sin^2 \theta \, dr d\theta = -\frac{a^6}{8} \pi,$$

where we changed the double integral to polar coordinates.

EXERCISES

1. Verify Stoke's theorem for the line integral

$$\int_C x^2 dx + yx dy,$$

where C is the boundary of the square $S = [0, a] \times [0, a]$ in the xy-plane.

2. Let $F(x, y, z) = (x^2 + y, yz, x - z^2)$ and C the boundary of the triangle S defined by the plane $2x + y + 2z = 2$ in the first octant.

   (a) Compute $\int_C F \cdot ds$ using Stokes' theorem.
   (b) Compute $\int_C F \cdot ds$ directly.

3. Let $F(x, y, z) = (x+y, x+z, z^2)$ and C the boundary of the surface S formed by the surface of cone $z^2 = x^2 + y^2$ cut off by the planes $z = 0$ and $z = 1$.

   (a) Compute $\int_C F \cdot ds$ using Stokes' theorem.
   (b) Compute $\int_C F \cdot ds$ directly.

4. Let $F(x, y, z) = (y, -x, x^2 y^2 z)$ and let S be the surface $x^2 + y^2 + 3z^2 = 1$, $z \le 0$. Compute $\int \int_S (\nabla \times F) \cdot dS$

(a) Directly.

(b) Using Stokes' theorem.

5. Show using Stoke's theorem, that

$$\int_C (y+z)dx + (z+x)dy + (x+z)dz = 0,$$

where C is the circle of intersection of the sphere $x^2 + y^2 + z^2 = a^2$ and the plane $x + y + z = 0$, oriented counterclockwise.

6. Use Stoke's theorem to find

$$\int_C (x-z)dx + (x+y)dy + (y+z)dz,$$

where C is the ellipse where the cylinder $x^2 + y^2 = 1$ is cut off by the plane $z = y$ oriented counterclockwise.

7. Using Stoke's theorem prove that,

$$\int_C ydx + zdy + xdz = -2\pi a^2 \sqrt{2},$$

where C is the curve of intersection of the plane $x + y = 2a$ and the sphere $x^2 + y^2 + z^2 - 2ax - 2ay = 0$.

8. Use Stoke's theorem to calculate

$$\int_C xzdx - ydy + x^2ydz,$$

where C consists of the boundary of the three faces not in the xz-plane of the tetrahedron bounded by the plane $3x + y + 3z = 6$ and the coordinate planes. The unit normal $\mathbf{n}$ is pointing out of the tetrahedron.

9. Compute

$$\int_C \sin yzdx + xz \cos yzdy + xy \cos ydz,$$

where C is the helix $\gamma(t) = (\cos t, \sin t, t)$, $t \in [0, 2\pi]$.

10. Let $F(x, y, z) = \frac{-y}{x^2+y^2}\mathbf{i} + \frac{x}{x^2+y^2}\mathbf{j}$ for $z \neq 0$.

   (a) Show that $\text{curl}(F) = \mathbf{0}$

   (b) Show by direct calculation $\int_C F \cdot d\mathbf{s} = 2\pi$ for any horizontal circle C centered at a point on the z-axis.

   (c) Explain why this does not contradict Stoke's theorem?

11. For a smooth oriented surface S and a fixed vector $\mathbf{v}$, show that

$$2 \int\int_S \mathbf{v} \cdot \mathbf{n} dS = \int_{\partial S} (\mathbf{v} \times \mathbf{r}) \cdot ds,$$

   where $\mathbf{r}(x, yz) = (x, y, z)$.

12. Let C be a properly oriented closed curve which is the boundary of a smooth oriented surface S in $\mathbb{R}^3$. Suppose f and g are C^2 functions on some open set containing S. Show that

   (a)

$$\int_C f\nabla g \cdot d\mathbf{s} = \int\int_S (\nabla f \times \nabla g) \cdot d\mathbf{S}.$$

   (b)

$$\int_C (f\nabla g + g\nabla f) \cdot d\mathbf{s} = 0.$$

Answers to selected Exercises

2. $-\frac{13}{6}$.   3. 0.   4. -2π.   6. 2π.   8. $\frac{4}{3}$.   9. 0.

6.7   The divergence theorem in $\mathbb{R}^3$

The *Divergence theorem* is the 3-dimensional analogue of Green's theorem in the form of Theorem 6.4.9. It relates the triple integral of the divergence of a vector field F over a solid in $\mathbb{R}^3$ to the surface integral of F over the surface bounding the solid. The Divergence theorem is also known as *Gauss's theorem*, or *Ostrogradsky's theorem*.[12]

---

[12]M. Ostrogradsky (1801-1862) published the divergence theorem in his article *On Heat Theory* in 1821. The theorem was also proved by Gauss and is often called Gauss' theorem.

Instead of reducing our result to Green's theorem, as we did in the case of Stokes' theorem, here we essentially use the method of proof of Green's theorem. Before doing so, for convenience, we introduce the following terminology. An elementary region V in $\mathbb{R}^3$ is said *xy-simple* (or *z-projectable*) if it has the form

$$V = \{(x, y, z) : (x, y) \in \Omega, \; \phi_1(x, y) \leq z \leq \psi_1(x, y)\},$$

where Ω is an elementary region in $\mathbb{R}^2$ and ϕ_1 and ψ_1 are continuous functions on Ω (see Definition 4.3.7 and Lemma 4.3.8). The notions of *yz-simple* and *xz-simple* are defined analogously. We say that V is a *symmetric elementary region* in $\mathbb{R}^3$ if it is xy-simple, yz-simple and xz-simple. An example of such a region is a rectangular parallelepiped, or a ball in $\mathbb{R}^3$.

Theorem 6.7.1. *(Gauss' or Divergence Theorem). Let $F = (F_1, F_2, F_3)$ be a C^1 vector field defined on a symmetric elementary region $V \subseteq \mathbb{R}^3$ with boundary ∂V. Assume the orientation on ∂V has been chosen so that the normal $\boldsymbol{n}$ varies piecewise smoothly and always points outward. Then*

$$\int \int \int_V div F \, dV = \int \int_{\partial V} \langle F, \boldsymbol{n} \rangle dS. \qquad (6.22)$$

Proof. Since

$$\int \int \int_V div F \, dV = \int \int \int_V \frac{\partial F_1}{\partial x} dV + \int \int \int_V \frac{\partial F_2}{\partial y} dV + \int \int \int_V \frac{\partial F_3}{\partial z} dV,$$

we would be done if we knew the following three equalities were satisfied.

$$\int \int \int_V \frac{\partial F_1}{\partial x} dV = \int \int_{\partial V} \langle F_1 \mathbf{i}, \mathbf{n} \rangle dS,$$

$$\int \int \int_V \frac{\partial F_2}{\partial y} dV = \int \int_{\partial V} \langle F_2 \mathbf{j}, \mathbf{n} \rangle dS,$$

and

$$\int \int \int_V \frac{\partial F_3}{\partial z} dV = \int \int_{\partial V} \langle F_3 \mathbf{k}, \mathbf{n} \rangle dS.$$

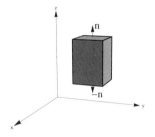

Figure 6.5: Elementary symmetric region

We shall prove only the last of these, the other two can be proved similarly. Since V is a symmetric elementary region, it can be written as xy-simple,

$$V = \{(x, y, z) : (x, y) \in \Omega,\ \phi_1(x, y) \le z \le \psi_1(x, y)\}.$$

Express $\int \int \int_V \frac{\partial F_3}{\partial z} dV$ as the iterated integral

$$\int \int_\Omega \left(\int_{\phi_1(x,y)}^{\psi_1(x,y)} \frac{\partial F_3}{\partial z} dz \right) dx dy,$$

where here Ω is the common domain of ϕ_1 and ψ_1. Then by the Fundamental Theorem of Calculus

$$\int \int \int_V \frac{\partial F_3}{\partial z} dV = \int \int_\Omega [F_3(x, y, \psi_1(x, y)) - F_3(x, y, \phi_1(x, y))] dx dy.$$

Now the boundary, ∂V, consists of six regions. The top and bottom S_1 and S_2, respectively, and the four sides S_3, S_4, S_5 and S_6. Hence

$$\int \int_{\partial V} \langle F_3 \mathbf{k}, \mathbf{n} \rangle dS = \sum_{i=1}^{6} \int \int_{S_i} \langle F_3 \mathbf{k}, \mathbf{n} \rangle dS.$$

The latter four have their normals perpendicular to the z-axis. Hence for each point (x, y, z) of such an S_i, $F_3(x, y, z)\mathbf{k}$ is perpendicular to $\mathbf{n}(x, y, z)$. It follows that for $i \ge 3$ each $\int \int_{S_i} \langle F_3 \mathbf{k}, \mathbf{n} \rangle dS = 0$. (Just as in the corresponding proof of Green's theorem, above, some of these sides may not really be surfaces, but curves. Nonetheless, the statement just made remains true. An example is provided by the northern half of the sphere together with the equator). What is left to show is that

$$\int\int_{S_1} \langle F_3\mathbf{k}, \mathbf{n}\rangle dS + \int\int_{S_2} \langle F_3\mathbf{k}, \mathbf{n}\rangle dS$$

$$= \int\int_{\Omega} [F_3(x, y, \psi_1(x, y)) - F_3(x, y, \phi_1(x, y))]dxdy.$$

Now for each point (x, y, z) of S_1 we have

$$\mathbf{n}(x, y, z) = \frac{(\frac{-\partial\psi_1}{\partial x}, \frac{-\partial\psi_1}{\partial y}, 1)}{\sqrt{1 + (\frac{\partial\psi_1}{\partial x})^2 + (\frac{\partial\psi_1}{\partial y})^2}}.$$

So that

$$\langle F_3(x, y, z)\mathbf{k}, \mathbf{n}(x, y, z)\rangle = \frac{F_3(x, y, z)}{\sqrt{1 + (\frac{\partial\psi_1}{\partial x})^2 + (\frac{\partial\psi_1}{\partial y})^2}}.$$

Hence

$$\int\int_{S_1} \langle F_3\mathbf{k}, \mathbf{n}\rangle dS = \int\int_{\Omega} F_3(x, y, \psi_1(x, y))dxdy.$$

Similarly,

$$\int\int_{S_2} \langle F_3\mathbf{k}, \mathbf{n}\rangle dS = -\int\int_{\Omega} F_3(x, y, \phi_1(x, y))dxdy.$$

The reason for the negative sign in this case is that the normal must be taken as the outward normal. For S_1 it is the surface normal, as defined above (which for such a surface always points *upward*). Therefore for S_2 it also points upwards, so we must take the negative of the surface normal.   □

Of course, the Divergence theorem is also valid for more general regions Ω in $\mathbb{R}^3$ that can be decomposed into finitely many symmetric elementary regions $\Omega_1, ..., \Omega_k$. The integrals of $divF$ over the regions $\Omega_1, ..., \Omega_k$ add up to the integral over Ω, and the surface integrals of $\langle F, \mathbf{n}\rangle$ over the boundaries $\partial\Omega_1, ..., \partial\Omega_k$ add up to the integral over $\partial\Omega$.

Notice that when we impose a positive orientation on the boundary $\partial\Omega$, the integrals over the portions of the boundaries $\Omega_1, ..., \Omega_k$ that are not part of $\partial\Omega$ all cancel. The reasoning here is the same as in the proof of Green's theorem

Example 6.7.2. Let $F(x, y, z) = (2x, -3y, z)$. Compute $\int \int_S F \cdot \mathbf{n}dS$, where S is the entire surface of the solid formed by the cylindrical surface $x^2 + y^2 = 1$, the planes $z = 0$, $z + x + 2$ and $\mathbf{n}$ is the outer unit normal.

Solution. This surface integral was computed directly in Example 6.3.21. Here we use the Divergence theorem. Let V be the solid in $\mathbb{R}^3$, bounded by the surface S. Since $div F = \frac{\partial}{\partial x}(2x) + \frac{\partial}{\partial y}(-3y) + \frac{\partial}{\partial z}(z) = 0$, it follows that $\int \int_S F \cdot \mathbf{n}dS = \int \int \int_V div F dV = 0$.

Example 6.7.3. Let $F(x, y, z) = (x^3, y^3, z^3)$. Compute $\int \int_S F \cdot \mathbf{n}dS$, where S is the unit sphere.

Solution. We have $div F = 3(x^2 + y^2 + z^2)$. Of course, $S = \partial B$, where B is the unit ball. By the divergence theorem we have

$$\int \int_S F \cdot \mathbf{n}dS = 3 \int \int \int_B (x^2 + y^2 + z^2)dxdydz.$$

Changing to spherical coordinates we get

$$\int_0^\pi \int_0^{2\pi} \int_0^1 r^4 \sin\phi dr d\theta d\phi = \frac{12\pi}{5}.$$

The following result is the 3-dimensional analogue of Corollary 6.4.3.

Corollary 6.7.4. *Let V be a symmetric elementary region in $\mathbb{R}^3$ and ∂V its boundary properly oriented. Then*

$$\nu(V) = \frac{1}{3} \int \int_{\partial V} xdydz + ydzdx + zdxdy.$$

Proof. Consider the vector field $F(x, y, z) = \frac{1}{3}(x, y, z)$. Then $divF = 1$ and the Divergence theorem gives

$$\nu(V) = \int\int\int_V 1dxdydz = \int\int_{\partial V} F \cdot d\mathbf{S}$$

$$= \frac{1}{3} \int\int_{\partial V} xdydz + ydzdx + zdxdy.$$

$\square$

With the help of the Divergence theorem we can obtain a better understanding of the meaning of $divF$. As in the planar case, we obtain a formula of the divergence that does not depend on the choice of the coordinates. In some books on vector analysis, Equation (6.23) below, is taken as the *definition* of divergence.

Corollary 6.7.5. *Let* B_r *be the ball of radius* $r > 0$ *centered at the point* $a \in \mathbb{R}^3$. *Let* F *be a* C^1 *vector field on* B_r. *Then*

$$divF(a) = \lim_{r \to 0} \frac{\int\int_{\partial B_r} \langle F, \mathbf{n} \rangle dS}{\nu(B_r)}. \tag{6.23}$$

Thus, $divF(a)$ *is the flow of* F *at the point* a *going out, in the normal direction, of the surface of a small ball centered at* a, *per unit volume.*

Proof. From Corollary 4.2.13, with $f = divF$, we have

$$divF(a) = \lim_{r \to 0} \frac{\int\int\int_{B_r} divF}{\nu(B_r)} = \lim_{r \to 0} \frac{\int\int_{\partial B_r} \langle F, \mathbf{n} \rangle dS}{\nu(B_r)}.$$

$\square$

Since for a smooth function f in a domain $V \subseteq \mathbb{R}^3$ we know that $div(\nabla f) = \Delta f$, applying the Divergence theorem here, just as in the planar case, yields the following corollaries.

Corollary 6.7.6. *Let* f *be a harmonic function on a bounded domain* V *in* $\mathbb{R}^3$. *Let* $B(r, p)$ *be a ball of radius* $r > 0$, *centered at* $p \in V$ *and contained in* V. *Then* $f(p)$ *is the mean value of* f *over the surface of* $B(r, p)$.

Corollary 6.7.7. *Let f be a harmonic function on V. Then any extreme value of f must occur on ∂V.*

Let V be a region in $\mathbb{R}^3$ of the type described in the Divergence theorem and let f be a C^2 function defined on an open set containing V. The directional derivative $\langle \nabla f \cdot \mathbf{n} \rangle$ in the direction of the unit outer normal $\mathbf{n}$ to the surface ∂V, is called the *outward normal derivative* of f on ∂V and is often denoted by $\frac{\partial f}{\partial \mathbf{n}}$, that is,

$$\frac{\partial f}{\partial \mathbf{n}} = \nabla f \cdot \mathbf{n}.$$

The following consequences of the Divergence theorem are known as Green's identities.

Corollary 6.7.8. *(Green's identities). Let f and g be C^2 functions on an open set containing the symmetric simple region V in $\mathbb{R}^3$ with boundary ∂V. Then*

1.
$$\int\!\!\int_{\partial V} f\frac{\partial g}{\partial \mathbf{n}}dS = \int\!\!\int\!\!\int_V (f\Delta g + \nabla f \cdot \nabla g)dV, \qquad (6.24)$$

(Green's first identity).

2.
$$\int\!\!\int_{\partial V} f(\frac{\partial g}{\partial \mathbf{n}} - g\frac{\partial f}{\partial \mathbf{n}})dS = \int\!\!\int\!\!\int_V (f\Delta g - g\Delta f)dV, \qquad (6.25)$$

(Green's second identity). In particular, if f and g are harmonic, then
$$\int\!\!\int_{\partial V} f\frac{\partial g}{\partial \mathbf{n}}dS = \int\!\!\int_{\partial V} g\frac{\partial f}{\partial \mathbf{n}}dS. \qquad (6.26)$$

Proof.   1. Applying the Divergence theorem to $F = f\nabla g$, we have

$$\int\!\!\int_{\partial V} f\frac{\partial g}{\partial \mathbf{n}}dS = \int\!\!\int_{\partial V} f\nabla g \cdot \mathbf{n}dS = \int\!\!\int\!\!\int_V div(f\nabla g)dV.$$

At the same time, by property (5),

$$div(f\nabla g) = f(div(\nabla g)) + \nabla f \cdot \nabla g = f\Delta g + +\nabla f \cdot \nabla g,$$

which yields (6.24).

2. Since $div(f\nabla g - g\nabla f) = f\Delta g - g\Delta f$ (by property 11), the Divergence theorem implies

$$\int\int_{\partial V} f\left(\frac{\partial g}{\partial \mathbf{n}} - g\frac{\partial f}{\partial \mathbf{n}}\right)dS = \int\int_{\partial V}(f\nabla g - g\nabla f)\cdot \mathbf{n}dS$$

$$= \int\int\int_V div(f\nabla g - g\nabla f)dV = \int\int\int_V (f\Delta g - g\Delta f)dV.$$

$\square$

EXERCISES

1. Let $F(x,y,z) = (x^2, y^2, z^2)$. Use the Divergence theorem to compute $\int\int_S F\cdot d\mathbf{S}$, where S is the boundary of the cube $[0,a]\times[0,a]\times[0,a]$.

2. Let $F(x,y,z) = (x^3, y^3, z^3)$. Use the Divergence theorem to compute $\int\int_S F\cdot d\mathbf{S}$, where S is the unit sphere.

3. Let $F(x,y,z) = (3x^2, xy, z)$. Use the Divergence theorem to compute $\int\int_S F\cdot d\mathbf{S}$, where S is the surface bounding the tetrahedron formed by the plane $x+y+z = 1$ and the coordinate planes.

4. Compute $\int\int_S F\cdot d\mathbf{S}$, where $F(x,y,z) = x\mathbf{i}+y\mathbf{j}+z\mathbf{k}$ and S is the surface bounding the region enclosed by the cylinder $x^2+y^2 = 9$, the paraboloid $z = x^2+y^2$ and the plane $z = 0$.

5. Use the Divergence theorem to show that

$$\int\int_S (x^2\cos\alpha + y^2\cos\beta + z^2\cos\gamma)dS = \frac{8a^4}{3}\pi,$$

where S is the surface $x^2+y^2+z^2 = 2az$ and $\cos\alpha$, $\cos\beta$, $\cos\gamma$ are the direction cosines of the outward normal $\mathbf{n}$ of S.

6. Let $F(x,y,z) = (x, -y, 2z)$. Use the Divergence theorem to compute $\int\int\int_\Omega divFdxdydz$, where Ω is the ball $x^2+y^2+(z-1)^2 < 1$.

7. Use the Divergence theorem to compute the surface integral of Exerxise 13 in Section 6.3.

8. Use the Divergence theorem to compute the surface integral

$$\int\int_S xz\,dydz + xy\,dzdx + yz\,dxdy,$$

where S is the surface of the cylinder in the first octant between the planes $z = 0$ and $z = h$.

9. Let V be a symmetric elementary region in $\mathbb{R}^3$ and ∂V its boundary. Let $\mathbf{r} = x\mathbf{i} + y\mathbf{j} + z\mathbf{k}$ and $r = \sqrt{x^2 + y^2 + z^2}$. Show that

$$\int\int\int_V \left(\frac{1}{r^2}\right) dxdydz = \int\int_{\partial V} \frac{1}{r^2}(\mathbf{r}\cdot\mathbf{n})dS,$$

where $\mathbf{n}$ is the outward unit normal to ∂V.

10. Let B_1 and B_2 be open balls in $\mathbb{R}^3$ such that $\overline{B_1} \subset B_2$. Prove that the Divergence theorem holds in the region $B_2 \setminus B_1$.

11. Suppose that the vector field F is tangent to the closed surface S of a symmetric elementary region Ω in $\mathbb{R}^3$. Show that

$$\int\int\int_\Omega divF\,dxdydz = 0.$$

12. Assume that F satisfies $divF = 0$ and $\operatorname{curl} F = \mathbf{0}$. Prove that we can find a harmonic function f such that $F = \nabla f$.

13. Let Ω be a symmetric elementary region in $\mathbb{R}^3$ with piecewise smooth boundary and let $f \in C^2(\overline{\Omega})$.

   (a) Show that

$$\int\int_{\partial\Omega} \frac{\partial f}{\partial \mathbf{n}}dS = \int\int\int_\Omega \Delta f\,dxdydz.$$

   (b) Show that if $\Delta f = 0$, then

$$\int\int_{\partial\Omega} f\frac{\partial f}{\partial \mathbf{n}}dS = \int\int\int_\Omega ||\nabla f||^2 dxdydz.$$

Answers to selected Exercises

1. $3a^4$.   2. $\frac{12}{5}\pi$.   3. $\frac{11}{24}$.   4. $\frac{243}{2}\pi$.   6. $\frac{8}{3}\pi$.   8. $a^2h\left(\frac{2a}{3} + \frac{\pi h}{8}\right)$.

6.8   Differential forms (*)

In this section, we shall formulate the classical theorems of vector analysis in terms of the so-called *differential forms*. Roughly speaking a differential k-form is an object whose mission in mathematics is to be integrated over k-dimensional sets. In the language of differential forms, classical results of vector analysis, in particular Stokes' theorem, can be generalized to higher dimensions.

In $\mathbb{R}^3$ there are 0-forms, 1-forms, 2-forms and 3-forms, while in $\mathbb{R}^2$ we deal with 0-forms, 1-forms and 2-forms. Here we discuss differential forms in $\mathbb{R}^3$.

Definition 6.8.1. Let U be an open set in $\mathbb{R}^3$.

(a) A 0-*form* on U is a function $f : U \to \mathbb{R}$.

(b) The *basic* 1-*forms* are the expressions

$$dx,\ dy,\ dz.$$

A 1-*form* ω on U is an expression of the form

$$\omega = F_1 dx + F_2 dy + F_3 dz,$$

where F_1, F_2 and F_3 are real-valued functions defined on U.

(c) The *basic* 2-*forms* are the expressions

$$dxdy,\ dydz,\ dzdx.$$

A 2-*form* ω on U is an expression of the form

$$\omega = F_1 dydz + F_2 dzdx + F_3 dxdy,$$

where the F_j's, as before, are functions defined on U.

(d) A *basic* 3-*form* is the expression $dxdydz$.
   A 3-*form* ω on U is an expression of the form

$$\omega = f(x, y, z)dxdydz,$$

where f is a real-valued function on U.

The differential form ω is called *continuous* or *differentiable* or C^1 etc., if its coefficient functions F_j's and f are continuous or differentiable or C^1 etc., respectively.

Definition 6.8.2. Let U be an open set in $\mathbb{R}^3$. For $k = 0, 1, 2, 3$ a *k-dimensional surface* or simply a *k-surface* in U is a smooth mapping $\Phi : \Omega \subset \mathbb{R}^k \to U$ which is one-to-one on the interior of Ω, where Ω is a compact set in $\mathbb{R}^k$. We think of S as the range of Φ, viz, $S = \Phi(\Omega)$.

A 0-*surface* is just a point x in $U \subseteq \mathbb{R}^3$.
A 1-*surface* is an oriented C^1 curve C in $U \subseteq \mathbb{R}^3$.
A 2-*surface* is an oriented C^1 surface S in $U \subseteq \mathbb{R}^3$.
A 3-*surface* is an elementary region V in $U \subseteq \mathbb{R}^3$. We think of differential forms as functions (functionals) that assign, via integration, to each k-dimensional surface a number:

(a) A 0-form f applied to a 0-surface $\{x\}$ is the *value* of f at the point $x \in U$, that is, $f(x)$.

(b) A 1-form
$$\omega = F_1 dx + F_2 dy + F_3 dz$$
applied to a 1-surface C is the *line integral* of the "associated" vector field $F = (F_1, F_2, F_3)$ along C, that is,
$$\int_C \omega = \int_C F_1 dx + F_2 dy + F_3 dz = \int_C F \cdot d\mathbf{s}.$$

(c) A 2-form
$$\omega = F_1 dy dz + F_2 dz dx + F_3 dx dy$$
applied to a 2-surface S is the *surface integral* of the "associated" vector field $F = (F_1, F_2, F_3)$ over S, that is,
$$\int\int_S \omega = \int\int_S F_1 dy dz + F_2 dz dx + F_3 dx dy = \int\int_S F \cdot d\mathbf{S}.$$

(d) A 3-form
$$\omega = f(x, y, z) dx dy dz$$
applied to a 3-surface V is the *triple (volume) integral*
$$\int\int\int_V \omega = \int\int\int_V f(x, y, z) dx dy dz = \int\int\int_V f dV.$$

Any two forms of the same class (order) can be added by combining coefficients of like terms. For example, if $\omega_1 = F_1 dx + F_2 dy + F_3 dz$ and $\omega_2 = G_1 dx + G_2 dy + G_3 dz$, then

$$\omega_1 + \omega_2 = (F_1 + G_1)dx + (F_2 + G_2)dy + (F_3 + G_3)dz.$$

For 2-forms when two terms contain the same differentials (that is, the same 2-basic forms), but in different orders, they must be brought into agreement before adding their coefficients, using the rules:

$$dxdy = -dydx, \quad dydz = -dzdy, \quad dzdx = -dxdz. \tag{6.27}$$

Multiplication of differential forms (called *exterior product* or *wedge product*) is defined in steps as follows:
1) Let g be a 0-form and ω a k-form.
For $k = 1$ the exterior product is

$$g \wedge \omega = g\omega = (gF_1)dx + (gF_2)dy + (gF_3)dz$$

For $k = 2$ the exterior product is

$$g \wedge \omega = (gF_1)dydz + (gF_2)ddzdx + (gF_3)dxdy$$

For $k = 3$ the exterior product is

$$g \wedge \omega = (gf)dxdydz.$$

2) For basic 1-forms we define

$$dx \wedge dy = dxdy = -(dy \wedge dx)$$

$$dy \wedge dz = dydz = -(dz \wedge dy)$$

$$dz \wedge dx = dzdx = -(dx \wedge dz)$$

$$dx \wedge (dy \wedge dz) = (dx \wedge dy) \wedge dz = dxdydz$$

3) Given a differential k-form ω and a differential l-form ψ their exterior product $\omega \wedge \psi$ is a $(k + l)$-form obtained using (1) and (2) and the *distributive* law

$$(f\omega_1 + \omega_2) \wedge \psi = f(\omega_1 \wedge \psi) + (\omega_2 \wedge \psi),$$

where f is a 0-form and ω_1, ω_2 and ψ are forms of the proper order.

Notice that from (2) $dx \wedge dx = -(dx \wedge dx)$, and hence $dx \wedge dx = 0$. Similarly, $dy \wedge dy = dz \wedge dz = 0$. Thus,

$$dxdx = dydy = dzdz = 0.$$

Furthermore note that the $\wedge$-product satisfies the *anticommutative*

$$\psi \wedge \omega = (-1)^{kl} \omega \wedge \psi,$$

and the *associative*

$$\omega_1 \wedge (\omega_2 \wedge \omega_3) = (\omega_1 \wedge \omega_2) \wedge \omega_3$$

laws.

Example 6.8.3. Let $\omega = dx + dz$ and $\psi = xy\,dydz + x^2 dxdy$. Find $\omega \wedge \psi$.

Solution.

$$\omega \wedge \psi = (dx + dz) \wedge (xy\,dydz) + x^2 dxdy$$

$$= (dx + dz) \wedge (xy\,dydz) + (dx + dz) \wedge (x^2 dxdy)$$

$$= dx \wedge (xy\,dydz) + dz \wedge (xy\,dydz) + dx \wedge (x^2 dxdy) + dz \wedge (x^2 dxdy)$$

$$= xy\,dxdydz + xy\,dzdydz + x^2 dxdxdy + x^2 dzdxdy$$

$$= xy\,dxdydz - 0 + 0 + x^2 dxdydz$$

$$= (x^2 + xy)dxdydz.$$

Next we define the notion of **exterior derivative** for k-forms. The differentiation operator d associates a $(k+1)$-form $d\omega$ to each k-form ω if $k < 3$. The form $d\omega$ is called the *exterior derivative* of ω.

(a) For a 0-form f (function) of class C^1 the exterior derivative is defined by

$$df = \frac{\partial f}{\partial x}dx + \frac{\partial f}{\partial y}dy + \frac{\partial f}{\partial z}dz,$$

called the *differential* of f. Identifying 1-forms with vector fields, this tells us that *the exterior deivative on 0-forms is the gradient.*

(b) If $\omega = F_1 dx + F_2 dy + F_3 dz$ is a 1-form with C^1 coefficients, we
define
$$d\omega = (dF_1)dx + (dF_2)dy + (dF_3)dz.$$

(c) If $\omega = F_1 dydz + F_2 dzdx + F_3 dxdy$ is a 2-form with C^1 coefficients,
we define

$$d\omega = (dF_1)dydz + (dF_2)dzdx + (dF_3)dxdy.$$

The exterior derivative of a 3-form is always zero.

Proposition 6.8.4. *Let $F = (F_1, F_2, F_3)$ be a C^1 vector field and let
$\omega = F_1 dx + F_2 dy + F_3 dz$ the 1-form associated with F.
Let $G = \nabla \times F = \operatorname{curl} F = (G_1, G_2, G_3)$. If ψ is the 2-form*

$$\psi = G_1 dydz + G_2 dzdx + G_3 dxdy,$$

then $d\omega = \psi$. Under the obvious identifications, this tells us the curl is
the exterior derivative on 1-forms in $\mathbb{R}^3$

Proof. We have

$$d\omega = (dF_1)dx + (dF_2)dy + (dF_3)dz$$

$$= \left(\frac{\partial F_1}{\partial x}dx + \frac{\partial F_1}{\partial y}dy + \frac{\partial F_1}{\partial z}dz \right) dx$$

$$+ \left(\frac{\partial F_2}{\partial x}dx + \frac{\partial F_2}{\partial y}dy + \frac{\partial F_2}{\partial z}dz \right) dy$$

$$+ \left(\frac{\partial F_3}{\partial x}dx + \frac{\partial F_3}{\partial y}dy + \frac{\partial F_3}{\partial z}dz \right) dz$$

$$= 0 - \frac{\partial F_1}{\partial y}dxdy + \frac{\partial F_1}{\partial z}dzdx + \frac{\partial F_2}{\partial x}dxdy + 0 - \frac{\partial F_2}{\partial z}dydz - \frac{\partial F_3}{\partial x}dzdx + \frac{\partial F_3}{\partial y}dydz + 0$$

$$= \left(\frac{\partial F_3}{\partial y} - \frac{\partial F_2}{\partial z} \right) dydz + \left(\frac{\partial F_1}{\partial z} - \frac{\partial F_3}{\partial x} \right) dzdx + \left(\frac{\partial F_2}{\partial x} - \frac{\partial F_1}{\partial y} \right) dxdy$$

$$= G_1 dydz + G_2 dzdx + G_3 dxdy = \psi.$$

$\square$

Proposition 6.8.5. Let $F = (F_1, F_2, F_3)$ be a C^1 vector field and let $\omega = F_1 dydz + F_2 dzdx + F_3 dxdy$ be the 2-form associated with F. Then

$$d\omega = (divF)dxdydz.$$

Under the obvious identifications, this tells us the divergence is the exterior derivative on 2-forms in $\mathbb{R}^3$

Proof.
$$d\omega = (dF_1)dydz + (dF_2)dzdx + (dF_3)dxdy$$

$$= \left(\frac{\partial F_1}{\partial x}dx + \frac{\partial F_1}{\partial y}dy + \frac{\partial F_1}{\partial z}dz \right) dydz$$

$$+ \left(\frac{\partial F_2}{\partial x}dx + \frac{\partial F_2}{\partial y}dy + \frac{\partial F_2}{\partial z}dz \right) dzdx$$

$$+ \left(\frac{\partial F_3}{\partial x}dx + \frac{\partial F_3}{\partial y}dy + \frac{\partial F_3}{\partial z}dz \right) dxdy$$

$$= \frac{\partial F_1}{\partial x}dxdydz + \frac{\partial F_2}{\partial y}dxdydz + \frac{\partial F_3}{\partial z}dxdydz = \left(\frac{\partial F_1}{\partial x} + \frac{\partial F_2}{\partial y} + \frac{\partial F_3}{\partial z} \right) dxdydz$$

$$= (divF)dxdydz.$$

$\square$

Remark 6.8.6. Note that for any 0-form f of class C^2, the equality of the mixed partials implies $d(df) = 0$. This is true in general. For any k-form ω of class C^2, we have $d(d\omega) = 0$. This captures the identities $\text{curl}(\nabla f) = \nabla \times (\nabla f) = \mathbf{0}$ and $div(\text{curl}\,F) = \nabla \cdot (\nabla \times F) = 0$ in one formula (see Proposition 6.5.4 and the basic identity (1)).

Now we formulate Green's, Stokes' and Gauss' theorems in the language of differential forms.

Theorem 6.8.7. *(Green's theorem).* Let Ω be a symmetric elementary region in $\mathbb{R}^2$ with $\partial\Omega$ given the counterclockwise orientation. If $\omega = F_1(x, y)dx + F_2(x, y)dy$ is a C^1 1-form on some open set U in $\mathbb{R}^3$ that contains Ω, then

$$\int_{\partial\Omega} \omega = \int\int_{\Omega} d\omega.$$

Proof. Here $d\omega$ is a 2-form on U and Ω is a 2-surface in $\mathbb{R}^3$ parametrized by $\Phi(x,y) = (x,y,0)$. Let $F = (F_1, F_2, 0)$. Then

$$G = \nabla \times F = \left(0, 0, \frac{\partial F_2}{\partial x} - \frac{\partial F_1}{\partial y}\right).$$

If ψ is the 2-form associated with G, then, from Proposition 6.8.4, $d\omega = \psi$. Now by the classical Green's theorem, we have

$$\int_{\partial\Omega} \omega = \int_{\partial\Omega} F_1 dx + F_2 dy = \int\int_{\Omega} \left(\frac{\partial F_2}{\partial x} - \frac{\partial F_1}{\partial y}\right) dxdy$$

$$= \int\int_{\Omega} \psi = \int\int_{\Omega} d\omega.$$

$\square$

Theorem 6.8.8. *(Stokes' theorem).* *Let S be a 2-surface in $\mathbb{R}^3$ with ∂S given the counterclockwise orientation. Let $\omega = F_1 dx + F_2 dy + F_3 dz$ be a 1-form of class C^1 defined on some open set U in $\mathbb{R}^3$ that contains S. Then*

$$\int_{\partial S} \omega = \int\int_S d\omega.$$

Proof. Let $F = (F_1, F_2, F_3)$ be the associated vector field. Let $G = \text{curl } F$ and let ψ be the 2-form associated with G. From Proposition 6.8.4, $d\omega = \psi$. By the classical Stoke's theorem

$$\int_{\partial S} \omega = \int_{\partial S} F_1 dx + F_2 dy + F_3 dz = \int_{\partial S} F \cdot ds = \int\int_S \text{curl } F \cdot d\mathbf{S}$$

$$= \int\int_S G \cdot d\mathbf{S} = \int\int_S \psi = \int\int_S d\omega.$$

$\square$

Theorem 6.8.9. *(Gauss' theorem).* *Let Ω be a 3-surface in $\mathbb{R}^3$ with $\partial\Omega$ given the outward orientation. Let $\omega = F_1 dydz + F_2 dzdx + F_3 dxdy$ be a 2-form of class C^1 defined on some open set U in $\mathbb{R}^3$ that contains Ω. Then*

$$\int\int_{\partial\Omega} \omega = \int\int\int_{\Omega} d\omega.$$

Proof. Let $F = (F_1, F_2, F_3)$ be the associated vector field. From Proposition 6.8.5, $d\omega = divF$. By the classical Divergence theorem theorem

$$\int\!\!\int_{\partial\Omega} \omega = \int\!\!\int_{\partial\Omega} F_1 dy dz + F_2 dz dx + F_3 dx dy$$

$$= \int\!\!\int_{\partial\Omega} F \cdot d\mathbf{S} = \int\!\!\int\!\!\int_\Omega divF \cdot d\mathbf{S} = \int\!\!\int\!\!\int_\Omega d\omega.$$

$\square$

These results generalize to $\mathbb{R}^n$. In $\mathbb{R}^n$ there are $n+1$ classes of differential forms which are called, *0-forms*, *1-forms*,...,*n-forms*, respectively.

Definition 6.8.10. Let $U \subseteq \mathbb{R}^n$ be an open set and $x \in U$. A *0-form* on an open set U in $\mathbb{R}^n$ is a function $f : U \to \mathbb{R}$. The *basic 1-forms* on U are the expressions $dx_1, ..., dx_n$. A *1-form* on U is an expression of the form[13]

$$\omega = F_1(x) dx_1 + ... + F_n(x) dx_n.$$

As before smoothness of a differential form ω means smoothness of its coefficient $F_1, ..., F_n$.

In general a *k-form* ω on an open set $U \subset \mathbb{R}^n$ is an alternating k-multilinear function

$$\omega : U \to \Lambda^k(\mathbb{R}^n),$$

where $\Lambda^k(\mathbb{R}^n)$ denotes the vector space of all alternating k-multilinear functions on $\mathbb{R}^n$. It turns out (as with 1-forms) that every alternating

---

[13]In rigorous terms, a *1-form* on U is a mapping $\omega : U \to L(\mathbb{R}^n, \mathbb{R})$ which assigns to each point $x \in U$ a linear functional $\omega_x : \mathbb{R}^n \to \mathbb{R}$. By Theorem 2.4.12, for each linear functional λ on $\mathbb{R}^n$ there is a unique $w = (w_1, ..., w_n) \in \mathbb{R}^n$ such that $\lambda(x) = \langle w, x \rangle = w_1 x_1 + ... + w_n x_n = w_1 p_1(x) + ... + w_n p_n(x)$, where $p_i(x_1, ..., x_n) = x_i$ is the projection (linear) map onto the ith coordinate axis. Since the derivative of p_i is itself, that is, since $D_{p_i}(x) = p_i(x) = x_i$, using the notation $p_i(x) = dx_i$, for $i = 1, ..., n$, we may write $\lambda = w_1 dx_1 + ... + w_n dx_n$. If $\lambda = \omega_x$ depends on $x \in U$, then so do the coefficients $w_1, ..., w_n$. Thus, ω_x has the form

$$\omega_x = w_1(x) dx_1 + ... + w_n(x) dx_n.$$

k-multilinear function on $\mathbb{R}^n$ is a (unique) linear combination of the multidimensional basic k-forms $dx_i = dx_{i_1} \cdots dx_{i_k}$, that is

$$\omega_x = \sum_{[i]} F_i(x)dx_i,$$

where $[i]$ denotes summation over all increasing k-tuples $i = (i_1, ..., i_k)$ with $1 \le i_s \le n$ and each F_i is a real-valued function on U. The *exterior derivative* of a k-form $\omega = \sum_{[i]} F_i dx_i$ in $\mathbb{R}^n$ is the $(k+1)$-form $d\omega$ defined similarly (as in the 3-dimensional case) by

$$d\omega = \sum_{[i]} d(F_i)dx_i.$$

Remark 6.8.11. (*Exact and Closed form*). A k-form ω defined on an open set U in $\mathbb{R}^n$ is called *exact* (on U) if there exist a $(k-1)$-form η on U such that $d\eta = \omega$. A k-form ω defined on an open set U in $\mathbb{R}^n$ is called *closed* (on U) if $d\omega = 0$.

Since $d(d\omega) = 0$, we see that every exact form is automatically closed. As we have seen in Example 6.2.11 the vector field $F(x,y) = (\frac{y}{x^2+y^2}, \frac{-x}{x^2+y^2})$ satisfies $\frac{\partial F_2}{\partial x} = \frac{\partial F_1}{\partial y}$, but F is not conservative on $\mathbb{R}^2 - \{(0,0)\}$. Therefore the associated 1-form

$$\omega = \frac{ydx - xdy}{x^2 + y^2}$$

is closed on $\mathbb{R}^2 - \{(0,0)\}$, but is not exact there. Hence, whether the converse is true or not depends on the topology of the set U. The Poincare lemma asserts that on a star-shaped (in particular, convex) open set U of $\mathbb{R}^n$ every closed C^1 k-form is exact (see [31]). This generalizes our Theorem 6.2.8 in $\mathbb{R}^n$.

Finally we state a result known as the *general Stokes' theorem*. It captures the classical theorems of vector analysis in one formula. For a discussion on differential forms in $\mathbb{R}^n$ and for its proof, we refer the interested reader to [31]. To do so we define a differentiable manifold.

Definition 6.8.12. Let M be a metric space with the property that each point $x \in M$ has a neighborhood U_x which is homeomorphic with some Euclidean space of a fixed dimension, say n. Moreover, we shall assume that whenever two of these neighborhoods, say U_x and U_y have a non empty intersection the resultant functions are smooth (C^k with $1 \leq k < \infty$. Then we say M is *differentiable manifold* of dimension n and degree k.

Theorem 6.8.13. *(Stokes' theorem in $\mathbb{R}^n$). Let M be a compact oriented k-dimensional manifold in $\mathbb{R}^n$ with boundary ∂M, a smooth $(k-1)$-dimensional manifold given the induced orientation from the orientation of M, and let ω be a $(k-1)$-form on M. Then*

$$\int_{\partial M} \omega = \int_M d\omega.$$

6.9   Vector fields on spheres and the Brouwer fixed point theorem (*)

In this section we deal with some questions at the interface between calculus in several variables and what is called *differential topology*. Let X be a compact subset of Euclidean space $\mathbb{R}^{n+1}$. (We choose the ambient space to have $\dim(n+1)$ for convenience). A continuous (resp. smooth) *tangential vector field F* on X is a continuous (resp. smooth) $\mathbb{R}^{n+1}$-valued function on X with the property that for each $x \in X$, $F(x)$ is tangent to X at x, that is, it lies in the tangent hyperplane, $T_x(X)$. We shall call a tangential vector field F *non singular* if $F(x)$ is never 0 at any point of X. If F is zero at some point (or several points) we say F is *singular*. For example, if $X = S^2$, the 2 sphere considered as the surface of the earth within $\mathbb{R}^3$, then the tangential component of the wind gives a tangential vector field (which one assumes is continuous, but *which may turn out to be singular*).

Notice that given a non singular smooth (or continuous) tangential vector field F on X, we can produce a unitary one G, that is, one whose vectors are all of unit length, which is also a smooth (or continuous) tangential vector field on X simply by normalizing F at each point: $G(x) = \frac{F(x)}{||F(x)||}$, $x \in X$. The reader should check that G is smooth (or continuous). An example of a non singular (unitary) tangent vector field on the circle, S^1, is given by choosing any point, x_0, and placing a unit vector v_0 at x_0 tangent to the circle there. Then, since any other point

$x \in S^1$ can be gotten from x_0 by a unique counterclockwise rotation we just rotate v_0 by the same amount and get a unit tangent vector at x. Doing this for all $x \in S^1$ gives a unitary smooth tangential vector field on S^1.

Here we are particularly interested in the unit sphere, $X = S^n \subseteq \mathbb{R}^{n+1}$. In this case each tangent vector $F(x)$ is perpendicular to x, for all $x \in S^n$. That is, $\langle x, F(x) \rangle = 0$ on all of S^n.

Exercise 6.9.1. Verify that the condition $\langle x, F(x) \rangle \equiv 0$ is both necessary and sufficient for tangency. That is, the tangent hyperplane at a point x on the unit sphere is the set of all vectors through x perpendicular to x.

The first question we want to address is: are there unitary smooth (or continuous) vector fields on S^n?

When $n = 2k - 1$ is odd there always are. We will show that for $x = (x_1, \ldots, x_{2k}) \in S^{2k-1}$,

$$F(x_1, \ldots, x_{2k}) = (x_2, -x_1, x_4, -x_3, \ldots x_{2k}, -x_{2k-1}) \tag{6.28}$$

is such a vector field. Here F is evidently a smooth function of x since each component is a polynomial. To see that it is unitary observe $F(x_1, \ldots, x_{2k}) = x_e - x_o$, where u_e is the vector in $\mathbb{R}^{2k}$ with even subscripts x_{2i} in the odd coordinates and zeros in the even ones, while x_o is the vector in $\mathbb{R}^{2k}$ with zeros in the odd coordinates and x_{2i-1} in the even ones. Hence

$$\|F(x)\|^2 = \langle x_e - x_o, x_e - x_o \rangle = \|x_e\|^2 + \|x_o\|^2 + 2\langle x_e, x_o \rangle.$$

However, $\langle x_e, x_o \rangle = 0$ and $\|x_e\|^2 + \|x_o\|^2 = \|x\|^2 = 1$. Thus $\|F(x)\|^2 = 1$. Moreover, one sees immediately using the fact that we are in an even dimensional space that $\langle F(x), x \rangle = 0$. Thus F is a smooth unitary tangential vector field on X.

Exercise 6.9.2. Show that, in the case of S^1, the vector field given in equation 6.28 is exactly the one gotten by rotation of a single unit tangent vector mentioned earlier.

Notice that one cannot even write down such a vector field when n is even. Indeed as we shall see, when n is even there are no smooth (or continuous) non singular vector fields on S^n. So for example, *at any moment there must be a point on the earth's surface where the wind doesn't blow tangentially.*

Our plan in this section, using an idea of J. Milnor, is to prove the following result Theorem 6.9.3 (sometimes called the hairy ball theorem). Then we shall extend Theorem 6.9.3 from the smooth to the continuous case. From this we will derive the Brouwer fixed point theorem, (Theorem 6.9.9). Finally, from (Theorem 6.9.9) as a further corollary we get the Invariance of Domain theorem, (Corollary 6.9.11).

6.9.1   Tangential vector fields on spheres

Theorem 6.9.3. S^n *possesses a non singular smooth vector field if and only if n is odd.*

Since as we saw when n is odd there is always such a vector field, we may assume $n = 2k$ is even and suppose there were a smooth unitary tangential vector field F on $X = S^n$ from which we shall get a contradiction. Put another way, when n is even any smooth tangential vector field on S^n must be zero at some point.

For $t \in \mathbb{R}$ and $x \in X$ let $G_t(x) = x + tF(x)$. Each such G_t is a vector field on S^n. But these are not tangent vector fields and they aren't of unit length. Let us first calculate $||x + tF(x)||$. Just as above,

$$\langle x + tF(x), x + tF(x) \rangle = \langle x, x \rangle + 2t\langle x, F(x) \rangle + t^2 \langle F(x), F(x) \rangle.$$

Since both $||x|| = ||F(x)|| = 1$ and $\langle x, F(x) \rangle = 0$, we see by the Pythagorean theorem that $||x + tF(x)|| = \sqrt{1 + t^2}$.

Since $G_t(x)$ is defined for every $x \in S^n$ we can regard $G_t : S^n \to \mathbb{R}^{n+1}$ as a vector valued function on the sphere with $G_t(S^n)$ contained in the sphere of radius $\sqrt{1 + t^2}$. To prove our theorem we shall need the following two lemmas. Let ν denote the n-dimensional volume on X (as opposed to the $n + 1$ dimensional volume in $\mathbb{R}^{n+1}$).

Lemma 6.9.4. *If $|t|$ is sufficiently small, G_t is an one-to-one mapping taking X onto a region whose volume can be expressed as a polynomial function of t.*

Lemma 6.9.5. *If $|t|$ is sufficiently small, G_t takes S^n onto the entire sphere of radius $\sqrt{1 + t^2}$ (and not just a subset).*

Proof of Lemma 6.9.4. Since X is compact and F is continuously differentiable on X we know (Corollary 3.4.8) there is some positive constant c so that

$$||F(u) - F(x)|| \leq c||u - x||,$$

for all $u, x \in X$.

We know this is true for u, x in a common neighborhood. Since X is compact we can cover it by a finite number of such neighborhoods and then choose c to be the minimum of the constants associated with each of these neighborhoods.

Now let $|t| < \frac{1}{c}$. If $G_t(x) = G_t(u)$, then $||u - x|| \leq |t|c||u - x||$. Hence $u = x$ and G_t is one-to-one. Taking $|t|$ small so this happens and observing that the matrix of $d(G_t)(x) = I + tdF(x)$, where I is the identity, we see the Jacobian, $\det(d(G_t)(x)) = \det(I + tdFv(x))$, is a polynomial function of t (of degree $n + 1$). Taking determinants $\det(d_{G_t})(x) = \det(I + td_F(x))$. Hence, by the Change of Variables formula, $\nu(G_t(S^n)) = \int_{S^n} |\det(I + tdF(x))|dx$. so $\nu(S^n)$ is a polynomial function of t (whose coefficients are the integrals over S^n of the coefficients of $\det(I + tdF(x))$. This completes the proof of Lemma 6.9.4.

We now turn to the *proof of Lemma 6.9.5.* Consider the region R between two concentric spheres defined by the inequalities $a \leq ||u|| \leq b$, where $a < 1 < b$. Extend the vector field F to R by $F(rx) = rF(x)$, where $a \leq r \leq b$ and $x \in S^n$. Then this also extends the map G_t to all of R because $G_t(rx) = rG_t(x)$ and moreover, using this last equation, by a calculation similar to one we have already made, G_t maps R into the region between the spheres of radii $a\sqrt{1 + t^2}$ and $b\sqrt{1 + t^2}$.

Now by the same argument as in Lemma 6.9.4 for $|t|$ sufficiently small G_t is one-to-one on R. Therefore by the Chain Rule its derivative, $d(G_t)(x)$, is invertible at every point. The Inverse Function Theorem tells us that G_t maps open sets in R to open sets in $G_t(R)$. In particular, $G_t(S^n)$ is an open set in the sphere of radius $\sqrt{1 + t^2}$. But this non empty open set is also compact and therefore closed. Since spheres are connected (Example 2.6.12) $G_t(S^n)$ must be the entire sphere of radius $\sqrt{1 + t^2}$, proving Lemma 6.9.5.

These two lemmas suffice to *prove Theorem 6.9.3.* By Lemma 6.9.4 for small t, $\nu(G_t(S^n)) = \nu(S^n) \cdot p(t)$, where $p(t)$ is a polynomial of degree

$n+1$ in t. On the other hand, in $n+1$ space, dilations change volume by a factor of r^{n+1}. Hence $\nu(G_t(S^n)) = \left(\sqrt{1+t^2}\right)^{n+1} \nu(S^n)$. Comparing these equations we see that for small $|\,t\,|$,

$$(\sqrt{1+t^2}\,)^{n+1} = p(t).$$

But since $n+1$ is odd, $\left(\sqrt{1+t^2}\right)^{n+1}$ is not a polynomial in t, a contradiction.   □

The reason for this is because in Section 3.3. we showed $\sqrt{1+t^2}$ is not a polynomial function of t. Let j be odd. Then $\left(\sqrt{1+t^2}\right)^{j}$ is a polynomial times $\sqrt{1+t^2}$. Hence $\left(\sqrt{1+t^2}\right)^{j}$ is also not a polynomial function of t. The reader should verify this is so by consulting the following worked exercise.

Exercise 6.9.6. Let $f(t) = q(t)\sqrt{1+t^2}$, where q is a polynomial. Then f is not a polynomial.

Proof. Calculating the derivatives of $g(t) = \sqrt{1+t^2}$ we see that the Taylor coefficients, $g^k(0)$, are 1 when k is odd and 0 when k is even, so that the Taylor expansion is $\sqrt{1+t^2} = 1 + t^2 + t^4 \ldots$. This means $q(t)\sqrt{1+t^2}$ has infinitely many of its Taylor coefficients nonzero so that f is not a polynomial. (Obviously the same argument works if $\sqrt{1+t^2}$ is replaced by any smooth $g(t)$ which is not a polynomial).   □

We remark that Theorem 6.9.3 is true for much more than the unit sphere. In fact, it's true for *anything diffeomorphic with the unit sphere.* For example it's true of an ellipsoid of any size, shape, or orientation in space. Another example to keep in mind is a partially inflated basketball. A non singular smooth vector field on such a surface will become a non singular smooth vector field on the sphere by simply blowing up the ball. More formally, if $\Phi : X \to S^n$ is the diffeomorphism, then for each $x \in X$, $d(\Phi)_x : T_x(X) \to T_{\Phi(x)}(S^n)$ is an invertable linear map between their tangent spaces. Hence it transports a non singular smooth vector field on X to one on S^n.

We now extend Theorem 6.9.3 from smooth to continuous vector fields.

Corollary 6.9.7. S^n *possesses a non singular continuous vector field if and only if n is odd.*

Proof. Suppose an even dimensional sphere X had a continuous unitary vector field, F. By the Weierstrass approximation (Theorem C.1.4), there is a polynomial map $p : X \to \mathbb{R}^{n+1}$ so that $||p(x) - F(x)|| < \frac{1}{2}$. Let $G(x) = p(x) - (\langle p(x), x \rangle)x$. Then G is a smooth vector field on X. But is it tangential?

$$\langle G(x), x \rangle = \langle p(x) - (\langle p(x), x \rangle)x, x \rangle = \langle p(x), x \rangle - \langle p(x), x \rangle ||x||^2 = 0$$

since $||x|| = 1$. Thus G is tangential.

Since both F and G are tangential, $\langle p(x) - G(x), x \rangle = \langle p(x), x \rangle = \langle p(x) - F(x), x \rangle$. By the Cauchy-Schwarz inequality the latter is $\leq \frac{1}{2}$. By the definition of G, $|\langle p(x), x \rangle| \leq \frac{1}{2}$ everywhere on X. Hence $||p(x) - G(x)|| < \frac{1}{2}$. Since also $||p(x) - F(x)|| < \frac{1}{2}$ we see $||F(x) - G(x)|| < 1$ and because F is unitary, G is never zero. Thus G is a *smooth* non singular vector field on an even dimensional sphere which is impossible by Theorem 6.9.3. $\qquad\square$

Exercise 6.9.8.   1. Let $a \leq r \leq b$. Show that we can extend the vector field F to R by $F(rx) = rF(x)$ and that this also extends the map G_t to all of R because $G_t(rx) = rG_t(x)$.

2. Prove G_t maps R onto the region between the spheres $a\sqrt{1 + t^2}$ and $b\sqrt{1 + t^2}$.

3. Prove if $\Phi : X \to S^n$ is a diffeomorphism, then for each $x \in X$, $d(\Phi)_x : T_x(X) \to T_{\Phi(x)}(S^n)$ is an invertible linear map between their tangent spaces. Hence it transports a non singular smooth vector field on X to one on S^n.

6.9.2   The Brouwer fixed point theorem

Let B^n stand for the closed unit ball in $\mathbb{R}^n$. That is, $B^n = \{x \in \mathbb{R}^n : ||x|| \leq 1\}$. Our purpose is to prove the following result.

Theorem 6.9.9. *(Brouwer[14]) Any continuous map $f : B^n \to B^n$ has a fixed point.*

---

[14]L. Brouwer (1881-1966). Mathematician and philoshoper. He workd in topology, set theory, measure theory and complex analysis.

The statement is similar to that of the Picard fixed point theorem (Theorem 7.3.5). However here the map is only required to be continuous and not a contraction map. To compensate, the space B^n is both compact and convex, not merely complete and there is no uniqueness. For example, the identity map has all of B^n as fixed points. Notice that if we replace the 2 ball by the 2 torus, together with its interior, then any rotation is a continuous map without fixed points. The Brouwer fixed point theorem can itself be generalized in a number of ways and one of these is a powerful tool in differential equations. Just as above, this theorem applies to any space homeomorphic with B^n.

As usual let $S^n = \{x \in \mathbb{R}^{n+1} : ||x|| = 1\}$ is the n-dimensional sphere which here we regard as the boundary of B^{n+1}.

Proof. We first remark that it is sufficient to consider prove the result when S^n has even dimension. For if we can prove Theorem 6.9.9 here, then when n is odd, say $n = 2k - 1$ is easily handled. For let $f : B^{2k} \rightarrow B^{2k}$ be a continuous map. Define $F : B^{2k+1} \rightarrow B^{2k+1}$ as follows: $F(x, x_{2k+1}) = (f(x), 0)$, where $x \in B^{2k}$. Evidently F is continuous and maps B^{2k+1} to itself. Since $2k + 1$ is odd F has a fixed point, (x, x_{2k+1}). This means $F(x, x_{2k+1}) = (x, x_{2k+1}) = (f(x), 0)$. Thus $x = f(x)$ and so x is a fixed point of f.

We now take n even and assume f is a continuous map from B^{n+1} to itself *without fixed points*. Since $f(x) \neq x$ the vector field $v(x) = x - f(x)$ is never zero on B^{n+1}.

We first show this vector field points *outward* in the sense that $\langle u, v(u) \rangle > 0$ everywhere on S^n. For

$$\langle u, u - f(u) \rangle = \langle u, u \rangle - \langle u, f(u) \rangle = 1 - \langle u, f(u) \rangle \geq 1 - |\langle u, f(u) \rangle|.$$

But since $f(u) \in B^{n+1}$, $|\langle u, f(u) \rangle| \leq 1$ by the Schwarz inequality. If we knew $|\langle u, f(u) \rangle| < 1$, then we could conclude $\langle u, v(u) \rangle > 0$ as desired.

Suppose $|\langle u, f(u) \rangle| = 1$. Then in particular, since $\| u \| = 1$ and $\| f(u) \| \leq 1$ we see by the Schwarz inequality that $\| f(u) \| = 1$. So by the equality part of Schwarz $f(u) = \lambda u$ for some real λ. On the other hand, as we just saw $f(u) \in S^n$. Since both u and $f(u)$ are in S^n, $f(u) = \pm u$. If $f(u) = u$ this violates our assumption that f has no fixed points and if $f(u) = -u$, then $1 - \langle u, f(u) \rangle = 2$, another contradiction. Thus $\langle u, v(u) \rangle > 0$ everywhere on S^n.

Next we modify v to get a nonzero vector field w on B^{n+1} pointing *directly outward* on the boundary, that is $w(u) = u$ everwhere on S^n. For $x \in B^{n+1}$ set

$$w(x) = x - \frac{f(x)(1 - ||x||^2)}{1 - \langle x, f(x) \rangle}.$$

Notice w is well defined since $1 - \langle x, f(x) \rangle$ is never zero. For if it were, then arguing as above, we would get $||x|| = 1 = ||f(x)||$ and $f(x) = \lambda x$. Hence $1 = \lambda$ so $f(x) = x$, a contradiction.

Since all functions involved in defining w are continuous, w depends continuously on x. Evidently $w(x) = x$ if and only if $f(x)(1 - ||x||^2) = 0$. In particular, when $x \in S^n$ since the second factor is zero we see that $w(u) = u$ on S^n.

We now check that w is never zero. If w were zero, then x and $f(x)$ would be linearly dependent. But then since $f(x) = \lambda x$, it follows that $x - \lambda \langle x, x \rangle x = \lambda x - \lambda \langle x, x \rangle x$. Hence $x = \lambda x$. If $x \neq 0$, then $\lambda = 0$ so $f(x) = 0$ and therefore $x = 0$, a contradiction. The only remaining possibility is $x = 0$. But then also $f(x) = 0$ so 0 is a fixed point of f, a contradiction.

Next we transplant w from B^{n+1} to the southern hemisphere of S^n by means of *stereographic projection* (or rather its inverse) σ from the north pole, $N = (0, 0, \ldots, 1)$. The southern hemisphere of $S^n \subseteq \mathbb{R}^{n+1}$ is identified as the points of S^n whose last coordinate $x_{n+1} = 0$.

Let σ be defined by

$$\sigma(x) = \frac{(2x_1, \ldots, 2x_n, ||x||^2 - 1)}{||x||^2 + 1},$$

where $x = (x_1, \ldots x_n, x_{n+1}) \in \mathbb{R}^{n+1} - N$. Since $||x||^2 \geq 0$, σ is a differentiable function of x.

Next we show $||\sigma(x)|| = 1$.

$$||\sigma(x)||^2 = \frac{4||x||^2 + (||x||^2 - 1)^2}{(||x||^2 + 1)^2}.$$

But this is just $\frac{||x||^4 + 2||x||^2 + 1}{||x||^4 + 2||x||^2 + 1} = 1$.

Also, σ maps into $S^n - N$. That is N is omitted. For suppose $\frac{(2x_1, \ldots, 2x_n, ||x||^2 - 1)}{||x||^2 + 1}$ were $(0, \ldots, 0, 1)$, then $(||x||^2 - 1) = ||x||^2 + 1$ (and each

$x_i = 0$, $i = 1, \ldots n$). The first statement is already tells us $1 = -1$, which is impossible.

Next let $\sigma(x) = u$, where u lies in the southern hemisphere. We calculate $d(\sigma)_x(w(x)) = W(u)$. Now $d(\sigma)_x(w(x)) = \frac{d(\sigma)}{dt}(x + tw(x))|_{t=0}$ so W is a nonzero continuous tangential vector field on the southern hemisphere of S^n. At every point $u = \sigma(u)$ on the equator $w(u) = u$ points directly outward and $W(u)$ points due north. Similarly stereographic projection from the south pole on the nonzero vector field $-w(x)$ gives rise to a nonzero continuous tangential vector field on the northern hemisphere which also points due north on the equator. Putting these together gives a smooth tangential vector field on the sphere which is certainly nonzero except perhaps at the equator. However, at the equator the vectors add up because they both point north. Hence we have a nonzero continuous tangential vector field on S^n. Since n is even this contradicts Corollary 6.9.7.   □

Exercise 6.9.10.   1. Prove that the formula given above is indeed stereographic projection from the north pole.

2. Check that $||\sigma(x)|| = 1$ for all $x \in S^n - N$.

3. Show that on the equator $u = \sigma(u)$.

4. At the equator $W(u)$ is a multiple of N (points due north).

The following result which is more or less equivalent to the Brouwer fixed point theorem is called *Invariance of Domain theorem* . Here S^{n-1} is the boundary of B^n.

Corollary 6.9.11. *There is no continuous map $g : B^n \to S^{n-1}$ leaving the boundary pointwise fixed.*

Proof. Indeed suppose g were such a map. Let i be the injection of S^{n-1} into B^n and ϕ the map of B^n onto itself sending each vector to its negative. Then ϕ preseves the boundary and the composition $\phi \cdot i \cdot g$ is a continuous map of B^n to itself. If $\phi \cdot i \cdot g$ had a fixed point x_0, then $g(x_0) = -x_0$ and $x_0 \in S^{n-1}$ violating the condition that g leaves S^{n-1} pointwise fixed. Since $\phi \cdot i \cdot g$ has no fixed point this contradicts Theorem 6.9.9.   □

Having defined stereographic projection, it now seems opportune to state and prove the simplest version of the *Whitney embedding theorem*. Here a smooth embedding means a C^∞ one-to-one map onto a closed set.

Theorem 6.9.12. *A compact manifold M can be embedded in some Euclidean space $\mathbb{R}^k$ of sufficiently high dimension.*

Proof. Let U_i be a family of open sets each diffeomorphic to $\mathbb{R}^n$ for fixed n. By compactness $M = \cup_{i=1}^s U_i$. Now by stereographic projection each U_i is diffeomorphic to $S^n \setminus \{p\}$ by say ϕ_i. Fix i and extend ϕ_i to all of X by making it take the constant value p on $X \setminus U_i$. Then these extensions are also differentiable and $\phi_i : M \to S^n$ (surjectively). If we compose this map with the natural embedding $S^n \to \mathbb{R}^{n+1}$ we finally have a family of smooth maps each taking M to $\mathbb{R}^{n+1}$.

Let $\Phi(x) = (\phi_1(x), \dots \phi_s(x))$, $x \in M$. Then Φ is a smooth map on X taking values in $\mathbb{R}^{s(n+1)}$. We show Φ is injective. Let x and $y \in M$ with $\Phi(x) = \Phi(y)$. Suppose x and y are in the same U_i. Then since $\phi_i(x) = \phi_i(y)$ and stereographic projection is one-to-one, $x = y$. Now let $x \in U_i$ and $y \in M$ with $\phi_i(x) = \phi_i(y)$. If $y \in M \setminus U_i$, $\phi_i(y) = p$. But then $\phi_i(x) \ne \phi_i(x)$, a contradiction. □

6.10   Solved problems for Chapter 6

Exercise 6.10.1. Find the centroid of the arc C of the cycloid

$$\gamma(t) = (t - \sin t, 1 - \cos t),\ t \in [0, \pi].$$

Solution. From Section 4.3.1 the coordinates of the centroid $(\overline{x}, \overline{y})$ are calculated by the formulas

$$\overline{x} = \tfrac{1}{L(C)} \int_C x\, ds, \quad \overline{y} = \tfrac{1}{L(C)} \int_C y\, ds,$$

where $L(C)$ is the arc length of C. From Example 6.1.5 $L(C) = 4$. We have

$$\overline{x} = \frac{1}{4} \int_C x\, ds = \frac{1}{4} \int_0^\pi (t - \sin t) 2\sin(\tfrac{t}{2}) dt = \frac{1}{2} \int_0^\pi [t \sin(\tfrac{t}{2}) - \sin t \sin(\tfrac{t}{2})] dt = \frac{8}{3}.$$

Similarly,

$$\overline{y} = \frac{1}{4} \int_C y\, ds = \frac{1}{4} \int_0^\pi (1 - \cos t) 2\sin(\tfrac{t}{2}) dt = \frac{1}{2} \int_0^\pi [\sin(\tfrac{t}{2}) - \cos t \sin(\tfrac{t}{2})] dt = \frac{4}{3}.$$

Thus $(\overline{x}, \overline{y}) = (\tfrac{8}{3}, \tfrac{4}{3})$.

Exercise 6.10.2. Let the surface S be parametrized by $\Phi(u,v) = (u^2, v^2, uv)$, where $(u,v) \in [0,2] \times [0,2]$.

1. Find the equation of the tangent plane of S at $\Phi((1,1))$.

2. Find the arc length of the curve $C = \Phi(1,v)$ on S.

Solution.

1. The normal vector to the surface at $\Phi(u,v)$ is

$$T_u \times T_v = \frac{\partial \Phi}{\partial u} \times \frac{\partial \Phi}{\partial v} = (-2v, -2u, 4uv).$$

At $\Phi(1,1) = (1,1,1)$ the vector $(-2,-2,4)$ is normal to the tangent plane and the plane has equation

$$(x-1, y-1, z-1) \cdot (-2,-2,4) = 0$$

or $x + y - 2z = 0$.

2. The curve C has parametrization $\gamma(v) = (1, v^2, v)$ with $v \in [0,2]$. Hence $\gamma'(v) = (0, 2v, 1)$, and

$$L(C) = \int_0^2 ||\gamma'(v)|| dv = \int_0^2 \sqrt{1 + 4v^2} dv = \frac{1}{2}[4\sqrt{17} + \log(4 + \sqrt{17})].$$

Exercise 6.10.3. Find the centroid of the plane $z = x$ bounded by the planes $x + y = 1$, $y = 0$ and $x = 0$.

Solution. Let S be the surface of the indicated plane. The coordinates of the centroid $(\overline{x}, \overline{y}, \overline{z})$ are given by

$$\overline{x} = \tfrac{1}{A(S)} \int \int_S x dS, \quad \overline{y} = \tfrac{1}{A(S)} \int \int_S y dS, \quad \overline{z} = \tfrac{1}{A(S)} \int \int_S z dS,$$

where $A(S)$ is the area of S. Let us find this area. The surface S is the graph of $z = f(x,y)$, where $f(x,y) = x$ defined on $\Omega = \{(x,y) : 0 \le x \le 1, 0 \le y \le 1 - x\}$. From Example 6.3.7

$$A(S) = \int \int_\Omega \sqrt{1 + \left(\frac{\partial f}{\partial x}\right)^2 + \left(\frac{\partial f}{\partial y}\right)^2} \, dx dy = \sqrt{2} \int_0^1 \left(\int_0^{1-x} dy \right) dx = \frac{\sqrt{2}}{2}.$$

Now

$$\bar{x} = \frac{1}{A(S)} \int \int_S x dS = \frac{2}{\sqrt{2}} \int \int_\Omega x\sqrt{2} dx dy = 2 \int_0^1 \left(x \int_0^{1-x} dy \right) dx = \frac{1}{3},$$

$$\bar{y} = \frac{1}{A(S)} \int \int_S y dS = \frac{2}{\sqrt{2}} \int \int_\Omega y\sqrt{2} dx dy = 2 \int_0^1 \left(\int_0^{1-x} y dy \right) dx = \frac{1}{3},$$

and since $z = x$ on S

$$\bar{z} = \frac{1}{A(S)} \int \int_S z dS = \frac{1}{A(S)} \int \int_S x dS = \frac{1}{3}.$$

Exercise 6.10.4. Let S be the spherical cap formed by cutting the unit sphere $x^2 + y^2 + z^2 = 1$ with the cone having vertex angle $\frac{\pi}{6}$ and with the vertex at the center of the sphere.

1. Find the area of S.

2. Let $F(x, y, z) = (z - y)\mathbf{i} + y\mathbf{k}$. Compute $\int_C F \cdot ds$, where C is the curve where the cone intersects the sphere.

   (a) Directly

   (b) Using Stoke's theorem.

Solution.

1. We may parametrize S by $\Phi(\theta, \phi) = (\cos\theta \sin\phi, \sin\theta \sin\phi, \cos\phi)$, where $0 \le \theta \le 2\pi$, $0 \le \phi \le \frac{\pi}{6}$. Then $||T_\theta \times T_\phi|| = \sin\phi$ and

$$A(S) = \int_0^{2\pi} \int_0^{\frac{\pi}{6}} \sin\phi d\theta d\phi = 2\pi \int_0^{\frac{\pi}{6}} \sin\phi d\phi = \pi(2 - \sqrt{3}).$$

2. (a) The intersection C of the cone with the sphere is the circle centered at the z-axis of radius $\frac{1}{2}$ on the plane $z = \frac{\sqrt{3}}{2}$. We parametrize C by $\gamma(t) = (\frac{1}{2}\cos t, \frac{1}{2}\sin t, \frac{\sqrt{3}}{2})$, where $t \in [0, 2\pi]$ and

$$\int_C F \cdot ds = \frac{1}{4} \int_0^{2\pi} [-\sqrt{3}\sin t + \sin^2 t] dt = \frac{\pi}{4}.$$

(b) A simple calculation gives $\operatorname{curl} F = \nabla \times F = (1,1,1)$. Taking the disc $D : x^2 + y^2 \le (\frac{1}{2})^2$ on the plane $z = \frac{\sqrt{3}}{2}$ as a cupping surface for the curve C, (i.e., $C = \partial D$), the unit normal to D is, of course, $\mathbf{n} = \mathbf{k} = (0,0,1)$. Stoke's theorem then yields

$$\int_C F \cdot d\mathbf{s} = \int\int_D (\operatorname{curl} F) \cdot \mathbf{k}\, dS = \int\int_D dx\, dy = A(D) = \pi \left(\frac{1}{2}\right)^2 = \frac{\pi}{4}.$$

Exercise 6.10.5. Let S be the unit sphere in $\mathbb{R}^3$. Compute the surface integral

$$\int\int_S (x + y^2 + z)\, dS.$$

Solution. We leave the direct computation of the integral to the reader as an exercise. Here one can use the Divergence theorem. Indeed, consider the vector field $F(x,y,z) = (1, y, 1)$, so that $div(F) = 1$. Since the outer unit normal to S is $\mathbf{n} = (x,y,z)$, we have $\langle F, \mathbf{n} \rangle = x + y^2 + z$. If B is the unit ball, so that $S = \partial B$, then by the Divergence theorem

$$\int\int_S (x + y^2 + z)\, dS = \int\int_S \langle F, \mathbf{n} \rangle\, dS = \int\int\int_B div(F)\, dV = \int\int\int_B dV = \frac{4}{3}\pi.$$

Exercise 6.10.6. Let Ω be the open unit ball in $\mathbb{R}^3$ and $f \in C^2(\overline{\Omega})$ a nonzero function such that $f \equiv 0$ on $\partial\Omega$ and $\Delta f = -\lambda f$ in Ω, where $\lambda \in \mathbb{R}$. (such a λ is called an *eigenvalue* for the Laplacian, and f a corresponding *eigenfunction*). Show that

$$\int\int\int_\Omega \|\nabla f\|^2 dx\, dy\, dz = \lambda \int\int\int_\Omega f^2 dx\, dy\, dz,$$

and conclude that $\lambda > 0$.

Solution. By the Divergence theorem

$$\int\int\int_\Omega div(f\nabla f)\, dx\, dy\, dz = \int\int_{\partial\Omega} \langle f\nabla f, \mathbf{n} \rangle\, dS = 0,$$

because $f \equiv 0$ on $\partial\Omega$. On the other hand the integrand of the left integral is

$$div(f\nabla f) = f\, div(\nabla f) + (\nabla f) \cdot (\nabla f) = f\Delta f + \|\nabla f\|^2 = -\lambda f^2 + \|\nabla f\|^2.$$

Exercise 6.10.7. (*Brouwer's fixed point theorem in $\mathbb{R}^3$*). Let $B = \{\mathbf{x} \in \mathbb{R}^3 : \|\mathbf{x}\| \le 1\}$ be the closed unit ball in $\mathbb{R}^3$. Prove that any continuous map $f : B \to B$ has a fixed point.

Proof. We first prove the theorem for a smooth $f : B \to B$. Suppose that f has no fixed points in B. So that $f(\mathbf{x}) \neq \mathbf{x}$ for all $\mathbf{x} \in B$. Since B is convex, for each $\mathbf{x} \in B$ the ray $\varphi_t(\mathbf{x}) = f(\mathbf{x}) + t(\mathbf{x} - f(\mathbf{x}))$ with initial point $f(\mathbf{x})$ passing from $\mathbf{x}$ meets the boundary, $\partial B = \{\mathbf{x} \in \mathbb{R}^3 : ||\mathbf{x}|| = 1\}$, at a point $\varphi_t(\mathbf{x}) \in \partial B$. If $\mathbf{x} \in \partial B$, then $\varphi_1(\mathbf{x}) = \mathbf{x}$. If $\mathbf{x} \in B - \partial B$, then $\varphi_t(\mathbf{x}) = f(\mathbf{x}) + t(\mathbf{x} - f(\mathbf{x}))$, where t is the positive root of the quadratic equation

$$||f(\mathbf{x}) - t(\mathbf{x} - f(\mathbf{x}))||^2 = 1.$$

The quadratic formula gives

$$t_{\mathbf{x}} = \frac{\langle f(\mathbf{x}), f(\mathbf{x}) - \mathbf{x} \rangle + \sqrt{\langle f(\mathbf{x}), \mathbf{x} - f(\mathbf{x}) \rangle^2 + ||\mathbf{x} - f(\mathbf{x})||^2 (1 - ||f(\mathbf{x})||^2)}}{||\mathbf{x} - f(\mathbf{x})||^2}.$$

If $||f(\mathbf{x})|| = 1$, then $\langle f(\mathbf{x}), f(\mathbf{x}) - \mathbf{x} \rangle \neq 0$; for otherwise, if $\langle f(\mathbf{x}), f(\mathbf{x}) - \mathbf{x} \rangle = 0$, then $\langle f(\mathbf{x}), \mathbf{x} \rangle = ||f(\mathbf{x})||^2 = 1$. Therefore, equality would hold in Cauchy-Schwarz inequality and so $f(\mathbf{x}) = \mathbf{x}$, which contradicts our assumption $f(\mathbf{x}) \neq \mathbf{x}$. Thus

$$t_{\mathbf{x}} = \frac{\langle f(\mathbf{x}), f(\mathbf{x}) - \mathbf{x} \rangle + |\langle f(\mathbf{x}), \mathbf{x} - f(\mathbf{x}) \rangle|}{||\mathbf{x} - f(\mathbf{x})||^2} > 0.$$

On the other hand if $||f(\mathbf{x})|| < 1$, then $||\mathbf{x} - f(\mathbf{x})||^2 (1 - ||f(\mathbf{x})||^2) > 0$ and therefore $t_{\mathbf{x}} > 0$. Thus, denoting φ_t simply by φ, we get a mapping $\varphi : B \to \partial B$ such that $\varphi(\mathbf{x}) = \mathbf{x}$ for all $\mathbf{x} \in \partial B$. Moreover φ is as smooth as f is.

Now consider the vector field $F(\mathbf{x}) = \frac{\mathbf{x}}{||\mathbf{x}||^3}$ in $\mathbb{R}^3 - \{\mathbf{0}\}$.

Since $\partial B \subseteq \mathbb{R}^3 \backslash \{\mathbf{0}\}$, the composition $F \circ \varphi$ gives a C^1 vector field $F \circ \varphi : B \to \partial B$. Moreover, since the restriction of φ to ∂B is the identity map and $F(\mathbf{x}) = \mathbf{x} = \mathbf{n}$ on ∂B, we have

$$\iint_{\partial B} (F \circ \varphi) \cdot d\mathbf{S} = \iint_{\partial B} F \cdot d\mathbf{S} =$$

$$\iint_{\partial B} \langle F, \mathbf{n} \rangle dS = \iint_{\partial B} \langle \mathbf{n}, \mathbf{n} \rangle dS = \iint_{\partial B} dS = 4\pi.$$

On the other hand, since $div F = 0$, the Divergence theorem gives

$$\iint_{\partial B} (F \circ \varphi) \cdot d\mathbf{S} = \iiint_B div(F \circ \varphi) dV = \iiint_B 0 = 0,$$

506 Chapter 6 Line and Surface Integrals

which is a contradiction. If f is merely continuous on B we use the Weierstrass approximation theorem to approximate f (uniformly) by C^1 functions f_k : $B \to B$ each having a fixed point x_k in B. Use compactness of B and uniform convergence to finish the proof. □

Miscellaneous Exercises

Exercise 6.10.8. Let C be a simple smooth curve in $\mathbb{R}^3$ parametrized by $\gamma(t)$ with $t \in [a,b]$ and $\rho(u) = (\gamma \circ s^{-1})(u)$ with $u \in [0,L]$ be its arc length parametrization. Show that the curvature is

$$\kappa(u) = ||\rho''(u)|| = \frac{||\gamma'(t) \times \gamma''(t)||}{||\gamma'(t)||^3}.$$

Find the curvature of the helix.

Exercise 6.10.9. Let the vector field $F(x,y) = (\frac{y}{x^2+y^2}, \frac{-x}{x^2+y^2})$ be defined in the plane minus the origin.

1. Use Green's theorem to show for any $\Omega \subseteq \mathbb{R}^2 - \{(0,0)\}$ that $\int_{\partial\Omega} F_1 dx + F_2 dy = 0$.

2. Let C any piecewise smooth closed curve in $\mathbb{R}^2 - \{(0,0)\}$ with a counter-clockwise orientation containing $(0,0)$ in its interior. If C goes around once, calculate $\int_C F \cdot d\mathbf{s}$.

3. What happens if C is a piecewise smooth closed curve in $\mathbb{R}^2 - \{(0,0)\}$ with a counterclockwise orientation, but $(0,0)$ is not in its interior?

Exercise 6.10.10. A particle of weight W descends from $(0,2,0)$ to $(4,0,0)$ along the parabola $8y = (x-4)^2$, $z = 0$. It is acted on both by gravity and a horizontal force $(y,0,0)$. Find the total work done by these forces. (Ans. $2W + \frac{8}{3}$).

Exercise 6.10.11. Let S be a parallelogram in $\mathbb{R}^3$ not parallel to any of the coordinate planes. Let S_{xy}, S_{xz}, and S_{yz} be the areas of the projections of S on the three coordinate planes respectively. Show that the area of S is $\sqrt{S_{xy}^2 + S_{xz}^2 + S_{yz}^2}$.

Exercise 6.10.12. Show that the area of the surface of the sphere $x^2 + y^2 + z^2 = a^2$ cut off by the cylinder $x^2 + y^2 = ax$ is $2(\pi - 2)a^2$.

Exercise 6.10.13. Find the surface area of the ellipsoid $\frac{x^2}{a^2} + \frac{y^2}{b^2} + \frac{z^2}{c^2} = 1$.

Exercise 6.10.14. Show that a curve lying on the unit sphere given by $\gamma(t) = (\theta(t), \phi(t))$ for $t \in [a, b]$ has length

$$L(\gamma) = \int_a^b \sqrt{(\sin^2 \phi)(\theta'(t))^2 + (\phi'(t))^2} \, dt.$$

Exercise 6.10.15. Let C be the curve in $\mathbb{R}^3$ parametrized by

$$\gamma(t) = (1 + \cos t, \sin t, 14 - 2\cos t), \text{ where } t \in [0, 2\pi].$$

1. Compute the line integral $\int_C x dx + z dy - y dy$ directly.

2. Use Stoke's theorem to express the integral in part (a) as a surface integral over a cupping surface S of C and evaluate the integral. (Ans. -4π).

Exercise 6.10.16. Let $F(x, y, z) = (xyz, y^2 + 1, z^3)$ and let S be the unit cube $S = [0, 1] \times [0, 1] \times [0, 1]$. Compute $\int \int_S (\nabla \times F) \cdot \mathbf{n} dS$ using

1. Stokes' theorem

2. the Divergence theorem

3. direct calculation

Exercise 6.10.17. Let $\mathbf{x} = (x, y, z) \in \mathbb{R}^3$. Compute

$$\int \int_S \frac{\mathbf{x}}{||\mathbf{x}||^2} \cdot \mathbf{n} dS,$$

where S is the sphere $x^2 + y^2 + z^2 = 4$. (Ans. 8π). Can you use the Divergence theorem?

Exercise 6.10.18. Let $S = \left\{ \mathbf{x} \in \mathbb{R}^3 : ||\mathbf{x}|| = r \right\}$ and $\mathbf{a} \notin S$. Show that

$$\int \int_S \frac{1}{||\mathbf{x} - \mathbf{a}||} dS = \begin{cases} 4\pi r & \text{if } \mathbf{a} \text{ is inside } S, \\ \frac{4\pi r^2}{||\mathbf{a}||} & \text{if } \mathbf{a} \text{ is outside } S. \end{cases}$$

Exercise 6.10.19. (*The mean value theorem for surface integrals*). Let F be a continuous vector field whose domain contains a smooth connected surface S. Prove that there exist $\mathbf{a} \in S$ such that

$$\frac{1}{A(S)} \int \int_S (F \cdot \mathbf{n}) dS = F(\mathbf{a}) \cdot \mathbf{n}(\mathbf{a}),$$

where $A(S)$ is the area of S.

Exercise 6.10.20. (*The change of variables formula in $\mathbb{R}^3$*). Suppose Ω and V are simple regions in $\mathbb{R}^3$ for which the divergence theorem holds and $\varphi : \Omega \to \mathbb{R}^3$ a C^2 change of variables given by $\varphi(u, v, w) = (x(u, v, w), y(u, v, w), z(u, v, w))$ with $\varphi(\Omega) = V$. Let f be a continuous function on V. Show that

$$\int \int \int_V f(x, y, z)dxdydz = \int \int \int_\Omega f(\varphi(u, v, w)) \left| \frac{\partial(x, y, z)}{\partial(u, v, w)} \right| dudvdw.$$

Hint. If $\Phi(s, t) = (u(s, t), v(s, t), w(s, t))$ is a parametrization of $\partial\Omega$, then $\varphi \circ \Phi$ is a parametrization of ∂V. Consider the function $G(x, y, z) = \int_0^z f(x, y, \xi)d\xi$. Note $\frac{\partial G}{\partial z} = f$ in $\mathbb{R}^3$. Apply the Divergence theorem twice; first in V and in the sequel in Ω.

Exercise 6.10.21.   1. Show that if f is harmonic $\nabla \cdot (f\nabla f) = ||\nabla f||^2$.

2. Let $f(x, y, z) = 3x + 2y + 4z$. Evaluate

$$\int \int_S f \frac{\partial f}{\partial \mathbf{n}} dS,$$

where is the sphere $x^2 + y^2 + z^2 = 4$. (Ans. $\frac{928}{3}\pi$).

Exercise 6.10.22. (*) (*Green's third identity*) Let f be a C^2 function on a set $\Omega \subseteq \mathbb{R}^3$ containing a ball B and let $\mathbf{a} \in B$. Use Green's second identity with $g = \frac{1}{r}$, where $r = ||\mathbf{x} - \mathbf{a}||$ ($\mathbf{x} \in B$) to show

$$f(\mathbf{a}) = -\frac{1}{4\pi} \int \int \int_B \frac{\Delta f}{r} dV + \frac{1}{4\pi} \int \int_{\partial B} \left[\frac{\nabla f}{r} - f\nabla \left(\frac{1}{r} \right) \right] \cdot \mathbf{n}dS.$$

Hint. Consider a small ball B_ϵ centered at $\mathbf{a}$ of radius $\epsilon > 0$. Set $D_\epsilon = B \backslash B_\epsilon$. Apply the Divergence theorem in D_ϵ and let $\epsilon \to 0$.

Chapter 7

Elements of Ordinary and Partial Differential Equations

In this chapter we present basic results on first, second, n-th order ordinary differential equations and linear systems of first order equations. Partial differential equations are also discussed. In particular, the classical second order equations of mathematical physics.

7.1 Introduction

In many problems of Geometry, Physics, Chemistry, Engineering, Economics and other disciplines a significant role is played by differential equations. A *differential equation* is an equation involving an unknown function, say $y = y(x)$, of a *single real* variable together with certain of its derivatives. Sometimes such an equation is called an *ordinary differential equation* (abbreviated ODE) the term ordinary referring to the fact that we are dealing with functions of just one variable. By the *order* of such an equation is meant the order of the highest derivative appearing in the equation. A *solution* of a differential equation on a finite or infinite interval $(\alpha, \beta) \subseteq \mathbb{R}$ is any function $y = y(x)$ on the interval (α, β) that satisfies the equation identically on (α, β). For example, the function $y(x) = 2\sin x - 5\cos x$ is a solution of the equation $y'' + y = 0$

on the interval $(-\infty, +\infty)$. The graph of a solution is called an *integral curve* of the equation. To solve the differential equation means to find *all* solutions. The simplest problem of this type is that of finding the indefinite integral of a given continuous function $f(x)$, that is, to find a function $y = y(x)$ for $x \in (\alpha, \beta)$ which satisfies the differential equation $y' = f(x)$. The Fundamental Theorem of Calculus tells us that such a function is given by

$$\varphi(x) = \int_\alpha^x f(t)dt$$

for all $x \in (\alpha, \beta)$. Moreover, the set of all solutions is $y(x) = \varphi(x) + C$, where C is an arbitrary constant. As a second simple example the differential equation $y'' = 0$ has the general solution $y(x) = C_1 x + C_2$, where C_1, C_2 are arbitrary constants. We shall assume that the functions involved in the differential equation have a common domain $(\alpha, \beta) \subseteq \mathbb{R}$ and we seek a solution defined on all (α, β) which statisfies the equation identically there. Such a function is called a *global solution*. When the solution y satisfies the ODE in some subinterval of (α, β) we say it is a *local solution*. One final introductory remark: Since we are interested in all possible solutions to the given equation and solving such an equation involves integrating, which perforce involves introducing an arbitrary constant of integration, as a rule of thumb we should expect that the number of arbitrary constants needed to describe the set of all solutions will be the same as the order of the equation.

In Section 7.8 we shall consider some basic and quite important *partial differential equations*, particularly of the second order. Here too the unknown is a function, but this time of several real variables and the equation involves its partial derivatives. Here again, the order of the equation is the order of the highest derivative occuring. In a partial differential equation, rather than the constant of integration that occurs in ordinary differential equations, instead here there will be an arbitrary function of all the variables except the one with respect to which differentiation takes place. Here we are also interested in all solutions and one should expect the number of arbitrary functions in the general solution to be the same as the order.

7.2   First order differential equations

A first order ordinary differential equation is an equation connecting an independent variable x, a sought for function $y = y(x)$ and its derivative $y' = y'(x) = \frac{dy}{dx}$. That is, an equation of the form

$$F(x, y, y') = 0. \tag{7.1}$$

We shall assume that it is always possible to solve a given ordinary differential equation for the highest derivative, here for y', obtaining

$$y' = f(x, y), \tag{7.2}$$

where the function $f = f(x, y)$ is defined in some domain in $\mathbb{R}^2$ and has continuous partial derivatives. By the Implicit Function theorem, this is possible locally whenever the function $F : \mathbb{R}^3 \to \mathbb{R}$ has continuous partial derivatives near a point (x_0, y_0, z_0) at which $F(x_0, y_0, z_0) = 0$ and $\frac{\partial F}{\partial z}(x_0, y_0, z_0) \neq 0$. Unfortunately even here, for an arbitrary function f there is no general method for solving the equation in terms of elementary functions (polynomial, trigonometric, exponential, logarithmic, hyperbolic functions.) However, a comprehensive mathematical theory exists for ordinary first order differential equations which are called *linear equations.*

7.2.1   Linear first order ODE

An equation of the form

$$\frac{dy}{dx} + p(x)y = q(x), \tag{7.3}$$

where $p(x)$ and $q(x)$ are given (continuous) functions is called a *linear first order differential equation.* When $q(x) \equiv 0$, the equation

$$\frac{dy}{dx} + p(x)y = 0, \tag{7.4}$$

is called the (corresponding) *homogeneous linear equation.*

We will solve equation (7.3) using a method due to Leibniz[1]. The idea is to multiply the equation by certain function $a(x) > 0$ (called an *integrating factor*) so that the equation is readily integrable. That is, we look for a function $a(x)$ with the property

$$a(x)[y' + p(x)y] = [a(x)y]'.$$

Multiplying out and using the product rule, this gives

$$a(x)p(x)y = a'(x)y.$$

Assuming $y(x) \neq 0$, we divide by $y(x)$ to get $p(x) = \frac{a'(x)}{a(x)}$. Now integrating we obtain

$$\int p(x)dx = \log(a(x)).$$

Hence the integrating factor is the form

$$a(x) = e^{\int p(x)dx}.$$

For a continuous function $p(x)$ we define $P(x) = \int p(x)dx$ to be any function whose derivative is $p(x)$, that is, $P'(x) = p(x)$, where we suppress the constant of integration. In general, the existence of such a function is guaranteed by the Fundamental Theorem of Calculus. However, in simple cases it is not necessary to appeal to general theorems, in fact it is *not even desirable since we are interested in explicit solutions.* For example, suppose $p(x) = \lambda$, a constant. Then $P(x) = \int p(x)dx = \lambda x$.

Theorem 7.2.1. *Suppose p and q are continuous functions on an interval (α, β). Let P be a function such that $P'(x) = p(x)$. Then the set of all solutions on (α, β) of the equation*

$$y' + p(x)y = q(x)$$

consists of functions of the form

$$y(x) = e^{-P(x)} \int q(x)e^{P(x)}dx + Ce^{-P(x)},$$

where C is an arbitrary constant.

---

[1]G. Leibniz (1646-1716). Mathematician and philosopher. To Liebniz belongs the honor, along with Newton, of having discovered the foundations of the infinitesimal calculus.

Proof. Let $a(x) = e^{\int p(x)dx} = e^{P(x)}$ be the integrating factor of the equation. Multiplying both sides of the equation $y' + p(x)y = q(x)$ by $a(x)$ yields $a(x)[y' + p(x)y] = a(x)q(x)$. That is,

$$\frac{d}{dx}[ye^{P(x)}] = q(x)e^{P(x)}.$$

Integrating we get $ye^{P(x)} = \int q(x)e^{P(x)}dx + C$, where C is a constant. Finally, since the exponential function is never zero, multiplying both sides of this last equation by $e^{-P(x)}$ yields

$$y = e^{-P(x)}\int q(x)e^{P(x)}dx + Ce^{-P(x)}.$$

Clearly all steps of our argument are reversible. Hence this shows that the set of all solutions of (7.3) consists of those of the form above.   □

Example 7.2.2. Solve $y' + 2xy = x^3 e^{-x^2}$.

Solution. In this case $p(x) = 2x$, $q(x) = x^3 e^{-x^2}$ are both continuous on $(-\infty, \infty)$. Here the integrating factor is $a(x) = e^{\int 2xdx} = e^{x^2}$. Hence

$$y'e^{x^2} + 2xe^{x^2}y = x^3,$$

$$\frac{d}{dx}[ye^{x^2}] = x^3,$$

and an integration gives

$$ye^{x^2} = \frac{x^4}{4} + C,$$

or

$$y(x) = \frac{x^4}{4}e^{-x^2} + Ce^{-x^2}.$$

Example 7.2.3. Solve $y' + \frac{1}{x}y = \frac{1}{x^2}$, $x > 0$.

Solution. Here $p(x) = \frac{1}{x}$, $q(x) = \frac{1}{x^2}$ are both continuous on $(0, \infty)$. The integrating factor is $a(x) = e^{\int \frac{1}{x}dx} = e^{\log x} = x$. So

$$x[y' + \frac{1}{x}y] = x\frac{1}{x^2},$$

$$xy' + y = \frac{1}{x},$$

or

$$\frac{d}{dx}[xy] = \frac{1}{x}.$$

Hence, integrating and solving for y we obtain

$$y(x) = \frac{\log x}{x} + \frac{C}{x}.$$

Let us see what Theorem 7.2.1 amounts to in some important special cases.

Special cases of the first order equation

1. Suppose, for example, that $p(x) = -\lambda$, a constant and $q(x) = 0$. Then the differential equation is simply

$$\frac{dy}{dx} = \lambda y.$$

This is the differential equation of *bacterial growth* $(\lambda > 0)$ or *radioactive decay* $(\lambda < 0)$. Of course, if $\lambda = 0$, then the equation is $\frac{dy}{dx} = 0$ which has as solutions only the constant functions. Notice this case is also covered by our general reasoning, according to which, the solutions are all functions of the form

$$y(x) = Ce^{\lambda x}.$$

2. Now suppose that $p(x) = -\lambda$ and $q(x) = \mu$, both nonzero constants. Thus we are solving the differential equation

$$\frac{dy}{dx} = \lambda y + \mu,$$

where λ and μ are constants. Clearly, if $\lambda = 0$, then the general solution is easily seen to be the family of linear functions $y(x) = \mu x + C$. Thus we may assume that $\lambda \neq 0$. After performing the requisite integration in the general solution, taking into account that $\lambda \neq 0$, we get as the set of all solutions

$$y(x) = -\frac{\mu}{\lambda} + Ce^{\lambda x}.$$

Of course, $\mu = 0$ we get the solutions $y = Ce^{\lambda x}$ of the *decay-growth equation* $y' = \lambda y$.

3. Finally, a special case which will play an important role in the next section is the differential equation

$$\frac{dy}{dx} = \lambda y + Ae^{\mu x},$$

where λ, μ and A are constants. Here $p(x) = -\lambda$ and $q(x) = Ae^{\mu x}$. Hence the general solution is $y(x) = Ae^{\lambda x} \int e^{(\mu-\lambda)x} dx + Ce^{\lambda x}$. It will now be important to distinguish the two cases, namely $\lambda = \mu$ and $\lambda \neq \mu$. For if $\lambda = \mu$, the general solution is

$$y(x) = e^{\lambda x}(Ax + C),$$

where C is a constant. Whereas if $\lambda \neq \mu$, the general solution is

$$y(x) = \frac{A}{\mu - \lambda}e^{\mu x} + Ce^{\lambda x}.$$

Frequently we are not interested in *all* solutions of an equation, but only those satisfying certain conditions. These conditions may take various forms. Two of the most important types are *initial conditions* and *boundary conditions*. An initial condition is a condition on the solution at one point, such as, $y(0) = 0$ or more generally $y(a) = b$ with $a, b \in \mathbb{R}$. A differential equation together with an initial condition is called an *initial-value problem* (or *Cauchy's problem*). In Section 7.3 we shall prove Picard's theorem on the local existence and uniqueness of the solution of an initial value problem for a first order differential equation. A boundary condition is a condition on the solution at two or more points. A differential equation together with a boundary condition is called *boundary-value problem*. Next we give two examples of initial-value problems.

Example 7.2.4. Solve the initial-value problem

$$y'(x) = -xy(x), \quad y(0) = 1.$$

Solution. The equation is $y'(x) + xy(x) = 0$, a first order linear equation with $p(x) = x$ and $q(x) = 0$. Hence, by Theorem 7.2.1, the general solution is $y(x) = Ce^{-P(x)}$, where $P(x) = \int x\,dx = \frac{x^2}{2}$. Therefore,

$y(x) = Ce^{-\frac{x^2}{2}}$. Now, using the initial condition $y(0) = 1$, we get $C = 1$. Thus,

$$y(x) = e^{-\frac{x^2}{2}}$$

is the unique solution of the problem.

Example 7.2.5. (*falling body with air resistance*). We now apply the solution of the first order linear equation to the situation of retarded fall, such as happens to a parachutist.

Solution. Here we consider a body falling along the vertical y-axis and we take into account the effect of the air resistance acting on the body. In this case, there is a retarding force proportional to the velocity $v = \frac{dy}{dt}$ of the form $-kv$. The acceleration is, of course, $a = \frac{dv}{dt}$. Thus Newton's law of motion $F = ma$ yields in this case

$$mg - kv = m\frac{dv}{dt},$$

where m is the mass of the falling body, $v(t)$ is its vertical velocity at time t, g is the constant of gravity, and k is a positive constant depending on the size of the parachute and perhaps its material. Dividing by m the equation becomes

$$\frac{dv}{dt} = -\frac{k}{m}v + g.$$

Thus here $\mu = g$ and λ is the negative number $-\frac{k}{m}$, so the general solution is $v(t) = \frac{W}{k} + Ce^{-\frac{k}{m}t}$, where $W = mg$ is the weight of the falling body. For example, if the parachutist leaves the plane from rest, i.e., with vertical velocity zero, then we can use this "initial condition" to determine C and thus get the *unique* solution to our problem; $v(0) = 0$ so $C = -\frac{W}{k}$ and the solution is

$$v(t) = \frac{W}{k}(1 - e^{-\frac{k}{m}t}).$$

This equation reveals an important property of the motion. The velocity does not increase with time beyond all bounds, but tends to a finite limit depending on the mass m. For $v(t) < \frac{W}{k}$ and letting $t \to \infty$ tells us that

$$v(t) \to \frac{W}{k}.$$

This number is called the terminal velocity and its existence is the reason that these people don't crash and die. Notice that, neglecting the weight of the parachute itself, a person of double the weight will have a terminal velocity double that of the lighter person. It is very typical in this subject for initial data to uniquely determine the solution to a problem.

Finally, note that integrating the expression for $v(t) = y'(t)$ yields

$$y(t) = \frac{mW}{k^2} e^{-\frac{k}{m}t} + \frac{W}{k}t + c_1,$$

where c_1 is a new constant of integration, which can be determined by knowing the initial height (at time $t = 0$) of the plane.

In the sequel, we consider a few other types of (nonlinear) differential equations of the first order which can be solved by integration (although in most cases the integration can not be performed explicitly in terms of elementary functions).

7.2.2   Equations with variables separated

If the differential equation $y' = f(x, y)$ can be written in the form

$$y' = \frac{a(x)}{b(y)},$$

where a, b are functions of a single argument, then the equation is said to have the *variables separated.* In this case we may write the equation as

$$b(y)\tfrac{dy}{dx} = a(x) \quad \text{or} \quad b(y)dy = a(x)dx,$$

that is, the variables are separated.

Theorem 7.2.6. *Let*
$$b(y)\frac{dy}{dx} = a(x) \tag{7.5}$$

with a, b continuous real-valued functions defined for $x \in [\alpha_1, \beta_1]$ and $y \in [\alpha_2, \beta_2]$ respectively. Then any solution $y = \varphi(x)$ of (7.5) satisfies the equation
$$B(y) = A(x) + C, \tag{7.6}$$

for some constant C, where $A(x) = \int a(x)dx$ and $B(y) = \int b(y)dy$.

Proof. Let $A(x) = \int a(x)dx$ and $B(y) = \int b(y)dy$ be any two functions such that $A'(x) = a(x)$ and $B'(y) = b(y)$. Then (7.5) becomes

$$B'(y)\frac{dy}{dx} = A'(x).$$

By the Chain Rule this gives

$$\frac{d}{dx}B(y) = \frac{d}{dx}A(x)$$

or

$$\frac{d}{dx}[B(y) - A(x)] = 0.$$

Therefore $B(y) - A(x) = C$, where C is a constant.   □

In practice the usual way of dealing with (7.5) is to write it as

$$b(y)dy - a(x)dx,$$

and then integrate

$$\int b(y)dy = \int a(x)dx + C,$$

where C is a constant, and thus obtain (7.6) which gives the solutions of the equation implicitly.

Example 7.2.7. Solve $y' = y^2$.

Solution. The equation is $\frac{dy}{dx} = y^2$. Separating the variables we have

$$\frac{1}{y^2}dy = dx.$$

Here $a(x) = 1$ and $b(y) = \frac{1}{y^2}$ (which is continuous for $y \neq 0$). Integrating we get $-\frac{1}{y} = x + C$. Hence $y = y(x) = -\frac{1}{x+C}$ is the solution[2], provided $x \neq -C$.

---

[2]Note that $y(x) \equiv 0$ is a particular solution of the equation. However, no constant C above will yield this solution. In other word, the separation of variables method of finding solutions may not yield *all* solutions of an equation. Attention to the possibilities of dividing by zero will often alert us to missing solutions. Note also that the equation $y' = y^2$ is *not* linear.

Example 7.2.8. Solve the initial-value problem

$$y' = y^5 \sin x, \quad y(0) = \tfrac{1}{2}.$$

Solution. Separating the variables, we have

$$\frac{dy}{y^5} = \sin x \, dx.$$

Integrating we get $-\tfrac{1}{4}y^{-4} = -\cos x + C$, or $y^4 = \frac{1}{4\cos x + c}$, where $c = 4C$. Using the initial condition $y(0) = \tfrac{1}{2}$, we find $c = 12$. Hence $y(x) = \pm \left(\frac{1}{4\cos x + 12} \right)^{\frac{1}{4}}$. Since $y(0) = \tfrac{1}{2} > 0$, the solution is $y(x) = \left(\frac{1}{4\cos x + 12} \right)^{\frac{1}{4}}$ for $x \in \mathbb{R}$.

Example 7.2.9. Solve the equation $2y' = x\sqrt{1 + y'^2}$

Solution. The variables can not be separated in the equation as it stands. We set[3] $y' = \tan\theta$ with $\theta \in (-\tfrac{\pi}{2}, \tfrac{\pi}{2})$. Then $x = \frac{2\tan\theta}{\sqrt{1+\tan^2\theta}} = 2\sin\theta$ and so $dx = 2\cos\theta d\theta$. Therefore

$$dy = \tan\theta dx = 2\sin\theta d\theta.$$

Integrating we get

$$y = -2\cos\theta + C.$$

From the system $x = 2\sin\theta$ and $y = -2\cos\theta + C$, we obtain

$$x^2 + (y - C)^2 = 4.$$

This equation defines the solutions of the differential equation implicitly. This is a family of circles centered at points $(0, C)$ of radius 2.

7.2.3   Homogeneous equations

Recall from Section 3.3 that a function $M(x, y)$ is called *homogeneous of degree* α if for any x, y and $t \in \mathbb{R}$ the following equality holds

$$M(tx, ty) = t^\alpha M(x, y).$$

---

[3]One could also set $y' = z$ and solving for z to get $y' = z = \pm\frac{x}{\sqrt{4-x^2}}$. Integrating $y = \pm\sqrt{4 - x^2} + C$ and so $x^2 + (y - C)^2 = 4$.

The differential equation

$$M(x,y)dx + N(x,y)dy = 0$$

is said to be *homogeneous* if M and N are homogeneous functions of the *same* degree. This equation can then be written in the form

$$\frac{dy}{dx} = f\left(\frac{y}{x}\right), \tag{7.7}$$

where f is a function of one variable. Indeed,

$$\frac{dy}{dx} = -\frac{M(x,y)}{N(x,y)} = -\frac{M(x, x\frac{y}{x})}{N(x, x\frac{y}{x}))} = -\frac{x^\alpha M(1, \frac{y}{x})}{x^\alpha N(1, \frac{y}{x}))} = f\left(\frac{y}{x}\right).$$

Now by means of the substitution $z = \frac{y}{x}$ the equation can *always* be reduced to an equation with separable variables.

In fact, since $y = zx$ and $\frac{dy}{dx} = z + x\frac{dz}{dx}$ equation (7.7) becomes

$$z + x\frac{dz}{dx} = f(z),$$

and separating the variables we obtain

$$\frac{dz}{f(z) - z} = \frac{dx}{x}.$$

Consequently $\int \frac{dz}{f(z)-z} = \log|x| + C$, and we complete the solution by integrating the first side and replacing z by $\frac{y}{x}$.

Example 7.2.10. Solve $ydx + (2\sqrt{xy} - x)dy = 0$, $x > 0$.

Solution. Here $M(x,y) = y$ and $N(x,y) = 2\sqrt{xy} - x$ are both homogeneous (the second for $t \geq 0$) of degree 1. We write the equation in the form

$$y + (2\sqrt{xy} - x)\frac{dy}{dx} = 0.$$

Letting $y = zx$, we have

$$zx + x(2z^{\frac{1}{2}} - 1)\left(z + x\frac{dz}{dx}\right) = 0$$

or

$$2z^{\frac{3}{2}} + x(2z^{\frac{1}{2}} - 1)\frac{dz}{dx} = 0,$$

and separating the variables we get

$$-\frac{(2z^{\frac{1}{2}} - 1)dz}{2z^{\frac{3}{2}}} = \frac{dx}{x}.$$

Integrating, we have

$$-\log z - z^{-\frac{1}{2}} = \log x + C$$

or

$$\log y + \sqrt{\frac{x}{y}} = C.$$

7.2.4   Exact equations

A differential equation of the form $M(x,y) + N(x,y)y' = 0$ or equivalently

$$M(x,y)dx + N(x,y)dy = 0$$

is said to be *exact* in a rectangle $R \subset \mathbb{R}^2$ if there exists a function $u = u(x,y)$ having continuous first partial derivatives there such that

$$\frac{\partial u}{\partial x} = M, \quad \frac{\partial u}{\partial y} = N$$

in R. Then the equation can be written in the form

$$\frac{\partial u}{\partial x}dx + \frac{\partial u}{\partial y}dy = 0 \quad \text{or} \quad du = 0,$$

where $du = \frac{\partial u}{\partial x}dx + \frac{\partial u}{\partial y}dy$ is the so-called differential of u. The next result characterizes the exact equations.

Theorem 7.2.11. *Let $M(x,y)$ and $N(x,y)$ have continuous first partial derivatives on a recatngle $R \subset \mathbb{R}^2$. The differential equation*

$$M(x,y)dx + N(x,y)dy = 0 \tag{7.8}$$

is exact in R if and only if

$$\frac{\partial M}{\partial y} = \frac{\partial N}{\partial x} \tag{7.9}$$

in R. Furthermore, every differentiable function $y = y(x)$ defined implicitly by $u(x,y) = C$, where C is a constant, is a solution of (7.8).

Proof. Suppose equation (7.8) is exact in R. Then there exists a function $u(x, y)$ on R such that $\frac{\partial u}{\partial x} = M$, $\frac{\partial u}{\partial y} = N$. Then $\frac{\partial M}{\partial y} = \frac{\partial^2 u}{\partial y \partial x} = \frac{\partial^2 u}{\partial x \partial y} = \frac{\partial N}{\partial x}$.

Conversely, suppose $\frac{\partial M}{\partial y} = \frac{\partial N}{\partial x}$ in R. Then from Poincare's lemma (Theorem 6.2.8), the vector field $F = (M, N)$ is conservative, ie , there exist a C^1 function $u = u(x, y)$ defined on R such that $F = \nabla u$. Hence equation (7.8) is exact.

Finally note that if $y = y(x)$ satisfies $M(x, y) + N(x, y)y' = 0$. Then $\frac{\partial u}{\partial x} + \frac{\partial u}{\partial y}\frac{dy}{dx} = 0$ or $\frac{d}{dx}u(x, y) = 0$ and so $u(x, y) = C$ for some constant C. $\square$

Remark 7.2.12. Note that *any* equation with variables separated is a special case of an exact equation. Indeed, given the equation

$$a(x)dx - b(y)dy = 0,$$

we can take $u(x, y) = A(x) - B(y)$, where $A' = a$ and $B' = b$.

Example 7.2.13. Solve $y' = -\frac{3x^2 + 6xy^2}{6x^2y + 4y^3}$.

Solution. We write the equation as $(3x^2 + 6xy^2)dx + (6x^2y + 4y^3)dy = 0$. Here $M(x, y) = 3x^2 + 6xy^2$ and $N(x, y) = 6x^2y + 4y^3$ and so $\frac{\partial M}{\partial y} = \frac{\partial N}{\partial x} = 12xy$. Hence our equation is exact. Theorem 7.2.11 tells us that there exists a function $u(x, y)$ such that

$$\frac{\partial u}{\partial x} = 3x^2 + 6xy^2, \quad \frac{\partial u}{\partial y} = 6x^2y + 4y^3.$$

Fixing y and integrating the first of these equations with respect to x yields $u(x, y) = x^3 + 3x^2y^2 + f(y)$, where f is independent of x. Now $\frac{\partial u}{\partial y} = N$ tells us $6x^2y + f'(y) = 6x^2y + 4y^3$, or that $f'(y) = 4y^3$. So $f(y) = y^4$ and therefore $u(x, y) = x^3 + 3x^2y^2 + y^4$. Thus the general solution of our differential equation is given implicitly by

$$x^3 + 3x^2y^2 + y^4 = C. \tag{7.10}$$

Note that at a given point (a, b) satisfying $u(a, b) = C$ and $\frac{\partial u}{\partial y}(a, b) \neq 0$ the Implicit Function theorem tells us that equation (7.10) will define a unique differentiable function φ in a neighborhood of the point (a, b). Notice also that the only points (x, y) satisfying (7.10) and $\frac{\partial u}{\partial y}(x, y) = 0$

are those for which $6x^2y + 4y^3 = 0$ and these are precisely the points for which our differential equation is not defined.

EXERCISES

1. Solve the linear differential equations.

   (a) $\frac{dy}{dx} + xy = x$.

   (b) $xy' - 3y = x^4$.

   (c) $y' + 2xy = xe^{-x^2}$.

   (d) $y' + (\cos x)y = e^{-\sin x}$.

   (e) $y' - (\tan x)y = e^{\sin x}$.

   (f) $(x \log x)y' + y = 3x^3$, $x > 0$.

   (g) $xy' - 2y = x^3 \cos x$.

2. Solve the differential equations.

   (a) $\frac{dy}{dx} - xy = x^3y^2$.

   (b) $\frac{dy}{dx} = \frac{1+y^2}{x}$.

   (c) $y' = \frac{x^2(y+1)}{y^2(1-x)}$.

   (d) $2x\sqrt{1-y^2} = y'(1+x^2)$.

   (e) $y' = \frac{e^{x-y}}{1+e^x}$.

   (f) $y' \sin x = y \log y = 0$, $y(\frac{\pi}{2}) = e$.

   (g) $(1 + e^x)yy' = e^x$, $y(0) = 1$.

3. Verify that the following equations are homogeneous and solve them:

   (a) $y' = \frac{xy}{x^2+y^2}$.

   (b) $x^2y' = 3(x^2 + y^2)\tan^{-1}(\frac{y}{x}) + xy$.

   (c) $y' = \frac{y + xe^{-\frac{2y}{x}}}{x}$.

   (d) $(x^3 + y^3)dx - xy^2dy = 0$.

   (e) $x\sin(\frac{y}{x})\frac{dy}{dx} = y\sin(\frac{y}{x}) + x$.

4. Determine which of the following equations are exact and solve these.

(a) $y' = -\frac{2xy+y}{x^2+x-1}$.

(b) $(2xy^4 + \sin y)dx + (4xy^3 + x \cos y)dy = 0$.

(c) $(x + \frac{2}{y})y' + y = 0$.

(d) $(e^y + \cos x \cos y)dx = (\sin x \sin y - xe^y)dy$.

(e) $(x \log y + xy)dx + (y \log x + xy)dy = 0$.

(f) $y' = \frac{2y \cos^2 x}{1+y^2 \sin 2x}$.

5. Solve the differential equations:

(a) $y' = (x + y)^2$.   *Hint.* Set $z = x + y$.

(b) $y' = \sin^2(x - y + 1)$.   *Hint.* Set $z = x - y + 1$.

(c) $y' = \frac{y-xy^2}{x+x^2y}$.   *Hint.* Reduce it to an equation with the variables separated, by setting $z = \frac{y}{x^n}$ for a convenient value of n.

(d) Let $k \neq 1$ be a constant. The equation $y' + a(x)y = b(x)y^k$, is called *Bernoulli's equation*. Show that the substitution $z = y^{1-k}$ transforms this into the linear equation

$$z' + (1 - k)a(x)z = (1 - k)b(x).$$

Apply this method to solve

$$y' - 2xy = xy^2.$$

6. By Newton's law the cooling rate of a body in air is proportional to the difference between the temperature T of the body and the air temperature T_0. If the air temperature is $20°C$ and in 20 minutes the body cools from $100°$ to $60°$, in what time will its temperature drop to $30°$?

7. The following equations illustrate one of the differences between linear and nonlinear equations.

(a) Find the solution ψ of the *linear* equation $y' = y+1$ satisfying $\psi(0) = 0$. Observe that this solution exists for all $x \in \mathbb{R}$.

(b) Find the (real) solution φ of the *nonlinear* equation $y' = y^2 + 1$ satisfying $\varphi(0) = 0$. Observe that this solution exist only for $x \in (-\frac{\pi}{2}, \frac{\pi}{2})$. *Hint.* For any t for which a φ exists we must have $\frac{\varphi'(t)}{1+\varphi(t)^2} = (\tan^{-1}\varphi)'(t) = 1$. Integrate from 0 to x to obtain $\varphi(x) = \tan x$. Check that this φ is the solution.

Answers to selected Exercises

1. (a) $y = 1 + Ce^{-\frac{x^2}{2}}$. (c) $y = \frac{1}{2}x^2e^{-x^2} + Ce^{-x^2}$.
   (e) $y = (\sec x)e^{\sin x} + C\sec x$. (f) $y = \frac{x^3+C}{\log x}$.
2. (b) $y = \tan(\log Cx)$. (d) $y = \sin[C + \log(1 + x^2)]$. (f) $y = e^{\tan(\frac{x}{2})}$.
   (g) $y = \sqrt{1 + \log(\frac{1+e^x}{2})^2}$.
3. (b) $y = x\tan Cx^3$. (d) $y = x^3 \log Cx^3$. (e) $\cos(\frac{y}{x}) + \log Cx = 0$.
4. (b) $x^2y^4 + x\sin y = C$. (c) $xy + \log y^2 + C$. (d) $xe^y + \sin x\sin y = C$.
   (e) not exact.
6. $\frac{dT}{dt} = k(T - T_0)$; $T = 20 + 80\left(\frac{1}{2}\right)^{\frac{t}{20}}$; $t = 60$ minutes.

7.3   Picard's theorem (*)

In this section, we consider the problem of existence of the solution of the *general* first order ODE

$$y' = f(x, y), \qquad\qquad (7.11)$$

where $f : \Omega \to \mathbb{R}$ is any continuous function defined on some domain (open connected set) $\Omega \subseteq \mathbb{R}^2$. As before, we view y as the dependent variable and x as the independent variable. That is, $y = y(x)$ is the unknown function we seek. Our main purpuse is to prove that a wide class of equations of the form (7.11) have local solutions, and that solutions to such initial value problems are unique. In Section 7.2, we saw that the *linear equation* $y' = -p(x)y + q(x)$, where $p(x)$ and $q(x)$ are continuous functions on some interval $(\alpha, \beta) \subseteq \mathbb{R}$ has global solutions (Theorem 7.2.1). However the reader should be aware that only in rather special cases it is possible to find explicit analytic expressions for the solutions of (7.11).

Here we prove that any initial value problem for equation (7.11) always has a unique solution which can be obtained by an *approximation process*, provided the function f satisfies an additional condition, known as the *Lipschitz condition*. This result due to E. Picard is of fundamental importance and has applications in a great variety of other mathematical problems. A simple extension to systems of first order ODE can also be used to establish existence and uniqueness of initial value problems for n^{th} order ODE (See, Theorems 7.5.10 and 7.5.11).

Definition 7.3.1. Let $(\alpha, \beta) \subseteq \mathbb{R}$ be a given interval, $a \in (\alpha, \beta)$ and $y(x)$ continuous on (α, β). Given any fixed constant b, an *initial value problem* for equation (7.11) is

$$y' = f(x, y), \quad y(a) = b. \tag{7.12}$$

The key idea for solving the initial value problem (7.12) lies in replacing it by the equivalent *integral equation* in y

$$y(x) = b + \int_a^x f(t, y(t))dt, \tag{7.13}$$

where $x \in (\alpha, \beta)$ (called so, because the unknown function appears in the integrand).

Proposition 7.3.2. *A function φ is a solution of the initial value problem (7.12) on an inteval $(\alpha, \beta) \subseteq \mathbb{R}$ if and only if φ is a solution of the integral equation (7.13) on (α, β).*

Proof. If φ is a solution of the initial value problem (7.12), then for $t \in (\alpha, \beta)$

$$\varphi'(t) = f(t, \varphi(t)), \tag{7.14}$$

and $\varphi(a) = b$. Since φ is continuous on (α, β), and f is continuous on Ω, the function $f(t, \varphi(t))$ is continuous on (α, β). Hence, integrating (7.14) from a to x yields

$$\varphi(x) = b + \int_a^x f(t, \varphi(t))dt,$$

and thus φ is a solution of (7.13).

Conversely, if φ satisfies the integral equation (7.13), then $\varphi(a) = b$, and differentiating (7.13) we find, using the Fundamental Theorem of Calculus, that

$$\varphi'(x) \equiv f(x, \varphi(x)) \text{ on } (\alpha, \beta).$$

Thus, φ is a solution of the initial value problem (7.12). $\square$

Definition 7.3.3. A function $f(x, y)$ defined on a set $S \subseteq \mathbb{R}^2$ is said to satisfy a *Lipschitz condition* in y on S if there is a constant $M > 0$ such that

$$|f(x, y_1) - f(x, y_2)| \leq M|y_1 - y_2|, \qquad (7.15)$$

for all (x, y_1), (x, y_2) in S.

Remark 7.3.4. A sufficient condition for a function $f(x, y)$ to satisfy a Lipschitz condition on a closed rectangle S in $\mathbb{R}^2$ is the continuity of $\frac{\partial f}{\partial y}(x, y)$ on S, (in this connection see also Corollary 3.4.8). Indeed, from Corollary 2.5.3, there exist $M > 0$ such that

$$\left| \frac{\partial f}{\partial y}(x, y) \right| \leq M \qquad (7.16)$$

for all $(x, y) \in S$. At the same time, for (x, y_1), $(x, y_2) \in S$ the Mean Value theorem implies

$$|f(x, y_1) - f(x, y_2)| = \left| \frac{\partial f}{\partial y}(x, \xi) \right| |y_1 - y_2|,$$

where $y_1 < \xi < y_2$. Now since $(x, \xi) \in S$, using (7.16) we see that

$$|f(x, y_1) - f(x, y_2)| \leq M|y_1 - y_2|.$$

Thus, in the following theorem it would suffice to assume f and $\frac{\partial f}{\partial y}$ are continuous on Ω.

Theorem 7.3.5. *(Picard[4]).* *Let Ω be a domain in $\mathbb{R}^2$ and let $f : \Omega \to \mathbb{R}$ be continuous. Let $(a, b) \in \Omega$, and consider the initial value problem*

$$y' = f(x, y), \quad y(a) = b. \qquad (7.17)$$

Suppose f satisfies a Lipschitz condition in y on Ω. Then there exists a $\delta > 0$ and a unique solution $\varphi = \varphi(x)$ to the initial value problem, for all $|x - a| \leq \delta$.

---

[4]E. Picard (1856-1914), made important contributions in analysis, differential equations and topology. Also known for his book *Traite' d'Analyse*.

Proof. Since Ω is open and $v = (a, b) \in \Omega$, there is $r_1 > 0$ such that the open disk $B_{r_1}(v) \subseteq \Omega$. Choose $0 < r < r_1$ so that the closed disk $\overline{B}_r(v) \subseteq B_{r_1}(v)$. Since f is continuous on Ω and so on $\overline{B}_r(v)$, there exists $K > 0$ such that

$$|f(x, y)| \leq K$$

for $(x, y) \in \overline{B}_r(v)$. In addition, since f satisfies a Lipschitz condition in y on Ω and hence on $\overline{B}_r(v)$, there exists $M > 0$ such that

$$|f(x, y_1) - f(x, y_2)| \leq M|y_1 - y_2|,$$

for all (x, y_1), (x, y_2) in $\overline{B}_r(v)$. We now choose $0 < \delta < \min\left\{\frac{r}{K+1}, \frac{1}{M}\right\}$ such that $\delta < \frac{1}{M}$ and the rectangle $\{(x, y) : |x - a| \leq \delta, \ |y - b| \leq K\delta\} \subseteq \overline{B}_r(v)$.

Let $J = [a - \delta, a + \delta]$ and consider the metric space $C(J)$ of all continuous real functions $\psi : J \to \mathbb{R}$ with the metric

$$d(\phi, \psi) = \max_{x \in J} |\phi(x) - \psi(x)|.$$

Let

$$\mathcal{C} = \{\psi \in C(J) : |\psi(x) - b| \leq K\delta\}.$$

The set $\mathcal{C}$ is a nonempty ($\mathcal{C}$ contains the constant function $\psi(x) = b$) closed subset of the complete metric space $(C(J), d)$ (see Exercise 2.7.2), and by Proposition 1.6.10, it is itself a complete metric space. In addition, note that if $\psi \in \mathcal{C}$, then $|\psi(x) - b| \leq K\delta$ and $(x, \psi(x)) \in \Omega$, for all $x \in J$.

Now for each $\psi \in \mathcal{C}$ consider the mapping F given by

$$F(\psi)(x) = b + \int_a^x f(t, \psi(t))dt,$$

where $x \in J$. First note that since ψ and f are continuous, all $F(\psi)$ are coninuous on J. Moreover, the estimate

$$|F(\psi)(x) - b| = \left|\int_a^x f(t, \psi(t))dt\right| \leq \int_a^x |f(t, \psi(t))|dt \leq K \int_a^x dt \leq K\delta,$$

shows that $F(\psi) \in \mathcal{C}$.

Next we show that $F : \mathcal{C} \to \mathcal{C}$ is a contraction mapping of this space. Let $\psi_1, \psi_2 \in \mathcal{C}$. Then

$$d(F(\psi_1), F(\psi_2)) = \max_{x \in J} |F(\psi_1)(x) - F(\psi_2)(x)|$$

$$\leq \max_{x \in J} \left| \int_a^x |f(t, \psi_1(t)) - f(t, \psi_2(t))| dt \right|$$

$$\leq \max_{x \in J} \left| \int_a^x M |\psi_1(t) - \psi_2(t)| dt \right|$$

$$\leq \max_{x \in J} \left| M \int_a^x d(\psi_1, \psi_2) dt \right|$$

$$\leq M d(\psi_1, \psi_2) \max_{x \in J} |x - a|$$

$$= M \delta d(\psi_1, \psi_2).$$

Since $M\delta < 1$, F is a contraction mapping. Therefore, from the Contraction Mapping principle (Theorem 2.3.20), it follows that there exists a unique $\varphi \in \mathcal{C}$ such that $F(\varphi) = \varphi$. That is,

$$\varphi(x) = b + \int_a^x f(t, \varphi(t)) dt.$$

In other words, φ is a solution of the integral equation (7.13). It follows from Proposition 7.3.2 that φ is a solution to the initial value problem (7.17). Moreover, φ is the only solution of (7.17). For if ψ were another solution of (7.17), ψ would also satisfy the integral equation

$$\psi(x) = b + \int_a^x f(t, \psi(t)) dt.$$

That is, $F(\psi) = \psi$ which contradicts the uniqueness of φ (i.e., the uniqueness of the fixed point). $\qquad \square$

Notice that we obtain a *sequence of successive approximations* $\varphi_0, \varphi_1, \varphi_2, \dots$ to the solution φ, as a by-product of the proof of the existence of this solution. According to the proof of Theorem 2.3.20,

$$\varphi_0(x) = b,$$

$$\varphi_{k+1}(x) = F(\varphi_k)(x) = b + \int_a^x f(t, \varphi_k(t))dt, \qquad (7.18)$$

for $k = 0, 1, 2, \dots$ and on taking the limit as $k \to \infty$ we obtain

$$\varphi_k(x) \to \varphi(x)$$

the unique solution φ to the initial value problem (7.17).

Example 7.3.6. The equation

$$y' = -y^2 \qquad (7.19)$$

is a particular case of differential equation (7.11). In this case the function $f(x, y) = y^2$. Since the function f is continuous together with its partial derivative $\frac{\partial f}{\partial y} = -2y$ in the entire xy-plane, we can conclude on the basis of Picard's (existence) theorem, without solving equation (7.19), that through any point (a, b) there passes a unique local integral curve of equation (7.19). As in Example 7.2.7 we see that the general solution in the upper half-plane $(y > 0)$ and in the lower half-plane $(y < 0)$ is $y = \frac{1}{x-C}$, together with the trivial solution $y \equiv 0$. At the point $(3, 1)$, for example, this gives $y = \frac{1}{x-2}$.

Example 7.3.7. Let $f(x, y) = x+y$. Use Picard's method of successive approximations to solve the initial value problem.

$$y' = f(x, y), \quad y(0) = 2. \qquad (7.20)$$

Solution. First note that this is a first order linear equation $y' - y = x$, with $p(x) = -1$ and $q(x) = x$. From Theorem 7.2.1, the general solution is $\psi(x) = -x-1+Ce^x$. The initial condition gives $C = 3$ and the unique solution of the initial value problem is

$$\varphi(x) = 3e^x - x - 1.$$

Now let us see what the successive approximations give. The equivalent integral equation is

$$y(x) = 2 + \int_0^x f(t, y(t))dt = 2 + \int_0^x [t + y(t)]dt.$$

Hence (7.18) gives

$$\varphi_{k+1}(x) = 2 + \int_0^x [t + \varphi_k(t)]dt, \tag{7.21}$$

$(k = 0, 1, 2, ...)$. Starting with $\varphi_0(x) = 2$ and using (7.21), the sequence of successive approximations $\varphi_0, \varphi_1, \varphi_2, ...$, is

$$\varphi_1(x) = 2 + \int_0^x (t + 2)dt = 2 + 2x + \frac{x^2}{2!},$$

$$\varphi_2(x) = 2 + \int_0^x \left(2 + 3t + \frac{t^2}{2!}\right)dt = 2 + 2x + 3\frac{x^2}{2!} + \frac{x^3}{3!},$$

$$\varphi_3(x) = 2 + \int_0^x \left(2 + 3t + 3\frac{t^2}{2!} + \frac{t^3}{3!}\right)dt = 2 + 2x + 3\frac{x^2}{2!} + 3\frac{x^3}{3!} + \frac{x^4}{4!},$$

$$= -1 - x + 3\left(1 + x + \frac{x^2}{2!} + \frac{x^3}{3!}\right) + \frac{x^4}{4!},$$

and so on ...

$$\varphi_k(x) = -1 - x + 3\left(1 + x + \frac{x^2}{2!} + \frac{x^3}{3!} + \frac{x^4}{4!} + ... + \frac{x^k}{k!}\right) + \frac{x^{k+1}}{(k+1)!}.$$

Thus

$$\lim_{k \to \infty} \varphi_k(x) = -1 - x + 3e^x$$

is a global solution in this case because the equation is linear.

Exercise 7.3.8. Let $y' = 3xy$ and $y(0) = 1$. Convert this initial vulue problem to an integral equation and use the method of successive approximations to solve it. Then solve it using Theorem 7.2.1.

Exercise 7.3.9. Let $y' = -y^2$ and $y(3) = 1$. Using the method of successive approximations show that the unique solution of this problem is $\varphi(x) = \frac{1}{x-2}$.

7.4   Second order differential equations

The general form of a second order differential equation is

$$F(x, y, y', y'') = 0. \tag{7.22}$$

It is assumed that $F(x, y, z, w)$ is a given continuously differentiable function defined on some domain in $\mathbb{R}^4$. As before, we assume that equation (7.22) can be solved for y'' obtaining

$$y'' = f(x, y, y'), \tag{7.23}$$

where the function $f = f(x, y, z)$ is defined in some domain in $\mathbb{R}^3$ and has continuous partial derivatives.

An *initial value problem* for a second order equation is the problem of solving (7.23) subject to the initial conditions $y(a) = b_0$ and $y'(a) = b_1$. This is also called *Cauchy's problem* for second order differential equations. The following result is a special case of Theorem 7.5.11 which we prove in the next section.

Theorem 7.4.1. *Let Ω be a domain in $\mathbb{R}^3$ and $f : \Omega \to \mathbb{R}$ be continuous and have continuous partial derivatives $\frac{\partial f}{\partial y}$, $\frac{\partial f}{\partial z}$ in Ω. Let $(a, b_0, b_1) \in \Omega$. Then there exist an interval J, ($a \in J$), and a unique twice continuously differentiable function $\varphi = \varphi(x)$ defined on J, satisfying the equation*

$$y'' = f(x, y, y')$$

and the initial conditions

$$y(a) = b_0, \quad y'(a) = b_1.$$

An important class of second order equations is the class of *linear equations*. These have a rich and far-reaching theory. The general *second order linear differential equation with variable coefficients* is an equation of the form

$$a(x)y'' + b(x)y' + c(x)y = G(x),$$

where the coefficient functions $a(x)$, $b(x)$, $c(x)$ and $G(x)$ are real-valued on some interval $(\alpha, \beta) \subseteq \mathbb{R}$. Points where $a(x) = 0$ are called *singular points*, and often the equation requires special consideration at such points. We shall assume that $a(x) \neq 0$ everywhere on (α, β) and we divide by it to obtain

$$y'' + p(x)y' + q(x)y = g(x),$$

where $p(x) = \frac{b(x)}{a(x)}$, $q(x) = \frac{c(x)}{a(x)}$ and $g(x) = \frac{G(x)}{a(x)}$.

Theorem 7.4.2. *Consider the linear equation*

$$y'' + p(x)y' + q(x)y = g(x), \qquad (7.24)$$

where the functions p, q, g are continuous on an interval (α, β). Let a, b_0, b_1 be fixed constants with $a \in (\alpha, \beta)$. Then there exist an interval $J \subset (\alpha, \beta)$, $(a \in J)$, and a unique twice continuously differentiable function $\varphi = \varphi(x)$ defined on the interval J, satisfying equation (7.24) and the initial conditions

$$y(a) = b_0, \ \ y'(a) = b_1.$$

Proof. Here $f(x, y, z) = -p(x)y - q(x)z + g(x)$, and so $\frac{\partial f}{\partial y} = -p(x)$, $\frac{\partial f}{\partial z} = -q(x)$. Since the coefficient functions p, q are continuous, the result follows from Theorem 7.4.1.   □

It will be convenient to denote the differential expression on the left of equation (7.24) by $L(y)$.
That is,

$$L(y) = \left(\frac{d^2}{dx^2} + p(x)\frac{d}{dx} + q(x) \right) y = y'' + p(x)y' + q(x)y,$$

and the equation is written

$$L(y) = g(x).$$

When $g(x) \equiv 0$

$$L(y) = 0$$

is the corresponding *homogeneous equation of second order.*
Let us introduce the notation

$$D = \frac{d}{dx}, \ \ D^2 = \frac{d^2}{dx^2}.$$

Then L may be written as

$$L = D^2 + p(x)D + q(x).$$

L is called *the differential operator of second order* and operates on functions having first and second derivatives on Ω. Since for $c \in \mathbb{R}$, $D(cy(x)) = cDy(x)$ and $D(y_1(x) + y_2(x)) = Dy_1(x) + Dy_2(x)$, it follows that L is linear, that is

$$L(c_1 y_1 + c_2 y_2) = c_1 L(y_1) + c_2 L(y_2),$$

with $c_1, c_2 \in \mathbb{R}$.

7.4.1   Linear second order ODE with constant coefficients

These are differential equations of the form

$$y'' + py' + qy = g(x),$$

where p and q are constants, but we allow g to be a function. The differential operator here is

$$L = D^2 + pD + q.$$

Note that $L(e^{rx}) = (r^2 + pr + q)e^{rx}$ and since $e^{rx} \neq 0$ the function $y = e^{rx}$ will be a solution of $L(y) = 0$, i.e.

$$L(e^{rx}) = 0,$$

if and only if r satisfies the quadratic equation $r^2 + pr + q = 0$. This equation is called the *characteristic equation* and the polynomial

$$\chi(r) = r^2 + pr + q,$$

is called the *characteristic polynomial* of L. There are three possibilities for the roots of $\chi(r)$:

1. *Real distinct roots* $r_1 \neq r_2$. Here the two corresponding solutions of the equation $L(y) = y'' + py' + qy = 0$ are

$$y_1 = e^{r_1 x}, \quad y_2 = e^{r_2 x}.$$

2. *Repeated roots* $r_1 = r_2 = r_0$ with corresponding solution $y_1 = e^{r_0 x}$. To find another solution note that $\chi(r) = (r - r_0)^2$, $\frac{d}{dr}\chi(r) = 2(r - r_0)$ and so $\frac{d}{dr}\chi(r)|_{r=r_0} = 0$. Therefore

$$L(xe^{rx}) = L\left(\frac{d}{dr}(e^{rx})\right) = \frac{d}{dr}L(e^{rx})$$

$$= \frac{d}{dr}[\chi(r)e^{rx}] = [\frac{d}{dr}\chi(r) + x\chi(r)]e^{rx}.$$

For $r = r_0$ this yields $L(xe^{r_0 x}) = 0$. Hence the two *linearly independent solutions*[5] are

$$y_1 = e^{r_0 x}, \quad y_2 = xe^{r_0 x}.$$

3. *Complex roots* $r_1 = \lambda + i\mu$ and $r_2 = \lambda - i\mu$. Here the two solutions $y_1 = e^{(\lambda + i\mu)x}$ and $y_2 = e^{(\lambda - i\mu)x}$ will be *complex-valued* functions. To obtain *real-valued* solutions set $\phi_1 = \frac{1}{2}(y_1 + y_2)$ and $\phi_2 = \frac{1}{2i}(y_1 - y_2)$. Then using Euler's identiy $e^{ix} = \cos x + i\sin x$ we find two (real-valued) independent solutions

$$\phi_1 = e^{\lambda x}\cos(\mu x), \quad \phi_2 = e^{\lambda x}\sin(\mu x).$$

Thus, in conclusion: The *general solution*[6] of the homogeneous second order linear equation with constant coefficients

$$y'' + py' + q = 0$$

is

$$y = c_1 y_1 + c_2 y_2,$$

where y_1 and y_2 are two linearly independent solutions given as above according to the nature of the roots of the characteristic polynomial and c_1, c_2 are arbitrary constants.

---

[5]Two solutions y_1, y_2 are said *linearly independent* on an interval (α, β) if for all $x \in (\alpha, \beta)$, the relation $c_1 y_1(x) + c_2 y_2(x) = 0$ implies $c_1 = c_2 = 0$, or equivalently, if the so-called *Wronskian* $W(y_1, y_2)(x) = \det\begin{bmatrix} y_1(x) & y_2(x) \\ y_1'(x) & y_2'(x) \end{bmatrix} \neq 0$, for all $x \in (\alpha, \beta)$.

[6]**Theorem**: If y_1, y_2, are linearly independent solutions of the equation $L(y) = 0$, then $c_1 y_1 + c_2 y_2$ is the general solution of the equation. Such a set $\{y_1, y_2\}$ of solutions is called a *fundamental set of solutions*. See, for example [17]. See also Exercise 7.10.2.

Example 7.4.3. Solve the equation $y'' - 3y' + 2y = 0$.

Solution. The characteristic polynomial is $\chi(r) = r^2 - 3r + 2 = (r - 2)(r - 1)$, and has two real roots $r_1 = 2$ and $r_2 = 1$. Hence the general solution is
$$y = c_1 e^{2x} + c_2 e^x.$$

Remark 7.4.4. Notice that $\chi(r)$ can be obtained from L by replacing D everywhere by r, where we use the convention that the zero-th derivative of y, $D^0 y = y$, i.e. we set $D^0 = I$ where I is the identity operator, and of course $r^0 = 1$. Moreover, if $\chi(r) = r^2 + pr + q = (r - r_1)(r - r_2)$, (with r_1, r_2 the roots of the characteristic equation), then

$$(D - r_1)(D - r_2)y = (D - r_1)(Dy - r_2 y) = D(Dy - r_2 y) - r_1(Dy - r_2 y)$$
$$= D^2 y - (r_1 + r_2)Dy + r_1 r_2 y$$
$$= [D^2 - (r_1 + r_2)D + r_1 r_2]y$$
$$= (D^2 + pD + qI)y = \chi(D)y = L(y).$$

Thus,
$$L = \chi(D) = (D - r_1 I)(D - r_2 I),$$

where here we *compose* these operators. *Note that these two first order operators commute, so the order of applying them is immaterial.*

We now turn to two rather simple, but quite important, second order equations. We shall see, that because *we are able to explicitly solve the general first order equations with variable coeffficients, we will be able to parlay this into the solution of second order linear equations with constant coefficients, and thus obtain an alternative method of solving them.*

The first equation we solve is the following:

Example 7.4.5. Solve the differential equation

$$y'' - \omega^2 y = 0,$$

where ω is a constant. That is, $(D^2 - \omega^2 I)y = 0$.

Solution. Here $\chi(r) = r^2 - \omega^2 = (r - \omega)(r + \omega)$. We shall assume $\omega \neq 0$, as the trivial case where $\omega = 0$ has for solutions, all linear functions and only those. Now suppose f is a solution to this equation. Then

$$(D - \omega I)(D + \omega I)f = 0.$$

Letting $g = (D + \omega I)f$, the equation says merely that $(D - \omega I)g = 0$. That is, $g' = \omega g$. This is the special case 1 of Theorem 7.2.1 and has solution $g(x) = Be^{\omega x}$ where B is some constant. But then, $f' = -\omega f + (-B)e^{\omega x}$. This is exactly the equation we have just solved, where here $\lambda = -\omega$, $\mu = \omega$ and $A = -B$. Since $\omega \neq 0$ we see that $\lambda \neq \mu$ and the general solution is $f(x) = \frac{-B}{2\omega}e^{\omega x} + Ce^{-\omega x}$. Since B and C are arbitrary constants we can, by renaming them, see that the general solution to $y'' - \omega^2 y = 0$ is

$$y(x) = ae^{\omega x} + be^{-\omega x}.$$

Conversely, by differentiation one sees easily that any such function is a solution.

Exercise 7.4.6. Show that the general solution to the above equation can also be expressed

$$f(x) = A\cosh(\omega x) + B\sinh(\omega x),$$

where A and B are arbitrary constants.

Exercise 7.4.7. Using the techniques just established solve the more general second order equation with constant coefficients a, b and c,

$$ay'' + by' + cy = 0,$$

under the assumption that the characteristic equation $ar^2 + br + c = 0$ has only real roots. (Observe that this assumption is a generalization of the one we have just dealt with.) Here there are two cases to consider: either the associated quadratic equation has equal or unequal roots. Then show, by differentiation, that conversely any such function is a solution.

Example 7.4.8. (*Simple harmonic oscillator equation*).

For completeness and beacause of its great importance, we also solve the second order equation

$$y'' + \omega^2 y = 0,$$

where ω is a nonzero constant. That is,

$$(D^2 + \omega^2 I)y = 0$$

Here, because of the plus sign, we shall see that the solutions are quite different. In fact they turn out to be *periodic* (and bounded).

Solution. We first consider the simplest case of the equation, namely when $\omega = 1$. If f is a solution to this equation, and we let $g = f'$, then $g' = -f$. But then, the derivative of $f^2 + g^2$ is $2ff' + 2gg' = 2fy - 2gf = 0$. Since $f^2 + g^2$ has an identically zero derivative it must be a constant; $f^2 + g^2 = C$. Clearly, unless f is identically zero, $C > 0$ so $C = \alpha^2$. Thus $f' = \sqrt{\alpha^2 - f^2}$, where $\alpha > 0$. But then, integrating, we get $x = \arcsin(\frac{f}{\alpha}) - C$, where C is some (other) constant. Applying the sin function tells us $f(x) = \alpha \sin(x + C)$. Finally, the addition formula for sin shows that $f(x) = \alpha(\sin x \cos C + \cos x \sin C)$ and taking $A = \alpha \cos C$ and $B = \alpha \sin C$, we see that $f(x) = A \sin x + B \cos x$.

Now consider the situation where ω is arbitrary, but nonzero. Let $g(x)$ be any solution to this equation and $f(x) = g(\frac{1}{\omega}x)$. Then a direct computation shows that $f'' = \frac{1}{\omega^2}g''(\frac{1}{\omega}x)$. Hence,

$$f''(x) + f(x) = \frac{1}{\omega^2}g''(\frac{1}{\omega}x) + g(\frac{1}{\omega}x) = \frac{1}{\omega^2}\left[-\omega^2 g(\frac{1}{\omega}x)\right] + g(\frac{1}{\omega}x) = 0.$$

But by the case $\omega = 1$ this means that there exist constants A and B such that for all real x, we have $f(x) = g(\frac{1}{\omega}x) = A \cos x + B \sin x$. Letting $u = \frac{1}{\omega}x$, i.e. $x = \omega u$, where u is arbitrary, we see that $g(u) = A \cos(\omega u) + B \sin(\omega u)$. Since we are now down to questions of notation any solution must be of the form

$$y(x) = A \cos(\omega x) + B \sin(\omega x).$$

Conversely, by differentiating, it follows immediately that any such function is a solution.

Figure 7.1: Spring

Definition 7.4.9. The *amplitute* and the *period* of $y(x) = A\cos(\omega x)$ and $y(x) = A\sin(\omega x)$ are $|A|$ and $T = \frac{2\pi}{\omega}$, respectively.

What is the amplitute and period of the function $y(x) = A\cos(\omega x) + B\sin(\omega x)$?

Exercise 7.4.10. Using the techniques established solve the second order equation with constant coefficients a, b and c,

$$ay'' + by' + cy = 0,$$

under the assumption that the roots of the associated quadratic equation $ar^2 + br + c = 0$ are *conjugate complex numbers*, say $z = \lambda + i\mu$, and $\bar{z} = \lambda - i\mu$. (Observe that this assumption is a generalization of the one we have just dealt with both in the harmonic oscillator and also the distinct (real) roots.)

Hint. Here the characteristic equation is $ar^2 + br + c = a(r-z)(r-\bar{z})$ and the differential equation $(D-z)(D-\bar{z})y = 0$. The solutions y_1 and y_2 will be *complex-valued* functions. To obtain *real-valued* solutions ϕ_1, ϕ_2, set $\phi_1 = \frac{1}{2}(y_1 + y_2)$ and $\phi_2 = \frac{1}{2i}(y_1 - y_2)$ and use the identiy $e^{ix} = \cos x + i\sin x$.

Applications

Here we give some applications to Physics. Generally speaking, vibrations occur whenever a physical system in stable equilibrium is disturbed. For then it is subject to forces tending to restore its equilibrium. We shall see below how situations of this kind can lead to second order differential equations.

Example 7.4.11. (*Undamped simple harmonic vibrations*).

Consider an object, which we assume to be concentrated at a point, hanging from a spring (Figure 1). First, we can locate the object by its center of gravity. Let $y(t)$ be the distance of the center of gravity above or below the equilibrium point at time t. If the weight is not negligible compared to the force exerted by the spring, then Hooke's law[7] tells us that the force the spring exerts on the point mass is proportional to the displacement, and opposite to the direction of motion. If k denotes the constant of proportionality then we see that the fundamental equation of motion $F = ma$ in this case says $mg - ky = m\frac{d^2y}{dt^2}$. Dividing by m we see that

$$\frac{d^2y}{dt^2} + \omega^2 y = g,$$

where $\omega = \sqrt{\frac{k}{m}}$. Now by inspection one sees that the constant function $y(t) = \frac{g}{\omega^2} = \frac{W}{k}$, where W is the weight of the object, is a solution to the non-homogeneous equation. Therefore

$$y(t) = A\cos(\omega t) + B\sin(\omega t) + \frac{W}{k}$$

is also such a solution. Of course, taking $W = 0$ gives the simple harmonic motion where $y(t) = A\cos(\omega t) + B\sin(\omega t)$, with A and B arbitrary constants, the *most general* solution to the homogeneous equation

$$\frac{d^2y}{dt^2} + \omega^2 y = 0.$$

To interpret the meaning of these constants let $t = 0$. Then $y(0) = A$. Having determined A, if we knew, say $y(t_0)$ at some other time t_0, or if we knew the initial velocity $v(0) = \frac{dy}{dt}(0)$, we could easily also determine B. In the latter case $B = \frac{v(0)}{\omega}$. This is another example of a specific solution being determined by initial conditions.

Without going into detail we mention that a slightly more general, second order differential equations can be used to solve other mechanics

---

[7]When a spring is stretched y inches beyond its natural length, *Hooke's law* states that the spring pulls back with a restoring force of $F = ky$ pounds, where k is a constant. The constant of proportionality k is called the *spring constant* can be thought of as a measure of the stiffness of the spring.

Figure 7.2: Pendulum

problems such as *forced oscillations* of a spring, with *damping*. *Characteristically of mathematics, those same principles apply equally well to seemingly unrelated problems in electric circuits* (see, for example, Exercise 7.10.16).

Example 7.4.12. (*The motion of a pendulum with small oscillations*).
Let the mass m of a small ball be concentrated at its center of gravity hanging from a weightless rod of length l be displaced through a small angle $\theta(0)$ and allowed to continue to swing in a vertical plane. We denote its angular position with respect to a vertical line through the point from which it is suspended by $\theta(t)$ and seek a differential equation for its position (at all other times) which we can solve to find $\theta(t)$. To do so we use Newton's Law of Motion, $F = ma$, where we consider forces and accelerations in the tangential direction only.

Now since the distance s traveled at time t is given by $s(t) = l\theta(t)$, the tangential velocity is $l\frac{d\theta}{dt}$ and the tangential acceleration is $l\frac{d^2\theta}{dt^2}$. On the other hand, using the geometry of the situation one sees easily that the tangential force is $-mg\sin\theta$, the minus sign because, just as in the Harmonic Oscillator, the force is in the direction opposite to the motion. Thus from Newton's law we get

$$ml\frac{d^2\theta}{dt^2} = -mg\sin\theta.$$

Then, just as in the falling body problem the m cancels so the situation is independant of the mass and we get $\frac{d^2\theta}{dt^2} + \omega^2\sin\theta = 0$, where $\omega = \sqrt{\frac{g}{l}} > 0$ (and we can take the square root since $\frac{g}{l}$ itself is positive). Finally, using the Taylor series for $\sin\theta$ and taking into account $\lim_{\theta\to 0}\frac{\sin\theta}{\theta} = 1$ and that we are dealing with *small* oscillations we can replace $\sin\theta$ by θ and get the familiar *Harmonic Oscillator Equation*,

$$\frac{d^2\theta}{dt^2} + \omega^2\theta = 0,$$

whose solutions we know to be $\theta(t) = A\cos(\omega t) + B\sin(\omega t)$.

We now turn to initial conditions: Taking $t = 0$, we see $\theta(0) = A$. On the other hand, the period $T = \frac{2\pi}{\omega}$ so when $t = \frac{1}{4}T = \frac{\pi}{2\omega}$, $\theta = 0$. Thus $0 = A\cos(\frac{\pi}{2}) + B\sin(\frac{\pi}{2})$ so $B = 0$ and the motion is given by,

$$\theta(t) = \theta(0)\cos(\sqrt{\frac{g}{l}}t).$$

Next we give an example illustrating that the techniques established here work for both *homogeneous* and *nonhomogeneous* second order linear differential equations with constant coefficients.

Example 7.4.13.   1 Solve the equation $y'' - 3y' + 2y = 0$.

2. Solve the equation $y'' - 3y' + 2y = xe^x$.

Solution.

1. The corresponding differential operator is $L = D^2 - 3D + 2I$. The characteristic polynomial is $\chi(r) = r^2 - 3r + 2 = (r-2)(r-1)$. Hence, we are solving

$$(D - 2I)(D - I)y = 0$$

Letting $g = (D - I)y$, the equation becomes $(D - 2I)g = 0$, that is,

$$g' - 2g = 0.$$

This is a first order equation in g with $p(x) = -2$ and $q(x) = 0$. From the special case 1 of Theorem 7.2.1, its solution is $g(x) = C_1 e^{2x}$. Now, we solve $(D - I)y = g(x)$ or

$$y' - y = g(x).$$

Here $p(x) = 1$ and the integrating factor is e^{-x}. Theorem 7.2.1, tells us $y(x) = e^x \int g(x)e^{-x}dx + C_2 e^x$. Substituting the formula of $g(x)$ and integrating we find

$$y = C_1 e^{2x} + C_2 e^x.$$

2. In this case we are solving

$$(D - 2I)(D - I)y = xe^x.$$

As above, letting $g = (D - I)y$, the equation becomes $(D - 2I)g = xe^x$. That is,

$$g' - 2g = xe^x.$$

Here $p(x) = -2$ and $q(x) = xe^x$. The integrating factor is e^{2x} and from Theorem 7.2.1, the general solution is

$$g(x) = e^{2x} \int xe^x e^{-2x} dx + C_1 e^{2x}.$$

Carrying out the integration we get $g(x) = -(x + 1)e^x + C_1 e^{2x}$. Now, we solve $(D - I)y = g(x)$. Again by Theorem 7.2.1, we obtain $y(x) = e^x \int g(x)e^{-x} dx + C_2 e^x$. Substituting the formula of $g(x)$ and integrating we get the general solution

$$y(x) = C_1 e^{2x} + C_2 e^x - (\frac{1}{2}x^2 + x)e^x.$$

Setting $\psi = -(\frac{1}{2}x^2 + x)e^x$, a glance at the solution, tells us that the *general solution* of the *nonhomogeneous* equation is the *sum* of the *general solution* of the corresponding *homogeneous* equation and a *particular solution* ψ of the nonhomogeneous equation. That is,

$$y(x) = C_1 e^{2x} + C_2 e^x + \psi.$$

This is not pure luck! This is an important result in the *general theory of linear differential equations*[8] on the structure of the general solution of a nonhomogeneous linear differential equations with *variable*, continuous, coefficients.

Next we solve a boundary value problem which will arise frequently in the subsequent discussion of second order partial differential equations.

---

[8]The usual two methods for obtaining a *particular solution* of a given nonhomogeneous linear equation with constant coefficients are either the *method of undetermined coefficients* or more generally the *method of variation of parameters*. For the method of variation of parameters see Exercise 7.10.3. For the general theory of linear equations, see, for example, [17].

Example 7.4.14. (*Eigenvalue problem*). Find all solutions of the differential equation

$$y'' + \lambda y = 0$$

on $[0, \pi]$, saisfying the boundary conditions $y(0) = 0$ and $y(\pi) = 0$, where λ is a real parameter. Those values of λ which permit nonzero solutions of this problem are called *eigenvalues* of the differential equation, while the corresponding nonzero solutions are called *eigenfunctions*. The problem itself is called an *eigenvalue problem*.

Solution. First note that, if $\lambda = 0$ then the general solution of the equation $y'' = 0$ is the linear function $y = Ax + B$, which can not vanish at two points unless it is identically zero. When $\lambda < 0$, then $\lambda = -\omega^2$ for some constant $\omega \neq 0$, and as we saw in Example 7.4.5 the equation $y'' - \omega^2 = 0$ has general solution $y(x) = Ae^{\omega x} + Be^{-\omega x}$. The only linear combination of this pair of exponentials which can satisfy the boundary conditions is the trivial one (prove!), that is $A = B = 0$, and the probem has only the zero solution.

Hence, the only interesting case is $\lambda > 0$. Letting $\lambda = \omega^2$, we are led to the differential equation $y'' + \omega^2 = 0$. In Example 7.4.8, we solved this equation and obtained the general solution

$$y(x) = A\cos(\omega x) + B\sin(\omega x).$$

Since $y(0) = 0$, it follows that $A = 0$ and the solution is

$$y(x) = B\sin(\omega x).$$

Now, since we are seeking nonzero solutions, applying the other boundary condition, $y(\pi) = 0$, we have $\omega\pi = n\pi$ for some positive integer n. Thus $\lambda = n^2$ are the eigenvalues of the problem, and the corresponding eigenfunctions (solutions) are

$$y(x) = B\sin(nx), \ n = 1, 2, 3, ...,$$

where B is an arbitrary nonzero constant.

7.4.2   Special types of second order ODE; reduction of order

In a number of problems the equation $y'' = f(x, y, y')$ is of the special type: (a) $y'' = f(x, y')$, or (b) $y'' = f(y, y')$.

   (a) When $y'' = f(x, y')$, we set $y' = z$, so that $y'' = z'$ and the equation reduces to the first order equation

$$\frac{dz}{dx} = f(x, z).$$

Solving this latter for z and then integrating the solution, we obtain the solution of the original equation.

Example 7.4.15. Solve the equation

$$xy'' - y' - y'^3 = 0, \quad (x > 0).$$

Solution. Setting $z = y'$, $y'' = z'$ we obtain the first order nonlinear equation

$$xz' - z - z^3 = 0.$$

Separating the variables we get $\frac{dz}{z + z^3} = \frac{dx}{x}$ and integrating (twice), we find

$$x^2 + (y - C_2)^2 = C_1^2.$$

   (b) When $y'' = f(y, y')$, as before, we set $y' = z$, but this time we express y'' in terms of a derivative with respect to y, that is using the Chain Rule,

$$y'' = \frac{dz}{dx} = \frac{dz}{dy}\frac{dy}{dx} = z\frac{dz}{dy}.$$

Hence the equation becomes

$$z\frac{dz}{dy} = f(y, z).$$

Example 7.4.16. Solve the equation

$$y'' + y'^2 = 2e^{-y}.$$

Solution. Setting $y' = z$, $y'' = z\frac{dz}{dy}$, the equation reduces to

$$z\frac{dz}{dy} + z^2 = 2e^{-y}.$$

Dividing both sides by z we get

$$\frac{dz}{dy} + y = 2e^{-y}z^{-1}.$$

This is a Bernoulli equation (with $k = -1$, see Section 7.2, Exercise 5(d)), and by the substitution $u = z^2$ reduces to the linear equation

$$\frac{du}{dy} + 2u = 4e^{-y},$$

whose general solution is $u = 4e^{-y} + C_1e^{-2y}$. Replacing u by $z^2 = y'^2$ we get

$$\frac{dy}{dx} = \pm\sqrt{4e^{-y} + C_1e^{\ 2y}}.$$

Separating the variables and integrating we get

$$\pm\frac{1}{2}\sqrt{4e^y + C_1} = x + C_2.$$

Thus, the general solution of the original equation is

$$e^y + \frac{1}{4}C_1 = (x + C_2)^2.$$

EXERCISES

1. Solve the linear differential equations.

   (a) $y'' - 2y' - 3y = 0$.
   (b) $5y'' + 3y' = 0$.
   (c) $y'' + 2y' + y = 0$.
   (d) $y'' + 2y' + 5y = 0$.
   (e) $y'' - 6y' + 25y = 0$.

2. Solve the nonhomogeneous linear differential equations.

(a) $y'' - 3y = x$.

(b) $y'' - 3y' = x$.

(c) $y'' + y' - 6y = 2$.

(d) $y'' - 4y' + 4y = x^2$.

(e) $y'' + y = 4x \cos x$.

(f) $y'' + y' - 2y = x^2 e^{4x}$.

(g) $y'' + 2y' - 3y = \sin x$.

(h) $y'' + 2y' + 5y = e^{-x} \cos 2x$.

(i) $y'' - 6y' + 9y = 4e^x - 16e^{3x}$.

(j) $y'' - 2y' + y = \frac{e^x}{x^2}, \quad x > 0$.

3. Solve the following initial-value problems.

(a) $y'' - 6y' + 9y = 0, \quad y(0) = 0, \, y'(0) = 5$.

(b) $y'' + 8y' - 9y = 0, \quad y(1) = 2, \, y'(1) = 0$.

(c) $y'' - y' = 4e^x, \quad y(0) = 0, \, y'(0) = 1$.

(d) $y'' + y' = e^{-x}, \quad y(0) = 1, \, y'(0) = -1$.

(e) $y'' + y = 2 \cos x, \quad y(0) = 1, \, y'(0) = 0$.

(f) $y'' - 4y' + 5y = 2x^2 e^x, \quad y(0) = 2, \, y'(0) = 3$.

4. Solve the differential equations.

(a) $y'' + (y')^2 = 1$.

(b) $y'' = 1 - (y')^2$.

(c) $xy'' + y' = 4x$.

(d) $xy'' - 2y' = x^3$.

(e) $x^2 y'' - 2xy' - (y')^2 = 0$.

(f) $y'' = yy'$.

(g) $y'' = 1 + (y')^2, \quad y(0) = 0, \, y'(0) = 0$.

(h) $y'' = y' e^y, \quad y(0) = 0, \, y'(0) = 2$.

5. Find the eigenvalues λ_n and the eigenfunctions $y_n(x)$ of $y'' + \lambda y = 0$ satisfying the following boundary conditions:

(a) $y(0) = 0$, $y(\frac{\pi}{2}) = 0$.

(b) $y(0) = 0$, $y(2\pi) = 0$.

(c) $y(0) = 0$, $y(L) = 0$, $(L > 0)$.

Answers to selected Exercises

1. (a) $y = C_1 e^{-x} + C_2 e^{3x}$. (b) $y = C_1 + C_2 e^{-\frac{3x}{5}}$. (3) $y = (C_1 + C_2 x)e^{-x}$.
   (d) $y = e^{-x}(C_1 \cos 2x + C_2 \sin 2x)$. (e) $y = e^{3x}(C_1 \cos 4x + C_2 \sin 4x)$.
2. (a) $y = C_1 e^{\sqrt{3}x} + C_2 e^{-\sqrt{3}x} - \frac{1}{3}x$. (b) $y = C_1 + C_2 e^{3x} - \frac{1}{6}x^2 - \frac{1}{9}x$.
   (d) $y = (C_1 + C_2 x)e^{2x} + \frac{1}{4}x^2 + \frac{1}{2}x + \frac{3}{8}$.
   (e) $y = C_1 \cos x + C_2 \sin x + x \cos x + x^2 \sin x$.
   (f) $y = C_1 e^x + C_2 e^{-2x} + \frac{1}{18}(x^2 - x + \frac{7}{18})e^{4x}$.
   (h) $y = (C_1 \cos 2x + C_2 \sin 2x)e^{-x} + \frac{1}{4}xe^{-x}\sin 2x$.
   (i) $y = (C_1 + C_2 x)e^{3x} + e^x - 8x^2 e^{3x}$.
   (j) $y = (C_1 + C_2 x)e^r - (\log x + 1)e^x$.
3. (a) $y = 5xe^{3x}$. (c) $y = 2xe^x - \sinh x$. (d) $y = 1 - xe^{-x}$.
   (f) $y = (\cos x - 2\sin x)e^{2x} + (x-1)^2 e^r$.
4. (a) $y = \log(C_1 e^x + e^{-x}) + C_2$. (c) $y = x^2 + C_1 \log x + C_2$.
   (d) $y = C_1 + C_2 x^3 + \frac{1}{4}x^4$.
   (e) $y = -\frac{1}{2}x^2 - C_1 x - C_1^2 \log(x - C_1) + C_2$.
   (f) $y = C$, or $y = -\frac{2}{x+C}$, or $y = 2C_1 \tan(C_1 x + C_2)$, $C_1 > 0$, $C_2 \in \mathbb{R}$,
   or $y = -2C_1 \coth(C_1 x + C_2)$, $C_1 > 0$, $C_2 \in \mathbb{R}$.
   (g) $y = -\log(\cos x)$, $x \in (-\frac{\pi}{2}, \frac{\pi}{2})$. (h) $y = -\log(2e^{-x} - 1)$.
5. (a) $\lambda = 4n^2$, $y(x) = \sin(2nx)$. (b) $\lambda = \frac{n^2}{4}$, $y(x) = \sin(\frac{nx}{2})$.
   (c) $\lambda = \frac{n^2\pi^2}{L^2}$, $y(x) = \sin(\frac{n\pi x}{L})$.

7.5   Higher order ODE and systems of ODE

The equation
$$F(x, y, y', ..., y^{(n)}) = 0$$

is called an *nth order differential equation*. Here $F(x, z_0, z_1, ..., z_n)$ is a continuously differentiable function in a certain domain in $\mathbb{R}^{n+2}$. Solving, as before, with respect to $y^{(n)}$, we may write

$$y^{(n)} = f(x, y, y', ..., y^{(n-1)}). \tag{7.25}$$

An *initial-value problem* here, is the equation (7.25) together with n initial conditions $y(a) = b_0$, $y'(a) = b_1$, ... , $y^{(n-1)}(a) = b_{n-1}$.

Everything we have done for the second order equation can be carried over to the case of the equation of order n. In particular, a *linear differential equation of order* n is an equation of the form

$$L(y) = y^{(n)} + p_1(x)y^{(n-1)} + p_2(x)y^{(n-2)} + ... + p_n(x)y = g(x), \quad (7.26)$$

where $p_1(x),...,p_n(x)$, $g(x)$ are continous functions on some interval (α, β). Briefly written

$$L(y) = g(x),$$

where

$$L = \frac{d^n}{dx^n} + p_1(x)\frac{d^{n-1}}{dx^{n-1}} + ... + p_{n-1}(x)\frac{d}{dx} + p_n(x)I$$

is the *differential operator of n-th order*. Again, when $g(x) \equiv 0$, the equation

$$L(y) = 0$$

is called the corresponding *homogeneuous linear equation of order* n.

As with the second order equations we confine ourselves to n-th order linear equations with *constant* coefficients. The differential operator here is

$$L = D^n + p_1 D^{n-1} + ... + p_{n-1}D + p_n I,$$

and the *characteristic polynomial* is

$$\chi(r) = r^n + p_1 r^{n-1} + ... + p_{n-1}r + p_n.$$

Next we consider some examples.

Example 7.5.1. Solve $y''' - 3y' + 2y = 0$.

Solution. The characteristic polynomial is $\chi(r) = r^3 - 3r + 2$ and factoring $\chi(r) = (r + 2)(r - 1)^2$. Hence we are solving

$$(D + 2I)(D - I)^2 y = 0.$$

Letting, $g = (D - I)^2 y$, the equation becomes $(D + 2I)g = 0$. That is,

$$g' + 2g = 0.$$

This is the special case 1 of Theorem 7.2.1 and its solution is $g(x) = c_1 e^{-2x}$. Now, we solve $(D - I)(D - I)y = g(x)$. Let $f = (D - I)y$. This amounts solving $(D - I)f = g(x)$. That is,

$$f' - f = g(x).$$

By Theorem 7.2.1, we have $f(x) = e^x \int g(x)e^{-x}dx + C_2 e^x$. Inserting $g(x)$ and integrating we find $f(x) = -\frac{1}{3}c_1 e^{-2x} + C_2 e^x$.

Finally, we solve $(D - I)y = f(x)$. Again, Theorem 7.2.1 yields $y(x) = e^x \int f(x)e^{-x}dx + C_3 e^x$. Substituting the formula for $f(x)$ and integrating we obtain the general solution of our equation $y(x) = \frac{1}{9}c_1 e^{-2x} + C_2 x e^x + C_3 e^x$, or

$$y(x) = C_1 e^{-2x} + (C_2 x + C_3)e^x,$$

where we set $C_1 = \frac{1}{9}c_1$. The last term is characteristic of repeated roots.

Example 7.5.2. Solve $y^{(4)} - y = 0$.

Solution. Here $\chi(r) = r^4 - 1 = (r^2 + 1)(r - 1)(r + 1)$. Hence we are solving
$$(D^2 + I)(D - I)(D + I)y = 0.$$

Letting $g = (D - I)(D + I)y$, we must solve $(D^2 + I)g = 0$ or $g'' + g = 0$. From Example 7.4.5, $g(x) = A \sin x + B \cos x$. Now

$$(D - I)(D + I)y = g(x).$$

Letting $f = (D + I)y$, this amounts to solving $(D - I)f = g(x)$. As before $f(x) = e^x \int g(x)e^{-x}dx + C_1 e^x$. Inserting $g(x)$ and integrating (by parts), we find $f(x) = C_3 \sin x + C_4 \cos x + C_1 e^x$. Finally, we solve $(D - I)y = f(x)$, and obtain

$$y(x) = C_1 e^x + C_2 e^{-x} + C_3 \sin x + C_4 \cos x.$$

We now turn to systems of ODE. Two or more differential equations that contain functions of the same independent variable and various of their derivatives form a *system* of differential equations.

Definition 7.5.3. A set of n differential equations

$$y_i' = f_i(x, y_1, y_2, ..., y_n), \quad i = 1, 2, ..., n \tag{7.27}$$

in the unknown functions $y_1(x)$, $y_2(x)$,...,$y_n(x)$ is called a *system of first-order differential equations.*

Proposition 7.5.4. *The differential equation*

$$y^{(n)} = f(x, y, y', ..., y^{(n-1)}). \tag{7.28}$$

is equivalent to the system of n first-odrer differential equations

$$y_1' = y_2,$$
$$y_2' = y_3,$$
$$...$$
$$y_{n-1}' = y_n,$$
$$y_n' = f(x, y_1, y_2, ..., y_n), \tag{7.29}$$

where $y_1(x) = y(x)$.

Proof. First consider the equation (7.28). Let $y = y(x)$ be a solution of (7.28) on some interval (α, β). Introducing the new notation

$$y_1 = y, \ y_2 = y', \ y_3 = y'', ..., \ y_n = y^{(n-1)},$$

equation (7.28) becomes the system

$$y_1' = y_2, \ y_2' = y_3, \ ..., \ y_{n-1}' = y_n, \ y_n' = f(x, y_1, y_2, ..., y_n),$$

which shows that the vector $Y = (y, y', y'', ..., y^{(n-1)})$ is a solution of the system (7.29) on (α, β).

Conversely, let $Y(x) = (y_1(x), y_2(x), ..., y_n(x)))$ be a solution of (7.29) on some interval (α, β). Then, since

$$y_1' = y_2, \ y_1'' = y_2' = y_3, \ ..., \ y_1^{n-1} = y_n, \ y_1^{(n)} = y_n' = f(x, y_1', y_1'', ..., y_1^{n-1})$$

the first component y_1 is a solution of (7.28) on (α, β), $\qquad \square$

Example 7.5.5. Reduce the system

$$y_1' = y_1 + y_2 + y_3,$$

$$y_2' = y_2 - 2y_3,$$

$$y_3' = y_1 + y_3$$

to a third-order differential equation.

Solution. Differentiating the first equation of the system and taking into account the two others, we have

$$y_1'' = y_1' + y_2' + y_3' = y_1' + y_1 + y_2 - y_3.$$

Differentiating once again,

$$y_1''' = y_1'' + y_1' + y_2' - y_3' = y_1'' + y_1' - y_1 + y_2 - 3y_3.$$

That is,

$$y_1''' = y_1'' + y_1' - y_1 + y_2 - 3y_3. \tag{7.30}$$

Now from the system

$$y_1' = y_1 + y_2 + y_3,$$

$$y_1'' = y_1' + y_1 + y_2 - y_3,$$

expressing y_2 and y_3 in terms of y_1, y_1' and y_1'', we obtain

$$y_2 = \frac{1}{2}(y_1'' - 2y_1)$$

$$y_3 = \frac{1}{2}(-y_1'' + 2y_1').$$

Substituting these expressions of y_2 and y_3 into (7.30) and setting $y = y_1$, we get the third-order equation

$$y''' - 3y'' + 2y' + 2y = 0.$$

Solving this (linear) equation, we find the function $y = y_1$, and then substituting into the above expressions of y_2 and y_3, we obtain the solution $Y = (y_1, y_2, y_3)$ of the system.

Example 7.5.6. Solve the initial-value problem for the system[9]

$$y_1' = 3y_1 + 8y_2,$$

$$y_2' = -y_1 - 3y_2,$$

$y_1(0) = 6, \quad y_2(0) = -2.$

Solution. Differentiating the first equation and taking into account the second we get $y_1'' = 3y_1' - 8y_1 - 24y_2$. From the first equation $y_2 = \frac{1}{8}(y_1' - 3y_1)$. Substituting this into the latter equation yields the second order equation

$$y_1'' - y_1 = 0,$$

whose general solution (from Example 7.4.5) is $y_1 = C_1 e^x + C_2 e^{-x}$. So that $y_1' = C_1 e^x - C_2 e^{-x}$, and $y_2 = \frac{1}{8}(y_1' - 3y_1) = -\frac{1}{4}C_1 e^x - \frac{1}{2}C_2 e^{-x}$.

Using the initial conditions we get a system of equations to determine the constants C_1, C_2:

$$C_1 + C_2 = 6$$

$$-C_1 - 2C_2 = -8.$$

Solving, we find $C_1 = 4$, $C_2 = 2$. Thus, $y_1 = 4e^x + 2e^{-x}$, $y_2 = -e^x - e^{-x}$.

As we just saw Proposition 7.5.4 connects first-order systems and n-th order equations of the form $y^{(n)} = f(x, y, y', ..., y^{(n-1)})$. Consequently, all results proved for first order systems may be applied to give results for n-th order equation of this form. Next we use this interplay to obtain existence and uniqueness of solutions for initial-value problems of n-th order equations.

The first order system

$$y_i' = f_i(x, y_1, y_2, ..., y_n), \quad i = 1, 2, ..., n$$

can be written briefly in vector notation as

$$Y'(x) = \mathbf{f}(x, Y(x)), \tag{7.31}$$

---

[9]The given system is a *linear system of first-order differential equations* and for such systems there is a comprehensive theory and method for solving them. In this connection see Theorem 7.5.12 and Example 7.5.14.

or $Y' = \mathbf{f}(x, Y)$, where $Y = (y_1, ..., y_n)$, $Y' = (y_1', ..., y_n')$ and $\mathbf{f} = (f_1, ..., f_n)$, the value of $\mathbf{f}$ at (x, Y) being given by $\mathbf{f}(x, Y) = (f_1(x, Y), ..., f_n(x, Y))$.

The reader will see this is just the case $n = 1$ in Section 7.3 where we replace y by Y.

Definition 7.5.7. Let $\mathbf{f} \colon \mathbb{R}^{n+1} \to \mathbb{R}^n$ be a continuous function defined on

$$R = \{(x, Y) : |x - a| \leq h, \; |y_1 - b_1| \leq h, ..., |y_n - b_n| \leq h\}, \; (h > 0).$$

An *initial value problem* for equation (7.31) is the problem of finding a solution of $Y' = \mathbf{f}(x, Y)$ satisfying $Y(a) = \mathbf{b}$, where $\mathbf{b} = (b_1, ..., b_n)$.

The definition of a Lipschitz condition is formally the same as before.

Definition 7.5.8. A function $\mathbf{f}(x, Y)$ defined on a set $S \subseteq \mathbb{R}^{n+1}$ is said to satisfy a *Lipschitz condition in Y* on S if there exists a constant $M > 0$ such that

$$\|\mathbf{f}(x, Y) - \mathbf{f}(x, Z)\|_1 \leq M\|Y - Z\|_1,$$

for all (x, Y), (x, Z) in S.

Remark 7.5.9. A sufficient condition for $\mathbf{f}(x, Y)$ to satisfy a Lipschitz condition on a closed ball (or parallelepiped) S in $\mathbb{R}^{n+1}$ is the continuity of $\frac{\partial \mathbf{f}}{\partial y_j}$ on S for all $j = 1, ..., n$, that is, the continuity of $\frac{\partial f_i}{\partial y_j}(x, y_1, ..., y_n)$ for all $i, j = 1, ..., n$. This is the analogue of Remark 7.3.4.[10]

By introducing the (vector) integral equation

$$Y(x) = \mathbf{b} + \int_a^x \mathbf{f}(t, Y(t))dt,$$

the following extension of Picard's theorem may be demonstrated exactly as in Theorem 7.3.5.

---

[10]The continuity of the partials and the compactness of S, imply that there is $M > 0$ such that $\|\frac{\partial \mathbf{f}}{\partial y_j}(x, Y)\|_1 \leq M$ for all $j = 1, ..., n$. Next for $t \in [0, 1]$, apply the Mean Value theorem to the function $\psi(t) = f_i(x, Z + t(Y - Z))$, for $i = 1, .., n$ (see, Theorem 3.4.1).

Theorem 7.5.10. *(Picard). Let Ω be a domain in $\mathbb{R}^{n+1}$ and let $\boldsymbol{f}$: $\Omega \to \mathbb{R}^n$ be continuous. Let $(a, \boldsymbol{b}) \in \Omega$, and consider the initial value problem*

$$Y' = \boldsymbol{f}(x, Y), \quad Y(a) = \boldsymbol{b}.$$

Suppose $\boldsymbol{f}$ satisfies a Lipschitz condition in Y on Ω. Then there exists a $\delta > 0$ and a unique solution $\Phi = \Phi(x)$ to the initial value problem, for all $|x - a| \leq \delta$.

Proof. Replace everywhere y, f, φ in Theorem 7.3.5 by Y, $\mathbf{f}$, Φ. $\square$

The *successive approximations* $\Phi_0, \Phi_1, ...$, here are given (in vector notation) by

$$\Phi_0(x) = \mathbf{b},$$

$$\Phi_{k+1}(x) = \mathbf{b} + \int_a^x \mathbf{f}(t, \Phi_k(t))dt, \quad (k = 0, 1, 2, ...).$$

The following theorem extends Theorem 7.3.5 to n-th order ODE.

Theorem 7.5.11. *(Existence and uniqueness for n-th order equations). Let Ω be a domain in $\mathbb{R}^{n+1}$ and let $f : \Omega \to \mathbb{R}$ be continuous. Let $(a, b_0, b_1, ..., b_{n-1}) \in \Omega$, and consider the initial value problem*

$$y^{(n)} = f(x, y, y', ..., y^{(n-1)}), \tag{7.32}$$

$$y(a) = b_0, \; y'(a) = b_1, \; ... \; , \; y^{(n-1)}(a) = b_{n-1}.$$

Suppose $f(x, y, y_1, ..., y_{n-1})$ satisfies a Lipschitz condition in $y, y_1, ..., y_{n-1}$ on Ω. Then there exists a $\delta > 0$ and a unique solution $\varphi = \varphi(x)$ to the initial value problem, for all $|x - a| \leq \delta$.

Proof. The proof follows readily by applying Proposition 7.5.4 to the equation (7.32) and by using Theorem 7.5.10. $\square$

We emphasize that in both Theorems 7.5.11 and 7.3.5 we only get *local* solutions.

An important application of these results is to the case of a *linear system*. This is a system

$$Y' = \mathbf{f}(x, Y),$$

where the component functions $f_1, ..., f_n$ of $\mathbf{f}$ have the form

$$f_i(x, Y) = a_{i1}(x)y_1 + a_{i2}(x)y_2 + ... + a_{in}(x)y_n + g_i(x), \quad (i = 1, 2, ..., n).$$

Here the functions $a_{i1}, ..., a_{in}, g_i$ are defined and continuous for x in some interval (α, β). Using matrix notation the system can be written as

$$Y' = AY + \mathbf{g},$$

where $A = (a_{ij})$ and $\mathbf{g} = (g_1, ..., g_n)$ (viewed as a column vector). When $G(x) \equiv 0$, the system is called *homogeneous*, and has the form

$$Y' = AY.$$

For *example*, the system of Example 7.5.6

$$y_1' = 3y_1 + 8y_2,$$

$$y_2' = -y_1 - 3y_2,$$

is written as $Y' = AY$ where

$$A = \begin{bmatrix} 3 & 8 \\ -1 & -3 \end{bmatrix},$$

$$Y = \begin{bmatrix} y_1 \\ y_2 \end{bmatrix} \quad \text{and} \quad Y' = \begin{bmatrix} y_1' \\ y_2' \end{bmatrix}.$$

With the above notation we have

Theorem 7.5.12. *(Existence and uniqueness for linear systems). Consider the linear system*

$$Y' = AY + \mathbf{g},$$

where all the functions $a_{ij}(x)$, $g_i(x)$ are continuous on an interval J containing a. Then for any vector $\mathbf{b}$ in $\mathbb{R}^n$ there exists a unique solution Φ of the initial value problem

$$Y' = AY + \mathbf{g}, \quad Y(a) = \mathbf{b}, \text{ on } J.$$

Proof. In the original notation the system is $Y' = \mathbf{f}(x, Y)$, with

$$f_i(x, Y) = \sum_{j=1}^{n} a_{ij}(x) y_j + g_i(x),$$

and, (of course), $\mathbf{f}$ is continuous. Since all $a_{ij}(x)$ are continuous, the partials $\frac{\partial f_i}{\partial y_j}(x, Y) = a_{ij}(x)$ are also continuous. Hence $\mathbf{f}$ satisfies a Lipschitz condition on the "strip" $S = \{(x, Y) : |x - a| \le h, \, ||Y||_1 < \infty\}$ (see Remark 7.5.9). Now, Theorem 7.5.10 applies giving the result. $\square$

Example 7.5.13. Use Picard's method of successive approximations to solve the problem.

$$y_1' = y_2,$$

$$y_2' = -y_1,$$

$$Y(0) = (0, 1).$$

Solution. Here $\mathbf{f}(x, Y) = (y_2, -y_1)$, $a = 0$, and $\mathbf{b} = (0, 1)$. The equivalent integral equation is

$$Y(x) = \mathbf{b} + \int_0^x \mathbf{f}(t, y_1(t), y_2(t)) dt = (0, 1) + \int_0^x (y_2(t), -y_1(t)) dt.$$

Now, in succession, we have $\Phi_0(x) = (0, 1)$,

$$\Phi_1(x) = (0, 1) + \int_0^x (1, 0) dt = (x, 1),$$

$$\Phi_2(x) = (0, 1) + \int_0^x (1, -t) dt = (0, 1) + \left(x, -\frac{x^2}{2} \right) = \left(x, 1 - \frac{x^2}{2} \right),$$

$$\Phi_3(x) = (0, 1) + \int_0^x \left(1 - \frac{t^2}{2}, -t \right) dt = \left(x - \frac{x^3}{3!}, 1 - \frac{x^2}{2} \right),$$

and so on ...

$$\Phi_k(x) = \left(x - \frac{x^3}{3!} + \frac{x^5}{5!} - \frac{x^7}{7!} + \dots, \, 1 - \frac{x^2}{2} + \frac{x^4}{4!} - \frac{x^6}{6!} + \dots \right).$$

Thus

$$\lim_{k \to \infty} \Phi_k(x) = \Phi(x) = (\sin x, \cos x).$$

As an exercise, we ask the reader to solve the problem of Example 7.5.13 by reducing it to an intitial value problem for the corresponding second order linear equation.

In the following example we describe a method for solving linear systems using techniques of linear algebra.

Example 7.5.14. Consider the linear system

$$y_1' = ay_1 + by_2,$$

$$y_2' = cy_1 + dy_2,$$

where a, b, c, d are *constants*. Show that this system always has a solution Φ of the form

$$\Phi(x) = e^{rx}\xi,$$

where $\xi = (\xi_1, \xi_2) \neq (0, 0)$ is a constant vector and r is a constant. In fact, show that r is an eigenvalue of the coefficient matrix $A = \begin{bmatrix} a & b \\ c & d \end{bmatrix}$, and ξ a corresponding eigenvector.

Solution. Substituting $\Phi(x) = (\xi_1 e^{rx}, \xi_2 e^{rx})$ and $\Phi'(x) = (r\xi_1 e^{rx}, r\xi_2 e^{rx})$ into the system and cancelling e^{rx}, we get the homogeneous algebraic linear system

$$(a - r)\xi_1 + b\xi_2 = 0$$

$$c\xi_1 + (d - r)\xi_2 = 0.$$

It is known from linear algebra that in order this system to have a nonzero solution, (ξ_1, ξ_2), the determinant of its coefficient matrix must be zero. That is,

$$\det \begin{bmatrix} a - r & b \\ c & d - r \end{bmatrix} = r^2 - (a + d)r + ad - bc = 0.$$

Hence any solution of the equation $r^2 - (a+d)r + ad - bc = 0$ can serve as the constant r and any nonzero solution $\xi = (\xi_1, \xi_2)$ of this (algebraic) system as the constant vector to construct the solution $\Phi(x) = e^{rx}\xi$.

Now, the given system in matrix form is

$$Y' = AY. \tag{7.33}$$

Assume a *nonzero* solution in the form $Y = e^{rx}\xi$, so that $Y' = re^{rx}\xi$. Substituting in (7.33) yields

$$A\xi = r\xi.$$

This tells us that r is an eigenvalue of A and ξ is a corresponding eigenvector of A. The characteristic equation, of course, is

$$r^2 - (a+d)r + ad - bc = 0.$$

There are three possibilities:

1. All eigenvalues are real and distinct,

2. Repeated eigenvalues,

3. Complex eigenvalues.

Note that when we reduce the system to a single ODE it is of the second order, viz,

$$y'' - (a+d)y + ad - bc = 0.$$

The following example gives an illustration of this method.

Example 7.5.15. Solve the system

$$y_1' = 2y_1 + y_2,$$

$$y_2' = -y_1 + 4y_2.$$

Solution. The system in matrix form is $Y' = AY$, that is,

$$Y' = \begin{bmatrix} 2 & 1 \\ -1 & 4 \end{bmatrix} Y.$$

Here the characteristic equation is $(r-3)^2 = 0$ with a double root $r_1 = r_2 = 3$. The eigenevectors belonging to the repeated eigenvalue

$r = 3$ are scalar multiples of the vector $\xi = (1,1)$. Hence a solution to the system is

$$\Phi_1(x) = e^{3x}\xi = (e^{3x}, e^{3x}).$$

To find another *linearly independent solution*[11] we look for a solution of the form $\Phi_2(x) = e^{3x}(x\xi + u)$, where the vector $u = (u_1, u_2)$ (called *generalized eigenvector*) satisfies

$$(A - 3I)u = \xi.$$

This algebraic system yields $-u_1 + u_2 = 1$. So $u = (c, 1 + c) = (0,1) + c(1,1)$, where $c \in \mathbb{R}$ is any constant. To obtain an independent solution Φ_2, we take $c = 0$, viz, $u = (0,1)$, and therefore

$$\Phi_2(x) = x\Phi_1(x) + ue^{3x} = (xe^{3x}, (x+1)e^{3x}).$$

The general theory of linear systems[12] now tells us that the general solution of our system is of the form $\Phi = c_1\Phi_1 + c_2\Phi_2$, where c_1, c_2 are arbitrary constants.

Remark 7.5.16. For a 3×3 system, in case the coefficient matrix A has a triple eigenvalue r, the three linearly independent solutions can be found by considering

$$\Phi_1(x) = e^{rx}\xi, \;\; \Phi_2(x) = e^{rx}(x\xi + u), \text{ and } \Phi_3(x) = e^{rx}(\tfrac{x^2}{2}\xi + xu + v),$$

where $\xi = (\xi_1, \xi_2, \xi_3) \neq (0,0,0)$ is an eigenvector belonging to r, and u, v are generalized eigenvectors of A.

Example 7.5.17. Consider the matrix ODE

$$\frac{dX(t)}{dt} = X(t)A,$$

---

[11]Two solutions Φ_1, Φ_2 are said *linearly independent* on an interval (α, β) if for all $x \in (\alpha, \beta)$, the relation $c_1\Phi_1(x) + c_2\Phi_2(x) = 0$ implies $c_1 = c_2 = 0$, or equivalently, if the so-called *Wronskian* $W(\Phi_1, \Phi_2)(x) = \det \begin{bmatrix} \Phi_1(x) & \Phi_2(x) \\ \Phi_1'(x) & \Phi_2'(x) \end{bmatrix} \neq 0$, for all $x \in (\alpha, \beta)$.

[12]**Theorem**: If $Y_1, Y_2, ..., Y_n$, are linearly independent solutions of an $n \times n$ linear system $Y' = AY$, then $c_1Y_1 + ... + c_2Y_n$ is the general solution of the system. Such a set $\{Y_1, Y_2, ..., Y_n\}$ of solutions is called a *fundamental set of solutions*. See, for example, [17].

where $X(t)$ is the unknown matrix function of t and A is a constant matrix (all real $n \times n$ matrices) and observe that this is a special case of the homogeneous system of 7.5.12 so there are solutions. Now consider the initial value problem where $X(0) = I$. Show that if A is skew symmetric (i.e., $A^t = -A$), then $X(t)$ is always orthogonal.

Solution. Denoting the transpose by ^{t} we have

$$\frac{d}{dt}(X(t)X(t)^t) = X(t)\frac{d}{dt}(X(t)^t) + \frac{dX(t)}{dt}X(t)^t$$

$$= X(t)\frac{d}{dt}(X(t))^t + \frac{dX(t)}{dt}X(t)^t$$

$$= X(t)(X(t)A)^t + X(t)AX(t)^t$$

$$= X(t)A^tX(t)^t + X(t)AX(t)^t$$

$$= X(t)(A^t + A)X(t)^t = 0.$$

Hence $X(t)X(t)^t$ is a constant. Since $X(0) = I$, this constant is I so $X(t)$ is orthogonal.

EXERCISES

1. Solve the differential equations.

   (a) $y''' - 5y'' - 8y' + 6y' = 0.$
   (b) $y''' - 5y'' - 8y' - 4y = 0.$
   (c) $y''' - y = 0.$
   (d) $y^{(4)} + 2y'' + y = 0.$
   (e) $y^{(6)} + 2y^{(5)} + y^{(4)} = 0.$

2. Solve the differential equations.

   (a) $y''' - y' = 1.$
   (b) $y''' - y'' + y' - y = x^2 + x.$
   (c) $y''' - 2y'' = y' = 2x + e^x.$
   (d) $y^{(4)} - 2y''' + 2y'' - 2y' + y = e^x.$

(e) $y^{(4)} - 2y'' + y = \cos x$.

(f) $y^{(5)} - y^{(4)} = xe^x - 1$.

(g) $y^{(5)} + y''' = x + 2e^{-x}$.

3. Solve the initial value problems.

(a) $y''' + y'' = 0$; $y(0) = 1$, $y'(0) = 0$, $y''(0) = 1$.

(b) $y''' - 3y'' + 3y' - y = 0$; $y(0) = 1$, $y'(0) = 2$, $y''(0) = 3$.

(c) $y''' - y = 2x$; $y(0) = y'(0) = 0$, $y''(0) = 2$.

(d) $y^{(4)} - y = 8e^x$; $y(0) = -1$, $y'(0) = 0$, $y''(0) = 1$, $y'''(0) = 0$.

4. (*Euler's equations*). Show that the change of the independent variable $x = e^t$ transforms the equation

$$x^n y^{(n)} + p_1 x^{n-1} y^{(n-1)} + \ldots + p_{n-1} x y' + p_n y = 0$$

into an n-th order homogeneous linear equation with constant coefficients. Apply this method to solve the equations.

(a) $xy'' + y' = 0$.

(b) $x^2 y'' + 2xy' - 6y = 0$.

(c) $x^2 y'' + 3xy' + y = 0$.

(d) $x^2 y'' - 2y = \sin(\log x)$.

(e) $x^2 y''' = 2y'$.

(f) $x^3 y''' + x^2 y'' - 2xy' + 2y = 0$.

5. Solve the following systems of differential equations.

(a)
$$y_1' = y_1 + y_2$$
$$y_2' = 4y_1 - 2y_2.$$

(b)
$$y_1' = -4y_1 - y_2$$
$$y_2' = y_1 - y_2.$$

(c)

$$Y' = \begin{bmatrix} 1 & -2 \\ 4 & 5 \end{bmatrix} Y$$

(d)

$$Y' = \begin{bmatrix} 1 & -5 \\ 2 & -1 \end{bmatrix} Y$$

6. Solve the system of differential equations.

(a)

$$Y' = \begin{bmatrix} 3 & 2 & 4 \\ 2 & 0 & 2 \\ 4 & 2 & 3 \end{bmatrix} Y.$$

(b)

$$Y' = \begin{bmatrix} 3 & 1 & -1 \\ -1 & 2 & 1 \\ 1 & 1 & 1 \end{bmatrix} Y.$$

(c)

$$Y' = \begin{bmatrix} 2 & -1 & 1 \\ 1 & 0 & 1 \\ 1 & -2 & -3 \end{bmatrix} Y,$$

$$y_1(0) = 0, \quad y_2(0) = 0, \quad y_3(0) = 1.$$

Answers to selected Exercises

1. (a) $y = C_1 + C_2 e^{2x} + C_3 e^{3x}$. (b) $y = C_1 e^x + (C_2 + C_3 x)e^{2x}$.
   (c) $y = C_1 e^x + e^{-\frac{x}{2}}\left(C_2 \cos \frac{\sqrt{3}}{2}x + C_3 \sin \frac{\sqrt{3}}{2}x\right)$.
   (d) $y = (C_1 x + C_2)\cos x + (C_3 x + C_4)\sin x$.
   (e) $y = C_1 + C_2 x + C_3 x^2 + C_4 x^3 + (C_5 + C_6)e^{-x}$.

2. (b) $y = C_1 e^x + C_2 \cos x + C_3 \sin x - (x^2 + 3x + 1)$.
   (c) $y = C_1 + (C + 2 + C_3 x)e^x + x^2 + 4x + \frac{1}{2}x^2 e^x$.
   (d) $y = (C_1 + C_2 x)e^x + C_3 \cos x + C_4 \sin x + \frac{1}{4}x^2 e^x$.
   (e) $y = (C_1 + C_2 x)e^x + (C_3 + C_4 x)e^{-x} + \frac{1}{4}\cos x$.
   (f) $y = C_1 x^3 + C_2 x^2 + C_3 x + C_4 + C_5 e^x + \frac{1}{24}(\frac{1}{3}x^2 - 4x)x^4 e^x$.
   (g) $y = C_1 + C_2 x + C_3 x^2 + C_4 \cos x + C_5 \sin x + \frac{1}{24}x^4 - e^{-x}$.

3. (a) $y = x + e^{-x}$. (b) $y = (1+x)e^x$. (c) $y = 2x - \frac{4}{\sqrt{3}}e^{-\frac{x}{2}}\sin\frac{\sqrt{3}}{2}x$.
   (d) $y = \cos x + 2\sin x + e^{-x} + (2x-3)e^x$.

4. (a) $y = C_1 + C_2\log x$. (b) $y = C_1 x^2 + C_2\frac{1}{x^3}$. (c) $y = \frac{1}{x}(C_1 + C_2\log x)$.
   (d) $y = C_1 x^2 + C_2\frac{1}{x} + \frac{1}{10}(\cos(\log x) - 3\sin(\log x))$.
   (e) $y = C_1 + C_2 x^3 + C_3\log x$. (f) $y = C_1 x + C_2 x^2 C_3\frac{1}{x}$.

5. (a) $y_1 = C_1 e^{-3x} + C_2 e^{2x}$, $y_2 = -4C_1 e^{-3x} + C_2 e^{2x}$.
   (b) $y_1 = C_1 e^{-xt} + C_2(1-x)e^{-3x}$, $y_2 = -C_1 e^{-3x} + C_2 x e^{-3x}$.
   (c) $y_1 = e^{3x}(C_1\cos 2x + C_2\sin 2x)$,
       $y_2 = e^{3x}[C_1(\sin 2x - \cos 2x) - C_2(\sin 2x + \cos 2x)]$.
   (d) $y_1 = 5C_1\cos 3x + 5C_2\sin 3x$,
       $y_2 = C_1(\cos 3x + 3\sin 3x) + C_2(\sin 3x - 3\cos 3x)$.

6. (a) $y_1 = -(C_1 + C_2)e^{-x} + 2C_3 e^{8x}$,
       $y_2 = 2C_1 e^{-x} + C_3 e^{8x}$, $y_3 = C_2 e^{-x} + 2C_3 e^{8x}$.
   (c) $y_1 = 1 - e^{-x}$, $y_2 = 1 - e^{-x}$, $y_3 = 2e^{-x} - 1$.

7.6   Some more advanced topics in ODE (*)

7.6.1   The method of Frobenius; second order equations with variable coefficients

In Section 7.4 we saw that a second order homogeneous linear differential equation with variable coefficients is of the form

$$a(x)y'' + b(x)y' + c(x)y = 0,$$

where the coefficient $a(x)$, $b(x)$ and $c(x)$ are continuous real-valued functions. In this section we will study one of these, namely the *Hermite equation* using the method of Frobenius[13]. Its solutions involve the so called *Hermite polynomials*, H_n, $(n = 0, 1, 2, ...)$.

Here we shall assume that the coefficient functions $a(x)$, $b(x)$ and $c(x)$ are all defined and real analytic on some interval (α, β) with

---

[13]G. Frobenius (1849-1917) taught in Berlin and Zurich. He worked on elliptic functions, differential equations, group representations and division algebras.

$a(x) \neq 0$ everywhere on (α, β). In fact, in many cases $a(x)$, $b(x)$ and $c(x)$ will be polynomials on $(-\infty, \infty)$, or an open interval (α, β) in $\mathbb{R}$.

As we observed since $a(x) \neq 0$ everywhere on J we can divide by it and write

$$y'' + p(x)y' + q(x)y = 0, \tag{7.34}$$

where $p(x) = \frac{b(x)}{a(x)}$ and $q(x) = \frac{c(x)}{a(x)}$ are themselves real analytic on (α, β) as well.[14]

As a result of this our next theorem will show that the solutions of equation (7.34) are also analytic on (α, β). We shall also see that, just as in the case of second order constant coefficient equations, the solutions depend on two arbitrary constants a_0 and a_1.

Theorem 7.6.1. *Let $x_0 \in (\alpha, \beta)$ and a_0 and a_1 be arbitrary constants. Then there is a unique solution $y(x)$ to the equation (7.34)*

$$y'' + p(x)y' + q(x)y = 0$$

in some neighborhood of x_0 in J which is real analytic there and satisfies the initial conditions $y(x_0) = a_0$ and $y'(x_0) = a_1$.

Proof. By translation we may assume $x_0 = 0$. This will make our notation simpler, so instead of power series in $x - x_0$ we just have to consider power series in x. Let $p(x) = \sum_{n=0}^{\infty} p_n x^n$ and $q(x) = \sum_{n=0}^{\infty} q_n x^n$ converge for $|x| < r$, where $r > 0$. If $y(x)$ is supposed to end up a power series about 0, then we write

$$y(x) = \sum_{n=0}^{\infty} a_n x^n$$

assuming it also converges for $|x| < r$, and try to find what requirements the a_n have to satisfy. Differentiating term by term and using multiplying and adding convergent power series substuting into equation (7.34)

---

[14]To see this, one extends both $a(x)$ and $b(x)$ as analytic functions into the complex domain (in a disk centered at the midpoint of J and of radius $\frac{1}{2}$ the length of (α, β)). Then by continuity, making the disk, say D, smaller if necessary to make sure that $a(z) \neq 0$ in this subdisk. Now we have two holomorphic functions $a(z)$ and $b(z)$ defined on D with $a(z)$ never vanishing. By the usual differentiation formula for a quotient we see that $\frac{b(z)}{a(z)}$ is holomorphic and therefore analytic. Hence so is its restriction, $\frac{b(x)}{a(x)}$. See for example, [25].

we get

$$\sum_{n=0}^{\infty}[(n+1)(n+2)a_{n+2} + \sum_{n=0}^{n} p_{n-k}(k+1)a_{k+1} + \sum_{n=0}^{n} q_{n-k}a_k]x^n = 0.$$

Now since a power series which is identically zero must have all its coefficients zero (and conversely), we obtain the recursion formula

$$(n+1)(n+2)a_{n+2} = -\sum_{n=0}^{n}[(k+1)p_{n-k}a_{k+1}+q_{n-k}a_k], \quad (n \geq 0). \quad (7.35)$$

To get a better feel for this let us write out the first few of these equations.

- $2a_2 = -(p_0a_1 + q_0a_0)$

- $2 \cdot 3a_3 = -(p_1a_1 + 2p_0a_2 + q_1a_0 + q_0a_1)$

- $3 \cdot 4a_4 = -(p_2a_1 + 2p_1a_2 + 3p_0a_3 + q_2a_0 + q_1a_1 + q_0a_2)$

These formulas determine the a_n for $n \geq 2$ in terms of a_0 and a_1. For instance the first formula determines a_2. Having done so, the second formula determines a_3 etc, and they do so in such a way that the resulting formal series satisfies equation (7.34). Furthermore from the formulas $y(x) = \sum_{n=0}^{\infty} a_n x^n$ and and $y'(x) = \sum_{n=0}^{\infty}(n + 1)a_{n+1}x^n$ it follows immediately that $y(0) = a_0$ and $y'(0) = a_1$. It only remains to show that the series for $y(x)$ converges in J.

We now complete the proof of Theorem 7.6.1 by showing that y is analytic. Let $0 < r_0 < r$. Since the series for $p(x)$ and $q(x)$ converge for $x = r_0$, there is a constant M for which both $|p_n r_0^n| \leq M$ and $|q_n r_0^n| \leq M$ for all $n \geq 0$. Hence $(n+1)(n+2)|a_{n+2}| \leq \frac{M}{r_0^n} \sum_{k=0}^{n}[(k+1)|a_{k+1}|+|a_k|]r_0^k$. Throwing in $M|a_{n+1}|r$ which can only make the inequality more so, we get

$$(n+1)(n+2)|a_{n+2}| \leq \frac{M}{r_0^n} \sum_{k=0}^{n}[(k+1)|a_{k+1}| + |a_k|]r_0^k + M|a_{n+1}|r_0.$$

For $n \geq 0$ we define a sequence of non negative terms b_n by $b_0 = |a_0|$, $b_1 = |a_1|$ and for $n \geq 2$ by

$$(n+1)(n+2)b_{n+2} = \frac{M}{r_0^n} \sum_{k=0}^{n} [(k+1)b_{k+1} + b_k] r_0^k + M b_{n+1} r_0. \quad (7.36)$$

Evidently $0 \leq |a_n| \leq b_n$ for all n. The question then is for which x does the series $\sum_{n=0}^{\infty} b_n x^n$ converge? Equation (7.36) tells us that

$$n(n+1)b_{n+1} = \frac{M}{r_0^{n-1}} \sum_{k=0}^{n-1} [(k+1)b_{k+1} + b_k] r_0^k + M b_n r_0.$$

and

$$(n-1)(n)b_n = \frac{M}{r_0^{n-2}} \sum_{k=0}^{n-2} [(k+1)b_{k+1} + b_k] r_0^k + M b_{n-1} r_0.$$

Multiplying the first of these by r_0 and substituting the second into it yields
$r_0 n(n+1)b_{n+1} = \frac{M}{r_0^{n-2}} \sum_{k=0}^{n-2} [(k+1)b_{k+1} + b_k] r_0^k + r_0 M(n b_n + b_{n-1}) + M b_n r_0^2$. But this is $[(n-1)n + r_0 M n + M r_0^2] b_n$. Thus

$$\frac{b_{n+1}}{b_n} = \frac{(n-1)n + r_0 M n + M r_0^2}{r_0 n(n+1)}.$$

From this we see easily that

$$\lim_{n \to \infty} \frac{b_{n+1}}{b_n} = \frac{1}{r_0},$$

and hence that

$$\lim_{n \to \infty} \frac{b_{n+1} x^{n+1}}{b_n x^n} = \frac{|x|}{r_0}.$$

The ratio test (see, Appendix B) tells us the series $\sum_{n=0}^{\infty} b_n x^n$ converges for $|x| < r_0$ and the comparison test tells us the same is true for $\sum_{n=0}^{\infty} a_n x^n$. Since this is so for all $r_0 < r$ we conclude $\sum_{n=0}^{\infty} a_n x^n$ converges whenever $|x| < r$. $\qquad \square$

Exercise 7.6.2. (*) Apply the method of Frobenius to solve the *harmonic oscillator* equation $\frac{d^2y}{dx^2} + \omega^2 y = 0$ where ω is a real parameter. Do the same with the *harmonic oscillator with drag* $\frac{d^2y}{dx^2} - k\frac{dy}{dx} + \omega^2 y = 0$, where k is a positive constant. How do these solutions compare with those of Section 7.4?

7.6.2   The Hermite equation

Here we solve the differential equation called the *Hermite equation*[15]

$$y'' - 2xy' + \lambda y = 0, \quad -\infty < x < \infty, \qquad (7.37)$$

where $\lambda \in \mathbb{R}$ is a parameter. Since in equation (7.37) the $a(x) \equiv 1$ the question of quotients of analytic functions doesn't arise at all. Here the interval $(\alpha, \beta) = \mathbb{R}$.

Solution. Theorem 7.6.1 tells us that all solutions are given by power series

$$y(x) = \sum_{n=0}^{\infty} a_n x^n$$

with $a_0 = y(0)$ and $a_1 = y'(0)$ arbitrary, and this series converges for $|x| < \infty$. We shall also shortly impose certain *boundary conditions* on the solutions.

Notice that if $y(x)$ is a solution to Hermite's equation, then so is $y(-x)$. Hence $y(x) + y(-x)$ and $y(x) - y(-x)$ are also solutions, so we can concern ourselves only with *even* and *odd* solutions. This means we consider power series of the form $y(x) = \sum_{n=0}^{\infty} a_{2n} x^{2n}$, or $y(x) = \sum_{n=0}^{\infty} a_{2n+1} x^{2n+1}$. The recursion formulas (7.35) then say

$$\frac{a_{n+2}}{a_n} = \frac{2n - \lambda}{(n+1)(n+2)}. \qquad (7.38)$$

The boundary conditions that we alluded to earlier are that $y(x) = O(x^k)$ as $x \to \pm\infty$ for a certain finite power k.[16] So taking either

---

[15]C. Hermite (1822-1901), a 19th century mathematician who made many contributions in number theory, quadratic forms, analysis and algebra. In 1873 he proved the transcendence of e.

[16]$y(x) = O(x^k)$ as $x \to \pm\infty$ means $\frac{y(x)}{x^k}$ is bounded near $\pm\infty$.

an even or odd solution we see that these boundary conditions force $\lambda = 2n$ for some n. For otherwise, if $a_0 \neq 0$ (resp. $a_1 \neq 0$), then in either case we have an infinite series and $y(x)$ can not be $O(x^k)$. Of course if either $a_0 = 0$ (resp. $a_1 = 0$) then the even (resp. odd) function is identically zero. Thus we know $\lambda = 2n$ and therefore our solutions, satisfying the boundary conditions, either are *polynomials consisting of all even degree terms, or those consisting of all odd degree terms*! These are called *Hermite polynomials* and the equation is actually

$$y'' - 2xy' + 2ny = 0.$$

The recursion formula (7.38) tells us these polynomials are given by:

$$H_n(x) = (2x)^n - \frac{n(n-1)}{1!}(2x)^{n-2} + \frac{n(n-1)(n-2)(n-3)}{2!}(2x)^{n-4} - ...,$$

where n is either odd or even and the last term is either $(-1)^{\frac{n}{2}}\frac{n!}{\frac{n}{2}!}$, if n is even, or $(-1)^{\frac{n-1}{2}}\frac{n!}{\frac{n-1}{2}!}2x$, if n is odd. In other words,

$$H_n(x) = \sum_{k=0}^{[\frac{n}{2}]}(-1)^k\frac{n!}{k!(n-2k)!}(2x)^{n-2k}, \tag{7.39}$$

where $[\frac{n}{2}]$ means the greatest integer $\leq \frac{n}{2}$.

Thus for example,

$$H_0(x) = 1, \quad H_1(x) = 2x, \quad H_2(x) = 4x^2 - 2, \quad H_3(x) = 8x^3 - 12x,$$

$$H_4(x) = 16x^4 - 48x^2 + 12, \quad H_5(x) = 32x^5 - 16x^3 + 120x, \ ....$$

Remark 7.6.3. The function $f(x) = e^{-x^2}$ has much to do with the Hermite equation and its solutions. In fact, the reader can easily check that Hermite's equation is equivalent with the equation

$$(e^{-x^2}y')' + \lambda e^{-x^2}y = 0. \tag{7.40}$$

This is the *Sturmian*[17] form of the Hermite equation.

---

[17]F. Sturm (1803-1855) whose main work was in what is now called the *Sturm-Liouville theory* of differential equations.

Thus Hermite's polynomials satisfy

$$H_n''(x) - 2xH_n'(x) + 2nH_n(x) = 0, \tag{7.41}$$

and

$$(e^{-x^2}H_n'(x))' + 2ne^{-x^2}H_n(x) = 0 \tag{7.42}$$

Corollary 7.6.4. *(Orthogonality of the Hermite polynomials). The Hermite polynomials are orthogonal with respect to the weight e^{-x^2} on the interval $(-\infty, \infty)$, that is,*

$$\int_{-\infty}^{\infty} H_n(x)H_m(x)e^{-x^2}\,dx = 0,$$

for $m \neq n$.

Proof. Let $m \neq n$. Multiplying (7.42) by H_m we get

$$(e^{-x^2}H_n'(x))'H_m(x) + 2ne^{-x^2}H_n(x)H_m(x) = 0,$$

which is symmetric with respect to n and m. So we also have

$$(e^{-x^2}H_m'(x))'H_n(x) + 2me^{-x^2}H_m(x)H_n(x) = 0.$$

Subtracting the second of these equations from the first, we get

$$\left[e^{-x^2}(H_mH_n' - H_nH_m')\right]' + 2(n - m)e^{-x^2}H_n(x)H_m(x) = 0.$$

Now integrating from $-\infty$ to ∞,

$$\left[e^{-x^2}(H_m(x)H_n'(x) - H_n(x)H_m'(x))\right]_{-\infty}^{\infty}$$

$$+ 2(n - m)\int_{-\infty}^{\infty} e^{-x^2}H_n(x)H_m(x) = 0$$

$$= 2(n - m)\int_{-\infty}^{\infty} e^{-x^2}H_n(x)H_m(x) = 0.$$

$\square$

Hermite's polynomials are an example of the wider class of *special functions* called *orthogonal polynomials*.[18] Orthogonal polynomials can also be defined as the coefficients in expansions in powers of t of suitably chosen functions $\psi(x,t)$, called *generating functions*.

Definition 7.6.5. A *generating function* for the H_n is a function $\psi(x,t)$ satisfying

$$\psi(x,t) = \sum_{n=0}^{\infty} H_n(x)\frac{t^n}{n!}, \quad |t| < \infty.$$

We shall now find a generating function for the H_n

Proposition 7.6.6.

$$\psi(x,t) = e^{-t^2+2tx} = \sum_{n=0}^{\infty} H_n(x)\frac{t^n}{n!}, \quad |t| < \infty.$$

Proof. Using equation (7.39) we have

$$\sum_{n=0}^{\infty} H_n(x)\frac{t^n}{n!} = \sum_{n=0}^{\infty}\left[\sum_{k=0}^{[\frac{n}{2}]}(-1)^k\frac{1}{k!(n-2k)!}(2x)^{n-2k}\right]t^n$$

$$= \left[\sum_{n=0}^{\infty}\frac{(-1)^n}{n!}t^{2n}\right]\left[\sum_{n=0}^{\infty}\frac{(2xt)^n}{n!}\right] = e^{-t^2}e^{2xt} = e^{-t^2+2xt} = \psi(x,t).$$

$\square$

---

[18] A system of real functions $\{\varphi_n(x)\}$, $n=0,1,2,\ldots$ is called *orthogonal with weight* $w(x) > 0$ on the interval $[a,b]$ if

$$\int_a^b \varphi_n(x)\varphi_m(x)w(x)dx = 0,$$

for every $n \neq m$, where $[a,b]$ could be a finite interval, a ray or $\mathbb{R}$. The weight function for Hermite's polynomials is $w(x) = e^{-x^2}$ and the interval $(-\infty, \infty)$. Other examples of such polynomials are the Legendre, Laguerre, Chebyshev, Jacobi and Gegenbauer polynomials. In addition to the orthogonality property, these polynomials are solutions of differential equations of a simple form and have other general properties with many applications in mathematical physics, approximation theory, partial differential equations, engineering and etc.

As a consequence we get *Rodrigues' Formula.*[19] It gives an efficient way to generate the Hermite polynomials.

Corollary 7.6.7.

$$H_n(x) = (-1)^n e^{x^2} \frac{d^n}{dx^n} e^{-x^2}, \quad n = 0, 1, 2, 3, \ldots \qquad (7.43)$$

Proof. From $\psi(x,t) = \sum_{n=0}^{\infty} H_n(x) \frac{t^n}{n!}$ and upon differentiating the power series term by term we get

$$\frac{\partial^n \psi(x,t)}{\partial t^n} \Big|_{t=0} = H_n(x).$$

On the other hand by induction one can easily show

$$\frac{\partial^n \psi(x,t)}{\partial t^n} \Big|_{t=0} = \frac{\partial^n}{\partial t^n} (e^{x^2} e^{-(t-x)^2}) \Big|_{t=0} = (-1)^n e^{x^2} \frac{d^n}{dx^n} e^{-x^2}.$$

$\square$

The following useful formula gives the derivative of a Hermite polynomial in terms of the next lower Hermite polynomial.

Corollary 7.6.8. *For* $n \geq 1$, $H_n'(x) = 2n H_{n-1}(x)$.

Proof. This follows from $\frac{\partial \psi(x,t)}{\partial x} = 2t\psi(x,t)$. $\square$

Corollary 7.6.9. *For* $n \geq 1$, $H_{n+1}(x) - 2x H_n(x) + 2n H_{n-1}(x) = 0$.

Proof. This follows from $\frac{\partial \psi(x,t)}{\partial t} = 2(x-t)\psi(x,t)$. $\square$

Final remark: By some miracle the Hermite equation and the resultant *Hermite functions,*

$$h_n(x) = e^{\frac{-x^2}{2}} H_n(x),$$

---

[19] O. Rodrigues (1794-1851) was a banker who supported Claude Henri Saint-Simon (the founder of socialism) in his destitute old age, and became one of his earliest disciples. He discovered the above formula in 1816, but soon thereafter became interested in the scientific organization of society and never returned to mathematics.

turn out to have a significance well beyond what Hermite envisaged in the mid 19th century. For example many decades later, Hermite functions played an important role on solving the Schroedinger equation of Quantum Mechanics. The Schroedinger equation which is the quantum mechanical analogue of the harmonic oscillator of classical mechanics is the following

$$\frac{d^2\psi}{dx^2} + (\frac{2E}{h\nu} - x^2)\psi = 0,$$

where E is the the total energy, ν is the frequency, h is Planck's constant and ψ is an unknown real valued function of x, the position. There are two additional conditions to be satisfied:

1. ψ vanishes at infinity. 2. $\int_{-\infty}^{+\infty} |\psi|^2 dx = 2\pi\sqrt{(\frac{\nu m}{h})}$, where m is the mass.

Leaving aside these two conditions, with a change of notation this second order differential equation is just the Hermite equation. Hence we know it has solutions satisfying $E = h\nu(n + \frac{1}{2})$ for some non negative integer n. Moreover the solutions are the Hermite functions $\psi(x) = ce^{-x^2}2H_n(x)$, where c is a constant and H_n is the nth Hermite polynomial. Of course the first condition is automatically satisfied (Prove!) while the second forces $c = \frac{4\pi\nu m}{2^{2n}(n!)^2 h}^{\frac{1}{4}}$.

7.7   Partial differential equations

Now that we have studied differential calculus in several variables and elements of ODE we can consider some elementary, but basic, partial differential equations. The usual abbreviation for *partial differential equation* is PDE. As mentioned in the introduction, a partial differential equation is an equation where the unknown is a numerical function u of several variables. Here we shall limit ourselves to two real variables and the equation involves partial derivatives of u. So, a partial differential equation in the dependent variable u and the independent variables x, y is an equation which is of the form

$$F(x, y, u, u_x, u_y, u_{xx}, u_{xy}, u_{yy}, u_{xxx}, u_{xxy}, ...) = 0,$$

where F is a function of the indicated quantities and at least one partial derivative occurs. As in ordinary differential equations the *order* of a

PDE means the order of the highest partial derivative occuring in the
equation. The simplest example of a first order PDE is $\frac{\partial u}{\partial y} = 0$, which,
clearly, is satisfied by any function of the form $u(x,y) = f(x)$, where
f is an arbitrary function of x alone. The class of partial differential
equations which will be of most interest is the class of second order
equations. We will be especially interested in certain *linear* second order
equations which often occur in physical applications. These are *linear
second-order partial differential equation* with *constant coefficients* in u
and independent variables $(x,y) \in \Omega$, where Ω is a domain in $\mathbb{R}^2$. They
can be put in the form

$$au_{xx} + 2au_{xy} + cu_{yy} + d_1 u_x + d_2 u_y + d_0 u = g(x,y), \qquad (7.44)$$

where a, b, c, d_1, d_2, d_0 are given real numbers where at least one of a, b, c
is different from zero, and $g(x,y)$ is a given smooth function defined in
Ω. When $g = 0$ on Ω, the equation (7.44) is called *homogeneuous*.

A *solution* of (7.44) on $\Omega \subseteq \mathbb{R}^2$ is a function $u(x,y)$ of class $C^2(\Omega)$
which satisfies the PDE identically in Ω. The *general solution* of (7.44)
is the set of all its solutions. To *solve* the PDE means to determine
its general solution. As above, in a PDE rather than the constant of
integration that occurs in ordinary differential equations here there will
be an arbitrary function of all the variables except the one with respect
to which differentiation takes place.

In general, it may not be possible to write the general solution of a
PDE in a closed form as we have done for many linear second order ODE.
For this reason doing this it is important to have methods for combining
known solutions when available. For homogeneous linear PDE we have
the following simple rule. An easy consequence of the linearity of the
PDE is the so-called *Superposition Principle*: If $u_1, ..., u_k$ are solutions of
some homogeneous linear PDE and $c_1, ..., c_k$ are constants, then $c_1 u_1 +
... + c_n u_k$ is also a solution.

A fundamental technique for obtaining solutions of partial differential equations is the method of *separation of variables*. This means we
look for particular solutions in the form $u(x,y) = X(x)Y(y)$ and then
obtain ordinary differential equations for $X(x)$ and $Y(y)$ which hope-

fully can be solved. The solution thus obtained is called a *separated solution*. When one independent variable represents time we denote the function u by $u = u(x,t) = X(x)T(t)$. The methods of separation of variables and the superposition principle enable us to solve boundary value problems and initial value problems for PDE in the form of a convergent infinite series of variable separable solutions (see subsections 7.8.2, 7.8.3, 7.8.4, below). Another major technique for solving boundary value problems and initial value problems for PDE (on unbounded domains) is the *Fourier transform* method presented in Section 7.9.

Some important examples of partial differential equations of the second order which we shall solve here are the following:

- The *wave equation* $c^2 u_{xx} = u_{tt}$,   or

$$c^2 \frac{\partial^2 u}{\partial x^2} = \frac{\partial^2 u}{\partial t^2},$$

  where c^2 is a positive constant. This occurs in problems involving propagation of sound such as in a vibrating string. Here $u(x,t)$ is the displacement of the sting at position x and time t.

- *Laplace's equation*[20] $u_{xx} + u_{yy} = 0$,   or

$$\frac{\partial^2 u}{\partial x^2} + \frac{\partial^2 u}{\partial y^2} = 0.$$

  This occurs in problems involving potentials and electrostatics. A solution to Laplace's equation is called a *harmonic* function.

- The *heat equation* $c^2 u_{xx} = u_t$,   or

$$c^2 \frac{\partial^2 u}{\partial x^2} = \frac{\partial u}{\partial t},$$

  with c^2 a positive constant. This occurs in problems involving heat flow, such as in a metal rod. Here $u(x,t)$ is the temperature of the rod at position x and time t.

---

[20]P. Laplace (1749-1827) made fundamental contributions to the development of celestial mechanics, the theory of probability and mathematical physics.

In the next section we shall solve boundary value and initial value problems for the wave, Laplace and heat equations using the method of separation of variables. In general it will reduce these partial differential equations in two variables to a pair of simultaneous ordinary differential equations which will then be solved by the methods of ordinary differential equations given earlier.

Although we shall not do so here, it is useful to consider analogues of these equations in more than two variables. These are:

Let $x = (x_1, \ldots, x_n) \in \Omega \subseteq \mathbb{R}^n$. Definition 3.5.6. tells us that the Laplacian, Δ, in these n variables is given by $\Delta u = \sum_{i=1}^{n} \frac{\partial^2 u}{\partial x_i^2}$. Here the *Laplace equation* is

$$\Delta u = 0.$$

The *wave equation* is

$$c^2 \Delta u = \frac{\partial^2 u}{\partial t^2},$$

while the *heat equation* is

$$c^2 \Delta u = \frac{\partial u}{\partial t}.$$

In the last two equations, $x = (x_1, \ldots, x_n)$ are called the *space* variables, t is called the *time* variable and in this case $(x, t) = (x_1, \ldots, x_n, t) \in \mathbb{R}^{n+1}$. Notice that in the *steady state*, i.e. when $\frac{\partial u}{\partial t} = 0$ the heat equation becomes Laplace's equation.

7.8   Second order PDE in two variables

In this section we give the classification of the second order linear PDE to canonical forms and find the general solution for each form. Then we use the methods of separation of variables and superposition to solve several problems involving the wave, Laplace's and heat equations.

7.8.1   Classification and general solutions

Here we find *general solutions* of second order linear homogeneous PDE with constant coefficients. We begin with a historical example, the wave equation which was first investigated and solved by D'Alembert and Euler in the mid eighteenth century.

Example 7.8.1. Find the general solution of the wave equation

$$c^2 \frac{\partial^2 u}{\partial x^2} = \frac{\partial^2 u}{\partial t^2}, \tag{7.45}$$

Solution. First note that, if φ and ψ are twice differentiable functions of one real variable, then a direct calculation using the chain rule shows that $u(x,t) = \varphi(x+ct) + \psi(x-ct)$ satisfies the wave equation. Proving the converse, is in effect, solving the wave equation. To do this we perform a change of variables $\xi = x + ct$ and $\eta = x - ct$ and see what the wave equation looks like in the new variables ξ and η. Solving for x and t we get

$$x = \tfrac{\xi+\eta}{2} \quad \text{and} \quad t = \tfrac{\xi-\eta}{2c}.$$

Moreover, $\frac{\partial x}{\partial \xi} = \frac{\partial x}{\partial \eta} = \frac{1}{2}$, $\frac{\partial t}{\partial \xi} = \frac{1}{2c} = -\frac{\partial t}{\partial \eta}$ and $\frac{\partial^2 x}{\partial \eta \partial \xi} = 0 = \frac{\partial^2 t}{\partial \eta \partial \xi}$.

By the Chain Rule,

$$\frac{\partial u}{\partial \xi} = \frac{\partial u}{\partial x}\frac{\partial x}{\partial \xi} + \frac{\partial u}{\partial t}\frac{\partial t}{\partial \xi}.$$

Hence

$$\frac{\partial^2 u}{\partial \eta \partial \xi} = \frac{\partial}{\partial \eta}\left(\frac{\partial u}{\partial x}\frac{\partial x}{\partial \xi}\right) + \frac{\partial}{\partial \eta}\left(\frac{\partial u}{\partial t}\frac{\partial t}{\partial \xi}\right)$$

$$= \frac{\partial u}{\partial x}\frac{\partial^2 x}{\partial \eta \partial \xi} + \frac{\partial^2 u}{\partial x^2}\frac{\partial x}{\partial \eta}\frac{\partial x}{\partial \xi} + \frac{\partial u}{\partial t}\frac{\partial^2 t}{\partial \eta \partial \xi} + \frac{\partial^2 u}{\partial t^2}\frac{\partial t}{\partial \eta}\frac{\partial t}{\partial \xi}$$

$$= \frac{1}{4}\left[\frac{\partial^2 u}{\partial x^2} - \frac{1}{c^2}\frac{\partial^2 u}{\partial t^2}\right] = 0.$$

Thus in the new variables (7.45) becomes

$$\frac{\partial^2 u}{\partial \eta \partial \xi} = 0.$$

As $\frac{\partial^2 u}{\partial \eta \partial \xi} = \frac{\partial}{\partial \xi}\left(\frac{\partial u}{\partial \eta}\right) = 0$, we have to solve

$$\frac{\partial w}{\partial \xi} = 0 \quad \text{and} \quad \frac{\partial u}{\partial \eta} = w.$$

Now, $\frac{\partial w}{\partial \xi} = 0$, implies $w = f(\eta)$, where f is an arbitrary smooth function on η alone. The equation $\frac{\partial u}{\partial \eta} = w$ in turn, implies that

$$u(\xi, \eta) = \int_0^\eta f(s)\,ds + \varphi(\xi),$$

where φ is a arbitrary smooth function of ξ alone. But then,

$$u(\xi, \eta) = \psi(\eta) + \varphi(\xi),$$

where ψ is (again) an arbitrary smooth function with $\psi' = f$. Thus, returning to the old variables we get

$$u(x, t) = \varphi(x + ct) + \psi(x - ct).$$

These solutions are known as *D'Alembert's solution*[21] of the wave equation.

Next we consider the general homogeneous linear second-order partial differential equations of the form (7.44). For simplicity we write

$$au_{xx} + 2bu_{xy} + cu_{yy} + \ldots = 0, \tag{7.46}$$

where $a, b, c, \in \mathbb{R}$ with $a^2 + b^2 + c^2 \neq 0$ and the dots represent terms with derivatives of lower orders. This, of course, includes the wave equation, Laplace's equation and the heat equation. Equation (7.46) will be solved by methods quite analogous to those of ordinary differential equations which were treated earlier.

The second order terms are called the *principal part* (or the *symbol*) of the equation and as we shall see they are decisive. The coefficient matrix of the principal part of the equation is

$$A = \begin{bmatrix} a & b \\ b & c \end{bmatrix},$$

---

[21] J. D'Alembert (1717-1783). Scholar and mathematician specializing in mechanics. He was the leading mathematician of the group of *philosophers* who wrote the *Encyclopedie* (28 vols., 1751-1772).

and the corresponding quadratic form of the symmetric matrix A is

$$Q(\xi, \eta) = a\xi^2 + 2b\xi\eta + c\eta.$$

The sign of the determinat $\det(A) = ac - b^2$ turns out to be invariant under smooth nonsingular transformations of coordinates. In addition, it is possible to show that there exists a linear transformation of variables x, y, which reduces equation (7.46) to one of the following forms, called *canonical forms*[22]. In this respect the role played by the sign of $\det(A)$ is decisive. Recalling Sylvester's criterion (Theorem 3.7.20) the sign of $\det(A)$ also determines the nature of the quadratic form Q.

1. If $\det(A) < 0$, that is, $ac - b^2 < 0$, the equation is reducible by a linear change of variables to the form

$$u_{xx} - u_{yy} + ... = 0,$$

    called the *hyperbolic form*. The quadratic form Q in this case is indefinite.

2. If $\det(A) > 0$, that is, $ac - b^2 > 0$, the equation is reducible to the form

$$u_{xx} + u_{yy} + ... = 0,$$

    called the *elliptic form*. In this case the quadratic form Q is strictly definite (positive or negative)

3. If $\det(A) = 0$, that is, $ac - b^2 = 0$, the equation is reducible to the form

$$u_{xx} + ... = 0, \text{ or } u_{yy} + ... = 0$$

    called the *parabolic form*. Here Q is degenerate.

In general, when the second order linear nonhomogeneous PDE involves more than two variables, it is of the form

$$\sum_{i,j=1}^{n} a_{ij} u_{x_i x_j} + ... = g(x)$$

---

[22]See for example, [36].

and the classification to canonical forms is based on the quadratic form $Q(v) = \langle Av, v \rangle$ associated to the coefficient (symmetric) matrix $A = (a_{ij})$. The equation is called *elliptic* if Q is strictly definite (positive or negative definite), *hyperbolic* if Q is indefinite but nondegenerate and *parabolic* if it is degenerate.[23] The significance of this classification is that properties and methods of solution for the three types of differential equations show considerable differences.

Now, if we want to solve equation

$$au_{xx} + 2bu_{xy} + cu_{yy} = 0, \qquad (7.47)$$

for $u = u(x, y)$, consider the function $u(x, y) = \phi(y + \lambda x)$, where λ is an unknown constant to be determined later and ϕ an arbitrary (but *non-linear*) smooth function of one variable. *When does this u satisfy the equation?*

A direct calculation using the Chain Rule shows $u(x, y) = \phi(y + \lambda x)$ satisfies the equation if and only if

$$\phi''(y + \lambda x)(a\lambda^2 + 2b\lambda + c) = 0.$$

Since this would have to hold for all x and y if the first factor were zero this would mean the function ϕ is linear. Hence the equation is satified for such a ϕ if and only if

$$a\lambda^2 + 2b\lambda + c = 0. \qquad (7.48)$$

Equation (7.48) is called the *auxilliary equation* and of course can be solved by the quadratic formula. We may as well have assumed $a \neq 0$, because if by chance $a = 0$ then we would merely reverse the roles of the variables so that c becomes a. The only problem here would be if $a = 0 = c$. Then our equation would be $bu_{xy} = 0$. Since if b is also 0 we have no equation at all, we can divide by b and get $u_{xy} = 0$, or

$$\frac{\partial^2 u}{\partial x \partial y} = 0,$$

---

[23]Actually, even when the coefficients of the PDE are *not* constant, $\sum_{i,j=1}^{n} a_{ij}(x)u_{x_i x_j} + \dots = g(x)$ is said to be *elliptic, hyperbolic* or *parabolic* in a domain Ω of $\mathbb{R}^n$ if it is respectively elliptic, hyperbolic or parabolic at every point x of Ω. However, partial differential equations may be of different type in different parts of the region in which they are to be solved. A typical example of this is the Tricomi equation $yu_{xx} + u_{yy} = 0$.

which we have already solved. The general solution is

$$u(x,y) = \varphi(x) + \psi(y),$$

where φ and ψ are arbitrary smooth functions in one variable.

Our objective is to reduce the general case to this one, where we are assured we have a second degree auxilliary equation (7.48). Now as above in the case of ordinary differential equations there are three possibilities depending on the *discriminant* $4(b^2 - ac)$ of the auxilliary equation $a\lambda^2 + 2b\lambda + c = 0$:

1. $b^2 - ac > 0$, that is, the roots are *real* and *distinct*. The *hyperbolic* case.

2. $b^2 - ac < 0$, that is, the roots are *complex conjugates* of one another. The *elliptic* case.

3. $b^2 - ac = 0$, that is, the roots are *real* and *equal*. The *parabolic* case.[24]

For example, the wave equation, $u_{xx} - c^2 u_{yy} = 0$, is hyperbolic since its auxillary equation is $\lambda^2 - c^2 = 0$ with roots $\lambda_1 = c$ and $\lambda_2 = -c$. Laplace's equation $u_{xx} + u_{yy} = 0$ is elliptic since its auxillary equation is $\lambda^2 + 1 = 0$ with roots $\lambda_1 = i$ and $\lambda_2 = -i$, while for the heat equation the principal part is $c^2 u_{xx} = 0$ and the equation is parabolic since its auxilliary equation is $c^2 \lambda^2 = 0$ has equal roots, $\lambda_1 = \lambda_2 = 0$.

Actually, just as in ODE we can deal with all three cases simultaneously. We solve the auxilliary equation for roots λ_1 and λ_2 which may be equal or not or real or not. Then the auxilliary equation becomes $(\lambda - \lambda_1)(\lambda - \lambda_2) = 0$, and our equation becomes

$$(D - \lambda_1)(D - \lambda_2)u = 0,$$

where $D = \frac{\partial}{\partial \xi}$ means differentiate with respect to $\xi = y + \lambda x$. Since these operators commute it makes no difference which factor comes first. So we have to solve $(D - \lambda_1)w = 0$ where $w = (D - \lambda_2)u$. Thus just

---

[24]The reader should note that this classification agrees with the one given earlier in terms of the sign of the $\det(A) = ac - b^2$ since $b^2 - ac = -\det(A)$.

as in Example 7.8.1, we have reduced our second order equation to two
simultaneous first order equations

$$(D - \lambda_1)w = 0,$$

and

$$w = (D - \lambda_2)u.$$

1. (*Hyperbolic case*). If the roots are real and distinct the equation
   $(D - \lambda_1)(D - \lambda_2)u = 0$ is exactly of the form $u_{\xi\eta} = 0$, where
   $\xi = y + \lambda_1 x$ and $\eta = y + \lambda_2 x$. Thus as above, the general solution
   is

   $$u = \varphi(y + \lambda_1 x) + \psi(y + \lambda_2 x),$$

   where φ and ψ are arbitrary smooth functions of one variable.

2. (*Elliptic case*). If the roots are of complex conjugates. Here just
   as in the case of distinct real roots we get as the general solution

   $$u = \varphi(y + \lambda_1 x) + \psi(y + \lambda_2 x),$$

   where φ and ψ are arbitrary smooth functions of one variable.
   Only this time $\lambda_1 = \alpha + i\beta$ and $\lambda_2 = \alpha - i\beta$, where $\beta \neq 0$.

Let $\Phi = Re(\varphi) + iRe(\psi)$ and $\Psi = Re(\varphi) - iRe(\psi)$. It is easy to
see, and we leave this as an exercise for the reader, that

$$u = u(x, y) = \varphi(y + \lambda_1 x) + \psi(y + \lambda_2 x)$$
$$= \Phi(y + (\alpha + i\beta)x) + \Phi(y + (\alpha - i\beta)x)$$
$$+ i(\Psi(y + (\alpha + i\beta)x) - \Psi(y + (\alpha - i\beta)x)),$$

and that, by calculating the conjugate, the right hand side is real.
Thus,

$$u = \Phi(y + (\alpha + i\beta)x) + \Phi(y + (\alpha - i\beta)x)$$
$$+ i(\Psi(y + (\alpha + i\beta)x) - \Psi(y + (\alpha - i\beta)x))$$

is the *general solution* in the *elliptic* case.

3. (*Parabolic case*). If the roots are real and equal we have

$$(D - \lambda_1)w = 0, \text{ where } w = (D - \lambda_1)u.$$

Hence from the first equation $w = f(y + \lambda_1 x)$ for some function f. So we have to solve $(D - \lambda_1)(f(y + \lambda_1 x)) = 0$ and just as in the ODE case we get the general solution

$$u = g(y + \lambda_1 x) + x f(y + \lambda_1 x),$$

where g and f are arbitrary smooth functions of one variable.

Because of its importance we write explicitly the general solution for the *Laplace equation* $\Delta u = u_{xx} + u_{yy} = 0$. As we saw the roots of the auxilliary equation are $\pm i$. Thus the general solution to Laplace's equation in two variables is

$$u(x, y) = \Phi(y + ix) + \Phi(y - ix) + i(\Psi(y + ix) - \Psi(y - ix)).$$

So, for *example*, if we choose $\Phi(t) = \sin(t)$ and $\Psi(t) = e^t$, we get the harmonic function

$$u(x, y) = 2(\sin(y)\cosh(x) - \sin(x)e^y).$$

7.8.2   Boundary value problems for the wave equation

Next we solve the wave equation satisfying *boundary* and *initial conditions* (such a problem is referred to as a *mixed-type problem*).

Small vibrations of a string

Example 7.8.2. *Solve the wave equation*

$$c^2 u_{xx} = u_{tt}, \quad 0 < x < \pi, \ t > 0 \tag{7.49}$$

with the boundary conditions

$$u(0, t) = u(\pi, t) = 0, \tag{7.50}$$

and the initial conditions

$$u(x,0) = f(x), \quad u_t(x,0) = g(x), \tag{7.51}$$

where $f(x)$ and $g(x)$ are functions given in advance.

Solution. We first seek a (nonzero) solution in the form

$$u(x,t) = X(x)T(t), \tag{7.52}$$

satisfying the boundary conditions

$u(0,t) = X(0)T(t) = 0$ and $u(\pi,t) = X(\pi)T(t) = 0$, for all $t > 0$.

Since we want a nonzero solution

$$X(0) = X(\pi) = 0. \tag{7.53}$$

Substitution of (7.52) into PDE (7.49) yields

$$c^2 X''(x)T(t) = X(x)T''(t),$$

or

$$\frac{X''(x)}{X(x)} = \frac{1}{c^2} \cdot \frac{T''(t)}{T(t)}.$$

But a function of x can be equal to a function of t only if both are equal to some constant number, say $-\lambda$ (no assumption is made at this point as to whether λ is positive, negative or zero). That is,

$$\frac{X''(x)}{X(x)} = \frac{1}{c^2} \cdot \frac{T''(t)}{T(t)} = -\lambda.$$

This gives rise to two second order linear ODE

$$X'' + \lambda X = 0 \tag{7.54}$$

$$T'' + c^2 \lambda T = 0. \tag{7.55}$$

The differential equation (7.54) together with conditions (7.53) is an eigenvalue problem and we saw in Example 7.4.14 it can have nontrivial solution only if $\lambda > 0$ and the eigenvalues are $\lambda = \lambda_n = n^2$. The corresponding eigenfunctions (nonzero solutions) are

$$X(x) = X_n(x) = \sin(nx) \text{ , for } n = 1, 2, 3, ...,$$

(we have set $B = 1$). We now turn to equation (7.55). Replacing λ in (7.55) by its value $\lambda_n = n^2$, we get

$$T'' + c^2 n^2 T = 0.$$

Setting $\omega = nc$ this is the equation $T'' + \omega^2 T = 0$ we solved in Example 7.4.8. Its solution is $T(t) = A\cos(\omega t) + B\sin(\omega t)$. That is,

$$T(t) = T_n(t) = A_n \cos(nct) + B_n \sin(nct),$$

where A_n and B_n are arbitrary constants. We have constructed a set of solutions to the wave equation satisfying the boundary conditions (7.50) given by

$$u_n(x, t) = X_n(x) \cdot T_n(t) = \sin(nx)[A_n \cos(nct) + B_n \sin(nct)], \quad (7.56)$$

for $n = 1, 2, 3, ...$ Now by the superposition principle any finite sum of these solutions

$$\sum_{n=1}^{N} u_n(x, t)$$

also satisfies equation (7.49) and the boundary conditions (7.50). The same is true for a convergent infinite sum of solutions

$$u(x, t) = \sum_{n=1}^{\infty} \sin(nx)[A_n \cos(nct) + B_n \sin(nct)], \quad (7.57)$$

if we can deal with the convergence problem, so that the series is twice differentiable termwise. For this it will be necessary that the series be uniformly and absolutely convergent.

We now try to determine the constants A_n and B_n so that the infinite series (7.57) satisfies the initial conditions (7.51). First

$$u(x, 0) = \sum_{n=1}^{\infty} A_n \sin(nx) = f(x)$$

must hold. Assuming termwise differentiation is valid we must also have

$$\frac{\partial u}{\partial t}(x, 0) = \sum_{n=1}^{\infty} cn B_n \sin(nx) = g(x).$$

Pursuing this further would take us to questions of *Fourier series*[25] which is another matter. However, for the reader who is familiar with Fourier series, we note that assuming the functions $f(x)$ and $g(x)$ have a Fourier series expansion, we see that the A_n's must be the *Fourier coefficients*[26] for $f(x)$, that is,

$$A_n = \frac{2}{\pi} \int_0^\pi f(x) \sin(nx) dx,$$

and the cnB_n must be the Fourier coefficients for $g(x)$, that is,

$$B_n = \frac{2}{cn\pi} \int_0^\pi g(x) \sin(nx) dx.$$

Remark. (*Principle of Conservation of Energy*). The quantity

$$E(t) = \int_0^\pi \left[(u_t)^2 + (u_x)^2 \right] dx$$

is called the *energy* of the solution for the problem. The first term of the inetgral denoted by $E_k(t) = \int_0^\pi (u_t)^2 dx$, (the integral of the square of the velocity of the vibrating string at the point x) represents the *kinetic energy*, while the second denoted by $E_p(t) = \int_0^\pi (u_x)^2 dx$, (the dilation of the string at the same point) represents the *potential energy* of the string. Now the *total energy* $E(t) = E_k(t) + E_p(t)$ is *constant*. Indeed,

$$\frac{dE}{dt} = \int_0^\pi \frac{\partial}{\partial t} \left[(u_t)^2 + (u_x)^2 \right] dx$$

$$= \int_0^\pi \left[2u_t u_{tt} + 2u_x u_{xt} \right] dx = 2c^2 \int_0^\pi \left[u_t u_{xx} + u_x u_{xt} \right] dx$$

$$= 2c^2 \int_0^\pi \left[\frac{\partial}{\partial x} (u_t u_x) \right] dx = 2c^2 u_t u_x \Big|_{x=0}^{x=\pi} = 0,$$

since $u_t(0,t) = u_t(\pi,t) = 0$.

---

[25] By a *Fourier series* is meant a series of the form

$$\frac{1}{2} a_0 + \sum_{n=1}^\infty (a_n \cos nx + b_b \sin nx),$$

where a_0, a_n, b_n , $n = 1, 2, 3, ...$, are constants.

[26] See, the note on Fourier series at the end of this section.

Vibrations of an infinite string. D'Alembert's formula

The boundary conditions $u(0,t) = u(\pi,t) = 0$ in the above problem of a vibrating string tell us that the string remains motionless at the points $x = 0$ and $x = \pi$ (such points are called *nodes*). If a string is very long, then its ends will exert little influence on the vibrations occuring near the middle. Hence, when considering free vibrations of an infinite string, the problem reduces to the following initial-value problem (also called the *Cauchy problem* for the two-dimensional wave equation) .

Example 7.8.3. *Find a twice continuously differentiable function* $u(x,t)$ *such that*

$$c^2 u_{xx} = u_{tt}, \quad -\infty < x < \infty, \ t > 0 \tag{7.58}$$

satisfying the initial conditions

$$u(x,0) = f(x), \quad u_t(x,0) = g(x), \tag{7.59}$$

where $f \in C^2(\mathbb{R})$ *and* $g \in C^1(\mathbb{R})$ *are given functions.*

Solution. The general solution (D'Alemberts's solution) of (7.58), was found in Example 7.8.1 to be

$$u(x,t) = \varphi(x + ct) + \psi(x - ct). \tag{7.60}$$

Now we use D'Alembert's solution (7.60) to solve the initial-value problem, that is, to determine the functions φ and ψ. Using the initial conditions (7.59) we see that

$$\varphi(x) + \psi(x) = f(x) \text{ and } c\phi'(x) - c\psi'(x) = g(x).$$

Differentiating the first equation and dividing by c in the second we obtain

$$\varphi'(x) + \psi'(x) = f'(x)$$
$$\phi'(x) - \psi'(x) = \tfrac{1}{c}g(x).$$

This is a linear system of two equations in φ' and ψ'. The solution is

$$\varphi'(x) = \tfrac{1}{2}f'(x) + \tfrac{1}{2c}g(x), \ \psi'(x) = \tfrac{1}{2}f'(x) - \tfrac{1}{2c}g(x),$$

which implies

$$\varphi(x) = \tfrac{1}{2}f(x) + \tfrac{1}{2c}\int_0^x g(s)ds + C_1 \text{ and } \psi(x) = \tfrac{1}{2}f(x) - \tfrac{1}{2c}\int_0^x g(s)ds + C_2.$$

Thus, D'Alembert solution

$$u(x,t) = \varphi(x + ct) + \psi(x - ct)$$

gives

$$u(x,t) = \frac{1}{2}[f(x + ct) + f(x - ct)] + \frac{1}{2c}\int_{x-ct}^{x+ct} g(s)ds, \qquad (7.61)$$

where $C_1 + C_2 = 0$, since $u(x,0) = f(x)$. Formula (7.61) is known as
D'Alembert's formula.

7.8.3   Boundary value problems for Laplace's equation

Dirichlet's problem

In this subsection we consider two-dimensional boundary value problems
for the Laplace equation. The typical boundary value problem for the
Laplace equation is the Dirichlet[27] problem.

Definition 7.8.4. (*Dirichlet's problem*). Let Ω be a bounded domain
(open connected set) in $\mathbb{R}^2$ with smooth boundary $\partial(\Omega)$. If f is a given
function which is defined and continuous on $\partial(\Omega)$, then the *Dirichlet
problem* is that of finding a function $u \in C^2(\Omega)$ which is:

1. defined and continuous on $\overline{\Omega} = \Omega \cup \partial(\Omega)$,

2. harmonic on Ω, i.e., $\Delta u = u_{xx} + u_{yy} = 0$ in Ω,

3. $u \equiv f$ on $\partial(\Omega)$.

The uniqueness of the solution was proved in Corollary 3.7.26. The
proof of existence of a solution of the Dirichlet problem in general is
beyond the scope of the present book[28]. Of particular importance are
the cases when the Dirichlet problem can be solved *explicitly*. Here we
solve the Dirichlet problem for a rectangle and for a disk.

---

[27]P. Dirichlet (1805-1859) made many contributions to analysis and number theory.
In 1855 he succeeded Gauss as professor at the University of Göttingen.

[28]See, for example, [27].

Example 7.8.5. (*Dirichlet's problem for a rectangle*). Let $\Omega = \{(x,y) : 0 < x < a,\ 0 < y < b\}$ be a rectangle in $\mathbb{R}^2$. We solve the problem

$$\Delta u = 0 \ \ in\ \Omega :\tag{7.62}$$

$$u(0,y) = u(a,y) = 0\tag{7.63}$$

$$u(x,0) = 0,\ \ u(x,b) = f(x)\tag{7.64}$$

Solution. We look for a variable separable (nonzero) solution in the form

$$u(x,y) = X(x)Y(y)\tag{7.65}$$

Substitution into the equation $\Delta u = 0$ and division by XY gives

$$\frac{X''}{X} = -\frac{Y''}{Y}.$$

Reasoning as in Example 7.8.2, there must exist a constant λ such that

$$X'' + \lambda X = 0$$

for $0 < x < a$, and

$$Y'' - \lambda Y = 0$$

for $0 < y < b$. Moreover, the homogeneous boundary conditions give $X(0) = X(a) = 0$ and $Y(0) = 0$. The eigenvalue problem

$$X'' + \lambda X = 0,\ \ X(0) = X(a) = 0$$

has eigenvalues $\lambda = \lambda_n = \left(\frac{n\pi}{a}\right)^2$, $n = 1, 2, 3, ...$, and eigenfunctions

$$X(x) = X_n(x) = \sin\left(\frac{n\pi x}{a}\right).$$

Now, we turn to the variable y. Setting $\beta_n^2 = \lambda_n$ and solving the ODE

$$Y'' - \beta_n^2 Y = 0,$$

(see, Example 7.4.5), we obtain

$$Y(y) = Y_n(y) = a_n e^{\beta_n y} + b_n e^{-\beta_n y}.$$

Here is preferable to get the general solution in terms of hyperbolic functions

$$Y_n(y) = A_n \cosh(\beta_n y) + B_n \sinh(\beta_n y).$$

Using the homogeneous boundary condition $Y(0) = 0$ (the nonhomogeneous boundary condition for $y = b$ will be considered in the last step), we get $A_n = 0$, and we obtain

$$Y_n(y) = B_n \sinh(\beta_n y) = B_n \sinh\left(\frac{n\pi y}{a}\right).$$

Thus, the functions

$$u_n(x,y) = X_n(x) \cdot Y_n(y) = B_n \sin\left(\frac{n\pi x}{a}\right) \sinh\left(\frac{n\pi y}{a}\right)$$

are solutions that satisfy the Laplace equation and all homogeneous boundary conditions. Superimposing (and assuming convergence) we arrive at the series solution

$$u(x,y) = \sum_{n=1}^{\infty} B_n \sin\left(\frac{n\pi x}{a}\right) \sinh\left(\frac{n\pi y}{a}\right)$$

that represents a harmonic function on the rectangle Ω and satisfies the homogeneous boundary conditions. Finally, we also require the solution to satisfy the boundary condition $u(x,b) = f(x)$, that is,

$$f(x) = u(x,b) = \sum_{n=1}^{\infty} B_n \sinh\left(\frac{n\pi b}{a}\right) \sin\left(\frac{n\pi x}{a}\right).$$

Setting $C_n = B_n \sinh(\frac{n\pi b}{a})$, we see that

$$f(x) = \sum_{n=1}^{\infty} C_n \sin\left(\frac{n\pi x}{a}\right). \tag{7.66}$$

Again pursuing this further would take us to questions of Fourier series. The reader who is familiar with the theory of Fourier series will recognize that expression (7.66) is the Fourier series of the function $f(x)$ (assuming that f permits such an expansion), and so C_n is the Fourier coefficient

$$C_n = \frac{2}{a} \int_0^a f(x) \sin\left(\frac{n\pi x}{a}\right) dx.$$

Hence, we obtain formulas for the unknown coefficients B_n, in fact

$$B_n = \frac{2}{a \sinh(\frac{n\pi b}{a})} \int_0^a f(x) \sin\left(\frac{n\pi x}{a}\right) dx.$$

Next we solve the Dirichlet problem in the case where Ω is a disk. This case has historic interest: Fourier's work on this problem was actually the origin of Fourier analysis.

Example 7.8.6. (*Dirichlet's problem for a disk*).   Let $\Omega = \{(x, y) : x^2 + y^2 < a^2\}$ *be a disk in* $\mathbb{R}^2$ *of radius a. The boundary of the disk is the circle* $\partial(\Omega) = \{(x, y) : x^2 + y^2 = a^2\}$. *We solve the problem*

$$\Delta u = 0 \quad in \ \ \Omega \tag{7.67}$$

$$u \equiv f \quad on \ \ \partial(\Omega), \tag{7.68}$$

where f is a continuous periodic function of period 2π with a piecewise continuous derivative defined on $\partial(\Omega)$.

Solution. As the geometry suggests here, we use polar coordinates $x = r\cos\theta$ and $y = r\sin\theta$. Now $\partial(\Omega) = \{(r, \theta) : r = a, \ -\pi \le \theta \le \pi\}$. From Example 3.5.8, the Laplace equation in polar cordinates is

$$\frac{\partial^2 u}{\partial r^2} + \frac{1}{r^2}\frac{\partial^2 u}{\partial \theta^2} + \frac{1}{r}\frac{\partial u}{\partial r} = 0, \tag{7.69}$$

and the boundary condition can be written in the form

$$u(a, \theta) = f(\theta), \ \ \theta \in (-\pi, \pi]. \tag{7.70}$$

Applying the method of separation of variables, we look for a nonzero solution of Laplace's equation in the form

$$u(r, \theta) = R(r) \cdot \Theta(\theta). \tag{7.71}$$

Substituting (7.71) into (7.69), multiplying through by r^2 and dividing by $R\Theta$, we obtain two ordinary differential equations

$$\Theta'' + \lambda\Theta = 0 \tag{7.72}$$

$$r^2 R'' + r R' - \lambda R = 0, \tag{7.73}$$

where λ is a constant.

Since $f(\theta) = u(a, \theta) = R(a)\Theta(\theta)$, the function $\Theta(\theta)$ must be continuous and periodic with period 2π. Hence, $\Theta(0) = \Theta(2\pi)$, $\Theta'(0) = \Theta'(2\pi)$. The eigenvalue problem of equation (7.72) with the above boundary conditions has eigenvalues $\lambda = n^2$ ($n = 0, 1, 2, 3, ...$) and corresponding eigenfunctions

$$\Theta_n(\theta) = a_n \cos(n\theta) + b_n \sin(n\theta).$$

Turning now to equation (7.73) we set $\lambda = n^2$ and we make a change of variable $r = e^t$, i.e. $\phi(t) = R(e^t)$. This has the effect of changing the equation (7.73) with variable coefficients to one with *constant coefficients*. The equation becomes $\phi''(t) - n^2\phi(t) = 0$. So the general solution is $\phi(t) = Ae^{nt} + Be^{-nt}$. Or, going back to $R(r)$ we get $R(r) = R_n(r) = Ar^n + Br^{-n}$, when $n = 1, 2, 3, ...$, and $R(r) = A + B\log r$ when $n = 0$. If we also want the solution to extend to the entire plane and say be bounded in a neighborhood of the origin, then we must have $B = 0$. Since the remaining constant A can be absorbed into the a_n and b_n we see that

$$u_n(r, \theta) = R_n(r)\Theta_n(\theta) = r^n(a_n \cos(n\theta) + b_n \sin(n\theta)).$$

Now by superposition we see that a very general solution of Laplace's equation is furnished by

$$u(r, \theta) = \sum_{n=0}^{\infty} r^n(a_n \cos(n\theta) + b_n \sin(n\theta)). \qquad (7.74)$$

For $n = 0$ (7.74) gives $u(r, \theta)$ is a constant, which we can take it to be $\frac{1}{2}a_0$, and so

$$u(r, \theta) = \frac{1}{2}a_0 + \sum_{n=1}^{\infty} r^n(a_n \cos(n\theta) + b_n \sin(n\theta)). \qquad (7.75)$$

The boundary condition $u(a, \theta) = f(\theta)$ is fulfilled provided the function f has a Fourier series expansion

$$f(\theta) = \frac{1}{2}a_0 + \sum_{n=1}^{\infty} a^n(a_n \cos(n\theta) + b_n \sin(n\theta)), \qquad (7.76)$$

and again the reader familiar with Fourier series will recognize that

$$a_n = \frac{1}{\pi a^n} \int_{-\pi}^{\pi} f(\phi) \cos(n\phi) d\phi \tag{7.77}$$

and

$$b_n = \frac{1}{\pi a^n} \int_{-\pi}^{\pi} f(\phi) \sin(n\phi) d\phi. \tag{7.78}$$

Remark 7.8.7. (*Poisson's integral formula*)[29] Using formulas (7.77) and (7.78) the sum of the series in (7.75) can be expressed explicitly by an integral formula. In fact,

$$u(r,\theta) = \frac{1}{2}a_0 + \sum_{n=1}^{\infty} \frac{r^n}{\pi a^n} \int_{-\pi}^{\pi} f(\phi)[\cos(n\phi)\cos(n\theta) + \sin(n\phi)\sin(n\theta)]d\phi$$

$$= \frac{1}{2}a_0 + \frac{1}{\pi}\sum_{n=1}^{\infty} (\frac{r}{a})^n \int_{-\pi}^{\pi} f(\phi)[\cos(n(\theta - \phi))]d\phi$$

$$= \frac{1}{\pi} \int_{-\pi}^{\pi} f(\phi) \left[\frac{1}{2} + \sum_{n=1}^{\infty} (\frac{r}{a})^n \cos(n(\theta - \phi)) \right] d\phi.$$

Setting $\alpha = \theta - \phi$ and using Euler's identity to express the cosine function $\cos(\alpha) = \frac{1}{2}(e^{i\alpha} + e^{-i\alpha})$, we can rewrite the expression in the brackets as:

$$\frac{1}{2} + \sum_{n=1}^{\infty} (\frac{r}{a})^n \cos(n\alpha)$$

$$= \frac{1}{2} \left[1 + \sum_{n=1}^{\infty} (\frac{r}{a})^n e^{in\alpha} + \sum_{n=1}^{\infty} (\frac{r}{a})^n e^{-in\alpha} \right].$$

Both these series are geometric. Since $|\frac{r}{a}e^{\pm i\alpha}| < 1$ for $r < a$. Summing them up yields

$$\frac{1}{2} + \sum_{n=1}^{\infty} (\frac{r}{a})^n \cos(n\alpha) = \frac{1}{2} \left[1 + \frac{re^{i\alpha}}{a - re^{i\alpha}} + \frac{re^{-i\alpha}}{a - re^{-i\alpha}} \right]$$

---

[29]S. Poisson (1781-1840), a student of Lagrange and Laplace at Ecole Polytechnique where he eventually taught for many years. Poisson made numerous important contributions in mathematical analysis, probability theory, applied mathematics and mathematical physics.

$$= \frac{1}{2} \left[\frac{a^2 - r^2}{a^2 - 2ar\cos(\alpha) + r^2} \right].$$

Putting this back to the integral, we finally obtain the solution u in the form

$$u(r, \theta) = \frac{a^2 - r^2}{2\pi} \int_{-\pi}^{\pi} \frac{f(\phi)}{a^2 - 2ar\cos(\theta - \phi) + r^2} d\phi, \qquad (7.79)$$

$0 \le r < a$, which is called the *Poisson integral formula*. The function

$$G(\theta, \phi) = \frac{1}{a^2 - 2ar\cos(\theta - \phi) + r^2}$$

is called the *Poisson kernel*.

The Poisson integral formula has some important theoretical consequences for the solution of Laplace's equation. As just one simple instance, we note that for $r = 0$ yields

$$u(0, \theta) = \frac{1}{2\pi} \int_{-\pi}^{\pi} f(\phi) d\phi,$$

which tells us that the value of the harmonic function u at the center of the disk Ω is equal to the mean value of u on the circle $\partial(\Omega)$, (this is known as the *Mean Value property*). Physically, this tells that in the steady-state the value of the temperature at the origin is the average value of the temperatures around the circular boundary.

7.8.4   Boundary value problems for the heat equation

Finally we solve the heat equation also by separation of variables.

Example 7.8.8. *Solve the heat equation*

$$u_t - c^2 u_{xx} = 0, \quad 0 \le x \le L, \ t \ge 0,$$

satisfying the boundary conditions $u(0, t) = u(L, t) = 0$.

Solution. We are interested in (nonzero) solutions of the form $u(x, t) = X(x)T(t)$. Separating variables, the equation takes the form

$$\frac{X''(x)}{X(x)} = c^{-2} \frac{T'(t)}{T(t)}.$$

As usual, this term is evidently independant of both x and t and so is constant. Calling this constant $-\lambda$ we get two ordinary differential equations

$$X''(x) + \lambda X(x) = 0 \text{ and } T'(t) + \lambda c^2 T(t) = 0.$$

The boundary conditions give $X(0) = X(L) = 0$. The eigenvalue problem for the first equation has eigenvalues $\lambda_n = \frac{n^2\pi^2}{L^2}$ and eigenfunctions

$$X_n(x) = A_n \sin\left(\frac{n\pi x}{L}\right) \quad (n = 1, 2, ...),$$

where A_n are arbitrary constants. Substituting λ_n in the second (first order) equation and solving (as per case 1 of Theorem 7.2.1) we find

$$T_n(t) = e^{-c^2 n^2 \pi^2 t/L^2}.$$

Thus

$$u_n(x,t) = A_n e^{-c^2 n^2 \pi^2 t/L^2} \sin\left(\frac{n\pi x}{L}\right).$$

Superimposing solutions as above leads to a (convergent) series

$$u(x,t) = \sum_{n=1}^{\infty} A_n e^{-c^2 n^2 \pi^2 t/L^2} \sin\left(\frac{n\pi x}{L}\right),$$

which satisfies both the heat equation (provided the series $\sum_{n=1}^{\infty} |A_n| < \infty$ converges), and the boundary conditions $u(0,t) = u(L,t) = 0$.

   Finally, if we also impose the initial condition $u(x,0) = f(x)$, under the assumption that the function f has the Fourier series expansion, then

$$u(x,0) = f(x) = \sum_{n=1}^{\infty} A_n \sin\left(\frac{n\pi x}{L}\right).$$

The numbers A_n can be determined by the formula

$$A_n = \frac{2}{L} \int_0^L f(x) \sin\left(\frac{n\pi x}{L}\right) dx.$$

We close this section with an example of the initial boundary value problem for a *nonhomogeneous* heat equation. The method of solving this problem is (in certain sense) analogous to the method of variation of parameters which is used for solving nonhomogeneous ODEs (see, Exercise 7.10.3).

Example 7.8.9. *Solve the nonhomogeneous heat equation*

$$u_t - c^2 u_{xx} = g(x, t), \quad 0 \le x \le 1, \ t \ge 0. \tag{7.80}$$

satisfying the initial condition $u(x, 0) = f(x)$ and the boundary conditions $u(0, t) = u(1, t) = 0$.

Solution. First we solve the corresponding homogeneous problem, as above. Here $L = 1$, and the eigenvalues are $\lambda_n = (n\pi)^2$ and the eigenfunctions $X_n(x) = \sin(n\pi x)$, $(n = 1, 2, ...)$. We assume that we can expand all data of the problem (viz, the functions f and g) in a Fourier series with respect to the eigenfunctions $\sin(n\pi x)$. That is,

$$f(x) = \sum_{n=1}^{\infty} A_n \sin(n\pi x),$$

where $A_n = 2 \int_0^1 f(x) \sin(n\pi x) dx$, and similarly

$$g(x, t) = \sum_{n=1}^{\infty} B_n(t) \sin(n\pi x),$$

with $B_n(t) = 2 \int_0^1 g(x, t) \sin(n\pi x) dx$.

The solution of corresponding homogeneous problem is

$$u(x, t) = \sum_{n=1}^{\infty} A_n e^{-c^2 n^2 \pi^2 t} \sin(n\pi x).$$

We seek a solution of equation (7.80) in the form of a series

$$u(x, t) = \sum_{n=1}^{\infty} T_n(t) X_n(x) = \sum_{n=1}^{\infty} T_n(t) \sin(\pi x). \tag{7.81}$$

Forcing the initial condition

$$u(x,0) = \sum_{n=1}^{\infty} T_n(0) \sin(n\pi x) = \sum_{n=1}^{\infty} A_n \sin(n\pi x) = f(x),$$

we see that $T_n(0) = A_n$.

Substituting the series expansion for $g(x,t)$ into equation (7.80), we have

$$\sum_{n=1}^{\infty} T_n'(t) \sin(n\pi x) + c^2 \sum_{n=1}^{\infty} (n\pi)^2 T_n(t) \sin(n\pi x) = \sum_{n=1}^{\infty} B_n(t) \sin(n\pi x).$$

Multiplying this equality by $\sin(\pi x)$ and integrating over $[0,1]$ in x, we obtain

$$T_n' + c^2 (n\pi)^2 T_n = B_n(t), \quad n = 1, 2, \dots.$$

Solving this first order ODE subject to the initial condition $T_n(0) = A_n$, we find

$$T_n(t) = A_n e^{-c^2 n^2 \pi^2 t} + \int_0^t e^{-c^2 n^2 \pi^2 (t-s)} B_n(s) ds.$$

Thus the solution (8.71) of the nonhomogeneous problem is

$$u(x,t) = \sum_{n=1}^{\infty} A_n e^{-c^2 n^2 \pi^2 t} \sin(n\pi x)$$

$$+ \sum_{n=1}^{\infty} \left(\int_0^t e^{-c^2 n^2 \pi^2 (t-s)} B_n(s) ds \right) \sin(n\pi x).$$

Note that the first sum on the right-hand side represents the solution of the corresponding homogeneous problem with the given initial condition, while the second sum describes the influence of the $g(x,t)$. This reasoning is also physically justified. Since the function $g(x,t)$ describes the density of distribution of heat sources inside the metal rod, we can expect that the development of temperature in time will not be expressed by terms $e^{-c^2 n^2 \pi^2 t}$ (as in the case of the homogeneous equation, when there is no source inside the rod), but by means of other functions $T_n(t)$ depending on $g(x,t)$.

7.8.5   A note on Fourier series

Definition 7.8.10. Let $f(x)$ be defined and integrable on $[-\pi, \pi]$. Then the trigonometric series

$$\frac{1}{2}a_0 + \sum_{n=1}^{\infty}(a_n \cos nx + b_b \sin nx),$$

where

$$a_n = \frac{1}{\pi}\int_{-\pi}^{\pi} f(x)\cos nx dx, \quad n = 0, 1, 2, ...,$$

$$b_n = \frac{1}{\pi}\int_{-\pi}^{\pi} f(x)\sin nx dx, \quad n = 1, 2, ...,$$

is said the *Fourier series* of f, and the numbers a_n and b_n are called the *Fourier coefficients* of f.

Recall that a function f is *piecewise continuous* on an interval $[a, b]$ if the graph of f consists of finitely many continuous curves.

Sufficient conditions that a function has (pointwise) a Fourier series representation are given in the following theorem[30].

Theorem 7.8.11. *Let f be a piecewise continuous function on $[-\pi, \pi]$. For points outside $[-\pi, \pi]$, let f be defined by $f(x + 2\pi) = f(x)$. Then for all x, except possibly points of discontinuity,*

$$f(x) = \frac{1}{2}a_0 + \sum_{n=1}^{\infty}(a_n \cos nx + b_b \sin nx),$$

where a_n, b_n are the Fourier coefficients. At a point of discontinuity x_0, the Fourier series converges to

$$\frac{1}{2}[f(x_0^+) + f(x_0^-)],$$

where $f(x_0^+) = \lim_{x \to x_0^+} f(x)$ and $f(x_0^-) = \lim_{x \to x_0^-} f(x)$).

---

[30]See, for example, [35]. Fourier series constitute one of the oldest parts of analysis. The existing theory of convergence for Fourier series is quite extensive. Parts of this theory are difficult and offer astonishing results. For example, a Fourier series can be divergent for every x, or it can diverge at a point of continuity.

Example 7.8.12. Expand the function in a Fourier series.

$$f(x) = \begin{cases} 1 \text{ for } -\pi \le x < 0, \\ 2 \text{ for } 0 \le x < \pi. \end{cases}$$

Solution. We calculate the Fourier coefficients.

$$a_0 = \frac{1}{\pi} \int_{-\pi}^{\pi} f(x) dx = \frac{1}{\pi} \int_{-\pi}^{0} dx + \frac{1}{\pi} \int_{0}^{\pi} 2 dx = 3,$$

and for $n = 1, 2, ...$

$$a_n = \frac{1}{\pi} \int_{-\pi}^{\pi} f(x) \cos nx dx = \frac{1}{\pi} \int_{-\pi}^{0} \cos nx dx + \frac{1}{\pi} \int_{0}^{\pi} 2 \cos nx dx = 0.$$

On the other hand,

$$b_n = \frac{1}{\pi} \int_{-\pi}^{\pi} f(x) \sin nx dx = \frac{1}{\pi} \int_{-\pi}^{0} \sin nx dx + \frac{1}{\pi} \int_{0}^{\pi} 2 \sin nx dx$$

and

$$b_n = \tfrac{2}{n\pi}, \text{ for } n \text{ odd, } \text{ and } \quad b_n = 0, \text{ for } n \text{ even.}$$

Now that all coefficients have been calculated we have

$$f(x) = \frac{3}{2} + \frac{2}{\pi} \left(\sin x + \frac{1}{3} \sin 3x + ... + \frac{1}{2n-1} \sin(2n-1)x + ... \right).$$

Example 7.8.13. Expand the following functions in a Fourier series.

1. $f(x) = \left(\frac{\pi - x}{2} \right)^2$, $0 \le x \le 2\pi$ and extend by periodicity to $\mathbb{R}$.

2. $g(x) = \left(\frac{\pi - x}{2} \right)$, $0 \le x \le 2\pi$ and extend by periodicity to $\mathbb{R}$.

Solution. We calculate the Fourier coefficients.

1. The function f is continuous and piecewise C^∞ on $\mathbb{R}$. Since f is even $b_n = 0$ for all $n = 1, 2, ...$

$$a_0 = \frac{1}{\pi} \int_0^{2\pi} \left(\frac{\pi - x}{2} \right)^2 dx = \frac{\pi^2}{6},$$

and for $n = 1, 2, ...$, integrating by parts we find

$$a_n = \frac{1}{\pi} \int_0^{2\pi} \left(\frac{\pi - x}{2} \right)^2 \cos nx \, dx = \frac{1}{n^2}.$$

Hence

$$f(x) = \frac{\pi^2}{12} + \sum_{n=1}^{\infty} \frac{1}{n^2} \cos nx.$$

Evaluated at $x = 0$ gives the famous formula

$$\frac{\pi^2}{6} = \sum_{n=1}^{\infty} \frac{1}{n^2}.$$

2. The function g is piecewise continuous $\mathbb{R}$. Since g is odd $a_n = 0$ for all $n = 0, 1, 2, ...$

$$b_n = \frac{1}{\pi} \int_0^{2\pi} \left(\frac{\pi - x}{2} \right) \sin nx = \frac{1}{n}.$$

Hence

$$g(x) = \sum_{n=1}^{\infty} \frac{1}{n} \sin nx.$$

Evaluated at $x = \frac{\pi}{2}$ gives

$$\frac{\pi}{4} = \sum_{n=1}^{\infty} \frac{\sin \frac{n\pi}{2}}{n}.$$

Since $\sin \frac{n\pi}{2} = 0$ when n is even and $\sin \frac{n\pi}{2} = \pm 1$ when n is odd, we see that

$$\frac{\pi}{4} = 1 - \frac{1}{3} + \frac{1}{5} - \frac{1}{7} + ...$$

We now come to the remarkable and versatile *Poisson summation formula*.

Theorem 7.8.14. *For any $f \in \mathcal{S}(\mathbb{R})$,*

$$\sum_{n \in \mathbb{Z}} f(n) = \sum_{n \in \mathbb{Z}} \widehat{f}(n),$$

where $\sum_{n \in \mathbb{Z}}$ means taking the limit as $k \to \infty$ of partial sums $s_k = \sum_{j=-k}^{j=k}$.

Proof. Let $F(x) = \sum_{n=-\infty}^{n=+\infty} f(x+n)$. This series is absolutely and uniformly convergent and gives a continuous periodic function of period 1. Therefore we can calculate its Fourier series expansion. $F(x) = \sum_{n=-\infty}^{n=+\infty} a_m e^{-2\pi i m x}$, where a_m, the m-th Fourier coefficient is given by $a_m = \int_0^1 F(x) e^{-2\pi i m x} dx$. But since $e^{2\pi i m x} = e^{2\pi i m (x+n)}$ for every $n \in \mathbb{Z}$ we get

$$a_m = \int_0^1 \sum_{n=-\infty}^{n=+\infty} f(x+n) e^{-2\pi i m(x+n)} dx = \int_0^1 \sum_{n=-\infty}^{n=+\infty} f(x) e^{-2\pi i m x} dx$$

$$= \int_{\mathbb{R}} f(x) e^{-2\pi i m x} dx = \widehat{f}(m).$$

But then $F(x) = \sum_{n=-\infty}^{n=+\infty} \widehat{f}(m) e^{-2\pi i m x}$ and letting $x = 0$ gives the Poisson summation formula. $\qquad \square$

EXERCISES

1. Solve the initial boundary value problem for the wave equation

$$u_{xx} - u_{tt} = 0, \quad 0 < x < \pi, \quad t > 0,$$

$u(0,t) = u(\pi, t) = 0$, for the initial conditions $u(x,0) = \pi x - x^2$, $u_t(x,0) = 0$.

2. Solve the mixed problem for the wave equation

$$u_{xx} - u_{tt} = 0, \quad 0 < x < 1, \quad t > 0,$$

$u_x(0,t) = u_x(1,t) = 0, \quad u(x,0) = 0, \quad u_t(x,0) = 2x - 1.$

3. Solve the Dirichlet problem

$$u_{xx} + u_{yy} = 0, \quad 0 < x < \pi, \quad 0 < y < 1,$$

$$u(0, y) = u(\pi, y) = 0, \quad u(x, 0) = 0, \quad u(x, 1) = f(x), \text{ where}$$

$$f(x) = \left\{ \begin{array}{ll} x & \text{for } 0 < x < \frac{\pi}{2}, \\ \pi - x & \text{for } \frac{\pi}{2} < x < \pi. \end{array} \right\}$$

4. Solve the Dirichlet problem $r^2 u_{rr} + r u_r + u_{\theta\theta} = 0$ in the unit disk Ω, with $u(1, \theta) = \theta$, and $u(r, -\pi) = u(r, \pi) = 0$.

5. Solve the initial boundary value problem for the heat equation

$$u_t = u_{xx}, \quad 0 < x < \pi, \quad t > 0,$$

$u(0, t) = u(\pi, t) = 0$, for the initial condition $u(x, 0) = x \sin x$.

6. Solve the initial boundary value problem for the heat equation

$$u_t = u_{xx}, \quad 0 < x < \pi, \quad t > 0,$$

$u_x(0, t) = u_x(\pi, t) = 0$, for the initial condition $u(x, 0) = \sin^2 x$.

7. Show that if $u(x, t)$ is a solution of the heat equation $u_{xx} - u_t = 0$, then so is $w(x, t) = \frac{1}{\sqrt{t}} e^{-\frac{x^2}{4t}} u\left(\frac{x}{t}, -\frac{1}{t}\right)$.

Answers to selected Exercises

1. $u(x, t) = \frac{8}{\pi} \sum_{n=0}^{\infty} \frac{1}{(2n+1)^3} \sin(2n+1)x \cos(2n+1)t$.

2. $u(x, t) = -\frac{8}{\pi^3} \sum_{n=0}^{\infty} \frac{1}{(2n+1)^3} \sin[(2n+1)\pi t] \cos[(2n+1)\pi x]$.

3. $u(x, y) = \frac{4}{\pi} \sum_{n=1}^{\infty} \frac{(-1)^{n-1}}{(2n-1)^2} \frac{\sinh(2n-1)y}{\sinh(2n-1)} \sin(2n-1)x$.

4. $u(r, \theta) = 2 \sum_{n=1}^{\infty} \frac{(-1)^{n-1}}{n} \frac{1}{r^n} \sin(n\theta)$.

5. $u(x, t) = \frac{\pi}{2} e^{-t} \sin x - \frac{16}{\pi} \sum_{n=1}^{\infty} \frac{n}{(4n^2-1)^2} e^{-4n^2 t} \sin(2nx)$.

6. $u(x, t) = \frac{1}{2} - \frac{1}{2} e^{-4t} \cos(2x)$.

7.9 The Fourier transform method (*)

In the previous section we used the method of separation of variables to solve initial value and boundary value problems for the wave, Laplace and heat equations. As we saw the method depends extensively on the theory of Fourier series. However, certain problems involving PDE on *unbounded* domains, require a continuous superposition of separated solutions. The Fourier tranform is an important tool that allows us to solve many such problems effectively.

The Fourier transform was introduced and its basic properties were developed in Section 5.5. For the convenience of the reader, we recall the definitions of the Fourier transform and its inverse. Let $x \in \mathbb{R}$ and $u = u(x)$ be an absolutely integrable function on $(-\infty, \infty)$. The *Fourier transform* $\mathcal{F}u$ or $\widehat{u}$ is defined by the formula

$$(\mathcal{F}u)(\xi) \equiv \widehat{u}(\xi) = \frac{1}{\sqrt{2\pi}} \int_{-\infty}^{\infty} u(x)e^{-i\xi x}dx.$$

Since $\int_{-\infty}^{\infty} |u(x)|dx < \infty$, the Fourier transform of u exists. We also saw in Section 5.5, that the theory of the Fourier transform works very well on the *Schwartz space* $\mathcal{S}$ where $\mathcal{F} : \mathcal{S} \to \mathcal{S}$ is a bijection (see, Theorem 5.5.14). The *inverse Fourier transform* is given by

$$(\mathcal{F}^{-1}\widehat{u})(x) = \frac{1}{\sqrt{2\pi}} \int_{-\infty}^{\infty} \widehat{u}(\xi)e^{i\xi x}d\xi.$$

The fundamental property of the Fourier transform is that it turns differentiation of the original function into multiplication of the image (or the transform side) by i times the independent variable. That is for $u \in \mathcal{S}$, we have

$$(\mathcal{F}u')(\xi) = i\xi\widehat{u}(\xi) \tag{7.82}$$

and hence

$$(\mathcal{F}u'')(\xi) = (i\xi)^2\widehat{u}(\xi) = -\xi^2\widehat{u}(\xi). \tag{7.83}$$

It is this property that makes the Fourier transform an indispensable tool in the study of partial differential equations. In the case of functions of two variables, $u = u(x, t)$, the Fourier transform with respect to x is given by

$$(\mathcal{F}u)(\xi,t) \equiv \widehat{u}(\xi,t) = \frac{1}{\sqrt{2\pi}} \int_{-\infty}^{\infty} u(x,t)e^{-i\xi x}dx,$$

where the variable t plays the role of a parameter. Thus, the derivatives with respect to x transform according to (7.82) and (7.83). However the derivatives with respect to t stay unchanged. That is,

$$\left(\mathcal{F}\frac{\partial u}{\partial t}\right)(\xi,t) = \frac{\widehat{\partial u}}{\partial t}(\xi,t) = \frac{\partial \widehat{u}}{\partial t}(\xi,t)$$

$$\left(\mathcal{F}\frac{\partial u}{\partial x}\right)(\xi,t) = i\xi\widehat{u}(\xi,t)$$

$$\left(\mathcal{F}\frac{\partial^2 u}{\partial x^2}\right)(\xi,t) = -\xi^2\widehat{u}(\xi,t).$$

We recall also the definition of the *convolution* $\varphi * \psi$ of two functions φ and ψ in $\mathcal{S}$ and its relation with the Fourier transform (see, Theorem 5.5.18)

$$(\varphi * \psi)(x) = \int_{-\infty}^{\infty} \varphi(x-y)\psi(y)dy.$$

$$\mathcal{F}(\varphi * \psi)(\xi) = \sqrt{2\pi}\widehat{\varphi}(\xi)\widehat{\psi}(\xi). \tag{7.84}$$

$$\frac{1}{\sqrt{2\pi}}(\varphi * \psi)(x) = \mathcal{F}^{-1}\left\{\widehat{\varphi}(\xi)\widehat{\psi}(\xi)\right\}. \tag{7.85}$$

In the sequel, we use the Fourier transform to solve the Cauchy problem for the two-dimensional *wave equation*, the *Dirichlet* and *Neumann* problems in the half-plane, and the Cauchy problem for the *heat equation*. We assume that the functions involved have Fourier transforms. This will be the case if they belong in $\mathcal{S}$ and will be our standing assumption.

Example 7.9.1. (*The Cauchy Problem for the wave equation*). *Solve the initial value problem*

$$c^2\frac{\partial^2 u}{\partial x^2} = \frac{\partial^2 u}{\partial t^2}, \quad -\infty < x < \infty, \ t > 0 \tag{7.86}$$

$$u(x,0) = f(x), \quad \frac{\partial u}{\partial t}(x,0) = g(x), \tag{7.87}$$

where f and g are given functions.

Solution. Applying the Fourier transform to the equation with respect to x we get

$$-c^2\xi^2\widehat{u}(\xi,t) = \frac{\partial^2 \widehat{u}}{\partial t^2}(\xi,t).$$

The initial conditions (7.87) become $\widehat{u}(\xi,0) = \widehat{f}(\xi)$ and $\frac{\partial \widehat{u}}{\partial t}(\xi,0) = \widehat{g}(\xi)$. To simplify the notation we set $(\mathcal{F}u)(\xi,t) = \widehat{u}(\xi,t) = U(\xi,t)$, $\widehat{f}(\xi) = F(\xi)$ and $\widehat{g}(\xi) = G(\xi)$. Hence,

$$\frac{d^2U}{dt^2} + c^2\xi^2 U = 0 \tag{7.88}$$

and

$$U(\xi,0) = F(\xi), \quad \frac{dU}{dt}(\xi,0) = G(\xi). \tag{7.89}$$

This is an initial value problem for the ODE (7.88). From Example 7.4.8, the general solution of (7.88) is

$$U(\xi,t) = A\cos(c\xi t) + B\sin(c\xi t).$$

Using the initial conditions (7.89) we obtain

$$U(\xi,t) = F(\xi)\cos(c\xi t) + \frac{1}{c\xi}G(\xi)\sin(c\xi t). \tag{7.90}$$

Equation (7.90) is the transformed solution. To find the original solution we apply the inverse Fourier tranform

$$u(x,t) = (\mathcal{F}^{-1}U)(x,t) = \frac{1}{\sqrt{2\pi}}\int_{-\infty}^{\infty} [F(\xi)\cos(c\xi t) + \frac{1}{c\xi}G(\xi)\sin(c\xi t)]e^{i\xi x}d\xi. \tag{7.91}$$

This integral representation (7.91) of the solution, though analogous to the Fourier series solution (7.57) derived in Example 7.8.2 is not very transparent. Nevertheless, it can be converted to D'Alemberts's solution (7.61) obtained in Example 7.8.3. To do this, recall that $\cos(\theta) = \frac{1}{2}(e^{i\theta} + e^{-i\theta})$ and $\sin(\theta) = \frac{1}{2i}(e^{i\theta} - e^{-i\theta})$.

Then the first integral in (7.91) becomes

$$\frac{1}{\sqrt{2\pi}} \int_{-\infty}^{\infty} F(\xi)\cos(c\xi t)e^{i\xi x}d\xi = \frac{1}{2\sqrt{2\pi}} \int_{-\infty}^{\infty} F(\xi)[e^{ic\xi t} + e^{-ic\xi t}]e^{i\xi x}d\xi$$

$$= \frac{1}{2\sqrt{2\pi}} \int_{-\infty}^{\infty} F(\xi)[e^{i\xi(x+ct)} + e^{i\xi(x-ct)}]d\xi = \frac{1}{2}[f(x+ct) + f(x-ct)].$$

Similarly for the second integral in (7.91) we have

$$\frac{1}{\sqrt{2\pi}} \int_{-\infty}^{\infty} \frac{1}{c\xi} G(\xi)\sin(c\xi t)e^{i\xi x}d\xi$$

$$= \frac{1}{2c\sqrt{2\pi}} \int_{-\infty}^{\infty} G(\xi)\frac{1}{i\xi}[e^{i\xi(x+ct)} - e^{i\xi(x-ct)}]d\xi$$

$$= \frac{1}{2c\sqrt{2\pi}} \int_{-\infty}^{\infty} G(\xi)\left[\int_{x-ct}^{x+ct} e^{i\xi s}ds\right]d\xi$$

$$= \frac{1}{2c} \int_{x-ct}^{x+ct} \left[\frac{1}{\sqrt{2\pi}} \int_{-\infty}^{\infty} G(\xi)e^{i\xi s}d\xi\right]ds$$

$$= \frac{1}{2c} \int_{x-ct}^{x+ct} g(s)ds.$$

Putting these together we get D'Alembert's formula

$$u(x,t) = \frac{1}{2}[f(x+ct) + f(x-ct)] + \frac{1}{2c} \int_{x-ct}^{x+ct} g(s)ds.$$

Now we turn to the Dirichlet problem for the half-plane.

Example 7.9.2. (*The Dirichlet problem for the upper half-plane*). *Let* $\Omega = \{(x,y) \in \mathbb{R}^2 : -\infty < x < \infty,\ y \geq 0\}$. *Solve the Laplace equation*

$$\frac{\partial^2 u}{\partial x^2} + \frac{\partial^2 u}{\partial y^2} = 0 \quad in \tag{7.92}$$

with the boundary conditions $u(x,0) = f(x)$, $\lim_{y\to\infty} |u(x,y)| < \infty$.

Solution. Applying the Fourier transform with respect to x in (7.92) we have

$$-c^2\xi^2\widehat{u}(\xi,y) + \frac{\partial^2\widehat{u}}{\partial y^2}(\xi,y) = 0.$$

Setting $U(\xi,y) = \widehat{u}(\xi,y)$ and $\widehat{f}(\xi) = F(\xi)$, as before, leads to the ODE

$$\frac{d^2U}{dy^2} - \xi^2 U = 0 \tag{7.93}$$

and

$$U(\xi,0) = F(\xi), \tag{7.94}$$

$$\lim_{y\to\infty} |U(\xi,y)| < \infty. \tag{7.95}$$

From Example 7.4.5 the general solution of (7.93) is

$$U(\xi,y) = ae^{\xi y} + be^{-\xi y},$$

where a, b are arbitrary constants. From condition (7.95) it follows that if $\xi > 0$, then $a = 0$. On the other hand, if $\xi < 0$, then $b = 0$. Hence,

$$U(\xi,y) = Ce^{-|\xi|y},$$

with C a constant (independent of y but depending on ξ). Since $U(\xi,0) = C$, using (7.94) we get $C = F(\xi)$. Thus,

$$U(\xi,y) = F(\xi)e^{-|\xi|y}.$$

This is the transformed solution. To find the original solution we apply the inverse Fourier transform. Hence, $u(x,y) = \mathcal{F}^{-1}\left\{F(\xi)e^{-|\xi|y}\right\}$. Since by Example 5.5.3

$$\mathcal{F}^{-1}\left\{e^{-|\xi|y}\right\} = \sqrt{\frac{2}{\pi}}\frac{y}{y^2 + x^2},$$

setting $P_y(x) = \sqrt{\frac{2}{\pi}}\frac{y}{y^2+x^2}$, the convolution property (7.85) yields

$$\mathcal{F}^{-1}\left\{F(\xi)e^{-|\xi|y}\right\} = (f * P_y)(x) = \frac{1}{\sqrt{2\pi}}\int_{-\infty}^{\infty} f(s)P_y(x - s)ds.$$

Thus,

$$u(x,y) = \frac{y}{\pi} \int_{-\infty}^{\infty} \frac{f(s)}{(x-s)^2 + y^2} ds. \tag{7.96}$$

This is closely related to Poisson's integral formula (7.79) obtained in Remark 7.8.7, and is called *Poisson's integral formula in the half-plane*. The function

$$P_y(x) = \sqrt{\frac{2}{\pi}} \frac{y}{y^2 + x^2}$$

is called the *Poisson kernel*. Note that

$$\widehat{P_y}(\xi) = e^{-|\xi| y}.$$

Next we deduce the solution of the *Neumann problem* in the half-plane from the solution of the Dirichlet problem.

Example 7.9.3. (*The Neumann problem in the half-plane*).[31] *Let* $\Omega = \{(x,y) \in \mathbb{R}^2 : -\infty < x < \infty, \ y \geq 0\}$. *Solve the Laplace equation*

$$\frac{\partial^2 u}{\partial x^2} + \frac{\partial^2 u}{\partial y^2} = 0 \quad in \ \Omega \tag{7.97}$$

with the boundary conditions $\frac{\partial u}{\partial y}(x,0) = f(x)$, $\lim_{y\to\infty} |\frac{\partial u}{\partial y}(x,y)| < \infty$.

Solution. Set $\frac{\partial u}{\partial y}(x,y) = w(x,y)$. So that

$$u(x,y) = \int_0^y w(x,\rho) d\rho + C, \tag{7.98}$$

where C is an arbitrary constant. Since we seek a smooth solution, the order of differentiation is immaterial and so

$$\frac{\partial^2 w}{\partial x^2} + \frac{\partial^2 w}{\partial y^2} = \Delta w = \frac{\partial}{\partial y}[\Delta u] = 0.$$

The boundary condition becomes

$$w(x,0) = \frac{\partial u}{\partial y}(x,0) = f(x).$$

---

[31]C. Neumann (1832-1925) is known for his work on integral equations, the Dirichlet problem and Neumann series.

That is, $w(x, y)$ satisfies the Laplace equation with the Dirichlet condition on the boundary of Ω. From the previous example, the solution is given by

$$w(x, y) = \frac{y}{\pi} \int_{-\infty}^{\infty} \frac{f(s)}{(x-s)^2 + y^2} ds. \tag{7.99}$$

Now, substituting (7.99) into (7.98), we get

$$u(x, y) = \int_0^y w(x, \rho) d\rho + C = \frac{1}{\pi} \int_0^y \left[\rho \int_{-\infty}^{\infty} \frac{f(s)}{(x-s)^2 + \rho^2} ds \right] d\rho + C$$

$$= \frac{1}{\pi} \int_{-\infty}^{\infty} f(s) \left[\int_0^y \frac{\rho}{(x-s)^2 + \rho^2} d\rho \right] ds + C$$

$$= \frac{1}{2\pi} \int_{-\infty}^{\infty} f(s) \log[(x-s)^2 + y^2] ds + C.$$

Thus,

$$u(x, y) = \frac{1}{2\pi} \int_{-\infty}^{\infty} f(s) \log[(x-s)^2 + y^2] ds + C,$$

is the solution of the Neumann problem in the half-plane uniquely determined up to an (arbitrary) additive constant.

Example 7.9.4. (*The Cauchy problem for the heat equation*). *Solve the heat equation*

$$\frac{\partial u}{\partial t} - c^2 \frac{\partial^2 u}{\partial x^2} = 0, \quad -\infty < x < \infty, \ t > 0 \tag{7.100}$$

satisfying the initial condition $u(x, 0) = f(x)$.

Solution. Applying the Fourier transform to the equation with respect to x, as before, we get

$$\frac{\partial \widehat{u}}{\partial t}(\xi, t) + c^2 \xi^2 \widehat{u}(\xi, t) = 0.$$

The initial condition transforms to $\widehat{u}(\xi, 0) = \widehat{f}(\xi)$.

Setting, as before, $U(\xi, t) = \widehat{u}(\xi, t)$ and $F(\xi) = \widehat{f}(\xi)$, we obtain the *first* order ODE in $U(\xi, t)$ (as a function of t)

$$\frac{dU}{dt} + \xi^2 c^2 U = 0 \tag{7.101}$$

with the initial condition $U(\xi, 0) = F(\xi)$. From Theorem 7.2.1 the solution of (7.101) is

$$U(\xi, t) = Ce^{-\xi^2 c^2 t}.$$

The initial condition implies $C = F(\xi)$ and hence

$$U(\xi, t) = F(\xi)e^{-\xi^2 c^2 t}.$$

The inverse Fourier transform gives the solution

$$u(x, t) = \mathcal{F}^{-1}\left\{F(\xi)e^{-\xi^2 c^2 t}\right\}. \tag{7.102}$$

From Proposition 5.5.12,

$$\mathcal{F}\left[\frac{1}{\sqrt{2c^2 t}}e^{-\frac{x^2}{4c^2 t}}\right] = e^{-c^2 \xi^2 t}.$$

Setting $G_t(x) = \frac{1}{\sqrt{2c^2 t}}e^{-\frac{x^2}{4c^2 t}}$, the convolution property (7.85), yields

$$\mathcal{F}^{-1}\left\{F(\xi)e^{-\xi^2 c^2 t}\right\} = (f * G_t)(x) = \frac{1}{\sqrt{2\pi}}\int_{-\infty}^{\infty} f(s)G_t(x - s)ds.$$

Thus, solution (7.102) becomes

$$u(x, t) = \frac{1}{\sqrt{4\pi c^2 t}}\int_{-\infty}^{\infty} f(s)e^{-\frac{(x-s)^2}{4c^2 t}}ds.$$

The integrand involved in the solution consists of the initial condition $f(x)$ and the function

$$G_t(x - s) = e^{-\frac{(x-s)^2}{4c^2 t}}.$$

The Gaussian function

$$G_t(x) = G(x, t) = e^{-\frac{x^2}{4c^2 t}}, \quad x \in \mathbb{R}, \ t > 0,$$

is called the *Gaussian kernel* or *heat kernel* and has interesting applications in analysis and probability.

EXERCISES

1. Solve the wave equation $u_{xx} - u_{tt} = 0$, $x \in \mathbb{R}$, $t > 0$, for the initial conditions $u(x,0) = \frac{1}{1+x^2}$ and $u_t(x,0) = 0$.

2. Use the Fourier transform to solve the wave equation

$$u_{xx} - u_{tt} = 0, \quad x \in \mathbb{R}, \ t > 0,$$

for the initial conditions

$$u(x,0) = \sqrt{\tfrac{2}{\pi}} \tfrac{\sin x}{x}, \quad u_t(x,0) = 0.$$

3. Use the Fourier transform to solve the Laplace equation $u_{xx} + u_{yy} = 0$ in the infinite strip $\Omega = \{(x,y) : x \in \mathbb{R}, \ 0 < y < 1\}$, together with the conditions $u(x,0) = 0$, $u(x,1) = f(x)$.

4. Use the Fourier transform to find the temperature distribution in an endless metal rod if the initial temperature distribution is $u(x,0) = e^{-\frac{x^2}{2}}$. (Take $c = 1$ in the heat equation).

5. Solve the heat equation $u_{xx} - u_t = 0$, $x > 0$, $t > 0$, satisfying the initial conditions

$$u(x,0) = \begin{cases} 1 \text{ for } 0 < x < 1, \\ 0 \text{ for } 1 < x. \end{cases}$$

and $u_x(0,t) = 0$.

Answers to selected Exercises

1. $u(x,t) = \frac{1}{2} \int_{-\infty}^{\infty} e^{-|\xi|} \cos(\xi t) e^{i\xi x} d\xi$.

2. $u(x,t) = \frac{50}{\sqrt{\pi t}} \int_{-\infty}^{\infty} \frac{1}{1+\xi^2} e^{-\frac{(x-\xi)^2}{4t}} d\xi$.

3. $u(x,y) = \int_0^{\infty} \int_{-\infty}^{\infty} \frac{1}{\pi \sinh(s)} f(\xi) \sinh(sy) \cos(sx - s\xi) ds d\xi$.

4. $u(x,t) = \frac{1}{\sqrt{1+2t}} e^{-\frac{x^2}{(2+4t)}}$.

7.10   Solved problems for Chapter 7

Exercise 7.10.1. (*Riccati's equation*). A first-order differential equation of the form

$$y' = c(x) + b(x)y + a(x)y^2 \qquad (7.103)$$

is called the *Riccati equation*. The Riccati[32] equation is a natural extension of the first order linear equation, for it reduces to a linear equation if the coefficient $a(x) \equiv 0$. If the coefficients a, b and c are constants, then the Riccati equation allows separation of the variables and we find the general solution

$$\int \frac{1}{ay^2 + by + c} dy = C - x.$$

In general this equation can not be solved by elementary methods. However, if a particular solution $y_1(x)$ is known, then the general solution has the form

$$y(x) = y_1(x) + z(x),$$

where $z(x)$ is the general solution of the *Bernoulli equation*

$$z' - (b - 2ay_1)z = az^2.$$

To see this, set $y(x) = y_1(x) + z(x)$, where $z(x)$ is a new unknown function. Substituting $y = y_1 + z$ into the Riccati equation (7.103) we find that

$$y_1' + z' = c(x) + b(x)(y_1 + z) + a(x)(y_1^2 + 2y_1z + z^2).$$

Since $y_1(x)$ is a solution of (7.103), we get $z' = b(x)z + a(x)(2y_1z + z^2)$, or

$$z' - [b(x) + 2a(x)y_1(x)]z = a(x)z^2. \qquad (7.104)$$

Equation (7.104) is a Bernoulli equation, which with the substitution $w = \frac{1}{z}$ reduces to a linear equation (see, Section 7.2, Exercise 5 (d)).

Example 1. Solve the Riccati equation $(1 - x^3)y' = y^2 - x^2y - 2x$, given the particular solution $y_1(x) = -x^2$.

Solution. Here $a(x) = \frac{1}{1-x^3}$, $b(x) = -\frac{x^2}{1-x^3}$ and $c(x) = -\frac{2x}{1-x^3}$. Setting

---

[32] J. Riccati (1676-1754) formulated and solved the Riccati equation.

$y = y_1 + z = -x^2 + z$ and substituting into the equation we get the Bernoulli equation

$$z' + \frac{3x^2}{1 - x^3} z = \frac{1}{1 - x^3} z^2.$$

Setting $z = \frac{1}{w}$, we obtain the linear equation

$$w' + \frac{3x^2}{x^3 - 1} w = \frac{1}{x^3 - 1}.$$

Solving this, we find $w = \frac{C+x}{x^3-1}$. Hence the general solution of the original equation is

$$y = y_1 + z = -x^2 + \frac{x^3 - 1}{C + x} = -\frac{Cx^2 + 1}{C + x}.$$

In some other special cases Riccati's equation can be solved using substitutions:

Example 2. Solve the Riccati equation $y' = \alpha y^2 + \beta x^{-2}$.

Solution. Here $a(x) = \alpha$, $b(x) = 0$, and $c(x) = \beta x^{-2}$, but we do not know a particular solution of the equation. However we can solve it by setting $y = \frac{1}{u}$. Hence $y' = -\frac{u'}{u^2}$, and substituting into the equation we get

$$u' = -\alpha - \beta \left(\frac{u}{x}\right)^2,$$

which is a homogeneous equation and can be solved by the methods of subsection 7.2.3.

Exercise 7.10.2. Consider the linear equation $L(y) = y'' + p(x)y' + q(x) = 0$, where p, q are continuous on some interval (α, β). Show that there exists a fundamental set of solutions of $L(y) = 0$.

Solution. Let $a \in (\alpha, \beta)$. From Theorem 7.4.2 each of following initial-value problems $L(y) = 0$ with $y(a) = 1$, $y'(a) = 0$, and $L(y) = 0$ with $y(a) = 0$, $y'(a) = 1$, has a unique solution ϕ_1 and ϕ_2 respectively. Since the *Wronskian*

$$W(\phi_1(a), \phi_2(a)) = \det \begin{bmatrix} \phi_1(a) & \phi_2(a) \\ \phi_1'(a) & \phi_2'(a) \end{bmatrix} = \det \begin{bmatrix} 1 & 0 \\ 0 & 1 \end{bmatrix} = 1 \neq 0.$$

Hence ϕ_1, ϕ_2 are linearly independent, and so $\{\phi_1, \phi_2\}$ is a fundamental set of solutions.

Exercise 7.10.3. (*Lagrange's method of variation of parameters*). Solve

$$L(y) = y'' + p(x)y' + q(x)y = g(x).$$

Solution. Solving the corresponding homogeneous equation $L(y) = 0$, we find the general solution $y = C_1 y_1 + C_2 y_2$. The Lagrange method of variation of parameters for obtaining a particular solution y_p of the nonhomogeneous equation $L(y) = g(x)$ involves the replacement of the constants C_1 and C_2 by, as yet, unkown functions $u_1 = u_1(x)$ and $u_2 = u_2(x)$ respectively, and we seek a particular solution $y_p = u_1 y_1 + u_2 y_2$. We argue in reverse and we force y_p to satisfy the equation. Let us calculate the derivatives of y_p. We have

$$y_p' = u_1' y_1 + u_1 y_1' + u_2' y_2 + u_2 y_2' = (u_1 y_1' + u_2 y_2') + (u_1' y_1 + u_2' y_2).$$

To simplify the calculations let us require that u_1 and u_2 satisfy $u_1' y_1 + u_2' y_2 = 0$. Then

$$y_p'' = (u_1 y_1' + u_2 y_2')' = u_1' y_1' + u_1 y_1'' + u_2' y_2' + u_2 y_2''.$$

Substituting we have $g = L(y_p) = u_1 L(y_1) + u_2 L(y_2) + u_1' y_1' + + u_2' y_2' = u_1' y_1' + + u_2' y_2'$, that is,

$$u_1' y_1' + + u_2' y_2' = g.$$

Now this latter equation together with $u_1' y_1 + u_2' y_2 = 0$ form an (algebraic) linear system in u_1' and u_2'. Its determinat is the Wronskian $W = W(y_1, y_2) \neq 0$. Therefore this system has a unique solution $u_1' = \frac{-y_2 g}{W}$ and $u_2' = \frac{y_1 g}{W}$. Integrating these we get $u_1(x) = - \int \frac{y_2(x)g(x)}{W(y_1(x), y_2(x))} dx$ and $u_2(x) = \int \frac{y_1(x)g(x)}{W(y_1(x), y_2(x))} dx$, where we ignored the constants of intgration. Now the general solution of the equation is

$$y = C_1 y_1 + C_2 y_2 + y_p.$$

The method generalizes, in a natural way, to higher order linear differential equations.

Example: For the equation $y'' + y = \tan x$, working as above, we find $u_1(x) = -\cos x$ and $u_2(x) = \sin x - \log|\sec x + \tan x|$. Hence a particular solution of the equation is $y_p(x) = -\cos x \sin x + (\sin x - \log|\sec x + \tan x|)\cos x$. Thus the general solution is

$$y = C_1 \sin x + C_2 \cos x - \cos x \sin x + (\sin x - \log|\sec x + \tan x|)\cos x.$$

Exercise 7.10.4. In this exercise we give an alternative proof of Rodrigues' formula

$$H_n(x) = (-1)^n e^{x^2} \frac{d^n}{dx^n} e^{-x^2},$$

for the Hermite polynomials H_n, $(n = 0, 1, 2, ...)$.

Solution. A generating function for H_n is (by definition) a function ψ satisfying $\psi(x, t) = \sum_{n=0}^{\infty} H_n(x) \frac{t^n}{n!}$. In Proposition 7.6.6, we found $\psi(x, t) = e^{-t^2 + 2xt}$. Now,

$$\psi(x, t) = e^{-t^2 + 2xt} = e^{x^2} e^{-(t-x)^2} = e^{x^2} \sum_{n=0}^{\infty} \left(\frac{\partial^n e^{-(t-x)^2}}{\partial t^n} \Big|_{t=0} \right) \frac{t^n}{n!}$$

$$= e^{x^2} \sum_{n=0}^{\infty} \left((-1)^n \frac{\partial^n e^{-s^2}}{\partial s^n} \Big|_{s=x} \right) \frac{t^n}{n!} = \sum_{n=0}^{\infty} \left((-1)^n e^{x^2} \frac{d^n e^{-x^2}}{dx^n} \right) \frac{t^n}{n!}.$$

Thus,

$$H_n(x) = (-1)^n e^{x^2} \frac{d^n}{dx^n} e^{-x^2}.$$

Exercise 7.10.5. Solve the heat equation $c^2 u_{xx} = u_t$, $0 < x < \pi$, $t > 0$, satisfying the boundary conditions $u(0, t) = u(\pi, t) = 0$, and the initial condition $u(x, 0) = \sin x$.

Solution. Solving by separating the variables, as in Subsection 7.8.4 (with $L = \pi$), the solution satisfying the boundary conditions is

$$u(x, t) = \sum_{n=1}^{\infty} A_n e^{-c^2 n^2 t} \sin(nx),$$

where, $A_n = \frac{2}{\pi} \int_0^\pi \sin x \sin(nx) dx$. Calculating the Fourier coefficients, we have $A_1 = \frac{2}{\pi} \int_0^\pi \sin^2 x\, dx = 1$, $A_n = \frac{2}{\pi} \int_0^\pi \sin x \sin(nx) dx = 0$, for $n > 1$. Hence

$$u(x, t) = e^{-c^2 t} \sin x.$$

Exercise 7.10.6. (*The transport equation*). Here we use the Fourier transform to solve the Cauchy problem for a first order PDE with *variable coefficients*. Solve the equation

$$t u_x - u_t = 0, \quad x \in \mathbb{R}, \ t > 0,$$

for the initial condition $u(x, 0) = f(x)$.

Solution. Note that the varying coefficient is a function of the time variable t alone. Applying the Fourier transform to the equation, we have

$$t\mathcal{F}(u_x) + \mathcal{F}(u_t) = 0, \text{ or } i\xi t\widehat{u} + \frac{\partial \widehat{u}}{\partial t} = 0.$$

Hence, it amounts to solve the Cauchy problem for the first order ODE

$$\frac{\partial \widehat{u}}{\partial t} + i\xi t\widehat{u} = 0, \quad \widehat{u}(\xi, 0) = \widehat{f}(\xi).$$

Solving we get, $\widehat{u}(\xi, t) = \widehat{f}(\xi)e^{-i\frac{t^2}{2}\xi}$. Applying the inverse Fourier transform we find

$$u(x, t) = f(x - \tfrac{t^2}{2}).$$

Miscellaneous Exercise

Exercise 7.10.7. Consider the equation $y' + (\cos x)y = e^{-\sin x}$.

1. Find the solution φ which satisfies $\varphi(\pi) = \pi$.

2. Show that any solution y has the property that $y(k\pi) - y(0) = k\pi$, $(k \in \mathbb{Z})$.

Exercise 7.10.8. Solve the equation $x^2 y' + 2xy = 1$ on $(0, \infty)$. Show that every solution tends to zero as $x \to \infty$. Find that solution φ which satisfies $\varphi(2) = 2\varphi(1)$.

Exercise 7.10.9. Suppose a continuously differentiable function φ on $[0, 1]$ satisfies there $\varphi'(x) - 2\varphi(x) \leq 1$, and $\varphi(0) = 1$. Prove that $\varphi(x) \leq \frac{3}{2}e^{2x} - \frac{1}{2}$.

Exercise 7.10.10. A spaceship is returning to earth. Assume that the only external force acting on it is the action of gravity, and that it falls along a straight line toward the center of the earth. The effect of gravity is partly overcome by firing a rocket directly downward. The rocket fuel is consumed at a constant rate of k pounds per second and the exhaust material has a constant speed of c feet per second relative to the rocket. Find a formula for the distance the spaceship falls in time t if it starts from rest at time $t = 0$ with an initial weight of W pounds.
Hint. Consult Example 7.2.5. (Ans. $y(t) = \frac{1}{2}gt^2 - ct\left(t - \frac{W}{k}\right)\log\left(1 - \frac{kt}{W}\right)$.)

Exercise 7.10.11. Solve the differential equations

1. $y'' - y' = e^x \sin x$.

2. $yy'' - (y')^2 = 0$.

Exercise 7.10.12. Let $\varphi : \mathbb{R} \to \mathbb{R}$ be a solution of the differential equation $5y'' + 10y' + 6y = 0$. Show that the function $f(x) = \frac{(\varphi(x))^2}{1+(\varphi(x))^4}$ attains a maximum value on $\mathbb{R}$.

Exercise 7.10.13. Find the values of $\lambda \in \mathbb{R}$ for which the differential equation $y'' - 2\lambda y' + y = 0$ has a solution satisfying $y(0) = y(2k\pi) = 0$, where $k \in \mathbb{Z}^+$.

Exercise 7.10.14. An integral curve $y = \phi(x)$ of the differential equation $y'' + 4y' + 13y = 0$ intersects an integral curve $y = \psi(x)$ of the differential equation $y'' - 4y' + 29y = 0$ at the origin. The two curves have equal slopes at the origin. Find ϕ and ψ if $\psi'(\frac{\pi}{2}) = 1$.

Exercise 7.10.15. Let $W = W(y_1, y_2) = y_1 y_2' - y_2 y_1'$ be the Wronskian of y_1, y_2.

1. Show that $W' = y_1 y_2'' - y_2 y_1''$.

2. If y_1 and y_2 are solutions of $y'' + p(x)y' + q(x) = 0$ on some interval (α, β), show that their Wronskian satisfies the first-order equation

$$W' + p(x)W = 0.$$

Solve this equation and deduce $W(x) = W(c)e^{-\int_c^x p(t)dt}$, for every choice of c in (α, β). This is *Abel's formula* for the Wronskian.

Exercise 7.10.16. The *current* $I = I(t)$ at time t flowing in an *electric circuit* obeys the differential equation

$$\frac{d^2 I}{dt^2} + R\frac{dI}{dt} + I = \sin \omega t,$$

where R and ω are positive constants. The solution can be expressed in the form $I(t) = F(t) + A\sin(\omega t + \alpha)$, where $F(t) \to 0$ as $t \to \infty$, and $A > 0$ and α are constants depending on R and ω. If there is a value of ω which makes A as large as possible, then $\frac{\omega}{2\pi}$ is called a *resonance frequency* of the circuit.

1. Find all resonance frequencies when $R = 1$. (Ans. $\frac{1}{\pi 2\sqrt{2}}$).

2. Find those values of R for which the circuit will have a resonance frequency. (Ans. $R < \sqrt{2}$).

Exercise 7.10.17. The function ϕ on $\mathbb{R}$ is defined by

$$\phi(x) = \int_0^\infty \frac{\sin xy}{y(1+y^2)} dy.$$

Prove that for $x > 0$, ϕ satisfies the differential equation $\phi''(x) - \phi(x) = -\frac{\pi}{2}$. Solve the equation and find $\phi(x)$ for all $x \in \mathbb{R}$. *Hint.* Differentiate under the integral sign. (Ans. $\phi(x) = -\frac{\pi}{2}(1 - e^{-x})$ for $x \geq 0$ and $\phi(x) = -\frac{\pi}{2}(1 - e^x)$ for $x < 0$.)

Exercise 7.10.18. (*). Show that the function $y(x) = \int_0^\pi \cos(n\theta - x\sin\theta)d\theta$ satisfies *Bessel's equation* $x^2 y'' + xxy' + (x^2 - n^2)y = 0$.

Exercise 7.10.19. (*). Prove that the *Legendre polynomials*

$$y = P_n(x) = \frac{1}{2^n n!} \frac{d^n}{dx^n} (x^2 - 1)^n, \ (n = 0, 1, 2, ...)$$

satisfy the differential equation $(1 - x^2)y'' - 2xy' + n(n+1)y = 0$.

Exercise 7.10.20. Solve the wave equation $u_{xx} - u_{tt} = 0$, $0 < x < \pi$, $t > 0$, satisfying the boundary conditions $u(0, t) = u(\pi, t) = 0$, for initial conditions $u(x, 0) = \sin x$, $u_t(x, 0) = 0$. Find the energy $E(t) = \int_0^\pi \left[(u_t)^2 + (u_x)^2 \right] dx$ of the solution for the problem. (Ans. $\frac{\pi}{2}$).

Chapter 8

An Introduction to the Calculus of Variations

In this final chapter we discuss a number of classical problems of the calculus of variations. These include, the brachistochrone problem, the problem of geometric optics, geodesics on a surface, the principle of least action, variational problems with constraints, isoperimetric problems, multiple integral variational problems, minimal surfaces, and the Dirichlet principle.

The calculus of variations is a beautiful and powerful subject invented by Euler and Lagrange in the eighteenth century. In a sense the subject is much older. A few of the problems of the calculus of variations (the isoperimetric problem) are very old, and were considered and partly solved by the ancient Greeks. Other important problems of the calculus of variations were studied in seventeenth century Europe, such as the work of Fermat on geometric optics (1662) or the famous brachistochrone problem which was formulated by Galileo in 1638 and was resolved by Giovanni Bernoulli in 1696 and almost immediately after by his brother Giacomo, Leibnitz and Newton. Nevertheless, it was Euler and Lagrange who discovered a systematic way of dealing with these problems by introducing what is now known as the *Euler-Lagrange equation*. Their work was then extended in many ways by a number of mathematicians such as, Bliss, Caratheodory, Du

Bois-Reymond, Hamilton, Hilbert, Jacobi, Kneser, Legendre and Weierstrass, to mention a few.

8.1   Simple variational problems

In its simplest form a typical problem in the calculus of variations is formulated as follows: Let $[a, b]$ be a closed inteval in $\mathbb{R}$ and $F(t, y, z)$ be a C^2 function $F : [a, b] \times \mathbb{R} \times \mathbb{R} \to \mathbb{R}$. Then among all continuously differentiable functions $y : [a, b] \to \mathbb{R}$, satisfying $y(a) = \alpha$ and $y(b) = \beta$ (called *admissible functions* or *admissible curves*), where α, β are given real numbers, find the function (or functions) for which the (well-defined) integral

$$\int_a^b F(t, y(t), y'(t)) dt, \tag{8.1}$$

has an *extremun*, that is, a *minimum* or a *maximum* value.

The set of admissible functions is a subset M of what is called a *function space*[1]. Each function is considered as an element or "point" of this space and the integral (8.1) assigns a definite number to each element $y = y(t)$ of the function space. In the language of functional analysis, the integral (8.1) is called a *functional* of y on the function space. We denote this functional by

$$J(y) = \int_a^b F(t, y(t), y'(t)) dt. \tag{8.2}$$

In contrast to differential calculus, the calculus of variations is concerned with minimizing or maximizing functionals rather than functions

---

[1]Here the function space is the normed linear space of all continuous functions $y(t)$ defined on $[a, b]$ having continuous first derivatives, and is denoted by $C^1([a, b])$. Addition and multiplication by scalars are defined pointwise and the norm

$$||y|| = \max_{t \in [a,b]} |y(t)| + \max_{t \in [a,b]} |y'(t)|.$$

The set of admissible functions $M = \{ y \in C^1([a, b]) : y(a) = \alpha,\ y(b) = \beta \}$. The normed linear space $C^1([a, b])$ turns out to be a complete space, i.e., is a Banach space. In general, the set of admissible functions is a subset of a function space of smooth functions satisfying certain pre-assigned boundary conditions.

of points on $\mathbb{R}$ or $\mathbb{R}^n$. Many problems in geometry, analysis or applied mathematics (in physics, optimal control theory or economics) can be formulated in this way.

We shall see that, just as in the case of minimum and maximum problems in one real variable, we will have necessary conditions and also sufficient conditions. The necessary condition in the case of extrema of a numerical function is a *numerical equation* expressing the fact that at an extreme point, or even a local extreme point, the derivative of the function must vanish, i.e., that the point is a *critical* point. In the present situation the numerical function will be replaced by the functional J, points will be replaced by smooth curves (admissible functions) and the numerical equation expressing the fact that the solution is a critical point will be replaced by a *differential equation* which expresses this vanishing, i.e., that the extreme "point" is a critical point of J. This differential equation is called the *Euler-Lagrange equation* and will be of fundamental importance in what follows.

We shall illustrate the use of this calculus with concrete problems both from geometry, mechanics and geometric optics (this list could be increased). As we shall see, this is an extremely fruitful viewpoint. It affords a consistent methodology to deal with many diverse problems with very little technical difficulty (at least at the early stages!). As we move along in this subject from time to time we shall introduce new problems whose solution will require various generalizations of the original formulation posed just above to vector-valued functions, whose core idea however will remain the same and the Euler-Lagrange equation then becomes a *system* of differential equations. These will be the exact analogues of going from one variable to several in the argument of F, but continuing to deal with curves, going to problems involving constraints, where we will employ an analogue of Lagrange multipliers, or passing to problems in several independent variables (where the Euler-Lagrange equations in this case are partial differential equations), or combinations of either of these types with one or several constraints.

In some cases we shall not concern ourselves with questions of the existence of an extremum for the functional involved. Instead, we will assume *a priori* the *existence*[2] of such an extremum. Problems concerning the existence and uniqeness of the extremum need methods of functional analysis (and as such are beyond the scope of this book). The uniqueness of the solution may also be a problem. This occurs, for example, in the case of geodesics on a manifold of positive curvature where conjugate points occur. In particular, this happens on the sphere (see Example 8.2.8). We do deal with sufficient conditions in 8.1.2.

Before we derive the Euler-Lagrange equation for the variational problem (8.1), we prove a preliminary result known as the *fundamental lemma of the calculus of variations* (Du Bois-Reymond lemma) which together with its *n*-dimensional analogue (Lemmas 8.2.1 and 8.4.1) will be used on several occasions when deriving the Euler-Lagrange equation associated to the various types of problem in the calculus of variations.

Lemma 8.1.1. *(Fundamental lemma). Let $\varphi : [a, b] \to \mathbb{R}$ be continuous and such that*

$$\int_a^b \varphi(t)\psi(t)dt = 0$$

for all functions $\psi \in C^1([a, b])$ satisfying $\psi(a) = \psi(b) = 0$. Then $\varphi \equiv 0$ on $[a, b]$.

Proof. Suppose not, that is, suppose that for some $t_0 \in (a, b)$ we have $\varphi(t_0) \neq 0$, say $\varphi(t_0) > 0$ (a similar argument works if $\varphi(t_0) < 0$). Since φ is continuous there is a small open interval $I = (t_1, t_2) \subset [a, b]$ about t_0 where $\varphi(t) \geq \frac{1}{2}\varphi(t_0) > 0$ for all $t \in I$. Choose a continuous ψ so that $\psi(t) = 0$ for t outside I, but with $\psi(t) > 0$ on a neighborhood of t_0 (for example, $\psi(t) = (t - t_1)(t_2 - t)$ for $t \in I$ and $\psi(t) = 0$ otherwise). Then

$$\int_a^b \varphi(t)\psi(t)dt = \int_I \varphi(t)\psi(t)dt \geq \frac{1}{2}\varphi(t_0) \int_I \psi(t)dt > 0,$$

---

[2] In 1899, David Hilbert (1862-1943), in a communication to the "*Deutsche Mathematiker-Vereinigung*" and in subsequent lectures (University of Göttingen, 1900) gave conditions upon the function F and the set of admissible curves for which the existence of an extremum for the more general functional

$$J = \int_a^b F(t, x(t), y(t), x'(t), y'(t))dt$$

can be ascertained *a priori*. See, [3].

a contradiction. Thus $\varphi \equiv 0$ on (a, b) and hence by continuity on $[a, b]$.

$\square$

Theorem 8.1.2. *Let*

$$J(y) = \int_a^b F(t, y(t), y'(t))dt$$

with $F(t, y, y')$ and y as in (8.1) above. If y minimizes or maximizes the functional J, then y satisfies the following differential equation called the Euler-Lagrange equation.

$$F_y - \frac{d}{dt}F_{y'} = 0. \tag{8.3}$$

Proof. We shall prove the theorem for a minimizer $y = y(t)(= \varphi(t))$ of J. The proof for a maximizer is completely similar. The central idea is that since $y(t)$ gives a minimum value to J, the functional J will increase if we make $y(t)$ to "vary" slightly. Consider a function $\eta(t)$ defined on $[a, b]$, which has a continuous second derivative and satisfies $\eta(a) = \eta(b) = 0$, but is otherwise arbitrary. Let $\epsilon > 0$ and for $|s| < \epsilon$ construct the function

$$\bar{y}(t) = y(t) + s\eta(t)$$

for $t \in [a, b]$. For each such s this function is as smooth as y and η are, satisfies $\bar{y}(a) = \alpha$ and $\bar{y}(b) = \beta$ and by taking ϵ small enough we can assume it is an admissible function in an open neighborhood (in the function space) of the function $y(t)$. The quantity $\delta y = s\eta(t)$ is known as the *variation* of the function $y = y(t)$. Since $J(y)$ is minimum, we have $J(y + s\eta) \geq J(y)$ for all $|s| < \epsilon$. Therefore the integral $J(\bar{y}) = J(y + s\eta)$, which may be regarded as a function $g(s)$ of s,

$$g(s) = J(y + s\eta) = \int_a^b F(t, y(t) + s\eta(t), y'(t) + s\eta'(t))dt, \tag{8.4}$$

must have a minimum at $s = 0$ in a sufficiently small neighborhood $(-\epsilon, \epsilon)$ of 0. By calculus of one real variable, $g'(0) = 0$.

Now differentiating (8.4) under the integral sign (which is permissible, see Theorem 5.3.2) we have

$$g'(s) = \int_a^b \frac{d}{ds}(F(t, \bar{y}(t), \bar{y}'(t))dt,$$

where $\overline{y}(t) = y(t) + s\eta(t)$ and $\overline{y}\,'(t) = y'(t) + s\eta'(t)$. By the Chain Rule

$$\frac{d}{ds}(F(t,\overline{y}(t),\overline{y}\,'(t))) = \frac{\partial F}{\partial t}\frac{\partial t}{\partial s} + \frac{\partial F}{\partial \overline{y}}\frac{\partial \overline{y}}{\partial s} + \frac{\partial F}{\partial \overline{y}\,'}\frac{\partial \overline{y}\,'}{\partial s} = \frac{\partial F}{\partial \overline{y}}\eta(t) + \frac{\partial F}{\partial \overline{y}\,'}\eta'(t).$$

Hence

$$g'(0) = \int_a^b \left(\frac{\partial F}{\partial y}\eta(t) + \frac{\partial F}{\partial y'}\eta'(t) \right) dt = 0.$$

By integration by parts we can eliminate $\eta'(t)$ from the second term

$$\int_a^b \frac{\partial F}{\partial y'}\eta'(t)dt = \eta(t)\frac{\partial F}{\partial y'}\Big|_a^b - \int_a^b \eta(t)\frac{d}{dt}\left(\frac{\partial F}{\partial y'}\right)dt = 0 - \int_a^b \eta(t)\frac{d}{dt}\left(\frac{\partial F}{\partial y'}\right)dt.$$

Thus,

$$\int_a^b \left[\frac{\partial F}{\partial y} + \frac{d}{dt}\left(\frac{\partial F}{\partial y'}\right) \right] \eta(t)dt = 0. \tag{8.5}$$

Up to this point our function η was fixed but arbitrary, except for the boundary conditions. Since (8.5) is valid for *all* such functions η, Lemma 8.1.1 implies

$$\frac{\partial F}{\partial y} + \frac{d}{dt}\left(\frac{\partial F}{\partial y'}\right) = 0.$$

Using subscript notation F_y and $F_{y'}$ for the partial derivatives of F this is $F_y - \frac{d}{dt}(F_{y'}) = 0$. $\square$

The quantity $\delta J = sg'(0)$ is called the *first variation*[3] of the functional J.

Note that the partial derivatives F_y and $F_{y'}$ in (8.3), are computed by treating t, y and y' as the three independent variables of the function F and are, of course, evaluated at $(t, y(t), y'(t)) \in \mathbb{R}^3$. At the same time, in general, $F_{y'}$ is a function of t explicitly and also implicitly via $y = y(t)$ and $y' = y'(t)$. Hence differentiating $\frac{d}{dt}(F_{y'}) = F_{ty'} + F_{yy'}y' + F_{y'y'}y''$. Expressed in these terms the Euler-Lagrange equation is

$$F_{y'y'}y'' + F_{y'y}y' + (F_{ty'} - F_y) = 0 \tag{8.6}$$

---

[3]This is the source of the name *calculus of variations*, which is meant to indicate that in this subject we are concerned with the behavior of functionals, J, when the argument function is made to vary by making small changes of a parameter s.

If $F_{y'y'} \neq 0$ is not identically zero this is a *second order* possibly non-linear (ordinary) differential equation for y, with variable but smooth coefficients defined on a closed interval with fixed values at the end points. In the general solution of (8.6) there occur two arbitrary constants, the number generally required in order to satisfy the boundary conditions. Solutions of (8.6) which are unrestricted by the boundary conditions are called *extremals*. When $F_{y'y'} \equiv 0$, the Euler-Lagrange equation degenerates either into a finite equation or into the identity $0 = 0$, but never into a *differential equation of the first order*, (see Problem 8.5.1).

Example 8.1.3. Among all continuously differentiable functions $y = y(t)$ defined on $[0, 1]$ satisfying $y(0) = 0$, $y(1) = \beta$, find those for which the functional

$$J(y) = \int_0^1 (y'(t))^2 dt$$

has a minimum value. Find the minimum.

Solution. Here $F(t, y, y') = (y')^2$. Hence not only is F independent of t, but it's also independent of y. Therefore $F_y = 0$, $F_t = 0$, and $F_{y'} = 2y'$. Hence the Euler-Lagrange equation (8.6) becomes $F_{y'y'}y'' = 0$. Since $F_{y'y'} = 2 \neq 0$, we obtain the one-dimensional Dirichlet problem

$$y'' = 0, \ on \ (0, 1)$$

$$y(0) = 0, \ y(1) = \beta.$$

The general solution is $y(t) = c_1 t + c_2$, so the extremals are all straight lines. Using the boundary condition, we see $y = \varphi(t) = \beta t$. Note that $F_{y'y'} > 0$ and as we shall see in Corollary 8.2.6, this is the condition for a minimum. Hence φ minimizes J and the minimum value is $J(\varphi) = \beta^2$. Note, for example, that for the neighboring admissible function $y(t) = \beta t^2$ we have $J(y) = \frac{4}{3}\beta^2 > \beta^2$.

In Section 8.4 we shall consider an analogue of this in $\mathbb{R}^2$. Namely, the variational formulation of the Dirichlet problem for the disk.

Example 8.1.4. Find the curves $y : [0, \frac{\pi}{2}] \to \mathbb{R}$ satisfying $y(0) = 0$, $y(\frac{\pi}{2}) = 1$ which extremize the functional

$$J(y) = \int_0^{\frac{\pi}{2}} \left[(y')^2 - y^2\right] dt.$$

Solution. Here $F(t, y, y') = (y')^2 - y^2$. So $F_y = -2y$, $F_t = 0$, $F_{y'} = 2y'$, $F_{y'y} = F_{ty'} = 0$, and $F_{y'y'} = 2$. Therefore the Euler-Lagrange equation (8.6) is $y'' + y = 0$. In Example 7.4.8 we solved this equation and found the general solution $y = A \sin t + B \cos t$. These are the extremals of the problem. Making use of the boundary conditions, $A = 1$ and $B = 0$. Thus $y = \varphi(t) = \sin t$ is the only extremum.

In the above examples, it was easy to solve the Euler-Lagrange equation. However in general it may not be possible to solve the Euler-Lagrange equation explicitly in terms of elementary functions or quadratures (simple integrals). On the other hand, in important special cases and, in fact, in most of the classical examples, the equation can be solved. We discuss these special cases below.

Special cases of Euler-Lagrange's equation

1. The function F does not contain the derivative $y' = y'(t)$, viz, $F = F(t, y)$. In this case (8.6) becomes

$$F_y(t, y) = 0,$$

   that is, it is no longer a differential equation at all but forms an implicit definition of the solution $y = f(t)$. Here of course no integration constants occur and the question of satisfying boundary conditions is impossible. In other words, the problem at hand simply does *not have solution.*

2. The function F does not contain y, viz, $F = F(t, y')$. Here (8.6) becomes

$$\frac{d}{dt}(F_{y'}) = 0.$$

   This can be integrated at once to give $F_{y'} = c_1$, where c_1 is an arbitrary constant. This latter equation may be used to express y' as a function, $\varphi(t, c_1)$, of t and c_1. That is, $y'(t) = \varphi(t, c_1)$. Then by a simple integration we get the general solution of (8.6) in the form

$$y(t) = \int_a^t \varphi(u, c_1) du + c_2.$$

3. The function F is independent of t, viz, $F = F(y, y')$. This is the most significant case in examples and applications. Here we have the following important result:

Proposition 8.1.5. *Suppose* $F = F(t, y, y')$ *is independent of* t *and satisfies the Euler-Lagrange equation* $F_y - \frac{d}{dt} F_{y'} = 0$. *Then*

$$F - y' F_{y'} = c,$$

where c *is a constant.*

Proof. Calculating $\frac{d}{dt}(F - y' F_{y'})$. We have

$$\frac{d}{dt}(F - y' F_{y'}) = F_y y' + F_{y'} y'' - \frac{d}{dt}(y' F_{y'}) = F_y y' + F_{y'} y'' - y' \frac{dF_{y'}}{dt} - y'' F_{y'}.$$

Canceling second derivatives gives

$$\frac{d}{dt}(F - y' F_{y'}) = F_y y' - y' \frac{dF_{y'}}{dt} = y' \left(F_y - \frac{dF_{y'}}{dt} \right) = 0$$

by the Euler-Lagrange equation. Hence $F - y' F_{y'}$ is constant. $\square$

The next subsection contains some classical variational problems and their solutions.

8.1.1   Some classical problems

Example 8.1.6. (*The problem of shortest distance in the Euclidean plane*). Let $P = (a, \alpha)$ and $Q = (b, \beta)$ be fixed points in the plane (with $a \neq b$) and consider all smooth curves $y = y(t)$ joining them. Find the curve of shortest length.

Solution.   From Chapter 6 the length of such a curve is given by $L = \int_a^b \sqrt{1 + (y'(t))^2} dt$. Thus we must minimize the functional

$$J(y) = \int_a^b \sqrt{1 + (y'(t))^2} dt$$

among all smooth curves satisfying $y(a) = \alpha$ and $y(b) = \beta$. Here $F(t, y, y') = \sqrt{1 + (y')^2}$ (and F is independent of both t and y). Therefore the Euler-Lagrange equation (8.6) becomes $F_{y'y'} y'' = 0$. Since

$F_{y'y'} = \frac{\partial^2 F}{\partial y'^2} = \frac{1}{(1+(y')^2)^{\frac{3}{2}}} \neq 0$, we have $y'' = 0$ and $y(t) = c_1 t + c_2$, so
the extremals are all straight lines. Taking into account the boundary
conditions we get the solution to the problem

$$y - \alpha = \frac{\beta - \alpha}{b - a}(t - a),$$

which is also clear from the geometry of the problem. If, by chance, the
points had the same first coordinates, then we just perform a rotation
of the plane. This preserves distance between points as well as lengths
of curves and will create new points with different first coordinates.

An interesting similar, but more difficult problem is that of finding
the shortest curve joining two fixed points on a given surface and lying
entirely on that surface. Such curves are called *geodesics*, and the study
of their properties is a one of the focal points of differential geometry.
In Section 8.2 we shall solve this problem in the case of certain surfaces
of constant curvature, such as the sphere.

Example 8.1.7. (*The Brachistochrone[4] problem*). Historically, this
was the first open problem posed in the calculus of variations and was
solved by Giovanni Bernoulli at the end of the seventeenth century. The
problem is: given two points P and Q lying in a vertical plane, with
different first and second coordinates, find the path joining them of
quickest descent.

Solution. Let us suppose that P is higher than Q and they are joined by
a greased wire with a hollow bead of mass m placed at P (starting from
rest, although this need not be so). The question is how should the wire
be shaped so that the bead slides down to Q in the shortest possible
time? We take coordinates $P = (0,0)$ and $Q = (a,b)$, where $a < 0$ and
$b < 0$. Since $\frac{ds}{dt} = v$, where s is the arc length along the curve and v is
the instantaneous velocity, we see that $dt = \frac{ds}{v}$ so that T, the total time
of descent, is $T = \int_0^a \frac{ds}{v}$. Now if the curve is given by $y = y(x)$, then
just as above $\frac{ds}{dx} = \sqrt{1 + (y'(x))^2} = \sqrt{1 + y'^2}$. So $ds = \sqrt{1 + y'^2}dx$. We
calculate v by conservation of energy (thus the greased wire). Since at
P the bead is at rest, the kinetic energy is zero. The potential energy is

---

[4] "Brachi-sto-chrone" in Greek "$B\rho\alpha\chi\upsilon s - \chi\rho o\nu os$" means "minimum-time".

also zero at P since there the height is zero. Hence $\frac{1}{2}mv^2 - mgy = 0$ so that $v = \sqrt{2gy}$. Since v and therefore the integral is independent of m, we see that the mass of the bead will be irrelevant to the question and

$$T = \frac{1}{\sqrt{2g}} \int_0^a \frac{\sqrt{1 + y'^2}}{\sqrt{y}} dx.$$

Thus, in the Brachistochrone problem we can take

$$F(x, y, y') = \frac{\sqrt{1 + y'^2}}{\sqrt{y}}$$

and we minimize the functional

$$J(y) = \int_0^a F(x, y, y') dx.$$

Since this is independent of x, applying Proposition 8.1.5 above tells us

$$\frac{y'^2}{\sqrt{y}\sqrt{1 + y'^2}} - \frac{\sqrt{1 + y'^2}}{\sqrt{y}} = c.$$

Multiplying this equation by $\sqrt{y}\sqrt{1 + y'^2}$ we see that $y'^2 - (1 + y'^2) = c\sqrt{y}\sqrt{1 + y'^2}$. (Here $c \neq 0$ since this would say $1 = 0$). That is, $\sqrt{y}\sqrt{1 + y'^2} = \frac{1}{c}$. This last constant is clearly positive so it is $\sqrt{2\alpha}$ for some $\alpha > 0$. Hence

$$y' = \sqrt{2\frac{\alpha}{y} - 1} \text{ and therefore } x = \int \frac{\sqrt{y}dy}{\sqrt{2\alpha - y}}.$$

To calculate this integral let $y = 2\alpha \sin^2 \frac{\theta}{2}$. Then we get

$$x = 2\alpha \int \sin^2 \frac{\theta}{2} = \alpha(\theta - \sin(\theta)) + x_0.$$

But then

$$y = 2\alpha \sin^2 \frac{\theta}{2} = \alpha(1 - \cos(\theta)).$$

So in parametric form $x = \alpha(\theta - \sin\theta) + x_0$ and $y = \alpha(1 - \cos\theta)$. Since the curve passes through $(0, 0)$ we see $x_0 = 0$. These are the parametric equations of a *cycloid*.

Finally, we deal with the question of fitting the cycloid to initial data. We shall show α can be adjusted to have the cycloid pass through $Q = (a, b)$ in the first loop and that this uniquely determines the cycloid. This gives an absolute minimum time of descent which can be calculated from the integral expressing T.

Lemma 8.1.8. *Let $x = \alpha(t - \sin t)$ and $y = \alpha(1 - \cos t)$ be the cycloid and $f(t) = \frac{t - \sin t}{1 - \cos t}$. Then this cycloid passes through (a, b) if and only if for some $t_0 \in [0, 2\pi)$, $f(t_0) = \frac{a}{b}$.*

Proof. If (a, b) lies on the cycloid in the first loop then this is clear. Conversely, suppose $f(t_0) = \frac{a}{b}$. Then $\frac{b}{1 - \cos t_0} = \frac{a}{t_0 - \sin t_0}$. Let α be this common value. Then $x(t_0) = \alpha(t_0 - \sin t_0) = a$ and $y(t_0) = \alpha(1 - \cos t_0) = b$. $\qquad\square$

Using this lemma we show that given any a and b both negative there is a unique cycloid passing through (a, b) in first loop. We calculate $\lim_{t \to 0+} f(t)$. By L'Hopital's rule differentiating numerator and denominator twice we see that this limit is zero. Hence by continuity $f(0) = 0$. On the other hand, $\lim_{t \uparrow 2\pi} f(t)$ is clearly $\frac{2\pi}{0+} = +\infty$. Since $\frac{a}{b} > 0$ it follows from the Intermediate Value theorem that for some $t_0 \in [0, 2\pi)$, $f(t_0) = \frac{a}{b}$. Clearly this can then be done in $[0, \pi)]$.

We remark that Bernoulli solved this problem, but by a different method. The solution we have presented here is due to L. Euler. We now explain how Bernoulli understood and solved the Brachistochrone problem at the end of the seventeenth century before the calculus of variations was actually invented.

Example 8.1.9. (*Giovanni Bernoulli's solution of the brachistochrone problem*). We first consider a related extremalization problem from calculus of one variable. Consider two materials of different uniform optical densities in the form of continuous strips, A and B, where v_A and v_B are the respective velocities of light in A and B. Let $P = (a, a')$ be a point of A and $Q = (b, b')$ be a point of B. Since in each media separately the velocity of light is constant, the problem of shortest time path has the same solution as the problem of shortest distance path. Thus the solution is a geodesic of the plane which we have already seen is a straight line. Hence within A and B light travels in straight lines.

Since this is so, if a path were to join P and Q the only possibility would be for it to change angle at the interface of A and B. Choose coordinates so the interface point on the trajectory is $(x,0)$, where x is the unknown. Then the total time of the trip is

$$T(x) = \frac{\sqrt{a'^2 + (a-x)^2}}{v_A} + \frac{\sqrt{b'^2 + (b-x)^2}}{v_B}$$

Calculating the derivative and setting this equal to zero we get

$$\frac{x-a}{v_A\sqrt{a'^2 + (a-x)^2}} = \frac{x-b}{v_B\sqrt{b'^2 + (b-x)^2}}.$$

That is to say

$$\frac{\sin(\theta_A)}{v_A} = \frac{\sin(\theta_B)}{v_B},$$

where θ_A and θ_B are the respective angles of incidence with the normal to the boundary at $(x,0)$. This is Snell's law. The quickest such path is the one where the ratio of sines of angles is equal to the ratios of the velocities. It says that as light enters a more dense optical medium and slows down it bends more towards the normal to make up for this.

Knowing this fact Bernoulli considered a nonhomogeneous media, but one in which the optical density depended only on the vertical coordinate y, where perhaps one doesn't know the exact nature of the dependance, only that it is smooth. He then divided this media (plane) into an infinite number of very thin horizontal strips of thickness dy. These strips being so thin that the velocity of light in each one was essentially constant, so that as the light passed from one of these strips to the next it would behave exactly as above and would obey Snell's law. Therefore, at the boundary we would have for all y, i.e. for all strips, $\frac{\sin(\theta(y))}{v(y)} = c$, a constant. However, since θ is the angle with the (each) normal if we let ϕ be the complementary angle, i.e., the angle with the x-axis, we see that $\sin(\theta(y)) = cv(y)$ and since this is the angle whose tangent is the derivative $\frac{dy}{dx}$ we get $\cot^2(\theta(y)) = (\frac{dy}{dx})^2$. Since $\csc^2(\theta(y)) = \cot^2(\theta(y)) + 1$, $\sin(\theta(y)) = \frac{1}{1+(\frac{dy}{dx})^2}$. Thus we have $cv(y) = \frac{1}{1+(\frac{dy}{dx})^2}$ and solving for $\frac{dy}{dx}$ we obtain a simple first order equation $y' = \pm\frac{\sqrt{1-c^2v(y)^2}}{cv(y)}$ in which the variables separate. Thus,

$$x = \pm c \int \frac{v(y)dy}{\sqrt{1-c^2v(y)^2}} + c'.$$

The constants c and c' being determined by P and Q. When $v(y) = \sqrt{y}$, this is the brachistochrone problem. Otherwise it is a generalization and an important one. We leave to the reader, as an exercise, to show that the solution is the same as the one in Example 8.1.7.

Thus, we have also solved the following problem of Geometric Optics: Given a smooth function $v(y)$ find the planar path joining the points P and Q which minimizes the functional $J(y) = \int \frac{\sqrt{1+y'^2}}{v(y)}dx$. The solution is the equation above. For example as a consequence, we see that when solar light enters the earth's atmosphere obliquely, since the optical density increases smoothly, the light bends smoothly more and more towards the normal.

We now turn to the so-called *isoperimetric problem*. This was proposed by the ancient Greeks. In its original formulation the problem is: *Among all the simple closed planar curves of given perimeter find the one that encloses the greatest area.* They called this the *isoperimetric*[5] problem and were able to show in a more or less rigorous manner that the obvious answer - a circle - is the correct answer.

Example 8.1.10. (*The Isoperimetric problem*). Consider all smooth simple closed planar curves $\gamma : [a,b] \to \mathbb{R}^2$ parametrized by $\gamma(t) = (x(t), y(t))$, with $\gamma(a) = \gamma(b)$ of a given length, say l. Find the curve which encloses the largest area and the relationship of this largest area to the length, l.

Solution. Suppose we had a curve which gave the largest area. Choose two points on the curve and in this way divide the curve into two subcurves. Clearly, if the original curve were to enclose the largest area, each of these subcurves would have to be convex in the sense that the original curve encloses a convex region (draw a picture). In particular, the line segment joining the two points would intersect the curve

---

[5]In Greek "$\iota\sigma os$" means "*equal*".

nowhere else. Now consider these points, as above, but with the additional requirement that the length along the curve joining them (in either direction) is $\frac{l}{2}$. Then the areas enclosed must be equal. For otherwise this would clearly violate the maximal area property which we are assuming. Thus we are reduced to considering the interval say $(0,0)$ to $(a,0)$ on the real axis and a convex curve defined on this interval lying in the first quadrant of length $\frac{l}{2}$ and enclosing an area, say A. We may also assume the curve $(x(s), y(s))$ is parametrized by arc length. Now since $A = \int_0^a y dx$ (see, Corollary 6.4.3) we see that $A = \int_0^{\frac{l}{2}} y(s)x'(s)ds$, where $y(0) = 0 = y(\frac{l}{2})$. Since we have chosen arc length as parameter we know $x'^2(s) + y'^2(s) \equiv 1$ for all s. Here of course A depends on the curve (i.e. in this case depends only on y) so we can write

$$A(y) = \int_0^{\frac{l}{2}} y(s)\sqrt{1 - y'(s)^2}ds,$$

where $y(0) = 0 = y(\frac{l}{2})$ and our task is to see what the condition that $A(y)$ be maximal forces on y.

Here $F(s, y, y') = y\sqrt{1 - y'^2}$ and since the integrand is independent of s Proposition 8.1.5 tells us that there is a constant, c where $-y = c\sqrt{1 - y'^2}$. If $c = 0$, then evidently $y = 0$ and the area enclosed is zero. This is the minimal area solution which doesn't interest us at all. So we may assume $c \neq 0$. Dividing by c and solving for y' we get $y' = \frac{\sqrt{c^2 - y^2}}{c}$. We can solve this differential equation by separation of variables and the substitution $y = c \sin \theta$ and get $y(s) = c \sin \frac{s+\alpha}{c}$. Using the boundary conditions and the fact that the curve is convex we see that $\alpha = 0$ and $\frac{l}{2c} = \pi$. Therefore

$$y(s) = \frac{l}{2\pi} \sin \frac{2\pi s}{l}.$$

Now since $x'^2(s) + y'^2(s) \equiv 1$ we can then also solve for $x(s)$ using the fact that $x(0) = 0$ and get

$$x(s) = \frac{l}{2\pi}\left(1 - \cos \frac{2\pi s}{l}\right).$$

Thus $(x(s) - \frac{l}{2\pi})^2 + (y(s))^2 \equiv \frac{l^2}{4\pi^2}$. This means our subcurve lies on a circle centered at $(\frac{l}{2\pi}, 0)$ and radius $\frac{l}{2\pi}$ and the area enclosed by the original curve is $\pi(\frac{l}{2\pi})^2 = \frac{l^2}{4\pi}$, thus proving the solution to the isoperimetric problem is a circle.

Note that at the same time this proves the *isoperimetric inequality* which states: If l is the length of a closed curve in $\mathbb{R}^2$ enclosing an area A, then

$$A \leq \frac{l^2}{4\pi},$$

with equality if and only if the curve is a circle.

We remark that the isoperimetric problem can also be solved using the so-called *Wirtinger's inequality* (See, Exercise 8.5.18) or with the use of Fourier series[6]. Also, it clearly could be considered a variational problem with a constraint, namely the arc length has constant value l. We will solve it on that basis in Example 8.3.3 and compare the result with what we have learned here.

The isoperimetric inequality has a generalization to $\mathbb{R}^n$, $n \geq 3$, as follows: If Ω is a bounded domain in $\mathbb{R}^n$ with smooth boundary $\partial\Omega$ and ν_n, ν_{n-1} are the n and $(n-1)$ volumes, respectively, then

$$[\nu_n(\Omega)]^{n-1} \leq \frac{[\nu_{n-1}(\partial\Omega)]^n}{n^n c_n},$$

where c_n is the volume of the unit ball of $\mathbb{R}^n$.

Example 8.1.11. (*The minimal surface problem for a surface of revolution*). Let P and Q be fixed points in the plane with different first coordinates and consider all smooth curves $y = f(x)$ joining them. Find a curve which minimizes the area of the resulting surface of revolution about the x-axis.

Solution. From Example 6.3.10 we want to minimize the functional

$$A(y) = 2\pi \int_a^b y\sqrt{1 + y'^2}dx.$$

Hence $F(x, y, y') = y\sqrt{1 + y'^2}$. Since F is independent of x we know if there were a solution $F - y'F_{y'} = c$. Thus $c\sqrt{1 + y'^2} = -y$. Squaring we

---

[6]See, Hurwitz's Solution of the Isoperimetric Problem, [6].

get $c^2(1+y'^2) = y^2$. Clearly $c \neq 0$ since then $y = 0$ and would not fit the initial data. So that $y' = \frac{\sqrt{y^2 - c^2}}{c}$. Separating variables and integrating gives $c \int \frac{dy}{\sqrt{y^2 - c^2}} = x - k$. To calculate this integral let $y = c \cosh t$. Then we get $y(x) = c \cosh(\frac{x-k}{c})$, a catenary. We leave it to the reader to show that this can be uniquely made to fit the initial data.

When we deal with problems involving several independent variables we will be able to drop the requirement that we have a surface of revolution (See, Example 8.4.4).

Example 8.1.12. (*The minimal volume problem for a surface of revolution*). Finally, although it seems similar to the above minimal surface area problem, here is an example of a variational problem with no solution. Let P and Q be fixed points of the plane with different first coordinates and consider all smooth curves $y = f(x)$ joining them. Find a curve which minimizes the volume of the resulting surface of revolution about the x-axis.

Solution. Here $V(y) = \pi \int_a^b y^2 dx$ and hence $F(t, y, y') = y^2$. Since F is independent of t we know if there were a solution $F - y'F_{y'}$ is constant. However, since here $F_{y'} = 0$ we see that F is constant. But $F = y^2$. Thus y is constant. But there is no constant passing through P and Q since they have different second coordinates, a contradiction. Actually, what goes wrong here is more fundamental. For even if the heights were equal our calculation shows that if there were a solution it would have to be the constant function. But this clearly doesn't minimize the volume of the revolved figure since we can get smooth curves y_n passing through the end points which are on the x axis for most of the curve. Hence $\inf_n V(y_n) = 0$. Therefore there is *no minimum*, since the zero function doesn't pass through the endpoints. Here there is an infimum of the volumes of all these curves, namely zero. But it is not achieved by any curve satisfying the required boundary conditions.

We conclude this section with a (continuous) generalization of Snell's law.

Example 8.1.13. (*The problem of geometric optics*). Given two points P and Q lying in a plane in which the optical density varies smoothly from point to point, our problem here is to find the path joining them along which the light takes the least time to travel.

Solution. Since just as in the brachistochrone problem $\frac{ds}{dt} = v$, where s is the arc length along the curve and v is the instantaneous velocity, we see that $dt = \frac{ds}{v}$ so that T, the total time, is given by $T = \int_0^a \frac{ds}{v}$. Now if the curve lying in this plane is $y = y(x)$, then $\frac{ds}{dx} = \sqrt{1 + y'^2}$. So $ds = \sqrt{1 + y'^2} dx$ and hence

$$T = \int_a^b \frac{\sqrt{1 + y'^2}}{v(x, y)} dx,$$

where $v(x, y)$ is the velocity of light in the medium at the point (x, y) in the plane. Now the reader will notice that this is very similar to the brachistochrone problem except here v is an unknown function of both x and y instead of a known function of y alone. In any case, the integrand of the functional we have to minimize is $F = \frac{\sqrt{1+y'^2}}{v(x,y)}$. If v is independent of x, then so is F and we could apply our usual methods.

8.1.2   Sufficient conditions

The Euler-Lagrange equation is a *necessary condition* for an extremum. In Examples 8.1.6, 8.1.7, 8.1.11, and 8.1.13 we have to prove our solution is actually a *minimum*. This means we must look at *sufficient conditions* for a minimum to occur. Here we shall not go into the problem of sufficient conditions in general, but confine ourselves to Corollary 8.2.6 below. In this situation it tells us that if F is independent of y and

$$F_{y'y'} \geq 0, \tag{8.7}$$

then each critical point (extremal) is actually a minimum. This inequality is known as *Legendre's condition*[7] and is of great importance in the problem of investigating whether an extremal actually gives an extremum. Note that it involves second order derivatives.

---

[7]A. Legendre (1752-1833) made important contributions to mathematical analysis, algebra, number theory and statistics. Other such criteria are, for example, Weierstrass, Weierstrass-Erdmann, Jacobi condition, or the so-called field theories. See [15].

Now in Examples 8.1.6 and 8.1.11 these conditions are easily verified and are left to the reader. However, in 8.1.7 and 8.1.13 the function F is of the form $F = p(x,y)\sqrt{1 + y'^2}$, where p is a positive smooth function and so although $F_{y'y'} = \frac{p(x,y)}{(1+y'^2)^{\frac{3}{2}}} \geq 0$ our criterion is *not* sufficient since F depends on y. Similarly in Example 8.1.10, Corollary 8.2.6 is not good enough to see that we actually get a maximum since although here $F_{y'y'}$ is positive, F again also depends on y. What is needed to deal with such situations to be sure that there is a solution to the extremal problem being considered are methods of functional analysis, especially *compactness* in certain *function spaces*. This will guarantee that the *sup* or *inf* of the functional involved is actually achieved. Once we know this is so and therefore that a solution exists we can use the Euler-Lagrange equations to find it. In this sense, here also, the situation very much resembles that of calculus in one or several real variables where compactness of the domain is what guarantees the *existence of extremum*[8], which is then found by sifting through the critical points.

EXERCISES

1. Find the extremals of the following functionals:

   (a) $J(y) = \int_0^1 yy'\,dt.$

   (b) $J(y) = \int_0^1 tyy'\,dt.$

   (c) $J(y) = \int_0^1 (\frac{1}{2}y'^2 + yy' + y' + y)\,dt.$

   (d) $J(y) = \int_a^b \frac{y'^2}{t^3}\,dt.$

2. Find the extremals of the functionals:

   (a) $J(y) = \int_a^b (ty' + y'^2)\,dt.$

   (b) $J(y) = \int_a^b y'(1 + t^2 y')\,dt.$

   (c) $J(y) = \int_a^b (y'^2 + 2yy' - 16y^2)\,dt.$

   (d) $J(y) = \int_a^b (y^2 + y'^2 - 2y\sin t)\,dt.$

   (e) $J(y) = \int_a^b \sqrt{1 + \frac{y'^2}{y}}\,dt.$

---

[8]See Weierstrass extreme-value theorem, viz, Corollary 2.5.3.

3. Show that the Euler-Lagrange equation of the functional

$$J(y) = \int_a^b [a(t)y'^2 + 2b(t)yy' + c(t)y^2]dt$$

is a second order linear differential equation.

4. Find the extremum and identify its nature (maximum or minimum) of the functional

$$J(y) = \int_0^1 \frac{\sqrt{1 + y'^2}}{t}dt, \quad y(0) = 1, \; y(1) = 2.$$

5. Consider the variational problem where the functional is

$$J(y) = \int_a^b p^2(x)y'^2 dx.$$

Show that if $y = y(x)$ is a critical point (extremal), then $y' = \frac{c}{p^2}$. Then show that the minimum value of J is $\frac{b-a}{\int_a^b p^{-2}dx}$.

Answers to selected Exercises

1. (a) $y(t) = c_1 t^4 + c_2$. (b) $y(t) = \frac{1}{2}te^t + c_1 e^t + c_2 e^{-t}$.
   (c) $y(t) = \frac{1}{2}(t^2 - 3t + 1)$. (d) $y(t) = c_1 t^4 + c_2$.

2. (a) $y(t) = -\frac{1}{4}t^2 + c_1 t + c_2$. (b) $y(t) = \frac{c_1}{t} + c_2$. (c) $y(t) = c_1 \sin(4t - c_2)$.
   (d) $y(t) = c_1 e^t + c_2 e - t + \frac{1}{2}\sin t$.

8.2   Generalizations

We now turn to the situation where our functional J is more elaborate and we will illustrate why this generalization is necessary with some significant examples. Since this is a mere introduction to the subject, all functions involved in this and the subsequent variational problems will be assumed to be as smooth (or piecewise smooth) as needed to make all arguments valid! In most cases, smoothness of order two, C^2, will suffice.

Let $F : [a,b] \times \mathbb{R}^n \times \mathbb{R}^n \to \mathbb{R}$ be a given smooth function of $2n+1$ real variables, $F(t, Y, Z)$, where $t \in [a, b]$, $Y : [a, b] \to \mathbb{R}^n$ and $Z : [a, b] \to \mathbb{R}^n$ its derivative. Form the functional

$$J(Y) = \int_a^b F(t, Y(t), Y'(t)) dt, \tag{8.8}$$

where $Y(t) = (y_1(t), ..., y_n(t))$ is a smooth vector valued curve parametrized by $t \in [a, b]$, $Y'(t) = (y_1'(t), ..., y_n'(t))$ is its derivative and $Y(a) = \mathbf{a}$ and $Y(b) = \mathbf{b}$ with $\mathbf{a}$, $\mathbf{b}$ given and fixed in $\mathbb{R}^n$. If it exists, our present objective is to find the curve $Y(t)$ in $\mathbb{R}^n$ which extremizes J subject to the boundary conditions, $Y(a) = \mathbf{a}$ and $Y(b) = \mathbf{b}$.

In this context in our original formulation, equation (8.2), was the case when $n = 1$. Here we shall see that, just as before, we have necessary conditions and also sufficient conditions. However, now the necessary condition expressing the fact that an extreme point is a critical point will not be a single differential equation, but rather a *system of n differential equations* (or a single vector equation), also called the Euler-Lagrange equations (Theorem 8.2.2). For the sufficient condition, instead of the second derivative $F_{y'y'}$ being ≥ 0 or ≤ 0, it will now state that the quadratic form associated to the $n \times n$ symmetric matrix of second derivatives $\left(F_{y_i' y_j'} \right)_{i,j=1,...,n}$ is positive semidefinite, or negative semidefinite (Corollary 8.2.6). The reader will note that, in addition to generalizing the calculus of variations in one variable, these results are exact analogues of the corresponding ones to those for minimum and maximum problems in calculus of several real variables in Chapter 3.

As usual, we denote by $\langle \, , \rangle$ the standard inner product on $\mathbb{R}^n$.

Lemma 8.2.1. *Let f be a continuous function $f : [a, b] \to \mathbb{R}^n$ and suppose*

$$\int_a^b \langle f(t), h(t) \rangle dt = 0$$

for all smooth functions $h : [a, b] \to \mathbb{R}^n$, satisfying $h(a) = 0 = h(b)$. Then $f \equiv 0$ on $[a, b]$.

Proof. Suppose $f(t_0) = (f_1(t_0), ..., f_n(t_0)) \neq 0$ for some $t_0 \in (a, b)$. Then it must be nonzero in some coordinate, say $f_i(t_0) \neq 0$. Since the lemma is true for $n = 1$, we can find a smooth function ψ_i : $[a,b] \to \mathbb{R}$ with $\psi_i(a) = 0 = \psi_i(b)$ and $\int_a^b f_i(t)\psi_i(t)dt \neq 0$. Take $h(t) = (0, ..., 0, \psi_i(t), 0, ..., 0)$ for all $t \in [a,b]$. Then $h(a) = 0 = h(b)$ and $\int_a^b \langle f(t), h(t) \rangle dt \neq 0$ a contradiction. Thus, $f \equiv 0$ on (a,b) and so by continuity also on $[a,b]$.   □

Theorem 8.2.2. *Let*

$$J(Y) = \int_a^b F(t, Y(t), Y'(t))dt$$

with F and Y as in (8.8) above. Suppose Y is an extremum of J. Then it satisfies the following system of differential equations for $i = 1, \ldots, n$,

$$F_{y_i} - \frac{d}{dt}F_{y_i'} = 0, \tag{8.9}$$

Written in vector form

$$F_Y - \frac{d}{dt}F_{Y'} = 0.$$

Proof. Let η be an arbitrary smooth function, $\eta : [a,b] \to \mathbb{R}^n$, satisfying $\eta(a) = 0 = \eta(b)$ and $\epsilon > 0$. Consider the function $\overline{Y}(t) = Y(t) + s\eta(t)$ on $[a,b]$, where $|s| < \epsilon$. For each such s this function is smooth, vector valued and takes the required values at the endpoints a and b. Assuming we are in an open set of functions by taking ϵ small enough, the function $\overline{Y} = Y + s\eta$ is in this set for all s, if Y itself is. Hence in the case of a minimizer, $J(Y + s\eta) \geq J(Y)$ for all $|s| < \epsilon$. Therefore by calculus of one real variable,

$$\frac{d}{ds}J(Y + s\eta)|_{s=0} = 0.$$

Set, as before, $g(s) = J(Y + s\eta)$. Then the necessary condition is $g'(0) = 0$. Differentiating under the integral sign and using the chain rule we get

$$g'(s) = \int_a^b \frac{d}{ds}F(t, \overline{Y}(t), \overline{Y}'(t))dt = \int_a^b \frac{d}{ds}F(t, \overline{y_1}, ..., \overline{y_1}, \overline{y_1}', ..., \overline{y_1}')dt$$

$$= \int_a^b \left[F_{\overline{y_1}}\eta_1 + ... + F_{\overline{y_n}}\eta_n + F_{\overline{y_1}'}\eta_1' + ... + F_{\overline{y_n}'}\eta_n' \right] dt.$$

In terms of the inner product on $\mathbb{R}^n$

$$g'(s) = \int_a^b \left[\langle F_{\overline{Y}}, \eta \rangle + \langle F_{\overline{Y}'}, \eta' \rangle \right] dt. \qquad (8.10)$$

At $s = 0$

$$g'(0) = \int_a^b \left[\langle F_Y, \eta \rangle + \langle F_{Y'}, \eta' \rangle \right] dt = 0.$$

Calculating each of the individual terms of the second term via integration by parts we get

$$\int_a^b \langle F_{Y'}, \eta' \rangle dt = \langle F_{Y'}, \eta \rangle \Big|_a^b - \int_a^b \langle \frac{d}{dt} F_{Y'}, \eta \rangle dt.$$

Taking into account the boundary values of η it follows the first of these terms is zero and so

$$\int_a^b \langle [F_Y - \frac{d}{dt} F_{Y'}], \eta \rangle dt = 0.$$

Finally, since this holds for *all* η satisfying the boundary conditions, an application of Lemma 8.2.1 completes the proof. $\qquad\square$

Thus in general, here, we get a system of second order ordinary differential equations with variable smooth coefficients. All solutions to this system of differential equations are called *extremals* of the variation problem. The possibility of giving a general solution of the system of the Euler-Lagrange differential equations is even more remote than in Section 8.1. It is only in special cases that we can find all the extremals explicitly.

Example 8.2.3. Consider the functional

$$J(Y) = \int_0^{\frac{\pi}{2}} (x'^{\,2} + y'^{\,2} + 2xy) dt,$$

where $Y(t) = (x(t), y(t))$ is a C^2 curve in $\mathbb{R}^2$ defined on $[0, 2\pi]$ and satisfies $Y(0) = (0,0)$, $Y(\frac{\pi}{2}) = (1, -1)$. Find the extremals of J.

Solution. Here $F(t, Y, Y') = F(t, x, y, x', y') = x'^{\,2} + y'^{\,2} + 2xy$, and so $F_x = 2y$, $F_{x'} = 2x'$, $F_y = 2x$ and $F_{y'} = 2y'$. The system of the two Euler-Lagrange equations is

$$F_x - \tfrac{d}{dt}F_{x'} = 0 \quad \text{and} \quad F_y - \tfrac{d}{dt}F_{y'} = 0.$$

That is,

$$x'' - y = 0 \quad \text{and} \quad y'' - x = 0.$$

Eliminating one of the unknown functions, say x, we have

$$y^{(4)} - y = 0.$$

This is a fourth order linear ODE with constant coefficients. In Example 7.5.2 we solved this equation and found its general solution $y = c_1 e^t + c_2 e^{-t} + c_3 \sin t + c_4 \cos t$. Therefore $x = y'' = c_1 e^t + c_2 e^{-t} - c_3 \sin t - c_4 \cos t$. These are the extremals of the problem. Using the boundary conditions we find $c_1 = c_2 = c_4 = 0$ and $c_3 = 1$, and hence the extermum is $Y(t) = (\sin t, -\sin t)$.

Next we look at the second variation of the functional J of Theorem 8.2.2. As before, the quantity $\delta J = sg'(0)$ is called the *first variation* of J. The *second variation* $\delta^2 J$ of J is defined by

$$\delta^2 J = \frac{s^2}{2} g''(0).$$

We now calculate the second variation.

Proposition 8.2.4. *Let* $g(s) = J(Y + s\eta)$, *where* η *is a smooth vector-valued function satisfying* $\eta(a) = 0 = \eta(b)$. *Then*

$$g''(0) = \int_a^b \sum_{i,j=1}^n [F_{y_i' y_j'} \eta_i' \eta_j' + 2F_{y_i y_j'} \eta_i \eta_j' + F_{y_i y_j} \eta_i \eta_j] dt. \qquad (8.11)$$

Proof. To simplify the notation we write $Z(t) = \overline{Y}(t) = Y(t) + s\eta(t)$ and $Z'(t) = \overline{Y}'(t) = Y'(t) + s\eta'(t)$. Now differentiating under the integral sign (8.10) we have

$$g''(s) = \int_a^b \frac{d}{ds}(\langle F_Z, \eta \rangle + \langle F_{Z'}, \eta' \rangle) dt$$

$$= \int_a^b \sum_{i,j=1}^n [F_{z_i' z_j'} \eta_i' \eta_j' + 2F_{z_i z_j'} \eta_i \eta_j' + F_{z_i z_j} \eta_i \eta_j] dt.$$

Setting $s = 0$ gives the result. $\qquad\qquad\square$

In the case $n = 1$ we considered in the previous section we have[9]

$$g''(0) = \int_a^b [F_{y'y'}\eta'^2 + 2F_{yy'}\eta\eta' + F_{yy}\eta^2]dt, \qquad (8.12)$$

From Proposition 8.2.4 we get a necessary condition for a minimum or a maximum.

Theorem 8.2.5. *If the functional J has minimum, (or a maximum) at Y respectively, then the second variation, $\delta^2 J \geq 0$, (or ≤ 0) respectively. Conversely, if $\delta^2 J \geq 0$ (or ≤ 0) and Y is a critical point (extremal), then it is a minimum (maximum) respectively.*

Proof. From Theorem 8.2.2 we see that if a mimimum occurs at Y, then for any η satisfying the boundary conditions and any small real s, since the Euler-Lagrange equations are satisfied by Y, the first variation is zero. By the second order Taylor expansion of g we have

$$g(s) = g(0) + sg'(0) + \frac{s^2}{2!}g''(0) + O(s^3).$$

Hence

$$J(Y + s\eta) = J(Y) + \delta^2 J + O(s^3).$$

Therefore Y is a minimum if and only if $\frac{s^2}{2!}g''(0) + O(s^3) \geq 0$. Since $\frac{s^2}{2!} \geq 0$, this occurs if and only if $g''(0) + O(s) \geq 0$. Taking the limit as $s \to 0$ shows this occurs if and only if $g''(0) \geq 0$. Equivalently, if and only if

$$\delta^2 J \geq 0.$$

The case of a maximum is similar. □

---

[9]Integrating the middle term by parts in (8.12) we get

$$\int_a^b 2F_{yy'}\eta\eta'\, dt = -\int_a^b \left(\frac{d}{dt}F_{yy'}\right)\eta^2\, dt.$$

Setting $P = P(t) = \frac{1}{2}F_{y'y'}$ and $Q = Q(t) = \frac{1}{2}\left(F_{yy} - \frac{d}{dt}F_{yy'}\right)$, (8.12) is usually transformed into a more convenient form

$$g''(0) = \int_a^b \left(P\eta'^2 + Q\eta^2\right)dt.$$

The derivation of sufficient condtitions for an extremum of the original functional (8.2) involves the analysis of this quadratic functional. See [15], Chapter 5.

An immediate consequence of this is the following *sufficiency condition*.[10]

Corollary 8.2.6. *Let J and F be as above with F independent of Y, viz $F = F(t, Y')$. Suppose the $n \times n$ symmetric matrix, $\left(F_{y_i' y_j'} \right)_{i,j=1,\ldots,n}$ is positive semi-definite everywhere (resp. negative semi-definite). If Y is a critical point of J, then Y is a global minimum (resp. maximum) for J.*

Proof. Since F is independent of Y we have $\frac{\partial F}{\partial y_i} = 0$ for all $i = 1, 2, \ldots, n$. It follows that $F_{y_i y_j} = \frac{\partial^2 F}{\partial y_i \partial y_j} = 0 = \frac{\partial^2 F}{\partial y_i \partial y_j'} = F_{y_i y_j'}$. Hence if $F_{y_i' y_j'} = \frac{\partial^2 F}{\partial y_i' \partial y_j'}$ is positive semi-definite everywhere, then by (8.11) so is the second variation. $\square$

Example 8.2.7. Consider the functional

$$J(Y) = \int_0^1 (x'^{\,2} + y'^{\,2} + x'y')dt,$$

where $Y(t) = (x(t), y(t))$ is a C^2 curve in $\mathbb{R}^2$ defined on $[0, 1]$ and satisfies $Y(0) = (-1, 1)$, $Y(1) = (2, 3)$. Find the extremals of J. Determine the nature of the extrema.

Solution. Here $F(t, Y, Y') = F(t, x, y, x', y') = x'^{\,2} + y'^{\,2} + x'y'$. The system of the two Euler-Lagrange equations is

$$F_x - \tfrac{d}{dt}F_{x'} = 0 \quad \text{and} \quad F_y - \tfrac{d}{dt}F_{y'} = 0.$$

---

[10]The sufficient conditions for the functional (8.8) with a general $F = F(t, Y, Y')$ to have a (weak) minimum read as follows: *Theorem.* Let the curve $Y(t) = y_1(t), \ldots, y_n(t))$ be an admissible function for the functional (8.8). Suppose Y is an external, i.e., satisfies the system of Euler-Lagrange equations $F_{y_i} - \frac{d}{dt}F_{y_i'} = 0$ ($i = 1, \ldots, n$). If the matrix

$$P(t) = \frac{1}{2}F_{y_i' y_j'}$$

is positive definite, and the interval $[a, b]$ contains no points conjugate to the point a, then the functional (8.8) has a (weak) minimum for the curve Y. See [15], Section 29.4, Theorem 6.

Since $F_x = 0 = F_y$, $F_{x'} = 2x' + y'$ and $F_{y'} = 2y' + x'$, we get the system

$$2x'' + y'' = 0$$

$$x'' + 2y'' = 0.$$

Eliminating x'', we get $-3y'' = 0$. Hence $y'' = 0$ and $x'' = 0$. It follows that the extremals are $x = C_1 t + C_2$, $y = C_3 t + C_4$, which is a family of straight lines in $\mathbb{R}^2$. Applying the boundary conditions we find $x = 3t - 1$ and $y = 2t + 1$. Hence $Y(t) = \varphi(t) = (3t - 1, 2t + 1)$ is the extremum.

To identify its nature, note that $F = F(t, Y')$ is independent of Y and we can apply the sufficient condition of Corollary 8.2.6. Since the 2×2 symmetric matrix

$$\begin{bmatrix} F_{x'x'} & F_{x'y'} \\ F_{y'x'} & F_{y'y'} \end{bmatrix} = \begin{bmatrix} 2 & 1 \\ 1 & 2 \end{bmatrix}$$

is positive definite, the extremum minimizes the functional J. The minimum is $J(\varphi) = 19$.

EXERCISES

Determine the extremals of the functionals

1. $J(x(t), y(t)) = \int_a^b (x'^2 + y'^2 + x'y')dt$.

2. $J(x(t), y(t)) = \int_a^b (2xy - 2x^2 + x'^2 - y'^2)dt$.

3. $J(x(t), y(t)) = \int_a^b x(x' + y' + t)dt$.

Answers to selected Exercises

1. $x(t) = c_1 t + c_2$, $y(t) = c_3 t + c_4$.
2. $x(t) = (c_1 t + c_2) \cos t + (c_3 t + c_4) \sin t$,
    $y(t) = 2x + x'' = (c_1 t + c_2 + 2c_3) \cos t + (c_3 t + c_4 - 2c_1) \sin t$.

We shall now illustrate the use of Theorem 8.2.2 with some applications from geometry and mechanics.

8.2.1   Geodesics on a Riemannian surface

Recall that a smooth manifold is a space which is locally diffeomorphic with an open ball in $\mathbb{R}^n$. Here n is fixed and is called the *dimension* of the manifold. When $n = 2$ this is called a *surface* and it will be surfaces with which we will be concerned. However, here is Riemann's idea in general: a *Riemannian manifold* M is a connected (so that any two points can be joined by some smooth curve) and complete (basically M has no holes) manifold with the property that at each point $p \in M$ there is a positive definite symmetric matrix (or a positive definite quadratic form) of order n, the dimension of the manifold, and these matrices vary smoothly from point to point in the sense that each of their $i, j = 1, \dots, n$ coordinates is a smooth function. The purpose of these positive definite quadratic forms is to measure infinitesimal distances, i.e. the length of a very small arc passing through a point p. This positive definite matrix can be constructed as follows:

Let $\sigma : [a, b] \to \mathbb{R}^n$ and $\phi : \mathbb{R}^n \to \mathbb{R}^n$ be C^1 mappings with

$$\sigma(t) = (u_1(t), ..., u_n(t))$$

and

$$\phi(u_1, ..., u_n) = (x_1(u_1, ..., u_n), ..., x_n(u_1, ..., u_n)).$$

Let $\gamma(t) = (\phi \circ \sigma)(t)$ be a small piece of curve passing through p at $t = a$. Then for $\epsilon > 0$ small its length from $p = \gamma(a)$ to $\gamma(a + \epsilon)$ should be roughly the length of its tangent vector to the curve at p. So if $(u_1, \dots, u_n)$ coordinatize a small neighborhood of p in which the curve lies, then by the chain rule

$$\gamma'(t) = \sum_{i=1}^{n} \frac{\partial \phi}{\partial u_i}(u_1(t), \dots, u_n(t)) \frac{du_i}{dt} = \sum_{i=1}^{n} T_{u_i} \frac{du_i}{dt},$$

where we denote $T_{u_i} = \frac{\partial \phi}{\partial u_i} = (\frac{\partial x_1}{\partial u_i}, ..., \frac{\partial x_n}{\partial u_i})$. So

$$ds = ||\gamma'(t)|| dt = \sqrt{\sum_{i=1,j}^{n} g_{ij} \frac{du_i}{dt} \frac{du_j}{dt}} \; dt,$$

where we define the positive definite matrix g_{ij} by

$$g_{ij}(p) = \langle T_{u_i}, T_{u_j} \rangle, \quad (i, j = 1, ..., n).$$

This is what the matrix $(g_{ij}(p))$ is for, and ds is called a *Riemannian metric*. We then integrate these infinitesimal lengths ds to calculate the length of an ordinary smooth arc joining two points on the manifold. That is,

$$s = s(\gamma) = \int_a^b \sqrt{\sum_{i=1,j}^n g_{ij} \frac{du_i}{dt} \frac{du_j}{dt}}\, dt. \tag{8.13}$$

The best ideas are often easily stated!

In the case of an *embedded surface* $S = \Phi(\Omega)$ in $\mathbb{R}^3$ parametrized by

$$\Phi(u,v) = (x(u,v), y(u,v), z(u,v)),$$

where $\Omega \subset \mathbb{R}^2$ is the parameter set and $\gamma = \Phi \circ \sigma$ a path on S, we know from Section 6.3.1 that the positive definite 2×2 symmetric matrix is

$$[g_{ij}(u,v)] = \begin{bmatrix} g_{11} & g_{12} \\ g_{21} & g_{22} \end{bmatrix} = \begin{bmatrix} E & F \\ F & G \end{bmatrix},$$

where $E = E(u,v) = ||T_u||^2$, $F = F(u,v) = \langle T_u, T_v \rangle$, and $G = G(u,v) = ||T_v||^2$ and therefore the metric (the positive definite quadratic form) is given by

$$ds^2 = E(u,v) \left(\frac{du}{dt} \right)^2 + 2F(u,v)\frac{du}{dt}\frac{dv}{dt} + G(u,v) \left(\frac{dv}{dt} \right)^2,$$

briefly written as

$$ds^2 = E du^2 + 2F du\, dv + G dv^2.$$

In order for γ to be a minimal-length path on S from $\gamma(a)$ to $\gamma(b)$, it must therefore be an extremal for the integral $s(\gamma)$. We say that γ is a *geodesic* on S if it is an extremal for the integral

$$s = \int_a^b \sqrt{E du^2 + 2F du\, dv + G dv^2}\, dt. \tag{8.14}$$

This is a functional of the form (8.8) with integrand

$$f(t, u, v, u', v') = \sqrt{E(u,v) u'^2 + 2F(u,v) u'v' + G(u,v) v'^2}.$$

Hence from Theorem 8.2.2 the Euler-Lagrange equation is the system of equations

$$f_u - \frac{d}{dt}f_{u'} = 0 \text{ and } f_v - \frac{d}{dt}f_{v'} = 0.$$

However, if in the parameter domain Ω we consider a curve of the form $v = v(u)$ or $u = u(v)$, then these give curves on the surface whose infinitesimal length is $ds = \sqrt{Eu'^2 + 2Fu' + G}$, where $u' = \frac{du}{dv}$ or $ds = \sqrt{E + 2Fv' + Gv'^2}$, where $v' = \frac{dv}{du}$ respectively and whose integral dv, respectively du gives the length function

$$s = \int_{v_1}^{v_2} \sqrt{Eu'^2 + 2Fu' + G}\,dv, \text{ respectively } s = \int_{u_1}^{u_2} \sqrt{E + 2Fv' + Gv'^2}\,du.$$

Therefore in order to minimize the length s, the Euler-Lagrange equation for one or the other of these functionals must be satisfied. In this way we have reduced the problem from having to deal with a system of second order equations to that of a single equation. Nevertheless, a moment's reflection tells us that this equation will not be easy to solve without some simplifying assumptions. For example, in general it undoubtably involves the independent variable with respect to which we are integrating, so the method which has worked so well for us by getting what is actually a first order equation will not work in general. But suppose, for example, that $g_{ij}(u,v)$ depends on u alone. Then we consider the second of these functionals $s = \int_{u_1}^{u_1} \sqrt{E + 2Fv' + Gv'^2}\,du$ and corresponding Euler-Lagrange equations

$$\frac{E_v + 2F_v v' + G_v v'^2}{\sqrt{E + 2Fv' + Gv'^2}} - \frac{d}{du}\left(\frac{F + Gv'}{\sqrt{E + 2Fv' + Gv'^2}}\right) = 0.$$

The situation is now much simpler. That is, $\frac{d}{du}\left(\frac{F+Gv'}{\sqrt{E+2Fv'+Gv'^2}}\right) = 0$. Hence for some constant c, we have $F + Gv' = c(\sqrt{E + 2Fv' + Gv'^2})$.

In the case that, in addition, the diagonal term F is identically zero (which means that up on the surface the curves $u = $ constant and $v = $ constant are orthogonal) the situation becomes even simpler and we have $Gv' = c(\sqrt{Gv'^2 + E})$. Solving for $v'^2 = \frac{c^2 E}{G^2 - c^2 G}$ and integrating in the usual manner gives

$$v = v(u) = \pm c \int \frac{\sqrt{E}}{\sqrt{G^2 - c^2 G}}\,du + c_1.$$

Of course, by symmetry if $g_{ij}(u,v)$ depended on v alone and $F = 0$ and we considered the first of these Euler-Lagrange equations where

$u = u(v)$, we would get

$$u = u(v) = \pm c \int \frac{\sqrt{G}}{\sqrt{E^2 - c^2 E}} dv + c_2.$$

Such things have been extensively investigated in the 19th century and are called Clairaut surfaces.

To illustrate the definition of a Riemannian surface we give two examples. The first example or rather class of examples are surfaces of revolution.

Example 8.2.8. (*Geodesics on surfaces of revolution*).   These surfaces are isometric to submanifolds of $\mathbb{R}^3$. As we have seen in Example 6.3.5, a sphere, a cylinder, a cone and a torus (a surface of a bagel) are particular cases of surfaces of revolution. We recall that these surfaces are constructed as follows:

For $a \leq z \leq b$, let $y = g(z)$ be a smooth curve in the yz-plane, where $g(z) \geq 0$ and consider the surface in $\mathbb{R}^3$ obtained by revolving the curve about the z-axis. The resulting surface of revolution is defined by the equation $x^2 + y^2 - g(z)^2 = 0$. We introduce surface parameters (which make it locally diffeomorphic to a ball in $\mathbb{R}^2$ at every point). Let $a \leq u \leq b$ and $0 \leq v \leq 2\pi$. In fact, $u = z$ and $v = \theta$ the polar angle. Then the surface of revolution consists of those points in $\mathbb{R}^3$ of the form

$$\Phi(u, v) = (g(u) \cos v, g(u) \sin v, u).$$

Then $T_u = (g'(u) \cos v, g'(u) \sin v, 1)$ and $T_v = (-g(u) \sin v, g(u) \cos v, 0)$. Hence $\langle T_u, T_u \rangle = 1 + g'(u)^2 = E$, $\langle T_u, T_v \rangle = F = 0$ (which is not usually the case, but when it is its helpful) and $\langle T_v, T_v \rangle = g(u)^2 = G$. So here the positive definite matrix is

$$(g_{ij}) = \begin{bmatrix} 1 + g'(u)^2 & 0 \\ 0 & g(u)^2 \end{bmatrix}$$

and gives a smoothly varying positive definite quadratic form at each point of the surface. (In the case of the sphere there are singularities at the two poles because the metric is not positive definite there, but nowhere else. This is just a technicality which can be dealt with in various ways.)

Here $ds^2 = (1 + g'(u)^2)du^2 + g(u)^2 dv^2$ and the matrix entries $g_{ij}(u, v)$ are independent of v. Hence, as above, using $s = \int_{u_1}^{u_2} \sqrt{E + 2Fv' + Gv'^2}\, du$, if we have a curve $v = v(u)$ in the parameter plane, the Euler-Lagrange equation becomes

$$v = \pm c \int \frac{\sqrt{1 + g'(u)^2}}{g(u)\sqrt{g(u)^2 - c^2}} du + c_1.$$

We may now suppose the generating curve has been parameterized by arc length, which we again call u. This is a normalization condition and every smooth curve can always be reparametrized in this way (See, 6.1.1). Then $\sqrt{1 + g'(u)^2} = 1$ for all u. Then

$$\frac{dv}{du} = \frac{c}{g(u)\sqrt{g(u)^2 - c^2}}.$$

There are two possibilities. If $c = 0$, then $\frac{dv}{du} = 0$ and so v is constant. These are exactly the plane sections through the z-axis of the surface and are always geodesics for any surface of revolution. The others require investigation.

1. (*Geodesics on a cylinder*). In the case of a *cylinder* we have $g(u) = c_0 > 0$. Hence

$$v = c \int \frac{1}{c_0 \sqrt{c_0^2 - c^2}} du + c_1.$$

Thus $v = v(u) = \alpha_1 u + c_1$ is a linear function of u. The only other possibility is u is itself constant. These give *circles* transverse to the axis of the cylinder and in this case these are also clearly geodesics. When $\alpha_1 = 0$, so v is constant we have *straight lines* parallel to the axis of the cylinder. When $\alpha_1 \neq 0$ we have other linear functions. These give a two-parameter family of *helical lines* on the cylinder. Thus these are also "geodesics". But given any two points on the cylinder which have different u and v coordinates we can find infinitely many helices joining them and exactly one of these will have minimal length.

2. (*Geodesics on a sphere*).   The surface of a *sphere* of radius
   $r > 0$, can be viewed as a surface of revolution by taking
   $g(u) = \sqrt{r^2 - u^2}$. We conclude from the above observations on
   surfaces of revolution in general that great circles through the
   north pole N (on the axis of rotation) are geodesics. But then
   since the group of rotations takes any point on the sphere to N,
   it follows that any great circle through any point is a geodesic.
   Moreover that is all. For if we had any geodesic it would satisfy

$$\frac{dv}{du} = \frac{c}{g(u)\sqrt{|g(u)|^2 - c^2}},$$

   where g is as above. Assume the geodesic passes through a point
   P.   By applying the rotation group always assume it passed
   through the N. Hence its derivative $\frac{dv}{du}(0) = 0$ since it would
   have to lie in the tangent plane at N. But $\frac{c}{g(u)\sqrt{|g(u)|^2-c^2}} \neq 0$,
   unless $c = 0$. Hence $c = 0$ and the geodesic is a great circle. Thus
   on the sphere the *geodesics* are exactly the *great circles*.

   The significance of knowing the geodesics here is that we now know
any two points can be joined by a geodesic and if the surface distance
is less then πr the geodesic is unique. This means a pilot flying from
New York to Paris knows exactly which path to take if it is to be the
shortest. However, if the distance is πr, then there are infinitely many
geodesics (a continuum) joining these points. This is because they are
then antipodal points and the great circle or plane is not uniquely determined, since the center of the earth lies on the line joining these points
and contributes nothing to determining the plane. The spheres of varius
radii are the compact *simply connected* surfaces of constant (positive)
curvature[11]. Any two geodesics ("straight lines") must meet. Hence,
here there are no parallel lines!

In the problem Section 8.5, we outline another way to deal with
geodesics on a surface using an analogue to Lagrange multipliers (see,
Exercise 8.5.5).

---

[11]See [33].

Our second example is an important Riemann surface. It is simply connected and has constant (negative) curvature.

Example 8.2.9. (*)(*Geodesics on the Poincare upper half plane*). The Poincare upper half plane, H^+, consists of points (x, y) in the plane where $y > 0$. The metric or positive definite symmetric matrix at (x, y) is a multiple of the identity, namely $\frac{1}{y^2}I$. It is clearly positive definite and smoothly varying over H^+. An important theorem of Hilbert proves that H^+ is not isometric to any surface in $\mathbb{R}^3$. We find the geodesics on H^+.

Solution. Here the hyperbolic metric is $ds = \frac{ds_{Euc}}{y}$, where $ds_{Euc} = \sqrt{dx^2 + dy^2}$. Thus if we are dealing with a curve joining points $P = (a, b)$ and $Q = (c, d)$ of H^+ given by a function $y = f(x)$, then its length is given by

$$s = \int_a^c \frac{\sqrt{1 + y'^2}}{y} dx.$$

Now it might be that P and Q are directly above one another. In this case the geodesic cannot be given by a function $y = f(x)$, nor can the formula just above give the length since here $a = c$ and so this formula gives zero. This means this formula is valid only when $a \neq c$. However, when P and Q are directly above one another it is very easy to see what the shortest length curve joining the points is. Namely, we consider curves of the form $x = \phi(y)$ joining these points. Then $\int_b^d \frac{\sqrt{1+x'^2}}{y} dy$ is the length of such a curve. But

$$\int_b^d \frac{\sqrt{1 + x'^2}}{y} dy \geq \int_b^d \frac{1}{y} dy$$

with equality only if $x'(y)$ is identically zero, since if it is nonzero at any point it is so in some neighborhood and then the integral on the left would definitely be bigger. Thus in this case, since $x'(y) \equiv 0$ we have $x(y)$ is constant, i.e. the curve is a vertical line (and its hyprbolic length is $\log d - \log b$).

We now consider the case where P and Q do not lie directly above one another. Then the formula above is the operative one and $F(x, y, y') = \frac{\sqrt{1+y'^2}}{y}$. Here since F is independent of x, by Proposition

8.1.5, we have $F - y'F_{y'} = c_0$ a constant. From this it easily follows that $1 = cy\sqrt{1 + y'^2}$. Solving for the derivative we get $\frac{dy}{dx} = \frac{\sqrt{c^2 - y^2}}{y}$, where $c = \frac{1}{c_0}$. We separate the variables and get $-2dx = \frac{-2ydy}{\sqrt{c^2 - y^2}}$. Hence integrating gives $-2x + 2\alpha = 2\sqrt{c^2 - y^2}$, where 2α is a constant of integration. Thus

$$(x - \alpha)^2 + y^2 = c^2.$$

This is a semi-circle whose center lies on the x-axis, or put another way it is a circle which meets the boundary of H^+ orthogonally. It is evident from Euclidean! geometry that there is a unique such circle on which both P and Q lie. The circular arc between them is therefore the *geodesic*.[12] We also remark that the problem of geodesics on H^+ can be regarded as falling under the heading of a problem in geometric optics mentioned above.

8.2.2   The principle of least action

The fundamental variational principle of mechanics is the *principle of least action or Hamilton's principle*,[13] which states that among all possible motions of a system of material points, i.e. among those compatible with the constraints, the actual motion traces out a curve that *mini-*

---

[12]In the upper half plane we have shown that, just as in the Euclidean plane itself, any two points can be joined by a unique geodesic and of minimal length. The significance of knowing all the geodesics ("straight lines") here is that we can see by inspection that given a geodesic γ and a point P not on it there are infinitely many geodesics σ passing through P which do not meet γ. Hence here there are infinitely many parallel lines! Now sometimes curves on a Riemannian manifold are called "geodesics" if they merely satisfy the Euler-Lagrange equations, *but may not have minimal length*. By contrast to the upper half plane, in the case of the sphere we saw an example of a surface where there are more than one (actually infinitely many) minimal length geodesic joining certain points and, in the case of the cylinder, a surface where there are infinitely many "geodesics" of larger and larger lengths joining a pair of fixed points. Of course this also occurs in the sphere when a geodesic goes around a great circle more than once.

[13]This is also known as the *Hamilton-Ostrogradski principle*. W. Hamilton (1805-1865). Mathematician, physicist and astronomer who made important contributions to classical mechanics, optics and algebra.

mizes the *action integral*

$$\int_a^b (T - U)dt, \tag{8.15}$$

where T is the kinetic energy and U the potential energy of the system (notice U appears with a minus sign). If the time interval $[a, b]$ is sufficiently small, it can be shown that the action is necessarily minimum. Hence the name "principle of least action", and can be loosely interpreted as saying that "nature tends to equalize the kinetic and potential energies throughout the motion", or expressed in even more philoshophical terms *nature pursues its diverse ends by the most efficient and economical means.*

We give some applications of this principle to mechanics. First we show that Hamilton's principle is equivalent to Newton's second law of motion.

Proposition 8.2.10. *Suppose that a particle of mass m moves in the force field $F : \mathbb{R}^3 \to \mathbb{R}^3$, where $F(x, y, z) = -\nabla U(x, y, z)$ with $U : \mathbb{R}^3 \to \mathbb{R}$ a given potential energy function depending on the coordinates only. Let $\gamma : [a, b] \to \mathbb{R}^3$ be the position vector $\gamma(t) = (x(t), y(t), z(t))$ of the particle, and consider the action integral*

$$J(\gamma) = \int_a^b (T - U)dt.$$

Then the path γ is an extremal for $J(\gamma)$ if and only if γ satisfies Newton's second law of motion

$$F(\gamma(t)) = m\gamma''(t), \quad t \in [a, b].$$

Proof. The velocity and speed of the particle are $\gamma'(t) = (x'(t), y'(t), z'(t))$ and $v = ||\gamma'(t)|| = \sqrt{x'(t)^2 + y'(t)^2 + z'(t)^2}$, respectively. Its kinetic energy is $T = \frac{1}{2}mv^2$. The function

$$L = T - U = \frac{1}{2}m||\gamma'(t)||^2 - U(\gamma(t))$$

is called the *Lagrangian*. The integrand of the action is therefore a function of the form $L(t, \gamma(t), \gamma'(t)) = L(x, y, z, x', y', z')$ and the action integral is a functional of the form (8.8). Suppose γ is an extremal for the action integral. The Euler-Largange equation in this case, in vector form, is

$$L_\gamma - \frac{d}{dt}L_{\gamma'} = 0.$$

That is,

$$-\frac{\partial U}{\partial x} - \frac{d}{dt}\left(\frac{\partial T}{\partial x'}\right) = 0, \ -\frac{\partial U}{\partial y} - \frac{d}{dt}\left(\frac{\partial T}{\partial y'}\right) = 0, \ -\frac{\partial U}{\partial z} - \frac{d}{dt}\left(\frac{\partial T}{\partial z'}\right) = 0,$$

or

$$F_1 - mx'' = 0, \ F_2 - my'' = 0, \ F_3 - mz'' = 0. \tag{8.16}$$

In vector form this is

$$F(\gamma(t)) = m\gamma''(t).$$

Conversely, rewrite (8.16) in the form

$$-\frac{\partial U}{\partial x} - \frac{d}{dt}\left(mx'\right) = 0, \ -\frac{\partial U}{\partial y} - \frac{d}{dt}\left(my'\right) = 0, \ -\frac{\partial U}{\partial z} - \frac{d}{dt}\left(mz'\right) = 0,$$

or

$$-\frac{\partial U}{\partial x} - \frac{d}{dt}\left[\frac{d}{dx'}\left(\tfrac{1}{2}mx'^2\right)\right] = 0, \ -\frac{\partial U}{\partial y} - \frac{d}{dt}\left[\frac{d}{dy'}\left(\tfrac{1}{2}my'^2\right)\right] = 0,$$

$$-\frac{\partial U}{\partial z} - \frac{d}{dt}\left[\frac{d}{dz'}\left(\tfrac{1}{2}mz'^2\right)\right] = 0.$$

Since U is independent of t, x', y', z' and T is independent of x, y, z, we can write

$$-\frac{\partial}{\partial x}\left[\tfrac{1}{2}mx'^2 - U\right] - \frac{d}{dt}\left\{\frac{d}{dx'}\left[\tfrac{1}{2}mx'^2 - U\right]\right\} = 0,$$
$$-\frac{\partial}{\partial y}\left[\tfrac{1}{2}my'^2 - U\right] - \frac{d}{dt}\left\{\frac{d}{dy'}\left[\tfrac{1}{2}my'^2 - U\right]\right\} = 0,$$
$$-\frac{\partial}{\partial z}\left[\tfrac{1}{2}mz'^2 - U\right] - \frac{d}{dt}\left\{\frac{d}{dz'}\left[\tfrac{1}{2}mz'^2 - U\right]\right\} = 0.$$

In this form these equations are the Euler-Lagrange equation of the curve $\gamma(t) = (x(t), y(t), z(t))$. The functional is

$$J(\gamma) = \int_a^b\left[\frac{1}{2}m||\gamma'(t)||^2 - U(\gamma(t))\right]dt = \int_a^b(T - U)dt.$$

$\square$

The argument given above works also for the motion a mechanical system consisting of n particles. Indeed we have:

Theorem 8.2.11. *Given a system of n particles with masses m_i, $i = 1, 2, ..., n$ and coordinates $\gamma_i(t) = (x_i(t), y_i(t), z_i(t))$, suppose that on each particle acts a corresponding conservative force $F_i = (F_{i1}, F_{i2}, F_{i3})$ where $F_i = -\nabla U$ with $U = U(x_i, y_i, z_i)$ a potential function depending on the coordinates only. Then the differential equation of motion of the system is Newton's law $F(\gamma(t)) = m\gamma''(t)$.*

Proof. Here $F_{i1} = -\frac{\partial U}{\partial x_i}$, $F_{i2} = -\frac{\partial U}{\partial y_i}$ and $F_{i3} = -\frac{\partial U}{\partial z_i}$. The kinetic energy is

$$T = \frac{1}{2} \sum_{i=1}^{n} m_i(x_i'^2 + y_i'^2 + z_i'^2), \tag{8.17}$$

and the potential energy of the system is U. In just the same way as above the Euler-Lagrange equations for the action integral $\int_a^b (T - U)dt$ are

$$-\frac{\partial U}{\partial x_i} - \frac{d}{dt}\left(\frac{\partial T}{\partial x_i'}\right) = 0,$$

$$-\frac{\partial U}{\partial y_i} - \frac{d}{dt}\left(\frac{\partial T}{\partial y_i'}\right) = 0,$$

$$-\frac{\partial U}{\partial z_i} - \frac{d}{dt}\left(\frac{\partial T}{\partial z_i'}\right) = 0.$$

Newton's equations of motion of the system are

$$F_i(\gamma_i(t)) = m\gamma_i''(t), \quad i = 1, 2, ..., n.$$

□

Thus, the *variational approach* (Hamilton's principle) and the *vectorial approach* (Newton's law) to the dynamics of a particle are equivalent to one another. This result emphasizes the essential characteristic of variational principles in physics: *they express the pertiment physical laws in terms of energy alone, without reference to any coordinate system.*

In classical mechanics, Hamilton's principle can be viewed as the source of Lagrange's equations of motion, which occupy a dominant position in the subject.

We now briefly trace this connection: A mechanical system has n degrees of freedom if its position is determined by n independent coordinates $q_1, q_2, ..., q_n$ (referred to as *Lagrange's generalized coordinates*). If for example, the system consists of a single particle, $n = 3$, since q_1, q_2, q_3 can be taken to be the three rectangular coordinates x, y, z (or cylindrical, spherical, or any other). By constraining the particle to move on a surface $G(x, y, z) = 0$, we reduce its degrees of freedom to $n = 2 = 3 - 1$. Similar considerations extend to a system of N particles in a space of dimension d, constrained by m consistent independent equations of the form

$$G_k(x_{11}, ..., x_{1d}, x_{21}, ..., x_{2d}, ..., ..., x_{N1}, ..., x_{Nd}) = 0, \quad (k = 1, ..., m).$$

Then the number of degrees of freedom is $n = dN - m$. In principle these m equations can be used to reduce the number of coordinates from dN to n by expressing the dN variables $x_{i1}, x_{i2}, ..., x_{id}$ (for $i = 1, 2, ..., N$) in terms of n of these variables.

A dynamical system of n degrees of freedom can be described with sufficient generality by means of two functions, the *kinetic energy, T*, and the *potential energy, U*. If we think of the system as moving in any way, the position coordinates $q_1, ..., q_n$ will be functions $q_1(t), ..., q_n(t)$ of the time t and their instantaneous velocities (or *momenta*) will be $q_1'(t), ..., q_n'(t)$. Then associated with the system there is a function called the kinetic energy, which is of the form

$$T = T(q_1, ..., q_n, q_1', ..., q_n') = \frac{1}{2} \sum_{i,j=1}^{n} a_{ij} q_i' q_j', \qquad (8.18)$$

where (a_{ij}) is a real symmetric matrix. Hence, the kinetic energy is a homogeneous quadratic function in the components of velocity. The coefficient matrix (a_{ij}) being taken as known functions *not* depending explicitly on the time or the coordinates $q_1, ..., q_n$ themselves. (Thinking of the position coordinates being the rectangular coordinates of the particles of the system, the kinetic energy has the form (8.17)).

The other function that is associated and which together with T characterize the dynamical system is the potential energy function $U = U(q_1, ... q_n)$, which depends on the position coordinates only and *not* on the time or the velocities.[14] The *Lagrangian* of the system, $L = T - U$, is a function of the form

$$L = L(q_1, ..., q_n, q_1', ..., q_n').$$

By Hamiltons' principle, the actual motion of the system from a given initial position A to a given final position B is along C^2 paths $\gamma(t) = (q_1(t), ..., q_n(t))$, with $\gamma(a) = A$ and $\gamma(b) = B$, which minimize the action integral

$$J(\gamma) = \int_a^b L dt = \int_a^b (T - U) dt.$$

From Theorem 8.2.2, the Euler-Lagrange equations of this functional are

$$L_{q_i} - \frac{d}{dt} L_{q_i'} = 0,$$

or

$$\frac{\partial T}{\partial q_i} - \frac{d}{dt}\left(\frac{\partial T}{\partial q_i'}\right) = \frac{\partial U}{\partial q_i} \quad (i = 1, ..., n), \tag{8.19}$$

which are called *Lagrange's equations* and are of fundamental importance in mechanics.

Using the form of the integrand in the action integral, we can find various functions which maintain constant values along each trajectory of the system, thereby obtaining so-called *conservation laws*[15]. In particular, when the given system is *conservative*, which, of course, means that the potential energy function U depends *only* on the position, we obtain the following generalization of Theorem 6.2.7 on the principle of conservation of energy.

---

[14]As is known from classical mechanics, this potential energy determines the effect of external forces acting on the system. In order the system to move from one position into another *mechanical work* is done. This work is equal to the difference between the corresponding values of U and does not depend on the trajectory in space as the system moves from one position to another.

[15]See, [15] Chapter 4, Sec. 22.

Theorem 8.2.12. *(The generalized principle of conservation of energy). When the potential energy of the system is conservative and the kinetic energy is quadratic, then the path of least action conserves the total energy of the system throughout its trajectory.*

Proof. Here $U = U(q_1, \ldots, q_n)$, the kinetic energy function is as in (8.18) and so the Lagrangian is independent of t. Hence $U_{q_i'} = 0$, $L_{q_i'} = T_{q_i'}$ and therefore by Euler's theorem on homogeneous functions (Theorem 3.3.2), we have $\sum_{i=1}^n q_i' L_{q_i'} = \sum_{i=1}^n q_i' T_{q_i'} = 2T$. Hence,

$$\sum_{i=1}^n q_i' L_{q_i'} - L = 2T - (T - U) = T + U = E.$$

We can now calculate

$$\frac{dE}{dt} = \frac{d}{dt}\left(\sum_{i=1}^n q_i' L_{q_i'} - L\right) = \sum_{i=1}^n q_i' \frac{d}{dt} L_{q_i'} + L_{q_i'} q_i'' - \frac{dL}{dt}.$$

By the Euler-Lagrange equations this is

$$\sum_{i=1}^n q_i' L_{q_i} + L_{q_i'} q_i'' - \frac{dL}{dt}.$$

Since $L = T - U$, $\frac{dL}{dt} = \sum_{i=1}^n q_i' L_{q_i} + L_{q_i'} q_i''$ and so $\frac{dE}{dt} = 0$. $\square$

Hamilton's principle can also be applied to yield the basic laws of electricity and magnetism, quantum theory and relativity. In fact, its influence is so profound and far-reaching that many scientists regard it as the most powerful single principle in mathematical physics.

8.3   Variational problems with constraints

In the variational problems we considered so far, the admissible functions were required (apart from certain smoothness requirements) to satisfy boundary conditions on the end points of the curves. We shall now consider problems in which additional conditions are imposed on the admissible functions. Such conditions are known as *constraints* or *side conditions.* These conditions[16] can express simple relation(s) among the variables, for example $G(x, y, z) = 0$ (known as *finite side conditions* and lead to problems related to *geodesics*, See Problem 8.5.5) or can involve integrals in which case the problem is called an *isoperimetric problem.* These problems are the analogues in the calculus of variations to the ordinary constraint extrema problems for functions in several variables studied in Section 3.9.

Here we present a theorem on extrema with integral constraint which is an analogue of the theorem on Lagrange multipliers of Chapter 3. In fact, its proof uses that result.

Theorem 8.3.1. *Given the functional*

$$J(y) = \int_a^b F(t, y(t), y'(t))dt,$$

suppose the admissible functions satisfy the conditions $y(a) = \alpha$, $y(b) = \beta$ *and*

$$K(y) = \int_a^b G(t, y(t), y'(t))dt = c, \tag{8.20}$$

where c is a constant. If $y = y(t)$ is a minimizer or maximizer of J satisfying the above conditions,[17] then there exist $\lambda \in \mathbb{R}$ such that $y = y(t)$ is a free extremal (without side conditions) for the functional

$$\int_a^b (F - \lambda G)dt,$$

i.e., $y = y(t)$ satisfies the Euler-Lagrange equation

$$F_y - \frac{d}{dt}F_{y'} - \lambda\left(G_y - \frac{d}{dt}G_{y'}\right) = 0. \tag{8.21}$$

---

[16]Side conditions can also involve a differential equation, such as, $G(x, y, z, y', z') = 0$, where the expression $G(x, y, z, y', z')$ can not be obtained by differentiating an expression $g(x, y, z)$ with respect to x (these are called *nonholonomic conditions*).

[17]We assume implicitly here that $y = y(t)$ is not an extremal of the functional $K(y)$, that is, y does not satisfies $G_y - \frac{d}{dt}G_{y'} = 0$. In this case there exist only one function satisfying the given side condition.

Proof. We shall prove the theorem for a minimizer. The proof for a maximizer is completely similar. Let the functional J have a minimum for the curve $y = y(t)(= \varphi(t))$, subject to the condition (8.20). As in the proof of Theorem 8.1.2, the idea is to make $y(t)$ to "vary" slightly. However, the neighboring curves we considered there of the form $\overline{y}(t) = y(t) + s\eta(t)$, can not be used here, for in general these will not satisfy (8.20). Instead we must consider a two-parameter family of neighboring curves of the form

$$\overline{y}(t) = y(t) + s_1\eta_1(t) + s_2\eta_2(t),$$

where η_1 and η_2 are functions defined on $[a, b]$, having continuous second derivatives and satisfy $\eta_1(a) = \eta_1(b) = \eta_2(a) = \eta_2(b) = 0$, but are otherwise arbitrary. The parameters s_1, s_2 are not independent, but are related by the condition

$$\Psi(s_1, s_2) = K(\overline{y}) = \int_a^b G(t, \overline{y}, \overline{y}')dt = c.$$

Setting

$$\Phi(s_1, s_2) = J(\overline{y}) = \int_a^b F(t, y(t) + s_1\eta_1(t) + s_2\eta_2(t), y'(t) + s_1\eta_1'(t) + s_2\eta_2'(t))dt,$$

the problem is now reduced to that the function $\Phi(s_1, s_2)$ must have a minimum at $(0, 0)$, for all sufficiently small values of s_1 and s_2 satisfying the condition $\Psi(s_1, s_2) = c$. According to Lagrange multiplier rule (Theorem 3.9.1), there exist $\lambda \in \mathbb{R}$ such that $\nabla\Phi(s_1, s_2) = \lambda\nabla\Psi(s_1, s_2)$. Introducing the Lagrangian function

$$L = F - \lambda G,$$

this amounts to the function

$$\Lambda(s_1, s_2) = \nabla\Phi(s_1, s_2) - \lambda\nabla\Psi(s_1, s_2) = \int_a^b L(t, \overline{y}, \overline{y}')dt$$

having an unconstrained (free) minimum at $(s_1, s_2) = (0, 0)$.

Now differentiating under the integral sign as before, (for $j = 1, 2$), we get

$$\frac{\partial\Lambda}{\partial s_j}(s_1, s_2) = \int_a^b \left[\frac{\partial L}{\partial \overline{y}}\eta_j(t) + \frac{\partial L}{\partial \overline{y}'}\eta_j'(t)\right]dt.$$

We thus have

$$\frac{\partial \Lambda}{\partial s_j}(0,0) = \int_a^b \left[\frac{\partial L}{\partial y}\eta_j(t) + \frac{\partial L}{\partial y'}\eta_j'(t)\right] dt = 0.$$

Integrating the second term by parts, we obtain

$$\int_a^b \left[\frac{\partial L}{\partial y} - \frac{d}{dt}\left(\frac{\partial L}{\partial y'}\right)\right] \eta_j(t) dt = 0.$$

Since this holds for all such η_j, the fundamental lemma implies

$$\frac{\partial L}{\partial y} - \frac{d}{dt}\left(\frac{\partial L}{\partial y'}\right) = 0.$$

Hence

$$F_y - \frac{d}{dt}F_{y'} - \lambda\left(G_y - \frac{d}{dt}G_{y'}\right) = 0.$$

$\square$

The general solution of (8.21) involves three undetermined parameters: two constants of integration and the Lagrange multiplier λ. The extremum is then selected from these solutions (extremals) by imposing the two boundary and giving the functional K its prescribed value c.

In the same way as in the previous section, Theorem 8.3.1 generalizes to functionals of the form

$$J(Y) = \int_a^b F(t, Y(t), Y'(t)) dt,$$

where the admissible functions $Y(t) = (y_1(t), ..., y_n(t))$ satisfy the conditions $Y(a) = \mathbf{a}$, $Y(b) = \mathbf{b}$ and are constrainted by

$$K(Y) = \int_a^b G(t, Y(t), Y'(t)) dt = c,$$

where c is a constant. In this case the extremals must satisfy the system of differential equations for $i = 1, ..., n$,

$$F_{y_i} - \frac{d}{dt}F_{y_i'} - \lambda\left(G_{y_i} - \frac{d}{dt}G_{y_i'}\right) = 0.$$

Example 8.3.2. (*The catenary*). Consider the problem of a hanging flexible cable of fixed length between two points. The physical principle here is that it hangs in such a way that its potential energy is minimal. However, we also have a constraint, namely that the length is fixed.

Solution. Let l be the length of the cable and δ be the constant linear density. A small piece of the cable of length ds, where s is the arc length has weight $(g\delta)ds$, where g is the constant of gravity. Therefore potential energy $y(g\delta)ds$. Notice that we have chosen the left end point to have height zero in calculating the potential energy.

Thus we want to minimize

$$U(y) = \int_0^l y\sqrt{1 + y'^2}dx,$$

while holding $\int_0^l \sqrt{1 + y'^2}dx$ constant.

Here $F(x, y, y') = y\sqrt{1 + y'^2}$, $G(x, y, y') = \sqrt{1 + y'^2}$ and

$$L = F - \lambda G = y\sqrt{1 + y'^2} - \lambda\sqrt{1 + y'^2}$$

Now L is independent of x. Applying Proposition 8.1.5, we see $L - y'L_{y'} = c$ for some constant c. Therefore

$$(y + \lambda)\left(\frac{y'^2}{\sqrt{1 + y'^2}} - \sqrt{1 + y'^2}\right) = c.$$

From this we get, by separation of variables, as before

$$y = \lambda - c\cosh\left(\frac{x - c_1}{c}\right),$$

where c_1 is a constant of integration. This is a family of *catenaries*. The three constants, λ, c and c_1 can then be used to fit the solution to the initial data.

Another example of an extremal problem with a constraint is provided by the isoperimetric problem we solved earlier. For instructional purposes, here we resolve it by Lagrange multipliers.

Example 8.3.3. We have $\int_a^b \sqrt{1 + y'^2} dx = l$, the fixed length, while we want to maximize the area

$$A(y) = \int_a^b y\, dx.$$

Solution. Here $F(x, y, y') = y$ and $G(x, y, y') = \sqrt{1 + y'^2}$. Therefore,

$$L = F - \lambda G = y - \lambda\sqrt{1 + y'^2},$$

where λ is the Lagrange multiplier. Since the function $F - \lambda G$ is independent of x and must satisfy the Euler-Lagrange equation, by Proposition 8.1.5 we get $L - y' L_{y'} = c$. Writing this out tells us

$$\frac{-\lambda y'^2}{\sqrt{1 + y'^2}} - y + \lambda\sqrt{1 + y'^2} = c.$$

This means $\lambda = (y + c)\sqrt{1 + y'^2}$. Calling $y + c = u$ so that $y' = u'$ we get $\lambda = u\sqrt{1 + u'^2}$. We solve for u' and get $\frac{du}{dx} = \frac{\sqrt{\lambda^2 - u^2}}{u}$. We now separate the variables $\int dx = \int \frac{u\, du}{\sqrt{\lambda^2 - u^2}}$. Integrating and squaring tells us that $(x + \alpha)^2 + u^2 = \lambda^2$, where α is a constant of integration. Thus in terms of the original parameters, we have a circle

$$(x + \alpha)^2 + (y + c)^2 = \lambda^2$$

of radius the Lagrange multiplier and so λ is determined by the initial data, namely the arc length of the curve.

The case of *several constraints* is treated similarly. In fact we have

Theorem 8.3.4. *Let*

$$J(Y) = \int_a^b F(t, Y, Y')\, dt$$

be a functional, where the admissible functions $Y(t) = (y_1(t), ..., y_n(t))$ satisfy the conditions $Y(a) = \mathbf{a}$, $Y(b) = \mathbf{b}$ and the constraints

$$K_i(Y) = \int_a^b G_i(t, Y, Y')\, dt = c_i,$$

where $i = 1, \ldots, k$, with $k \leq n$. If Y extremizes J, then there exist real numbers $\lambda_1, \ldots, \lambda_k$ such that Y satisfies the Euler-Lagrange equation

$$L_Y - \frac{d}{dt} L_{Y'} = 0,$$

where L is the Lagrangian, $L = F - \sum_{i=1}^{k} \lambda_i G_i$.

Proof. As above let $L = F - \sum_{i=1}^{k} \lambda_i G_i$, where the Lagrange multipliers $\lambda_1, \ldots, \lambda_k$ are to be determined later. For $j = 1, 2$ let $\eta_j(t) = (\eta_{j,1}(t), \ldots, \eta_{j,n}(t))$ be arbitrary smooth functions on $[a, b]$ with values in $\mathbb{R}^n$ satisfying $\eta_j(a) = 0 = \eta_j(b)$ and let $s_j = (s_{j,1}, \ldots, s_{j,n}) \in \mathbb{R}^n$. Holding η_1 and η_2 fixed consider

$$\Lambda(s_1, s_2) = \int_a^b L(t, Y + s_1\eta_1 + s_2\eta_2, Y' + s_1\eta_1' + s_2\eta_2')dt.$$

Thus Λ is a function of $2n$ real variables, $s_{1,1}, \ldots, s_{1,n}, s_{2,1}, \ldots, s_{2,n}$. We want to calculate its partial derivatives, $\frac{\partial \Lambda(s_1, s_2)}{\partial s_{j,l}}$ at $(s_1, s_2) = (0, 0)$. However, these variables $s_{j,l}$ are not all independent. There are relations among them because, throughout this process (i.e. for all s_1 and s_2), we insist that the $K_i(\overline{Y}) = K_i(Y + s_1\eta_1 + s_2\eta_2)$ be held at the constant value c_i and this must hold for all $i = 1, \ldots k$. Thus, the number of independent variables is actually only $2n - k$. This is where the k Lagrange multipliers come in. They boost the number of independent variables back to $2n$.

The proof now follows the general pattern of that of Theorem 8.2.2. $\qquad\square$

EXERCISES

1. Complete the proof of Theorem 8.3.4.

2. Find the Euler-Lagrange equation for the isoperimetric problem in which

$$J(y) = \int_a^b [a(t)y'^2 + 2b(t)yy' + c(t)y^2]dt$$

is to be extremized subject to the condition $K(y) = \int_a^b y^2 dt = 1$.

3. Find the extremals of the functional $J(y) = \int_a^b y'^2 dt$ subject to the constraint $K(y) = \int_a^b y dt = c$, where c is a constant.

4. Find the extremals of

$$J(y) = \int_0^1 (y'^2 + t^2) dt$$

subject to $\int_0^1 y^2 dt = 2$, $y(0) = 0$ and $y(1) = 0$.

5. A curve in the first quadrant joins the points $(0,0)$ and $(1,0)$ and has a given area under it. Prove that the shortest such curve is an arc of a circle.

6. Find the extremals of

$$J(x(t), y(t)) = \int_0^1 (x'^2 + y^2 - 4ty' - 4y) dt$$

subject to $K(x(t), y(t)) = \int_0^1 (x'^2 - tx' - y'^2) dt = 2$, $x(0) = 0$, $y(0) = 0$, $x(1) = 1$ and $y(1) = 1$.

Answers to selected Exercises

2. $\frac{ay'' + a'y' + y(b' - c)}{y} = \lambda$, where λ is a constant.
3. $y = \lambda t^2 + c_1 t + c_2$, where the three constants c_1, c_2 and λ, can be determined by the boundary conditions and the constraint condition.
4. $y = \pm 2\sin(n\pi t)$, with $n \in \mathbb{Z}$.   6. $x(t) = -\frac{5}{2}t^2 + \frac{7}{2}t$, $y(t) = t$.

8.4   Multiple integral variational problems

In this final section we deal with variational problems involving functionals depending on functions of several independent real variables. In this case the functional J is defined by a multiple integral. We shall see that the general method of finding necessary conditions for it to have an extreme value can be applied equally well to multiple integrals. To keep the notation simple, in 8.4.1 we first confine ourselves to the case of two independent variables, but as we shall see in 8.4.2, all considerations remain valid when there are n independent variables.

8.4.1   Variations of double integrals

Let $\Omega \subset \mathbb{R}^2$ be a bounded domain whose boundary is a closed smooth (or piecewise smooth) positively oriented curve $C = \partial\Omega$. Let $u = u(x, y)$ be a C^2 function defined on Ω and its boundary, assuming prescribed boundary values on $\partial\Omega$, viz, $u \equiv f$ on $\partial\Omega$ for some given C^1 function f, but being otherwise arbitrary. This function can be thought of as defining a variable surface fixed along its boundary in space. Furthermore, let $F : \overline{\Omega} \times \mathbb{R} \times \mathbb{R}^2 \to \mathbb{R}$ be a C^2 function in five variables $F(x, y, u, u_x, u_y)$, where $u_x = \frac{\partial u}{\partial x}$ and $u_y = \frac{\partial u}{\partial y}$, evaluated at $(x, y) \in \overline{\Omega}$.

The problem we consider here is to find necessary conditions so that the *admissible functions* (or *admissible surfaces*) $u = u(x, y)$ minimize or maximize the functional

$$J(u) = \int\!\!\int_{\Omega} F(x, y, u, u_x, u_y) dx dy \tag{8.22}$$

We shall see that in this case the necessary condition, viz, the Euler-Lagrange equation is a *partial differential equation*. Such multidimensional variational problems arise in several applications in mechanics (*strings, membranes*), in geometry (*minimal surfaces*) and in partial differential equations (*existence* and *uniqueness* of *solutions*).

Before we derive the Euler-Lagrange equation, we need the multivariable analogue of the fundamental lemma.

Lemma 8.4.1. *Let* $\varphi : \Omega \subset \mathbb{R}^2 \to \mathbb{R}$ *be continuous and such that*

$$\int\!\!\int_{\Omega} \varphi(x, y)\psi(x, y) dx dy = 0,$$

for all functions $\psi \in C^1(\Omega)$ *satisfying* $\psi \equiv 0$ *on* $\partial\Omega$*. Then* $\varphi \equiv 0$ *on* Ω*.*

Proof. Suppose the function $\varphi(x, y)$ is nonzero, say $\varphi(x_0, y_0) > 0$, at some point $(x_0, y_0) \in \Omega$. Continuity implies that $\varphi(x, y) > 0$ in an open disk B_0 of sufficiently small radius with center (x_0, y_0). Choose a continuous ψ on Ω such that $\psi \equiv 0$ outside B_0 but $\psi(x, y) > 0$ on a ϵ-neighborhood B_ϵ of (x_0, y_0) (for example, $\psi(x, y) = [\epsilon^2 - (x - x_0)^2 - (y - y_0)^2]^k$ if $(x, y) \in B_\epsilon$ with $(k > 1)$, and $\psi(x, y) = 0$ otherwise). Then

$$\int\!\!\int_{\Omega} \varphi(x, y)\psi(x, y) dx dy = \int\!\!\int_{B_\epsilon} \varphi(x, y)\psi(x, y) dx dy > 0.$$

This is a contradiction.   □

We remark that the fundamental Lemma 8.4.1 (proved above for $n = 2$) actually holds for all n. Simply replace the 2-dimensional neighborhoods B_0 and B_ϵ by n-dimensional ones.

Theorem 8.4.2. *Let*

$$J(u) = \int\int_\Omega F(x, y, u, u_x, u_y)\, dx\, dy$$

with F and u as in (8.22) above. If u minimizes or maximizes the functional J, then u satisfies the following partial differential equation, called the Euler-Lagrange equation

$$F_u - \frac{\partial}{\partial x} F_{u_x} - \frac{\partial}{\partial y} F_{u_y} = 0. \tag{8.23}$$

Proof. Suppose the function $u = u(x, y)(= \varphi(x, y))$ is an extremum of J. Let $\epsilon > 0$ and for $|s| < \epsilon$, form the function

$$\overline{u}(x, y) = u(x, y) + s\eta(x, y),$$

where $\eta(x, y)$ is a smooth function on Ω which vanishes on $\partial\Omega$. When $\overline{u}$ is substituted into the integral, the functional $J(\overline{u}) = J(u + s\eta)$ again becomes a function $g(s)$ of s, and a necessary condition for an extremum is

$$g'(0) = 0.$$

As before (differentiating under the integral sign) this condition takes the form

$$\int\int_\Omega \left(F_u \eta + F_{u_x} \eta_x + F_{u_y} \eta_y \right) dx\, dy = 0.$$

To eliminate η_x and η_y from the integral notice that

$$F_{u_x}\eta_x + F_{u_y}\eta_y = \frac{\partial}{\partial x}\left(F_{u_x}\eta \right) + \frac{\partial}{\partial y}\left(F_{u_y}\eta \right) - \left(\frac{\partial}{\partial x} F_{u_x} + \frac{\partial}{\partial y} F_{u_y} \right)\eta.$$

Then according to Green's theorem (Theorem 6.4.1)

$$\int\int_\Omega \left(F_{u_x}\eta_x + F_{u_y}\eta_y \right) dx\, dy$$

$$= \int_{\partial\Omega} \eta(F_{u_x}\, dy - F_{u_y}\, dx) - \int\int_\Omega \left(\frac{\partial}{\partial x} F_{u_x} + \frac{\partial}{\partial y} F_{u_y} \right)\eta\, dx\, dy.$$

Since $\eta = 0$ on $\partial\Omega$, we find

$$\int\int_\Omega \left(F_u - \frac{\partial}{\partial x}F_{u_x} - \frac{\partial}{\partial y}F_{u_y}\right)\eta\, dx dy = 0.$$

An application of Lemma 8.4.1 completes the proof. □

Written out in detail the Euler-Lagrange equation (8.23) is the second order partial differential equation

$$F_{u_x u_x}u_{xx} + 2F_{u_x u_y}u_{xy} + F_{u_y u_y}u_{yy} + F_{u_x u}u_x + F_{u_y u}u_y + F_{u_x x} + F_{u_y y} - F_u = 0.$$

From the infinite set of solutions of this PDE a particular solution must be determined by means of the given boundary condition, viz, by solving the corresponding *boundary value problem*. Our first example is the analogue of Example 8.1.3 in $\mathbb{R}^2$.

Example 8.4.3. (*Dirichlet's energy integral*). Let $\Omega = \{(x,y) : x^2 + y^2 < \alpha\}$ be a disk in $\mathbb{R}^2$ and $u \in C^2(\overline{\Omega})$ with $u \equiv f$ on $\partial\Omega$. Consider the so-called *Dirichlet functional*

$$J(u) = \int\int_\Omega \frac{1}{2}(u_x^2 + u_y^2)dx dy = \int\int_\Omega \frac{1}{2}||\nabla u(x,y)||^2 dx dy \qquad (8.24)$$

Find the minimum[18] value of J.

Solution. Here $F(x,y,u,u_x,u_y) = \frac{1}{2}(u_x^2+u_y^2)$. So that, $F_u = 0$, $F_{u_x} = u_x$ and $F_{u_y} = u_y$. Hence the Euler-Lagrange equation (8.23) here is the Laplace equation

$$\Delta u = u_{xx} + u_{yy} = 0,$$

and the corresponding boundary-value problem is the Dirichlet problem for the disk Ω

$$\Delta u = 0 \ in \ \Omega$$

$$u \equiv f \ on \ \partial\Omega.$$

---

[18]The existence of a minimizer of the Dirichlet functional is not an easy matter. Such concerns we will take us well beyond the subject of this book. The interested reader can consult [6].

In Example 7.8.6 we solved this problem and found the solution

$$\varphi = u(r, \theta) = \frac{1}{2}a_0 + \sum_{n=1}^{\infty} r^n (a_n \sin(n\theta) + b_n \cos(n\theta)). \qquad (8.25)$$

Thus, φ is the extremal of the functional J. Changing the integral (8.24) into polar coordinates we get

$$J(u) = \int_0^{2\pi} \int_0^{\alpha} \left(u_r^2 + \frac{1}{r^2} u_\theta^2 \right) r \, dr \, d\theta. \qquad (8.26)$$

Calculating the partial derivatives φ_r and φ_θ (by term-by-term differentiation) and substituting in (8.26) we obtain the minimum

$$J(\varphi) = \pi \sum_{n=1}^{\infty} n(a_n^2 + b_n^2),$$

where a_n and b_n are given by (7.77) and (7.78) respectively.

Next we consider the problem of minimizing the area of a surface subject to fixed boundary conditions. Namely, that all such surfaces have the same boundary curve. Earlier we dealt with a simpler version of this problem when we considered surfaces of revolution. Also, notice that the problem of geodesics joining two fixed points is a lower dimensional analogue.

Example 8.4.4. (*minimal surfaces*). Among the surfaces enclosed by a given boundary curve, find the surface of minimum area.

Solution. Let Ω be a bounded region in $\mathbb{R}^2$ with smooth boundary $\partial\Omega$, $u \in C^2(\overline{\Omega})$ and S the embedded surface $S = \Phi(\Omega)$, where $\Phi(x, y) = (x, y, u(x, y))$. From Section 6.3 we know $T_x = (1, 0, u_x)$ and $T_y = (0, 1, u_y)$ and so $E = 1 + u_x^2$, $F = u_x u_y$ and $G = 1 + u_y^2$. Since $EG - F^2 = 1 + u_x^2 + u_y^2$ we see that the infinitesimal area of S is $dS = \sqrt{1 + u_x^2 + u_y^2}$. Hence the area of S is given by the functional

$$J(u) = A(S) = \int \int_{\Omega} \sqrt{1 + u_x^2 + u_y^2} \, dx \, dy.$$

The problem reduces to minimizing the functional J. Here $F(x, y, u, u_x, u_y) = \sqrt{1 + u_x^2 + u_y^2}$ and since $F_u = 0$, $F_{u_x} = \frac{u_x}{\sqrt{1+u_x^2+u_y^2}}$ and $F_{u_y} = \frac{u_y}{\sqrt{1+u_x^2+u_y^2}}$, the Euler-Lagrange equation (8.24) is $\frac{\partial}{\partial x}\left(\frac{u_x}{\sqrt{1+u_x^2+u_y^2}}\right) + \frac{\partial}{\partial y}\left(\frac{u_y}{\sqrt{1+u_x^2+u_y^2}}\right) = 0$. From which we obtain the PDE[19]

$$u_{xx}(1 + u_y^2) + u_{yy}(1 + u_x^2) - 2u_x u_y u_{xy} = 0. \tag{8.27}$$

Now the mean curvature[20] $H(x, y)$ at a point of such a surface is given by

$$H(x, y) = \frac{Eg - 2Ff + Ge}{2(EG - F^2)},$$

where $e = \frac{u_{xx}}{\sqrt{1+u_x^2+u_y^2}}$, $f = \frac{u_{xy}}{\sqrt{1+u_x^2+u_y^2}}$ and $g = \frac{u_{yy}}{\sqrt{1+u_x^2+u_y^2}}$. Hence

$$H(x, y) = \frac{u_{yy}(1 + u_x^2) - 2u_x u_y u_{xy} + u_{xx}(1 + u_y^2)}{\sqrt{1 + u_x^2 + u_y^2}}.$$

It follows that *minimal surfaces* (ones where the area is minimal) are exactly those with *mean curvature identically zero*.

Example 8.4.5. (*Vibrating string*). Derive and solve the partial differential equation describing the free vibrations of a string.

Solution. This problem consists of a string which is clamped at both ends and then plucked (such as with a violin string). Here we assume that the string is perfectly elastic (so there is no potential energy stored in the string other than due to its height, i.e. to gravity) and that it is rigidly clamped at each end (as a violin string is). Let τ denote the tension in the string and l be its fixed equilibrium length at that tension. When plucked the string vibrates in a vertical plane in such a way that each particle in the string moves in a straight vertical line. We

---

[19]This second order partial differential equation we just derived is an elliptic equation. As such it has a unique global solution, if Ω is simply connected. However, in general the solution cannot be written explicitly since it depends on Ω and the given curve lying above the boundary. Even when these are given it is not easy to write down the solution. Doing so is solving the so-called *Plateau problem*.

[20]See [33].

assume the amplitude of this vibration is so small that at every point and at all times the slope of the tangent line to it is much less than 1. We assume that the elongation is so slight as to have no effect on the tension τ. That is to say τ is constant. We also assume there are no losses due to friction with the air, or anything else. Finally, we suppose uniform linear density, δ of the string. (Otherwise, instead of equations with constant coefficients we would have to deal with the more difficult subject of equations with smooth, but variable coefficients.)

We are looking for the shape of this vibrating string which is a function of both position and time. Thus the two independent variables, x, t. Let the height above the x-axis be $u = u(x, t)$ where $0 \le x \le l$ and $0 \le t$. We know, since the string is clamped at both ends, that $u(0, t) = 0 = u(l, t)$ for all t. Let dx be a small portion of the interval $[0, l]$ on the x-axis. The length of the corresponding piece of the string is $ds = \sqrt{1 + u_x^2}\,dx$ and the corresponding potential energy due to stretching the string is τds. Thus the total potential energy of the configuration is the difference of all this with the initial or rest potential energy, τl. Hence the potential energy of the whole string is $U = \tau(\int_0^l \sqrt{1 + u_x^2}\,dx - l)$. Now our assumption on $|u_x|$ being small throughout $[0, l]$ allows us to use the Taylor series of $\sqrt{1 + u_x^2}$ to approximate it by the first two polynomials terms (in much the same way as we did when we studied the pendulum undergoing small vibrations in Chapter 7). Thus when we replace $\sqrt{1 + u_x^2}$ by $1 + \frac{1}{2}u_x^2$ we see that the total potential energy of the configuration takes the form

$$U = \frac{1}{2}\tau \int_0^l u_x^2\,dx$$

On the other hand, since δdx represents the mass of this same small piece of string, the kinetic energy of this piece is $\frac{1}{2}\delta dx u_t^2$ and so the total kinetic energy is $T = \int_0^l \frac{1}{2}\delta u_t^2\,dx$. Hence for a particular time interval $[t_1, t_2]$ the action is

$$\int_{t_1}^{t_2}(T - U)dt = \frac{1}{2}\int_{t_1}^{t_2}\int_0^l (\delta u_t^2 - \tau u_x^2)dxdt.$$

By the Principle of Least Action we must minimize this double integral over the product space $[0, l] \times [t_1, t_2]$, subject to the boundary conditions on the edges of this rectangle that $u(x,t) = 0$ for all $t \in [t_1, t_2]$ when $x \in [0, l]$ and in particular at t_1 and t_2 and that $u(0, t) = u(l, t) = 0$ and in particular at t_1 and t_2. Thus $u(x,t)$ is identically zero on the boundary. Accordingly, taking F in (8.23) to be

$$F(t, x, u_t, u_x) = \frac{1}{2}(\delta u_t^2 - \tau u_x^2),$$

we get for the Euler-Lagrange equation what is usually called the one (space) dimensional wave equation.

$$u_{xx}^2 = \frac{\delta}{\tau} u_{tt}^2.$$

Now since the interval $[t_1, t_2]$ is arbitrary this holds for all t and because the constant $\frac{\tau}{\delta} > 0$ one often calls this c^2. Then the wave equation takes the form $c^2 u_{xx}^2 = u_{tt}^2$.

To complete our analysis of the vibrating string we must solve the equation

$$c^2 u_{xx}^2 = u_{tt}^2,$$

subject to the boundary conditions $u(0, t) = 0 = u(l, t)$ and the initial condition $u(x, 0) = 0$ for all $x \in [0, l]$. This was done in Example 7.8.2, with $l = \pi$, by the method of separation of variables. In our case the solution is

$$u(x, t) = \sum_{n=1}^{\infty} B_n \sin\left(\frac{\sqrt{\lambda_n}}{c} x\right) \sin(\sqrt{\lambda_n} t),$$

where each $\lambda_n = n^2 c^2 \frac{\pi^2}{l^2}$ and each B_n is a constant.

8.4.2   The case of n variables

Let $u = u(x_1, ..., x_n)$ be a C^2 function[21] in n independent variables $x = (x_1, ..., x_n)$, defined on a bounded domain $\Omega \subset \mathbb{R}^n$ and its boundary. We

---

[21] The function space is $C^1(\overline{\Omega})$ with norm

$$\|u\| = \max_{x \in \overline{\Omega}} |u(x)| + \max_{x \in \overline{\Omega}} |\nabla u(x)|.$$

Recall from Section 3.5 that $C^2(\overline{\Omega}) \subseteq C^1(\overline{\Omega})$. However, if we were also concerned with the question of existence of the extremum, a better choice would be the larger

assume that the domain has a smooth (or perhaps a piecewise smooth) boundary $\partial\Omega$. (Whatever it takes to make the integral converge and Stokes' theorem in $\mathbb{R}^n$ to be valid). Let $F : \overline{\Omega} \times \mathbb{R} \times \mathbb{R}^n \to \mathbb{R}$ be a C^2 function of $2n + 1$ variables

$$F = F(x_1, \ldots, x_n, u, u_{x_1}, \ldots, u_{x_n}).$$

Consider the functional

$$J(u) = \int \ldots \int_\Omega F(x, u, \nabla u) dx_1 \ldots dx_n, \qquad (8.28)$$

where $u = u(x) = u(x_1, \ldots, x_n)$ and $\nabla u = \nabla u(x) = (u_{x_i}(x), \ldots, u_{x_n}(x))$ and partial derivatives are denoted by $u_{x_i} = \frac{\partial u}{\partial x_i}$.

We want to extremize the functional J subject to the condition that $u \equiv f$ on $\partial\Omega$ with f a given C^1 function. In order to do this we proceed just as in previous case for $n = 2$. In fact we have

Theorem 8.4.6. *Let $n \geq 3$. For F and u as above, if the functional*

$$J(u) = \int \ldots \int_\Omega F(x, u, \nabla u) dx_1 \ldots dx_n,$$

has an extremum at $u = \varphi(x)$, then u satisfies the Euler-Lagrange equation

$$F_u - \sum_{i=1}^n \frac{\partial}{\partial x_i} F_{u_{x_i}} = 0 \qquad (8.29)$$

Proof. As before we consider a function $\eta = \eta(x_1, \ldots, x_n)$ which is defined on the domain and its boundary, it is smooth on the domain and $\eta(x) \equiv 0$ for $x \in \partial\Omega$. The varied function is $\overline{u}(x) = u(x) + s\eta(x)$, and so

$$J(\overline{u}) = J(u + s\eta) = g(s).$$

---

class of (generalized) functions

$$W^{1,2}(\Omega) = \left\{ u : \ \|u\|_{1,2} = \left[\int_\Omega (|u(x)|^2 + |\nabla u(x)|^2) dx \right]^{\frac{1}{2}} < \infty \right\}.$$

This class includes, for example $C^1(\overline{\Omega})$, and is called a *Sobolev space*.

Now the necessary condition $g'(0) = 0$ takes the form

$$\int \cdots \int_\Omega \left(F_u \eta - \sum_{i=1}^n F_{u_{x_i}} \eta_{x_i} \right) dx_1 \ldots dx_n = 0. \qquad (8.30)$$

To eliminate η_{x_i} from the integral, for $n = 3$, we work as in Theorem 8.4.2 and we use the Divergence theorem (Theorem 6.7.1). In the general case of n, we work with differential forms and use Stoke's theorem in $\mathbb{R}^n$. This is done as follows. Consider the $(n-1)$-form

$$\omega = \sum_{i=1}^n (-1)^{i+1} (F_{u_{x_i}} \eta) dx_1 \cdot \hat{\cdots} \cdot dx_n,$$

where $dx_1 \cdot \hat{\cdots} \cdot dx_n$ denotes the basic $(n-1)$ forms (leaving out the term below the ˆ). Then

$$d\omega = \sum_{i=1}^n (-1)^{i+1} d(F_{u_{x_i}} \eta) dx_1 \ldots dx_n$$

$$= \left[\sum_{i=1}^n F_{u_{x_i}} \eta_{x_i} + \frac{\partial}{\partial x_i} \left(F_{u_{x_i}} \right) \eta \right] dx_1 \ldots dx_n.$$

Therefore

$$\sum_{i=1}^n F_{u_{x_i}} \eta_{x_i} dx_1 \ldots dx_n = d\omega - \left(\sum_{i=1}^n \frac{\partial}{\partial x_i} \left(F_{u_{x_i}} \right) \eta \right) dx_1 \ldots dx_n.$$

Hence the integral in (8.30) becomes

$$\int \cdots \int_\Omega \left(F_u - \sum_{i=1}^n \frac{\partial}{\partial x_i} \left(F_{u_{x_i}} \right) \right) \eta dx_1 \ldots dx_n + \int \cdots \int_\Omega d\omega.$$

By Stokes' theorem (Theorem 6.8.13) $\int_\Omega d\omega = \int_{\partial\Omega} \omega = 0$ because $\omega = 0$ on $\partial\Omega$ (since $\eta \equiv 0$ on $\partial\Omega$). Thus,

$$\int \cdots \int_\Omega \left(F_u - \sum_{i=1}^n \frac{\partial}{\partial x_i} \left(F_{u_{x_i}} \right) \right) \eta dx_1 \ldots dx_n = 0,$$

and the fundamental lemma completes the proof.   □

The next result shows the close relationship and interplay between partial differential equations and the calculus of variations. In case $\rho = 0$ this is the so-called *Dirichlet principle*[22].

Theorem 8.4.7. *Let* $\Omega = \{(x, y, z) : x^2 + y^2 + z^2 < \alpha\}$ *be the open ball of radius* $\alpha > 0$ *in* $\mathbb{R}^3$. *Let* $u = u(x, y, z)$ *and* $\rho = \rho(x, y, z)$ *where* $u \in C^2(\overline{\Omega})$ *and* $\rho \in C(\Omega)$. *The variational problem of minimizing the functional*

$$J(u) = \int \int \int_{\Omega} \left(\frac{1}{2}||\nabla u||^2 - \rho u \right) dx dy dz, \qquad (8.31)$$

subject to $u \equiv f$ *on* $\partial\Omega$ *where* $f \in C(\partial\Omega)$ *is a given function, is equivalent to the boundary value problem for the Poisson equation*

$$\Delta u = -\rho \ in \ \Omega \qquad (8.32)$$

$$u \equiv f \ on \ \partial\Omega.$$

Proof. Here $F = F(x, y, z, u, u_x, u_y, u_z) = \frac{1}{2}(u_x^2 + u_y^2 + u_z^2) - \rho u$. Hence $F_u = -\rho$, $F_{u_x} = u_x$, $F_{u_y} = u_y$, $F_{u_z} = u_z$ and the Euler-Lagrange equation here is $\Delta u(x, y, z) = -\rho(x, y, z)$. However, since we did not work the case $n = 3$ in Theorem 8.4.6, we give a detailed proof here. For ϕ of class C^1 and ψ of class C^2 recall the identity

$$div(\psi \nabla \phi) = \nabla \psi \cdot \nabla \phi + \psi \Delta \phi. \qquad (8.33)$$

Let $\eta = \eta(x, y, z)$ be a smooth function on $\overline{\Omega}$ and $\eta \equiv 0$ on $\partial\Omega$. Consider the function $\overline{u} = u + s\eta$ with s a small real parameter. As before $J(\overline{u}) = g(s)$ and the necessary condition $g'(0) = 0$ gives

$$\int \int \int_{\Omega} \left(F_u \eta + F_{u_x} \eta_x + F_{u_y} \eta_y + F_{u_z} \eta_z \right) dx dy dz = 0.$$

That is,

$$\int \int \int_{\Omega} (-\rho \eta + \nabla \eta \cdot \nabla u) \, dV = 0.$$

---

[22]Dirichlet's principle asserts that *the solution of the Dirichlet problem is realized by obtaining from all functions satisfying given boundary conditions the one that minimizes the Dirichlet functional.* Bernhard Riemann attributed this "principle" to his teacher, Dirichlet.

Using (8.33) and the Divergence theorem we have

$$\int\int\int_\Omega (\nabla\eta \cdot \nabla u)\, dV = \int\int\int_\Omega div(\eta\nabla u)dV - \int\int\int_\Omega (\eta\Delta u)dV$$

$$= \int\int_{\partial\Omega} \langle\eta\nabla u, \mathbf{n}\rangle dS - \int\int\int_\Omega (\eta\Delta u)dV = -\int\int\int_\Omega (\eta\Delta u)dV,$$

since $\eta = 0$ on $\partial\Omega$. Thus,

$$\int\int\int_\Omega (\rho + \Delta u)\,\eta dV = 0,$$

and the fundamental lemma gives $\Delta u + \rho = 0$.

Conversely, suppose $u \in C^2(\overline{\Omega})$ is a solution of $\Delta u = -\rho$ with $u = f$ on $\partial\Omega$. Take $\eta = 0$ on $\partial\Omega$ and let $\overline{u} = u + s\eta$, so that $\nabla\overline{u} = \nabla u + s\nabla\eta$. Now we have

$$J(\overline{u}) = \int\int\int_\Omega \left(\frac{1}{2}||\nabla\overline{u}||^2 - \rho\overline{u}\right) dV$$

$$= \int\int\int_\Omega \left[\frac{1}{2}\langle\nabla u + s\nabla\eta, \nabla u + s\nabla\eta\rangle\right] dV - \int\int\int_\Omega \rho(u + s\eta)dV$$

$$= \int\int\int_\Omega \frac{1}{2}\left[||\nabla u||^2 + 2s\langle\nabla u, \nabla\eta\rangle + s^2||\nabla\eta||^2\right] dV$$

$$- \int\int\int_\Omega (\rho u + s\rho\eta)dV$$

$$= s^2 \int\int\int_\Omega \frac{1}{2}||\nabla\eta||^2 dV + s \int\int\int_\Omega (\langle\nabla u, \nabla\eta\rangle - \rho\eta)dV + J(u)$$

Since $\Delta u = -\rho$, this is equal to

$$= s^2 \int\int\int_\Omega \frac{1}{2}||\nabla\eta||^2 dV + s \int\int\int_\Omega (\langle\nabla u, \nabla\eta\rangle - \eta\Delta u)dV + J(u).$$

By the identity (8.33) and the Divergence theorem the second integral yields

$$\int\int\int_\Omega div(\eta\nabla u)dV = \int\int_{\partial\Omega} \langle\eta\nabla u, \mathbf{n}\rangle dS = 0.$$

Hence

$$J(\overline{u}) = s^2 \int \int \int_\Omega \frac{1}{2}||\nabla \eta||^2 dV + J(u) \geq J(u).$$

$\square$

Remark 8.4.8. It is true quite generally that a smooth minimum of a functional $J(u)$ is indeed a solution of the corresponding Euler-Lagrange equation. However, if we try to replicate Theorem 8.4.7 for other variational problems we encounter several difficulties. It may be difficult to show that a solution of a Euler-Lagrange equation is necessarily a minimum of the associated functional $J(u)$. For example, the functional $J(u) = \int \int_\Omega (c^2 u_x^2 - u_t^2) dx dt$ has for the Euler-Lagrange equation the *wave equation* $c^2 u_{xx} = u_{tt}$. A minimum (or maximum) of $J(u)$ is necessarily a solution of the wave equation, but the converse is not true (see, Exercise 8.5.25).

We close this section by looking at the connection between *eigenvalue problems* and the *calculus of variations*. In our discussion of ODE in Example 7.4.14 we solved the eigenvalue problem

$$-u'' = \lambda u, \ 0 < x < \pi, \tag{8.34}$$

$$u(0) = u(\pi) = 0,$$

for which the eigenvalues were $\lambda_n = n^2$ and the corresponding eigenfunctions were $u_n(x) = \sin(nx), \ n = 1, 2, 3, ...$, where we take the constant $B = 1$.

Since $\frac{d}{dx}[u(x)u'(x)] = u(x)u''(x) + u'^2(x)$ we see that

$$\int_0^\pi u(x)u''(x)dx + \int_0^\pi u'^2(x)dx = u(x)u(x)'|_0^\pi.$$

Now by the boundary conditions u is zero at both 0 and π hence this last term is zero. On the other hand, for a solution of (8.34) the first term is $-\lambda \int_0^\pi u(x)^2 dx$ so we see

$$\lambda = \frac{\int_0^\pi u'^2(x)dx}{\int_0^\pi u^2(x)dx} > 0.$$

This quotient is called *Rayleigh's quotient*[23]. In particular for the eigenfunctions $u_n(x) = \sin(nx)$ we have

$$\lambda_n = \frac{\int_0^\pi u_n'^2(x)dx}{\int_0^\pi u_n^2(x)dx} = \frac{n^2 \int_0^\pi \cos^2(nx)dx}{\int_0^\pi \sin^2(nx)dx} = n^2.$$

Moreover, (from Calculus)

$$\int_0^\pi u_n(x)u_m(x)dx = \begin{cases} \frac{\pi}{2} \text{ when } m = n, \\ 0 \text{ when } m \neq n. \end{cases}$$

That is, the eigenfunctions form an *orthogonal set*[24] of functions.

As we shall see in the next example, the variational characterization of the solution for this eigenvalue problem is given by minimizing the functional

$$J(u) = \frac{\int_0^\pi u'^2(x)dx}{\int_0^\pi u^2(x)dx}, \quad u(0) = u(\pi) = 0.$$

Example 8.4.9. We consider the problem of determining the *vibrating frequencies* of a 3-dimensional region Ω. These are the numbers λ_n that are obtained as the *eigenvalues* of the eigenvalue problem

$$-\Delta u = \lambda u \ in \ \Omega \qquad (8.35)$$

$$u \equiv 0 \ on \ \partial\Omega.$$

Consider the functional

$$J(u) = \frac{\int \int \int_\Omega ||\nabla u||^2 dV}{\int \int \int_\Omega |u|^2 dV}, \qquad (8.36)$$

where $u \in C^2(\overline{\Omega})$ with $u \equiv 0$ on $\partial\Omega$. If $u = \varphi$ minimizes J, then $\Delta u = \lambda u$, where

$$\lambda = \min_u J(u) = J(\varphi).$$

---

[23]L. Rayleigh worked in diverse areas of classical spectral analysis of light, sound, color and electromagmetism that led to his book *"Theory of sound"* in 1877.

[24]In fact, the set of the eigenfunctions of the problem form an *orthonormal basis* for the underlying Hilbert space $L^2(0,\pi)$. (A *Hilbert space* is a complete inner-product space). Such concerns are again beyond the scope of this book.

Solution. Let $\overline{u} = u + s\eta$, with $\eta = 0$ on $\partial\Omega$, be a neighboring function. Then $\lambda = J(u) \leq J(\overline{u})$ implies

$$\int\int\int_\Omega ||\nabla u + s\nabla\eta||^2 dV \geq \lambda \int\int\int_\Omega |u + s\eta|^2 dV.$$

Expanding as before and simplifying we have,

$$s^2 \int\int\int_\Omega (||\nabla\eta||^2 - \lambda\eta^2)dV + 2s \int\int\int_\Omega (\langle\nabla u, \nabla\eta\rangle - \lambda u\eta)dV \geq 0.$$

This quadratic polynomial in s has minimum at $s = 0$. Therefore

$$\int\int\int_\Omega (\langle\nabla u, \nabla\eta\rangle - \lambda u\eta)dV = 0.$$

As in the proof of Theorem 8.4.7, the Divergence theorem and the boundary condition for η give $\int\int\int_\Omega \langle\nabla u, \nabla\eta\rangle dV = -\int\int\int_\Omega \eta\Delta u dV$. Hence

$$\int\int\int_\Omega (\Delta u - \lambda u)\eta dV,$$

and the fundamental lemma implies

$$\Delta u - \lambda u = 0.$$

The minimum just obtained is the *smallest frequency* of the 3-dimensional region Ω (in other words, the *smallest eigenvalue*[25] of the Laplace operator Δ on Ω with the Dirichlet boundary conditions). Indeed, if μ is another eigenvalue corresponding to the solution $w = \psi$ of $\Delta w = \mu w$, then by the Divergence theorem and the boundary condition we must have

$$\mu = \frac{\int\int\int_\Omega ||\nabla\psi||^2 dV}{\int\int\int_\Omega |\psi|^2 dV}.$$

However this quotient is greater or equal to the global minimum λ found above, viz, $\mu \geq \lambda$.

---

[25]The next eigenvalue is obtained as the minimum of the Rayleigh quotient but now the minimum is taken over the functions w which are "orthogonal" to the eigenfunction φ. For the very next eigenvalue, similarly, we minimize the Rayleigh quotient but this time over the functions which are orthogonal to both eigenfunctions found at the previous steps, etc.

EXERCISES

1. Let Ω be a bounded domain in $\mathbb{R}^2$ whose boundary is a closed smooth curve and $u = u(x, y)$, with $u \in C^2(\overline{\Omega})$. Consider the functional

$$J(u) = \int\int_\Omega \left(\frac{1}{2}(u_x^2 + u_y^2) + auu_x + buu_y + \frac{1}{2}cu^2 \right) dxdy,$$

where $a = a(x, y)$, $b = b(x, y)$, $c = c(x, y)$ are C^1 functions in Ω. Find the Euler-Lagrange equation.
(Ans. $\Delta u + (a_x + b_y - c)u = 0$).

2. Let Ω be an open disk in $\mathbb{R}^2$, and $u = u(x, y)$ and $\rho = \rho(x, y)$, with $u \in C^2(\overline{\Omega})$ and $\rho \in C(\Omega)$. Consider the functional

$$J(u) = \int\int_\Omega \left(\frac{1}{2}(u_{xx} + u_{yy})^2 + \rho(x, y)u \right) dxdy.$$

Suppose that $u = \varphi(x, y)$ minimizes the functional J among all smooth functions u satisfying $u = 0$ and $\frac{\partial u}{\partial \mathbf{n}} = 0$ on $\partial\Omega$, where $\frac{\partial u}{\partial \mathbf{n}} = \langle \nabla u, \mathbf{n} \rangle$ is the outward normal derivative of u. Prove that φ satisfies, in Ω, the equation[26]

$$\Delta\Delta u = u_{xxxx} + 2u_{xxyy} + u_{yyyy} = -\rho.$$

3. Let Ω be an open disk in $\mathbb{R}^2$, and $u = u(x, y)$ and $\rho = \rho(x, y)$, with $u \in C^2(\overline{\Omega})$ and $\rho \in C(\Omega)$. Consider the functional

$$J(u) = \int\int_\Omega \left(\frac{1}{2}||\nabla u(x, y)||^2 - \rho(x, y)u \right) dxdy.$$

Suppose that $u = \varphi(x, y)$ minimizes the functional J among all smooth functions u (assuming no boundary conditions).

(a) Prove that φ satisfies the equation $\Delta u = -\rho$ in Ω together with the boundary condition $\frac{\partial u}{\partial \mathbf{n}} = 0$ on $\partial\Omega$.

(b) Conclude that $\int\int_\Omega \rho(x, y)dxdy = 0$.

---

[26]When $\rho = 0$, the equation $\Delta\Delta u = 0$ is the so-called *biharmonic equation*.

8.5   Solved problems for Chapter 8

Exercise 8.5.1. Show that if $F_{y'y'} \equiv 0$, then the Euler-Lagrange equation $F_{y'y'}y'' + F_{y'y}y' + F_{ty} - F_y = 0$ degenerates into an identity.

Solution. From $F_{y'y'} \equiv 0$ it follows that $F(t, y, y') = M(t, y) + N(t, y)y'$. Substituting into the Euler-Lagrange equation $F_{y'y}y' + F_{ty} - F_y = 0$, we get $\frac{\partial M}{\partial y} = \frac{\partial N}{\partial t}$. This is only a functional identity, not a differential equation. In fact, by Chapter 7 (exact equations) this is the exactness condition, viz, the equation $M(t, y)dt + N(t, y)dy = 0$ is exact. Therefore there exist a function $f(t, y)$ such that $\frac{\partial f}{\partial t} = M$ and $\frac{\partial f}{\partial y} = N$. Hence, $F(t, y, y') = \frac{\partial f}{\partial t} + \frac{\partial f}{\partial y}y' = \frac{d}{dt}f(t, y)$. Thus

$$J(y) = \int_a^b F(t, y, y')dt = \int_a^b \left(\frac{d}{dt}f(t, y) \right) dt = f(b, y(b)) - f(a, y(a)),$$

so that the functional has the same constant value along every admissible curve. In this case, there is no variational problem.

Exercise 8.5.2. For $n \geq 2$, consider the functional

$$J(u) = \int_a^b F(t, y, y', y'', \ldots, y^{(n)})dt,$$

where the admissible functions $y = y(t)$ have continuous derivatives up to the $2n$-th order and for which the derivatives up to order $(n - 1)$ have prescribed values at the end points a and b. Show that the corresponding Euler-Lagrange equation is the $2n$-th order differential equation

$$F_y - \frac{d}{dt}(F_{y'}) + (-1)^2 \frac{d^2}{dt^2}(F_{y''}) + \ldots + (-1)^n \frac{d^n}{dt^n}(F_{y^{(n)}}) = 0.$$

Solution. As in Theorem 8.1.2, consider the function $\overline{y}(t) = y(t) + s\eta(t)$ on $[a, b]$, where $|s| < \epsilon$ and η is an arbitrary function having continuous derivatives up to the $2n$-th order satisfying $\eta(a) = \eta'(a) = \ldots = \eta^{(n-1)}(a) = 0$ and $\eta(b) = \eta'(b) = \ldots = \eta^{(n-1)}(b) = 0$. The integral then takes the form $g(s) = J(u + s\eta)$, and the necessary condition $g'(0) = 0$ must be satisfied for all functions $\eta(t)$. Proceeding in a similar way as in Theorem 8.1.2, we differentiate under the integral sign and we obtain the above condition in the form

$$\int_a^b \left(F_y\eta + F_{y'}\eta' + F_{y''}\eta'' + \ldots + F_{y^{(n)}}\eta^{(n)} \right) dt = 0.$$

By repeated integration by parts we can eliminate all the derivatives of the function η from the integral, and thus obtain

$$\int_a^b \left(F_y - \frac{d}{dt} + F_{y'} + \frac{d^2}{dt^2} F_{y''} + \ldots + (-1)^n \frac{d^n}{dt^n} F_{y^{(n)}} \right) \eta \, dt = 0.$$

The fundamental lemma (Lemma 8.1.1) completes the proof.

Exercise 8.5.3. Find the extremals of the functional

$$J(y) = \int_0^{\frac{\pi}{2}} \left(t^2 - y^2 + 3y' + y''^{\,2} \right) dt,$$

where the curve $y = y(t)$ satisfies $y(0) = 1$, $y'(0) = 0$, $y(\frac{\pi}{2}) = 0$, $y'(\frac{\pi}{2}) = -1$.

Solution. Here $F(t, y, y', y'') = t^2 - y^2 + 3y' + y''^{\,2}$ and the Euler-Lagrange equation $F_y - \frac{d}{dt}(F_{y'}) + \frac{d^2}{dt^2}(F_{y''}) = 0$ becomes $y^{(4)} - y = 0$. By Example 7.5.2, the general solution is $y = c_1 e^t + c_2 e^{-t} + c_3 \sin t + c_4 \cos t$. Using the boundary conditions we find the extremum $y = \cos t$.

Exercise 8.5.4. Consider the problem of extremizing $J(y) = \int_a^b (y')^2 dx$, where $y(a) = \alpha$ and $y(b) = \beta$ with α, β given fixed values subject to the constraint $K(y) = \int_a^b y^2 dx = c$. Show that such a function $y = y(x)$ must be a solution to $y'' - \lambda y = 0$. Thus here the Lagrange multiplier is an eigenvalue of the operator $\frac{d^2}{dx^2}$ with the given boundary conditions.

Solution. Here $L(x, y, y') = F(x, y, y') - \lambda G(x, y, y') = y'^2 - \lambda y^2$, where λ is the Lagrange multiplier. The Euler-Lagrange equation becomes $L_{y'y'}y'' - L_y = 0$, that is, $y'' - \lambda y = 0$. In other words, the extremals here are solutions of the eigenvalue problem $y'' - \lambda y = 0$, $a < x < b$, and $y(a) = \alpha$, $y(b) = \beta$.

Exercise 8.5.5. Here we outline an approach of dealing with a constraint on the manifold rather than on the functional. We want to solve variational problems where we restrict the locus to a submanifold of $\mathbb{R}^n$ which is the zero set of a smooth function G defined on $\mathbb{R}^n$ or on an open subset of $\mathbb{R}^n$. Let $x = (x_1, \ldots, x_n) \in \mathbb{R}^n$ and consider the submanifold $S = \{x \in \mathbb{R}^n : G(x) = 0\}$. For example, the sphere of radius $r > 0$ is the zero set of $G(x) = x_1^2 + \ldots + x_n^2 - r^2$. In other words, we want to solve variational problems with a finite side condition. Here we have the following theorem which we state without proof.

Theorem 8.5.6. . *Let $Y(t) = (y_1(t), \ldots, y_n(t))$ be a curve in $\mathbb{R}^n$ which makes the functional $J(Y) = \int_a^b F(t, y_1, \ldots, y_n, y_1', \ldots, y_n') dt$ have an extremum subject* to the constraint $G(y_1, ..., y_n) = 0$. *Then there is a smooth function $\lambda(t)$ defined on $[a, b]$ such that Y satisfies the Euler-Lagrange equation for*

$$L(t, Y, Y') = F(t, Y, Y') - \lambda(t)G(Y).$$

Our purpose is to indicate how this result gives another approach to the problem of geodesics on a 2-surface S in $\mathbb{R}^3$. Let $L = F - \lambda(t)G$, where $\lambda(t)$ is an unknown smooth function and $F = F(x'(t), y'(t), z'(t))$. Since we are in $\mathbb{R}^3$, the Euler-Lagrange equation is the system of 3 equations

$$\lambda(t)G_x + \frac{d}{dt}F_{x'} = 0, \quad \lambda(t)G_y + \frac{d}{dt}F_{y'} = 0, \quad \lambda(t)G_z + \frac{d}{dt}F_{z'} = 0.$$

Eliminating $\lambda(t)$ we have

$$\frac{\frac{d}{dt}F_{x'}}{G_x} = \frac{\frac{d}{dt}F_{y'}}{G_y} = \frac{\frac{d}{dt}F_{z'}}{G_z}. \tag{8.37}$$

When we specialize this result to the problem of finding *geodesics* on S, we have

$$F = \sqrt{x'^2 + y'^2 + z'^2}$$

and equations (8.37) become

$$\frac{\frac{d}{dt}\left(\frac{x'}{F}\right)}{G_x} = \frac{\frac{d}{dt}\left(\frac{y'}{F}\right)}{G_y} = \frac{\frac{d}{dt}\left(\frac{z'}{F}\right)}{G_z} \tag{8.38}$$

and the problem reduces to extract information from this system. Note that if we use arc length s instead of t to parametrize the curve, then $F = 1$ and equations (8.38) become

$$\frac{x''(s)}{G_x} = \frac{y''(s)}{G_y} = \frac{z''(s)}{G_z}. \tag{8.39}$$

Example. If S is the *sphere*, then $G(x, y, z) = x^2 + y^2 + z^2 - r^2$ and equations (8.38) become

$$\frac{Fx'' - x'F'}{2xF^2} = \frac{Fy'' - y'F'}{2yF^2} = \frac{Fz'' - z'F'}{2zF^2}.$$

Put another way

$$\frac{yx'' - xy''}{yx' - xy'} = \frac{F'}{F} = \frac{zy'' - yz''}{zy' - yz'}$$

Equating the first and last of these tells us

$$\frac{\frac{d}{dt}(yx' - xy')}{yx' - xy'} = \frac{\frac{d}{dt}(zy' - yz')}{zy' - yz'}.$$

This integrates to give

$$\log(yx' - xy') = \log(zy' - yz') + \log c.$$

Therefore, $yx' - xy' = c(zy' - yz')$. But then, dividing by y and solving for $\frac{y'}{y}$ yields $\frac{x' + cz'}{x + cz} = \frac{y'}{y}$. Integrating again gives $\log(x + cz) = \log(y) + \log c_1$ and so

$$x + cz - c_1 y = 0.$$

This is a plane through the origin, i.e. through the center of the sphere and therefore intersects the sphere in a great circle. Which planes, i.e. which great circles are unaccounted for? The only ones unaccounted for are planes of the form $ay + bz = 0$, the ones passing through the x-axis. Hence we don't get the geodesics joining $(r, 0, 0)$ with $(-r, 0, 0)$. But since, as discussed above, the rotation group carries such a pair of points to different pair of antipodal points and preserves spherical distance and geodesics we see that the geodesics on a sphere are precisely the great circles.

We remark that this method's success depends on a certain symmetry between F and G. For example, the reader should verify that there is trouble if we simply replace the sphere by an ellipsoid. In general is not easy to solve equations (8.38). The main significance of these equations lies in their connection with the following important result in mathematical physics.

Exercise 8.5.7. If a particle of mass m is constrained to move on a given surface $S = \{(x, y, x) \in \mathbb{R}^3 : G(x, y, z) = 0\}$, free from the action of any external force, then its path is a geodesic on S.

Solution. Let $\gamma(t) = (x(t), y(t), z(t))$ be the path describing the trajectory of the particle on S. Form Section 8.2.2, we know that the motion of the particle is described by the Lagrangian $L = T - U$, where T and U are the kinetic and potential energy respectively. Since no force is acting on the particle its potential energy is zero, viz, $U = 0$, and so the Lagrangian $L = T$. The kinetic energy is $T = \frac{1}{2}mv^2$, where $v = \sqrt{x'(t)^2 + y'(t)^2 + x'(t)^2}$ is the speed of the particle. According to Hamilton's principle, the actual path of motion extremizes the action integral

$$\int_{t_0}^{t_1} L\, dt = \int_{t_0}^{t_1} T\, dt,$$

subject to the constraint $G(x, y, z) = 0$. From Theorem 8.3.1 we must extremize the functional

$$J(\gamma) = \int_{t_0}^{t_1} [T - \lambda(t)G]\, dt$$

free of the side condition, where $\lambda(t)$ is a smooth function of t. The Euler-Lagrange equations for this functional are

$$\lambda(t)G_x + mx''(t) = 0, \quad \lambda(t)G_y + my'(t)' = 0, \quad \lambda(t)G_z + mz''(t) = 0.$$

Eliminating m and $\lambda(t)$, we have

$$\frac{x''(t)}{G_x} = \frac{y''(t)}{G_y} = \frac{z''(t)}{G_z}. \tag{8.40}$$

By the principle of conservation of energy (Theorem 8.2.12), the total energy $T + U = T$ of the particle is constant, and therefore its speed is also constant. That is, $v = c$ for some constant c. Hence the arc length $s = ct$, when s is measured from the point $\gamma(t_0)$. Thus we can write equations (8.40) in the form

$$\frac{x''(s)}{G_x} = \frac{y''(s)}{G_y} = \frac{z''(s)}{G_z},$$

which are the equations (8.39) of a geodesic on the surface S.

Exercise 8.5.8. Let M be a Riemannian surface with metric $ds^2 = E\,du^2 + 2F\,du\,dv + G\,dv^2$. Thus if $\gamma(t) = \Phi(u(t), v(t))$, where $a \leq t \leq b$ is a path in M, its length $s(\gamma)$ is given by $s(\gamma) = \int_a^b \sqrt{Eu'^2 + 2Fu'v' + Gv'^2}\,dt$. We also define its energy by $E(\gamma) = \frac{1}{2}\int_a^b (Eu'^2 + 2Fu'v' + Gv'^2)\,dt$. Prove that

$$s(\gamma) \leq 2(b - a)E(\gamma),$$

with equality if and only if the path is parametrized by arc length. Thus geodesics are exactly the energy minimizers.

Solution. Let $\varphi(t) = \|\gamma'(t)\| = \sqrt{Eu'^2 + 2Fu'v' + Gv'^2}$. Then $s(\gamma) = \int_a^b \varphi(t)\,dt$ and $E(\gamma) = \frac{1}{2}\int_a^b \varphi(t)^2\,dt$. Now the Cauchy-Schwartz inequality gives

$$s(\gamma) = \int_a^b \varphi(t)\,dt \leq \left(\int_a^b \varphi(t)^2\,dt\right)\left(\int_a^b 1\,dt\right) = 2(b - a)E(\gamma).$$

The path γ is parametrized by arc length if and only if $\|\gamma'(t)\| = 1$. That is, if and only if $ds = dt$ or equivalently $s(\gamma) = \int_a^b dt = b - a = 2(b - a)E(\gamma)$.

Miscellaneous Exercises

Exercise 8.5.9. Consider the variational problem of Geometric Optics $J(y) = \int \frac{\sqrt{1+y'^2}}{v(x,y)} dx$, where $v(x,y)$ is the velocity of light at the point (x,y) of the medium. Let $p(x,y) = \frac{1}{v(x,y)}$. Show the Euler-Lagrange equation for $J(y) = \int p(x,y)\sqrt{1+y'^2}dx$ is a first order equation with variable coefficients.

Exercise 8.5.10. Let $J(y) = \int_0^a \frac{1}{y'^2} dt$, where $y \in C^1([0,a])$ with $y(0) = 0$, $y(a) = b$, $a > 0$ and $b > 0$. Show that the curve $y = y(t) = \frac{bt}{a}$ minimizes the functional J.

Exercise 8.5.11. Let $J(y) = \int_1^2 \frac{t^3}{y'^2} dt$, where $y \in C^1([1,2])$ with $y(1) = 1$, $y(2) = 4$. Show that the curve $y = y(t) = t^2$ minimizes J.

Exercise 8.5.12. Let $J(y) = \int_0^1 ty'^2 dt$, where $y \in C^1([0,1])$ with $y(0) = 1$, $y(1) = 0$. Investigate the functional for extrema.

Exercise 8.5.13. Let $J(y) = \int_a^b (y^2 + 2y'^2 + y''^2)dt$. Show that the extremals of J are the curves $y = (c_1 t + c_2)e^t + (c_3 t + c_4)e^{-t}$, where c_1, c_2, c_3, c_4 are arbitrary constants.

Exercise 8.5.14. Determine the extremals of the functional

$$J(x(t), y(t)) = \int_a^b (x' + y)(x + y')dt.$$

Exercise 8.5.15. Find the geodesics of a paraboloid of revolution $z = x^2 + y^2$. Show that each geodesic on the paraboloid which is not a plane section through the axis of rotation intersects itself infinitely often.

Exercise 8.5.16. Let the surface of revolution S be the right circular cone $z^2 = (x^2 + y^2)$ with $z \geq 0$, show that any geodesic has the following property: if the cone is cut along a generator[27] and flattened into a plane, then the geodesic becomes a straight line.
Hint. Parametrize the cone by $x = \frac{r\cos(\theta\sqrt{2})}{\sqrt{2}}$, $y = \frac{r\sin(\theta\sqrt{2})}{\sqrt{2}}$, $z = \frac{r}{\sqrt{2}}$. Show that the parameters r and θ represent ordinary polar coordinates on the flattened cone; and show that a geodesic $r = r(\theta)$ is a straight line in these polar coordinates.

Exercise 8.5.17. Let u be a C^2 function on $[a,b]$ with $u(a) = 0$. Show

$$\int_a^b u(x)^2 dx \leq \frac{(b-a)^2}{2} \int_a^b u'(x)^2 dx.$$

Hint. Since $u(a) = 0$ by the fundamental theorem of calculus for all $x \in [a,b]$, $u(x) = \int_a^x u'(t)dt$. Squaring, integrating and applying the Schwarz inequality yields $\int_a^b u(x)^2 dx \leq \int_a^b u'(t)^2 dt \int_a^b (x-a)dx$. Then note that $\frac{d}{dx}\frac{(x-a)^2}{2} = x - a$.

---

[27] A generator is a line formed by the intersection of the surface and a plane passing through the axis of revolution.

Exercise 8.5.18. (*Wirtinger's inequality*)[28]

(a) Let u be a C^2 function on $[a, b]$ with $u(a) = u(b)$ and $\int_a^b u(x)dt = 0$. Show

$$\int_a^b u'(t)^2 dt \geq \left(\frac{2\pi}{b-a}\right)^2 \int_a^b u(t)^2 dt.$$

Equality holds if and only if $u(t) = A \sin \pi t + B \cos \pi t$.

(b) Let $u, v \in C^2([a, b])$ with $u(a) = u(b)$, $v(a) = v(b)$, and $\int_a^b u(t)dx = 0$. Then

$$\int_a^b (u'(t)^2 + v'(t)^2)dt \geq \left(\frac{2\pi}{b-a}\right)^2 \int_a^b u(t)v'(t)dt.$$

Hint. Show that the functional $J(u) = \int_{-1}^1 (u'^2 - \pi^2 u^2)dt \geq 0$.

Exercise 8.5.19. (*The isoperimetric inequality*). Let C be a smooth closed curve enclosing a bounded open set Ω in $\mathbb{R}^2$. Let $L = \int_a^b \sqrt{u'(t) + v'(t)}dt$ be the length of C, and $A = \frac{1}{2}\int_a^b (u(t)v'(t) - v(t)u'(t))dt = \int_a^b u(t)v'(t)dt$ be the area of Ω, where C is parametrized by $\gamma(t) = (u(t), v(t))$ with u, v as in Exercise 8.5.8. Use Exercise 8.5.8 to show that

$$A \leq \frac{L^2}{4\pi}.$$

Exercise 8.5.20. Consider two functionals $J(y) = \int_a^b F(x, y, z)dx$ and $K(y) = \int_a^b G(x, y, z)dx$. Show the y which extremize J subject to the constraint $K = c$ also extremize K subject to the constraint $J = c$.

Exercise 8.5.21. Formulate the variational problem of maximizing the volume of a figure enclosed in a surface of fixed area. What isoperimetric inequality does one get?

Exercise 8.5.22. Suppose in a variational problem with a constraint the Lagrange multiplier turns out to be zero. What conclusion can you draw?

Exercise 8.5.23. Formulate the corresponding result of Exercise 8.5.2 when

$$J(y) = \int F(t, y, y', y'')dt$$

and there is an integral constraint. Then prove your statement.

Exercise 8.5.24. Show that $Y(t) = (x(t), y(t)) = (-\frac{5}{2}t^2 + \frac{7}{2}t, t)$ is the extremal of the functional $J(x(t), y(t)) = \int_0^1 (x'^2 + y'^2 - 4ty' - 4y)dt$ subject to the conditions:

$$\int_0^1 (x'^2 - tx' - y'^2)dt = 2, \quad x(0) = y(0) = 0, \quad x(1) = y(1) = 1$$

---

[28]W. Wirtinger (1865-1945), proposed a generalization of eigenvalues, the spectrum of an operator, in an 1897 paper; the concept was extended by David Hilbert into spectral theory.

Exercise 8.5.25. (*). Let

$$J(u) = \int_0^T \int_0^\pi (u_t^2 - u_x^2)dxdt,$$

with the boundary conditions $u(0,t) = 0$, $u(\pi,t) = 0$, $u(x,0) = 0$ and $u(x,T) = \beta \sin x$.

(a) Show that the Euler-Lagrange equation is the wave equation $u_{xx} = u_{tt}$, and solve it if $T \neq \pi, 2\pi, 3\pi, ...$

(b) Compute the value of $J(u)$ for a function of the form

$$u(x,t) = \left(\frac{\beta t}{T}\right) \sin x + \alpha \sin(nx) \sin \left(\frac{k\pi t}{T}\right),$$

where n, k are integers and α is a constant.

(c) By suitable choice of n, k, α, show that we have

$$\inf_u J(u) = -\infty \quad \text{and} \quad \sup_u J(u) = +\infty,$$

where both the inf and sup are taken over the class of functions in (2) for *any* $T > 0$, *no matter how small*. Thus, we have no minimum or maximum in this case.

Exercise 8.5.26. (*) Consider the interior of the unit disk D with the hyperbolic metric

$$ds^2 = \frac{ds_{euc}^2}{(1 - (x^2 + y^2))^2} = \frac{ds_{euc}^2}{(1 - r^2)^2}.$$

Find all its geodesics. In particular, show geodesics always intersect the boundary orthogonally. Show the ones passing through 0 are straight lines. Show the sum of the angles of a geodesic triangle is strictly less than π. Show there are infinitely many lines parallel to a given line through a given point. This Riemannian surface is actually isometric to H^+.

Hint. Consider $z \mapsto \frac{z-i}{z+i}$ which maps $H^+ \to D$ isometrically[29].

---

[29]See [25].

This page intentionally left blank

Appendix A

Countability and Decimal Expansions

Definition A.0.1. A set S is called *countable* if there is a bijection $f : \mathbb{N} \to S$.

A set which is not countable is called *uncountable*. The set $\mathbb{R}$ of real numbers is uncountable.

Proposition A.0.2. *A countable union of countable sets is itself countable.*

In particular, the set $\mathbb{Q}$ of rational numbers is countable.

Any $x \in [0, 1]$ can be written in decimal form as

$$x = 0.a_1 a_2 \ldots ,$$

where $a_i \in \{0, 1, \ldots, 9\}$. More generally, if n is any integer ≥ 2, then any $x \in [0, 1]$ can be written as

$$x = 0.a_1 a_2 \ldots ,$$

where $a_i \in \{0, 1, \ldots, n - 1\}$.

These facts can be found in [18] pages 12, 15.

This page intentionally left blank

Appendix B

Calculus in One Variable

B.1 Differential calculus

In what follows each result of the three is a special case of the next. The reader should check that this is so.

Theorem B.1.1. *(Rolle's Theorem). Let $f : [a, b] \to \mathbb{R}$ be continuous on the closed interval $[a, b]$ and differentiable on (a, b). If $f(a) = f(b)$, then there is at least one point $\xi \in (a, b)$ such that $f'(\xi) = 0$.*

Theorem B.1.2. *(Mean Value Theorem). Suppose $f : [a, b] \to \mathbb{R}$ is continuous on $[a, b]$ and differentiable on (a, b). Then there is at least one $\xi \in (a, b)$ such that*

$$f(b) - f(a) = f'(\xi)(b - a).$$

Theorem B.1.3. *(Cauchy's Mean Value Theorem). Suppose f and g are continuous on $[a, b]$ and differentiable on (a, b), and $g'(x) \neq 0$ for all $x \in (a, b)$. Then there exist $\xi \in (a, b)$ such that*

$$\frac{f(b) - f(a)}{g(b) - g(a)} = \frac{f'(\xi)}{g'(\xi)}.$$

A consequence of the Cauchy Mean value theorem is L'Hopital's Rule.

Theorem B.1.4. *(L'Hopital's Rule). Let f and g be differentiable functions on an open (finite or infinite) interval I with $g'(x) \neq 0$ for all $x \in I$. Let c be either one of the endpoints of I. Suppose that either*

$$\lim_{x \to c} f(x) = 0 = \lim_{x \to c} g(x), \quad or \quad \lim_{x \to c} f(x) = \infty = \lim_{x \to c} g(x).$$

If

$$\lim_{x \to c} \frac{f'(x)}{g'(x)} = l$$

exists, where l is either a real number, or $l = \infty$, or $l = -\infty$, then $g(x) \neq 0$ for all $x \in I$ and

$$\lim_{x \to c} \frac{f(x)}{g(x)} = \lim_{x \to c} \frac{f'(x)}{g'(x)}.$$

B.2 Integral calculus

We now come to basic results in Integral Calculus of one variable.

Theorem B.2.1. *(Fundamental Theorem of Calculus).*

1. *Let $f : [a, b] \to \mathbb{R}$ be continuous on $[a, b]$. For $x \in [a, b]$, let*

$$F(x) = \int_a^x f(t)dt.$$

Then F is continuous on $[a, b]$, and $F'(x) \equiv f(x)$. (Such an F is called a primitive *of f.)*

2. *Let f be continuous on $[a, b]$ and let F be a primitive of f. Then*

$$\int_a^b f(t)dt = F(b) - F(a). \tag{B.1}$$

Often the right side of (A.1) is written in the form $F(x)\big|_a^b$.

Theorem B.2.2. *(First Mean Value Theorem for Integrals). Let f and g be continuous functions on $[a, b]$ with $g \geq 0$. Then there exist $\xi \in [a, b]$ such that*

$$\int_a^b f(x)g(x)dx = f(\xi) \int_a^b g(x)dx.$$

Theorem B.2.3. *(Second Mean Value Theorem for Integrals). Let f be continuous on $[a,b]$ and g have continuous derivative with $g'(x) \neq 0$ for all $x \in [a,b]$. Then there exist $\xi \in (a,b)$ such that*

$$\int_a^b f(x)g(x)dx = g(a) \int_a^\xi f(x)dx + g(b) \int_\xi^b f(x)dx.$$

Theorem B.2.4. *(Integration by parts). Let f and g be functions having continuous derivatives on a closed interval $[a,b]$. Then*

$$\int_a^b f(x)g'(x)dx = f(x)g(x)\big|_a^b - \int_a^b f'(x)g(x)gx.$$

B.2.1 Complex-valued functions

For a complex number $z = x + iy$ the complex conjugate is $\bar{z} = x - iy$ and so

$$z\bar{z} = |z|^2 = x^2 + y^2.$$

Similarly, a *complex-valued* function $f : [a,b] \to \mathbb{C}$ is decomposed as $f(x) = u(x) + iv(x)$ with $u, v : [a,b] \to \mathbb{R}$ and $\overline{f(x)} = \overline{u(x) + iv(x)} = u(x) - iv(x)$. Briefly written $f = u + iv$ and $\bar{f} = u - iv$. Moreover,

$$\int_a^b f(x)dx = \int_a^b u(x)dx + i \int_a^b v(x)dx,$$

and

$$\overline{\int_a^b f(x)dx} = \int_a^b \overline{f(x)}dx.$$

B.3 Series

Results concerning infinite series also played a role so we have the comparison and ratio tests.

Theorem B.3.1. *(Comparison and Ratio Tests).*

1. *(The Comparison Test). Let $a_n \geq 0$ and $b_n \geq 0$ for all $n \geq 1$. Suppose there exists a positive constant c such that $a_n \leq cb_n$ for all n sufficiently large. If $\sum b_n$ converges, then so does $\sum a_n$. If $\sum a_n$ diverges, then so does $\sum b_n$.*

2. *(The Ratio Test). Let $\sum a_n$ be a series of positive terms such that*

$$\lim_{n \to \infty} \frac{a_{n+1}}{a_n} = L.$$

(a) If $L < 1$, the series converges.

(b) If $L > 1$, the series diverges.

(c) If $L = 1$, the test is inconclusive.

Appendix C

Uniform Convergence

The results on uniform convergence can be found, for example, in [13].

Definition C.0.1. (Uniform Convergence). Let $\{f_n\}$ be a sequence of functions defined on an interval I. Suppose that there is a function f defined on I such that

$$\lim_{n \to \infty} f_n(x) = f(x) \text{ for each } x \in I.$$

We say that $\{f_n\}$ *converges uniformly* to f on I if for any $\epsilon > 0$ there is an $N = N(\epsilon)$ such that, for $n \geq N$

$$|f_n(x) - f(x)| < \epsilon \text{ for all } x \in I.$$

Equivalently, if for $n \geq N$

$$\sup_{x \in I} |f_n(x) - f(x)| < \epsilon.$$

Theorem C.0.2. *Let $\{f_n\}$ be a sequence of continuous functions on $[a, b]$.*

1. *If $f_n \to f$ uniformly on $[a, b]$, then f is continuous.*

2. *If $f_n \to f$ uniformly on $[a, b]$, then*

$$\lim_{n \to \infty} \int_a^b f_n(x)dx = \int_a^b f(x)dx.$$

This tells us that $J(f) = \int_a^b f(x)dx$ is a continuous linear functional on $C[a, b]$.

Definition C.0.3. Let $\sum_{n=1}^{\infty} f_n(x)$ be a series of functions defined on I and let $s_n(x) = \sum_{k=1}^{n} f_k(x)$ be the sequence of partial sums. We say that the series $\sum_{n=1}^{\infty} f_n(x)$ *converges uniformly to f on I if* $\{s_n\}$ *converges uniformly to f on I*

Theorem C.0.4. *(Weierstrass M-test). Let $\sum_{n=1}^{\infty} f_n(x)$ be a series of functions defined on an interval I. Let $\{M_n\}$ be a sequence of positive constants, such that $|f_n(x)| \leq M_n$, for all n and $x \in I$. If $\sum_{n=1}^{\infty} M_n < \infty$, then $\sum_{n=1}^{\infty} f_n(x)$ is absolutely and uniformly convergent on I.*

Theorem C.0.5. *(Term-by-term differentiation). Let $f_n(x)$ be a sequence of functions with continuous derivatives in $[a, b]$. Suppose $\sum_{n=1}^{\infty} f'_n(x)$ is uniformly convergent on $[a, b]$, and that $\sum_{n=1}^{\infty} f_n(x)$ is pointwise convergent to $f(x)$ for each $x \in [a, b]$. Then f is differentiable and for each $x \in [a, b]$*

$$f'(x) = \sum_{n=1}^{\infty} f'_n(x).$$

C.1   The Stone-Weierstrass theorem

Here we deal with the *real* Stone-Weierstrass theorem[30]. We first define the term *algebra* as follows:

Definition C.1.1. A commutative algebra with identity $\mathcal{A}$ is a real vector space consisting of elements f, g, h, etc. together with a multiplication satisfying for all $f, g, h \in \mathcal{A}$ and $t \in \mathbb{R}$:

1. $f(gh) = fg(h)$

2. $f(g + h) = fg + fh$

3. $f1 = 1f = f$, where $1 \in \mathcal{A}$

4. $(tf)g = t(fg) = f(tg)$

5. $fg = gf$

---

[30]M. Stone (1903-1989) made contributions to functional analysis. In 1946, he became the chairman of the Mathematics Department at the University of Chicago and made it into a world class department. He is best known for Stone-Weierstrass theorem, the spectral theorem and Stone-von Neumann theorem.

For example, if (X, d) is a compact metric space, $C(X)$ with point-wise operations is a commutative algebra with identity. (The reader should verify this). The algebras we shall be interested in here are sub-algebras of $C(X)$. We need one more definition. Let $\mathcal{A}$ be a subalgebra of $C(X)$. We say $\mathcal{A}$ *separates points* of X if for any two points of X there is a function $f \in \mathcal{A}$ taking different values at these two points. The continuous Urysohn lemma tells us this is true of $C(X)$ itself. The reader should also verify this. Alternatively, let $a \in X$ be fixed and for each $x \in X$ define $f(x) = d(x, a)$. Then since f is (uniformly) continuous, $C(X)$ separates points of X.

We now state the Stone-Weierstrass theorem (proved in, for example, [12]) and then apply it to various problems in functions of several real variables.

Theorem C.1.2. *(Stone-Weierstrass).   Let X be a compact metric space and $C(X)$ be the algebra of continuous real valued functions on X. Let $\mathcal{A}$ be a subalgebra of $C(X)$ which contains the constants and separates points of X. Then $\mathcal{A}$ is dense in $C(X)$. That is, any real valued continuous function f on X can be uniformly approximated on X by a function $g \in \mathcal{A}$.*

Now let $f : C \to \mathbb{R}$ where C is a closed cube in $\mathbb{R}^n$. We now define what are known as variable separable functions. Suppose $f(x) = f_1(x_1) \cdots f_n(x_n)$, where each of the f_i is continuous. Then of course f is continuous. Such a function, or even a finite linear combination of such functions is called a *variable seperable function*. More generally, let X and Y be compact. Then a variable seperable function on $X \times Y$ is

$$\sum_{i=1}^{k} f_i(x)g_i(y),$$

where each $f_i \in C(X)$ and $g_i \in C(Y)$.

Corollary C.1.3. *Let X and Y be compact.   Then any continuous function $f : X \times Y \to \mathbb{R}$ can be uniformly approximated on $X \times Y$ by a variable seperable function.*

Proof. Consider $\mathcal{A}$, the set of variable seperable functions on $X \times Y$. $\mathcal{A}$ is a subspace of $C(X \times Y)$ and since the pointwise product of such functions is again in $\mathcal{A}$ the latter is a subalgebra of $C(X \times Y)$. $\mathcal{A}$ evidently contains the constants. Now let (x, y) and $(x', y') \in X \times Y$ be distinct points. Then either $x \neq x'$ or $y \neq y'$ (or both). If $x \neq x'$ then choose $f \in C(X)$ with $f(x) \neq f(x')$. Let $g = 1$ and $k = 1$. Then we have separated (x, y) from (x', y'). If $y \neq y'$ we can similarly separate. Therefore by the Stone-Weierstrass theorem $\mathcal{A}$ is dense in $C(X \times Y)$. $\square$

We now come to the Weierstrass Approximation theorem in several variables.

Corollary C.1.4. *Let K be a compact set in $\mathbb{R}^n$ and $f : K \to \mathbb{R}$ be a continuous function. Then f can be uniformly approximated on K by polynomials in n variables.*

The reader will notice that we have already proved this when $n = 1$ in Theorem 5.4.2.

Proof. Since K is contained in a closed cube and such things are compact, we may assume K is a closed cube B in $\mathbb{R}^n$. Let $\mathcal{A}$ be the polynomials in n variables restricted to B. Then $\mathcal{A}$ is a subalgebra of $C(B)$ which certainly contains the constants. $\mathcal{A}$ also separates the points. For if we had two distinct points they would have to differ in some coordinate. So it would be sufficient to show that the polynomials in one variable x separate the points. But we know that $C(X)$ separates the points and by the Weierstrass approximation theorem in one variable any $f \in C(X)$ can be uniformly approximated on X by a polynomial in $P(X)$, where $P(X)$ is the subalgebra of polynomials on X. Therefore $P(X)$ also separates the points of X. $\square$

Appendix D

Linear Algebra

We first mention some important properties of the determinant.

Theorem D.0.1. *Let A and B be $n \times n$ real matrices. Then*

1. $\det(AB) = \det(A)\det(B)$.

2. $\det(A) = \det(A^t)$, *where A^t is the transpose of $A = (a_{ij})$, i.e.,* $A^t = (a_{ji})$.

A linear transformation gives rise to a matrix as follows:

Definition D.0.2. *Let $T : \mathbb{R}^n \to \mathbb{R}^m$ be a linear transformation, and let $\{e_1 \dots e_n\}$ and $\{u_1 \dots u_m\}$ be the standard bases of $\mathbb{R}^n$ and $\mathbb{R}^m$, respectively. For each $i = 1, ..., n$ write*

$$T(e_i) = \sum_{j=1}^{m} c_{ji} u_j.$$

The $m \times n$ matrix

$$[T] = (c_{ji}) = \begin{pmatrix} c_{11} & \dots & c_{1n} \\ . & \dots & . \\ . & \dots & . \\ . & \dots & . \\ c_{m1} & \dots & c_{mn} \end{pmatrix}$$

is called the (standard) matrix representation of T with respect to the two given bases.

Given $x \in \mathbb{R}^n$ and regarding x as an $n \times 1$ column matrix (a column vector in $\mathbb{R}^n$), we recover T from its matrix representation $[T]$ by multiplication, that is,

$$T(x) = [T] \, x.$$

Furthermore, the mapping $T \mapsto [T]$ is an *isomorphism* from $L(\mathbb{R}^n, \mathbb{R}^m)$, the vector space of linear transformation from $\mathbb{R}^n$ to $\mathbb{R}^m$, onto $M_{m \times n}(\mathbb{R})$, the vectror space of $m \times n$ real matrices. Under this isomprhism we *identify* T with $[T]$ and we just denote both by T.

Theorem D.0.3. *Let* $T : \mathbb{R}^n \to \mathbb{R}^n$ *be a linear transformation. The following are all equivalent:*

- T *is invertible.*

- $\ker(T) = \{0\}$, *where* $\ker(T) = \{v \in \mathbb{R}^n : T(v) = 0\}$ *is the kernel of* T.

- $\mathrm{range}(T) = \mathbb{R}^n$, *that is,* $\mathrm{rank}(T) = n$.

- *The columns of the matrix of* T *are linearly independent.*

- *The rows of the matrix of* T *are linearly independent.*

- $\det(T) \neq 0$.

Theorem D.0.4. *Let* V *be a finite dimensional real vector space and* $\{x_1, \ldots x_n\}$ *and* $\{y_1 \ldots y_n\}$ *be bases. Let* $T : V \to V$ *be a linear transformation. If* X *and* Y *are respectively the matrix of* T *which goes with each of these bases, then there exists an invertible* $n \times n$ *matrix* P *so that*

$$PXP^{-1} = Y,$$

that is, the matrices X *and* Y *are similar. Furthermore, if the basis are each orthonormal, then* P *can be taken to be orthogonal.*

A real $n \times n$ matrix is called *symmetric* if $A^t = A$.

Theorem D.0.5. *(Spectral Theorem). Let* A *be an* $n \times n$ *real symmetric matrix. Then* $\mathbb{R}^n$ *has an orthonormal basis each vector of which is an* *eigenvector of A. That is,*

$$A = ODO^{-1},$$

where O is an orthogonal $n \times n$ martix and D is a diagonal $n \times n$ martix with diagonal entries the eigenvalues of A.

This page intentionally left blank

Bibliography

[1] T. Apostol, *Calculus*, Vol. I. John Wiley and Sons, second edition, 1967.

[2] T. Apostol, *Calculus*, Vol. II. John Wiley and Sons, second edition, 1969.

[3] O. Bolza, *Lectures in the Calculus of Variations*. University of Chicago, 1904.

[4] C. Buck, *Advanced Calculus*. Waveland Press. Inc., third edition, 2003.

[5] E. Coddington, *An Introduction to Ordinary Differential Equations*. Dover Publications, 1989.

[6] R. Courant and D. Hilbert, *Methods of Mathematical Physics*, Vol I, II. Wiley Classics Publications, 1989.

[7] R. Courant and F. John, *Introduction to Calculus and Analysis*, Vol II/1, II/2. Sringer Verlag, Berlin Heidelberg, 2000.

[8] R. Dorfman, P. Samuelson and R. Solow, *Linear Programming and Economic Analysis*. Dover Publications, 1987.

[9] J. Dugunji, *Topology*. McGraw-Hill Companies, 1966.

[10] W. Fleming, *Functions of Several Variables*. Springer-Verlag, New York, 1977.

[11] G. Folland, *Advanced Calculus*. Prentice-Hall, New Jersey, 2002.

[12] G. Folland, *Real Analysis: Modern Techniques and Their Applications.* John Wiley and Sons, New York, 1984.

[13] A. Friedman, *Advanced Calculus.* Dover Publications, 2007.

[14] K. Hofmann and R. Kunze, *Linear Algebra.* Prentice Hall, second edition, 1971.

[15] I. Gelfand and S. Fomin, *Calculus of Variations.* Dover Publications, 2000.

[16] W. Kaplan, *Advanced Calculus.* Addison Wesley, fifth edition, 2002.

[17] W. Kaplan, *Ordinary Differential Equations.* Addison Wesley, 1962.

[18] A. Kolmogorov and S. Fomin, *Introductory Real Analysis*, Dover Publications, 1970.

[19] S. Lang, *Analysis I.* Addison-Wesley Publising Company, fifth edition, 1976.

[20] P. Lax, "Change of variables in multiple integrals", *Amer. Math. Monthly*, Vol. 106, 497-501 (1999).

[21] P. Lax, "Change of variables in multiple integrals II", *Amer. Math. Monthly*, Vol. 108, 115-119 (2001).

[22] L. Maligranda, "Simple norm inequalities", *Amer. Math. Monthly*, Vol. 113, 256-260 (2006).

[23] J. Marsden and A. Tromba, *Vector Calculus.* W. H. Freeman and Company, third edition, New York 1988.

[24] J. Milnor, "Analytic proofs of the "Hairy Ball Theorem" and the Brouwer Fixed Point Theorem", *Amer. Math. Monthly*, Vol. 85, 521-524 (1978).

[25] M. Moskowitz, *A Course in Complex Analysis in One Variable.* World Scientific Publishing, 2002.

[26] S. Nikolsky, *A Course of Mathematical Analysis*, Vol I, II. Mir Publishers, Moscow, 1977.

[27] I. Petrovsky, *Lectures on Partial Differential Equations*, Dover Publications, 1991.

[28] H. Royden, *Real Analysis*. Macmillan, New York, third edition, 1988.

[29] J. Schwartz, "The formula of change of variables in multiple intgrals", *Amer. Math. Monthly*, Vol. 61, 81-85 (1954).

[30] A. Schep, "A Simple Complex Analysis and an Advanced Calculus Proof of the Fundamental Theorem of Algebra". *Amer. Math. Monthly*, Vol. 116, No. 1, (2009).

[31] M. Spivak, *Calculus on Manifolds*. Westview Press, fifth edition, 1971.

[32] D. Struik, *A Concise History of Mathematics*. Dover Publications, 1987.

[33] D. Struik, *Lectures on Classical Differential Geometry*. Dover Publications, 1988.

[34] A. Taylor and R. Mann, *Advanced Calculus*. John Wiley & Sons, third edition, New York, 1983.

[35] G. Tolstov, *Fourier Series*. Dover Publications, 1976.

[36] E. Zachmanoglou and D. Thoe, *Introduction to Partial Differential Equations with Applications*, Dover Publications, 1986.

[37] J. Webb, *Functions of Several Real Variables*. Ellis Horwood Limited, 1991.

[38] D. Widder, *Advanced Calculus*, BN Publishing, 2009.

This page intentionally left blank

Index

Abel's formula, 617
absolute value, 5
absolutely integrable, 369
accelaration, 115
accumulation point, 5
action integral, 654
admissible surface, 667
affine group of $\mathbb{R}$, 324
affine transformation, 82
almost everywhere, 254
amplitude, 539
arc-length, 406
Archimedean property, 3
area, 453
arithmetic mean, 210

Banach's fixed point theorem, 76
basis, 17
Bernoulli, 630
Bernoulli equation, 524, 613
Bernoulli inequality, 7
Bessel equation, 618
Bessel's inequality, 55
Beta function, 365
bijection, 65
Bolzano-Weierstrass, 6
Bolzano-Weiertsrass theorem in $\mathbb{R}^n$, 47

boundary, 39
boundary condition, 515, 568, 625
boundary value problem, 583
Brachistochrone problem, 628
Brouwer fixed point theorem, 497, 504

calculus of variations, 619
canonical form, 579
Cantor set, 242
capping surface, 470
catenary, 663
Cauchy problem, 515
Cauchy problem for the heat equation, 609
Cauchy sequence, 13, 45
Cauchy's Mean Value Theorem, 693
Cauchy-Schwarz inequality, 20, 257, 370, 686
Cavalieri principle, 271, 277, 284, 288
center of mass, 273
centroid, 274
Chain Rule, 108, 116
Change of Variable, 304, 335, 456, 508
characteristic equation, 534

characteristic function, 247
characteristic polynomial, 534
circulation, 412
closed form, 491
closed surface, 439
Cobb-Douglas production func-
    tion, 207
commutativity of diff.   with
    conv., 371
compact support, 307
compactness, 48
Comparison test, 333
complete metric space, 45
completeness, 13
Completeness Axiom, 2
component functions, 61
composition, 73
cone, 432
conformal, 434
connectedness, 88
Conservation of Energy,   419,
    586, 658
conservative vector field,   417,
    470
constrained extrema, 194
continuous function, 71
Contraction Mapping principle,
    76
convergent sequences, 44
convex hull, 102
convex set, 95
convolution, 368, 604
cosine law, 25
critical point, 161, 200, 643
cross product, 32
curl, 462
curvature, 409

curve, 62
cycloid, 408, 629
cylinder, 433, 650
cylindrical coordinates, 315

D'Alembert formula, 588
D'Alembert solution, 578
decay-growth equation, 514
degenerate critical point, 172
dense, 59
denseness of $\mathbb{Q}$, 4
derivative, 104
determinant, 298
diagonal matrix, 298
diff. under int. sign, 347, 355,
    362
diffeomorphism, 182, 294
Differentiability criterion, 119
differentiable function, 104
differentiable manifold, 492
differential equation, 509
differential form, 483
differential topology, 492
dimension, 17
directional derivative, 110
Dirichlet energy integral, 669
Dirichlet integral, 360
Dirichlet principle, 676
Dirichlet problem,   588,   591,
    606, 669
divergence, 462
Divergence theorem,   458,   474,
    677
domain, 61, 94
Du-Bois-Reymond lemma, 622

Economics, applications to, 205
eigenfunction, 679

eigenvalue, 168, 197
eigenvalue problem, 544, 678
electric circuit, 617
elementary region, 266
ellipsoid, 99
elliptic form, 579
energy minimizer, 686
equality of mixed partials, 135, 289
equivalent norms, 53
escape velocity, 419
Euler, 14, 630
Euler equation in ODE, 562
Euler's equation, 129
Euler-Lagrange differential equations, 641
Euler-Lagrange equation, 624, 667
Euler-Poisson integral, 332, 340
exact equations, 521
exact form, 491
exhaustion, 329
existence and uniqueness, 527, 555, 556
exterior derivative, 486
exterior product, 485
extrema over compact sets, 203
extremals, 625, 643
Extreme Value theorem, 86

falling body with air resistance, 516
first fundamental form, 434
First Mean Value Th. for Integrals, 694
first order differential equation, 511

first order linear ODE, 511
first variation, 642
flux, 445
Fourier coefficient, 586, 598
Fourier inversion formula, 386
Fourier series, 586, 598, 634
Fourier transform, 376, 603
Fresnel integrals, 398
Fubini theorem, 261, 267, 289, 340, 354
functional dependence, 211
Fundamental lemma, 622, 639, 667
fundamental set of solutions, 535, 560
Fundamental Theorem of Algebra, 349
Fundamental Theorem of Calculus, 293, 694
Fundamental Theorem of line integrals, 417

Gamma function, 361, 364
Gauss theorem, 475, 489
Gaussian function, 384, 390
general solution, 574
generating function, 571
geodesic, 646, 649, 652
geodesic triangle, 689
geometric mean, 210
geometric series, 242
global solution, 510
gradient, 110, 462
graph, 62
gravitational attraction, 418
great circle, 651
Green identities, 480, 508

Green theorem, 450, 488, 668

Hamilton principle, 654
harmonic function, 137, 175,
    454, 459, 479
harmonic oscillator with drag,
    568
heat equation, 575, 594, 615
Heine-Borel theorem, 49, 58
helical lines, 650
Hermite equation, 568
Hermite functions, 572
Hermite polynomials, 564, 569
Hessian, 139
higher order differential equa-
    tions, 548
Holder's iequality, 260
homeomorphism, 76, 303
homogeneous equations, 519
homogeneous function, 129
homogeneous PDE, 574
hyperbolic form, 579
hypersurface, 123

implicit differentiation, 188
Implicit Function theorem, 185
improper multiple integral, 330
infimum, 2
initial condition, 515
inner product, 19
integers, 1
integrable function over a set,
    248
integral curve, 509, 617
integral form of the remainder,
    150
integrating factor, 512

integration by parts, 363, 624,
    695
Intermediate Value theorem, 91
Invariance of Domain theorem,
    500
invariant integrals on groups,
    324
inverse Fourier transform, 387,
    392
Inverse Function theorem, 177
irrationality of e, 153
isometry, 76
isoperimetric inequality, 634,
    688
isoperimetric problem, 632, 663
iterated integral, 262

Jacobian, 114

kinetic energy, 418, 654

L'Hopital's Rule, 694
Lagrange equations, 658
Lagrange generalized coordi-
    nates, 657
Lagrange multipliers, 194, 660
Lagrange rule, several con-
    straints, 197
Lagrangian, 665, 685
Lagrangian function, 200
Laplace equation, 137, 575, 592,
    669
Laplace tranform, 373
Laplace's eq. in spherical coor-
    dinates, 144
Laplacian, 137, 403, 459, 463
Laplacian in polar coordinates,
    137

least upper bound axiom, 1
Lebesgue criterion, 245
Legendre condition, 636
Legendre polynomials, 618
level sets, 62, 123
limit point, 5
line integral, 410, 411
line segment, 62
Linear Approximation theorem, 105, 301
linear functional, 84
linear programming, 97
linear second order PDE, 574
linear span, 16
linear transformation, 80
linearly independent, 17, 301
Lipschitz condition, 133, 296, 527, 554
local extrema, 160
local solution, 510

matrix ODE, 560
matrix representation, 112, 298, 701
maximum-minimum principle, 175
mean value property, 594
Mean Value theorem, 131, 693
Mean value theorem for integrals, 257
measure zero, 241
method of Frobenius, 564
method of least squares, 166
metric space, 38
Milnor, J., 494
minima and maxima in n variables, 161

minimal surface problem, 634
minimal surfaces, 670
minimal volume problem, 635
Minkowski's inequality, 60, 260
Mobius, 439
moment of inertia, 274
momenta, 657
monotone sequence, 9
Morse's lemma, 218
multiple integral, 233

nested intervals property, 11
nested sets, 57
Neumann problem in the half-plane, 608
Newton's law, 418, 419, 516, 541, 654
Newtonian potential, 448
nondegenerate, 580
nonhomogeneous, 542
nonhomogeneous heat equation, 596
nonuniform convergence, 356
norm, 19
normal derivative, 480, 681
normal vector, 430
nowhwere dense, 57

ODE with variables separated, 517
orthogonal set, 679
orthogonal transformation, 82
orthogonality, 24
orthogonality of Hermite's polynomials, 570
orthogonalization process, 27
orthogonally conjugate, 298
oscillation, 244

Pappus theorem, 277, 437
parabolic form, 579
parallelepiped, 36, 301
parallelogram law, 21
parameter curves, 430
parametrization, 405
parametrized surface, 429
Parseval's equality, 28, 387
partial derivative, 111
partial differential equation, 573, 667
partition, 233
partition of unity, 282
path, 405
path connectedness, 91
path independence, 426
pendulum with small oscillations, 541
period, 539
Picard theorem, 527, 555
Poincare lemma, 417, 421
Poincare upper half plane, 652
Poisson integral formula, 593
Poisson kernel, 594
Poisson summation formula, 600
polar coordinates, 117, 313
Polar decomposition, 83, 298
polynomial, 72
positive definite symmetric matrix, 298
positive orientation, 405
positive side, 441
potential energy, 418, 654
Principal axis theorem, 168, 340
Principle of least action, 654

Principle of mathematical induction, 1
problem of geometric optics, 636
problem of shortest distance, 627
Projection theorem, 29

quadratic form, 141, 162

radial function, 339
range, 61
rational numbers, 1
Rayleigh quotient, 679
real analytic, 152, 565
real vector space, 15
reduction of order, 545
refinement, 233
regular value, 283
regularity condition, 430
resonance frequency, 617
Riccati equation, 612
Riemann condition, 236
Riemann integral, 233
Riemann-Lebesgue lemma, 379
Riemannian metric, 647
Riemannian surface, 646, 686
Riesz lemma, 58
Rodrigues' formula, 572, 615
Rolle's theorem, 693

saddle point, 161
Sard's theorem, 283, 307
scalar field, 410
Schroedinger equation, 573
Schwartz space, 603
second derivative test, 162, 200
Second MVT for integrals, 695
second order linear ODE, 534

second variation, 642
separation of variables, 574, 594
sequential compactness, 49
several constraints, 664
simple closed path, 405
simple harmonic oscillator, 538
simple set, 247, 297, 300
simply connected, 423
small vibrations of a string, 583
Snell's law, 631
special cases of Euler-Lagrange
    equation, 626
special types of second order
    ODE, 545
Spectral theorem, 702
speed, 115, 406
sphere, 651
spherical coordinates, 317, 321
spherical shell, 332
squeezing theorem, 14
standard basis, 17
Steiner theorem, 292
stereographic projection, 86,
    499
Stokes theorem, 467, 489, 492,
    675
Stone-Weierstrass theorem, 698
subsequence, 12
successive approximation, 78,
    529
sufficiency condition, 644
Superposition principle, 574
support of a function, 280
supremum, 2
surface area, 433, 436
surface area of the unit ball, 399
surface integral, 441

surface of revolution, 431, 649
Sylvester criterion, 170, 579
symbol, 578
systems of ODE, 548

tangent hyperplane, 124
tangent vector, 105, 406, 429
tangential component, 412
tangential vector field, 492
Taylor's theorem, 145
Taylor's theorem in several vari-
    ables, 154
terminal velocity, 517
torus, 277, 312, 436
transport equation, 615
triangle inequality, 5, 19
triple scalar product, 35
Tychonoff theorem, 52

undamped simple harmonic vi-
    brations, 540
uniform continuity, 88
uniform convergence, 351, 697
uniformly continuous, 378
unit cube, 301
unit normal, 430
Urysohn lemma, 280

variable coefficients, 564
variable seperable, 699
variation of parameters, 614
vector field, 410
vector space, 15
velocity, 115
vibrating frequencies, 679
vibrating string, 671
volume, 249, 478
volume of revolution, 273

volume of unit ball, 299, 322, 364, 399
volume zero, 241, 255, 297

wave equation, 575, 604, 673, 689
Weierstrass M-test, 353
Weierstrass approximation theorem, 374

Whitney embedding theorem, 501
Wirtinger inequality, 634, 688
Wronskian, 535, 560

zero curvature, 409

世界著名数学史专家 E. T. Bell 曾指出:

> 常常说,单单 19 世纪贡献的数学就是以往时代全部贡献的五倍之多,这个断言是有其可靠根据的.这不仅是指数量上的,而且还包括质量方面的,后者有着不可比拟的重要性.

本书就是一部以 19 世纪产生并完善的多元微积分为主要内容的英文数学教程,中文书名或可译为《多元实函数教程》.

本书的作者有两位,一位为马丁·莫斯科维茨(Martin Moskowitz),美国数学家,美国纽约城市大学教授,另一位为福蒂奥斯·帕里奥詹尼斯(Fotios Paliogiannis),也是美国数学家,美国圣弗朗西斯学院教授.

正如作者在前言中所介绍:

> 这本书源于我们每个人都教授过多年的多变量微积分课程.我们的书具有双重好处,既强调了该主题的概念和计算内容,又拥有现代观点.其主

要目标是以清晰和系统的方式教授该主题. 尽管完成该书的要求比预期的要高, 但在完成后, 我们认为这是一项值得付出努力的工作. 前面的章节对经典主题进行了成熟的介绍, 包括多变量中的微积分、高级微积分或向量分析, 这些主题通常在本科数学课程的三年级或四年级进行讲授. 然后我们转向常微分方程以及二阶经典偏微分方程, 这些内容通常可以在高级微积分或本科生的数学物理学书籍中找到. 最后, 我们对变分法这一强大而重要的主题进行了基本介绍.

本书的主题是众多学科的基础, 因此对低年级数学研究生, 以及数学、物理学、化学、生物学、工程学甚至经济学专业的高年级本科生都有用. 本书包括 8 章, 第 1 章和第 2 章处理了欧几里得空间的基本的几何与拓扑内容; 第 3 章处理了微分学内容; 第 4,5 和 6 章是关于多变量的积分学的内容. 本书有两个新奇的特征: 第 7 章是基本但非常重要的常微分方程和二阶经典偏微分方程的相关内容; 第 8 章深入介绍了变分法, 它被视为类似于多变量微积分中常见的极值问题. 出于这两个原因, 我们似乎很好地安排了其余的材料, 这些材料也非常适合成为本书的一部分.

当我们挑选背景材料的附录时, 将这些材料分为四类: 可数性与小数展开式、一元微积分、一致收敛和线性代数. 可数性的结果可以在 [18] 中找到, 一元微积分的结果在 [1] 中找到, 与一致收敛相关的, 特别是斯通–魏尔斯特拉斯定理在 [13] 和 [12] 中, 线性代数的相关内容在 [14] 中.

教师可以根据课程的主题进行多项选择. 例如, 第 3,4,5 和 6 章制定了多变量微积分的标准年课程. 第 3 章和第 4 章是第一学期的课程, 而第 5 章和第 6 章是第二学期的课程, 第 7 章可作为一学期的常微分方程和偏微分方程课程的教材, 对于实力较强的学生, 第 7 章和第 8 章可用于常微分方程、偏微分方程和变分法的一个学期的课程. 通过集中讲授标星的部分, 本书还可用于分析学和几何学中的选定主题.

读者将会找到每部分的许多例子. 每个部分包含了与该部分直接相关的练习. 这些练习的答案在每部分的结尾给出. 除此之外, 每章以已解决的练习题和建议的练习题结束, 其中一些带有答案或提示.整本书都有更具挑战性的标星的练习题来吸引水平更高的读者.星号(∗)表示第一次阅读时可能会省略的部分、例子或练习题.然而,我们希望对主题有更深入了解的雄心勃勃的读者可以去掌握这些材料.

我们要感谢佛瑞德·格林利夫(Fred Greenleaf)为我们提供了一组他在纽约大学使用的分析学笔记.感谢斯坦利·卡普兰(Stanley Kaplan),他阅读了本书的大部分内容,并给出了评论,大大改进了本书的阐述方式.当然,任何错误或错误陈述都是我们的责任.我们还要感谢埃雷兹·秀沙(Erez Shochat)提供的数据,这对我们准备本书的最终版本非常有帮助.最后,感谢安妮塔(Anita)和卡西亚尼(Kassiani)在我们开始这个项目的三年中给予我们的鼓励和耐心.

本书的版权编辑李丹女士为了能使读者快速了解这部 700 多页巨著的基本内容.特翻译了本书的目录如下:

1. 欧几里得空间 $\mathbf{R}^n$ 的基本特征

1.1 实数
1.2 作为向量空间的 $\mathbf{R}^n$
1.3 作为内积空间的 $\mathbf{R}^n$
1.4 作为度量空间的 $\mathbf{R}^n$
1.5 $\mathbf{R}^n$ 中序列的收敛性
1.6 紧性
1.7 等价范数(∗)
1.8 第 1 章的已解问题

2. 欧几里得空间的函数

2.1 从 $\mathbf{R}^n$ 到 $\mathbf{R}^m$ 的函数
2.2 函数的极限
2.3 连续函数
2.4 线性变换
2.5 紧集的连续函数

2.6　连通性与凸性

2.7　第2章的已解问题

3. 多变量的微分学

3.1　可微函数

3.2　偏导数与方向导数、切空间

3.3　齐次函数与欧拉方程

3.4　均值定理

3.5　高阶导数

3.6　泰勒定理

3.7　多变量中的极大值和极小值

3.8　逆函数定理与隐函数定理

3.9　约束极值,拉格朗日乘子

3.10　函数依赖

3.11　莫尔斯引理

3.12　第3章的已解问题

4. 多变量中的积分学

4.1　$\mathbf{R}^n$ 中的积分

4.2　多重积分的性质

4.3　富比尼定理

4.4　光滑乌雷松引理与单位(∗)的划分

4.5　萨德定理(∗)

4.6　第4章的已解问题

5. 变量代换公式,反常多重积分

5.1　变量代换公式

5.2　反常多重积分

5.3　用积分定义的函数

5.4　魏尔斯特拉斯逼近定理(∗)

5.5　傅里叶变换(∗)

5.6　第5章的已解问题

6. 线积分与面积分

6.1　弧长和线积分

6.2　守恒向量场和庞加莱引理

6.3 表面积与面积分

6.4 格林定理与 $\mathbf{R}^2$ 中的散度定理

6.5 散度与旋度

6.6 斯托克斯定理

6.7 $\mathbf{R}^3$ 中的散度定理

6.8 微分形式(∗)

6.9 球体上的向量场与布劳威尔不动点定理(∗)

6.10 第 6 章的已解问题

7. 常微分函数与偏微分函数基础

7.1 介绍

7.2 一阶微分函数

7.3 毕卡定理(∗)

7.4 二阶微分函数

7.5 高阶常微分函数与常微分函数的系统

7.6 常微分函数中若干更高级的主题(∗)

7.7 偏微分函数

7.8 二元二阶偏微分函数

7.9 傅里叶变换方法(∗)

7.10 第 7 章的已解问题

8. 变分法介绍

8.1 简单的变分问题

8.2 概述

8.3 带约束的变分问题

8.4 多重积分变分问题

8.5 第 8 章的已解问题

本书的学习是后续许多课程的基础,比如在学习数学物理方程时有:

拉普拉斯算符在直角坐标系中的表达式为

$$\Delta U = \frac{\partial^2 U}{\partial x^2} + \frac{\partial^2 U}{\partial y^2} + \frac{\partial^2 U}{\partial z^2}$$

在许多数学物理方程(偏微分方程)中,常要讨论含 ΔU 的方程,在有些情况下,在球坐标系或柱坐标系下进行讨论比较方便,因此,需要用到 ΔU 在球坐标系和柱坐标系下的表达式.有了本书的准备我们就可以很方便的推导它.

先推导 ΔU 在球坐标系下的表达式,令 $A = \operatorname{grad} U$,我们知道

$$\operatorname{grad} U = \frac{\partial U}{\partial x}\boldsymbol{i} + \frac{\partial U}{\partial y}\boldsymbol{j} + \frac{\partial U}{\partial z}\boldsymbol{k}$$

故

$$A_x = \frac{\partial U}{\partial x}, A_y = \frac{\partial U}{\partial y}, A_z = \frac{\partial U}{\partial z}$$

于是

$$\operatorname{div} \boldsymbol{A} = \frac{\partial A_x}{\partial x} + \frac{\partial A_y}{\partial y} + \frac{\partial A_z}{\partial z} = \frac{\partial^2 U}{\partial x^2} + \frac{\partial^2 U}{\partial y^2} + \frac{\partial^2 U}{\partial z^2} = \Delta U$$

由此,得 ΔU 的一个表达式

$$\Delta U = \operatorname{div}(\operatorname{grad} U) \tag{1}$$

我们就根据公式(1)来求出 ΔU 在球坐标系下的表达式,即先求出 $A = \operatorname{grad} U$ 在球坐标系下的表达式,再根据

$$\operatorname{div} \boldsymbol{A} \bigg|_M = \lim_{V \to M} \frac{\oiint_S \boldsymbol{A} \cdot \mathrm{d}\boldsymbol{s}}{V} \tag{2}$$

求出 $\operatorname{div} \boldsymbol{A}$ 的表达式.

下面求 $A = \operatorname{grad} U$ 在球坐标系下的表达式,如图1,令点 $M(r, \theta, \varphi)$ 处沿 r 增加方向的单位向量为 $\boldsymbol{n}_r$,沿 θ 增加方向的单位向量为 $\boldsymbol{n}_\theta$,沿 φ 增加方向的单位向量为 $\boldsymbol{n}_\varphi$,于是 $\boldsymbol{n}_r, \boldsymbol{n}_\theta, \boldsymbol{n}_\varphi$ 是互相垂直的三个单位向量.

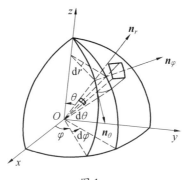

图 1

令 $A = \operatorname{grad} U$ 在 $\boldsymbol{n}_r, \boldsymbol{n}_\theta, \boldsymbol{n}_\varphi$ 三个方向上的投影分别为 A_r, A_θ, A_φ, 于是

$$A = A_r \boldsymbol{n}_r + A_\theta \boldsymbol{n}_\theta + A_\varphi \boldsymbol{n}_\varphi \tag{3}$$

另外,由梯度的性质,知 $A = \operatorname{grad} U$ 在 $\boldsymbol{n}_r$ 上的投影等于 U 沿 $\boldsymbol{n}_r$ 方向的方向导数,即

$$A_r = \frac{\partial U}{\partial r}$$

同样, $A = \operatorname{grad} U$ 在 $\boldsymbol{n}_\theta$ 上的投影等于 U 沿 $\boldsymbol{n}_\theta$ 方向的方向导数,即(注意, $\boldsymbol{n}_\theta$ 方向的微弧长 $\mathrm{d}s_1 = r\mathrm{d}\theta$)

$$A_\theta = \frac{\partial U}{\partial s_1} = \frac{1}{r} \frac{\partial U}{\partial \theta} \tag{5}$$

同样, $A = \operatorname{grad} U$ 在 $\boldsymbol{n}_\varphi$ 上的投影等于 U 沿 $\boldsymbol{n}_\varphi$ 方向的方向导数,即(注意, $\boldsymbol{n}_\varphi$ 方向的微弧长 $\mathrm{d}s_2 = r\sin\theta\mathrm{d}\varphi$)

$$A_\varphi = \frac{\partial U}{\partial s_2} = \frac{1}{r\sin\theta} \frac{\partial U}{\partial \varphi} \tag{6}$$

由此,式(3)成为

$$A = \frac{\partial U}{\partial r}\boldsymbol{n}_r + \frac{1}{r}\frac{\partial U}{\partial \theta}\boldsymbol{n}_\theta + \frac{1}{r\sin\theta}\frac{\partial U}{\partial \varphi}\boldsymbol{n}_\varphi \tag{7}$$

现根据式(2)来求 $\operatorname{div} A$,将 V 取成图 1 中所示的小体积则

$$V \approx \mathrm{d}r \cdot r\mathrm{d}\theta \cdot r\sin\theta\mathrm{d}\varphi = r^2\sin\theta\mathrm{d}r\mathrm{d}\theta\mathrm{d}\varphi \tag{8}$$

$\oiint_S A \cdot \mathrm{d}s$ 中的 s 即 V 的边界,共包括六块曲面,记垂直于 $\boldsymbol{n}_r$ 方向的两块为 S_1 与 S_2,垂直于 $\boldsymbol{n}_\theta$ 的两块为 S_3 与 S_4,垂直于 $\boldsymbol{n}_\varphi$ 的两块为 S_5, S_6,由于 V 很小,显然,此六块小曲面都近似于矩形,于是(注意到式(4))

$$\iint_{S_1} A\mathrm{d}s + \iint_{S_2} A\mathrm{d}s \approx A_r r\mathrm{d}\theta \cdot r\sin\theta\mathrm{d}\varphi \bigg|_{r+\mathrm{d}r,\theta,\varphi} - $$

$$A_r r\mathrm{d}\theta \cdot r\sin\theta\mathrm{d}\varphi \bigg|_{r,\theta,\varphi} \approx $$

$$\frac{\partial(A_r r\mathrm{d}\theta \cdot r\sin\theta\mathrm{d}\varphi)}{\partial r}\mathrm{d}r = $$

$$\frac{\partial}{\partial r}\left(r^2 \frac{\partial U}{\partial r}\right)\sin\theta\mathrm{d}r\mathrm{d}\theta\mathrm{d}\varphi$$

同样,有(注意到式(5)与(6))

$$\iint_{S_3} \boldsymbol{A}\mathrm{d}s + \iint_{S_4} \boldsymbol{A}\mathrm{d}s \approx A_\theta \mathrm{d}r \cdot r\sin\theta\mathrm{d}\varphi \Big|_{r,\theta+\mathrm{d}\theta,\varphi} -$$

$$A_\theta \mathrm{d}r \cdot r\sin\theta\mathrm{d}\varphi \Big|_{r,\theta,\varphi} \approx$$

$$\frac{\partial(A_\theta \mathrm{d}r \cdot r\sin\theta\mathrm{d}\varphi)}{\partial\theta}\mathrm{d}\theta =$$

$$\frac{\partial}{\partial\theta}\Big(\sin\theta\,\frac{\partial U}{\partial\theta}\Big)\mathrm{d}r\mathrm{d}\theta\mathrm{d}\varphi$$

$$\iint_{S_5} \boldsymbol{A}\mathrm{d}s + \iint_{S_6} \boldsymbol{A}\mathrm{d}s \approx A_\varphi \mathrm{d}r \cdot r\mathrm{d}\theta \Big|_{r,\theta,\varphi+\mathrm{d}\varphi} - A_\varphi \mathrm{d}r \cdot r\mathrm{d}\theta \Big|_{r,\theta,\varphi} =$$

$$\frac{\partial(A_\varphi \mathrm{d}r \cdot r\mathrm{d}\theta)}{\partial\varphi}\mathrm{d}\varphi =$$

$$\frac{1}{\sin\theta}\frac{\partial^2 U}{\partial\varphi^2}\mathrm{d}r\mathrm{d}\theta\mathrm{d}\varphi$$

故

$$\oiint_S \boldsymbol{A}\cdot\mathrm{d}\boldsymbol{s} =$$

$$\Big\{\sin\theta\,\frac{\partial}{\partial r}\Big(r^2\,\frac{\partial U}{\partial r}\Big) + \frac{\partial}{\partial\theta}\Big(\sin\theta\,\frac{\partial U}{\partial\theta}\Big) + \frac{1}{\sin\theta}\frac{\partial^2 U}{\partial\varphi^2}\Big\}\mathrm{d}r\mathrm{d}\theta\mathrm{d}\varphi \qquad (9)$$

由式(8)和(9),得

$$\frac{\oiint_S \boldsymbol{A}\cdot\mathrm{d}\boldsymbol{s}}{V} \approx \frac{1}{r^2}\frac{\partial}{\partial r}\Big(r^2\,\frac{\partial U}{\partial r}\Big) + \frac{1}{r^2\sin\theta}\frac{\partial}{\partial\theta}\Big(\sin\theta\,\frac{\partial U}{\partial\theta}\Big) + \frac{1}{r^2\sin^2\theta}\frac{\partial^2 U}{\partial\varphi^2}$$

并且 V 越小时,此近似等式越精确,故令 $V \to M$ 取极限时,即成为等式

$$\lim_{V\to M}\frac{\oiint_S \boldsymbol{A}\cdot\mathrm{d}\boldsymbol{s}}{V} \approx \frac{1}{r^2}\frac{\partial}{\partial r}\Big(r^2\,\frac{\partial U}{\partial r}\Big) + \frac{1}{r^2\sin\theta}\frac{\partial}{\partial\theta}\Big(\sin\theta\,\frac{\partial U}{\partial\theta}\Big) + \frac{1}{r^2\sin^2\theta}\frac{\partial^2 U}{\partial\varphi^2}$$

由此,根据式(1)及式(2),即得球坐标系下拉普拉斯算符 ΔU 的表达式

$$\Delta U = \frac{1}{r^2}\frac{\partial}{\partial r}\Big(r^2\,\frac{\partial U}{\partial r}\Big) + \frac{1}{r^2\sin\theta}\frac{\partial}{\partial\theta}\Big(\sin\theta\,\frac{\partial U}{\partial\theta}\Big) + \frac{1}{r^2\sin^2\theta}\frac{\partial^2 U}{\partial\varphi^2}\ (10)$$

类似地,可得柱坐标系下拉普拉斯算符 ΔU 的表达式

$$\Delta U = \frac{1}{r}\frac{\partial}{\partial r}\Big(r\,\frac{\partial U}{\partial r}\Big) + \frac{1}{r^2}\frac{\partial^2 U}{\partial\theta^2} + \frac{\partial^2 U}{\partial z^2} \qquad (11)$$

再比如许多学生在学习调和方程及其边值问题时都会遇到这样的问题：

试证明：在调和方程的诺伊曼问题

$$\begin{cases} \Delta u = 0, 在 \varOmega 中 \\ \dfrac{\partial u}{\partial n} = g, 在边界 \partial\varOmega 上 \end{cases}$$

中，出现在边界条件右端的函数 g 不能随意给出，它必须满足条件

$$\int_{\partial\varOmega} g \mathrm{d}s = 0$$

我们将方程在区域 $\varOmega$ 上积分，可得

$$\iint_{\varOmega} \Delta u \mathrm{d}x\mathrm{d}y = 0$$

再利用本书 6.4 中的格林定理即可得

$$\int_{\partial\varOmega} g \mathrm{d}s = 0$$

顺便说说为什么要引进这本书. 就图书市场而言，微积分教程的出版是空前内卷化的. 以微积分为主要内容的教科书多达几百种，几乎每所高校，每个教授微积分的教师都有自己编写的课本. 按理说"百花齐放，百家争鸣"是好事，是会从中诞生出精品，但遗憾的是并没有出现这种局面，而是高度重复，平庸单调，缺乏个性化，其背后的原因是我们的教授们并没有主人翁精神，把自己定位于一个"打工人"，以完成 KPI 的计时工心态去写，这也是这么多年来高校行政化带来的隐性恶果，正严重腐蚀着我们大学质量. 而国外正相反，曾任北京大学校长的林建华教授在一次会议中讲过一个故事. 说是美国的麦克阿瑟将军，退役之后到哥伦比亚大学做校长，有一次不小心对学校的教授说，教授是学校的雇员，教授当即回怼道：你才是学校的雇员，而我就是哥伦比亚大学.

主仆关系，一目了然！只有主人才会出精品，奴隶永远在应付！

刘培杰
2022 年 9 月 23 日
于哈工大

刘培杰数学工作室
已出版(即将出版)图书目录——原版影印

书　　名	出版时间	定　价	编号
数学物理大百科全书.第1卷(英文)	2016—01	418.00	508
数学物理大百科全书.第2卷(英文)	2016—01	408.00	509
数学物理大百科全书.第3卷(英文)	2016—01	396.00	510
数学物理大百科全书.第4卷(英文)	2016—01	408.00	511
数学物理大百科全书.第5卷(英文)	2016—01	368.00	512
zeta函数,q-zeta函数,相伴级数与积分(英文)	2015—08	88.00	513
微分形式:理论与练习(英文)	2015—08	58.00	514
离散与微分包含的逼近和优化(英文)	2015—08	58.00	515
艾伦·图灵:他的工作与影响(英文)	2016—01	98.00	560
测度理论概率导论,第2版(英文)	2016—01	88.00	561
带有潜在故障恢复系统的半马尔柯夫模型控制(英文)	2016—01	98.00	562
数学分析原理(英文)	2016—01	88.00	563
随机偏微分方程的有效动力学(英文)	2016—01	88.00	564
图的谱半径(英文)	2016—01	58.00	565
量子机器学习中数据挖掘的量子计算方法(英文)	2016—01	98.00	566
量子物理的非常规方法(英文)	2016—01	118.00	567
运输过程的统一非局部理论:广义波尔兹曼物理动力学,第2版(英文)	2016—01	198.00	568
量子力学与经典力学之间的联系在原子、分子及电动力学系统建模中的应用(英文)	2016—01	58.00	569
算术域(英文)	2018—01	158.00	821
高等数学竞赛:1962—1991年的米洛克斯·史怀哲竞赛(英文)	2018—01	128.00	822
用数学奥林匹克精神解决数论问题(英文)	2018—01	108.00	823
代数几何(德文)	2018—04	68.00	824
丢番图逼近论(英文)	2018—01	78.00	825
代数几何学基础教程(英文)	2018—01	98.00	826
解析数论入门课程(英文)	2018—01	78.00	827
数论中的丢番图问题(英文)	2018—01	78.00	829
数论(梦幻之旅):第五届中日数论研讨会演讲集(英文)	2018—01	68.00	830
数论新应用(英文)	2018—01	68.00	831
数论(英文)	2018—01	78.00	832

刘培杰数学工作室
已出版(即将出版)图书目录——原版影印

书　　名	出版时间	定　价	编号
湍流十讲(英文)	2018—04	108.00	886
无穷维李代数:第3版(英文)	2018—04	98.00	887
等值、不变量和对称性(英文)	2018—04	78.00	888
解析数论(英文)	2018—09	78.00	889
《数学原理》的演化:伯特兰·罗素撰写第二版时的手稿与笔记(英文)	2018—04	108.00	890
哈密尔顿数学论文集(第4卷):几何学、分析学、天文学、概率和有限差分等(英文)	2019—05	108.00	891
偏微分方程全局吸引子的特性(英文)	2018—09	108.00	979
整函数与下调和函数(英文)	2018—09	118.00	980
幂等分析(英文)	2018—09	118.00	981
李群、离散子群与不变量理论(英文)	2018—09	108.00	982
动力系统与统计力学(英文)	2018—09	118.00	983
表示论与动力系统(英文)	2018—09	118.00	984
分析学练习.第1部分(英文)	2021—01	88.00	1247
分析学练习.第2部分,非线性分析(英文)	2021—01	88.00	1248
初级统计学:循序渐进的方法:第10版(英文)	2019—05	68.00	1067
工程师与科学家微分方程用书:第4版(英文)	2019—07	58.00	1068
大学代数与三角学(英文)	2019—06	78.00	1069
培养数学能力的途径(英文)	2019—07	38.00	1070
工程师与科学家统计学:第4版(英文)	2019—06	58.00	1071
贸易与经济中的应用统计学:第6版(英文)	2019—06	58.00	1072
傅立叶级数和边值问题:第8版(英文)	2019—05	48.00	1073
通往天文学的途径:第5版(英文)	2019—05	58.00	1074
拉马努金笔记.第1卷(英文)	2019—06	165.00	1078
拉马努金笔记.第2卷(英文)	2019—06	165.00	1079
拉马努金笔记.第3卷(英文)	2019—06	165.00	1080
拉马努金笔记.第4卷(英文)	2019—06	165.00	1081
拉马努金笔记.第5卷(英文)	2019—06	165.00	1082
拉马努金遗失笔记.第1卷(英文)	2019—06	109.00	1083
拉马努金遗失笔记.第2卷(英文)	2019—06	109.00	1084
拉马努金遗失笔记.第3卷(英文)	2019—06	109.00	1085
拉马努金遗失笔记.第4卷(英文)	2019—06	109.00	1086
数论:1976年纽约洛克菲勒大学数论会议记录(英文)	2020—06	68.00	1145
数论:卡本代尔1979:1979年在南伊利诺伊卡本代尔大学举行的数论会议记录(英文)	2020—06	78.00	1146
数论:诺德韦克豪特1983:1983年在诺德韦克豪特举行的Journees Arithmetiques数论大会会议记录(英文)	2020—06	68.00	1147
数论:1985—1988年在纽约城市大学研究生院和大学中心举办的研讨会(英文)	2020—06	68.00	1148

书　名	出版时间	定　价	编号
数论:1987年在乌尔姆举行的Journees Arithmetiques数论大会会议记录(英文)	2020-06	68.00	1149
数论:马德拉斯1987:1987年在马德拉斯安娜大学举行的国际拉马努金百年纪念大会会议记录(英文)	2020-06	68.00	1150
解析数论:1988年在东京举行的日法研讨会会议记录(英文)	2020-06	68.00	1151
解析数论:2002年在意大利切特拉罗举行的C.I.M.E.暑期班演讲集(英文)	2020-06	68.00	1152
量子世界中的蝴蝶:最迷人的量子分形故事(英文)	2020-06	118.00	1157
走进量子力学(英文)	2020-06	118.00	1158
计算物理学概论(英文)	2020-06	48.00	1159
物质,空间和时间的理论:量子理论(英文)	2020-10	48.00	1160
物质,空间和时间的理论:经典理论(英文)	2020-10	48.00	1161
量子场理论:解释世界的神秘背景(英文)	2020-07	38.00	1162
计算物理学概论(英文)	2020-06	48.00	1163
行星状星云(英文)	2020-10	38.00	1164
基本宇宙学:从亚里士多德的宇宙到大爆炸(英文)	2020-08	58.00	1165
数学磁流体力学(英文)	2020-07	58.00	1166
计算科学:第1卷,计算的科学(日文)	2020-07	88.00	1167
计算科学:第2卷,计算与宇宙(日文)	2020-07	88.00	1168
计算科学:第3卷,计算与物质(日文)	2020-07	88.00	1169
计算科学:第4卷,计算与生命(日文)	2020-07	88.00	1170
计算科学:第5卷,计算与地球环境(日文)	2020-07	88.00	1171
计算科学:第6卷,计算与社会(日文)	2020-07	88.00	1172
计算科学.别卷,超级计算机(日文)	2020-07	88.00	1173
多复变函数论(日文)	2022-06	78.00	1518
复变函数入门(日文)	2022-06	78.00	1523
代数与数论:综合方法(英文)	2020-10	78.00	1185
复分析:现代函数理论第一课(英文)	2020-07	58.00	1186
斐波那契数列和卡特兰数:导论(英文)	2020-10	68.00	1187
组合推理:计数艺术介绍(英文)	2020-07	88.00	1188
二次互反律的傅里叶分析证明(英文)	2020-07	48.00	1189
旋瓦兹分布的希尔伯特变换与应用(英文)	2020-07	58.00	1190
泛函分析:巴拿赫空间理论入门(英文)	2020-07	48.00	1191
卡塔兰数入门(英文)	2019-05	68.00	1060
测度与积分(英文)	2019-04	68.00	1059
组合学手册.第一卷(英文)	2020-06	128.00	1153
-代数、局部紧群和巴拿赫-代数丛的表示.第一卷,群和代数的基本表示理论(英文)	2020-05	148.00	1154
电磁理论(英文)	2020-08	48.00	1193
连续介质力学中的非线性问题(英文)	2020-09	78.00	1195
多变量数学入门(英文)	2021-05	68.00	1317
偏微分方程入门(英文)	2021-05	88.00	1318
若尔当典范性:理论与实践(英文)	2021-07	68.00	1366
伽罗瓦理论.第4版(英文)	2021-08	88.00	1408

刘培杰数学工作室
已出版(即将出版)图书目录——原版影印

书　名	出版时间	定　价	编号
典型群,错排与素数(英文)	2020—11	58.00	1204
李代数的表示:通过 gln 进行介绍(英文)	2020—10	38.00	1205
实分析演讲集(英文)	2020—10	38.00	1206
现代分析及其应用的课程(英文)	2020—10	58.00	1207
运动中的抛射物数学(英文)	2020—10	38.00	1208
2—纽结与它们的群(英文)	2020—10	38.00	1209
概率,策略和选择:博弈与选举中的数学(英文)	2020—11	58.00	1210
分析学引论(英文)	2020—11	58.00	1211
量子群:通往流代数的路径(英文)	2020—11	38.00	1212
集合论入门(英文)	2020—10	48.00	1213
酉反射群(英文)	2020—11	58.00	1214
探索数学:吸引人的证明方式(英文)	2020—11	58.00	1215
微分拓扑短期课程(英文)	2020—10	48.00	1216
抽象凸分析(英文)	2020—11	68.00	1222
费马大定理笔记(英文)	2021—03	48.00	1223
高斯与雅可比和(英文)	2021—03	78.00	1224
π 与算术几何平均:关于解析数论和计算复杂性的研究(英文)	2021—01	58.00	1225
复分析入门(英文)	2021—03	48.00	1226
爱德华·卢卡斯与素性测定(英文)	2021—03	78.00	1227
通往凸分析及其应用的简单路径(英文)	2021—01	68.00	1229
微分几何的各个方面.第一卷(英文)	2021—01	58.00	1230
微分几何的各个方面.第二卷(英文)	2020—12	58.00	1231
微分几何的各个方面.第三卷(英文)	2020—12	58.00	1232
沃克流形几何学(英文)	2020—11	58.00	1233
彷射和韦尔几何应用(英文)	2020—12	58.00	1234
双曲几何学的旋转向量空间方法(英文)	2021—02	58.00	1235
积分:分析学的关键(英文)	2020—12	48.00	1236
为有天分的新生准备的分析学基础教材(英文)	2020—11	48.00	1237
数学不等式.第一卷.对称多项式不等式(英文)	2021—03	108.00	1273
数学不等式.第二卷.对称有理不等式与对称无理不等式(英文)	2021—03	108.00	1274
数学不等式.第三卷.循环不等式与非循环不等式(英文)	2021—03	108.00	1275
数学不等式.第四卷.Jensen 不等式的扩展与加细(英文)	2021—03	108.00	1276
数学不等式.第五卷.创建不等式与解不等式的其他方法(英文)	2021—04	108.00	1277

刘培杰数学工作室
已出版(即将出版)图书目录——原版影印

书　名	出版时间	定　价	编号
冯·诺依曼代数中的谱位移函数:半有限冯·诺依曼代数中的谱位移函数与谱流(英文)	2021—06	98.00	1308
链接结构:关于嵌入完全图的直线中链接单形的组合结构(英文)	2021—05	58.00	1309
代数几何方法.第1卷(英文)	2021—06	68.00	1310
代数几何方法.第2卷(英文)	2021—06	68.00	1311
代数几何方法.第3卷(英文)	2021—06	58.00	1312
代数、生物信息和机器人技术的算法问题.第四卷,独立恒等式系统(俄文)	2020—08	118.00	1199
代数、生物信息和机器人技术的算法问题.第五卷,相对覆盖性和独立可拆分恒等式系统(俄文)	2020—08	118.00	1200
代数、生物信息和机器人技术的算法问题.第六卷,恒等式和准恒等式的相等 问题、可推导性和可实现性(俄文)	2020—08	128.00	1201
分数阶微积分的应用:非局部动态过程,分数阶导热系数(俄文)	2021—01	68.00	1241
泛函分析问题与练习:第2版(俄文)	2021—01	98.00	1242
集合论、数学逻辑和算法论问题:第5版(俄文)	2021—01	98.00	1243
微分几何和拓扑短期课程(俄文)	2021—01	98.00	1244
素数规律(俄文)	2021—01	88.00	1245
无穷边值问题解的递减:无界域中的拟线性椭圆和抛物方程(俄文)	2021—01	48.00	1246
微分几何讲义(俄文)	2020—12	98.00	1253
二次型和矩阵(俄文)	2021—01	98.00	1255
积分和级数.第2卷,特殊函数(俄文)	2021—01	168.00	1258
积分和级数.第3卷,特殊函数补充:第2版(俄文)	2021—01	178.00	1264
几何图上的微分方程(俄文)	2021—01	138.00	1259
数论教程:第2版(俄文)	2021—01	98.00	1260
非阿基米德分析及其应用(俄文)	2021—03	98.00	1261
古典群和量子群的压缩(俄文)	2021—03	98.00	1263
数学分析问题集.第3卷,多元函数:第3版(俄文)	2021—03	98.00	1266
数学习题:乌拉尔国立大学数学力学系大学生奥林匹克(俄文)	2021—03	98.00	1267
柯西定理和微分方程的特解(俄文)	2021—03	98.00	1268
组合极值问题及其应用:第3版(俄文)	2021—03	98.00	1269
数学词典(俄文)	2021—01	98.00	1271
确定性混沌分析模型(俄文)	2021—06	168.00	1307
精选初等数学习题和定理.立体几何.第3版(俄文)	2021—03	68.00	1316
微分几何习题:第3版(俄文)	2021—05	98.00	1336
精选初等数学习题和定理.平面几何.第4版(俄文)	2021—05	68.00	1335
曲面理论在欧氏空间 E_n 中的直接表示(俄文)	2022—01	68.00	1444
维纳—霍普夫离散算子和托普利兹算子:某些可数赋范空间中的诺特性和可逆性(俄文)	2022—03	108.00	1496
Maple 中的数论:数论中的计算机计算(俄文)	2022—03	88.00	1497
贝尔曼和克努特问题及其概括:加法运算的复杂性(俄文)	2022—03	138.00	1498

刘培杰数学工作室
已出版（即将出版）图书目录——原版影印

书　　名	出 版 时 间	定　价	编号
复分析：共形映射(俄文)	2022－07	48.00	1542
微积分代数样条和多项式及其在数值方法中的应用(俄文)	2022－08	128.00	1543
蒙特卡罗方法中的随机过程和场模型：算法和应用(俄文)	2022－08	88.00	1544
线性椭圆型方程组：论二阶椭圆型方程的迪利克雷问题(俄文)	2022－08	98.00	1561
动态系统解的增长特性：估值、稳定性、应用(俄文)	2022－08	118.00	1565
群的自由积分解：建立和应用(俄文)	2022－08	78.00	1570
狭义相对论与广义相对论：时空与引力导论(英文)	2021－07	88.00	1319
束流物理学和粒子加速器的实践介绍：第2版(英文)	2021－07	88.00	1320
凝聚态物理中的拓扑和微分几何简介(英文)	2021－05	88.00	1321
混沌映射：动力学、分形学和快速涨落(英文)	2021－05	128.00	1322
广义相对论：黑洞、引力波和宇宙学介绍(英文)	2021－06	68.00	1323
现代分析电磁均质化(英文)	2021－06	68.00	1324
为科学家提供的基本流体动力学(英文)	2021－06	88.00	1325
视觉天文学：理解夜空的指南(英文)	2021－06	68.00	1326
物理学中的计算方法(英文)	2021－06	68.00	1327
单星的结构与演化：导论(英文)	2021－06	108.00	1328
超越居里：1903年至1963年物理界四位女性及其著名发现(英文)	2021－06	68.00	1329
范德瓦尔斯流体热力学的进展(英文)	2021－06	68.00	1330
先进的托卡马克稳定性理论(英文)	2021－06	88.00	1331
经典场论导论：基本相互作用的过程(英文)	2021－07	88.00	1332
光致电离量子动力学方法原理(英文)	2021－07	108.00	1333
经典域论和应力：能量张量(英文)	2021－05	88.00	1334
非线性太赫兹光谱的概念与应用(英文)	2021－06	68.00	1337
电磁学中的无穷空间并矢格林函数(英文)	2021－06	88.00	1338
物理科学基础数学.第1卷,齐次边值问题、傅里叶方法和特殊函数(英文)	2021－07	108.00	1339
离散量子力学(英文)	2021－07	68.00	1340
核磁共振的物理学和数学(英文)	2021－07	108.00	1341
分子水平的静电学(英文)	2021－08	68.00	1342
非线性波：理论、计算机模拟、实验(英文)	2021－06	108.00	1343
石墨烯光学：经典问题的电解解决方案(英文)	2021－06	68.00	1344
超材料多元宇宙(英文)	2021－07	68.00	1345
银河系外的天体物理学(英文)	2021－07	68.00	1346
原子物理学(英文)	2021－07	68.00	1347
将光打结：将拓扑学应用于光学(英文)	2021－07	68.00	1348
电磁学：问题与解法(英文)	2021－07	88.00	1364
海浪的原理：介绍量子力学的技巧与应用(英文)	2021－07	108.00	1365
多孔介质中的流体：输运与相变(英文)	2021－07	68.00	1372
洛伦兹群的物理学(英文)	2021－08	68.00	1373
物理导论的数学方法和解决方法手册(英文)	2021－08	68.00	1374
非线性波数学物理入门(英文)	2021－08	88.00	1376
波：基本原理和动力学(英文)	2021－07	68.00	1377
光电子量子计量学.第1卷,基础(英文)	2021－07	88.00	1383
光电子量子计量学.第2卷,应用与进展(英文)	2021－07	68.00	1384
复杂流的格子玻尔兹曼建模的工程应用(英文)	2021－08	68.00	1393

刘培杰数学工作室
已出版(即将出版)图书目录——原版影印

书　名	出版时间	定　价	编号
电偶极矩挑战(英文)	2021—08	108.00	1394
电动力学:问题与解法(英文)	2021—09	68.00	1395
自由电子激光的经典理论(英文)	2021—08	68.00	1397
曼哈顿计划——核武器物理学简介(英文)	2021—09	68.00	1401
粒子物理学(英文)	2021—09	68.00	1402
引力场中的量子信息(英文)	2021—09	128.00	1403
器件物理学的基本经典力学(英文)	2021—09	68.00	1404
等离子体物理及其空间应用导论.第1卷,基本原理和初步过程(英文)	2021—09	68.00	1405
拓扑与超弦理论焦点问题(英文)	2021—07	58.00	1349
应用数学:理论、方法与实践(英文)	2021—07	78.00	1350
非线性特征值问题:牛顿型方法与非线性瑞利函数(英文)	2021—07	58.00	1351
广义膨胀和齐性:利用齐性构造齐次系统的李雅普诺夫函数和控制律(英文)	2021—06	48.00	1352
解析数论焦点问题(英文)	2021—07	58.00	1353
随机微分方程:动态系统方法(英文)	2021—07	58.00	1354
经典力学与微分几何(英文)	2021—07	58.00	1355
负定相交形式流形上的瞬子模空间几何(英文)	2021—07	68.00	1356
广义卡塔兰轨道分析:广义卡塔兰轨道计算数字的方法(英文)	2021—07	48.00	1367
洛伦兹方法的变分:二维与三维洛伦兹方法(英文)	2021—08	38.00	1378
几何、分析和数论精编(英文)	2021—08	68.00	1380
从一个新角度看数论:通过遗传方法引入现实的概念(英文)	2021—07	58.00	1387
动力系统:短期课程(英文)	2021—08	68.00	1382
几何路径:理论与实践(英文)	2021—08	48.00	1385
论天体力学中某些问题的不可积性(英文)	2021—07	88.00	1396
广义斐波那契数列及其性质(英文)	2021—08	38.00	1386
对称函数和麦克唐纳多项式:余代数结构与Kawanaka恒等式(英文)	2021—09	38.00	1400
杰弗里·英格拉姆·泰勒科学论文集:第1卷.固体力学(英文)	2021—05	78.00	1360
杰弗里·英格拉姆·泰勒科学论文集:第2卷.气象学、海洋学和湍流(英文)	2021—05	68.00	1361
杰弗里·英格拉姆·泰勒科学论文集:第3卷.空气动力学以及落弹数和爆炸的力学(英文)	2021—05	68.00	1362
杰弗里·英格拉姆·泰勒科学论文集:第4卷.有关流体力学(英文)	2021—05	58.00	1363

刘培杰数学工作室
已出版(即将出版)图书目录——原版影印

书　名	出版时间	定　价	编号
非局域泛函演化方程:积分与分数阶(英文)	2021－08	48.00	1390
理论工作者的高等微分几何:纤维丛、射流流形和拉格朗日理论(英文)	2021－08	68.00	1391
半线性退化椭圆微分方程:局部定理与整体定理(英文)	2021－07	48.00	1392
非交换几何、规范理论和重整化:一般简介与非交换量子场论的重整化(英文)	2021－09	78.00	1406
数论论文集:拉普拉斯变换和带有数论系数的幂级数(俄文)	2021－09	48.00	1407
挠理论专题:相对极大值,单射与扩充模(英文)	2021－09	88.00	1410
强正则图与欧几里得若尔当代数:非通常关系中的启示(英文)	2021－10	48.00	1411
拉格朗日几何和哈密顿几何:力学的应用(英文)	2021－10	48.00	1412

时滞微分方程与差分方程的振动理论:二阶与三阶(英文)	2021－10	98.00	1417
卷积结构与几何函数理论:用以研究特定几何函数理论方向的分数阶微积分算子与卷积结构(英文)	2021－10	48.00	1418
经典数学物理的历史发展(英文)	2021－10	78.00	1419
扩展线性丢番图问题(英文)	2021－10	38.00	1420
一类混沌动力系统的分歧分析与控制:分歧分析与控制(英文)	2021－11	38.00	1421
伽利略空间和伪伽利略空间中一些特殊曲线的几何性质(英文)	2022－01	68.00	1422
一阶偏微分方程:哈密尔顿—雅可比理论(英文)	2021－11	48.00	1424
各向异性黎曼多面体的反问题:分段光滑的各向异性黎曼多面体反边界谱问题:唯一性(英文)	2021－11	38.00	1425

项目反应理论手册.第一卷,模型(英文)	2021－11	138.00	1431
项目反应理论手册.第二卷,统计工具(英文)	2021－11	118.00	1432
项目反应理论手册.第三卷,应用(英文)	2021－11	138.00	1433
二次无理数:经典数论入门(英文)	2022－05	138.00	1434
数,形与对称性:数论,几何和群论导论(英文)	2022－05	128.00	1435
有限域手册(英文)	2021－11	178.00	1436
计算数论(英文)	2021－11	148.00	1437
拟群与其表示简介(英文)	2021－11	88.00	1438
数论与密码学导论:第二版(英文)	2022－01	148.00	1423

刘培杰数学工作室
已出版(即将出版)图书目录——原版影印

书　名	出版时间	定　价	编号
几何分析中的柯西变换与黎兹变换:解析调和容量和李普希兹调和容量、变化和振荡以及一致可求长性(英文)	2021-12	38.00	1465
近似不动点定理及其应用(英文)	2022-05	28.00	1466
局部域的相关内容解析:对局部域的扩展及其伽罗瓦群的研究(英文)	2022-01	38.00	1467
反问题的二进制恢复方法(英文)	2022-03	28.00	1468
对几何函数中某些类的各个方面的研究:复变量理论(英文)	2022-01	38.00	1469
覆盖、对应和非交换几何(英文)	2022-01	28.00	1470
最优控制理论中的随机线性调节器问题:随机最优线性调节器问题(英文)	2022-01	38.00	1473
正交分解法:涡流流体动力学应用的正交分解法(英文)	2022-01	38.00	1475
芬斯勒几何的某些问题(英文)	2022-03	38.00	1476
受限三体问题(英文)	2022-05	38.00	1477
利用马利亚万微积分进行 Greeks 的计算:连续过程、跳跃过程中的马利亚万微积分和金融领域中的 Greeks(英文)	2022-05	48.00	1478
经典分析和泛函分析的应用:分析学的应用(英文)	2022-03	38.00	1479
特殊芬斯勒空间的探究(英文)	2022-03	48.00	1480
某些图形的施泰纳距离的细谷多项式:细谷多项式与图的维纳指数(英文)	2022-05	38.00	1481
图论问题的遗传算法:在新鲜与模糊的环境中(英文)	2022-05	48.00	1482
多项式映射的渐近簇(英文)	2022-05	38.00	1483
一维系统中的混沌:符号动力学,映射序列,一致收敛和沙可夫斯基定理(英文)	2022-05	38.00	1509
多维边界层流动与传热分析:粘性流体流动的数学建模与分析(英文)	2022-05	38.00	1510
演绎理论物理学的原理:一种基于量子力学波函数的逐次置信估计的一般理论的提议(英文)	2022-05	38.00	1511
R^2 和 R^3 中的仿射弹性曲线:概念和方法(英文)	2022-08	38.00	1512
算术数列中除数函数的分布:基本内容、调查、方法、第二矩、新结果(英文)	2022-05	28.00	1513
抛物型狄拉克算子和薛定谔方程:不定常薛定谔方程的抛物型狄拉克算子及其应用(英文)	2022-07	28.00	1514
黎曼-希尔伯特问题与量子场论:可积重正化、戴森-施温格方程(英文)	2022-08	38.00	1515
代数结构和几何结构的形变理论(英文)	2022-08	48.00	1516
概率结构和模糊结构上的不动点:概率结构和直觉模糊度量空间的不动点定理(英文)	2022-08	38.00	1517

刘培杰数学工作室
已出版(即将出版)图书目录——原版影印

书　名	出版时间	定　价	编号
反若尔当对:简单反若尔当对的自同构(英文)	2022－07	28.00	1533
对某些黎曼－芬斯勒空间变换的研究:芬斯勒几何中的某些变换(英文)	2022－07	38.00	1534
内诣零流形映射的尼尔森数的阿诺索夫关系(英文)	即将出版		1535
与广义积分变换有关的分数次演算:对分数次演算的研究(英文)	即将出版		1536
强子的芬斯勒几何和吕拉几何(宇宙学方面):强子结构的芬斯勒几何和吕拉几何(拓扑缺陷)(英文)	2022－08	38.00	1537
一种基于混沌的非线性最优化问题:作业调度问题(英文)	即将出版		1538
广义概率论发展前景:关于趣味数学与置信函数实际应用的一些原创观点(英文)	即将出版		1539
纽结与物理学:第二版(英文)	2022－09	118.00	1547
正交多项式和 q－级数的前沿(英文)	2022－09	98.00	1548
算子理论问题集(英文)	即将出版		1549
抽象代数:群、环与域的应用导论:第二版(英文)	即将出版		1550
菲尔兹奖得主演讲集:第三版(英文)	即将出版		1551
多元实函数教程(英文)	2022－09	118.00	1552
球面空间形式群的几何学:第二版(英文)	2022－09	98.00	1566

联系地址:哈尔滨市南岗区复华四道街 10 号　哈尔滨工业大学出版社刘培杰数学工作室
网　　址:http://lpj.hit.edu.cn/
邮　　编:150006
联系电话:0451－86281378　　13904613167
E-mail:lpj1378@163.com